MOBILE AD-HOC NETWORKS: APPLICATIONS

Edited by **Xin Wang**

Mobile Ad-Hoc Networks: Applications
Edited by Xin Wang

Published by InTech
Janeza Trdine 9, 51000 Rijeka, Croatia

Publishing Process Manager Vedran Greblo

Technical Editor Teodora Smiljanic

Cover Designer InTech Design Team

Additional hard copies can be obtained from orders@intechopen.com

Mobile Ad-Hoc Networks: Applications, Edited by Xin Wang
p. cm.
ISBN 978-953-307-416-0

For EU product safety concerns:
IN TECH d.o.o., Prolaz Marije Krucifikse Kozulić 3, 51000 Rijeka, Croatia,
info@intechopen.com or visit our website at intechopen.com.

Meet the editor

Xin Wang received the B.S. and M.S. degrees in electronic engineering from Tsinghua University, Beijing, China. He received PhD degree in computer engineering from University of California, Santa Cruz, U.S. He currently works at Cisco Systems, San Jose, CA, U.S. His research interests include protocol design and performance evaluation for computer networks, especially for cross layer optimization in wireless networks, server access virtualization and data center networks. He is a member of IEEE and ACM.

Contents

Preface XI

Part 1 Vehicular Ad Hoc Networks 1

Chapter 1 **Survey on Multi-hop Vehicular Ad Hoc Networks under IEEE 802.16 Technology 3**
Gabriel Alejandro Galaviz Mosqueda, Raúl Aquino Santos, Luis A. Villaseñor González, Víctor Rangel Licea and Arthur Edwards Block

Chapter 2 **Communications in Vehicular Networks 19**
Zaydoun Yahya Rawashdeh and Syed Masud Mahmud

Chapter 3 **Modeling and Simulation of Vehicular Networks: Towards Realistic and Efficient Models 41**
Mate Boban and Tiago T. V. Vinhoza

Chapter 4 **Security Issues in Vehicular Ad Hoc Networks 67**
P. Caballero-Gil

Chapter 5 **Routing in Vehicular Ad Hoc Networks: Towards Road-Connectivity Based Routing 89**
Nadia Brahmi, Mounir Boussedjra and Jospeh Mouzna

Chapter 6 **Traffic Information Dissemination in Vehicular Ad Hoc Networks 107**
Attila Török, Balázs Mezny and Péter Laborczi

Chapter 7 **CARAVAN: Context-AwaRe Architecture for VANET 125**
Sławomir Kukliński and Grzegorz Wolny

Part 2 Security and Caching in Ad Hoc Networks 149

Chapter 8 **Trust Establishment in Mobile Ad Hoc Networks: Key Management 151**
Dawoud D.S., Richard L. Gordon, Ashraph Suliman and Kasmir Raja S.V.

Chapter 9 **Grouping-Enabled and Privacy-Enhancing Communications Schemes for VANETs 193**
T.W. Chim, S.M. Yiu, Lucas C.K. Hui and Victor O.K. Li

Chapter 10 **APALLS: A Secure MANET Routing Protocol 221**
Sivakumar Kulasekaran and Mahalingam Ramkumar

Chapter 11 **Meta-heuristic Techniques and Swarm Intelligence in Mobile Ad Hoc Networks 245**
Floriano De Rango and Annalisa Socievole

Chapter 12 **Impact of the Mobility Model on a Cooperative Caching Scheme for Mobile Ad Hoc Networks 265**
F.J. Gonzalez-Cañete and E. Casilari

Part 3 Applications of Ad Hoc Networks 287

Chapter 13 **Ad Hoc Networks for Cooperative Mobile Positioning 289**
Francescantonio Della Rosa, Helena Leppäkoski, Ata-ul Ghalib, Leyla Ghazanfari, Oscar Garcia, Simone Frattasi and Jari Nurmi

Chapter 14 **Ad-hoc Networks As an Enabler of Brain Spectroscopy 305**
Salah Sharieh

Chapter 15 **MANET Mining: Mining Association Rules 323**
Ahmad Jabas

Chapter 16 **Wired/Wireless Compound Networking 349**
Juan Antonio Cordero, Emmanuel Baccelli, Philippe Jacquet and Thomas Clausen

Chapter 17 **Multiple Multicast Tree Construction and Multiple Description Video Assignment Algorithms 375**
Osamah Badarneh and Michel Kadoch

Part 4 TCP in Ad Hoc Networks 399

Chapter 18 **TCP-MAC Interaction in Multi-hop Ad-hoc Networks 401**
Farzaneh R. Armaghani and Sudhanshu S. Jamuar

Chapter 19 **The Effect of Packet Losses and Delay on TCP Traffic over Wireless Ad Hoc Networks 427**
May Zin Oo and Mazliza Othman

Part 5 Other Topics 451

Chapter 20 **A Survey on The Characterization of the Capacity of Ad Hoc Wireless Networks 453**
Paulo Cardieri and Pedro Henrique Juliano Nardelli

Chapter 21 **Design and Analysis of a Multi-level Location Information Based Routing Scheme for Mobile Ad hoc Networks 473**
Koushik Majumder, Sudhabindu Ray and Subir Kumar Sarkar

Chapter 22 **Power Control in Ad Hoc Networks 489**
Muhammad Mazhar Abbas and Hasan Mahmood

Preface

Mobile Ad hoc Networks (MANETs) are a fundamental element of pervasive networks, where user can communicate anywhere, any time and on-the-fly. MANETs introduce a new communication paradigm, which does not require a fixed infrastructure - they rely on wireless terminals for routing and transport services. This edited volume covers the most advanced research and development in MANET. It seeks to provide an opportunity for readers to explore the emerging fields about MANET.

It includes five parts in total. Part 1 discusses the emerging vehicular ad-hoc networks. Part 2 focuses on the security and caching protocols. Part 3 introduces some new applications for MANET. Part 4 presents novel approaches in transport-layer protocol design. Some interesting topics about network capacity, power control, etc. are discussed in Part 5.

Acknowledgements
The editors are particularly grateful to the authors who present their work in this book. They would also like to express their sincere thanks to all the reviewers, who help to maintain the high quality of this book. We hope that the readers will share our excitement to present this volume on ad-hoc networks and will find it useful.

Prof. Xin Wang
University of California, Santa Cruz,
USA

Part 1

Vehicular Ad Hoc Networks

Survey on Multi-hop Vehicular Ad Hoc Networks under IEEE 802.16 Technology

Gabriel Alejandro Galaviz Mosqueda[1], Raúl Aquino Santos[2], Luis A. Villaseñor González[1], Víctor Rangel Licea[3] and Arthur Edwards Block[2]
[1]Centro de Investigación Científica y Educación Superiore de Ensenada. Carretera Ensenada-Tijuana, núm. 3918, Zona playitas, C. P. 22860, Ensenada, Baja California,
[2]Facultad de Telemática, Avenida Universidad 333, C. P. 28040, Colima, Col.,
[3]Facultad de Ingeniería, Edificio Valdez Vallejo, 3er piso, Circuito Interior, Ciudad Universitaria, Delegación Coyoacán, C. P. 04510,
México

1. Introduction

Today, there are many existing technologies designed to make vehicular road travel safer, easier and more enjoyable, using geographical positioning system, proximity sensors, multimedia communication, etc. The current data transmission requirements of these technologies, unfortunately, place great demand on both the algorithms and equipment, which often perform less than optimally, especially when having to interact with other vehicles. For example, GPS can trace a route to a specific location, but does so without taking into account some very important variables such as congestion caused by road conditions, high traffic volume and traffic accidents, which can entirely block one-lane traffic and affect two-lane traffic by almost 65% [1].

Presently, GPS permits users to obtain real-time location information. However, expanded communications among vehicles and with roadside infrastructure can substantially expand services drivers currently enjoy in the areas of traffic flow, safety, information (Internet), communications (VoIP) and comfort applications, among others [2].

According to Sichitiu et al. applications for vehicular communications include the following:

- Proactive safety applications: geared primarily to improve driver reaction and decision making to avoid possible accidents (e.g. broadcast warnings from a vehicle that has ignored red stop light) or minimize the impacts of an imminent crash (automated braking systems).
- Traffic management applications: mainly implemented to improve traffic flow and reduce travel time, which is particularly useful for emergency vehicles.
- Traffic coordination and traffic assistance: principally concerned with improving the distribution and flow of vehicles by helping drivers pass, change lanes, merge and form columns of vehicles that maintain constant relative speeds and distances (platooning).
- Traveler Information Support: mainly focused on providing specific information about available resources and assistance persons require, making their traveling experience less stressful and more efficient.

- Comfort Applications: primarily designed to improve the travel experience of the passengers and the driver (e.g. gaming, internet, automatic tolls, etc.)

Figure 1 shows some potential applications.

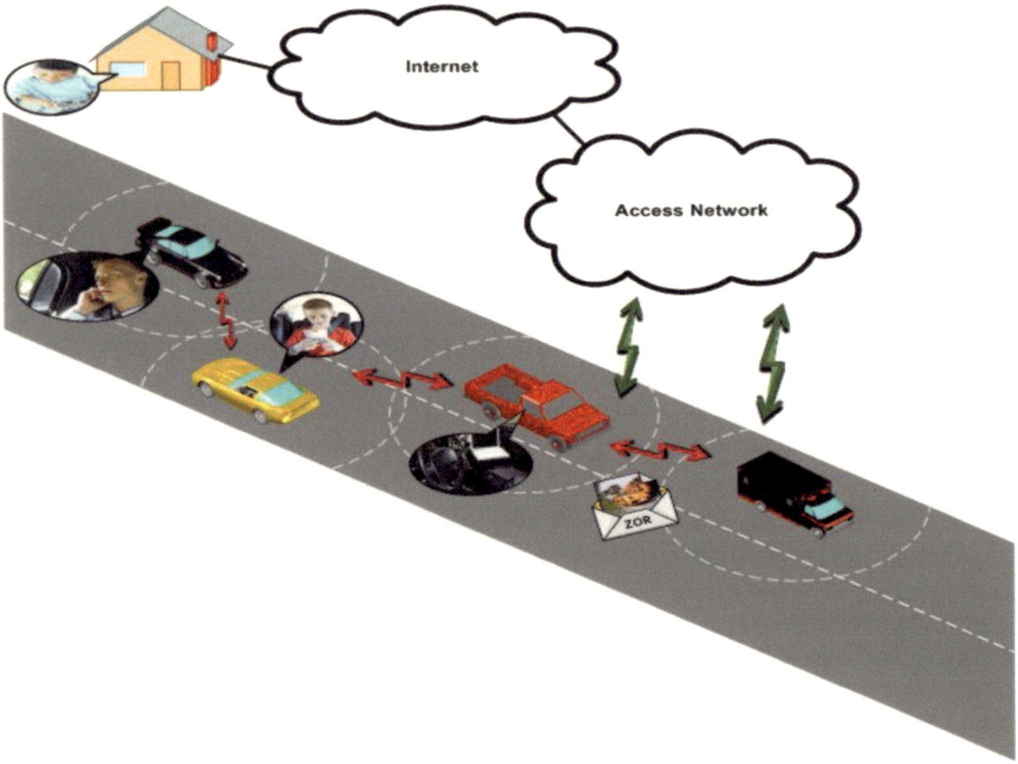

Fig. 1. Some potential services to be offered by vehicular communication networks

In order to provide greater passenger safety, convenience and comfort, protocols and equipment must provide more timely and reliable data transfer between network nodes for them to effectively share vital information. In the case of WiMAX, network nodes must efficiently transmit and receive data in a instantaneously changing network environment, characterized by the constant entry and exit of nodes. In addition, mobile nodes must handle handoffs between different clusters, all while functioning within very strict technical parameters regarding packet loss, delay, latency, and throughput, among others.

Sichitiu and Kihl in [3] construct a taxonomy based on the way nodes exchange data. Their work involves two forms of vehicular communication: vehicle to vehicle (IVC) and vehicle to roadside (RVC). IVC can employ either a one hop (SICV) or multi-hop (MIVC) strategy. On the other hand, RVC can be ubiquitous (URVC) or scarce (SRVC). Figure 2 schematizes these authors' taxonomy [3]. The following three figures explain this taxonomy and provide examples of IVC, RVC and HVC.

Communications within VANETs can be either inter-vehicular or vehicle to roadside and each type of communication imposes its specific requirements. For example, highway collision warning systems can more easily be implemented using multi-hop communications between vehicles (without infrastructure). On the other hand, traveller information requires fixed infrastructure to provide connectivity between the vehicles and

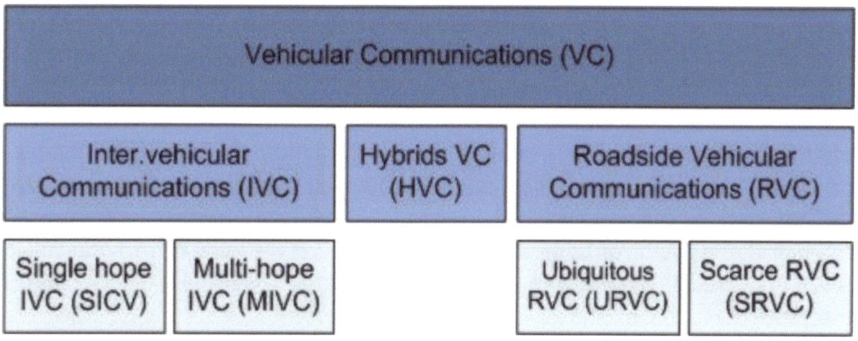

Fig. 2. Vehicular communications Taxonomy

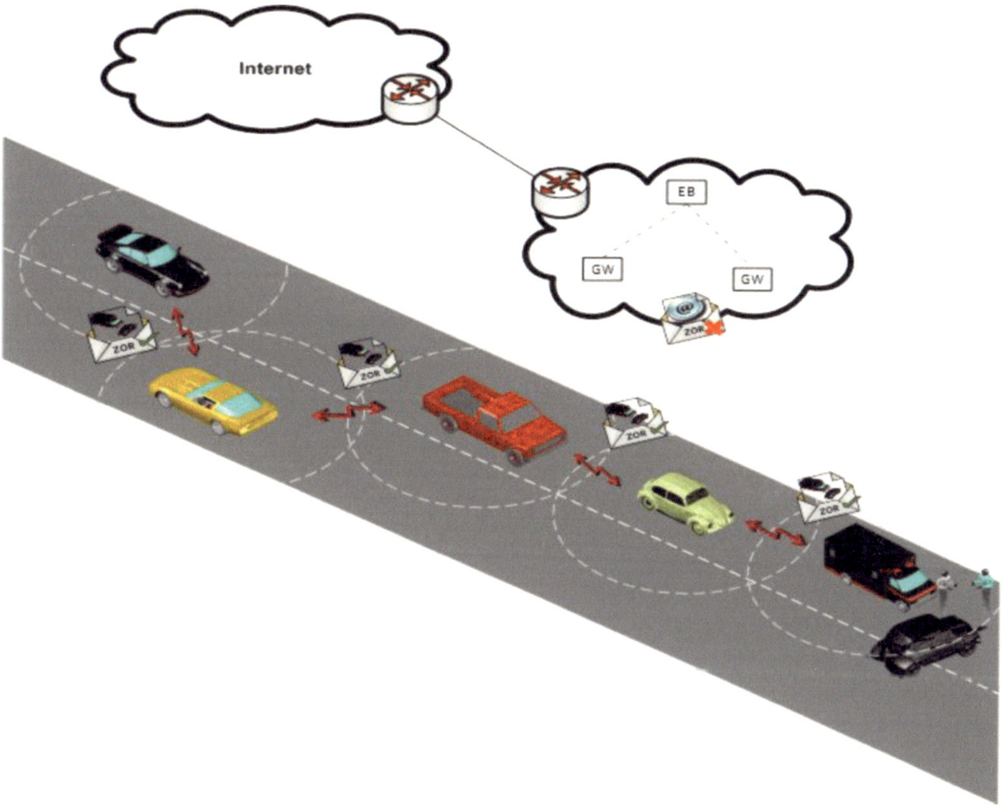

Fig. 3. An IVC example

an information center. IVC deployment is significantly less expensive than RVC because it is infrastructureless. This kind of architecture allows vehicles to send information between each other via multi-hop communication, even with vehicles that are beyond their immediate radio coverage area. IVC internet access is much more complicated than with RVC. As a result, IVC can only provide a reduced number of applications. However, IVC is better suited for safety applications because the vehicles can almost immediately detect collision or congestion warning that is transmitted within the affected area. Figure 3 provides an example of inter vehicular communication, where a vehicle approaching an accident detects the crash and informs the vehicles behind it that it is about to brake suddenly. This forewarning could help avoid other accidents caused by drivers who cannot apply their brakes opportunely and allows vehicles further behind to change lanes to lessen traffic congestion.

RVC can offer a wider range of applications because of its more stable and robust access to the Internet, which allows ready availability of information about specific places and the services they provide. RVC, however, has two important drawbacks when considered for safety applications:

- the cost of deployment of base stations (BS) makes it difficult to provide full coverage for so many vehicles over such a large area as vehicles leaving the BS coverage area lose connectivity.
- the delay caused by sending packets through a base station can prove disastrous in time sensitive safety applications.

Different technologies have been tested to enable RVC, including cellular, WiFi (IEEE 802.11p) and WiMAX (IEEE 802.16e), but no standard has been established as of yet. Presently, authors believe that WiMAX best fits VCN requirements because of its high bandwidth, robust medium access control (MAC), versatility (i.e. wide range of compatible standards) and QoS support. Importantly, it meets the already existing standard for mobile nodes (IEEE 802.16e). Figure 4 illustrates examples of some RVC applications, which include broadcasting the location of specific businesses and providing information about goods and services offered by them.

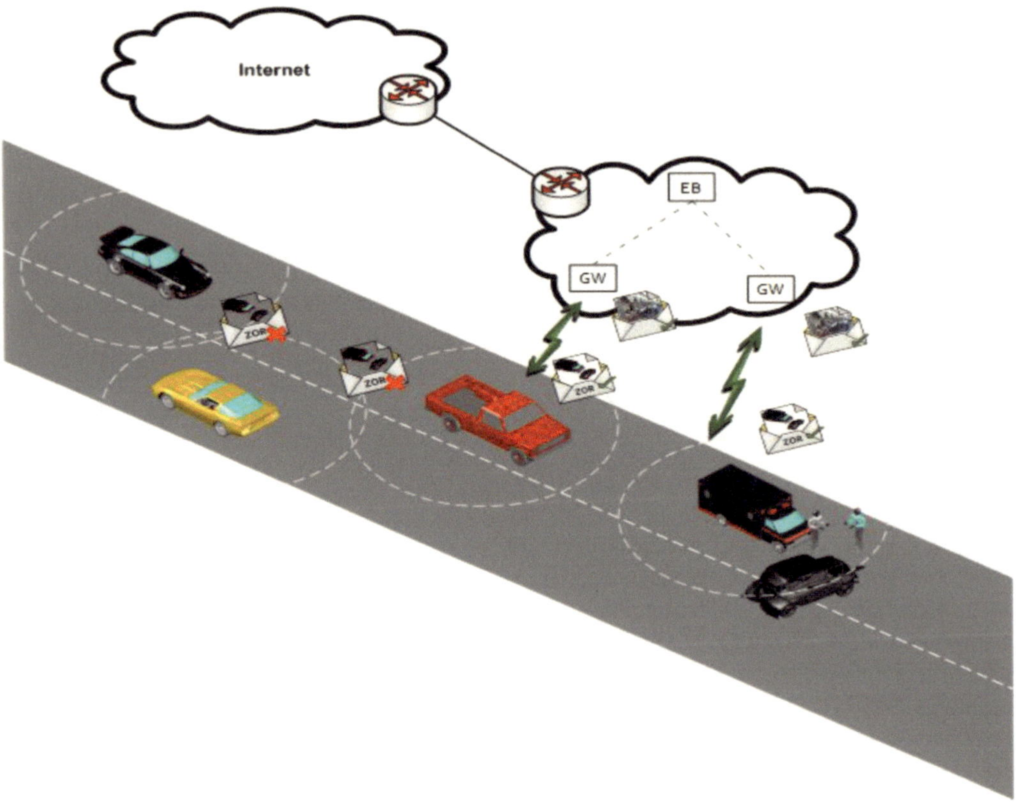

Fig. 4. A RVC network example

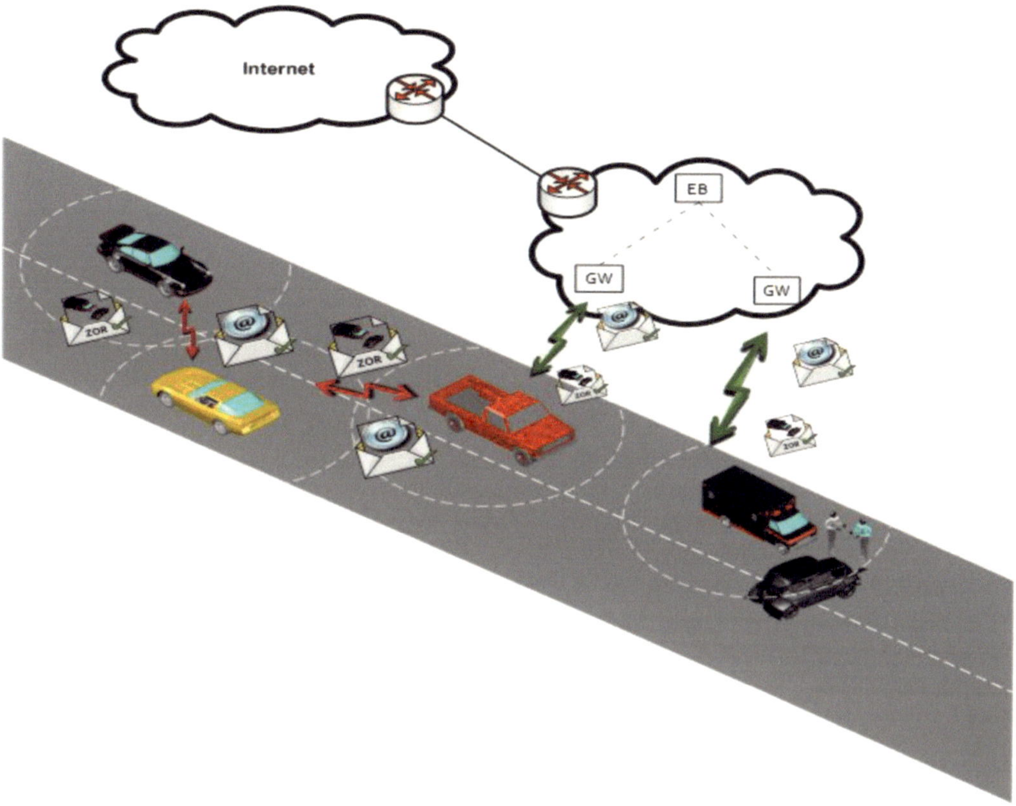

Fig. 5. A mixture of IVC and RVC (HVC)

Both IVC and RVC have desirable benefits; while with IVC users can form groups practically anywhere, with RVC persons can have access to internet and extend the vehicular applications. Importantly, combining both of these architectures into a hybrid vehicular communications (HVC) network can maximize benefits. HVC, however, is more complex in various aspects: HVC need more complex routing protocols, a robust physical layer and a medium access layer that is sufficiently dynamic to fully exploit the short duration of links and organized enough to minimize interference.

Figure 5 illustrates a hybrid vehicular communication network where vehicles inside the coverage area of a RVC can act as gateways for vehicles outside the coverage area. HVC networks are very desirables because they can provide virtually any kind of service. Importantly, however, as previously mentioned, research must first overcome many technical challenges before HVC networks can be implemented in real-world applications. This is primarily because of the incompatibility of technologies (e.g. WiFi was developed for WLANs, while cellular communications were designed for WANs).

As previously mentioned, each type of vehicular communications (IVC, RVC or HVC) has different technological requirements, although they all must meet several common demands inherent in VCN (see Table 1 and Figure 6). Three of these network requirements include [4]:

- radio transceiver technology that provides omni-directional coverage
- rapid vehicle-to-vehicle communications to keep track of dynamic topology changes
- highly efficient routing algorithms that fully exploit network bandwidth

Fig. 1. Types of scenarios for VCN

	Rural	Urban	City	Highway
Speed	Low	Medium/High	Low/Very Low	Very high
Vehicles Density	Low	Medium	Very high	Med/Low
Interference	Low	Medium	Very high	Low
Infrastructure	Low	Medium	Very high	Med/Low

Table 1. Features of Vehicular Scenarios

Numerous researchers have worked to overcome issues related to vehicular communications (e.g. [5-9, 10-12]). In 2004, the IEEE group created the IEEE 802.11p (wireless access in vehicular environments-WAVE) task force [13]. The workforce established a new standard that essentially employs the same PHY layer of the IEEE 802.11a standard, but uses a 10 MHz channel bandwidth instead of the 20 MHz used in IEEE 802.11a. With respect to the MAC layer, WAVE is based on a contention method (i.e. CSMA/CA), similar to other standards in this group.

The MAC layer in IEEE 802.11p has several significant drawbacks. For example, in vehicular scenarios, WAVE drops over 53% of packets sent according to simulation results [14].

WAVE also has a limited transmission range; simulations carried out by [15] show that only 1% of communication attempts at 750m are successful in a highway scenario presenting multipath shadowing. Furthermore, results in [16] show that throughput decays as the number of vehicles increases. In fact, throughput decreases to almost zero with 20 concurrent transmissions. The authors thus conclude that WAVE is not scalable. Additionally, IEEE 802.11p does not support QoS, which is essential in Vehicular Ad hoc Networks (VANETs). Importantly, safety applications using VCNs require not only expanded radio coverage, but also demand minimal delay, robust bandwidth, negligible packet loss and reduced jitter, among others (see Table 2).

Recently, the IEEE 802.16 taskforce [17, 18] actualized this standard to support QoS, mobility, and multihop relay communications. Networks using the IEEE 802.16 MAC layer now can potentially meet a wider range of demands, including VCN.

Worldwide Interoperability for Microwave Access (WiMAX) is a nonprofit consortium supported by over 400 companies dedicated to creating profiles based on the IEEE 802.16 standard.

Application	Maximum Required Range (m)
Approaching emergency vehicle warning	~1000
Emergency vehicle signal preemption	~1000
SOS services	~400
Postcrash Warning	~300

Table 2. Maximum required range for some applications in VCN

The first IEEE 802.16 standard considers fixed nodes with line of sight (LOS) between the base station and each fixed remote node [19]. Later, the IEEE 802.16 task force amended the original standard to provide mobility to end users (Mobile WiMAX[17]) in non-line-of-sight (NLOS) conditions. The most recent modification to IEEE 802.16e was in March, 2007, which later resulted in the IEEE802.16j multihop relay communications protocol, approved in 2009 [18].

IEEE 802.16j operates in both transparent and non-transparent modes. In transparent mode, mobile stations (MS) must decode the control messages relayed from the base station (BS). In other words, they must operate within the physical coverage radius of the BS because relay stations (RS) do not retransmit control information. In non-transparent mode, one of the RS provides the control messages to the MS. The main difference between transparent and non-transparent mode architecture is that in transparent mode, RS increase network capacity while in non-transparent mode, RS extend the BS range. Additionally, RS can be classified according to their mobility and can be fixed (FRS), nomadic (NRS) or mobile (MRS) [20].

Despite recent progress in implementing VCN with WiMAX, much work still has to be done. This work presents proposals that employ IEEE 802.16 as their underlying technology for multi-hop vehicular communication networks.

This paper is organized as follows: Section II analyzes various proposals suggested by researchers for VCN using WiMAX networks; Section III presents challenges of using WiMAX in VCN and Section IV presents conclusions.

2. State of the art of WiMAX in multi-hop vehicular communication networks

The authors in [21] propose a routing protocol called Coordinated External Peer Communications (CEPEC), whose cross-layer protocol is designed for multi-hop vehicular networks. They obtained their simulation results using a proprietary development tool which guaranteed all vehicles fair access to the Internet, even over nodes that were several hops distant from the BS. Their proposal includes organize the OSI model into three layers: PHY, MAC and Network. However, the authors do not specify the modifications they made to the IEEE 802.16-2004 standard that permitted the increased mobility and quicker registration of the MS. The authors employ TDMA to assign channels, exploiting TDMA's centralized scheduler and time division duplexing. Finally, and very importantly, CEPEC needs to determine the geographic position of every vehicle. To do this, all vehicles must be equipped with GPS.

An important disadvantage of CEPEC is that it only allows data communication from vehicles to the BS and vice versa; therefore, it does not provide for vehicle-to-vehicle data exchange. Additionally, CEPEC's centralized scheduling mechanism reduces its scalability. Since, as previously mentioned, the authors of [21] do not specify the changes they made to the IEEE 802.16 standard, we must assume that vehicles enter the network according to standard specifications for nodes in mesh mode. Of course, this implies that network performance suffers significant deterioration. Also, the authors fail to detail the modifications they made to the standard that permitted increased mobility and topology control.

Figure 7 shows the segment configuration of a CEPEC simulation in which the green vehicles are segment subscriber stations (SSSs) and the red ones are segment heads (SH).

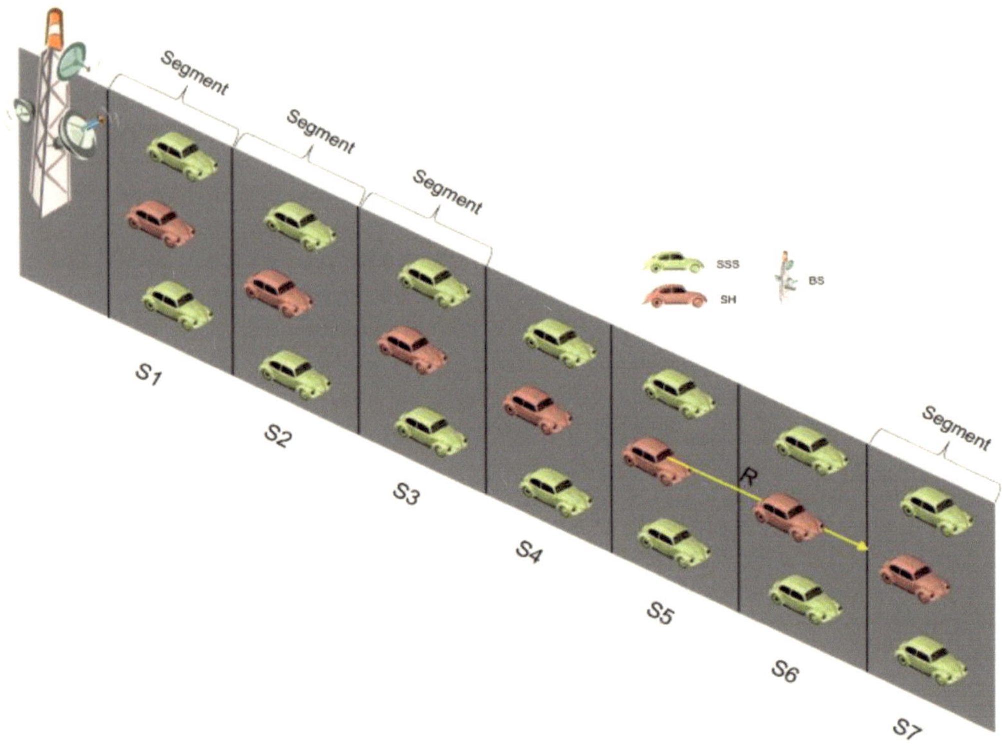

Fig. 7. CEPEC Topology

The authors in [22] do not provide simulation or test bed results and limit their work to making suggestions at a conceptual level about how to apply a hierarchical topology using WiFi hotspots (i.e. IEEE 802.11p) as access points for vehicles and WiMAX mesh stations as access points for WiFi hotspots. One major issue concerning this topology is that the IEEE 802.11p standard does not support QoS and the MAC contention-based method represents a significant disadvantage.

The topology in [22] is comprised of a point of access (PoA) consisting of a WiMAX mesh point (MP) and at least one access point (AP). The clusters are formed by several PoAs, one of which serves as a cluster head (CH) and domain, which is formed by a group of clusters. Figure 8 schematizes the described topology.

The authors in [23] propose a handoff mechanism called SWIFT, which includes modification in the MAC and network layers.

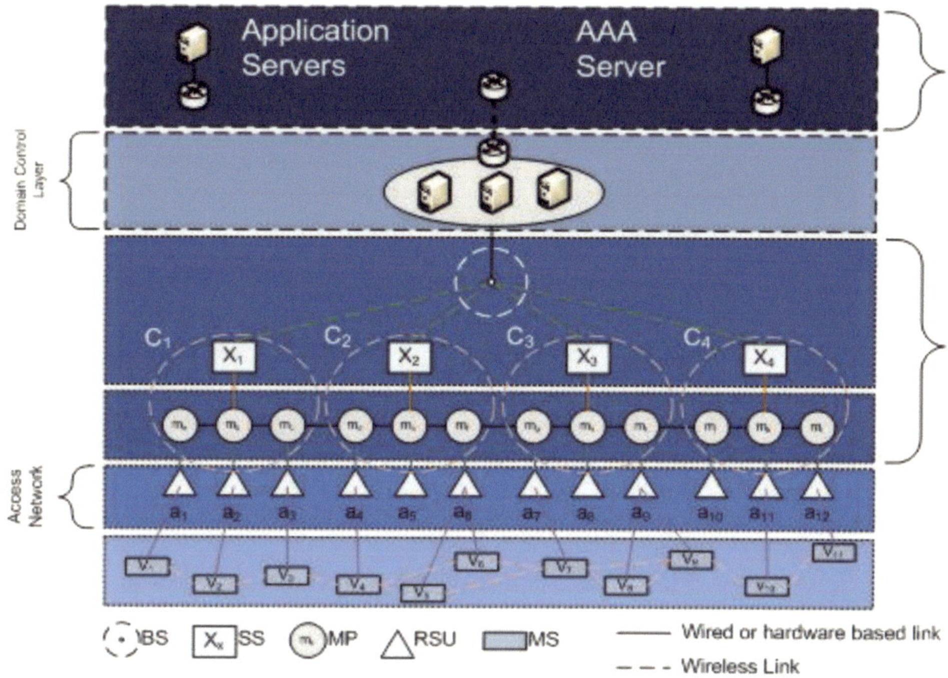

Fig. 8. Conceptual Architecture of [22]

The objective of the architecture is to provide high speed internet access in trains with a soft handoff, and having a minimum of connectivity interruptions. This proposal consists of a three layer topology: Level 0 is an access point functioning under the IEEE 802.11e standard; Level 1 uses base stations (BS) that work in conjunction with the IEEE 802.16m standard and Level 2 enables an optical backbone to interconnect with base stations located alongside the train tracks. Each train possesses two gateway interfaces that serve both as WLAN access points (i.e. IEEE 802.11e) and IEEE 802.16m subscriber stations. Results obtained using the popular NS-2 simulator show that the handoff latency of SWiFT is 52% less than with ipV6 mobile.

The SWiFT protocol can be seen as having a vehicle-to-roadside architecture where, as in [22], there is no possibility of inter-vehicular communications to cause a reduction in network services. Figure 9 shows the architecture of the SWiFT proposal.

In [24], the authors develop a handoff mechanism with a hybrid architecture using the IEEE 802.16e and IEEE 802.16j standards, which also includes control information of the vehicles via V2V. In this handoff mechanism, vehicles leaving their relay vehicle coverage area, called oncoming small size vehicles-OSV, directly transmit the information maintained in layers 2 and 3 to the vehicles outside the coverage area (called broken vehicles) of the relay vehicle. The information passed from OSV to BV is necessary to synchronize communications between the oncoming vehicle and the network.

The NS-2 simulator tool was used in this work and results show that the handoff mechanism developed helped reduce the handoff latency between relay vehicles. Figure 10 shows the topology described in [24] where the relay vehicles, in this case public buses, are equipped with IEEE 802.16j, which is used to register the buses at a base station that

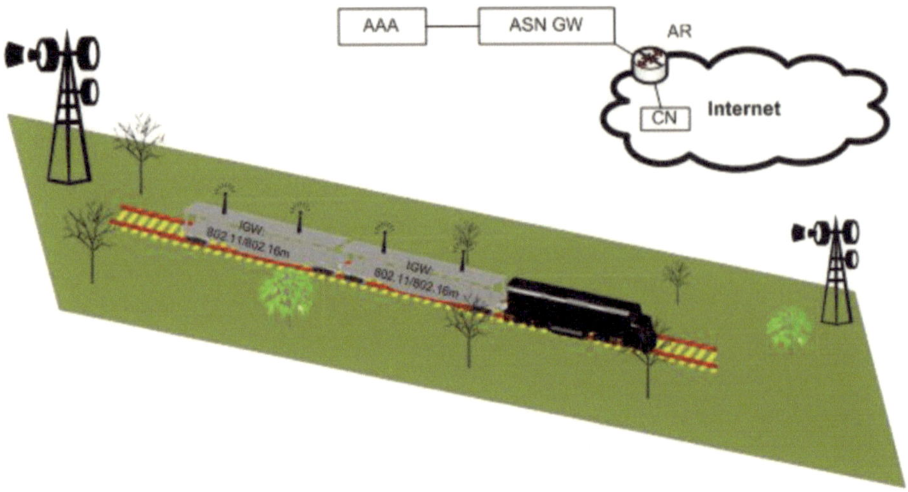

Fig. 9. SWiFT architecture

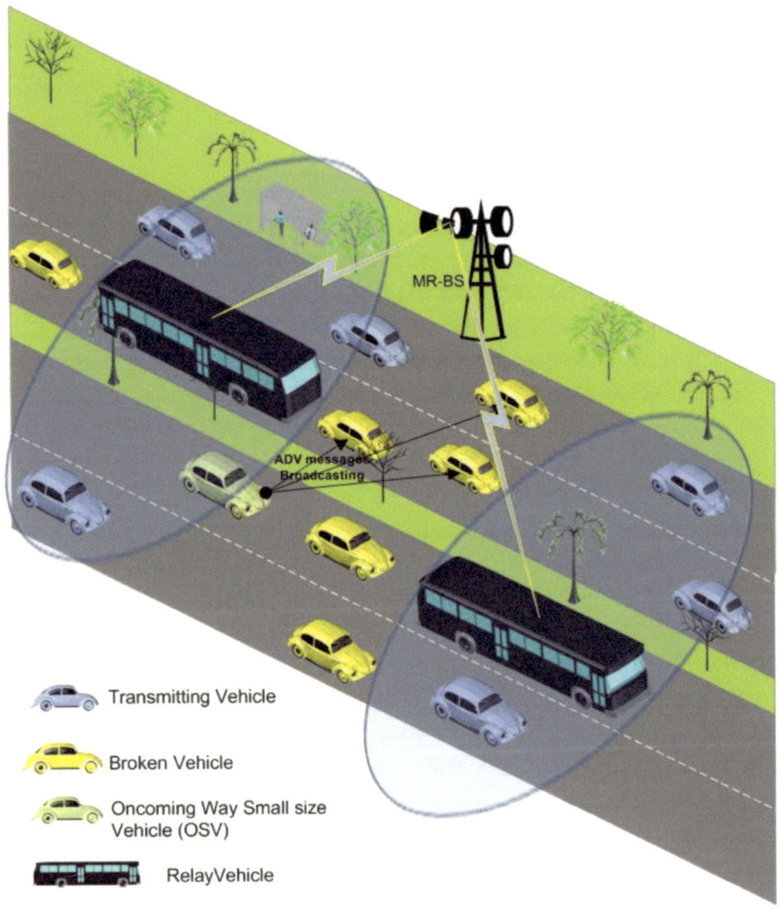

Fig. 10. A handoff with the VFHS mechanism.

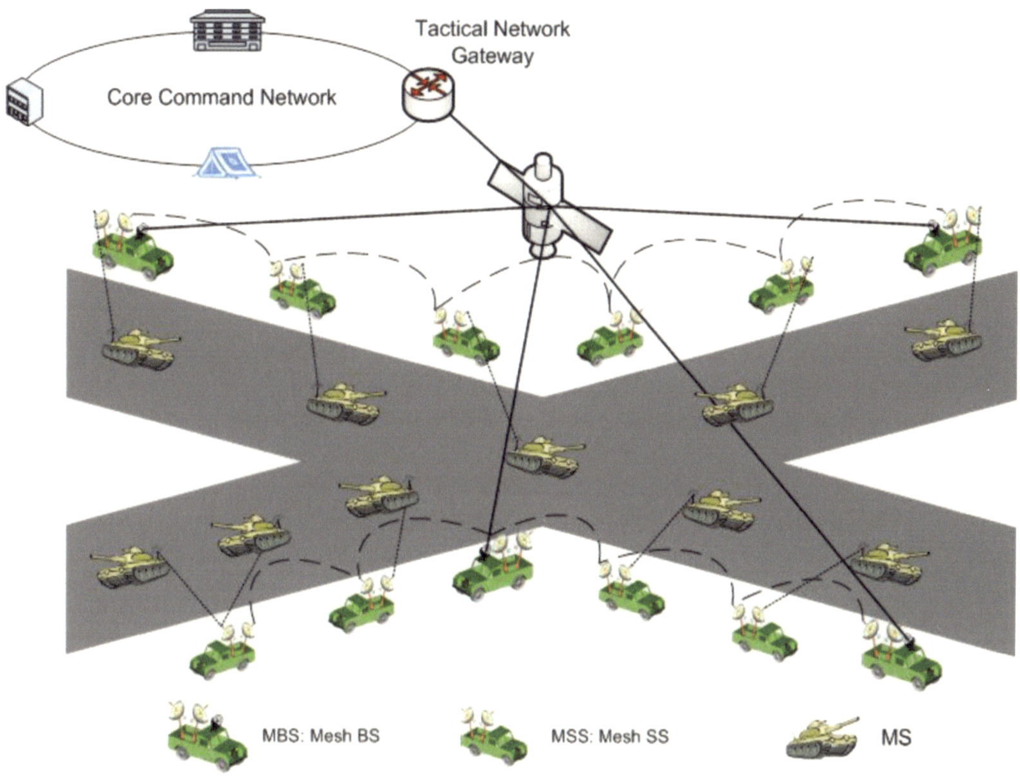

Fig. 11. A possible network deployment using the scheduler mechanism proposed in [21].

functions according to IEEE 802.16e. This proposal does not provide a communications solution for vehicles beyond the RS or BS coverage. Additionally, it does not recommend a routing mechanism to assist nodes select the optimal RV for overlapping coverage areas.

The authors in [25] design a scheduling mechanism called "An interference and QOS aware distributed scheduling approach for hybrid IEEE 802.16e mesh networks," which was obtained using the NS-2 simulator. Their results show that the developed scheduling mechanism facilitates efficient spectral reuse by permitting the deployment of base stations under the IEEE 802.16-2004 mesh standard. Each BS also has an IEEE 802.16e interface that provides access to mobile subscribers. Importantly, the backbone is enabled by satellite communications and their proposal does not provide a routing mechanism to improve network performance. Finally, vehicles outside the coverage area of the BS cannot access network services. Figure 11 shows the topology suggested by [25].

The authors in [26] propose a routing mechanism for Mobile Ad-hoc networks (MANET). This mechanism uses a WiMAX architecture to relay routing information. After the route is enabled by a WiMAX BS, the data is sent through participating nodes.

The researchers in [26] implement their routing mechanism simulating speeds of up to 108 km/h. Their results show that packet delivery is good, but they do not mention the method used to combine the MANET and WiMAX architectures. Also, the simulations varied node densities at a speed of 18 km/h, which is an insufficient velocity for their results to be conclusive. Another important issue concerns nodes leaving the BS coverage area, because network performance can be compromised by node mobility.

In [27], the authors use a roadside architecture based on IEEE 802.16j as shown in Figure 12. They suggest a method to select an optimal relay station. The method proposed is numerical and based on non-linear optimization. Their results show that network capacity can significantly increase; however, the authors do not validate their results via benchmarking or simulation.

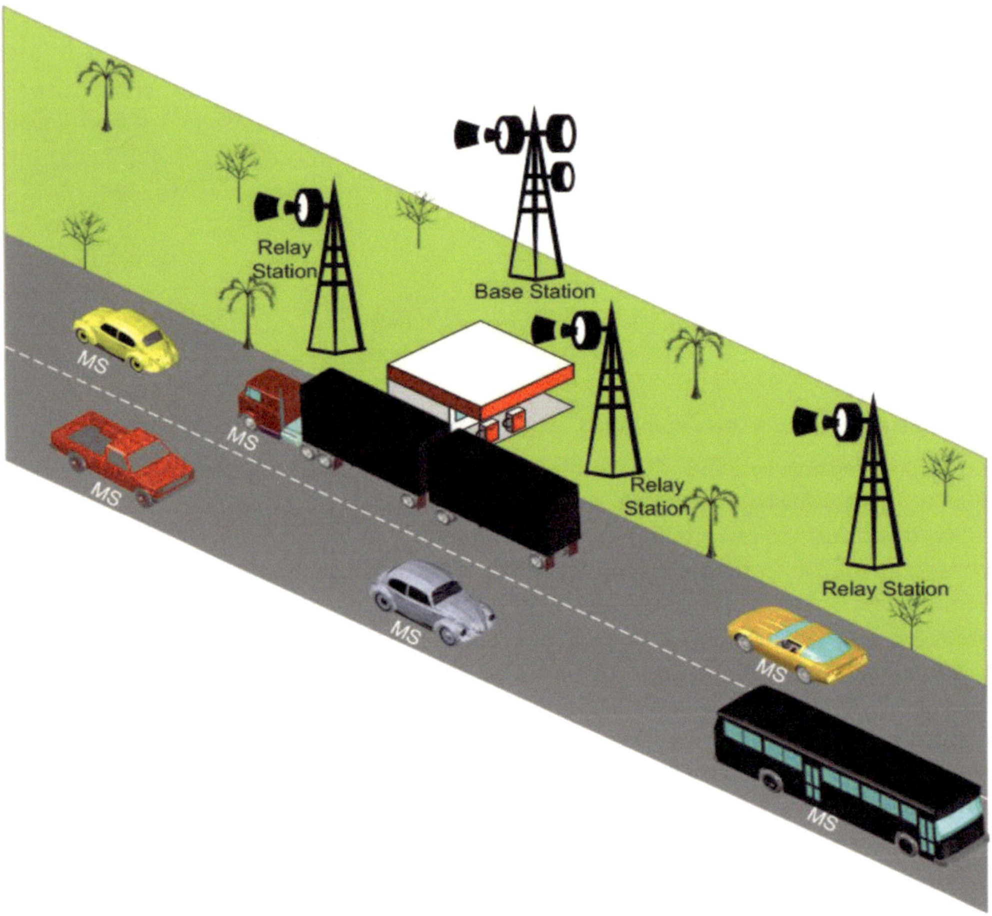

Fig. 12. Underlying Architecture with IEEE 802.16j for [27].

In [28], the authors propose a dynamic bandwidth allocation algorithm with a QoS guarantee for IEEE 802.16j-enabled vehicular networks. They suggest employing an optimization application that uses Lagrange multipliers. Their simulation results, obtained by Mathlab simulation, minimized queue delay and maximized network utilization. They assume that the primary link is the downlink and overlook that vehicular networks consist of both a downlink and uplink, both of which are equally important.

3. Actual challenges and conclusions

Although the proposals reported in this work show that WiMAX is suitable for multi-hop VCN because of its versatility and robustness in the physical and medium access layers, there still remain several technical challenges that must be overcome before deploying WiMAX as the underlying medium access and physical layer in VCN.

This current state of the art reveals that dynamic fully distributed routing mechanism which satisfies the demands of VCN, have not been proposed. Only the proposal by [21] includes a cross-layer hierarchical routing protocol; however, the scheduling mechanism is centralized and is based on the infrastructure, instead of a distributed algorithm. How to best permit intervehicular communication in VCN is still a contentious topic. Equally important are issues related to architecture; more precisely, how to form groups in the absence of the BS (ad-hoc domain) that can still interact with the BS upon demand. Equally important is how to provide control access in the boundaries of the coverage areas and determine which routing mechanism best optimizes network bandwidth. To the best of our knowledge, any proposal that involves WiMAX and VCN can form an ad-hoc domain without a BS or RS. The authors in [21] and [24] suggest a cooperative ad-hoc environment and infrastructure domain; however, the ad-hoc domain must exist with at least one node within the BS coverage area. Another important issue to be resolved is how to allocate bandwidth resource in ad-hoc networks while optimizing their performance. Research carried out by [25] and [28] provides a resource allocation mechanism for multi-hop networks; however, only in presence of a roadside BS.

4. Conclusions

The proposals analyzed in this work suggest that WiMAX can represent a viable alternative for roadside communication using present standards. Importantly, it also has the potential to be used in conjunction with radio technology for inter-vehicular communications because its strong PHY and QoS support. However, there are still significant technical challenges to be overcome before WiMAX can be implemented as radio technology for inter-vehicular communications networks.

Research provided in this chapter shows that integrating WiMAX technology into vehicular ad hoc networks is a very rich area of inquiry, although current research is somewhat limited. We believe that this is because standards for VCN are still in their infancy or have only very recently been published (i.e. IEEE 802.16j/June 2009, and IEEE 802.16m/February 2010[1]). Consequently, we predict there will be much more research carried out in the future as these standards are more fully exploited.

Table 3 shows the most outstanding features of the proposals included in this work [21-28].

Ref	VRC	IVC	PHY	MAC	Red	Sim	802.2	App	Otras. Tec	Año
CEPEC	✓	✓	✓	✓	✓	Prop.	2004	Inter.	⊘	2007
Aspects of roadside backbone	✓	✓	⊘	⊘	✓	⊘	N/A	Back.	⊘	2009
SWIFT	✓	⊘	⊘	✓	⊘	NS2	e,m	Inter.	IEEE 802.11e	2008
A cross layer Fast Handover	✓	⊘	✓	✓	⊘	NS2/NIST	e, j	Hand	⊘	2009
An interference and QoS aware distributed sched.	✓	⊘	⊘	✓	✓	NS2/NIST	E,2004	Schd	⊘	2008
Position based conectionless	⊘	✓	⊘	✓	✓	NS2	2004	Capa.	MANET	2008
Optimal Relay Selection in 16j MHR Vehicular Networks	✓	⊘	⊘	✓	✓	Mod. Mat.	j	Inter.	⊘	2010
Bandwidth Allocation Algorithm with QoS Guarantee for 16j vehicular Networks	✓	⊘	⊘	✓	⊘	Matl.	j	Inter.	⊘	2009

*CEPEC, vehicle-to-vehicle exchange is for control packages only.

Table 3. Outstanding features of the proposals discussed in this paper.

5. References

[1] http://hcmguide.com/

[2] VANET Vehicular Applications and Inter-Networking Technologies (Intelligent Transport Systems) [Hardcover]

[3] M. L. Sichitiu and M. Kihl. Inter-vehicle communication systems: a survey. Communications Surveys & Tutorials, IEEE, 10(2):88-105, 2008.

[4] L. Briesemeister, L. Schäfers, and G. Hommel, "Disseminating Messages Among Highly Mobile Hosts Based on Inter-Vehicle Communication," Proc. IEEE Intelligent Vehicle Symp., 2000, pp. 522–27.

[5] X. Yang, J. Liu, F. Zhao, and N. Vaidya, "A Vehicle-to-Vehicle Communication Protocol for Cooperative Collision Warning," Proc. 1st Annual Int'l. Conf. Mobile and Ubiquitous Sys.: Networking and Services, 2004, pp. 1–4.

[6] J. Yin et al., "Performance Evaluation of Safety Applications over DSRC Vehicular Ad Hoc Networks," Proc. 1st ACM Wksp. Vehic. Ad Hoc Networks, 2004, pp. 1–9.

[7] R. Rajamani and S. Shladover, "An Experimental Comparative Study of Autonomous and Co-Operative Vehicle-Follower Control Systems," Transportation Research Part C, vol. 9, 2001, pp. 15–31.

[8] P. Varayia, "Smart Cars on Smart Roads: Problems of Control," IEEE Trans. Automatic Control, vol. 38, no. 2, 1993, pp. 195–207.

[9] A. Brown et al., "Vehicle to Vehicle Communication Outage and Its Impact on Convoy Driving," Proc. IEEE Intelligent Vehicle Symp., 2000, pp. 528–33

[10] SAFESPOT, http://www.safespot-eu.org

[11]COMeSafety, http://www.comesafety.org

[12] Car2car Communication Consortium, http://www.car-tocar.org

[13] Jiang, D.; Delgrossi, L.; , "IEEE 802.11p: Towards an International Standard for Wireless Access in Vehicular Environments," Vehicular Technology Conference, 2008. VTC Spring 2008. IEEE , vol., no., pp.2036-2040, 11-14 May 2008

[14] Evaluation of the IEEE 802.11p MAC Method for Vehicle-to-Vehicle Communication, 2008. [Online].Available: http://dx.doi.org/10.1109/VETECF.2008.446

[15] Yi Wang; Ahmed, A.; Krishnamachari, B.; Psounis, K., "IEEE 802.11p performance evaluation and protocol enhancement", Vehicular Electronics and Safety, 2008. ICVES 2008. IEEE International Conference on, pp. 317 - 322, 2008

[16] L. Stibor, Y. Zang, and H.-J. Reumerman, "Evaluation of communication distance of broadcast messages in a vehicular ad-hoc network using IEEE 802.11p," in Proceedings of the IEEE Wireless Communications and Networking Conference (WCNC '07), pp. 254-257, Kowloon, China, March 2007.

[17] IEEE. Standard 802.16e-2005. Part16: Air interface for fixed and mobile broadband wireless access systems—Amendment for physical and medium access control layers for combined fixed and mobile operation in licensed band. December 2005.

[18] 'IEEE 802.16j Mobile Multi-hop Relay Project Authorization Request (PAR)', Official IEEE 802.16j Website: http://standards.ieee.org/board/nes/projects/802-16j.pdf, March 2006.

[19] IEEE. Standard 802.16-2004. Part16: Air interface for fixed broadband wireless access systems. October 2004.

[20] S. W. Peters and R. W. Heath, \The future of wimax: Multihop relaying with ieee 802.16j," Communications Magazine, IEEE, vol. 47, no. 1, pp. 104{111, 2009. [Online]. Available: http://dx.doi.org/10.1109/MCOM.2009.4752686

[21] K. Yang, S. Ou, H. Chen, J. He, "A Multihop Peer Communication Protocol With Fairness Guarantee for IEEE 802.16 Based Vehicular Networks", IEEE Trans. Veh. Technol., Vol. 56, No. 6, Nov. 2007. Page(s): 3358-3370.

[22] Krohn, M.; Daher, R.; Arndt, M.; Tavangarian, D.; , "Aspects of roadside backbone networks," Wireless Communication, Vehicular Technology, Information Theory and Aerospace & Electronic Systems Technology, 2009. Wireless VITAE 2009. 1st International Conference on , vol., no., pp.788-792, 17-20 May 2009

[23] K. R. Kumar, P. Angolkar, D. Das, R. Ramalingam, "SWiFT A Novel Architecture for Seamless Wireless Internet for Fast Trains",IEEE In Vehicular Technology Conference, 2008. VTC Spring 2008. IEEE (2008), pp. 3011-3015.

[24] Kuan-Lin Chiu, Ren-Hung Hwang, Yuh-Shyan Chen, "A Cross Layer Fast Handover Scheme in VANET", IEEE International Conference on Communications (IEEE ICC 2009), Dresden, Germany, June 14-18, 2009.

[25] Amin, R. Kuang-Ching Wang Ramanathan, "An interference and QOS aware distributed scheduling approach for hybrid IEEE 802.16E mesh networks", Military Communications Conference, 2008. MILCOM 2008. IEEE,

[26] Hsien- Chou Liao and Cheng- Jung Lin, "A Position-Based Connectionless Routing Algorithm for MANET and WiMAX under High Mobility and Various

Node Densities", Information Technology Journal, ANSI Journal, pp. 458-465, 2008

[27] Ge, Y. Wen, S. Ang, Y.-N. Liang, Y.-C., "Optimal Relay Selection in IEEE 802.16j Multihop Relay Vehicular Networks", Vehicular Technology, IEEE Transactions on, 04 marzo 2010

[28] Ridong Fei Kun Yang Shumao Ou Shaochun Zhong Lixun Gao, "A Utility-Based Dynamic Bandwidth Allocation Algorithm with QoS Guarantee for IEEE 802.16j-Enabled Vehicular Networks", Scalable Computing and Communications; Eighth International Conference on Embedded Computing, 2009. SCALCOM-EMBEDDEDCOM'09. International Conference on, pp. 200-205, 30 noviembre 2009

2

Communications in Vehicular Networks

Zaydoun Yahya Rawashdeh and Syed Masud Mahmud
Wayne State University, Detroit, Michigan, USA

1. Introduction

Recent advances in wireless networks have led to the introduction of a new type of networks called Vehicular Networks. Vehicular Ad Hoc Network (VANET) is a form of Mobile Ad Hoc Networks (MANET). VANETs provide us with the infrastructure for developing new systems to enhance drivers' and passengers' safety and comfort. VANETs are distributed self organizing networks formed between moving vehicles equipped with wireless communication devices. This type of networks is developed as part of the Intelligent Transportation Systems (ITS) to bring significant improvement to the transportation systems performance. One of the main goals of the ITS is to improve safety on the roads, and reduce traffic congestion, waiting times, and fuel consumptions. The integration of the embedded computers, sensing devices, navigation systems (GPS), digital maps, and the wireless communication devices along with intelligent algorithms will help to develop numerous types of applications for the ITS to improve safety on the roads. The up to date information provided by the integration of all these systems helps drivers to acquire real-time information about road conditions allowing them to react on time. For example, warning messages sent by vehicles involved in an accident enhances traffic safety by helping the approaching drivers to take proper decisions before entering the crash dangerous zone (ElBatt et al., 2006) (Xu et al., 2007). And Information about the current transportation conditions facilitate driving by taking new routes in case of congestion, thus saving time and adjusting fuel consumption (Dashtinezhad et al., 2004) (Nadeem et al., 2004). In addition to safety concerns, VANET can also support other non-safety applications that require a Quality of Service (QoS) guarantee. This includes Multimedia (e.g., audio/video) and data (e.g., toll collection, internet access, weather/maps/ information) applications.

Vehicular networks are composed of mobile nodes, vehicles equipped with On Board Units (OBU), and stationary nodes called Road Side Units (RSU) attached to infrastructure that will be deployed along the roads. Both OBU and RSU devices have wireless/wired communications capabilities. OBUs communicate with each other and with the RSUs in ad hoc manner. There are mainly two types of communications scenarios in vehicular networks: Vehicle-to-Vehicle (V2V) and Vehicle-to-RSU (V2R). The RSUs can also communicate with each other and with other networks like the internet as shown in Figure 1.

Vehicular Networks are expected to employ variety of advanced wireless technologies such as Dedicated Short Range Communications (DSRC), which is an enhanced version of the WiFi technology suitable for VANET environments. The DSRC is developed to support the data transfer in rapidly changing communication environments, like VANET, where time-critical responses and high data rates are required.

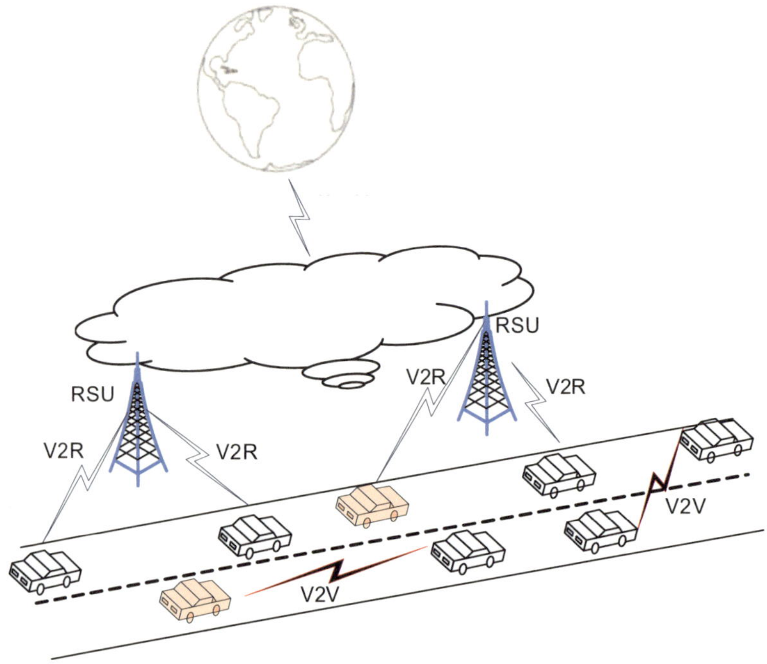

Fig. 1. VANET architecture

A number of technical challenges need to be addressed to make Vehicular networks more efficient to provide services to drivers and passengers (Torrent-Moreno et al., 2005). These challenges came from the unique features and characteristics (like frequent topology change, abundant of nodes, etc) of the vehicular networks. However, due to these unique characteristics, the standard MANET communication protocols are inefficient in the VANET environment. Therefore, the new communication mechanisms, like media access, data dissemination, routing, etc., designed for VANET should consider these unique characteristics to provide reliable communications. The Media Access Control (MAC) mechanisms should support fast link establishment and low latency communications to ensure the service reliability for safety applications considering the time constraints required by this type of applications. The data dissemination techniques should be designed to efficiently deliver the safety data to the intended receivers on time. Safety messages are of a broadcast nature targeting vehicles in a certain geographic area. Therefore, safety message dissemination mechanisms should deal with different types of network densities to eliminate the redundant rebroadcasted data, especially in very high network density scenarios. The frequent topology change characteristics pose another challenge for routing methods in VANET. In addition to the traditional routing challenges like broadcast problems, the VANET routing algorithms should be designed to ensure the quality and continuity of services for non-safety applications with high probability. Other challenges related to security and data managements should also be studied in depth in VANET.

The main objective of this book chapter is to introduce the reader to the main applications used in vehicular networks, the main characteristics of VANETs, and the challenges associated with the designing of new VANET communication protocols. All these will be covered in the context of the MAC, data dissemination, and routing mechanisms in VANET.

2. VANET characteristics and challenges

VANETs are characterized by their unique characteristics that distinguish them from MANET. These special characteristics can be summarized as follows:

1. **High mobility**: VANET nodes are characterized by their high relative speed which makes VANET environment high dynamic.
2. **Predictable and restricted mobility patterns**: Unlike the random mobility of MANET, VANET node movements are governed by restricted rules (traffic flow theory rules), which make them predictable at least on the short run.
3. **Rapid topology change**: VANET nodes are characterized by their high speed. This leads to frequent network topology changes, which introduces high communication overhead for exchanging new topology information.
4. **No power constraints**: Each vehicle is equipped with a battery that is used as an infinite power supply for all communications and computation tasks.
5. **Localization**: Vehicles can use the Global Positioning System (GPS) to identify their locations with high accuracy.
6. **Abundant network nodes**: Unlike MANETs that are characterized by a small network sizes, VANET networks can be very large due to high density of the vehicles.
7. **Hard delay constraints**: Safety messages are the main goal of VANETs. Therefore, safety messages should be given high priority and must be delivered on time.

The above unique characteristics create new challenges that need to be resolved in the vehicular network environments. According to (Torrent-Moreno et al., 2005), the main challenges of the vehicular networks can be summarized as follows:

- Frequent neighbourhood change due to high mobility.
- Increasing channel load (high density environment).
- Irregular connectivity due to the variation of the received signal power.
- Packet loss due to exposed and hidden terminal problems.

However, lots of efforts have been made to resolve these issues. The literature contains a huge amount of studies addressing these challenges in all aspects. The studies tried to address all layers related issues ranging from lower layers (physical and MAC layers) enhancement to upper layers (application) developments.

2.1 DSRC technology

DSRC is an emerging technology developed based on the WiFi standards. The DSRC technology will be used in the ITS domain to provide secure and reliable communication links among vehicles and between vehicles and infrastructure. These communication links allow the transfer of data that are necessary for the operation of different ITS applications. The DSRC is developed to work in very high dynamic networks to support fast link establishments and to minimize communication latency. Mainly, the DSRC is designed to ensure the service reliability for safety applications taking into account the time constraints for this type of applications. It can also support other non-safety applications that require a Quality of Service (QoS) guarantee. DSRC is developed for the environments where short time response (less than 50 msec.) and/or high data rates are required in high dynamic networks.

2.2 Characteristics of DSRC spectrum and data rates

In the United States, the Federal Communications Commission has allocated the 5.9 GHz Dedicated Short Range Communications (DSRC) (Xu et al., 2004) technique to support

public safety and commercial applications in V2V and V2R communication environments. The 5.9 GHz (5.850-5.925) band is divided into seven non-overlapping 10 MHz channels as shown in Figure. 2. One channel is called the control channel, and the other six are called service channels. The channels at the edges are reserved for future use. The control channel is used to broadcast safety data like warning messages to alert drivers of potential dangerous conditions. It can also be used to send advertisements about the available services, which can be transferred over the service channels. The service channels are used to exchange safety and non-safety data like announcements about the sales in nearby malls, video/audio download, digital maps, etc. Vehicles, using service channels, can relay the received data to other vehicles in other regions or/and to the roadside units.
The DSRC supports different data transfer rates: 6, 9, 12, 18, 24, and 27 Mbps with 10 MHz channels. The data rate can be increased to 54 Mbps with 20 MHz channels. Switching between the different data transfer rates can be achieved by changing the modulation schemes and channel code rate.

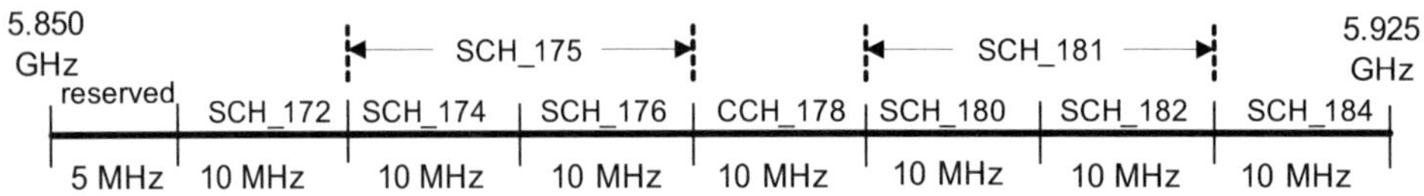

Fig. 2. DSRC channels

2.3 Applications and DSRC data traffic requirements

Numerous applications enabled by the DSRC technology have been proposed for VANET. These applications are categorized as Safety and Non-safety applications and installed on the OBUs and RSUs to process the safety and non-safety data. Different applications have different requirements. Safety messages are given higher priority over the non-safety data. Safety messages are time sensitive and should be disseminated to the vehicles in the surrounding area of the event within a bounded time. Safety messages are of a broadcast nature, therefore smart dissemination strategies should be employed to ensure the fast delivery of these messages. Safety messages in DSRC are either event-driven or periodic-based. For example, event-based safety messages are high priority messages generated and sent by vehicles involved in an accident to warn the vehicles approaching the accident area. On the other hand, periodic safety messages are considered preventive safety methods sent at specific intervals (e.g., every T second). The periodic messages carry the current status information like velocity, acceleration, direction, etc of the vehicles. These information are used by vehicles in the neighbourhood to update the status of their neighbouring vehicles. Periodic safety messages can also be sent by the RSU e.g., the RSU installed at the intersection periodically sends messages about the intersection conditions. The non-safety applications have different goals and can be used to provide a number of services ranging from transportation management, toll collection, infotainment, music download, to commercial advertisements. The non-safety data have low priority compared to the safety data. Table I shows requirements for various DSRC applications.
The DSRC also supports a number of different network protocols, which gives it the ability to interact with different types of networks. The DSRC supports the TCP/IP as well as IPv6 protocols. Thus, the internet applications/services can also be available in the VANET. Moreover, IP-based routing can also be enabled in VANET.

Applications	Packet Size (bytes)/ Bandwidth	Latency (ms)	Network Data Type	Application Range (m)	Priority
Intersection Collision Warning / Avoidance	~100	~100	Event	300	Safety of life
Cooperative Collision Warning	~100/ 10Kbps	~100	Periodic	50 - 300	Safety of life
Work Zone Warning	~100/ 1Kbps	~1000	Periodic	300	Safety
Transit Vehicle Signal Priority	~100	~1000	Event	300 - 1000	Safety
Toll Collections	~100	~50	Event	15	Non - Safety
Service Announcments	~100/ 2Kbps	~500	Periodic	0 - 90	Non - Safety
Movie Download (2 hours of MPEG 1)	>20 Mbps	NA	NA	0 - 90	Non - Safety

Table 1. DSRC Application Requirements (Xu et al., 2004)

2.4 VANET applications enabled by DSRC

In the context of the VANET applications, new types of applications enabled by the DSRC have been developed. These applications will benefit from the integration of different hardware components (CPUs, Wireless transceivers, Sensors, Navigations Systems, and input and output devices) that will be embedded in future vehicles. In the followings, we summarize some of the different VANET applications:

2.4.1 Safety applications:

The main goal of the safety applications is to increase public safety and protect the loss of life. The main characteristic of these applications is that the safety data should be delivered to the intended receivers (vehicles approaching the dangerous area) within a bounded time. The Vehicles Safety Communication (VSC) project has defined 34 different safety applications to work under the DSRC technology (Xu et al., 2004). These applications were studied in depth to determine the potential benefit provided by them. In the following, we present some of the applications that, according to the VSC, provide the greatest benefit in terms of safety of life.

2.4.1.1 Cooperative Collision Avoidance (CCA):

The main goal of this application is to prevent collisions. This type of safety applications will be triggered automatically when there is a possibility of collisions between vehicles. Vehicles, upon detecting a possible collision situation, send warning messages to alert the drivers approaching the collision area. The drivers can take the proper actions or the vehicle itself can stop or decrease the speed automatically. Another scenario where the CCA are of great importance is to avoid crashes during lane change. The CCA messages are disseminated to vehicles approaching the collision area. One of the proposed techniques to disseminate CCA messages on a highway was presented in (Biswas et al., 2006).

2.4.1.2 Emergency Warning Messages (EWM):

This type of applications is similar to the CCA. However, depending on the type of the emergency event, the EWMs either vanish once they are disseminated or may reside in the relevant area for longer period of time. For example, when vehicles detect an accident they start to send EWMs to warn vehicles that are close to the accident area. Another example is when vehicles sense a dangerous road conditions they send EWMs to other vehicles in a certain area, and these vehicles disseminate the EWMs to the new vehicles entering that area. Some of the proposed techniques are presented in (Yang et al., 2004) (Maih¨ofer, 2003) (Yu & Heijenk, 2008).

2.4.1.3 Cooperative Intersection Collision Avoidance (CICA):

This type of applications will be used to avoid collisions at the intersections (signalized or non-signalized). Mainly, an RSU installed at the intersection periodically distributes the state of the intersection to the approaching vehicles. The distributed information includes: 1) Traffic Signal State (e.g., red, green, yellow, and time remaining until the traffic switches to a new state). 2) State of the vehicles approaching the intersection that are within a relevant distance/time from the intersection (e.g. location, speed, and so on). 3) Intersection environmental conditions (e.g., weather, visibility, road surface at the intersection, and so on). Several of the works about the intersection collision avoidance can be found in (Tong et al., 2009) (Benmimoun et al., 2005).

2.4.2 Traffic managements:

This type of applications is used to facilitate traffic flow, thus reducing traffic congestion, fuel consumption, and travel time. This type of applications is less strict on real-time constraints. This means that if the messages are delayed, there is no real threat to life (no collision to occur), as opposed to the safety messages where a real threat to the life may occur if the messages are delayed. The information provided by these applications mainly describes the status of the traffic in a certain areas like intersection or road constructions. In this kind of applications, vehicles cooperate to generate messages;. These messages are aggregated and sent, using inter-vehicle communications, in a multi-hop manner to other vehicles in other geographic areas. Some of the papers that discuss the aggregation and forwarding are (Kihl et al., 2008).

2.4.3 Advertisements, entertainment and comfort applications:

The goal of this type of applications is to provide comfort and entertainments to the passengers. The advertisement applications have commercial purposes. The data of this type of applications should not consume the bandwidth on the count of the safety data. The priority should always be given to the safety data. Some of these applications are:

2.4.3.1 Electronic toll collection:

Using this service, the drivers don't need to stop and make the payment; instead the payment is done electronically through the network.

2.4.3.2 Entertainment Applications:

Multimedia files (music, movies, news, e-books, and so on) can be uploaded to vehicles. These data can also be transferred from one vehicle to another. Information about local restaurants, hotels, malls, gas stations can be uploaded to the vehicles and can also be exchanged among vehicles using the inter-vehicular networks to facilitate travelling.

2.4.3.3 Internet Access:

Passengers can browse the internet and send/receive emails (Zhang et al., 2007). Most of these applications will be downloaded from other networks (like internet). However, vehicles use the inter-vehicle networks to distribute these information to reduce the cost associated with the installation of the infrastructure along the roads (Zhang et al., 2007) (Wang, 2007).

2.5 Wireless Access in Vehicular Environment (WAVE) stack

Lots of efforts have been made to design new standards for the services and the interfaces for VANET. These standards form the basis for wide range of applications in the vehicular network environments. Recently, a trial of a set of standardized services and interfaces defined under WAVE stack has been released. These services and interfaces cooperatively enable a secure V2V and V2R communications in a rapidly changing communications environment, where communications and transactions need to be completed in a short time frame. The WAVE architecture is developed based on the IEEE 802.11p and the IEEE P1609 standards (Nadeem, 2004). The IEEE 802.11p deals with the physical and Media Access Control layers, whereas the IEEE 1609 deals with the higher-layer protocols. In this section we try to give a background on these standards especially the MAC protocol.

2.5.1 The IEEE 1609 family of standards for WAVE

The IEEE has defined four standards and released them for a trial use (IEEE, 2007). Figure 3 shows the architecture of the WAVE family of standards, while Figure 4 shows the IEEE protocol architecture for vehicular communications (IEEE, 2007). These standards can be defined as follows:

2.5.1.1 IEEE 1609.1: Resource Manager

This standard defines the services and the interfaces of the WAVE Resource Manager applications. It describes the message formats and the response to those messages. It also describes data storage format that is used by applications to access other architectures.

2.5.1.2 IEEE 1609.2: Security Services

This standard defines security and secure message formatting and processing. It also defines how secure messages are exchanged.

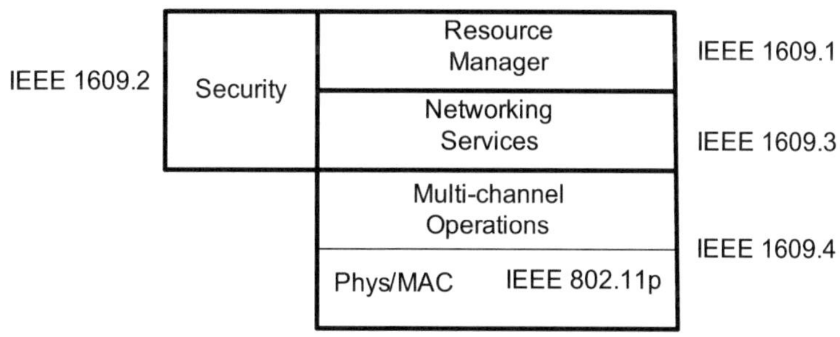

Fig. 3. IEEE WAVE Stack for trial use

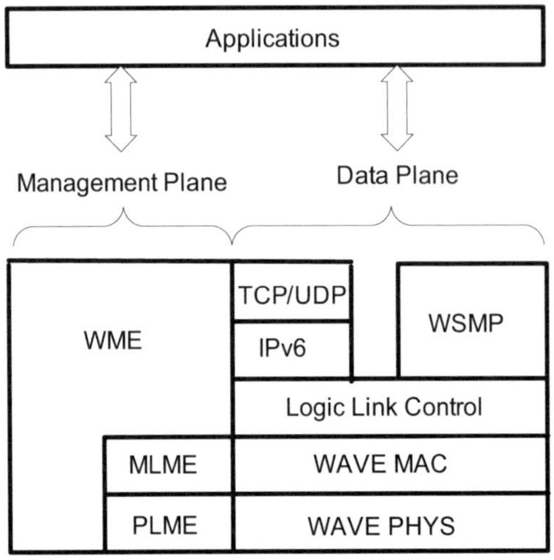

Fig. 4. IEEE protocol architecture for vehicular communications (IEEE, 2007).

2.5.1.3 IEEE 1609.3: Networking Services

This standard defines routing and transport layer services. It also defines a WAVE-specific messages alternative to IPv6 that can be supported by the applications. This standard also defines the Management Information Base (MIB) for the protocol stack.

2.5.1.4 IEEE 1609.4: Multi-Channel Operations

Multi-Channel Operations: This standard defines the specifications of the multi-channel in the DSRC. This is basically an enhancement to the IEEE 802.11a Media Access Control (MAC) standard.

2.5.2 The IEEE 802.11p MAC protocol for VANET

A new MAC protocol known as the IEEE 802.11p is used by the WAVE stack. The IEEE 802.11p basic MAC protocol is the same as IEEE 802.11 Distributed Coordination Function (DCF), which uses the Carrier Sense Multiple Access/Collision Avoidance (CSMA/CA) method for accessing the shared medium. The IEEE 802.11p MAC extension layer is based on the IEEE 802.11e (IEEE, 2003) that uses the Enhanced Distributed Channel Access (EDCA) like Access Category (AC), virtual station, and Arbitration Inter-Frame Space (AIFS). Using EDCA, the Quality of Service (QoS) in the IEEE 802.11p can be obtained by classifying the data traffic into different classes with different priorities.

The basic communication modes in the IEEE 802.11p can be implemented either using broadcast, where the control channel (CCH) is used to broadcast safety critical and control messages to neighbouring vehicles, or using the multi-channel operation mode where the service channel (SCH) and the CCH are used. The later mode is called the WAVE Basic Service Set (WBSS). In the WBSS mode, stations (STAs) become members of the WBSS in one of two ways, a WBSS provider or a WBSS user. Stations in the WAVE move very fast and it's very important that these stations establish communications and start transmitting data very fast. Therefore, the WBSSs don't require MAC sub-layer authentication and association (IEEE, 2007). The provider forms a WBSS by broadcasting a WAVE Service advertisement (WSA) on the CCH. The WSA frame contains all information including the service channels

(SCH) that will be used for the next SCH interval. After receiving the WBS advertisement, the user joins the WBSS, and at the beginning of the next SCH interval, both the provider and the user switch to the chosen SCH to start data exchange. Since the provider and the user keep jumping between CCH and SCH, the provider can send a WSA frames during the CCH to let other users detect and join the WBSS. The users have the option to join the WBSS. The user can also receive other WBS frames while listening to the CCH to update the operational parameters of existing WBSSs. Once the provider and the user finish sending out all data frames, the provider ends the WBSS and the user also leaves the WBSS when no more data frames are received from the provider.

2.5.3 Media access control in VANET

Different MAC schemes targeting VANET have been proposed in the literature. Mainly, these schemes are classified as probability based and time based.

2.5.3.1 Probability-based MAC schemes

This type of media access control uses CSMA/CA technique to access the media. The advantage of this method is that vehicle movements don't cause any protocol reconfiguration. However, using this type of media access doesn't provide guarantee on a bounded access delay. Therefore, one of the main challenges of this method is to limit the access delay. The rest of this section presents a summary of three MAC schemes developed based on CSMA method.

The authors of (Zang et al., 2007) proposed a congestion detection and control architecture for VANET. The authors divided the messages into beacons (background data) having lower priority, and event driven alert messages with higher priority. One of the congested control methods is the adaptive QoS that deals with traffic of different types. The main goal of this work is to prevent the channel from being exhausted by the lower priority traffic (e.g., background beacon messages). The paper presented a congestion detection method called measurement based congestion detection, where nodes sense the usage level of the channel. The authors adopted a technique similar to the IEEE 802.11e to prioritize the traffic. In this technique the transmission queues are mapped to traffic with different priorities (access categories). The basic concept of the QoS adaptive method is to reserve a fraction of the bandwidth for safety applications. The authors defined three thresholds for the channel usage value.

1. If 95% of the total channel usage has been exceeded, then all output queues, except the safety message queue, are closed.
2. If 70% of the total channel usage has been exceeded, then the contention window size is doubled for all queues except for the safety message queue.
3. If the total channel usage becomes less the 30%, then the contention window of all queues is halved.

This work mainly uses the access category concept that is considered the core of the IEEE 802.11e. The work was implemented using one type of safety messages. It didn't show how to prioritize safety messages among themselves (which safety messages have higher priority than others when they attempt to access the media at the same time).

Another media access method called Distributed Fair Transmit Power Adjustment for Vehicular Ad hoc Networks (D-FPAV) was proposed in (Torrent-Moreno, 2006). The authors focused on adjusting the transmission power of periodic messages, and tried to keep the transmission power under a certain predefined threshold called Maximum

Beaconing Load (MBL). Thus, using this technique a certain amount of the overall bandwidth can be kept to handle unexpected situations. The authors tried to compromise between increasing the transmission power to ensure safety (increasing power means increasing transmission range, which means more receivers can be reached), and reducing it to avoid packet collisions. The authors used the centralized approach algorithm presented in (Moreno et al., 2005) to build the D-FPAV presented in (Torrent-Moreno, 2006). The algorithm in (Moreno et al., 2005) works as follows: every node in the network starts an initial minimum transmit power, then during every step, all nodes in the network start increasing their transmission power by an increment ε as long as MBL is not exceeded. Then, after this phase, each node finds the optimal transmit power value. Based on this, the authors proposed the D-FPAV that works for node *u* as follows:

- Based on the current state of the vehicles in the Carrier Sense (CS) range, use the FPAV to calculate the transmission power level P_i such that the MBL is not violated at any node.
- Send P_i to all vehicles in the transmit range.
- Receive messages and collect the power level calculated by all vehicles.
- Assign the final power level according to the following equation:

$$PA_i = \min\{P_i, min_{j:u_i \in CS_{MAx}(j)}\{P_j\}\} \quad (1)$$

Whereas $CS_{MAx}(j)$ is the carrier sense range of node *j* at the max power. The proposed work relies on adjusting the transmission power of the periodic messages. However, reducing the transmission power makes the coverage area small, which reduces the probability of receiving periodic messages by distant nodes.

In (Yang et al., 2005), the authors proposed a CSMA-based protocol, which gives different priority levels to different data types. The authors use different back-off time spacing (TBS) to allow the higher priority traffic to access the media faster than those with lower priorities. The TBS is inversely proportional to the priority such that high priority packets are given shorter back-off time before a channel access attempt is made. However, this type of prioritization mechanism was implemented in the IEEE 802.11e (IEEE, 2003). The paper also proposes another feature in which a receiving vehicle polls vehicles in its proximity. If a polled vehicle's data is ready for transmission, then the vehicle generates a tone indicating that state. Upon receiving the tone, the receiving vehicle clears it to transmit the packets (Yang et al., 2005). However, even with the use of busy tones, there is no upper bound on which channel access can take place.

2.5.3.2 Time-based MAC schemes

The time-based scheme is another approach to control the media access. In this approach, the time is divided into frames, which are divided into time slots. This approach is called Time Division Multiple Access (TDMA). The TDMA mechanism is a contention free method that relies on a slotted frame structure that allows high communication reliability, avoids the hidden terminal problem, and ensures, with high probability, the QoS of real-time applications. The TDMA technique can guarantee an upper limit on the message dissemination delay, the delay is deterministic (the access delay of messages is bounded) even in saturated environments. However, this technique needs a complex synchronization procedure (e.g., central point to distribute resources fairly among nodes). Some of the time-based methods use distributed TDMA for media access (Yu & Biswas, 2007), while most of the others use centralized structure like the clustering techniques (Su & Zhang, 2007) (Rawashdeh & Mahmud, 2008). Some of the time-based approaches used in VANET are summarized as follows:

The authors of (Yu & Biswas, 2007) proposed a distributed TDMA approach called Vehicular Self-Organizing MAC (VeSOMAC) that doesn't need virtual schedulers such as leader vehicle. The time is divided into transmission slots of constant duration τ, and the frame is of duration T_{frame} sec. Each vehicle must send at least one packet per frame, which is necessary for time slot allocation. Vehicles use the bitmap vector included in the packet header for exchanging slot timing information. Each bit in the bitmap vector represents a single slot inside the frame (1 means the slot is in use, 0 means it's free). Vehicles continuously inform their one-hop neighbours about the slot occupied by their one-hop neighbours. Vehicles upon receiving the bitmap vector can detect the slot locations in the bitmap vector for their one-and two-hop neighbours, and based on this they can choose the transmission slots such that no two one-hop or two-hop neighbours' slot can overlap. The authors proposed an iterative approach, using acknowledgments through the bitmaps, to resolve the slot collision problem. The idea is to have each vehicle move its slot until no collision is detected. The vehicles detect the collision as follows: each vehicle upon joining the network marks its slot reservation and inform its neighbours. Upon receiving a packet from a neighbouring node, the vehicle looks at its time slot. If the time slot is marked, by the neighbouring node, as occupied, then the vehicle knows that the reservation was successful. If the time slot is marked as free, then this means a collision occurred and the reservation was not successful. However, this approach is inefficient when the number of the vehicles exceeds the number of time slots in a certain area.

In (Su & Zhang, 2007), the authors try to make best use of the DSRC channels by proposing a cluster-based multi-channel communication scheme. The proposed scheme integrates clustering with contention-free and/or -based MAC protocols. The authors assumed that each vehicle is equipped with two DSRC transceivers that can work simultaneously on two different channels. They also redefined the functionality of the DSRC channels. In their work, the time is divided into periods that are repeated every T msec. Each period is divided into two sub-periods to upload and exchange data with the cluster-head. After the cluster-head is elected by nearby nodes, the cluster-head uses one of its transceivers, using the contention free TDMA-based MAC protocol, to collect safety data from its cluster members during the first sub-period, and deliver safety messages as well as control packets to its cluster members in the second sub-period. The cluster-head uses the other transceiver to exchange the consolidated safety messages among nearby cluster-head vehicles via the contention-based MAC protocol. However, this method is based on the assumption that each vehicle is equipped with two transceivers. The authors also redefined the functionality of all DSRC channels such that each channel is used for a specific task.

In (Rawashdeh & Mahmud, 2008), the authors proposed a hybrid media access technique for cluster-based vehicular networks. The proposed method uses scheduled-based approach (TDMA) for intra-cluster communications and managements, and contention-based approach for inter-cluster communications, respectively. In the proposed scheme, the control channel (CTRL) is used to deliver safety data and advertisements to nearby clusters, and one service channel (SRV) is used to exchange safety and non- safety data within the cluster. The authors introduced the so called system cycle that is divided into Scheduled-Based (SBP) and Contention-Based (CBP) sub-periods and repeated every T msec. The system cycle is shared between the SRV channel and CTRL channels as shown in Figure. 5.

The SRV channel consists of Cluster Members Period (CMP) and Cluster Head Period (CHP). CMP is divided into time slots. Each time slot can be owned by only one cluster member. The end of the CHP period is followed by the CBP period during which CRL is used. At the beginning of each cycle, all vehicles switch to the SRV channel. During CMP, each cluster member uses its time slot to send its status, safety messages and advertisements. The CHP period follows the CMP and is allocated to the cluster-head to process all received messages and to respond to all cluster members' requests. Vehicles remain listening to the SRV channel until the end of the SBP. After that they have the option to stay on the SRV channel or to switch to any other service channel. By default, vehicles switch to the CTRL channel. Through analysis and simulation, the authors studied the delay of the safety messages. They focused on informing cluster members and informing neighbouring cluster members. The analysis showed that the maximum delay to inform cluster members is less than T, and to inform neighbouring cluster-members is less that 2T in the worst Case scenarios (depending on when the message is generated and when the message is sent). The authors showed the delay to deliver safety messages between two clusters.

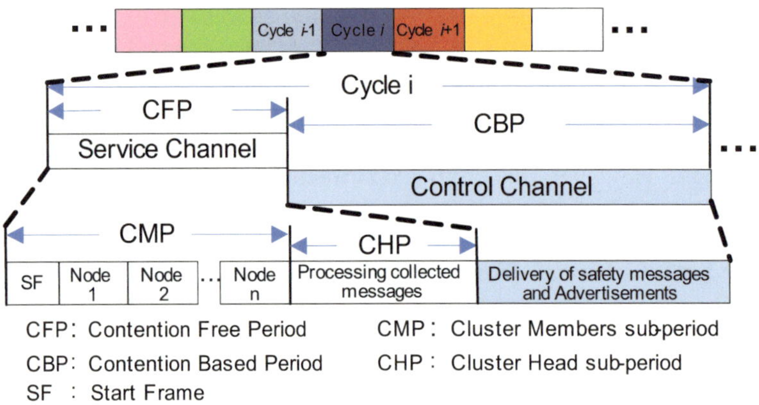

Fig. 5. System Cycle (Rawashdeh & Mahmud, 2008)

3. Data disseminations in VANET

In the context of the vehicular ad hoc networks data can be exchanged among vehicles to support safe and comfort driving. Several applications that rely on distributing data in a geographic region or over long distances have been developed. Different from routing that is concerned with the delivery of data packets from source to destination via multi-hop steps (intermediate nodes) over long distance, data dissemination refers to distributing information to all nodes in a certain geographic region. Its key focus is on conveying data related to safety applications particularly real-time collision avoidance and warning. While one of dissemination's main goals is to reduce the overload of the network; guaranteeing the exchange of information between all necessary recipients without noticeable delay, is also of great importance. Dissemination in VANET can also be seen as a type of controlled flooding in the network. Consider a scenario of a high density network, assume that vehicles detect an event and try to distribute the information about this event to other vehicles. The shared wireless channel will be overloaded when the number of forwarders that are trying to relay this data increases. Therefore, a smart forwarding strategy should be adopted to avoid having the wireless channel congested. Moreover, safety messages are of a broadcast nature, and they should be available to all vehicles on time. Therefore, the dissemination techniques should minimize the number of unnecessary retransmissions to avoid overloading the channel. The data dissemination methods can be categorized as flooding-based where each node rebroadcasts the received message, and relay-based where smart flooding techniques are used to select a set of nodes to relay received messages.

3.1 Flooding-based method

Flooding is the process of diffusion the information generated and received by a node to other approaching vehicles. In this approach, each node participates in dissemination. The flooding can be suitable for delay sensitive applications and also for sparsely connected network. The main problem of this approach is that rebroadcasting each received message leads to network congestions, especially when the network is dense. The flooding of data is also limited by the ability of the system to handle properly new arrivals and dealing with the scalability issues (network size).

3.2 Relay-based method

In this approach, smart flooding algorithms are used to eliminate unnecessary data retransmissions. Instead of having all nodes disseminate the information to all neighbors, a relay node or a set of nodes are selected to forward the data packet further in an effort to maximize the number of reachable nodes. The relay-based methods have the ability to handle the scalability problem (increasing number of nodes in the network) of the high density nodes. However the main challenge of these approaches is how to select the suitable relaying node in the algorithm. Different algorithms were developed under the smart flooding techniques as follows: the time-based algorithms, the location-based algorithms.

3.2.1 Time-based algorithms

This type of dissemination algorithms is designed to eliminate unnecessary retransmissions caused by classical flooding. This mechanism gives the nodes that cover more area and maximizes the number of new receivers the chance (high priority) to forward the received message. In (Briesemeister, 2000), nodes calculate the distance between themselves and the sender of the message. If the message is received for the first time, each node sets a countdown timer and starts decrementing until a duplicate message is overheard or the timer is expired. The value of the timer is proportional to the distance from the sender. The higher the distance, the lower the timer value as shown in the following equation.

$$WT(d) = -\frac{MaxWT}{Range} \cdot \hat{d} + MaxWT$$
$$\hat{d} = \min\{d, Range\} \tag{2}$$

Where *Range* is the transmission range, *MaxWT* is the maximum waiting time, and $\hat{d}$ is the distance to the sender.

The node whose timer expires first (timer value reaches zero), forwards the received message. The other nodes, upon receiving the same message more than once, stop their countdown timer. The same process is repeated until the maximum number of forwarding hops is reached; in this case the packet is discarded.

3.2.2 Location-based algorithm

This approach relies on the location of the nodes with respect to the sender node. The node that reaches a large number of new receivers in the direction of the dissemination is selected to forward the messages. The goal is to reach as many new receivers as possible with less number of resources. The authors of (Korkmaz et al., 2004) proposed a new dissemination approach called Urban Multi-hop Broadcast for inter-vehicle communications systems (UMB). The algorithm is composed of two phases, the directional broadcast and the intersection broadcast. In this protocol, the road portion within the transmission range of the sender node is divided into segments of equal lengths. Only the road portion in the direction of the dissemination is divided into segments. The vehicle from the farthest segment is assigned the task of forwarding and acknowledging the broadcast without any apriori knowledge of the topology information. However, in dense scenarios more than one vehicle might exist in the farthest segment. In this case, the farthest segment is divided into sub-segments with smaller width, and a new iteration to select a vehicle in the farthest sub-segment begins. If these sub-segments are small and insufficient to pick only one vehicle, then the vehicles in the last subs-segment enter a random phase. When vehicles in the direction of the dissemination receive a request form the sender to forward the received data, each vehicle calculates its distance to the source node. Based on the distance, each vehicle sends a black-burst signal (jamming signal) in the Shortest Inter Frame Space (SIFS) period. The length of the black-burst signal is proportional to the distance from the sender. The equation below shows the length of the black-burst in the first iteration.

$$L_1 = \left\lfloor \frac{\hat{d}}{R} \cdot N_{max} \right\rfloor * SlotTime \qquad (3)$$

Where L_1 is the length of the black-burst signal, $\hat{d}$ is the distance from the sender, R is the transmission range, N_{max} is the number of segments in the transmission range, and *SlotTime* is the length of a time slot.

As shown in Equ. (3), the farther the node, the longer the black-burst signal period. Nodes, at the end of the black-burst signal, listen to the channel. If the channel is found empty, then they know that their black-burst signal was the longest, and thus, they are the suitable nodes to forward the message.

In the intersection phase, repeaters are assumed to be installed at the intersections to disseminate the packets in all directions. The node that is located inside the transmission range of the repeater sends the packet to the repeater and the repeater takes the responsibility of forwarding the packet further to its destination. To avoid looping between intersections, the UMB uses a caching mechanism. The vehicles and the repeaters record the ID's of the packets. The repeaters will not forward the packet if they have already received it. However, having the vehicle record the ID's of the packets will be associated with a high cost in terms of memory usage. Moreover, the packet might traverse the same road segment more than one time in some scenarios, which increases the bandwidth usage.

4. Routing in VANET

Routing is the process of forwarding data from source to destination via multi-hop steps. Specifically, routing protocols are responsible for determining how to relay the packet to its destination, how to adjust the path in case of failure, and how to log connectivity data. A good routing protocol is one that is able to deliver a packet in a short amount of time, and consuming minimal bandwidth. Different from routing protocols implemented in MANETs, routing protocols in VANET environment must cope with the following challenges:

- Highly dynamic topology: VANETs are formed and sustained in an ad hoc manner with vehicles joining and leaving the network all the time, sometimes only being in the range for a few seconds.
- Network partitions: In rural areas traffic may become so sparse that networks separate creating partitions.
- Time sensitive transmissions: Safety warnings must be relayed as quickly as possible and must be given high priority over regular data.

Applying traditional MANET's routing protocols directly in the VANET environment is inefficient since these methods don't take VANET's characteristics into consideration. Therefore, modifying MANET routing protocols or developing new routing protocols specific for VANET are the practical approaches to efficiently use routing methods in VANET. One example of modifying MANET's protocols to work in the VANET environment is modifying the Ad hoc On Demand Distance Vector (AODV) with Preferred Group Broadcasting (PGB). On the other hand, new routing protocols were developed specifically for VANET (Lochert et al., 2003) (Lochert et al., 2005) (Tian et al., 2003) (Seet et al., 2004) (Tee & Lee, 2010). These protocols are position-based that take advantage of the knowledge of road maps and vehicle's current speed and position. Mainly, most of VANET's routing protocols can be split into two categories: topology-based routing and position-based routing. In the following sections, we will further define these two types of routing protocols. But, we will focus on the position-based type since it is more suitable for VANET environments.

4.1 Topology based routing

Topology-based routing protocols rely on the topology of the network. Most of the topology-based routing algorithms try to balance between being aware of the potential routes and keeping overhead at the minimum level. The overhead here refers to the bandwidth and computing time used to route a packet. Protocols that keep a table of information about neighbouring nodes are called proactive protocols; while reactive protocols route a packet on the fly.

4.1.1 Reactive topology based protocols

This type of protocols relies on flooding the network with query packets to find the path to the destination nodes. The Dynamic Source Routing (DSR) (Johnson & Maltz, 1996) is one of the reactive topology-based routing protocols. In the DSR, a node sends out a flood of query packets that are forwarded until they reach their destination. Each node along the path to the destination adds its address to the list of relay nodes carried in the packet. When the destination is reached, it responds to the source listing the path taken. After waiting a set amount of time, the source node then sends the packet from node to node along the shortest path.

The Ad Hoc On-Demand Distance Vector (AODV) (Perkins & Royer) is another reactive topology-based routing protocol developed for MANETs. The AODV routing protocol works similar to DSR in that when a packet must be sent routing requests flood the network, and the destination confirms a route. However unlike the DSR, in AODV the source node is not aware of the exact path that the packet must take, the intermediate nodes store the connectivity information. AODV-PGB (Preferred Group Broadcasting) is a modified version of AODV that reduces overhead by only asking one member in a group to forward the routing query.

4.1.2 Proactive topology based protocols

This type of protocols builds routing tables based on the current connectivity information of the nodes. The nodes continuously try to keep up to date routing information. Proactive-topology based Routing protocols are developed to work in low mobility environments (like MANET). However, some of these protocols were modified to work in high mobility environment (Benzaid et al., 2002). In (Benzaid et al., 2002), the authors proposed a fast Optimized Link State Routing (OLSR), where nodes exchange the topology information using beacons to build routing paths. The exchange of beacon messages is optimized such that the frequency of sending these messages is adapted to the network dynamics. Mainly, the proactive routing protocols consume a considerable amount of bandwidth. This is because a large amount of data is exchanged for routing maintenance, especially in very high dynamic networks where the neighbourhood of nodes is always changing. The high dynamics of the network leads to frequent change in the neighbourhood, which increases the overhead needed to maintain the routing table, and consume more bandwidth.

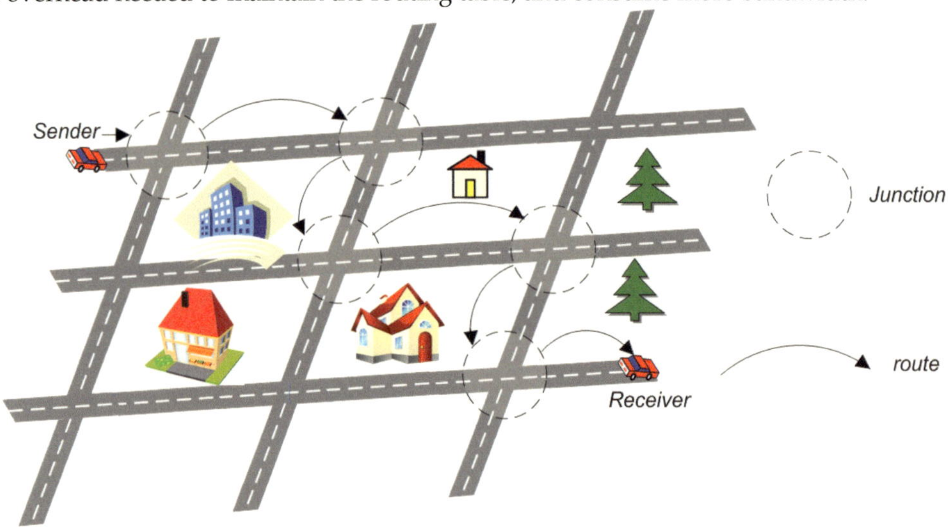

Fig. 6. Paths and junctions to route the packet

4.2 Position based routing

Position-based routing protocols or geographic routing protocols rely on the actual real world locations to determine the optimal path for a packet. The nodes are assumed to be equipped with device, like GPSs, allowing them to record their locations. Position-based protocols usually perform better in VANET than topology-based protocols because overhead is low, and node connectivity is so dynamic that sending a packet in the general direction of its destination is the most effective method.

In (Lochert et al., 2003), the authors proposed a position-based routing protocol for VANET called Geographic Source Routing (GSR). GSR relies on the maps of the cities and the locations of the source and destination nodes. The nodes use Dijkstra's algorithm to compute the shortest path between source and destination nodes. In GSR, intersections can be seen as junctions that represent the path that packets have to pass through to reach their destination as shown in Figure 6. The GSR uses the greedy forwarding technique to determine the location of the next junctions on the path. The greedy destination is the location of the next junction on the path. A received packet is forwarded to the node that is closer to the next junction. This process is repeated until the packet is delivered to its final destination. Two approaches were proposed to deal with the sequence of junctions: the first approach requires that the whole list of junctions is included in the packet header. In this approach, the computation complexity and overhead is reduced, but bandwidth usage is increased. The second approach requires that each forwarding node computes the list of junctions. In this approach, bandwidth consumption is reduced, but computation overhead is increased. Finally, there are some issues that are not clear in GSR implementation, for example it is not clear how GSR deals with low connectivity scenarios and what happens when the forwarding node can't find another node closer to the next junction.

Lochert et al. (Lochert et al., 2005) proposed a position-based routing protocol suitable for urban scenarios. The routing protocols called Greedy Perimeter Coordinator Routing (GPCR). Similar to GSR, the proposed algorithm considers intersections as junctions and streets as paths. One of the main ideas implemented in the algorithm is restricted greedy forwarding. In the restricted greedy forwarding, the junctions play very important role in routing. Therefore, instead of forwarding packets as close as possible to the destination, restricted greedy routing forwards packets to a node in the junction as shown in Figure 7.

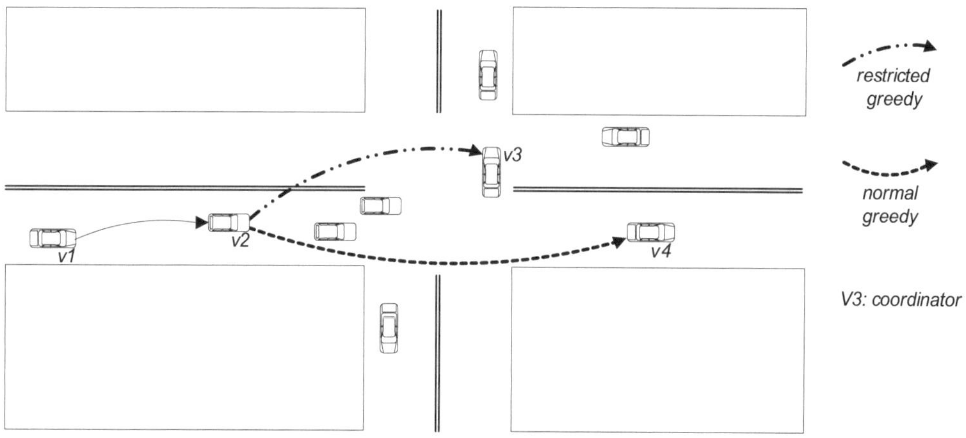

Fig. 7. Restricted greedy in GPCR

This is because the node on the junction has more options to route packets. In addition to that, the local optimum can be avoided (local optimum happens when a forwarding vehicle can't find a node closer to the destination than itself). The nodes close to the junction are called Coordinators. Coordinators announce their role via beacons to let neighbouring nodes know about them. Two approaches were proposed for the node to know whether its role is a coordinator or not. The first approach requires that nodes include their neighbours in the beacons, so that nodes can have information about their 2-hop neighbours. Based on this, the node is considered a coordinator if it has two neighbours that are within direct communication range with respect to each other, but don't list each other as neighbours. This means that nodes are separated by obstacles. The second approach requires each node calculate the correlation coefficient with respect to its neighbours. Assume that x_i and y_i represent the coordinates for node i. Assume also that $\hat{x}$ and $\hat{y}$ are the means for x-coordinate and y-coordinate respectively. Let σ_{xy} represents the covariance of x and y, σ_x and σ_y indicate the standard deviation of x and y respectively. The correlation coefficient can be calculated as follows:

$$\sigma_{xy} = \left|\frac{\sigma_{xy}}{\sigma_x \sigma_x}\right| = \left|\frac{\sum_{i=1}^{n}(x_i - \hat{x})(y_i - \hat{y})}{\sqrt{(\sum_{i=1}^{n}(x_i - \hat{x}))^2 (\sum_{i=1}^{n}(y_i - \hat{y}))^2}}\right| \tag{4}$$

The value of σ_{xy} is in the range [0,1]. If the value is close to 1, then it indicates linear coherence, which is found when a vehicle is located in the middle of the street. A value close to 0 shows no linear relationship between the positions of the nodes indicating that a node is located on the junction. The authors used a threshold ϵ such that, if $\sigma_{xy} \geq \epsilon$ then the node is located on the street, and if $\sigma_{xy} < \epsilon$, then the node is close to the junction.

Packets are forwarded along the street. The farthest node is a candidate to forward the packets until they reach the intersection. Once a packet is delivered to a coordinator on the junction, a decision about which road the packet should traverse is made. Mainly, a neighbor that has the highest progress toward destination is selected.

The Spatially Aware Routing (SAR) (Tian et al., 2003) is a position based routing protocol that is more relevant to an urban setting. SAR takes into account that packets cannot be forwarded through the dense buildings in urban areas, so they must be forwarded through the streets and intersections (similar to GSR). SAR uses the maps of the cities such that the roads and intersections are represented as paths and junctions on a graph. The nodes select the junctions that the packet has to go through to reach its destination. Nodes use Dijkstra's algorithm to compute the shortest path on the graph. Then, this path is included in the header of the messages. The source node routes the packet using the shortest path algorithm on that graph. Upon receiving a packet, the forwarding node chooses the neighbor that is closer to the first junction in the GSR. The packet is forwarded to the next junction in the path until it gets delivered. The SAR algorithm uses different approaches to deal with the scenario when the forwarding node can't find another node closer to the next junction on the path. The first option is storing the packet and periodically trying to forward it. The packet will be discarded if the time limit is passed or the buffer becomes full. The second option is forwarding the packet, using the traditional greedy forwarding routing, toward the destination instead of the next junction. The third option is recalculating new path based on the current situation after discarding the path computed by the source node.

Anchor Based Street and Traffic Aware Routing (A-STAR) (Seet et al., 2004) is similar to SAR in that it also routes along streets and intersections. The packet is routed along a directional vector that contains anchors or fixed geographic points that the packet must go through. When A-STAR calculates the best path it prefers, streets with higher vehicle density, making the protocol traffic aware. Higher vehicle density in a street provides better transmission and less delay for a packet traveling along it. Traffic information is taken into consideration when the routing protocol uses the shortest path algorithm to determine the best path for the packet. Traffic information can be determined by the number of bus stops on a street, or by actual real-time measurements of traffic density. The first method is called the statistically rated map and the second is called the dynamically rated map. A-STAR also has a novel way to deal with local maximums. When a packet reaches a void, the anchor path is recalculated and the surrounding nodes are notified that particular path is out of service.

Junction Based Adaptive Reactive Routing (JARR) (Tee & Lee, 2010) is a new routing protocol designed specifically to deal with urban environments. It uses different algorithms for when the packet is traveling to a junction, and when it has reached a junction. First the packet is forwarded down an optimal path to a junction. At that point a different algorithm takes over that determines the next optimal path and auxiliary routes. JARR takes into consideration velocity, direction, current position, and density when determining the path for a packet. In order for nodes to gather that information, a beacon regularly informs neighboring nodes of its position and velocity. JARR is able to reap the benefits of the beacon without paying the full price in overhead by adapting the frequency of the beacon as vehicle density increases. The higher the density, the less frequently the beacon is used to disseminate information. JARR also increases its throughput by allowing for some delay tolerance. For example, if a packet is transferred to a node that loses connectivity with the network, the packet will be carried until it can be forwarded.

5. Conclusion

This book chapter presented an overview and tutorial of various issues related to communications in vehicular networks. Various types of challenges in vehicular communications have been identified and addressed. A number of media access and routing techniques are also clearly presented. This book chapter will allow readers to get an understanding about what a vehicular network is and what type of challenges are associated with vehicular networks.

6. References

Benmimoun, A.; Chen, J.; Neunzig, D.; Suzuki, T. and Kato, Y. (2005). Communication-based intersection assistance, *Proceedings of the IEEE Intelligent Vehicle Symposium*, Las Vegas, NV, 2005.

Benzaid, M.; Minet, P. and Agha, K. (2002). Integrating fast mobility in the OLSR routing protocol, *Proceedings of IEEE Conference on Mobile and Wireless Communications Networks, Stockholm*, 2002.

Biswas, S.; Tachikou, R.; and Dion, F. (2006). Vehicle-to-Vehicle wireless communication protocols for enhancing highway traffic safety, *IEEE communications Magazine*. 44(1), 2006.

Briesemeister, Linda; Sch¨afers, Lorenz and Hommel, G¨unter (2000). Disseminating Messages among Highly Mobile Hosts based on Inter-Vehicle Communication., *Proceedings of the IEEE Intelligent Vehicles Symposium*, Detroit, USA 2000.

Clausen, T.; Jacquet, P.; Laouiti, A.; Muhlethaler, P.; Qayyum, A.; and Viennot, L. (2001). Optimized link state routing protocol, *Proceedings of IEEE International Multitopic Conference INMIC*, Pakistan, 28–30 December, 2001.

Dashtinezhad, S.; Nadeem, T.; Dorohonceanu, B.; Borcea, C. (2004); Kang, P.; Iftode, L.; TrafficView: A Driver Assistant Device for Traffic Monitoring based on Car-to-Car

Communication, *Proceedings of the IEEE Semiannual Vehicular Technology Conference*, Milan, Italy, May 2004.

ElBatt, T.; Goel, S. K.; Holland, G.; Krishnan, H.; Parikh, J. (2006). Cooperative collision warning using dedicated short range wireless communications, *Proceedings of ACM VANET* 2006, Page(s): 1-9

IEEE (2003). IEEE 802.11e/D4.4, Draft Supplement to Part 11: Wireless Medium Access Control (MAC) and physical layer (PHY) specifications: Medium Access Control (MAC) Enhancements for Quality of Service (QoS), June 2003

IEEE (2007). IEEE Draft, "Trial Use Standard for Wireless Access in Vehicular Environments (WAVE)—Architecture," P1609.0/D01, February 2007.

IEEE (2007). IEEE WG, IEEE 802.11p/D2.01, Draft Amendment to Part 11: Wireless Medium Access Control (MAC) and Physical Layer (PHY) specifications: Wireless Access in Vehicular Environments, March 2007.

IEEE (2007). IEEE WG, IEEE 802.1lp/D2.01, Draft Amendement to Part 11: Wireless Medium Access Control (MAC) and Physical Layer (PHY) specifications: Wireless Access in Vehicular Environments March 2007.

Johnson, D. B. and Maltz, D. A. (1996). Dynamic Source Routing in Ad Hoc Wireless Networks. In *Mobile Computing, T. Imielinski and H. Korth, Eds.*, Kluwer Academic Publisher, 1996, ch.5, pp. 153–81.

Kihl, M.; Sichitiu, M. and Joshi, H. P. (2008). Design and evaluation of two Geocast protocols for vehicular ad-hoc networks, *Journal of Internet Engineering* 2(1), 2008.

Korkmaz, G. et al. (2004). Urban Multi-Hop Broadcast Protocol for Inter-Vehicle Communication Systems, *Proceedings ACM Int'l. Wksp. Vehic. Ad Hoc Networks, Philadelphia*, PA, Oct. 2004.

Lochert, C.; Hartenstein, H.; Tian, J.; Fler, H.; Herrmann, D. and Mauve, M. (2003). A routing strategy for vehicular ad hoc networks in city environments, *Proceedings of IEEE Intelligent Vehicles Symposium*, Columbus, OH, 2003.

Lochert, C.; Mauve, M.; Fusler, H. and Hartenstein, H. (2005). Geographic routing in city scenarios, *ACM SIGMOBILE Mobile Computing and Communications Review*, 2005:69–72.

Maih¨ofer, C.; Cseh, C.; Franz, W.; and Eberhardt, R. (2003). Performance evaluation of stored geocast, *Proceedings of the IEEE 58th Vehicular Technology Conference*, Orlando, FL, 2003.

Moreno, T.; Santi, P.; and Hartnestien, H. (2005). Fair Sharing of Bandwidth in VANETS. *Proceedings of the 2nd ACM international workshop on Vehicular Ad Hoc Networks (VANET)*, pages 49-58, NewYork, USA, 2005.

Nadeem, T.; Dashtinezhad, S.; Liao, C.; Iftode, L. (2004) TrafficView: Traffic Data Dissemination using Car-to-Car Communication, *ACM Mobile Computing and Communications Review (MC2R)*, Vol. 8, No. 3, pp. 6-19, July 2004.

Perkins, C. E. and Royer, E. M. (1999). Ad-Hoc On-Demand Distance Vector Routing, *Proceedings of the IEEE WMCSA '99*, New Orleans, LA, Feb. 1999, pp. 90–100.

Rawashdeh, Z. Y. and Mahmud, S. M. (2008). Media Access Technique for Cluster-Based Vehicular Ad Hoc Networks, *Proceedings of the 2nd IEEE International Symposium on Wireless Vehicular Communications*, Calgary, Canada, September 21 - 22, 2008.

Seet, B. C.; Liu, G.; Lee, B. S.; Foh, C. H.; Wong, K. J. and Lee, K. K. (2004). A-STAR: A mobile ad hoc routing strategy for metropolis vehicular communications,

Proceedings of 3rd International Networking Conference IFIP-TC6 (IFIP '04), Athens, Greece, Dec 2004. Lecture Notes in Computer Science 3042:989–999.

Su, Hang and Zhang, Xi (2007). Clustering-based multichannel MAC protocols for QoS provisionings over vehicular ad hoc networks, *IEEE Transactions on Vehicular Technology* 56(6):3309–3323, November 2007.

Tee, C. A. T. H.; Lee, A. (2010). A novel routing protocol – Junction based Adaptive Reactive Routing (JARR) for VANET in city environments, *Proceeding of the 12th European Wireless Conference (EW 2010)*, vol., no., pp.1-6, 12-15, L ucca (Tuscany), Italy Apr. 2010.

Tian, J.; Han, L.; Rothermel, K.; and Cseh, C. (2003). Spatially aware packet routing for mobile ad hoc inter-vehicle radio networks, *Proceedings of IEEE Intelligent Transportation System Conference (ITSC '03)*, Shanghai, China, October, 2003:1546–1551.

Tong, Zhu; Jian, Xu; Yu, Bai and Xiaoguang, Yang (2009). A research on risk assessment and warning strategy for intersection collision avoidance system, *Proceedings of the 12th Intelligent Transportations Systems*. (ITSC'09), pages 1-6, St. Louis, Missouri, U.S.A. 3-7 Oct. 2009.

Torrent-Moreno, M.; Santi, P.; and Hartenstien, H. (2005). The challenges of robust inter-vehicle communications, *Proceeding of IEEE 62nd Semiannual Vehicular Technology Conference*, VTC 2005-Fall, Dallas, Texas, Sep. 2005.

Torrent-Moreno, M.; Santi, P. and Hartnestien, H. (2006). Distributed Fair Transmit Power Adjustment for Vehicular Ad hoc Networks. *Proceedings of 3rd IEEE Sensor and Ad Hoc Communications and Networks*. SECON '06. Sep 2006.

Wang, S. Y. (2007). The potential of using inter-vehicle communication to extend the coverage area of roadside wireless access points on the highway, *Proceedings of the IEEE International Conference on Communications*, Glasgow, UK, 2007.

Xu, Q.; Mark, T.; Ko, J.; and Sengupta, R. (2004). Vehicle-to-Vehicle Safety Messaging in DSRC, *Proceedings of VANET*, October 2004.

Xu, Q.; Mark, T.; Ko, J.; Sengupta, R. (2007). Medium Access Control Protocol Design for Vehicle-Vehicle Safety Messages, *IEEE Transactions on Vehicular Technology*, Vol. 56, N. 2, pp.499-518, March 2007.

Yang, S.; Refai, H.; and Ma, X. (2005). CSMA based inter-vehicle communication using distributed and polling coordination, *Proceedings IEEE Int. Conf. on ITS*, Vienna, Austria, Sept. 2005, pp. 167-171.

Yang, X.; Liu, J.; Zhao, F.; and Vaidya, N. H.; (2004). A vehicle-to-vehicle communication protocol for cooperative collision warning, *Proceedings of the First Annual International Conference on Mobile and Ubiquitous Systems: Networking and Services*, Boston, MA, 2004.

Yu, B.; Gong, J.; and Xu, C.-Z. (2008). Catch-up: A data aggregation scheme for VANETs, *Proceedings of ACM VANET*, San Francisco, CA, Sep. 2008, pp. 49–57.

Yu, Fan F. and Biswas, S. (2007). A Self-Organizing MAC Protocol for DSRC based Vehicular Ad Hoc Networks, *ICDCS Workshops 2007*.

Yu, Q. and Heijenk, G. (2008). Abiding geocast for warning message dissemination in vehicular ad hoc networks, *Proceedings of the IEEE Vehicular Networks and Applications Workshop 2008*, 2008.

Zhang, Y.; Weiss, E.; Chen, L.; and Cheng, X. (2007). Opportunistic wireless internet access in vehicular environments using enhanced WAVE devices, *Proceedings of the International Conference on Future Generation Communication and Networking*, Jeju, South Korea, 2007.

Zang, Y.; Stibor, L.; Cheng, Xi; Reumerman, H.-J.; Paruzel, A. and Barroso, A. (2007). Congestion Control in Wireless Networks for Vehicular Safety Applications,. *Proceeding of the 8th European Wireless Conference*, Paris, France, Apr. 2007.

3

Modeling and Simulation of Vehicular Networks: Towards Realistic and Efficient Models

Mate Boban[1,2] and Tiago T. V. Vinhoza[2]
[1]*Department of Electrical and Computer Engineering, Carnegie Mellon University 5000 Forbes Avenue, Pittsburgh, PA, 15213*
[2]*Instituto de Telecomunicações, Departamento de Engenharia Electrotécnica e de Computadores Faculdade de Engenharia da Universidade do Porto, 4200-465, Porto, Portugal*
USA and Portugal

1. Introduction

Vehicular Ad Hoc Networks (VANETs) have been envisioned with three types of applications in mind: safety, traffic management, and commercial applications. By using wireless interfaces to form an ad hoc network, vehicles will be able to inform other vehicles about traffic accidents, hazardous road conditions and traffic congestion. Commercial applications (e.g., data exchange, audio/video communication) are envisioned to provide incentive for faster adoption of the technology.

To date, the majority of VANET research efforts have relied heavily on simulations, due to prohibitive costs of employing real world testbeds. Current VANET simulators have gone a long way from the early VANET simulation environments, which often assumed unrealistic models such as random waypoint mobility, circular transmission range, or interference-free environment Kotz et al. (2004). However, significant efforts still remain in order to enhance the realism of VANET simulators, at the same time providing a computationally inexpensive and efficient platform for performance evaluation of VANETs. In this work, we distinguish three key building blocks of VANET simulators:

- Mobility models,
- Networking (data exchange) models,
- Signal propagation (radio) models.

Mobility models deal with realistic representation of vehicular movement, including mobility patterns (i.e., constraining vehicular mobility to the actual roadway), interactions between the vehicles (e.g., speed adjustment based on the traffic conditions) and traffic rule enforcement (e.g., intersection control through traffic lights and/or road signs). Networking models are designed to provide realistic data exchange, including simulating the medium access control (MAC) mechanisms, routing, and upper layer protocols. Signal propagation models aim at realistically modeling the complex environment surrounding the communicating vehicles, including both static objects (e.g., buildings, overpasses, hills), as well as mobile objects (other vehicles on the road).

We first present the state-of-the art in vehicular mobility models and networking models and describe the most important proponents for these two aspects of VANET simulators.

Then, we describe the existing signal propagation models and motivate the need for more accurate models that are able to capture the behavior of the signal on a per-link basis, rather than relying solely on the overall statistical properties of the environment. More specifically, as shown in Koberstein et al. (2009), simplified stochastic radio models (e.g., free space Goldsmith (2006), log-distance path loss Rappaport (1996), two-ray ground reflection Goldsmith (2006), etc.), which are based on the statistical properties of the chosen environment and do not account for the specific obstacles in the region of interest, are unable to provide satisfactory accuracy for typical VANET scenarios. Contrary to this, topography-specific, highly realistic channel models (e.g., based on ray tracing Maurer et al. (2004)) yield results that are in very good agreement with the real world. However, these models are computationally too expensive and usually bound to a specific location (e.g., a particular neighborhood in a city), thus making them impractical for extensive simulation studies. For these reasons, such models are not implemented in VANET simulators. Based on the experimental assessment of the impact of mobile obstacles on vehicle-to-vehicle communication, we point out the importance of the realistic modeling of mobile obstacles and the inconsistencies that arise in VANET simulation results in case these obstacles are omitted from the model. Motivated by this finding, we developed a novel model for incorporating the mobile obstacles (i.e., vehicles) in VANET channel modeling. A useful model that accounts for mobile obstacles must satisfy a number of requirements: accurate vehicle positioning, realistic underlying mobility model, realistic propagation characterization, and manageable complexity. The model we developed satisfies all of these requirements Boban et al. (2010). The proposed model accounts for vehicles as three-dimensional obstacles and takes into account their impact on the LOS obstruction, received signal power, and the packet reception rate. The algorithm behind the model allows for computationally efficient implementation in VANET simulators. Furthermore, the proposed model can easily be used in conjunction with the existing models for static obstacles to accurately simulate the entire spectrum of VANET environments with regards to both road conditions (e.g., sparse or dense vehicular networks), as well as various surroundings (including highway, suburban, and urban environments).

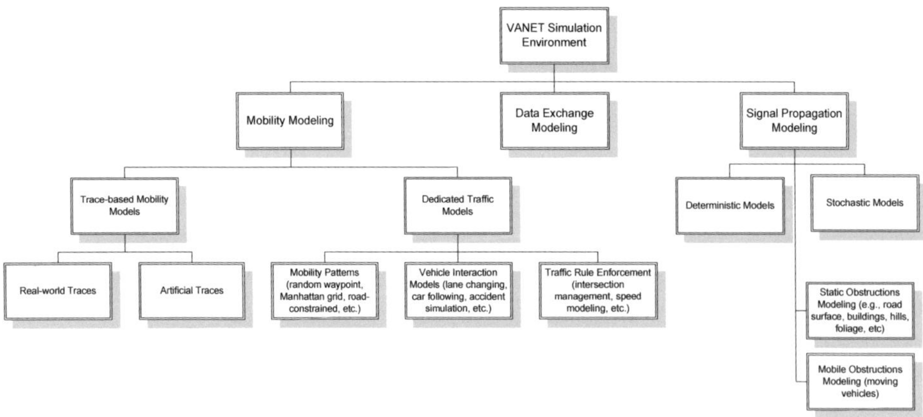

Fig. 1. Structure of VANET simulation environment

2. Mobility models

Mobility models can be roughly divided in trace-based models and dedicated traffic models (Fig. 1). Trace-based models are based on a set of generated vehicular traces which are then used as an underlying mobility pattern over which the data communication is carried over. The traces can be either real world (i.e., based on mapping of the positions of vehicles) Ferreira et al. (2009) and Ho et al. (2007), or artificially generated using the dedicated traffic engineering tools Naumov et al. (2006). The advantage of trace-based models is they provide the highest level of realism achievable in VANET simulations. However, there are also several important shortcomings. Firstly, in order to collect the real world mobility traces, significant time and cost are involved. This often makes the traces collected limited with respect to both the number of the vehicles that are recorded and the region over which the recording has been made. Further this implies that there is rarely a chance to record the mobility of all the vehicles in a certain region (as it would often involve equipping each vehicle with the location devices), thus leading to a need for compensating algorithms for the non-recorded vehicles. Finally, since the traces are collected/recorded beforehand, the feedback connection from the networking model to the mobility model is not available. This is a very important shortcoming, given that a large number of proposed Intelligent Transportation System (ITS) applications carried over VANETs can affect the movement of the vehicles (this is especially the case with traffic management applications), thus rendering the trace-based models inadequate for any application with the feedback loop between the traffic and networking models. A vivid example of such application is Congested Road Notification Bai et al. (2006), which aids the vehicles in circumventing congested roads, thus directly affecting the mobility of the vehicles through the network communication.

A characteristic that distinguishes the dedicated traffic models from the trace-based ones, capability to support the feedback loop between the mobility model and the networking model, is an important reason for adopting the more flexible dedicated traffic models. This way, the information from the networking model (e.g., a vehicle receives a traffic update advising the circumvention of a certain road) can affect the behavior of the mobility model (e.g., the vehicle takes a different route than the one initially planned). Early VANET mobility models were characterized by their simplicity and ease of implementation. For quite some time, the random waypoint mobility model Saha & Johnson (2004), where the vehicles move over a plane from one randomly chosen location to another, was the de facto standard for VANET simulations. However, it was shown that the overly simplified mobility models such as random waypoint are not able to model the vehicular mobility adequately Choffnes & Bustamante (2005). A significant step towards the realism were the simple one-dimensional freeway model and the so-called Manhattan grid model Bai et al. (2003), where the mobility is constrained to a set of grid-like streets which represent an urban area. Further elaboration of the mobility models was achieved by using map generation techniques, such as Voronoi graphs Davies et al. (2006), which constrain the movement of the vehicles to a network of artificially generated irregular streets. Recently, the most prominent mobility models (e.g., Choffnes & Bustamante (2005), Conceição et al. (2008), and Mangharam et al. (2005)) started utilizing real world maps in order to constrain the vehicle movement to real streets based on some of the available geospatial databases (e.g., the U.S. Census Bureau's TIGER data *U.S. Census Bureau TIGER system database* (n.d.) or the data collected in the OpenStreetMap project *Open Street Map Project* (n.d.)). Furthermore, the distinction can be made with regards to the coupling between the mobility and networking and signal propagation models. On one side of the spectrum are the mobility models embedded with the networking model (e.g., Choffnes &

Bustamante (2005) and Mangharam et al. (2005)), thus allowing for a more efficient execution of the simulation. On the other side, there are mobility models which are based on the dedicated traffic simulators stemming from the traffic engineering community (e.g., *SUMO - Simulation of Urban MObility* (n.d.) and *CORSIM: Microscopic Traffic Simulation Model* (n.d.)), which are then bidirectionally coupled with the networking model (e.g., Sommer et al. (2008) and Piórkowski et al. (2008)). These types of environments are characterized by a high level of traffic simulation credibility, but often suffer from inefficiencies caused by the integration of two separate systems Harri (2010).

Vehicle interaction (Fig. 1) includes modeling the behavior of a vehicle that is a direct consequence of the interaction with the other vehicles on the road. This includes the microscopic aspects of the impact of other vehicles, such as lane changing Gipps (1986) and decreasing/increasing the speed due to the surrounding traffic, as well as the macroscopic aspects, such as taking a different route due to the traffic conditions (e.g., congestion). Another important aspect of mobility modeling is traffic rule enforcement, which includes intersection management, changing the vehicle speed based on the speed limits of the roads, and generally making the vehicle obey any other traffic rules set forth on a certain highway. Even though the vehicle interaction and the enforcement of traffic rules were shown to be essential for accurate modeling of vehicular traffic Helbing (2001), as noted in Harri et al. (2009), many of VANET mobility models have scarce support for these microscopic aspects of vehicular mobility. For this reason, significant research efforts remain in order to make these aspects of mobility models more credible, and for the research community to strive for the simulation environments that realistically model these components.

With regards to the implementation approaches for the mobility models, the most prolific proponents are Helbing (2001): the cellular automata models (e.g., Nagel & Schreckenberg (1992)), the follow-the-leader models (e.g., car-following Rothery (1992) and intelligent driver model Treiber et al. (2000)), the gas-kinetic models (e.g., Hoogendoorn & Bovy (2001)), and the macroscopic models (e.g., Lighthill-Whitham model Lighthill & Whitham (1955)). Further classification of mobility models can be made with respect to the granularity at which the mobility is simulated, categorizing the mobility models as microscopic, mesoscopic, and macroscopic. Microscopic models are simulating the mobility at the per-vehicle level (i.e., each vehicle's motion is simulated separately). Prominent examples of such models are the car following model Rothery (1992) and cellular automata models Nagel & Schreckenberg (1992), Tonguz et al. (2009). Macroscopic models simulate the entire vehicular network as an entity that possesses certain physical properties. Such models can give insights into the overall statistical properties of vehicular networks (e.g., the average vehicular density, average speer, or the flow/density relationship of a given vehicular network). Examples of such models are kinematic wave models Jin (2003) and fluid percolation Cheng & Robertazzi (Jul. 1989). Mesoscopic models are simulating the mobility at the flow level, where a number of vehicles is characterized by certain averaging properties (e.g., arrival time, average speed, etc.), but the flows are distinguishable. Gas-kinetic model Hoogendoorn & Bovy (2001) is an example of mesoscopic models. For an extensive treatment focusing on modeling the vehicular traffic in general, we refer the reader to Helbing (2001), and for the overview of the mobility models used in VANET research, we refer the reader to Harri (2010).

3. Networking models

Unlike the mobility models or signal propagation models for VANETs, which have significant differences when compared to models used in other types of mobile ad hoc networks

(MANETs) Murthy & Manoj (2004), the networking models for VANETs are quite similar to those used in other fields of MANET research. The data models used in the current simulators, such as NS-2 *Network Simulator 2* (n.d.), JiST/SWANS/STRAW Choffnes & Bustamante (2005), and NCTU-NS Wang et al. (2003), rely on discrete event simulation, where different protocols of the network stack are executed based on the events triggered either by upper layer (e.g., an application sends a message to the networking protocol) or by lower layer (e.g., the link layer protocol notifies the network layer protocol about the correct reception of the message). The main difference arises in the use of a dedicated VANET protocol stack called Wireless Access in Vehicular Environments (WAVE), standardized under the IEEE 1609 working group *IEEE Trial-Use Standard for Wireless Access in Vehicular Environments (WAVE) - Networking Services* (Apr. 2007).

In 1999, the U.S. Federal Communications Commission (FCC) allocated 75 MHz of spectrum between 5850 - 5925 MHz for WAVE systems operating in the Intelligent Transportation System (ITS) radio service for vehicle-to-vehicle (V2V) and infrastructure-to-vehicle (V2I) communications. Similarly, the European Telecommunications Standards Institute (ETSI) has allocated 30 MHz of spectrum in the 5.9 GHz band for ITS services in August 2008, and many other countries are actively working towards standardizing the 5.9 GHz spectrum, thus allowing worldwide compatibility of WAVE devices in the future. WAVE provisions for public safety and traffic management applications. Commercial (tolling, comfort Bai et al. (2006), entertainment Tonguz & Boban (2010), etc.) services are also envisioned, creating incentive for faster adoption of the technology. The lower layers of the WAVE protocol stack are being standardized under the Dedicated short-range communications (DSRC) set of protocols *IEEE Draft Standard IEEE P802.11p/D9.0* (July 2009). DSRC is based on IEEE 802.11 technology and is proceeding towards standardization as IEEE 802.11p. Fig. 2 shows the WAVE protocol stack. On the network layer, WAVE Short Message Protocol (WSMP) is being developed for fast and efficient message exchange in VANETs. It is planned to support both safety as well as for non-safety applications. Applications running over WSMP will directly control the physical layer characteristics (e.g., channel number and transmitter power) on a per message basis. As seen in Fig. 2, applications running over the standard TCP/IP protocol stack are also supported. Their operation is restricted to the predefined underlying physical layer characteristics, based on the application type. The applications will be divided in up to eight levels of priority, with the safety applications having the highest level of priority. The multi-channel operation *IEEE Trial-Use Standard for Wireless Access in Vehicular Environments (WAVE) - Multi-channel Operation* (2006) is aimed at providing higher availability and managing contention. Channels are divided into Control Channel (CCH) and Service Channels (SCH). WAVE devices must monitor the Control Channel (CCH) for safety application advertisements during specific intervals known as CCH intervals. CCH intervals and are specified to provide a mechanism that allows WAVE devices to operate on multiple channels while ensuring all WAVE devices are capable of receiving high-priority safety messages with high probability *IEEE Trial-Use Standard for Wireless Access in Vehicular Environments (WAVE) - Multi-channel Operation* (2006). For a tutorial on WAVE protocol stack, we refer the reader to Uzcátegui & Acosta-Marum (2009).

Due to the relative novelty of DSRC and WAVE protocols, the majority of the widely used VANET simulators do not implement the DSRC and WAVE protocols. One exception is the NCTUNS simulation environment Wang et al. (2003), which implements both DSRC (IEEE 802.11p) and WAVE (IEEE 1609 set of standards) in its current version. Modeling the networking stack realistically is important for the credibility of the results obtained at each

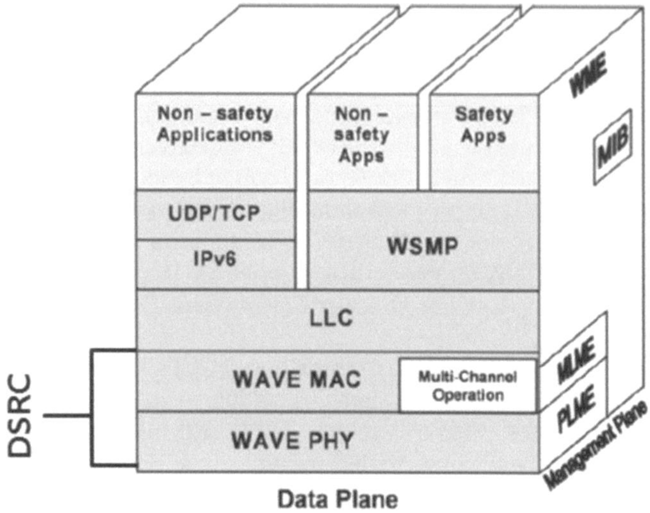

Fig. 2. WAVE protocol stack.

level of the protocol stack, and especially for the application level, since all the potential simulation errors from the lower layers are reflected at the application layer. To this end, it was recently shown that several stringent constraints exist in VANETs for applications Boban et al. (2009), and even with the optimal settings with regards to the networking model, some of the results reported with simplified models, especially with regards to connectivity and message reachability (e.g., Palazzi et al. (2007)) are unachievable.

4. Signal propagation models

In order to adequately model the signal propagation in VANETs, appropriate models need to be developed that take into account the unique characteristics of VANET environment (e.g., high speed of the vehicles, obstruction-rich setting, specific location of the antennas, etc.). In the early days of VANET research, simple signal propagation models were utilized (e.g., unit area disk model Gupta & Kumar (2000), free-space path loss Goldsmith (2006), among others), which were carried over from MANET research. Due to the significantly different environment, these models do not provide satisfying accuracy for typical VANET scenarios Koberstein et al. (2009). Based on whether the model is accounting for a specific location of the objects or generalized distribution of objects in the environment, we can distinguish deterministic and stochastic models (Fig. 1). Deterministic models attempt to model the signal behavior based on the exact environment in which the vehicle is currently located, and with specific locations of the objects surrounding the vehicle (e.g., Maurer et al. (2004).) Stochastic models, on the other hand, assume a location of the surrounding objects based on a certain (often pre-defined) statistical distribution (e.g., Acosta & Ingram (2006)). Based on the approach of modeling the environment, we distinguish geometrical or non-geometrical models. Geometrical models use the concepts of computational geometry to characterize the environment by generating the possible paths or rays between the transmitting and receiving vehicle. Non-geometrical models use the higher level properties of the environment (e.g., path-loss exponent Wang et al. (2004)) to approximate the signal power at the receiver. Furthermore, geometrical signal propagation models have to account for two types of obstructions that affect the signal: static obstructions (e.g., road surface, buildings, overpasses,

Fig. 3. Experiment setup.

hills, etc.) and mobile obstructions (moving vehicles). Numerous studies have dealt with static obstacles as the key factors affecting signal propagation (e.g., Nagel & Eichler (2008) and Giordano et al. (2010)) and proposed models for accurately quantifying the impact of static obstacles. However, due to the nature of VANETs, where communication is often performed in V2V fashion, it is reasonable to expect that the moving vehicles will act as obstacles to the signal, often affecting the signal propagation even more than static obstacles (e.g., in case of an open road).

Furthermore, the fact that the communicating entities in VANETs are vehicles exchanging data in a V2V fashion raises new challenges in signal modeling. We observe, for example, that antenna heights of both transmitter and receiver are relatively low (on top of the vehicles at best), such that other vehicles can act as obstacles for signal propagation by obstructing the LOS between the communicating vehicles. The natural conclusion is that analyzing static obstacles only is not sufficient; vehicles as moving obstacles have to be taken into account. These assumptions have been confirmed in several previous studied. Specifically, Otto et al. in Otto et al. (2009) performed V2V experiments at 2.4 GHz frequency band in an open road environment and pointed out a significantly worse signal reception on the same road during the traffic heavy, rush hour period in comparison to a no traffic, late night period. A similar experimental V2V study presented in Takahashi et al. (2003) analyzed the signal propagation in "crowded" and "uncrowded" highway scenarios (depending on the number of cars currently on the road) for the 60 GHz frequency band, and reported significantly higher path loss for the crowded scenarios. Several other studies (Jerbi et al. (2007), Wu et al. (2005), Matolak et al. (2005), and *Vehicle Safety Communications Project, Final Report* (2006)) hint that other vehicles apart from the transmitter and receiver could be an important factor in modeling the signal propagation by obstructing the LOS between the communicating vehicles. Despite this, virtually all of the state-of-the-art VANET simulators consider the vehicles as dimensionless entities that have no influence on signal propagation Martinez et al. (2009).

This motivated our study on the impact of vehicles as obstacles on V2V communication described in Boban et al. (2010) and presented in the next section.

5. Model for incorporating vehicles as obstacles in VANET simulation environments

5.1 Empirical measurements

In order to quantify the impact that the vehicles have on the received signal strength, we performed experimental measurements. To isolate the effect of the obstructing vehicles, we aimed at setting up a controlled environment without other obstructions and with minimum

	Dimensions (m)		
Vehicle	Height	Width	Length
2002 Lincoln LS (TX)	1.453	1.859	4.925
2009 Pontiac Vibe (RX)	1.547	1.763	4.371
2010 Ford E-250 (Obstruction)	2.085	2.029	5.504

Table 1. Dimensions of Vehicles

impact of other variables (e.g., other moving objects, electromagnetic radiation, etc). For this reason, we performed experiments in an empty parking lot in Pittsburgh, PA (Fig. 3). We analyzed the received signal strength for the no obstruction, LOS case, and the non-LOS case where we introduced an obstructing vehicle (the van shown in Fig. 3) between the transmitter (Tx) and the receiver (Rx) vehicles. The received signal strength was measured for the distances of 10, 50, and 100 m between the Tx and the Rx. In case of the non-LOS experiments, the obstructing van was placed in the middle between the Tx and the Rx. We performed experiments at two frequency bands: 2.4 GHz (used by the majority of commercial WiFi devices) and 5.9 GHz (the band at which spectrum has been allocated for automotive use worldwide *IEEE Draft Standard IEEE P802.11p/D9.0* (July 2009)). For 2.4 GHz experiments, we equipped the Tx and Rx vehicles with laptops that had Atheros 802.11b/g wireless cards installed and we used 3 dBi gain omnidirectional antennas. For 5.9 GHz experiments, we equipped the Tx and Rx vehicles with NEC Linkbird-MX devices Festag et al. (2008), which communicate via IEEE 802.11p *IEEE Draft Standard IEEE P802.11p/D9.0* (July 2009) wireless interfaces and we used 5 dBi gain omnidirectional antennas. In both cases, antennas were mounted on the rooftops of the Tx and Rx vehicles (Fig. 3). The dimensions of the vehicles are shown in Table 1, and the height of the antennas used in both experiments was 260 mm. The transmission power was set to 18 dBm. The Atheros wireless cards in laptops as well as IEEE 802.11p radios in LinkBird-MXs were evaluated beforehand using a real time spectrum analyzer and no significant power fluctuations were observed. The central frequency was set to 2.412 GHz and 5.9 GHz, respectively, and the channel width was 20 MHz. The data rate for 2.4 GHz experiments was 1 Mb/s, with 10 messages (140 bytes in size) sent per second using the ping command, whereas for 5.9 GHz experiments the data rate was 6 Mb/s (the lowest data rate in 802.11p for 20 MHz channel width) with 10 beacons Festag et al. (2008) (36 bytes in size) sent per second. Each measurement was performed for at least 120 seconds, thus resulting in a minimum of 1200 data packets transmitted per measurement. We collected the per-packet Received Signal Strength Indication (RSSI) information.

Figures 4a and 4b show the RSSI for the LOS (no obstruction) and non-LOS (van obstructing the LOS) measurements at 2.4 GHz and 5.9 GHz, respectively. The additional attenuation at both central frequencies ranges from approx. 20 dB at 10 m distance between Tx and Rx to 4 dB at 100 m. Even though the absolute values for the two frequencies differ (resulting mainly from the different quality radios used for 2.4 GHz and 5.9 GHz experiments), the relative trends indicate that the obstructing vehicles attenuate the signal more significantly the closer the Tx and Rx are. To provide more insight into the distribution of the received signal strength for LOS and non-LOS measurements, Fig. 5 shows the cumulative distribution function (CDF) of the RSSI measurements for 100 m in case of LOS and non-LOS at 2.4 GHz. The non-LOS case exhibits a larger variation and the two distributions are overall significantly different, thus clearly showing the impact of the obstructing van. Similar distributions were observed for other distances between the Tx and the Rx.

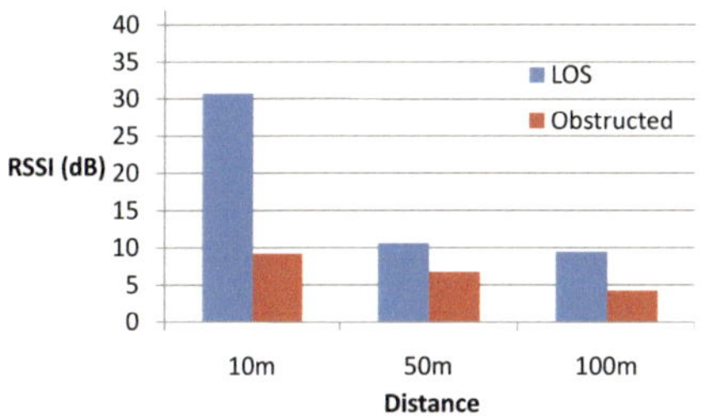

(a) 2.4 GHz

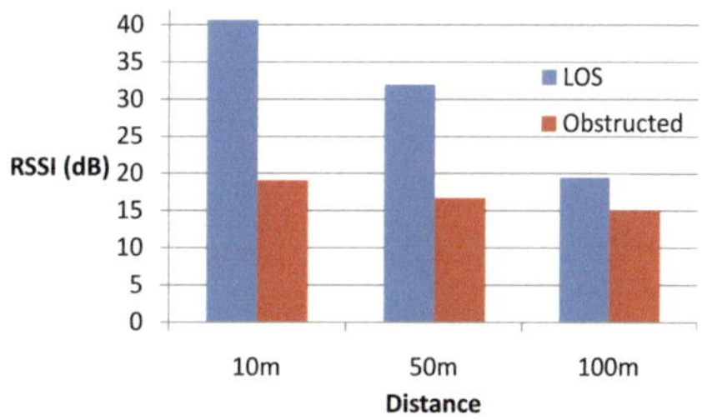

(b) 5.9 GHz

Fig. 4. RSSI measurements: average RSSI with and without the obstructing vehicle.

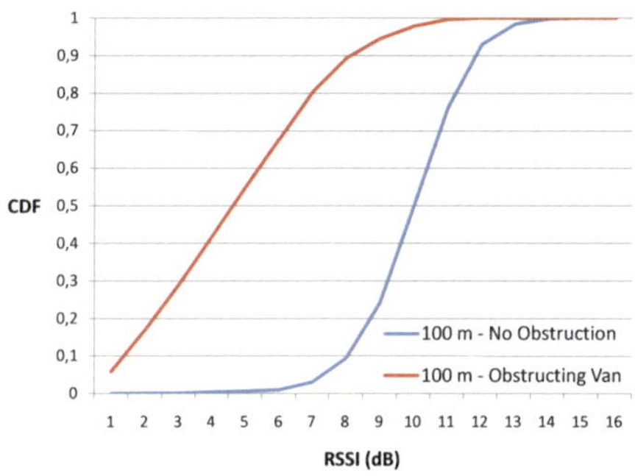

Fig. 5. Distribution of the RSSI for 100 m in case of LOS (no obstruction) and non-LOS (obstructing van) at 2.4 GHz.

5.2 Model analysis

5.2.1 The impact of vehicles on line of sight

In order to isolate and quantify the effect of vehicles as obstacles on signal propagation, we do not consider the effect of other obstacles such as buildings, overpasses, vegetation, or other roadside objects on the analyzed highways. Since those obstacles can only further reduce the probability of LOS, our approach leads to a best case analysis for probability of LOS.

Figure 6 describes the methodology we use to quantify the impact of vehicles as obstacles on LOS in a V2V environment. Using aerial imagery (Fig. 6a) to obtain the location and length of vehicles, we devise a model that is able to analyze all possible connections between vehicles within a given range (Fig. 6b). For each link – such as the one between the vehicles designated as transmitter (Tx) and receiver (Rx) in Fig. 6b – the model determines the existence or non-existence of the LOS based on the number and dimensions of vehicles potentially obstructing the LOS (in case of the aforementioned vehicles designated as Tx and Rx, the vehicles potentially obstructing the LOS are those designated as Obstacle 1 and Obstacle 2 in Fig. 6b).

The proposed model calculates the (non-)existence of the LOS for each link (i.e., between all communicating pairs) in a deterministic fashion, based on the dimensions of the vehicles and their locations. However, in order to make the model mathematically tractable, we derive the expressions for the microscopic (i.e., per-link and per-node) and macroscopic (i.e., system-wide) probability of LOS. It has to be noted that, from the electromagnetic wave propagation perspective, the LOS is not guaranteed with the existence of the visual sight line

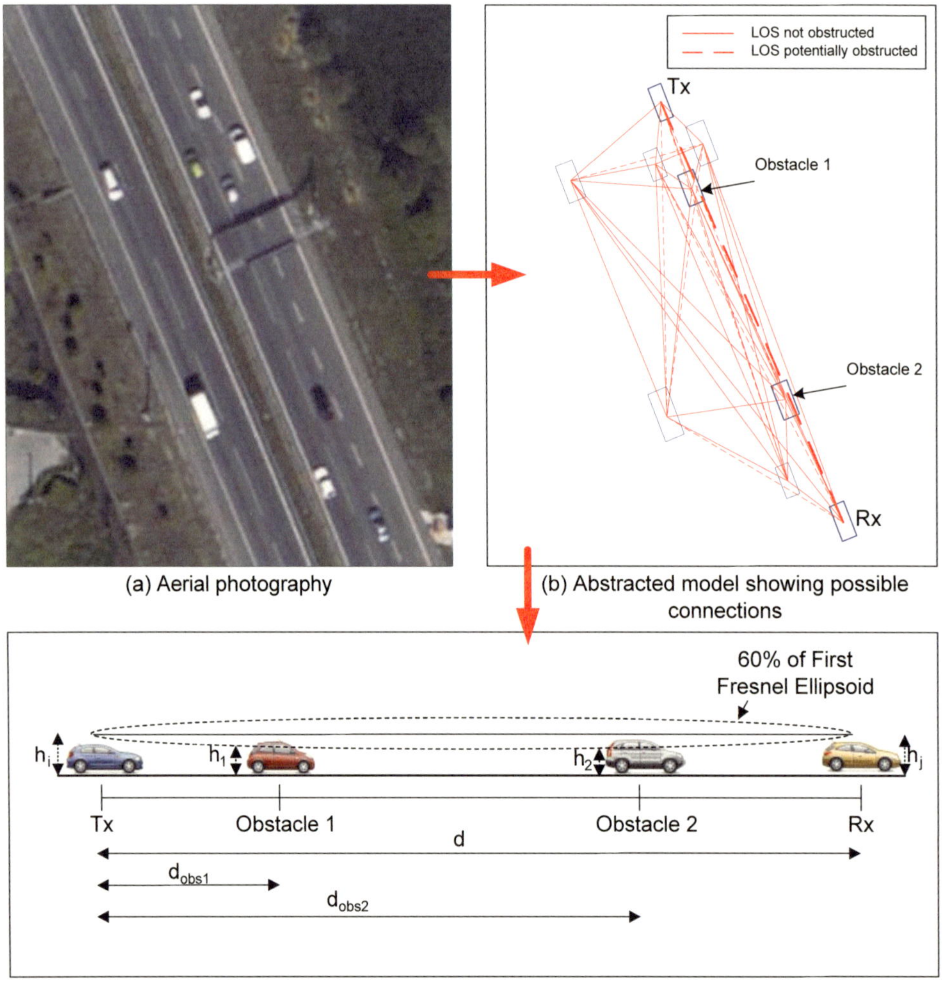

(a) Aerial photography

(b) Abstracted model showing possible connections

(c) P(LOS) calculation for a given link

Fig. 6. Model for evaluating the impact of vehicles as obstacles on LOS (for simplicity, vehicle antenna heights (h_a) are not shown in subfigure (c)).

between the Tx and Rx. It is also required that the Fresnel ellipsoid is free of obstructions (Rappaport, 1996, Chap. 3). Any obstacle that obstructs the Fresnel ellipsoid might affect the transmitted signal. As the distance between the transmitter and receiver increases, the diameter of the Fresnel ellipsoid increases accordingly. Besides the distance between the Tx and Rx, the Fresnel ellipsoid diameter is also a function of the wavelength.

As we will show later in Section 5.3, the vehicle heights follow a normal distribution. To calculate $P(LOS)_{ij}$, i.e., the probability of LOS for the link between vehicles i and j, with one vehicle as a potential obstacle between Tx and Rx (of height h_i and h_j, respectively), we have:

$$P(LOS|h_i,h_j) = 1 - Q\left(\frac{h-\mu}{\sigma}\right) \tag{1}$$

and

$$h = (h_j - h_i)\frac{d_{obs}}{d} + h_i - 0.6r_f + h_a, \tag{2}$$

where the i,j subscripts are dropped for clarity, and h denotes the effective height of the straight line that connects Tx and Rx at the obstacle location when we consider the first Fresnel ellipsoid. Furthermore, $Q(\cdot)$ represents the Q-function, μ is the mean height of the obstacle, σ is the standard deviation of the obstacle's height, d is the distance between the transmitter and receiver, d_{obs} is the distance between the transmitter and the obstacle, h_a is the height of the antenna, and r_f is the radius of the first Fresnel zone ellipsoid which is given by

$$r_f = \sqrt{\frac{\lambda d_{obs}(d - d_{obs})}{d}},$$

with λ denoting the wavelength. We use the appropriate λ for the proposed standard for VANET communication (DSRC), which operates in the 5.9 GHz frequency band. In our studies, we assume that the antennas are located on top of the vehicles in the middle of the roof (which was experimentally shown to be the overall optimum placement of the antenna Kaul et al. (2007)), and we set the h_a to 10 cm. As a general rule commonly used in literature, LOS is considered to be unobstructed if intermediate vehicles obstruct the first Fresnel ellipsoid by less than 40% (Rappaport, 1996, Chap. 3). Furthermore, for N_o vehicles as potential obstacles between the Tx and Rx, we get (see Fig. 6c)

$$P(LOS|h_i,h_j) = \prod_{k=1}^{N_o}\left[1 - Q\left(\frac{h_k-\mu_k}{\sigma_k}\right)\right], \tag{3}$$

where h_k is the effective height of the straight line that connects Tx and Rx at the location of the k-th obstacle considering the first Fresnel ellipsoid, μ_k is the mean height of the k-th obstacle, and σ_k is the standard deviation of the height of the k-th obstacle.

Averaging over the transmitter and receiver antenna heights with respect to the road, we obtain the unconditional $P(LOS)_{ij}$

$$P(LOS)_{ij} = \int\int P(LOS|h_i,h_j)p(h_i)p(h_j)dh_i dh_j, \tag{4}$$

where $p(h_i)$ and $p(h_j)$ are the probability density functions for the transmitter and receiver antenna heights with respect to the road, respectively.

The average probability of LOS for a given vehicle i, $P(LOS)_i$, and all its N_i neighbors is defined as

$$P(LOS)_i = \frac{1}{N_i}\sum_{j=1}^{N_i} P(LOS)_{ij} \tag{5}$$

To determine the system-wide ratio of LOS paths blocked by other vehicles, we average $P(LOS)_i$ over all N_v vehicles in the system, yielding

$$\overline{P(LOS)} = \frac{1}{N_v}\sum_{i=1}^{N_v} P(LOS)_i. \tag{6}$$

Furthermore, we analyze the behavior of the probability of LOS for a given vehicle i over time. Let us denote the i-th vehicle probability of LOS at a given time t as $P(LOS)_i^t$. We define the change in the probability of LOS for the i-th vehicle over two snapshots at times t_1 and t_2 as

$$\Delta P(LOS)_i = |P(LOS)_i^{t_2} - P(LOS)_i^{t_1}|, \tag{7}$$

where $P(LOS)_i^{t_1}$ and $P(LOS)_i^{t_2}$ are obtained using (5).
It is important to note that equations (1) to (7) depend on the distance between the node i and the node j (i.e., transmitter and receiver) in a *deterministic manner*. More specifically, the snapshot obtained from aerial photography provides the exact distance d (Fig. 6c) between the nodes i and j. While in our study we used aerial photography to get this information, any VANET simulator would also provide the exact location of vehicles based on the assumed mobility model (e.g., car-following Rothery (1992), cellular automata Tonguz et al. (2009), etc.), hence the distance d between the nodes i and j would still be available. This also explains why the proposed model is independent of the simulator used, since it can be incorporated into any VANET simulator, regardless of the underlying mobility model, as long as the locations of the vehicles are available. Furthermore, even though we used the highway environment for testing, the proposed model can be used for evaluating the impact of obstructing vehicles on any type of road, irrespective of the shape of the road (e.g., single or multiple lanes, straight or curvy) or location (e.g., highway, suburban, or urban[1]).

5.2.2 The impact of vehicles on signal propagation

The attenuation on a radio link increases if one or more vehicles intersect the ellipsoid corresponding to 60% of the radius of the first Fresnel zone, independent of their positions on the Tx-Rx link (Fig. 6c). This increase in attenuation is due to the diffraction of the electromagnetic waves. The additional attenuation due to diffraction depends on a variety of factors: the obstruction level, the carrier frequency, the electrical characteristics, the shape of the obstacles, and the amount of obstructions in the path between transmitter and receiver. To model vehicles obstructing the LOS, we use the knife-edge attenuation model. It is reasonable to expect that more than one vehicle can be located between transmitter (Tx) and receiver (Rx). Thus, we employ the multiple knife-edge model described in ITU-R recommendation ITU-R (2007). When there are no vehicles obstructing the LOS between the Tx and Rx, we use

[1]However, to precisely quantify the impact of obstructing vehicles in complex urban environments, further research is needed to determine the interplay between the vehicle-induced obstruction and the obstruction caused by other objects (e.g., buildings, overpasses, etc.).

the free space path loss model Goldsmith (2006)[2].

Single Knife-Edge
The simplest obstacle model is the knife-edge model, which is a reference case for more complex obstacle models (e.g., cylinder and convex obstacles). Since the frequency of DSRC radios is 5.9 GHz, the knife-edge model theoretically presents an adequate approximation for the obstacles at hand (vehicles), as the prerequisite for the applicability of the model, namely a significantly smaller wavelength than the size of the obstacles ITU-R (2007), is fulfilled (the wavelength of the DSRC is approximately 5 cm, which is significantly smaller than the size of the vehicles).
The obstacle is seen as a semi-infinite perfectly absorbing plane that is placed perpendicular to the radio link between the Tx and Rx. Based on the Huygens principle, the electric field is the sum of Huygens sources located in the plane above the obstruction and can be computed by solving the Fresnel integrals Parsons (2000). A good approximation for the additional attenuation (in dB) due to a single knife-edge obstacle A_{sk} can be obtained using the following equation ITU-R (2007):

$$A_{sk} = \begin{cases} 6.9 + 20\log_{10}\left[\sqrt{(v-0.1)^2+1}+v-0.1\right]; \\ \qquad \text{for } v > -0.7 \\ 0; \text{ otherwise,} \end{cases} \tag{8}$$

where $v = \sqrt{2}H/r_f$, H is the difference between the height of the obstacle and the height of the straight line that connects Tx and Rx, and r_f is the Fresnel ellipsoid radius.

Multiple Knife-Edge
The extension of the single knife-edge obstacle case to the multiple knife-edge is not immediate. All of the existing methods in the literature are empirical and the results vary from optimistic to pessimistic approximations Parsons (2000). The method in Epstein & Peterson (1953) presents a more optimistic view, whereas the methods in Deygout (1966) and Giovaneli (1984) are more pessimistic approximations of the real world. Usually, the pessimistic methods are employed when it is desirable to guarantee that the system will be functional with very high probability. On the other hand, the more optimistic methods are used when analyzing the effect of interfering sources in the communications between transmitter and receiver. To calculate the additional attenuation due to vehicles, we employ the ITU-R method ITU-R (2007), which can be seen as a modified version of the Epstein-Patterson method, where correcting factors are added to the attenuation in order to better approximate reality.

5.3 Model requirements

The model proposed in the previous section is aimed at evaluating the impact of vehicles as obstacles using geometry concepts and relies heavily on realistic modeling of the physical environment. In order to employ the proposed model accurately, realistic modeling of the following physical properties is necessary: determining the exact position of vehicles and the inter-vehicle spacing; determining the speed of vehicles; and determining the vehicle dimensions.

[2]We acknowledge the fact that the free space model might not be the best approximation of the LOS communication on the road. However, due to its tractability, it allows us to analyze the relationship between the LOS and non-LOS conditions in a deterministic manner.

Dataset	Size	# vehicles	# large vehicles	Veh. density
A28	12.5 km	404	58 (14.36%)	32.3 veh/km
A3	7.5 km	55	10 (18.18%)	7.3 veh/km

Table 2. Analyzed highway datasets

Determining the exact position of vehicles and the inter-vehicle spacing

The position and the speed of vehicles can easily be obtained from any currently available VANET mobility model. However, in order to test our methodology with the most realistic parameters available, we used *aerial photography*. This technique is used by the traffic engineering community as an alternative to ground-based traffic monitoring McCasland, W T (1965), and was recently applied to VANET connectivity analysis Ferreira et al. (2009). It is well suited to characterize the physical interdependencies of signal propagation and vehicle location, because it gives the exact position of each vehicle. We analyzed two distinct data sets, namely two Portuguese highways near the city of Porto, A28 and A3, both with four lanes (two per direction). Detailed parameters for the two datasets are presented in Table 2. For an extensive description of the method used for data collection and analysis, we refer the reader to Ferreira et al. (2009).

Determining the speed of vehicles

For the observed datasets, besides the exact location of vehicles and the inter-vehicle distances, stereoscopic imagery was once again used to determine the speed and heading of vehicles. Since the successive photographs were taken with a fixed time interval (5 seconds), by marking the vehicles on successive photographs we were able to measure the distance the vehicle traversed, and thus infer the speed and heading of the vehicle. The measured speed and inter-vehicle spacing is used to analyze the behavior of vehicles as obstacles while they are moving.

The distribution of inter-vehicle spacing for both cases can be well fitted with an *exponential probability distribution*. This agrees with the empirical measurements made on the I-80 interstate in California reported in Wisitpongphan et al. (Oct. 2007). The speed distribution on both highways is well approximated by a normal probability distribution. Table 3 shows the parameters of best fits for inter-vehicle distances and speeds.

Determining the vehicle dimensions

From the photographs, we were also able to obtain the length of each vehicle accurately, however the width and height could not be determined with satisfactory accuracy due to resolution constraints and vehicle mobility. To assign proper widths and heights to vehicles, we use the data made available by the Automotive Association of Portugal *Associação Automóvel de Portugal* (n.d.), which issued an official report about all vehicles currently in circulation in Portugal. From the report we extracted the eighteen most popular personal vehicle brands which comprise 92% of all personal vehicles circulating on Portuguese roads, and consulted an online database of vehicle dimensions *Automotive Technical Data and Specifications* (n.d.) to arrive at the distribution of height and width required for our analysis. The dimensions of the most popular personal vehicles showed that both the vehicle widths and heights can be modeled as a normal random variable. Detailed parameters for the fitting process for both personal and large vehicles are presented in Table 4. For both width and height of personal vehicles, the standard error for the fitting process remained below 0.33% for both the mean and the standard deviation. The data regarding the specific

Data for A28		
Parameter	**Estimate**	**Std. Error**
Speed: normal fit		
mean (km/h)	106.98	1.05
std. deviation (km/h)	21.09	0.74
Inter-vehicle spacing : exponential fit		
mean (m)	51.58	2.57
Data for A3		
Parameter	**Estimate**	**Std. Error**
Speed: normal fit		
mean (km/h)	122.11	3.97
std. deviation (km/h)	28.95	2.85
Inter-vehicle spacing : exponential fit		
mean (m)	215.78	29.92

Table 3. Parameters of the Best Fit Distributions for Vehicle Speed and Inter-vehicle spacing

types of large vehicles (e.g., trucks, vans, or buses) currently in circulation was not available. Consequently, the precise dimension distributions of the most representative models could not be obtained. For this reason, we infer large vehicle height and width values from the data available on manufacturers' websites, which can serve as rough dimension guidelines that show significantly different height and width in comparison to personal vehicles.

5.4 Computational complexity of the proposed model

In order to determine the LOS conditions between two neighboring nodes, we analyzed the existence of LOS in a three dimensional space, as shown in Fig. 6 and explained in the previous sections. Our model for determining the existence of LOS between vehicles and, in case of obstruction, obtaining the number and location of the obstructions, belongs to a class of computational geometry problems known as geometric intersection problems de Berg et al. (1997), which deal with pairwise intersections between line segments in an n-dimensional space. These problems occur in various contexts, such as computer graphics (object occlusion) and circuit design (crossing conductors), amongst others Bentley & Ottmann (1979).
Specifically, for a given number of line segments N, we are interested in determining, reporting, and counting the pairwise intersections between the line segments. For our specific application, the line segments of interest are of two kinds: a) the LOS rays between the communicating vehicles (lines colored red in Fig. 6b); and b) the lines that compose the bounding rectangle representing the vehicles (lines colored blue in Fig. 6b). It has to be noted that the intersections of interest are only those between the LOS rays and the bounding rectangle lines, and not between the lines of the same type. Therefore, we arrive at a special case of the segment intersection problem, namely the so-called "red-blue" intersection problem. Given a set of red line segments r and a set of blue line segments b, with a total of $N = r + b$ segments, the goal is to report all K intersections between red and blue segments, for which Agarwal in Agarwal (1991) presented an efficient algorithm. The time-complexity of the algorithm proposed in Agarwal (1991), using the randomized approach of Clarkson (1987), is $\mathcal{O}(N^{4/3} \log N + K)$, where K is the number of red-blue intersections, with space complexity of $\mathcal{O}(N^{4/3})$. This algorithm fits our purposes perfectly, as the red segments correspond to the LOS rays between the communicating vehicles and blue segments are the lines of the

Personal vehicles	
Parameter	**Estimate**
Width: normal fit	
mean (cm)	175
std. deviation (cm)	8.3
Height: normal fit	
mean (cm)	150
std. deviation (cm)	8.4
Large vehicles	
Parameter	**Estimate**
Width: constant	
mean (cm)	250
Height: normal fit	
mean (cm)	335
std. deviation (cm)	8.4

Table 4. Parameters of the Best Fit Distributions for Vehicle Width and Height

bounding rectangles representing the vehicles (see Fig. 6b).
To assign physical values to r and b, we denote v as the number of vehicles in the system and v' as the number of transmitting vehicles. The number of LOS rays results in $r = Cv'$, where the average number of neighbors C is an increasing function of the vehicle density and transmission range. The number of lines composing the bounding vehicle rectangles can be expressed as $b = 4v$, since each vehicle is represented by four lines forming a rectangle (see Fig. 6b). Therefore, a more specific time-complexity bound can be written as $\mathcal{O}((Cv' + 4v)^{4/3}\log(Cv' + 4v) + K)$.
Apart from the algorithm for determining the red-blue intersections, the rest of the proposed model consists in calculating the additional signal attenuation due to vehicles for each communicating pair. In the case of non-obstructed LOS the algorithm terminates, whereas for obstructed LOS, the red-blue intersection algorithm is used to store the number and location of intersecting blue lines (representing obstacles). The total number of intersections is given by $K = gr$, where g is the number of obstacles (i.e., vehicles) in the LOS path and is a subset of C. The complete algorithm for additional attenuation due to vehicles is implemented as follows.

Algorithm 1 Calculate additional attenuation due to vehicles

for $i = 1$ to r **do**
  $[coord] = getIntersect(i)$ {For each LOS ray in r, obtain the location of intersections as per Agarwal (1991)}
  if $size([coord]) \neq 0$ **then**
    $att = calcAddAtten([coord])$ {Calculate the additional attenuation due to vehicles as per ITU-R (2007)}
  else
    $att = 0$ dB {Additional attenuation due to vehicles equals zero.}
  end if
end for

Highways			
	Transmission Range (m)		
Highway	100	250	500
A3 $\overline{P(LOS)}$	0.8445	0.6839	0.6597
A28 $\overline{P(LOS)}$	0.8213	0.6605	0.6149

Table 5. $\overline{P(LOS)}$ for A3 and A28

The function $getIntersect(\cdot)$ is based on the aforementioned red-blue line intersection algorithm Agarwal (1991), and has complexity $\mathcal{O}((Cv' + 4v)^{4/3}\log(Cv' + 4v) + gr)$, whereas the function $calcAddAtten(\cdot)$ is based on multiple knife-edge attenuation model described in ITU-R (2007) with time-complexity of $\mathcal{O}(g^2)$ for each LOS ray r. It follows that the time-complexity of the entire algorithm is given by $\mathcal{O}((Cv' + 4v)^{4/3}\log(Cv' + 4v) + g^2 r)$.
In order to implement the aforementioned algorithm in VANET simulators, apart from the information available in the current VANET simulators, very few additional pieces of information are necessary. Specifically, the required information pertains to the physical dimensions of the vehicles. Apart from this, the model only requires the information on the position of the vehicles at each simulation time step. This information is available in any vehicular mobility model currently in use in VANET simulators.

5.5 Results

We implemented the model described in previous sections in Matlab. In this section we present the results based on testing the model using the A3 and A28 datasets. We also present the results of the empirical measurements that we performed in order to characterize the impact of the obstructing vehicles on the received signal strength. We emphasize that the model developed in the paper is not dependent on these datasets, but can be used in any environment by applying the analysis presented in Section 5.2. Furthermore, the observations pertaining to the inter-vehicle and speed distributions on A3 and A28 are used only to characterize the behavior of the highway environment over time. We do not use these distributions in our model; rather, we use actual positions of the vehicles. Since the model developed in Section 5.2 is intended to be utilized by VANET simulators, the positions of the vehicles can easily be obtained through the employed vehicular mobility model.
We first give evidence that vehicles as obstacles have a significant impact on LOS communication in both sparse (A3) and more dense (A28) networks. Next, we analyze the microscopic probability of LOS to determine the variation of the LOS conditions over time for a given vehicle. Then, we used the speed and heading information to characterize both the microscopic and macroscopic behavior of the probability of LOS on highways over time in order to determine how often the proposed model needs to be recalculated in the simulators, and to infer the stationarity of the system-wide probability of LOS. Using the employed multiple knife-edge model, we present the results pertaining to the decrease of the received power and packet loss for DSRC due to vehicles. Finally, we corroborate our findings on the impact of the obstructing vehicles and discuss the appropriateness of the knife-edge model by performing empirical measurements of the received signal strength in LOS and non-LOS conditions.

5.5.1 Probability of line of sight

Macroscopic probability of line of sight. Table 5 presents the values of $\overline{P(LOS)}$ with respect to the observed range on highways. The highway results show that even for the sparsely populated

A3 highway the impact of vehicles on $\overline{P(LOS)}$ is significant. This can be explained by the exponential inter-vehicle spacing, which makes it more probable that the vehicles are located close to each other, thus increasing the probability of having an obstructed link between two vehicles. For both highways, it is clear that the impact of other vehicles as obstacles can not be neglected even for vehicles that are relatively close to each other (for the observed range of 100 m, $\overline{P(LOS)}$ is under 85% for both highways, which means that there is a non-negligible 15% probability that the vehicles will not have LOS while communicating). To confirm these results, Fig. 7 shows the average number of neighbors with obstructed and unobstructed LOS for the A28 highway. The increase of obstructed vehicles in both absolute and relative sense is evident.

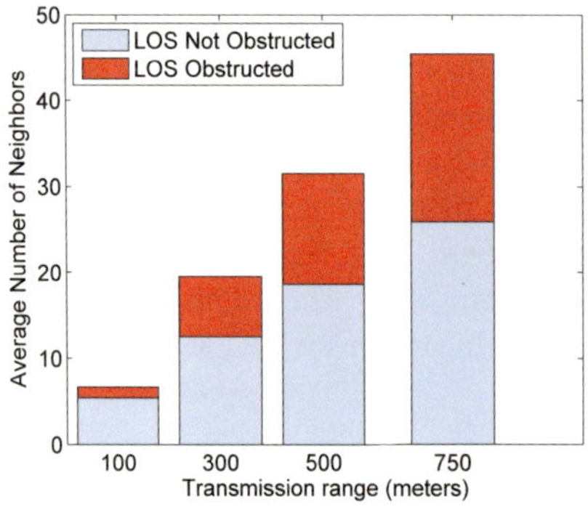

Fig. 7. Average number of neighbors with unobstructed and obstructed LOS on A28 highway.

Microscopic probability of line of sight. In order to analyze the variation of the probability of LOS for a vehicle and its neighbors over time, we observe the $\Delta P(LOS)_i$ (as defined in equation (7)) on A28 highway for the maximum communication range of 750 m. Table 6 shows the $\Delta P(LOS)_i$. The variation of probability of LOS is moderate for periods of seconds (even for the largest offset of 2 seconds, only 15% of the nodes have the $\Delta P(LOS)_i$ greater than 20%). This result suggests that the LOS conditions between a vehicle and its neighbors will remain largely unchanged for a period of seconds. Therefore, a simulation time-step of the order of seconds can be used for calculations of the impact of vehicles as obstacles. From a simulation execution standpoint, the time-step of the order of seconds is quite a long time when compared with the rate of message transmission, measured in milliseconds; this enables a more efficient and scalable design and modeling of vehicles as obstacles on a microscopic, per-vehicle level. With the proper implementation of the LOS intersection model discussed in Sections 5.2 and 5.4 , the modeling of vehicles as obstacles should not induce a large overhead in the simulation execution time.

	$\Delta P(LOS)_i$ in %			
Time offset	$< 5\%$	5-10%	10-20%	>20%
1ms	100%	0%	0%	0%
10ms	99%	1%	0%	0%
100ms	82%	15%	3%	0%
1s	35%	33%	22%	10%
2s	31%	25%	29%	15%

Table 6. Variation of $P(LOS)_i$ over time for the observed range of 750 m on A28.

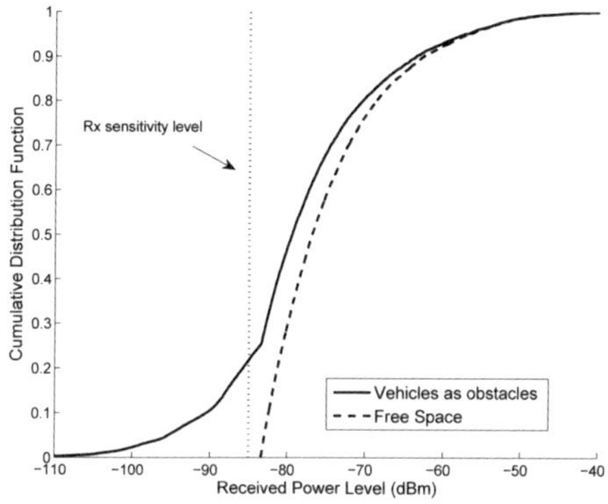

Fig. 8. The impact of vehicles as obstacles on the received signal power on highway A28.

Data Rate (Mb/s)	Modulation	Minimum sensitivity (dBm)
3	BPSK	−85
4.5	BPSK	−84
6	QPSK	−82
9	QPSK	−80
12	QAM-16	−77
18	QAM-16	−70
24	QAM-64	−69
27	QAM-64	−67

Table 7. Requirements for DSRC Receiver Performance

5.5.2 Received power

Based on the methodology developed in Section 5.2, we utilize the multiple knife-edge model to calculate the additional attenuation due to vehicles. We use the obtained attenuation to calculate the received signal power for the DSRC. We employed the knife-edge model for its simplicity and the fact that it is well studied and often used in the literature. However, we point out that the LOS analysis and the methodology developed in Section 5.2 can be used in conjunction with any channel model that relies on the distinction between the LOS and NLOS communication (e.g., Zang et al. (2005) or Wang et al. (2004)).

For the A28 highway and the observed range of 750 m, with the transmit power set to 18 dBm, 3 dBi antenna gain for both transmitters and receivers, at the 5.9 GHz frequency band, the results for the free space path loss model Goldsmith (2006) (i.e., not including vehicles as obstacles) and our model that accounts for vehicles as obstacles are shown in Fig. 8. The average additional attenuation due to vehicles was 9.2 dB for the observed highway.

Using the minimum sensitivity thresholds as defined in the DSRC standard (see Table 7) *Standard Specification for Telecommunications and Information Exchange Between Roadside and Vehicle Systems - 5GHz Band Dedicated Short Range Communications (DSRC) Medium Access Control (MAC) and Physical Layer (PHY) Specifications* (Sep. 2003), we calculate the packet success rate (PSR, defined as the ratio of received messages to sent messages) as follows. We analyze all of the communicating pairs within an observed range, and calculate the received signal power for each message. Based on the sensitivity thresholds presented in Table 7, we determine whether a message is successfully received. For the A28 highway, Fig. 9 shows the PSR difference between the free space path loss and the implemented model with vehicles as

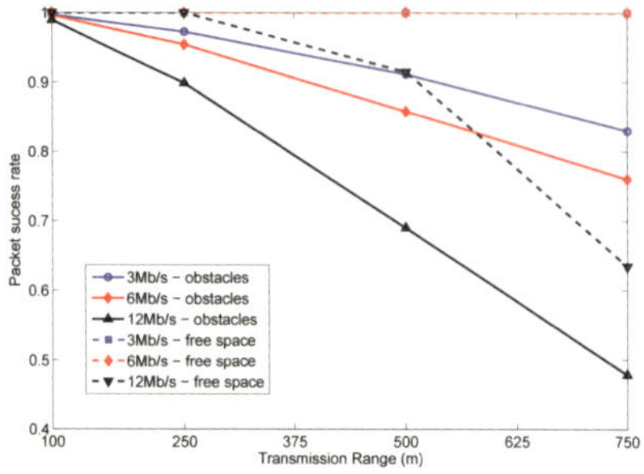

Fig. 9. The impact of vehicles as obstacles on packet success rate for various DSRC data rates on A28 highway.

obstacles for rates of 3, 6, and 12 Mb/s. The results show that the difference is significant, as the percentage of lost packets can be up to 25% higher when vehicles are accounted for.
These results show that not only do the vehicles significantly decrease the received signal power, but the resulting received power is highly variable even for relatively short distances between the communicating vehicles, thus calling for a microscopic, per-vehicle analysis of the impact of obstructing vehicles. Models that try to average the additional attenuation due to vehicles could fail to describe the complexity of the environment, thus yielding unrealistic results. Furthermore, the results show that the distance itself can not be solely used for determining the received power, since even the vehicles close by can have a number of other vehicles obstructing the communication path and therefore the received signal power becomes worse than for vehicles further apart that do not have obstructing vehicles between them.

5.6 Discussion

In this section, we describe the impact that the obtained results have on various aspects of V2V communication modeling, ranging from physical to application layer to the realism of VANET simulators.

5.6.1 Impact on signal propagation modeling

The results presented in this paper clearly indicate that vehicles as obstacles have a significant impact on signal propagation; therefore, in order to properly model V2V communication, it is imperative that vehicles as obstacles are accounted for. Furthermore, the effect of vehicles as obstacles cannot be neglected even in the case of relatively sparse vehicular networks, as one of the two analyzed highway datasets showed (namely, the dataset collected on the A3 highway). Therefore, previous efforts pertaining to signal propagation modeling in V2V communication which do not account for vehicles as obstacles, can be deemed as optimistic in overestimating the received signal power level.

5.6.2 Impact on data link layer

Neglecting vehicles as obstacles on the physical layer has profound effects on the performance of upper layers of the communication stack. The effects on the data link layer are twofold: a) the medium contention is overestimated in models that do not include vehicles as obstacles in the calculation, thus potentially representing a more pessimistic situation than the real-world with regards to contention and collision; and b) the network reachability is bound to be overestimated, due to the fact that the signal is considered to reach more neighbors and at a higher power than in the real world. These results have important implications for vehicular

Medium Access Control (MAC) protocol design; MAC protocols will have to cope with an increased number of hidden vehicles due to other vehicles obstructing them.

5.6.3 Impact on the design of routing protocols

If vehicles as obstacles are not accounted for, the impact on routing protocols is represented by an overly optimistic hop count; in the process of routing, next hop neighbors are selected that are actually not within the reach of the current transmitter, thus inducing an unrealistic behavior of the routing protocol, as the message is considered to reach the destination with a smaller number of hops than it is actually required.

As an especially important class of routing protocols, safety messaging protocols, are often modeled and evaluated using distance information only. As our results have shown, not accounting for vehicles as obstacles in such calculations results in the overestimation of the number of reachable neighbors, which yields unrealistic results with regards to network reachability and message penetration rate. Therefore, it is extremely important to account for vehicles as obstacles in V2V, especially since safety applications running over such protocols require that practically all vehicles receive the message, thus posing very stringent requirements on the routing protocols.

For these reasons, it is more beneficial to design routing protocols that rely primarily on the received signal strength instead of the geographical location of vehicles, since this would ensure that the designated recipient is actually able to receive the message. However, even with smart protocols that are able to properly evaluate the channel characteristics between the vehicles, in case of lower market penetration rates of the communicating equipment, the vehicles that are not equipped could significantly hinder the communication between the equipped vehicles; this is another aspect of routing protocol design that is significantly affected by the impact of vehicles as obstacles in V2V communication.

Similarly, the results suggest that, where available, vehicle-to-infrastructure (V2I) communication (where vehicles are communicating with road side equipment) should be favored instead of V2V communication; since the road side equipment is supposed to be placed in lamp posts, traffic lights, or on the gantries above the highways such as the one in the Fig. 6a), all of which are located 3-6 meters above ground level, other vehicles as obstacles would impact the LOS much less than in the case of V2V communication. Therefore, similarly to differentiating vehicles with regards to their dimensions, routing protocols would benefit from being able to differentiate between the road side equipment and vehicles.

5.6.4 Impact on VANET simulations

VANET simulation environments have largely neglected the modeling of vehicles as obstacles in V2V communication. Results presented in this paper showed that the vehicles have a significant impact on the LOS, and in order to realistically model the V2V communication in simulation environments, vehicles as obstacles have to be accounted for. This implies that the models that relied on the simulation results that did not account for vehicles as obstacles have been at best producing an optimistic upper bound of the results that can be expected in the real world.

In order to improve the realism of the simulators and to enable the implementation of a scalable and realistic framework for describing the vehicles as obstacles in V2V communication, we proposed a simple yet realistic model for determining the probability of LOS on both macroscopic and microscopic level. Using the results that proved the stationarity of the probability of LOS, we showed that the average probability of LOS does not change over time if the vehicle arrival rate remains constant. Furthermore, over a period of seconds, the LOS conditions remain mostly constant even for the microscopic, per-vehicle case. This implies that the modeling of the impact of vehicles as obstacles can be performed at the rate of seconds, which is two to three orders of magnitude less frequent than the rate of message exchange (most often, messages are exchanged on a millisecond basis). Therefore, with the proper implementation of the proposed model, the calculation of the impact of vehicles on LOS should not induce a large overhead in the simulation execution time.

6. Conclusions

We discussed the state-of-the-art in VANET modeling and simulation, and described the building blocks of VANET simulation environments, namely the mobility, networking and signal propagation models. We described the most important models for each of these categories, and we emphasized that several areas are not optimally represented in state-of-the-art VANET simulators. Namely, the vehicle interaction and traffic rule enforcement models in most current simulators leave a lot to be desired, and the lack of WAVE and DSRC protocol implementation in the simulators is also a fact for most simulators. Finally, we pointed out that the models for moving obstacles are lacking in modern simulators, and we described our proposed model for vehicles as physical obstacles in VANETs as follows. First, using the experimental data collected in a measurement campaign, and by utilizing the real world data collected by means of stereoscopic aerial photography, we showed that vehicles as obstacles have a significant impact on signal propagation in V2V communication; in order to realistically model the communication, it is imperative that vehicles as obstacles are accounted for. The obtained results point out that vehicles are an important factor in both highway and urban, as well as in sparse and dense networks. Next, we characterized the vehicles as three-dimensional objects that can obstruct the LOS between the communicating pair. Then, we modeled the vehicles as physical obstacles that attenuate the signal, which allowed us to determine their impact on the received signal power, and consequently on the packet error rate. The presented model is computationally efficient and, as the results showed, can be updated at a rate much lower than the message exchange rate in VANETs. Therefore, it can easily be implemented in any VANET simulation environment to increase the realism.

7. References

Acosta, G. & Ingram, M. (2006). Model development for the wideband expressway vehicle-to-vehicle 2.4 ghz channel, *IEEE Wireless Communications and Networking Conference, 2006. WCNC 2006.*, Vol. 3, pp. 1283–1288.

Agarwal, P. K. (1991). *Intersection and Decomposition Algorithms for Planar Arrangements*, Cambridge University Press.

Associação Automóvel de Portugal (n.d.). URL: *http://www.acap.pt/*

Automotive Technical Data and Specifications (n.d.). URL: *http://www.carfolio.com/*

Bai, F., Elbatt, T., Hollan, G., Krishnan, H. & Sadekar, V. (2006). Towards characterizing and classifying communication-based automotive applications from a wireless networking perspective, *1st IEEE Workshop on Automotive Networking and Applications (AutoNet)*.

Bai, F., Sadagopan, N. & Helmy, A. (2003). IMPORTANT: a framework to systematically analyze the impact of mobility on performance of routing protocols for adhoc networks, *INFOCOM 2003. Twenty-Second Annual Joint Conference of the IEEE Computer and Communications Societies. IEEE* 2: 825–835 vol.2.

Bentley, J. & Ottmann, T. (1979). Algorithms for reporting and counting geometric intersections, *Computers, IEEE Transactions on* C-28(9): 643–647.

Boban, M., Tonguz, O. & Barros, J. (2009). Unicast communication in vehicular ad hoc networks: a reality check, *IEEE Communications Letters* 13(12): 995–997.

Boban, M., Vinhoza, T. T. V., Barros, J., Ferreira, M. & Tonguz, O. K. (2010). Impact of vehicles as obstacles in vehicular ad hoc networks, *IEEE Journal on Selected Areas in Communications (to appear)* .

Cheng, Y. & Robertazzi, T. (Jul. 1989). Critical Connectivity Phenomena in Multihop Radio Models, *IEEE Transactions on Communications* 37(7): 770–777.

Choffnes, D. R. & Bustamante, F. E. (2005). An integrated mobility and traffic model for vehicular wireless networks, *VANET '05: Proceedings of the 2nd ACM international workshop on Vehicular ad hoc networks*, ACM, New York, NY, USA, pp. 69–78.

Clarkson, K. (1987). New applications of random sampling in computational geometry, *Discrete and Computational Geometry* 2: 195–222.

Conceição, H., Damas, L., Ferreira, M. & Barros, Jo a. (2008). Large-scale simulation of v2v environments, *SAC '08: Proceedings of the 2008 ACM symposium on Applied computing*, ACM, New York, NY, USA, pp. 28–33.

CORSIM: Microscopic Traffic Simulation Model (n.d.).
URL: *http://www-mctrans.ce.ufl.edu/featured/TSIS/*

Davies, J. J., Beresford, A. R. & Hopper, A. (2006). Scalable, distributed, real-time map generation, *IEEE Pervasive Computing* 5(4): 47–54.

de Berg, M., van Kreveld, M., Overmars, M. & Schwarzkopf, O. (1997). *Computational Geometry Algorithms and Applications*, Springer-Verlag.

Deygout, J. (1966). Multiple knife-edge diffraction of microwaves, *IEEE Transactions on Antennas and Propagation* 14(4): 480–489.

Epstein, J. & Peterson, D. W. (1953). An experimental study of wave propagatin at 850MC, *Proceedings of the IRE* 41(5): 595–611.

Ferreira, M., Conceição, H., Fernandes, R. & Tonguz, O. K. (2009). Stereoscopic Aerial Photography: An Alternative to Model-Based Urban Mobility Approaches, *Proceedings of the Sixth ACM International Workshop on VehiculAr Inter-NETworking (VANET 2009)*, ACM New York, NY, USA.

Festag, A., Baldessari, R., Zhang, W., Le, L., Sarma, A. & Fukukawa, M. (2008). Car-2-x communication for safety and infotainment in europe, *NEC Technical Journal* 3(1).

Giordano, E., Frank, R., Pau, G. & Gerla, M. (2010). Corner: a realistic urban propagation model for vanet, *WONS'10: Proceedings of the 7th international conference on Wireless on-demand network systems and services*, IEEE Press, Piscataway, NJ, USA, pp. 57–60.

Giovaneli, C. L. (1984). An analysis of simplified solutions for multiple knife-edge diffraction, *IEEE Transactions on Antennas and Propagation* 32(3): 297–301.

Gipps, P. G. (1986). A model for the structure of lane-changing decisions, *Transportation Research Part B: Methodological* 20(5): 403–414.

Goldsmith, A. J. (2006). *Wireless Communications*, Cambridge University Press.

Gupta, P. & Kumar, P. (2000). The Capacity of Wireless Networks, *IEEE Transactions on Information Theory* 46(2): 388–404.

Harri, J. (2010). Vehicular mobility modeling for vanet, *VANET Vehicular Applications and*

Inter-Networking Technologies, Wiley, pp. 107–152.
Harri, J., Filali, F. & Bonnet, C. (2009). Mobility models for vehicular ad hoc networks: a survey and taxonomy, *IEEE Communications Surveys & Tutorials* 11(4): 19–41.
Helbing, D. (2001). Traffic and related self-driven many-particle systems, *Rev. Mod. Phys.* 73(4): 1067–1141.
Ho, I. W., Leung, K. K., Polak, J. W. & Mangharam, R. (2007). Node connectivity in vehicular ad hoc networks with structured mobility, *32nd IEEE Conference on Local Computer Networks, LCN 2007.* pp. 635–642.
Hoogendoorn, S. & Bovy, P. (2001). Generic gas-kinetic traffic systems modeling with applications to vehicular traffic flow.
IEEE Draft Standard IEEE P802.11p/D9.0 (July 2009). *Technical report.*
IEEE Trial-Use Standard for Wireless Access in Vehicular Environments (WAVE) - Multi-channel Operation (2006). *IEEE Std 1609.4-2006* pp. c1–74.
IEEE Trial-Use Standard for Wireless Access in Vehicular Environments (WAVE) - Networking Services (Apr. 2007). *IEEE Std 1609.3-2007* pp. c1–87.
ITU-R (2007). Propagation by diffraction, *Recommendation P.526*, International Telecommunication Union Radiocommunication Sector, Geneva.
Jerbi, M., Marlier, P. & Senouci, S. M. (2007). Experimental assessment of V2V and I2V communications, *Proc. IEEE Internatonal Conference on Mobile Adhoc and Sensor Systems (MASS 2007)*, pp. 1–6.
Jin, W. (2003). Kinematic wave models of network vehicular traffic. Ph.D. Dissertation, UC Davis.
Kaul, S., Ramachandran, K., Shankar, P., Oh, S., Gruteser, M., Seskar, I. & Nadeem, T. (2007). Effect of antenna placement and diversity on vehicular network communications, *Proc. IEEE SECON.*, pp. 112–121.
Koberstein, J., Witt, S. & Luttenberger, N. (2009). Model complexity vs. better parameter value estimation: comparing four topography-independent radio models, *Simutools '09: Proceedings of the 2nd International Conference on Simulation Tools and Techniques*, ICST (Institute for Computer Sciences, Social-Informatics and Telecommunications Engineering), ICST, Brussels, Belgium, Belgium, pp. 1–8.
Kotz, D., Newport, C., Gray, R. S., Liu, J., Yuan, Y. & Elliott, C. (2004). Experimental evaluation of wireless simulation assumptions, *Proc. ACM MSWiM '04*, ACM, New York, NY, USA, pp. 78–82.
Lighthill, M. J. & Whitham, G. B. (1955). On Kinematic Waves. II. A Theory of Traffic Flow on Long Crowded Roads, *Royal Society of London Proceedings Series A* 229: 317–345.
Mangharam, R., Weller, D. S., Stancil, D. D., Rajkumar, R. & Parikh, J. S. (2005). Groovesim: a topography-accurate simulator for geographic routing in vehicular networks, *VANET '05: Proceedings of the 2nd ACM international workshop on Vehicular ad hoc networks*, ACM, New York, NY, USA, pp. 59–68.
Martinez, F. J., Toh, C. K., Cano, J.-C., Calafate, C. T. & Manzoni, P. (2009). A survey and comparative study of simulators for vehicular ad hoc networks (VANETs), *Wireless Communications and Mobile Computing* .
Matolak, D., Sen, I., Xiong, W. & Yaskoff, N. (2005). 5 ghz wireless channel characterization for vehicle to vehicle communications, *Proc. IEEE Military Communications Conference (MILCOM 2005)*, Vol. 5, pp. 3016–3022.
Maurer, J., Fugen, T., Schafer, T. & Wiesbeck, W. (2004). A new inter-vehicle communications (ivc) channel model, *Vehicular Technology Conference, 2004. VTC2004-Fall. 2004 IEEE*

60th, Vol. 1, pp. 9–13 Vol. 1.

McCasland, W T (1965). Comparison of Two Techniques of Aerial Photography for Application in Freeway Traffic Operations Studies, *Photogrammetry and Aerial Surveys* .

Murthy, C. S. R. & Manoj, B. (2004). *Ad Hoc Wireless Networks: Architectures and Protocols*, Prentice Hall PTR, Upper Saddle River, NJ, USA.

Nagel, K. & Schreckenberg, M. (1992). A cellular automaton model for freeway traffic, *J. de Physique* 2: 2221.

Nagel, R. & Eichler, S. (2008). Efficient and realistic mobility and channel modeling for vanet scenarios using omnet++ and inet-framework, *Simutools '08: Proceedings of the 1st international conference on Simulation tools and techniques for communications, networks and systems & workshops*, ICST, Brussels, Belgium, Belgium, pp. 1–8.

Naumov, V., Baumann, R. & Gross, T. (2006). An evaluation of inter-vehicle ad hoc networks based on realistic vehicular traces, *MobiHoc '06: Proceedings of the 7th ACM international symposium on Mobile ad hoc networking and computing*, New York, NY, USA, pp. 108–119.

Network Simulator 2 (n.d.).
URL: *http://www.isi.edu/nsnam/ns/*

Open Street Map Project (n.d.).
URL: *http://www.openstreetmap.org*

Otto, J. S., Bustamante, F. E. & Berry, R. A. (2009). Down the block and around the corner – the impact of radio propagation on inter-vehicle wireless communication, *Proc. of IEEE International Conference on Distributed Computing Systems (ICDCS)*.

Palazzi, C. E., Ferretti, S., Roccetti, M., Pau, G. & Gerla, M. (2007). How do you quickly choreograph inter-vehicular communications? a fast vehicle-to-vehicle multihop broadcast algorithm, explained, *in Proc. of the 3rd IEEE CCNC International Workshop on Networking Issues in Multimedia Entertainment (CCNC/NIME 2007), Las Vegas, NV, USA, IEEE Communications Society*.

Parsons, J. D. (2000). *The Mobile Radio Propagation Channel*, John Wiley & Sons.

Piórkowski, M., Raya, M., Lugo, A. L., Papadimitratos, P., Grossglauser, M. & Hubaux, J.-P. (2008). Trans: realistic joint traffic and network simulator for vanets, *SIGMOBILE Mob. Comput. Commun. Rev.* 12(1): 31–33.

Rappaport, T. S. (1996). *Wireless Communications: Principles and Practice*, Prentice Hall.

Rothery, R. W. (1992). Car following models, *In Trac Flow Theory* .

Saha, A. K. & Johnson, D. B. (2004). Modeling mobility for vehicular ad-hoc networks, *in* K. P. Laberteaux, R. Sengupta, C.-N. Chuah & D. Jiang (eds), *Vehicular Ad Hoc Networks*, ACM, pp. 91–92.

Sommer, C., Yao, Z., German, R. & Dressler, F. (2008). Simulating the influence of IVC on road traffic using bidirectionally coupled simulators, *27th IEEE Conference on Computer Communications (IEEE INFOCOM 2008): Mobile Networking for Vehicular Environments (IEEE MOVE 2008)*, IEEE, Phoenix, AZ, pp. 1–6.

Standard Specification for Telecommunications and Information Exchange Between Roadside and Vehicle Systems - 5GHz Band Dedicated Short Range Communications (DSRC) Medium Access Control (MAC) and Physical Layer (PHY) Specifications (Sep. 2003). *ASTM E2213-03* .

SUMO - Simulation of Urban MObility (n.d.).
URL: *http://sumo.sourceforge.net*

Takahashi, S., Kato, A., Sato, K. & Fujise, M. (2003). Distance dependence of path loss for millimeter wave inter-vehicle communications, *Proc. IEEE 58th Vehicular Technology Conference (VTC 2003-Fall)*, Vol. 1, pp. 26–30.

Tonguz, O. K. & Boban, M. (2010). Multiplayer games over vehicular ad hoc networks: A new application, *Ad Hoc Networks* 8(5): 531 – 543.

Tonguz, O. K., Viriyasitavat, W. & Bai, F. (2009). Modeling urban traffic: a cellular automata approach, *Comm. Mag.* 47(5): 142–150.

Treiber, M., Hennecke, A. & Helbing, D. (2000). Congested traffic states in empirical observations and microscopic simulations, *Phys. Rev. E* 62(2): 1805–1824.

U.S. Census Bureau TIGER system database (n.d.).
URL: *http://www.census.gov/geo/www/tiger*

Uzcátegui, R. A. & Acosta-Marum, G. (2009). Wave: a tutorial, *Comm. Mag.* 47(5): 126–133.

Vehicle Safety Communications Project, Final Report (2006). *Technical Report DOT HS 810 591*, U.S. Department of Transportation, NHTSA, Crash Avoidance Metrics Partnership.

Wang, S. Y., Chou, C. L., Huang, C. H., Hwang, C. C., Yang, Z. M., Chiou, C. C. & Lin, C. C. (2003). The design and implementation of the nctuns 1.0 network simulator, *Computer Networks* 42(2): 175 - 197.

Wang, Z., Tameh, E. & Nix, A. (2004). Statistical peer-to-peer channel models for outdoor urban environments at 2 ghz and 5 ghz, *Vehicular Technology Conference, 2004. VTC2004-Fall. 2004 IEEE 60th*, Vol. 7, pp. 5101–5105 Vol. 7.

Wisitpongphan, N., Bai, F., Mudalige, P., Sadekar, V. & Tonguz, O. (Oct. 2007). Routing in Sparse Vehicular Ad Hoc Wireless Networks, *IEEE Journal on Selected Areas in Communications* 25(8): 1538–1556.

Wu, H., Palekar, M., Fujimoto, R., Guensler, R., Hunter, M., Lee, J. & Ko, J. (2005). An empirical study of short range communications for vehicles, *Proc. of the 2nd ACM International workshop on Vehicular ad hoc networks*, pp. 83–84.

Zang, Y., Stibor, L., Orfanos, G., Guo, S. & Reumerman, H.-J. (2005). An error model for inter-vehicle communications in highway scenarios at 5.9ghz, *PE-WASUN '05: Proceedings of the 2nd ACM international workshop on Performance evaluation of wireless ad hoc, sensor, and ubiquitous networks*, ACM, New York, NY, USA, pp. 49–56.

4

Security Issues in Vehicular Ad Hoc Networks

P. Caballero-Gil
University of La Laguna
Spain

1. Introduction

Communications are becoming more wireless and mobile than ever. Thus, in the near future, we can expect that vehicles will be equipped with wireless devices, which will enable the formation of Vehicular Ad Hoc NETworks (VANETs). The main goal of these wireless networks will consist in providing safety and comfort to passengers, but their structure will be also taken advantage with many different aims, such as commercial, access to Internet, notification, etc.

From a general point of view, the basic idea of a VANET is straightforward as it can be seen as a particular form of Mobile Ad hoc NETwork (MANET). Consequently, in a first approach we could think on considering well-known and widely adopted solutions for MANETs and install them on VANETs. However, as explained in this chapter, that proposal would not work properly.

A VANET is a wireless network that does not rely on any central administration for providing communication among the so-called On Board Units (OBUs) in nearby vehicles, and between OBUs and nearby fixed infrastructure usually named Road Side Unit (RSU). In this way, VANETs combine Vehicle TO Vehicle (V2V) also known as Inter-Vehicle Communication (IVC) with Vehicle TO Infrastructure (V2I) and Infrastructure TO Vehicle (I2V) communications (see Figure 1).

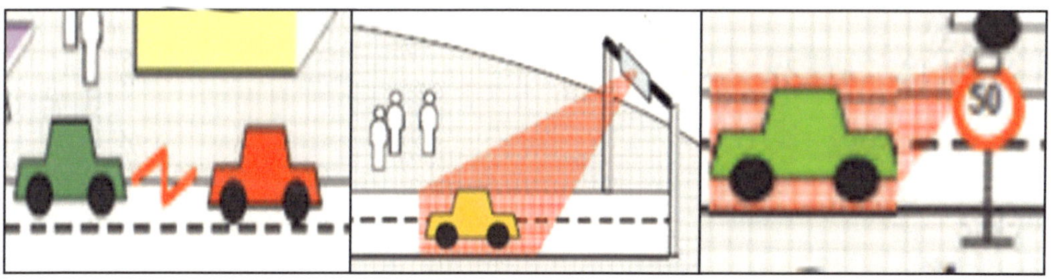

Fig. 1. V2V, V2I & I2V Communications

On the one hand, OBUs in vehicles will broadcast periodic messages with the information about their position, time, direction, speed, etc., and also warnings in case of emergency. On the other hand, RSUs on the roads will broadcast traffic related messages.

Additional communications can be also useful depending on the specific application. Among all these messages, routine traffic-related will be one hop broadcast, while emergency warnings will be transmitted through a multi hop path where the receiver of each warning will continue broadcasting it to other vehicles. In this way, drivers are expected to get a better awareness of their driving environment so that in case of an abnormal situation they will be able to take early action in order to avoid any possible damage or to follow a better route.

VANETs are expected to support a wide variety of applications, ranging from safety-related to notification and other value-added services. However, before putting such applications into practice, different security issues such as authenticity and integrity must be solved because any malicious behaviour of users, such as modification and replay attacks with respect to disseminated traffic-related messages, could be fatal to other users.

Moreover, privacy-regarding user information such as driver's name, license plate, model, and travelling route must also be protected. On the other hand, in the case of a dispute such as an accident scene investigation, the authorities should be able to trace the identities of the senders to discover the reason of the accident or look for witnesses. Therefore, specific security mechanisms for VANETs must be developed (Hubaux et al., 2004).

Great attention both from industry and academia has been received to this promising network scenario, and standards for wireless communications in VANETs are nowadays under preparation. In particular, IEEE 802.11p is a draft standard for Wireless Access in Vehicular Environment (WAVE), and IEEE 1609 is a higher layer standard on which IEEE 802.11p is based. At a superior level, Communications, Air-interface, Long and Medium (CALM) range is an initiative to define a set of wireless communication protocols and air interfaces for the so-called Intelligent Transportation System (ITS).

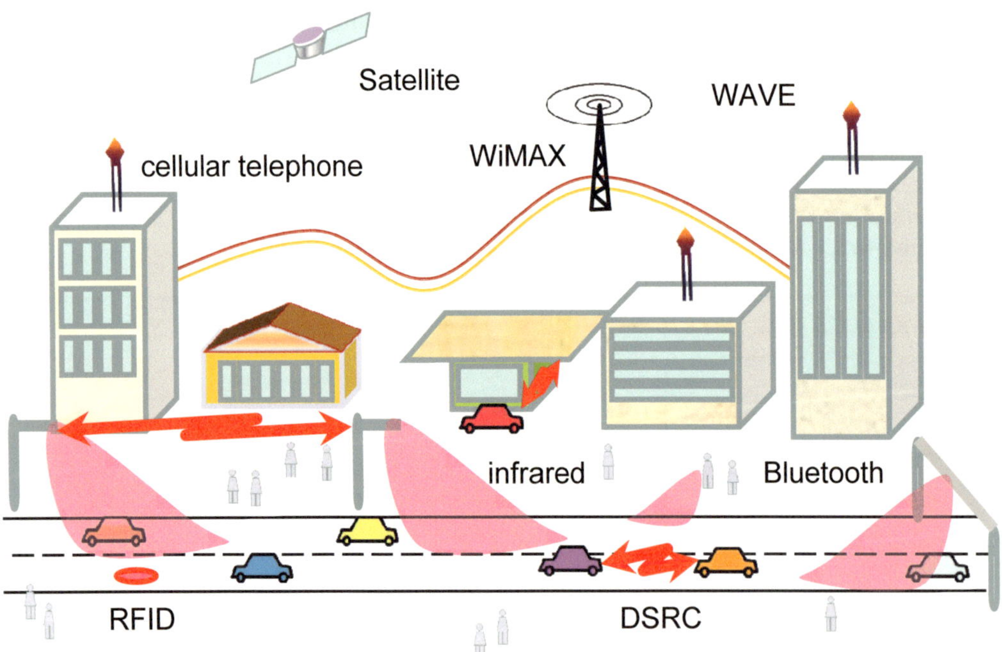

Fig. 2. Convergence of technologies

It is foreseeable that VANETs will combine a variety of wireless methods of transmission used by CALM and based on different types of communication media such as WAVE, infrared, cellular telephone, 5.9 GHz Dedicated Short-Range Communication (DSRC), WiMAX, Satellite, Bluetooth, RFID, etc. The current state of all these standards is trial use (see Figure 2).

In this way, the field of vehicular applications and technologies will be based on an interdisciplinary effort from the sectors of communication and networking, automotive electronics, road operation and management, and information and service provisioning. Without cooperation among the different participants, practical and wide deployment of VANETs will be difficult, if not impossible.

In the future it could be expected that each vehicle will have as part of its equipment: a black box (EDR, Event Data Recorder), a registered identity (ELP, Electronic License Plate), a receiver of a Global Navigation Satellite System like GPS (Global Positioning System) or Galileo, sensors to detect obstacles at a distance lesser than 200 ms, and some special device that provides it with connectivity to an ad hoc network formed by the vehicles, allowing the node to receive and send messages through the network (see Figure 3). One of the most interesting components of this future vehicle is the ELP, which would securely broadcast the identity of the vehicle.

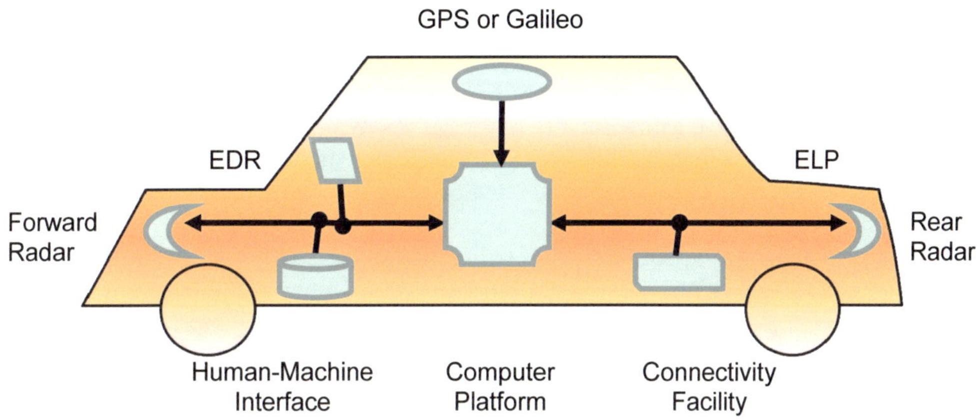

Fig. 3. Components of a future vehicle

Two hypotheses that are necessary to guarantee the protection of a VANET are that security devices are reliable and tamper-proof, and that the information received through sensors is also trustworthy. It is generally assumed by most authors that messages sent through the VANET may be digitally signed by the sender with a public-key certificate.

This certificate is assumed to be emitted by a Certification Authority (CA) that is admitted as reliable by the whole network. The moments corresponding to the vehicle purchase and to the periodic technical inspections are proposed to be respectively associated to the emission and renovation of its public-key certificate. In general, symmetric authentication is acknowledged by most authors as not a valid option due to important factors in VANETs such as time and scalability (Raya & Hubaux, 2005).

Different security challenges of vehicular networks are here addressed, paying special attention to the application of several known security primitives such as symmetric and asymmetric cryptography, strong authentication, data aggregation and cooperation enforcement.

In particular, the chapter is organized as follows. A brief summary of the main characteristics of VANETs is included in Section 2. Section 3 classifies their most important applications while Section 4 describes several security threats and challenges in VANETs. The following section introduces definitions of basic cryptographic requirements and drafts of several solutions that other researchers have proposed to provide these networks with security. Section 6 briefly describes some security schemes here proposed to protect VANET authenticity, privacy and integrity. Finally, Section 7 concludes the chapter by highlighting conclusions and open problems.

2. Characteristics

There are several general security requirements, such as authenticity, scalability, privacy, anonymity, cooperation, stability and low delay of communications, which must be considered in any wireless network, and which in VANETs are even more challenging because of their specific characteristics such as high mobility, no fixed infrastructure and frequently changing topology that range from rural road scenarios with little traffic to cities or highways with a huge number of communications.
Consequently, VANET security may be considered one of the most difficult and technically challenging research topics that need to be taken into account before the design and wide deployment of VANETs (Caballero-Gil, Hernández-Goya & Fúster-Sabater, 2009).
Among the main key technical challenges the following issues can be remarked:

- The lack of a centralized infrastructure in charge of synchronization and coordination of transmissions makes that one of the hardest tasks in the resulting decentralized and self-organizing VANETs is the management of the wireless channel to reach an efficient use of its bandwidth.
- High node mobility, solution scalability requirements and wide variety of environmental conditions are three of the most important challenges of these decentralized self-organizing networks. A particular problem that has to be faced comes from the high speeds of vehicles in some scenarios such as highways. These characteristics collude with most iterative algorithms intended to optimize the use of the channel bandwidth or of predefined routes.
- Security and privacy requirements in VANETs have to be balanced. On the one hand, receivers want to make sure that they can trust the source of information but on the other hand, this might disagree with privacy requirements of the sender.
- The radio channel in VANET scenarios present critical features for developing wireless communications, which degrade strength and quality of signals.
- The need for standardization of VANET communications should allow flexibility as these networks have to operate with many different brands of equipment and vehicle manufacturers.
- Real-time communication is a necessary condition because no delay can exist in the transmission of safety-related information. This implies that VANET communication requires fast processing and exchange of information.
- The existence of a central registry of vehicles, possible periodic contact with it, and qualified mechanisms for the exigency of fulfilment of the law are three usual assumptions that are necessary for some proposed solutions.
- Communication for information exchange is based on node-to-node connections. This distributed nature of the network implies that nodes have to relay on other nodes to

make decisions, for instance about route choice, and also that any node in a VANET can act either as a host requesting information or a router distributing data, depending on the circumstances.

Another interesting characteristic is the dependency of confidentiality requirements on specific applications. On the one hand, secret is not needed when the transmitted information is related to road safety, but on the other hand, it is an important requirement in some commercial applications (Caballero-Gil et al., 2010).

As aforementioned, VANETs can be seen as a specific type of MANET. However, the usual assumption of these latter networks about that nodes have strict restrictions on their power, processing and storage capacities does not appear in VANETs. Another difference with respect to pure MANETs is that in vehicular networks, we can consider that access to a fixed infrastructure along the roadside is possible when RSU is available either directly or through routing.

When developing a simulation of a VANET (see Figure 4), some special features have to be considered:

- Each vehicle generally moves according to a road network pattern and not at random like in MANETs.
- The movement patterns of vehicles are normally occasional, that is to say, they stop, move, park, etc.
- Vehicles must respect speed limitations and traffic signals.
- The behaviour of each vehicle depends on the behaviour of its neighbour vehicles as well as on the road type.
- VANETs can provide communication over 5-10 Km.
- Two nodes cannot exist in the same location at the same time.
- Nodes usually travel at an average speed lower than 120 Km/h.

Fig. 4. Example of simulation

Despite the aforementioned differences between MANETs and VANETs, some security tools designed for their use in MANETs have been evaluated for their possible application in VANETs (Füßler et al., 2007).

Such as it happens in MANETs, in VANETs the nodes are in charge of package routing. Up to now, several routing protocols originally defined for MANETs have been adapted to VANETs following different approaches.

Reactive protocols designed for MANETs such as Ad hoc On-demand Distance Vector (AODV) and Dynamic Source Routing (DSR) have been modified to be used in VANETs. Nevertheless, simulation results do not indicate a good performance due to the highly unstable routes. Consequently, we can conclude that those adaptations might be successfully used only in small VANETs.

In other routing protocols based on geographic location of nodes, the decisions related to package routing are taken based on street guides, traffic models and data collected with global positioning systems available in the vehicles.
According to simulations, this type of protocols based on geographic information seems to be the most promising for its use in different types of sceneries such as cities and highways. In particular, in VANETs it might be useful to send messages only to nodes in a precise geographic zone. Specific routing protocols with this characteristic have been designed, and mentioned in the bibliography as geocast routing. This way to proceed allows disseminating information only to interested nodes (for instance, in case of an accident, only to proximal vehicles, and in case of an advertisement, only to nodes that are in the zone of the advertised service). In (Li & Wang, 2007) a comparative study among different routing schemes is presented.
Also like in MANETs, routing in VANETs basically follows two ways of action:

- Proactive: All vehicles periodically broadcast messages on their present states (beacons) containing their ELP, position, timestamp, speed, etc., and resend such messages if it is necessary.
- Reactive: Each vehicle sends messages only after it detects an incident, generates a request, or must resend a received message.

We have an example of how to take advantage of the proactive mode when a parked vehicle is witness of an accident thanks to its sensors, and stores the corresponding data in its EDR, so that they could be later used to determine liabilities.
In the proactive mode, the frequent beacons are very costly. Furthermore, they imply the possibility of their use to track vehicles. This fact leads to the necessity of a solution that might consist in encrypted beacons. The high frequency of those beacons combined with the higher computational cost of asymmetric cryptography suggests the application of a hybrid solution combining it with symmetrical cryptography. This hybrid solution also seems the best option, independently of the routing protocol, for some specific applications.

3. Applications

After full deployment of VANETs, when vehicles can directly communicate with other vehicles and with the road side infrastructure, several safety and non-safety applications will be developed. Although less important, non-safety applications can greatly enhance road and vehicle efficiency and comfort.

3.1 Safety-Related

A possible application of VANETs for road safety, besides the warning dissemination of accidents or traffic jumps that constitute their main application, is the warning dissemination of danger before any accident or traffic jump has taken place. This would be the case for example of a high speed excess or a violation of a traffic signal (such as a traffic light or a stop sign). In these cases, when some vehicle detects a violation through its sensors, it must activate the automatic dissemination of warning messages communicating the fact to all neighbour vehicles in order to warn them about the danger.
An additional difficulty of this application is due to the fact that the dangerous vehicle is in motion. This implies that it is not clear what any vehicle that receives the message can do to avoid the danger without being able to identify the actual location of the guilty vehicle.

Another related application of VANETs in road safety is the warning dissemination of emergency vehicle approach.
The situations of vehicles that have suffered an accident or have met a traffic jump can be dealt in the same way as any other detection of anything that might be classified as an obstacle, such as extremely slow vehicles, results of possible natural phenomena on the road, stones, bad conditions of the pavement due to works on the road, or bad meteorological conditions like low visibility. In all these cases we have that the corresponding information is important for road safety, and that the incident can be characterized by a certain location and moment.
Consequently, in these cases of applications for driver assistance, the aforementioned hypothesis referring to the existence of a Global Navigation Satellite System in vehicles is fundamental because it allows locating both the own location and that of the detected incident (see Figure 5).

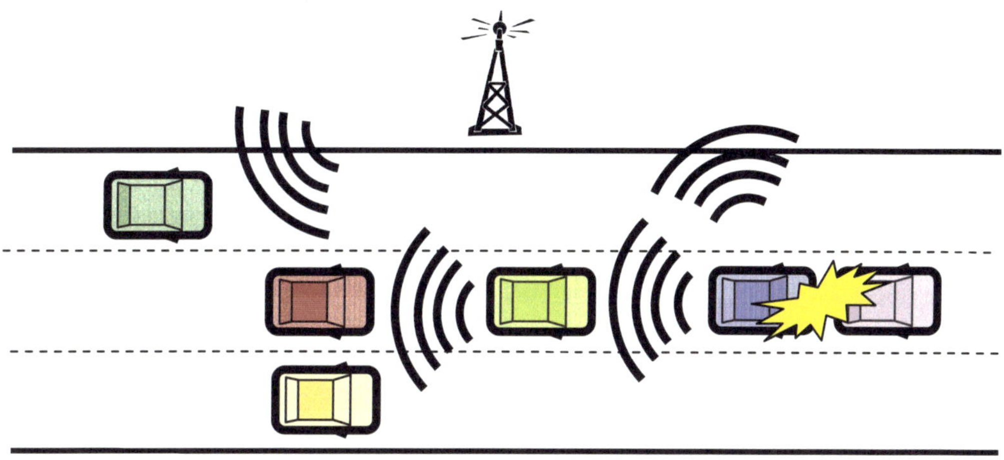

Fig. 5. Accident warning

Given the importance of the warnings of incidents for road safety, in these cases it would be advisable the use of an evaluation system of messages previous to their massive dissemination. For example, we could stipulate that in the scenery of the incident at least a minimum number of vehicles higher than a pre-established threshold activates or signs the same warning. This can be implemented for example by means of a voting scheme among the vehicles in the area nearby the incident.
In addition, note that with this proposal, possible Denegation of Service (DoS) attacks and sending of false warnings are prevented. In this sense, note that, although privacy is an important aspect in VANETs, its protection cannot stop the use of information by the authorities in order to establish responsibility in case of accident (Caballero-Gil et al., 2010).
On the other hand, it is foreseeable that the reception of a warning of abnormal and/or potentially dangerous incident will have influence in the behaviour of the other drivers. For that reason, in these schemes it is necessary to consider possible attacks based on trying to inject or to modify messages in order to obtain an effect like for example a road free of vehicles.
In order to inform cars in their vicinity to warn their drivers earlier of potential hazards, so that they have more time to react and avoid accidents, vehicles exhibiting abnormal driving

patterns, such as a dramatic change of direction, send messages including information derived from many sources like sensors, devices ABS, ESP, etc., use of airbags, speed, acceleration or deceleration of vehicle, as well as information originating from other sources like radars or video monitors, and SOS telephones or traffic lights used as repeaters to extend the dissemination rank of warnings.

From the combination of all these data, neighbouring vehicles can directly identify in many cases the type of incident by means of the interpretation of this information. A similar approach can be applied at intersections where cars communicate their current position and speed, making it possible to predict possible collisions between cars.

There is another important case that does not correspond exactly to a warning of an incident with a determined location and moment, but has also important implications in road safety. That is the case of a warning of the presence of an emergency vehicle like police, ambulance, fire-fighters, etc. In this case, the warning should include location, moment and foreseeable destiny or route of the emergency vehicle, and the objective is that the other vehicles can receive this information with enough time to clear the path of emergency vehicles in real-time, hence saving crucial time.

3.2 Non-safety-related

There is a whole variety of non-safety applications included in Value-Added Services (VASs), which can be provided through a VANET. Passengers in vehicles who spend a very long period in transit might be interested in certain application domain for vehicular networks consisting in the provision of many different types of information. Such information could be data about the surrounding area such as nearby businesses, services, facilities or road conditions, different entertainment-oriented services like Internet access (see Figure 6) or sharing multimedia contents with neighbours (Franz et al., 2005), and advertisement services (Lee et al., 2007). This diversity of possible applications comes from the fact that vehicular networks can be considered a form of pervasive network, that is to say, they operate anywhere and at any time.

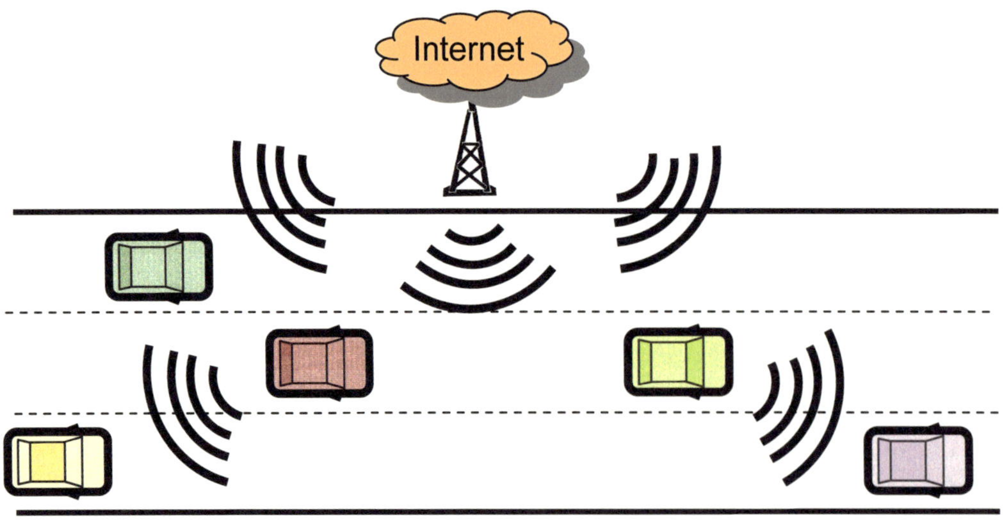

Fig. 6. Internet access

Vehicular networks could be also used for traffic monitoring. In particular, traffic authorities might be interested in obtaining information about road users so that for example they could get traffic flows to deduce current congestion levels and detect potential traffic jams.

In general, dissemination of that type of information among nodes can be used to manage traffic, not only in the aforementioned cases when an incident occurs, but also in normal conditions, when it can be used for the optimization of traffic flow.

Therefore, on the one hand, VANETs could be used for traffic management by extending drivers' horizons and supporting driving manoeuvres so that they provide drivers with information they might have missed or might not yet be able to see, in order to help them in decision making. A special traffic management application is a lane positioning system that uses inter-vehicle communication to improve GPS accuracy and provide lane-level positioning. Such detailed positioning allows the provision of services such as lane departure warning, as well as lane-level navigation systems.

On the other hand, if junctions are equipped with a controller that can either listen to communication between vehicles or receive messages from arriving vehicles, then the controller would be able to build an accurate view of the traffic at the junction through the aggregation of the received data corresponding to traffic conditions in the area, and could therefore adapt its behaviour to optimize the throughput. Traffic management applications could be also used to allow emergency vehicles to change traffic lights at signalized intersection in order to synchronize adequately to the objective of clearing the path.

An approach similar to the general case of traffic monitoring could be extended by the use of audio and video devices, which could be used for terrorist activities monitoring.

Closely related to traffic monitoring and a current particularly useful application of VANETs is traffic management. For instance, V2I solutions for road tolling are already deployed in certain places in the world to allow paying for road usage on congested roads, with prices depending on congestion levels. In the future, vehicular networks could enable that drivers are charged for their specific usage of the road network (Cottingham et al., 2007).

The idea of autonomous vehicles that are able to operate in urban areas while obeying traffic regulations is part of a collection of revolutionary applications called coordinated driving applications. This special type of safety-related applications improves performance and safety of participant vehicles through their collaboration with each other. Proposed coordinated driving applications focus mainly on three scenarios: adaptive cruise control, platooning and intersection management.

The simplest coordination application is adaptive cruise control, which performs control manoeuvres in order to maintain a safe distance for each vehicle to the vehicle in front by using forward sensors, wireless communication and cooperation among vehicles.

In a platoon, V2V communication is used to coordinate platoon members through a leader or a teamwork model in which autonomous vehicles follow a decentralized management scheme. The main benefits of platoon applications are: increase of road capacity and efficiency, reduction in congestion, energy consumption and pollution, and enhancement of safety and comfort. Demonstrations of cars travelling in platoons have already proven the feasibility of such a radical approach in certain protected settings. In particular, (Hedrick et al., 1994) and (Gehring & Fritz, 1997) have demonstrated the technique of coupling two or more vehicles together electronically to form a train.

Finally, the third mentioned coordinated driving application is intersection management for collaborative collision avoidance of autonomous vehicles while reducing delay in comparison to traffic lights or stop signs. This interesting application allows improving road safety through cooperative driving in dangerous road points where certain circumstances exist according to which several vehicles compete for a common critical point that all have to go over so that the VANET can offer support for certain driving manoeuvres. That is the case for example of the access to a highway or a road intersection without visibility or traffic lights, where it is convenient that vehicles act co-ordinately through group communications in order to avoid accidents.

Each application implies several important differences in the security schemes that are used. In order to use VANETs as practical support for advertisement dissemination, a system of incentives must be defined both for the advertiser and for the nodes of the VANET, so that both gain when disseminating the advertisements through the network (Caballero-Gil et al., 2009). In this sense, the VANET can offer several advantages because the driver would be aimed to listen to advertisements, and even to help in their dissemination, if it obtains something in return, for example, some valuable good as gasoline. Obviously, in these cases it is necessary to define measures to prevent possible frauds of those who try to gain without receiving/redistributing the advertisements.

A similar incentive-based approach might be used for other Value-Added Services, like for example, the supply and demand of useful information like alternative routes, near parking zones, gas stations, hotels, restaurants, access points to Internet, etc. In all these cases it is fundamental that the information is encrypted in order to prevent access to non-authorized users who have not paid for the service. These other VAS applications have some similarities and differences with respect to the described advertising support service.

Both in the case when the information is a warning of incident or emergency vehicle, and in the case of dissemination of publicity or other VAS, it is remarkable that the messages have a definite origin (crashed/in traffic jump/emergency/VAS applicant vehicle, or advertiser business) but do not have a unique and definite destiny, what has clear implications in security issues. In fact, in all those cases the objective is to disseminate the message to the largest number of nodes but with different optimization criteria. In order to achieve such a goal the origin broadcasts the message to all the vehicles within its neighbourhood.

There are several authors (Dousse et al., 2002); (Wischof et al., 2005) who have proposed different algorithms to optimize the propagation of information through a VANET depending on the road type, traffic density, vehicles speed, etc. For example, in highways, the authors of (Little & Agarwal, 2005) consider the possible formation of vehicle blocks, with more or less frequent gaps between blocks. Since these gaps could cause a temporal fragmentation of the network, in order to solve the problem, the authors propose the use of vehicles against the sense of the march for spreading communications.

4. Threats

VANETs represent a challenge in the field of communication security, as well as a revolution for vehicular safety and comfort in road transport. In some of the aforementioned applications, messages can influence on driver behaviour, and consequently on road safety. In other cases like certain VASs, they can have economic consequences. In any of these cases, VANET deployment must consider the possible existence of adversaries or attackers who try to exploit the different situations, for example by injecting false, modified or repeated messages or by impersonating vehicles. Therefore, the security of communications in VANETs is an essential factor to preventing all these threats.

Even though some physical security measures can help to defend certain vehicular components against manipulations, tamper-protection instruments rarely can help to identify attacks or threats. Hence, even perfect tamper-proof components like ELPs could be stolen and installed into another vehicle to carry out impersonation attacks. Consequently, it is necessary to develop security algorithms that help to guarantee the correct and secure operation of VANETs.

An attacker can be seen as an entity who wants to spread false information, interrupt communications, impersonate legitimate nodes, compromise their privacy, or take advantage of the network without cooperating in its normal operation.

Attacks can be categorized on the basis of the attackers, into internal or external. Also they can be classified according to their behaviour, into passive or active attackers.

External attackers are mainly nodes outside the network who want to get illegitimate access mostly to inject erroneous information and cause the network to stop functioning properly. Internal attackers are legitimate nodes that have been compromised, so that they launch attacks from inside the network mostly to feed other nodes with incorrect information. In general, internal attacks are more severe then external attacks.

On the other hand, most passive attackers are illegitimate eavesdroppers, or selfish nodes that do not cooperate with the purpose of energy saving. In contrast to active attacks, in general passive attackers do not try to actively interfere with communications. In active attacks, misbehaving nodes spend some energy to perform a harmful action.

Most usual active attacks are malicious attempts to introduce invalid data into the network or to produce communication failures. Both types of attackers can have a direct influence on the correct functioning of the network. On the one hand, active malicious nodes can directly cause network traffic to be dropped, redirected to a different destination or to take a longer route to the destination by increasing communication delays. On the other hand, selfish nodes can severely degrade it by simply not participating in the network operations.

Malicious nodes can execute two of the most harmful actions in VANETs: DoS and integrity attacks.

DoS attacks, and especially jamming, are relatively simple to launch yet their effects can be devastating, bringing down the whole VANET. Jammers deliberately generate interfering transmissions to prevent communication in the VANET. Since the network coverage area, e.g., along a highway, is well-defined, jamming is a low-effort exploit opportunity because such an attacker can easily, without compromising cryptographic mechanisms and with limited transmission power, partition the vehicular network.

With respect to integrity attacks, especially interesting are spoofing where malicious nodes impersonate legitimate nodes, and transmission of false information to contaminate the communication network.

Consider, for example, an attacker that masquerades an emergency vehicle to mislead other vehicles, or impersonates RSU to spoof false service advertisements or safety hazard warnings. In conclusion, fundamental security functions in vehicular networks should always include correct authentication of the origin of data packets and of their integrity (Caballero-Gil et al., 2009); (Caballero-Gil & Hernández-Goya, 2009). To achieve this, most authors assume that vehicles will in general sign each message with their private key and attach the corresponding certificate. Thus, when another vehicle receives this message, it verifies the key used to sign the message and the message.

Selfish behaviour of any node acting as a relay forwarding other nodes traffic can also seriously impair communications in the network because it can drop messages that might be valuable or even critical traffic notifications or safety messages.
There exists a different type of attacks whose main objective is the privacy of nodes. In this case, the attacker either passively or actively, and internally or externally, tries to extract data such as time, location, vehicle identifier, technical descriptions, or trip details. Afterwards, based on those data, the attacker tries to derive private information about the attacked node.

5. Security background

Among the main cryptographic requirements to solve security issues in VANETs are:

- Availability: The network must be available at all times in order to send and receive messages. Two possible threats to availability are for example DoS and jamming attacks. Another availability problem might be caused by selfish nodes that do not provide their services for the benefit of other nodes in order to save their own resources like battery power.
- Confidentiality: Secrecy must be provided to sensitive material being sent over the VANET, like in certain commercial applications.
- Integrity: Messages sent over the network should not be corrupted. Possible attacks that would compromise their integrity are malicious attacks or signal failures producing errors in the transmission.
- Authenticity: The identity of the nodes in the network must be ensured. Otherwise, it would be possible for an attacker to masquerade a legitimate node in order to send and receive messages on its behalf.
- Non-Repudiation: A sender node might try to deny having sent the message in order to avoid its responsibility for its contents. Non-repudiation is particularly useful to detect compromised nodes.

It is almost impossible to protect all the aforementioned characteristics against the wide variety of existing threats. Furthermore, different applications have specific security requirements to take into consideration. As a result of this diversity, many different approaches exist that focus on different properties.
Authentication is a must in order to achieve the necessary trust in vehicular ad hoc networks. The existence of an authentication service makes it more difficult for attackers to join the network in the first place and thus increases the cost of misbehaviour. Hence, by verifying the authenticity of any node before exchanging information, mobile nodes reduce the amount of undesired data. For example, users of many VAS applications should obtain authentication credentials by subscribing to the service.
According to the DSRC protocol, the security overhead of this type of schemes is usually bigger than the message contents. Consequently, such an issue has to be well addressed due to the limited wireless channel bandwidth available in VANETs. Symmetric cryptography usually implies less communication overhead than asymmetric cryptography. Consequently, we might think that symmetric cryptography is a good solution, but due to the huge amount of network members in VANETs, it seems not appropriate as a generalized solution for all communications.
For comfort/commercial-related packets, sent information should be encrypted. However, safety-related messages have a different management due to their strict requirements on

delay, reliability and dissemination. In fact, urgent safety-related messages must be automatically sent and checked through tamper-proof devices so that they are not encrypted/decrypted them. What is really important for such type of information is that it must be truly reliable, what implies the need of aggregation schemes for checking not only possible unintentional transmission errors but also probable intentional fraud attempts.
There are safety related events that can be detected by a single vehicle's sensors. In that case local sensor information is aggregated and if there is a matching event, a message is sent out (Doetzer et al., 2005).
Most researchers in security of VANETs (Parno & Perrig, 2005); (Raya & Hubaux, 2007) propose a Public Key Infrastructure (PKI) solution, with anonymous or pseudonymous certificates issued by a CA. This solution assumes that each vehicle is assigned a public/private key pair that is stored in a tamper-proof device.
Every time a vehicle sends a message, it includes its signature produced with its private key together with the public-key certificate signed by the CA. So, digital signatures are added to each message, and messages are not always encrypted. Its main drawback is the big computational need and bandwidth overhead of all communications. Furthermore, since messages are not always encrypted, even outsiders can eavesdrop and possibly create movement profiles. In this way, the receiver can verify the integrity and authenticity of each message and signer.
In order to reduce overhead, some authors have proposed to attach certificates only if new neighbours are discovered (Papadimitratos et al., 2008). Also to meet the overhead requirements in terms of either processing or bandwidth, Elliptic Curve Cryptography has been chosen for the IEEE 1609 trial standard. On the other hand, the authors in (Choi et al., 2005) suggest a system based exclusively on symmetric cryptography. The main problem of their proposal is that vehicles have to contact always the base station to decrypt and verify messages.
Some other authors (Zarki et al., 2002) outline security and privacy issues in VANETs but do not present a security infrastructure. Regarding routing protocols, authors of (Rudack et al., 2002) focus on the impacts of vehicular traffic dynamics on them. With respect to node authentication, (Caballero-Gil et al., 2009) proposes differentiated services according to privacy and efficiency needs. Finally, the first to investigate the potential of ring signatures to achieve anonymity and untraceability in mobile networks were the authors of (Freudiger et al., 2008).

6. Security proposal

In this section group formation is proposed as a valid strategy to strengthen privacy and provide authenticity, privacy and integrity protection, while reducing communications in VANETs. To make it possible, group management within the network must be very fast to minimize time lost in that task (Johansson, 2004).
In particular, we propose location-based group formation according to dynamic cells dependent on the characteristics of the road, and especially on the average speed. In this way, any vehicle that circulates at such a speed will belong to the same group within its trajectory. It is also proposed here that the leader of each group be the vehicle that has belonged to the same group for the longest time (see Figure 5).
According to our proposal, V2V between groups will imply package routing from the receiving vehicle towards the leader of the receiving group, who is in charge of broadcasting

it to the whole group if necessary. If the cells have a radio that is greater than the wireless coverage of the OBU, the group communication may be carried out by proactive Optimized Link State Routing (OLSR).

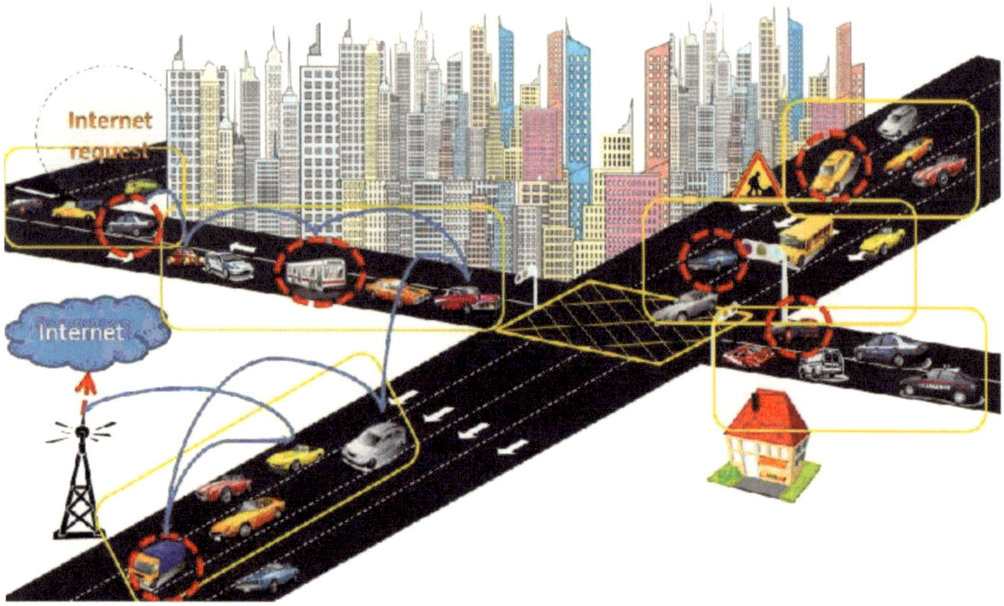

Fig. 5. Proposal structure

In the two phases corresponding to group formation and node joining, each new node has to authenticate itself to the leader through asymmetric authentication. Later, the leader sends a shared secret key to it, encrypted with the public key of the new node. In particular, this secret key is shared among all the members of the group, and used both for V2V within the group and for V2V between groups, as it is explained in the following sections.

We propose the application of different cryptographic primitives for node authentication, while paying special attention to the efficiency of communications and to the need of privacy. In this way, we distinguish four different ways of authentication, which are analyzed in the following subsections.

6.1 I2V authentication

Since privacy-preserving authentication is not necessary in I2V, we propose for such a case the use of Identity-Based Cryptography because it provides a way to avoid the difficult public-key certificate management problem.

Identity-Based Cryptography is a type of public-key cryptography in which the public key of a user is some unique information about the identity of the user (e.g. the ELP in VANETs).

The first implementation of an Identity-Based scheme was developed in (Shamir, 1984), which allowed verifying digital signatures by using only public information such as the users' identifier. A possible choice for VANETs could be based on the modern schemes that include Boneh/Franklin's pairing-based encryption scheme (Boneh & Franklin, 2001), which is an application of Weil pairing over elliptic curves and finite fields.

6.2 V2I authentication

Unlike I2V communication, in V2I communications privacy is an essential ingredient. Here we propose a challenge-response authentication protocol based on a secret-key approach where each valid user is assigned a random key-ring with k keys drawn without replacement from a central key pool of n keys (Xi et al., 2007).

According to the proposed scheme, during authentication each user chooses at random a subset with c keys from its key-ring, and uses them in a challenge-response scheme to authenticate itself to the RSU in order to establish a session key, which is sent encrypted under the RSU's public key.

This scheme preserves user privacy due to the feature that each symmetric key is with a high probability (related to the birthday paradox and dependent on the specific choice of parameters) shared by several vehicles.

When a vehicle wants to communicate with the RSU, it sends an authentication request together with a set of c keys taken at random from its key-ring and a timestamp. All this information is then encrypted by the established session key. Note that a set of keys, instead of only one key, is proposed for authentication, because there is a high probability for the OBU to have one key shared by a large amount of vehicles. This makes it difficult to identify a possible malicious vehicle if just one key is used. However, there is a much lower probability that a set of keys be shared by a large number of vehicles, and so it is much easier to catch a malicious vehicle in the proposal.

After the RSU gets the authentication request from the vehicle, it creates a challenge message by encrypting a random secret with the set of keys indicated in the request, by using Cipher-Block Chaining (CBC) mode. Upon receiving the challenge, the vehicle decrypts the challenge with the chosen keys and creates a response by encrypting the random secret with the session key. Finally, the RSU verifies the response and accepts the session key for the next communications with the vehicle.

In the first step, in order to make easier the task of checking the key subset indicated in the request by the RSU, we propose a tree-based version where the central key pool of n keys may be represented by a tree with c levels (Buttyán et al., 2006). Each user is associated to k/c leaves, and each edge represents a secret key.

In this way, the key-ring of each user is formed by several paths from the root to the leaves linked to it. During each authentication process the user chooses at random one of its paths, which may be shared by several users. In this way, to check the keys, the RSU has to determine which first-level key was used, then, it continues by determining which second-level key was used but by searching only through those second-level keys below the identified first-level key.

This process continues until all c keys are identified, what at the end implies a positive and anonymous verification. The key point of this proposal is that it implies that the RSU reduces considerably the search space each time a vehicle is authenticated.

6.3 V2V authentication inside groups

At the stages of group formation and group joining, each new node has to authenticate itself to the group leader by using public-key signatures (Sampigethava et al., 2006).

After group formation or group joining, the group leader sends a secret shared key to every new member of the group, encrypted with the public key of this new node (see Figure 6). Such a secret group key is afterwards used for any communication within the group both for node authentication and for secret-key encryption if necessary (e.g. for commercial applications).

In this way, the efficiency of communications inside the group is maximized because on the one hand certificate management is avoided, and on the other hand, secret-key cryptography is in general more efficient than public-key. Note that the use of a shared secret key also contributes to the protection of privacy.

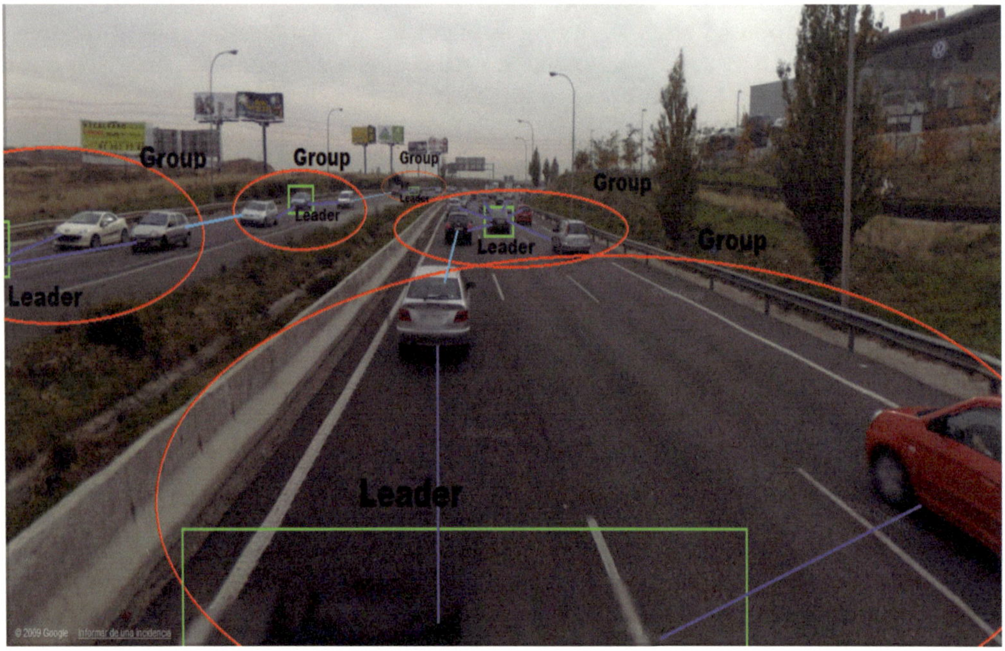

Fig. 6. Group-based organization

6.4 V2V authentication between groups

In order to protect privacy, group signatures might be proposed for node authentication between groups. A group signature scheme is a method for allowing a member of a group to anonymously sign a message on behalf of the group so that everybody can verify such a signature with the public key of the group. This group signature identifies the signer as a valid member of the group and does not allow distinguishing among different group members. This concept was first introduced in (Chaum & van Heyst, 1991).

Essential for a group signature scheme is the group leader, who is in charge of adding group members and has the ability to reveal the original signer in the event of disputes. In this proposal, the group leader issues a private key to each vehicle within the group, which uniquely identifies each vehicle, and at the same time allows it to compute a group signature and prove its validity without revealing its identity.

In this way, any vehicle from any group will be able to communicate with any vehicle belonging to other group anonymously. In particular, a proposal for group signature might be based on the cryptographic primitive of bilinear pairings, which was also proposed for I2V authentication.

6.5 Privacy

In order to guarantee the privacy of mobile nodes, they must be both anonymous and untraceable. Our proposal allows both saving communications and preserving privacy mainly thanks to the management of communications through groups. Group keys and symmetric cryptography are used for one-hop communications inside groups.

In order to protect privacy of group members and to avoid the need of group managers, ring signatures might be used in communications between groups. Since each node of a VANET is assumed to have a public/private key pair, the knowledge of the public keys of the other nodes in the group is sufficient to create a ring signature without any interaction, so it can be performed by any member of any group (Rivest et al., 2001). Hence, unlike group signatures, ring signatures have no group managers and do not require any coordination among ring members. In this way, it would be difficult to determine which of the group members' keys was used to produce the ring signature. Also, it would be impossible to revoke the anonymity of an individual signature, and any group of nodes might behave as a group without any additional setup. Furthermore, ring signatures can be constructed with any public-key cryptographic scheme, and are usually based on combining functions.

Both membership management and group support in the absence of any infrastructure are complex research issues. Consequently, a model to describe group membership dynamics is essential. In particular it is necessary to provide efficient and flexible mechanisms for group formation within the highly dynamic scenario of the VANETs. The capability of creating and dynamically manage the membership of groups in such a mobile scenario is, at the same time, a critical issue and a challenging research area.

The problem of group key establishment can be dealt in different ways. A first solution that might be considered is key transport, which consists in allowing a group leader to create a group key and multicast it to all members. This solution involves just one round but focuses most computational load on the group leader, which is also a possible point of failure. As a second possible solution, key agreement might also be considered but in general it involves several operations and rounds of multicasts or anycasts among all group members (Rafaeli, & Hutchison, 2003). A third interesting solution (Boyd, 1997) is the combination of key transport and key agreement where the leader plays a special role but it is not exactly who chooses the group key. In such a protocol, the group key is generated with a combining function on some number contributed by the leader together with the outputs of a one-way function over the contribution of each other node. First, all members except the leader multicast their contributions, then the group leader sends its contribution encrypted with the public key of each group member, and finally each member decrypts such a contribution and generates the group key.

Group memberships in VANETs are likely to change very fast. Hence another challenge in secure group management is the efficient handling of join and leave operations of members. The simplest approach for a join operation would be based on a key transport process to transfer the existing group key to the new member. Also, if the key must be changed during each joining operation, the necessary process is not too complex since it is possible to send the new group key through multicast to the old group members encrypted with the old group key. However, changing the group key after a member leaves is far more complicated since the old key cannot be used to distribute a new one, because the leaving member knows the old key. Therefore, for the sake of simplicity, it can be assumed that when a member leaves a group it is not necessary to update the group key.

Another critical problem of group management is the definition of group memberships. Regarding this issue, group formation will take place in the VANET as soon as vehicle density exceeds a threshold. Two other characteristics of the proposal are that the cell size depends on the transmission range of vehicles (around 300m), and the closest vehicle to the cell centre is considered the group leader.

6.6 Integrity

The trustworthiness of messages sent by a node is determined by the trustworthiness of the sender because messages from any node are trusted if and only if node authentication is valid. Apart from checking node authenticity, in vehicular networks it is extremely important to validate also the trustworthiness of data since, although in most cases identities of the nodes are irrelevant, correctness of the data they send is fundamental. For example, a simple attack based on transmitting fraudulent data about road congestion or vehicle position can be quite damaging and hence must be avoided.

In our proposal, a pervasive communication system is assumed in which mobile nodes automatically exchange information upon meeting. However, instead of doing message dissemination in VANETs through direct flooding, an approach based on location-based data aggregation is assumed so that message dissemination is delegated only to selected vehicles, which in our proposal are the group leaders. The data that group leaders disseminate are computed through a data aggregation scheme using those data received from members of its group that share a similar view of their environment. In this way, data aggregation helps both to improve security and efficiency of VANETs.

The data aggregation scheme here proposed is based on the most consistent version of data with respect to the collected information. In order to obtain such a version, one solution might be based on that versions of data obtained from other nodes receive scores according to nodes trustworthiness, and in this case the collector node accepts just those data with the highest scoring. However, due to the large size and mobility of VANETs, such reputation schemes carried out by nodes are not appropriate. We only consider a type of reputation scheme where nodes that are the source of incorrect information are detected by the RSU, which stores such information and scores nodes trustworthiness.

Data aggregation requires that group leaders crosscheck information concerning an event by comparing messages received from several sources with the data obtained from their own sensors, which are always considered trustworthy. After this step, instead of independent safety-related messages reporting the same event and sent by individual nodes, aggregated messages signed by a group with a ring signature are sent by the group leader. Thus, all the overhead will be grouped in one message as an alternative to be spread over several messages, resulting in a more efficient channel usage. In addition, once a vehicle receives such a combined message, it can trust data after the ring signature verification because the combined signature implies that all the involved signers agree on the content of the message.

Another possible useful application of data aggregation schemes in VANETs is the exploitation of data exchanged among vehicles in order to produce knowledge that can be used later by the nodes. For example, such data might allow detecting potentially dangerous road segments or determining the areas with a higher probability to find an available parking space. Furthermore, within this secondary application of data aggregation schemes it would be possible to exchange aggregated data between vehicles in order to improve their respective knowledge. According to this idea, each node should collect aggregated data, according to a map concept named Local Dynamic Map (LDM), which must reflect all relevant static and dynamic information in the vicinity, organized as a four layer structure with increasing dynamics. Furthermore, every time a vehicle moves towards some place, it should merge its LDM with the LDM of its group neighbours in order to try to build an LDM containing information about its destination and route.

7. Conclusion

VANETs represent a challenge in the field of communications security, as well as a revolution for vehicular safety, comfort and efficiency in road transport. In this chapter we have briefly described different security characteristics and services for VANETs.

Some basic ideas of some tools that can be used to improve communication security in VANETs have been here presented. We have addressed several important security issues with a special focus on efficiency and self-organization in our proposal.

The main goals of any design for VANETs should be: wide applicability, node privacy, efficient group management, strong authentication, and data verification. In order to reach them, any solution has to combine well-known building blocks (e.g. PKI, ring signatures, identity-based schemes) according to a modular design that includes several components specifically devoted to authentication, encryption, group management, data aggregation, simulation, safety-related/value-added applications, etc.

A brief description of several proposed security schemes has been given. In particular, for I2V authentication, since there is no need of privacy, Identity-Based cryptography seems the best option to avoid certificates management. In the remaining cases, privacy is a must. In V2I a challenge-response authentication protocol using a secret-key approach based on random key-trees might be a good scheme as it provides an efficient solution for anonymous authentication. In this chapter, groups have been proposed as the most efficient way to save communications. On the one hand, in order to provide privacy between groups, we proposed group or ring signatures. On the other hand, for V2V inside groups, secret-key authentication is the basis of the proposed solution.

Since security in VANETs is yet a work in progress, many questions are open. Some of those questions are the concrete definitions of proposals, the analysis of interactions among existing schemes, and the implementation of the different proposed algorithms in order to be able to compare different possible solutions to choose the best option for a wide practical deployment of VANETs.

8. Acknowledgment

This research has been supported by the Spanish Ministry of Education and Science and the European FEDER Fund under TIN2008-02236/TSI Project, and by the Agencia Canaria de Investigación, Innovación y Sociedad de la Información under PI2007/005 Project.

9. References

Boneh D. & Franklin M. K., (2001), Identity-Based Encryption from the Weil Pairing. *Proceedings of CRYPTO 2001*, Advances in Cryptology: Lecture Notes in Computer Science Vol. 2139, pp. 213-229, California, USA, August 2001

Boyd, C., (1997), On key agreement and conference key agreement, *Proceedings of the Information Security and Privacy: Australasian Conference*. Lecture Notes in Computer Science, Vol. 1270. Springer, pp. 294–302

Buttyán L.; Holczer T. & Vajda I., (2006), Optimal Key-Trees for Tree-Based Private Authentication, *Proceedings of the 6th International Workshop Privacy Enhancing Technologies- PET*, Lecture Notes in Computer Science Vol. 4258 Springer, pp. 332-350, Cambridge, UK, June 2006

Caballero-Gil, P. & Hernández-Goya, C. (2009), Designing Communication-Oriented Node Authentication for VANETs, *Proceedings of Mobiquitous - International Conference on Mobile and Ubiquitous Systems: Networks and Services*, Toronto, Canada, July 2009

Caballero-Gil, P.; Caballero-Gil, C.; Molina-Gil, J. & Fúster-Sabater, A., (2010), On Privacy and Integrity in Vehicular Ad Hoc Networks, *Proceedings of the International Conference on Wireless Networks (ICWN'10)*, Las Vegas, USA, July 2010

Caballero-Gil, P.; Caballero-Gil, C.; Molina-Gil, J. & Hernández-Goya, C. (2009), Flexible Authentication in Vehicular Ad hoc Networks, *Proceedings of APCC IEEE Asia Pacific Conference on Communications*, Vol. 208, pp. 876-879, Shanghai, China, October 2009

Caballero-Gil, P.; Hernández-Goya, C. & Fúster-Sabater, A., (2009), Securing Vehicular Ad-Hoc Networks, *International Journal on Information Technologies & Security*, Vol. 1, (25-36)

Caballero-Gil, P.; Hernández-Goya, C. & Fúster-Sabater, A., (2009), Differentiated Services to Provide Efficient Node Authentication in VANETs, *Proceedings of the International Conference on Security and Management SAM-WorldComp2009*, pp. 184-187, Las Vegas, Nevada, USA, July 2009

Caballero-Gil, P.; Molina-Gil, J., Caballero-Gil, C. & Hernández-Goya, C., (2010), Security in Commercial Applications of Vehicular Ad-Hoc Networks, *Proceedings of Financial Cryptography and Data Security '10*, Lecture Notes in Computer Science, Vol. 6052, pp. 427, Springer-Verlag, Tenerife, Spain, January 2010

Caballero-Gil, P.; Molina-Gil, J., Hernández-Goya, C. & Caballero-Gil, C., (2009), Stimulating Cooperation in Self-Organized Vehicular Networks, *Proceedings of APCC IEEE Asia Pacific Conference on Communications*, Vol. 82, pp. 346-349, Shanghai, China, October 2009

Chaum D. & van Heyst E., (1991), Group signatures, *Proceedings of EUROCRYPT '91*, Advances in Cryptology, Lecture Notes in Computer Science Vol. 547, pp. 257-265, Brighton, UK, April 1991

Choi, J.Y.; Jakobsson, M. & Wetzel, S., (2005), Balancing auditability and privacy in vehicular networks, *Proceedings of the 1st ACM international workshop on quality of service and security in wireless and mobile networks Q2SWinet*, Montreal, Canada, October 2005

Cottingham, D.; Beresford, A. & Harle, R., (2007), A Survey of Technologies for the Implementation of National-Scale Road User Charging, *Transport Reviews*, Vol. 27, No. 4, (July 2007) (499-523)

Doetzer, F.; Kosch, T. & Strassberger, M., (2005), Classification for traffic related intervehicle messaging. *Proceedings of the 5th IEEE International Conference on ITS Telecommunications*, Brest, France, June 2005

Dousse, O.; Thiran, P. & Hasler, M., (2002), Connectivity in ad-hoc and hybrid networks, *Proceedings of Infocom*, pp. 1079-1088, New York, USA, June 2002

Franz, W.; Hartenstein, H. & Mauve, M. (2005), *InterVehicle-Communications Based on Ad Hoc Networking Principles - The FleetNet Project*, Universitätsverlag Karlsruhe, ISBN 3-937300-88-0

Freudiger, J.; Raya, M. & Hubaux, J.-P., (2009), Self-organized Anonymous Authentication in Mobile Ad Hoc Networks, *Proceedings of the Conference on Security and Privacy in Communication Networks (Securecomm)*, pp. 350-372, Athens, Greece, September 2009

Füßler, H.; Schnaufer, S.; Transier, M. & Effelsberg W. (2007). Vehicular Ad-Hoc Networks: From Vision to Reality and Back. *Proceedings of the Fourth IEEE/IFIP Annual Conference on Wireless On demand Network Systems and Services (WONS)*, Obergurgl, Austria, January 2007

Gehring O. & Fritz, H., (1997), Practical results of a longitudinal control concept for truck platooning with vehicle to vehicle communication, *Proceedings of the 1st IEEE Conference on Intelligent Transportation System (ITSC'97)*, pp. 117–122, Boston, USA, November 1997

Hedrick, J.K.; Tomizuka, M. & Varaiya, P., (1994), Control issues in automated highway systems, *IEEE Control Systems Magazine*, Vol. 14, No. 6, (December 1994) (21–32)

Hernández-Goya, C.; Caballero-Gil, P.; Molina-Gil, J. & Caballero-Gil, C. (2009). Cooperation Enforcement Schemes in Vehicular Ad-Hoc Networks, *Proceedings of the 11th International Conference on Computer Aided Systems Theory EUROCAST 2009*, Lecture Notes in Computer Science, No. 5717, pp. 429-436, Springer-Verlag, Las Palmas de Gran Canaria, Spain, February 2009

Hubaux, J.P.; Capkun, S. & Luo, J., (2004), Security and privacy of smart vehicles. IEEE Security & Privacy, Vol. 2, No. 3, (May 2004) (49–55)

Johansson, T. & Carr-Motyckova, L., (2004), Bandwidth-constrained Clustering in Ad Hoc Networks, *Proceedings of the Third Annual Mediterranean Ad Hoc Networking Workshop*, Turkcell, Turkey, June 2004

Lee, S.; Pan, G.; Park, J.; Gerla, M. & Lu, S. (2007). Secure incentives for commercial ad dissemination in vehicular networks. *Proceedings of ACM International Symposium on Mobile Ad Hoc Networking and Computing (MOBIHOC)*, Montreal, Canada, September 2007

Li, F. & Wang, Y., (2007), Routing in vehicular ad hoc networks: A survey. *IEEE Vehicular Technology Magazine*, Vol. 2, No. 2 (June 2007) (12–22)

Little, T.D.C. & Agarwal, A., (2005), An Information Propagation Scheme for VANETs. *Proceedings of the 8th Intl. IEEE Conf. on Intelligent Transportation Systems (ITSC2005)*, Vienna Austria, September 2005

Molina-Gil, J.; Caballero-Gil, P. & Caballero-Gil, C., (2010), Group Proposal to Secure Vehicular Ad-Hoc networks, *Proceedings of the International Conference on Security and Management SAM*, Las Vegas, USA, July 2010

Papadimitratos, P.; Calandriello, G.; Hubaux, J.-P. & Lioy, A., (2008) Impact of Vehicular Communication Security on Transportation Safety, *Proceedings of the IEEE INFOCOM. Mobile Networking for Vehicular Environments*, pp. 1-6, Phoenix, USA, April 2008

Parno, B. & Perrig, A., (2005), Challenges in securing vehicular networks, *Proceedings of the ACM Workshop on Hot Topics in Networks (HotNets-IV)*, Maryland, USA, November 2005

Rafaeli, S. & Hutchison, D., (2003), A survey of key management for secure group communication, *ACM Computing Surveys*, Vol. 35, No. 3, (September 2003) (309–329)

Raya, M. & Hubaux, J.-P., (2005). The security of vehicular ad hoc networks. *Proceedings of the ACM Workshop on Security of Ad Hoc and Sensor Networks*, pp. 11–21

Raya, M. & Hubaux, J.-P., (2007), Securing vehicular ad hoc networks, *Journal of Computer Security*, Special Issue on Security of Ad Hoc and Sensor Networks, Vol. 15, No. 1, (39–68)

Rivest, R.L.; Shamir, A. & Tauman, Y., (2001), How to leak a secret, *Proceedings of the Asiacrypt*, Lecture Notes in Computer Science, Vol. 2248, Springer, pp. 552, Queensland, Australia, December 2001

Rudack, M.; Meincke, M. & Lott, M., (2002), On the Dynamics of Ad Hoc Networks for Inter Vehicle Communications (IVC), *Proceedings of the International Conference on Wireless Networks, WORLDCOMP*, Las Vegas, USA, July 2002

Sampigethava, K.; Huang, L.; Li, M.; Poovendran, R.; Matsuura, K. & Sezaki, K., (2006), CARAVAN: Providing Location Privacy for VANET, *Proceedings of the 3rd ACM International workshop on Vehicular ad hoc networks (VANET)*, California, USA, September 2006

Shamir A., (1984), Identity-Based Cryptosystems and Signature Schemes, *Proceedings of CRYPTO 84*, Advances in Cryptology, Lecture Notes in Computer Science Vol. 7, pp. 47-53, California, USA, August 1984

Wischof, L.; Ebner, A. & Rohling, H., (2005), Information dissemination in self-organizing intervehicle networks, *IEEE Transactions on intelligent transportation systems*, Vol. 6, No. 1, (March 2005) (90–101)

Xi Y.; Sha K.; Shi W.; Scnwiebert, L. & Zhang T., (2007), Enforcing Privacy Using Symmetric Random Key-Set in Vehicular Networks, *Proceedings of the Eighth International Symposium on Autonomous Decentralized Systems ISADS*, pp. 344-351, Arizona, USA, March 2007

Zarki, M.E.; Mehrotra, S.; Tsudik, G. & Venkatasubramanian, N., (2002), Security issues in a future vehicular network, *Proceedings of the European Wireless 2002 Conference*, Florence, Italy, February 2002

5

Routing in Vehicular Ad Hoc Networks: Towards Road-Connectivity Based Routing

Nadia Brahmi, Mounir Boussedjra and Jospeh Mouzna
Department of Intelligent Transportation System,
IRSEEM-ESIGELEC
France

1. Introduction

The proliferation of wireless technologies has inspired researchers from both academia and automotive industry to integrate advanced capabilities to the vehicles and provide new services and mobile applications. In particular, vehicular networks have emerged as a novel class of Mobile Ad Hoc Networks (MANETs) formed between moving vehicles equipped with wireless devices. Based on multi-hop communications, these self-organizing networks enable data exchanges among nearby vehicles and between vehicles and the road side infrastructure.

Driven by the transportation safety and efficiency issues, Vehicle-to-Vehicle (V2V) and Vehicle-to-Infrastructure (V2I) communications are attracting considerable attention in providing Intelligent Transportation Systems (ITS). In this context, a variety of services are offered to road users for improving their security and comfort. These emerging applications include among others safety applications for traffic monitoring and collision prevention, road information services, and infotainment and so on.

However, unlike other ad hoc networks, Vehicular ad hoc Networks (VANETs) have their unique characteristics which give rise to many challenging issues. One of the most salient features is the high mobility of vehicles resulting in dynamic topology changes. Accordingly, data routing remains a key networking issue that needs to be addressed in order to support the emerging applications. Over the last decades, many efforts have been concerted to design efficient routing protocols after recognizing the inefficiency of traditional MANET protocols to meet the requirements of vehicular environments.

This chapter presents an analysis of the routing problem in vehicular ad hoc networks. First, it discusses the main characteristics and challenges of VANETs that distinguish them from the traditional MANETs. Then, it reviews the most relevant routing strategies proposed in the research community highlighting their advantages and disadvantages.

Based on these considerations, we introduce a new class of geographic routing protocols called RCBR, Road Connectivity-based Routing for vehicular networks. The proposed approach exploits information about road connectivity and vehicles distribution to find stable routes and reduce the probability of links breakage. Simulations results are used to show how traffic awareness combined with a spatial knowledge of the environment can optimize the routing decisions in high dynamic networks.

2. Routing protocols in vehicular networks

This section presents a brief overview of routing protocols proposed or adapted for vehicular ad hoc networks. According to the type of information used to make the routing decisions, these protocols can be classified into 5 categories as shown in figure 2.1. In the following subsections, we describe the principal protocols in each group and analyses their adaptability for VANETs scenarios.

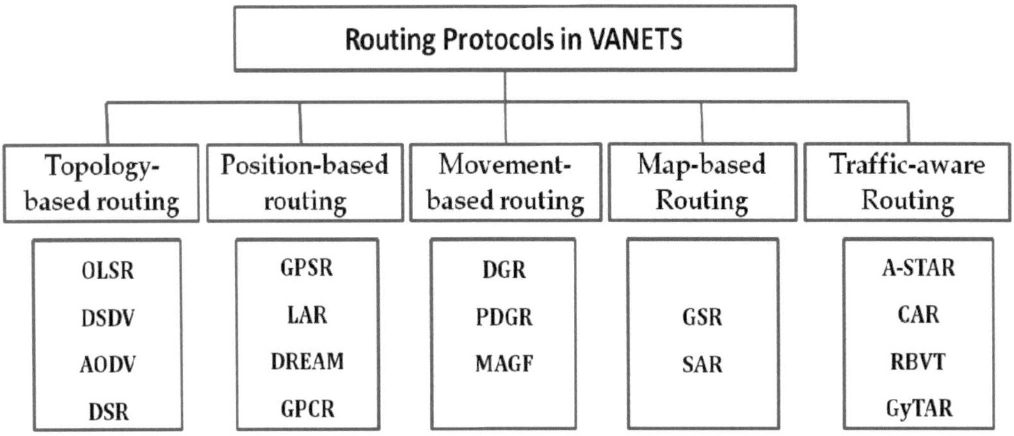

Fig. 1. Taxonomy of existing routing protocols in vehicular ad hoc networks

2.1 Topology-based routing

The *topology-based* protocols use the information about the network topology and the state of communication links between nodes to perform the routing decisions. They can be further categorized into proactive and reactive approaches.

The proactive protocols, such as Optimized Link State Routing (OLSR) (Clausen et al., 2001) and Destination-Sequenced Distance-Vector Routing (DSDV) (Perkins & Bhagwat, 1994) compute and maintain routing information about all available paths in the networks even if no data traffic is exchanged. For instance, in DSDV, every node maintains a vector of distances to every known destination. Therefore, frequent broadcast messages are issued by all nodes to learn periodically about their 1-hop neighbors or to advertize topology changes (e.g. link breakages). Similarly, OLSR floods the network by the topology control messages in order to disseminate the link states information throughout the entire network showing which nodes are connected to which other nodes.

This additional traffic used in proactive approaches for the maintenance of unused paths has several drawbacks. First, it consumes the networks resources and wastes a part of the bandwidth for control messages that increase with rapid changes. Moreover, the use of flooding increases the network congestion and leads to the loss of messages because of collision. They face a trade-off between the freshness of the routing information and the control overhead. Clearly, proactive solutions do not scale well in very large networks with a high number of nodes joining and leaving the network over a short time, which is the case for VANETs.

On the contrary, reactive protocols such as AODV (Perkins & Royer, 1999) and DSR (Johnson & Maltz, 1996) determine a route to a given destination only on-demand. They reduce the overhead by restricting the route maintenance only between nodes that need to communicate. In other words, route discovery is only initiated when a sending node has to set up a valid path towards a given destination. Obviously, this extends the delay before that the packets could be actually sent through the network. In addition, most of reactive protocols use the flooding technique to establish the communication between the source and destination and consequently, consume a lot of the available bandwidth.

Because of the high mobility of vehicles, the topology-based algorithms fail to handle frequent broken routes usually constructed as a succession of vehicles between the source and the destination. Moreover, the route instability and frequent topology changes increase the overhead for path repairs or change notifications and thus, degrade the routing performances.

Generally, in topology-based approaches, routing paths are built as successions of mobile nodes, and hence the chance of losing the connectivity is higher. Therefore, they are not suitable for vehicular scenarios and no further investigation will be done on their applications in these extremely dynamic environments.

2.2 Position-based routing

Position-based protocols perform the routing decisions based on the geographic information of the nodes. This class offers an alternative approach known to be more robust to face the mobility issues (Giordano & Stojmenovic, 2003). Indeed, no global knowledge of the network topology is required; a purely local decision is made by each node to make a better progress towards the destination. Therefore, they require all nodes to be aware of their physical positions as well as their neighbours' positions. They also assume that the sending node knows the position of the destination. Typically, a location management service is responsible for querying this information (Li et al., 2000).

2.2.1 Greedy perimeter stateless routing

As a representative example of the position-based algorithms, Greedy Perimeter Stateless Routing (GPSR) seems to be the most popular candidate for dynamic networks (Karp & Kung, 2000). Typically, there are several requirements on the availability of position information: GPSR requires that each node is able to obtain its current location e.g., through a GPS receiver as it is becoming standard equipment in vehicles. Furthermore, it assumes that each node learns about the existence of its direct neighbors and their current positions through the exchange of periodic HELLO messages. To make the routing decisions, a source node needs to know the position of the destination. The source node forwards the packets to its neighbor which is geographically closest to the destination. This procedure, known as Greedy Forwarding, is recursively applied by intermediate nodes until the final destination is reached. However, packets can reach node that has no neighbor which is closer to the destination than itself. This problem known as local maximum is likely to happen in case of sparse networks.

In such a case, GPSR switches to a recovery strategy called Perimeter Mode using the right hand rule algorithm of planer graph traversal to route the packets out of the local maximum region. Being expensive this recovery procedure is abandoned as soon as possible to go back to the greedy strategy since it can decrease the performance when used often.

Nevertheless, using only the position information may lead packets to be forwarded a wrong direction and looses consequently good candidates that ensure its delivery. As shown in figure 2, following a position-based approach the packets take the direction of the node A instead of the destination D facing so a local maximum problem.

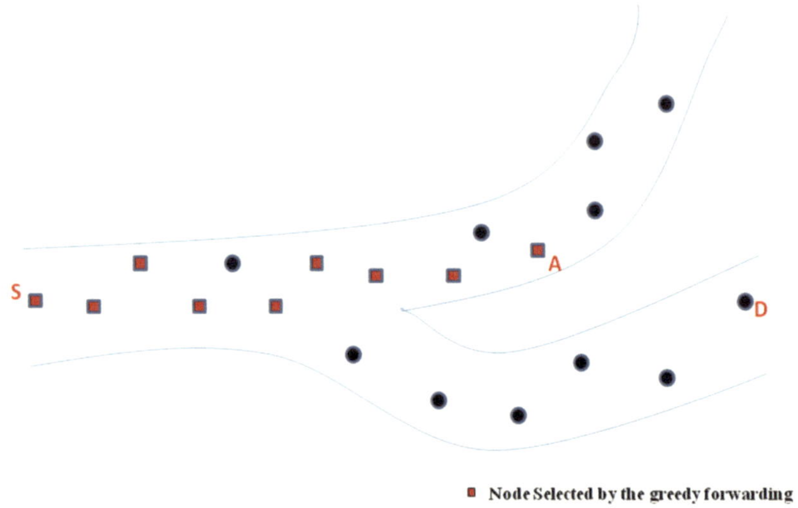

Fig. 2. Failure of Greedy Forwarding by selecting wrong direction

2.3 Movement-based routing

Numerous protocols enhance the basic position-based scheme to optimize the routing decisions. Indeed, due to the local maximum problem packet drops still occur using only position information. To address this shortcoming, some approaches like Directional Greedy Forwarding (DGR) (Gong et al., 2007) and Movement Aware Greedy Forwarding (MAGF) (Brahmi et al., 2009) suggest making use of additional information about vehicles movement such as direction and speed. The basic idea is to compute a weighted score W_i as a function of different factors (position, direction, speed) for assigning priority between neighboring nodes while selecting the next forwarder. This enhancement of the pure position-based scheme reduces the number of encountered local maximum by avoiding sending packets away from the destination while selecting a wrong direction.

Considering the fact that vehicles follow a predicable mobility pattern, the authors of (Gong et al., 2007) propose Predictive Directional Greedy Routing (PDGR) to forward packet to the most suitable next hop based on both current and predicable future locations. The mobility prediction scheme allows a packet relay to ensure the validity of a selected neighbor.

These enhanced solutions are likely to fit more to highway scenarios than to city-scenario where the topology of road determines the movement and the behaviours of cars.

2.4 Map-based routing

The Map-based routing protocols combine the position information with topological knowledge about the road and the surroundings. The idea is to build a spatial model representing the underlying road topology and select a routing path that overlaps with the streets. For this purpose, the road maps are represented by graphs where vertices are crossroads and edges are road segments. Commonly, the edges of the graph are weighted with static data extracted from the street maps. Examples of these static data could be distance, travel time or speed limits. Accordingly, the routing path is selected based on the new constructed graph and the data packets are only forwarded respecting the particular mobility pattern restricted by the road topology.

These approaches vary from source routing approaches, where the entire path towards the destination is pre-computed by the data source, to the dynamic routing where decisions are made only at road intersections based on various parameters.

As example of protocols that belong to this class we present in the following sub-sections Geographic Source Routing (GSR) (Lochert et al., 2003) and Spatial Aware Routing (SAR) (Tian et al., 2003).

2.4.1 Geographic source routing

The first protocol to use the knowledge of the underlying map of the streets was Geographic Source Routing (GSR) which is mainly proposed for urban environments. (Lochert et al., 2007). Assuming the availability of such information through a navigation system, a given source computes the shortest path to an intended destination using Dijkstra's algorithm based on the distance metric. The computed path consists of a sequence of junctions IDs known as Anchor Points (AP), along which packets should be forwarded to reach the destination. These anchors, obtained from the streets map, reflect the underlying road topology and usually represent the road intersections where decisions are made. The list of junctions is then inserted into the header of each data packet sent by the source.

The packets are forwarded over the selected path successively from one AP to the next AP using the greedy forwarding scheme. Moreover, it is important to note that the authors make use of a reactive location service to retrieve the current position of a desired destination. Concretely, the source node floods the network to query the location of a specific distant node and thus, leads to bandwidth wastes.

The studies conducted to compare GSR with topology-based protocols show the advantage of this map-based approach in realistic vehicular environments. However, it should be noted that the insertion of the entire path in the packet's header cannot be preferred in case of a long route between the source and the destination since it causes an additional packet overhead. Furthermore, assuming the connectivity of the shortest path is not realistic since it does not consider situations where there is no sufficient number of vehicles on the road between two involved junctions to ensure the road connectivity. That is, if along one road segment the packets face a local maximum situation that prevent them from progressing towards the next AP, they are directly discarded although an alternative longer path may exist.

2.4.2 Spatially aware packet routing

The authors of (Tian et al., 2003) introduce the Spatially-Aware Packet Routing (SAR) protocol to improve the basic GSR with a recovery procedure to avoid a local maximum. As we have already noted, the greedy routing used to forward packets along the shortest path may fail if there are no vehicles ensuring connectivity to the next intersection. In such situations, GSR drops the packets although a valid path may exist. On the contrary, SAR suggests finding an alternative path from the current location where the local maximum occurs and then replaces the original route with the new one. The new path is computed again using Dijkstra algorithm after removing the current road segment where the local maximum is detected. According to the authors, another option would be to store the packet

in a suspension buffer and wait for an incoming neighbor that provides positive progress towards the next intersection. The suspended packets will be dropped if the buffer is full or if they cannot be forwarded during a predefined interval depending on the application requirements. Although these recovery procedures are defined to be used separately, it seems advantageous to combine both mechanisms to decrease the risk of packet drop.
The performances evaluation has shown that SAR is more robust to the mobility than topology-based routing protocols (DSR) since the routing path is computed independently of specific mobile nodes.
Although knowing the road topology represents a big advantage, this approach fails in the case where the algorithm tries to forward packets over streets where no vehicles are moving. Moreover, frequent network partitions can cause path disconnections and prevent packets from progressing towards their destinations. These problems were addressed by extending the road topology knowledge with the vehicular traffic awareness. The following sub-section presents some representative examples of such protocols that include information about vehicular traffic and density in addition to basic street-level data extracted from digital maps.

2.5 Traffic-aware routing

The traffic-aware routing protocols suggest the use of available data about vehicular traffic density and flows in addition to spatial information. Thus, only streets where vehicles are moving will be used for packet forwarding. The following sub-section examines examples of such routing solution which are designed using traffic information.

2.5.1 Anchor-based street and traffic aware routing

One of the protocols that exploit the idea of traffic awareness in designing a routing scheme is the Anchor-based Street and Traffic-Aware Routing (A-STAR) (Lee et al., 2004). Generally, in city environments, vehicles are concentrated more in some areas than in others and hence the connectivity is higher in these roads. A-STAR added a new feature to the basic GSR using historical information on bus traffic to identify anchor-based paths with the highest connectivity according to the bus traffic regularity. It builds a weighted graph where edge's weight is assigned based on the number of bus lines that traverse the road segment. The more the bus lines are, the more stable the traffic is and so the less weight attributed to the road is. Then, the anchor route is constructed using the Dijkstra's algorithm applied to the produced graph. The bus traffic information can be extracted from statically rated maps with preconfigured routes or from dynamically rated ones where the street traffic is updated periodically based on road-side units. In A-STAR, the packets are forwarded along the defined route on the same way defined in SAR. Besides, the authors define a recovery procedure similar to the one used by SAR to counter local maximum. It consists of computing an alternative anchor route from the local maximum to the destination. To prevent other packets from traversing the same area, a road segment where a local maximum is detected is market as out of service (OFF). This information is afterwards disseminated in the network to update the graph so that these routes will not be used for new paths computation.
The study comparing the protocol to other existing protocols like GPSR and GSR show that the traffic information are useful for routing in VANETs. Nevertheless, the authors do not give any indication about the network overhead generated in order to monitor the city traffic condition and distribute such information to every vehicle.

In addition, it seems worthy to observe that historical data, such as bus traffic, cannot always accurately describe the current road traffic conditions since road congestion and events like road constructions cannot be detected. This makes the protocol inappropriate for highly dynamic environments. Based on this observation, new approaches propose to investigate more the use of real traffic information.

2.5.2 Connectivity aware routing

Yang et al. propose Connectivity Aware Routing protocol (Yang et al., 2008) which uses the statistical data collected by different vehicles to estimate the probability of connectivity of each road segment. In their model, the authors consider also the clustering phenomenon resulting from vehicles movement affected by the traffic light. The connectivity information is disseminated in the entire network to provide a global vision about the network connectivity. Based on that, a connectivity graph is defined from a combination between road topology information extracted from digital maps and the gathered connectivity information. For packet routing, CAR uses Dijkstra's algorithm to compute the optimal path along which packets will face the least probability of network disconnection. In other words, the route with the highest probability of connectivity is selected as the routing path.
The simulations studies have shown that the real-time information used in CAR improves the performances of routing in VANETs compared to GPSR and GSR. However, no indication was given about the overhead generated by collecting and exchanging connectivity information about the entire network especially that this information is volatile due to vehicles mobility. Since CAR relies on the statistic traffic data, the authors propose to investigate in their future work how they can further improve the protocol's performances by exploiting real-time traffic information.

2.5.3 Road-based with vehicular traffic

Recently, a group of Road-Based with Vehicular Traffic protocols (Nzouonta et al., 2009) has been proposed for VANETs. These protocols incorporate real-time vehicular traffic to compute road-based paths consisting of successions of road intersections connected among them through vehicular communications. Two variants of RBVT are presented: a reactive protocol, RBVT-R, and a proactive protocol, RBVT-P. In RBVT-R, only source nodes discover the connected road segments on demand by initiating route discovery packets which traverse the network towards the destination. Being a source routing, RBVT-R includes the discovered routes in the packets headers and utilizes a greedy forwarding procedure to transmit packets along road segments forming the selected path. On the other hand, RBVT-P maintains a graph of all connected road segments. To discover the network topology, connectivity packets (CP) are generated periodically by multiple vehicles randomly selected in the network. Each node decides independently whether to generate a new CP based on the estimated current number of vehicles in the networks, the historic hourly traffic information and the time interval since it last received a CP update.
These packets traverse the road map and record the road segments with enough vehicular traffic before returning to the generator segment. Using the collected information, any vehicle belonging to that segment which receives the CP after all intersections marked are visited will construct the connectivity graph and disseminate it in the entire network. Then, the shortest path to the destination is computed only from the road segments that are marked as connected. The evaluation study comparing RBVT to topology-based routing,

position-based routing and map-based routing show an improvement in its performances when used for dynamic vehicular networks. This improvement is due to real-time traffic consideration that makes routing decisions adapted to network conditions. Nevertheless, this procedure generates an additional overhead to maintain the freshness of the topology information. More adapted and suitable schemes for providing the connectivity information should be used to improve the scalability of RBVT protocols.

In the rest of this chapter, we introduce a new routing approach which is well adapted to vehicular ad hoc networks called Road Connectivity-based Routing (RCBR). Based on the fact that the density of vehicles moving along one road is not an accurate indicator of its connectivity, RCBR defines the concept of road connectivity to provide real-time view of the network topology. In addition to providing a good support for delay sensitive applications, RCBR has the advantage of performing well under sparse networks. A detailed description of the proposed scheme is given on the following section.

4. New approach: road connectivity-based routing protocols

RCBR routing approach combines information about the real-time vehicular traffic and the road-topology to select more stable routing paths. The idea is mainly based on the concept of road connectivity describing the state of each road segment whether it is connected or disconnected. In this context, a road is defined as connected if it has enough vehicular traffic which allows the transmission of the packet through multi-hop communications between its two adjacent intersections. For that, we define an algorithm predicting the connectivity duration over each road segment.

We designed two variants of RCBR protocols: a source routing protocol S-RCBR and a dynamic version of D-RCBR. S-RCBR computes the route using a global connectivity graph of the real-time state of the road segments and includes them in the packets. In D-RCBR, dynamic routing decisions are executed only in the proximity of road intersections to select a next segment through which data packets will be forwarded.

This class of protocols assumes that each car is equipped with a Global Positioning System (GPS) to get its own position and a navigation system that provides information about the local road map. In addition, the current position of a destination node is discovered by mean of location service. The road topology is mapped into a graph, G (V, E) where V is the set of vertices representing the road intersections and E is the set of edges representing the segments of road connecting adjacent vertices.

4.1 Road-connectivity model

In this subsection, we present the mathematical model used by RCBR routing protocols to estimate the connectivity of each road segment. First, we introduce some definitions that serve to this illustration and will be used throughout this chapiter. Then we describe the prediction model.

1. **Intersection virtual range:** in this context, the range of a road intersection is defined as the area within the circle centred on it and which radius is half of the wireless communication range. This value is delimited to the half of the transmission radius to ensure that the distance between any two vehicles in this area is less than the radio range and hence they can communicate.
2. **Link duration (LD):** the link duration between two vehicles represents the period during which they remain within the transmission range of each other. It can be

estimated by applying the mobility prediction method presented in [Su et al.,]. If we consider, two vehicles N_i and N_j, with a transmission range R, speeds vi and vj, coordinates (x_i, y_i) and (x_j, y_j), and velocity angles θ_i and θ_j, respectively, the Link Duration $LD_{i,j}$ is predicted by:

$$LD_{i,j} = \frac{-(ab+cd)+\sqrt{(a^2+c^2)r^2-(ad-bc)^2}}{a^2+c^2} \quad (1)$$

Where

$$a = v_i \cos\theta_i - v_j \cos\theta_j$$
$$b = x_i - x_j$$
$$c = v_i \sin\theta_i - v_j \sin\theta_j$$
$$d = y_i - y_j$$

Through the beacon messages periodically exchanged between neighboring nodes, each vehicle maintains a table of its neighbours' information which uses to compute their corresponding link durations [17].

3. **Path Connectivity (PC):** the path connectivity CP_i of a multi-hop path P_i formed by n-1 links connecting n neighboring vehicles N_1, N_2, ...,N_n is defined as the duration for which all the links are still available. It is called also lifetime and can be formulated as:

$$CPi = \min(LD_{Nj,Nj+1}) \quad (2)$$

Where N_i and N_{j+1} are two successive nodes of P_i.

4. **Road Connectivity (RC):** A road segment is said to be connected if there is at least one multi-hop path connecting its two adjacent intersections. To estimate the connectivity over one road, we exploit the concept of path connectivity. In this context, a path between two adjacent intersections I_i and I_j is defined as a multi-hop path formed by links between neighbor vehicles moving on the road segment delimited by these intersections and connecting two vehicles situated on virtual range of I_i and I_j respectively. Figure.3 shows an example of a path between two adjacent intersections I_1 and I_2.

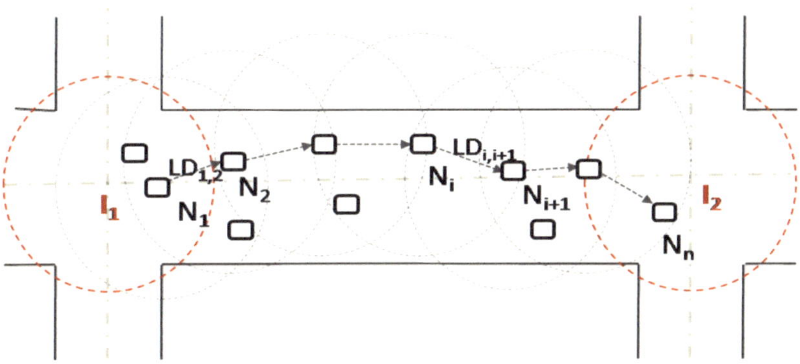

Fig. 3. A connected road segment delimited by intersections I_1

As a consequence, the Road Connectivity of a segment $[I_i, I_j]$ can be formulated as the highest Path Connectivity of all the paths P_i between the adjacent intersections I_i and I_j. It is computed by the following formula:

$$CPi = \max(CPi)\ \{\forall Pi \text{ path connecting } I_i \text{ and } I_j\} \quad (3)$$

In practice, a vehicle directly connected to one intersection computes the period during which it remains in its virtual range and inserts it in its hello message. Through the propagation of the beaconing messages, all vehicles in this road are then able to estimate their connectivity to both intersections delimiting the road segment. Only the vehicles in the proximity of the intersection maintain a connectivity table containing the information about all the adjacent intersections. This table is updated based on the information exchanged between different vehicles in the proximity of the intersection.

4.2 S-RCBR: source routing protocol

RCBR is a source routing protocol that proactively computes paths between the source and the destination using the connected road segments. Based on the road connectivity model described above, it defines a global real-time graph called "Connectivity Graph" to maintain a consistent view of the network connectivity. The connectivity information is exchanged between vehicles and a server deployed on the roadside infrastructure using V2I communications. Each source uses the road segments marked as connected to compute an optimal stable path which is then stored in the header of data packets to be used for geographic forwarding.

4.2.1 Network connectivity discovery

To optimize the routing decisions using the support of the infrastructure, we suggest deploying a Connectivity Server (CS) integrated to the roadside infrastructure and able to communicate with the vehicles through V2I communications. The CS aggregates all the connectivity information received from different vehicles in order to build a Connectivity Graph describing the state of all the road segments in the nearby zone.
Therefore it maintains a table with entries of the form

$$<Ibegin, Iend, Duration, Ts> \quad (4)$$

where *Ibegin* and *Iend* indicate the two adjacent intersections limiting the road segment, *Duration* represents the connectivity period calculated at the instant *Ts*.
In order to reduce the data traffic managed by the server, only some particular vehicles transmit Connectivity Packets (*CP*) to the server. In fact, after predicting the connectivity of the road segment using the model described below, the nearest vehicle to the intersection sends a CP to the server and notifies its neighbors by adding into the next hello message.
Further, the CP initiation time is known by all the vehicles located on the range of the intersection and only one CP is sent per intersection. As a consequence, the server receives a connectivity packet from each intersection; note that it is possible to receive multiple CP related to the same road from different nodes present in both intersections defining the segment.
On the reception of each CP, the server updates the corresponding entry in the connectivity graph. Once the graph is rebuilt, it can be transmitted on-demand to different nodes present in the zone. To give an overview of the above process, figure 4 illustrates an example of the server updates and the form of connectivity graph created for the road structure.

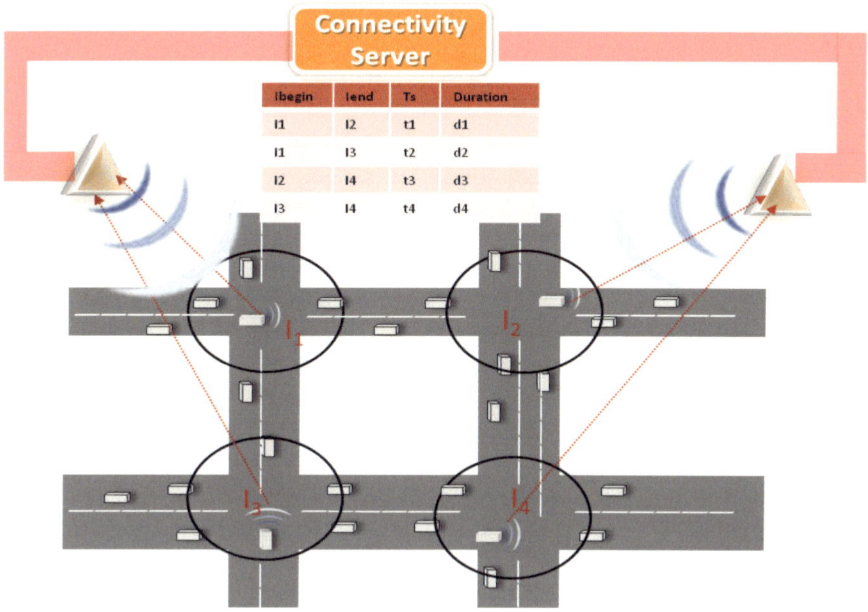

Fig. 4. The Connectivity Graph (CG) is constructed using connectivity packets (CP) sent by the nearest vehicle to each intersection.

4.2.2 The routing algorithm

In S-RCBR, the routing process consists of two main tasks: 1) defining the routing paths through which the packets will be forwarded and 2) forwarding data packets along the selected path using the greedy forwarding.

S-RCBR uses road-based paths consisting of sequence of intersections to transmit the data packet through connected road segments. When a source node needs to send information to a given destination, it initiates a CRequest to obtain the connectivity graph from the server. Based on the newly received graph, a routing path with most stable routes is constructed along the segments with the highest connectivity. These routes are stored in the headers of the data packet to be used by intermediate nodes while transferring packets between intersections denoting the defined path. In between intersections, the greedy forwarding is used.

To maintain fresh information about the network connectivity, a data source periodically generates a CUpdate to get the latest information from the server. The routing paths are updated accordingly using fresher information.

Finally, since network partitions cannot be avoided in highly dynamic environment like VANET, S-RCBR uses the Carry-and-forward strategy. Indeed, to handle network disconnections, packets are buffered and later forwarded when an available next hop is found to restore the connection.

4.3 D-RCBR: dynamic routing protocol

D-RCBR is a dynamic variant of RCBR that only requires a local view of the road connectivity, since collecting global real-time information about the entire network can be expensive especially with the mobility of vehicles. The new protocol performs local routing decisions only near road intersections. It uses the road connectivity prediction model described in the section above to estimate the connectivity over each road segment. Through the propagation of the beaconing messages, all vehicles moving along a given road are able to estimate the expected time for which they remain connected to both intersections delimiting the road segment. Then, this connectivity information is gathered near each intersection thanks to the dissemination mechanism based on the exchange of HELLO messages between different vehicles in the proximity of the intersection. Therefore, each vehicle located inside the virtual range of an intersection maintains a local connectivity table with entries about all the adjacent intersections. Based on this local connectivity information, the vehicles make the routing decisions and select the next vertex towards the destination. The idea of the greedy scheme is applied to select the closest intersection to the destination only among the adjacent connected intersections. However, the packets can reach an intersection which has no adjacent intersection closer to the destination. This situation known also as a local maximum is likely to happen considering only a greedy selection of vertex. To address this problem, D-RCBR defines a recovery procedure inspired from the right hand rule (Karp & Kung, 2000).

The routing process includes two main tasks: 1) Select the next intersections towards the destination using one of the two strategies: Greedy or right-hand rule for the vertex selection 2) forward data packets hop by hop towards the selected intersection.

1. Greedy Vertex-Selection: In this mode the idea of the greedy scheme is applied to select the closest intersection to the destination among all the adjacent connected intersections. When a packet reaches a vehicle in the range of an intersection, the vehicle selects the next intersection towards the destination. Only a connected adjacent vertex can be selected to ensure the delivery of the packet along the forwarding road. However, to minimize the networking delays, the closest intersection to the destination is chosen. To do so, all the neighbor vertices which are disconnected from the current vertex are removed from the road graph G and then the shortest path between the current vertex and the destination is computed using Dijkstra algorithm. The next intersection in the determined path is inserted into the packet header. Between two intersections the greedy forwarding scheme is used to forward the packet. An example of packet routing with the proposed D-RCBR is shown in Figure 5 where a source node S has a packet addressed to the destination D. S is in the proximity of the intersection I_1 so the shortest path should be computed from intersection I_1 to the destination near the intersection I_6. By exploiting the local connectivity information gathered by the nodes near I_1, the intersection I_2 is marked as unreachable and is not considered for the shortest path computation. As a consequence, the closest vertex to the destination among all the adjacent connected vertices is selected as the next intersection. The greedy vertex selection is repeated until the packet reached the intersection I_6 as one of the destination's road. In the figure, the disconnected roads are marked by a cross.

2. Right-Hand rule for Vertex Selection: Using the greedy selection of vertex, D-RCBR helps reducing the overhead needed by a global knowledge of the network connectivity. However, there is no guarantee for the packets to be delivered until the destination. An example is shown in Figure 6 when a packet reaches the range of intersection I_5 and the adjacent intersection I_6 which represents the destination vertex is disconnected. As a consequence, the greedy selection fails although a possible path exists between I_1 and I_6. To address the aforementioned problem, we suggest using the idea of the right hand rule to select an intersection in counter clockwise. This idea was previously adopted by GPSR, but contrary to GPSR, in D-RCBR the right hand rule is applied to the road graph where vertices are intersections instead of the network graph where vertices are mobile nodes.

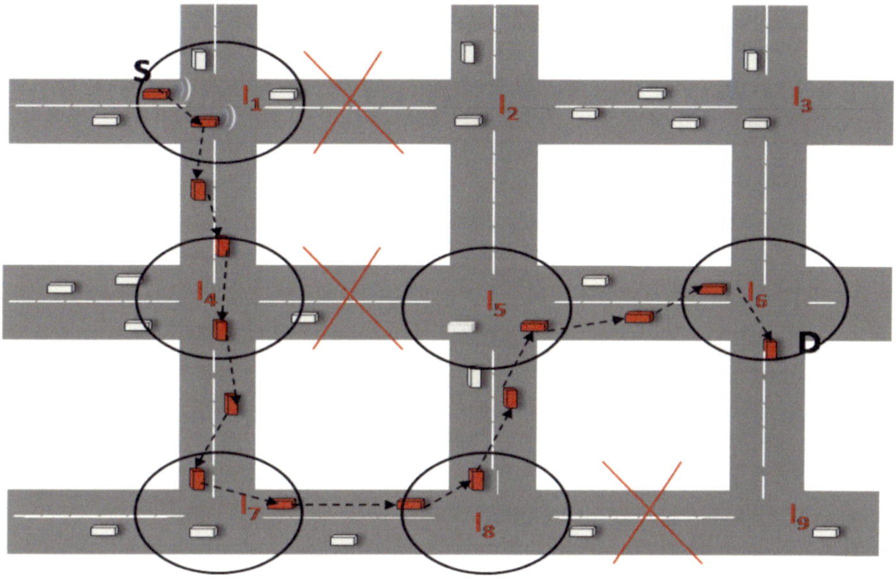

Fig. 5. The greedy strategy applied for the vertex selection in D-RCBR

Hence, if the greedy selection of intersection fails, the forwarding node in a range of an intersection selects, following the right hand rule, a next vertex among the connected neighbor vertices. The protocol should returns back to the greedy selection of vertex as soon as the packet escapes from the local maximum. With this procedure, D-RCBR can ensure finding a possible path to destination if any exists.

To illustrate the recovery procedure described above, a scenario of the failure of greedy selection is described in figure 6 using the same road topology. A data packet reaches the

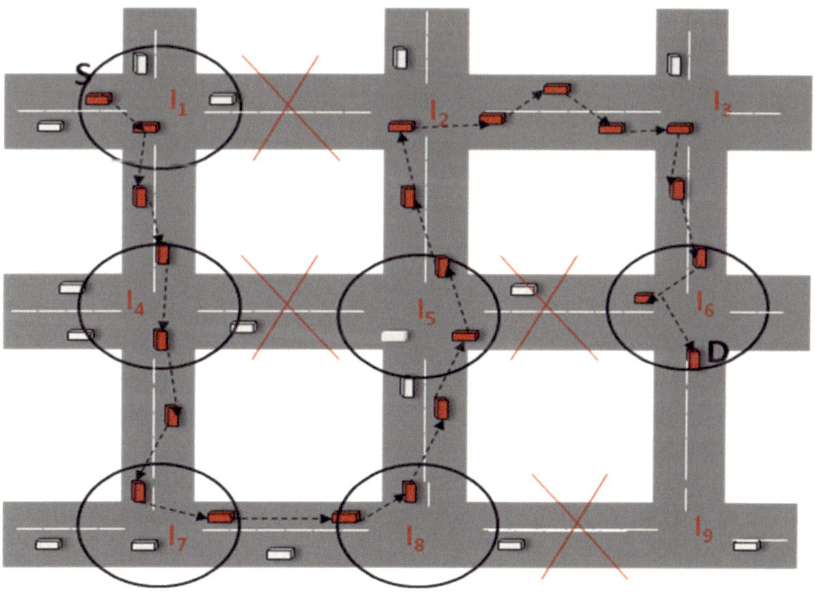

Fig. 6. The right hand rule for vertex selection

range of intersection I_5 where a local maximum occurs since no adjacent connected intersection is closer to the destination. D-RCBR switches to the recovery mode and selects according to the right hand rule the vertex I_2 as next vertex. The packet is the sequentially forwarded through the intersections I_3 and I_6 where it can be delivered to the destination.

4.4 Simulation and analysis

In order to evaluate the proposed solution, an implementation two variants of RCBR protocols has been developed under Network Simulator (NS2). The simulations were carried out with different nodes densities and velocities. The results were then compared with those achieved by three other existing protocols: GPSR, GSR and CAR.
In particular, we were interested in comparing two main metrics: the packet delivery ratio and the average end-to-end delay.
In the following subsections, we describe the simulation environment and present a detailed analysis of the results.

4.4.1 Simulation environment and setup

The simulations have been performed for a vehicular mobility scenario in city environment. The road topology is based on a real map extracted from TIGER (Topologically Integrated Geographic Encoding and Referencing) database. The mobility traces of vehicle movement were generated using a realistic vehicular traffic generator VanetMobiSim (Härri et al., 2006). Vehicles move along the streets with speed limits equal to 50km/h and they change their directions at road intersections. The key parameters of the simulation are summarized in table1.

Simulation parameter	Value
Simulation time	600s
Map size	2500 x 2500 m2
Number of roads	39
Number of road-intersections	33
Number of vehicles	150
Vehicle velocity	15-50km/m
Wireless transmission range	250m
Beacon interval	1s
Data packet size	512bytes

Table 1. The simulation parameters

4.4.2 Packet Delivery Ratio

One of the metric used to evaluate the performance of a routing protocol is the packet delivery ratio (PDR). It is computed as the ratio of the total number of packets received by the total number of packets transmitted by different source nodes.
The graph in Figure 7 shows the average delivery ratio varies as a function of the packet generation rate obtained by varying the sending interval for the different studied protocols.
GPSR considers neither the road topology nor the vehicular traffic and hence packets are more likely to encounter a local maximum which explain the low delivery ratio. On the other hand, GSR improved the forwarding decision with spatial awareness as the sequence

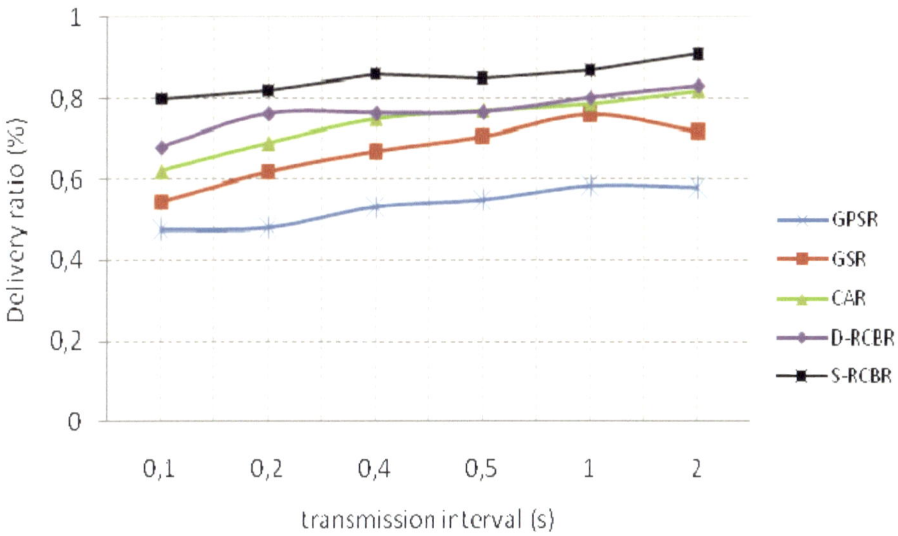

Fig. 7. Packet Delivery ratio Vs Packet sending interval

of junctions is computed before data forwarding. However, since the path is determined without considering real-time traffic, some packets fail to reach their destination when being forwarding along non connected streets which explain the obtained success delivery rate.

The proposed S-RCBR protocol demonstrates the highest delivery ratio than other protocols. This is because the real time traffic information guaranties the connectivity of the entire selected path. Hence, packets are forwarded along connected paths. Moreover, networks partitions are avoided and fewer packets are suspended. Nevertheless a disadvantage that can be noted in S-RCBR is the need for roadside infrastructure which can be costly and not always possible.

The figure depicts also that the number of successfully received packets in D-RCBR are comparable with CAR and even with a relative improvement. The advantage of D-RCBR is that, contrary to S-RCBR and CAR no global knowledge of the network traffic density or real-time connectivity is assumed. The path is dynamically determined following the local connectivity information available in crossroads. So, a packet is only forwarded along connected roads that successfully lead to the destination. Hence, D-RCBR adapts to frequent networks changes.

4.4.3 End-To-end Delay

The results presented in Figure 8 show that S-RCBR achieves a lower end-to-end delay compared to the rest of the protocols (GPSR, GSR, D-RCBR and CAR). The main reason is that S-RCBR offers an accurate view of the network that helps a source node to select a connected path reducing so the chance of facing network disconnections. The packets are simply forwarded along a pre-computed path following the greedy scheme which decreases the networking delays.

GSR does not consider the vehicular traffic to guarantee the connectivity of the shortest path and that is why more packets are likely to be suspended and buffered. CAR also may select

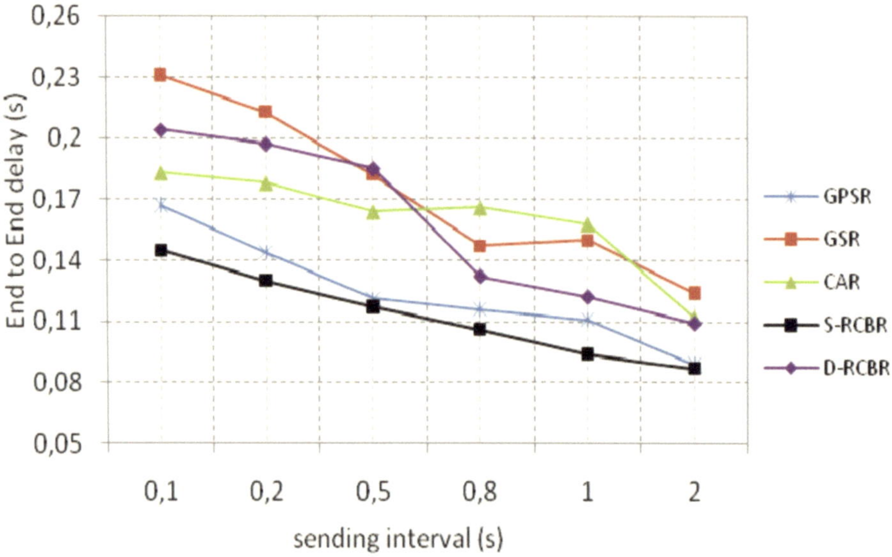

Fig. 8. The end-to-end delay of different protocols

a non optimal path due to the error in the road density information that affects the estimation of the probability of connectivity.

In its turn, D-RCBR achieves a lower end-to-end delay compared to GSR and its performances are as good as CAR. In D-RCBR approach, the routes are discovered while relaying the packet so that the probability of route breaks is much reduced during the forwarding delay. However, CAR uses a source routing approach and generates an additional overhead for the density estimation.

The delay remains higher in D-RCBR than in GPSR because the packets which are usually dropped in GPSR when the perimeter mode fails to handle the local maximum frequently encountered in city environments; they are successfully delivered with D-RCBR mechanism. Note that both D-RCBR and S-RCBR provide an average latency less than 240 ms which proves that the proposed scheme meets the requirements of delay sensitive applications with a good tradeoff between the delivery ratio and the end-to-end delay.

5. Conclusion

Throughout this chapter, we have analyzed the routing problem in vehicular ad hoc networks and presented a taxonomy of existing protocols.

Several routing protocols have been proposed or adapted for the vehicular applications. Nevertheless, the geographic routing has become the trends taking advantages of the availability of navigation system that makes the vehicle aware of its own location as well as its surrounding. Many studies showed that the exploitation of the road-topology improves the routing performances especially with complex mobility patterns of vehicular environments. Also the use of traffic information is proved to be of a great importance and demonstrated better performances. Different ways are used to model this traffic awareness through the historical density data or the real-time traffic information.

In this chapter, we proposed two routing protocols S-RCBR and D-RCBR that combine both the road topology and the real-time traffic. RCBR protocols define a prediction model to estimate the connectivity along the road segments. Then based on this connectivity information either a source route is computed as a sequences of intersection along the connected roads or the path is dynamically adjusted at each intersection. Geographical forwarding is used to transfer the data packets between the vehicles along the road segments that form these paths. The simulation results showed that the proposed protocols outperforms existing approaches and provide a good support for vehicular scenarios. In particular, D-RCBR can be used for vehicular applications where throughput is the main requirement while S-RCBR is suitable for delay-sensitive applications.

6. References

B. Karp and H. T. Kung, "Gpsr: greedy perimeter stateless routing for wireless networks," in MobiCom '00: Proceedings of the 6th annual international conference on Mobile computing and networking, New York, NY, USA, 2000, pp. 243–254.

B.-S. Lee, B.-C. Seet, C.-H. Foh, K.-J. Wong, and K.-K. Lee, "A routing strategy for metropolis vehicular communications," in Proceedings of the International Conference on Networking Technologies for Broadband and Mobile Networks (ICOIN '04), pp. 134–143, Busan, Korea, February 2004.

Charles E. Perkins and Pravin Bhagwat, "Highly dynamic destination-sequenced distance vector routing (DSDV)," in Proceedings of ACM SIGCOMM'94 Conference on Communications Architectures, Protocols and Applications, 1994.

Charles E. Perkins and Elizabeth M. Royer, "Adhoc on-demand distance vector routing," in Proceedings of the 2nd IEEE Workshop on Mobile Computing Systems and Applications, February 1999, pp. 1405–1413.

C. Lochert, H. Hartenstein, J. Tian, H. Fussler, D. Hermann, and M. Mauve. "A routing strategy for vehicular ad hoc networks in city environments". In Proceedings of the IEEE Intelligent Vehicles Symposium, pages 156-161, June 2003.

D. B. Johnson and D. A. Maltz, "Dynamic source routing in ad hoc wireless networks," in Mobile Computing, 1996, pp. 153–181.

Giordano S, Stojmenovic I. Position based routing algorithms for ad hoc networks: A taxonomy. In: Cheng X, Huang X, Du D Z, Kluwer. Ad Hoc Wireless Networking. Holland: Kluwer Academic Publishers, 2003. 103-136

J. Gong, C.-Z. Xu, and J. Holle, "Predictive Directional Greedy Routing in Vehicular Ad hoc Networks," Proc. of Intl. Conf. on Distributed Computing Systems Workshops (ICDCSW), pp. 2-10, June 2007.

J. Härri, F. Filali, C. Bonnet, and M. Fiore, .VanetMobiSim: Generating realistic mobility patterns for VANETs, in VANET '06: Proceedings of the 3rd international workshop on Vehicular ad hoc networks. ACM Press, 2006, pp. 96.97.

J. Li, J. Jannotti, D. De Couto, D. Karger, and R. Morris, "A scalable location service for geographic ad hoc routing", ACM/IEEE MOBICOM'2000, pp. 120.130, 2000.

J. Tian, L. Han, K. Rothermel, and C. Cseh, "Spatially aware packet routing for mobile ad hoc inter- vehicle radio networks,". In Proceedings of the IEEE Intelligent Transportation Systems, Volume: 2, pages 1546-1551, Oct. 2003.

Josiane Nzouonta, Neeraj Rajgure, Guiling Wang, and Cristian Borcea, "VANET Routing on City Roads using Real-Time Vehicular Traffic Information", IEEE Transactions on Vehicular Technology, Vol 58, No. 7, 2009.

N. Brahmi, M. Boussedjra, J. Mouzna, and B. Mireille, "Adaptative movement aware routing for vehicular ad hoc networks," in The 5th International Wireless Communications and Mobile Computing Conference IWCMC09, Leipzig , Germany, Jun 2009.

Q. Yang, A. Lim, and P. Agrawal, "Connectivity aware routing in vehicular networks", In Wireless Communications and Networking Conference, 2008. WCNC 2008. IEEE, 2008, pp. 2218.2223.

Thomas Clausen , Philippe Jacquet , Anis Laouiti , Pascale Minet , Paul Muhlethaler , Amir Quayyum , and Laurent Viennot , "Optimized link state routing protocol," Internet Draft, draftietf-manet-olsr-05.txt, work in progress, October 2001.

W. Su, S.-J. Lee, and M. Gerla, "Mobility prediction and routing in ad hoc wireless networks," in International Journal of Network Management, Wiley and Sons, Eds., 2000.

6

Traffic Information Dissemination in Vehicular Ad Hoc Networks

Attila Török, Balázs Mezny and Péter Laborczi
Bay Zoltán Foundation For Applied Research,
Institute for Applied Telecommunication Technologies
Hungary

1. Introduction

Today, cars are equipped with all kind of on-board sensors and microcomputers that are able to measure geolocation, speed, tire pressure, raindrops on the windshield, etc., and based on these information Intelligent Transportation Systems (ITS) are built. The ITS applications are intended to ease the everyday life of drivers by reducing the risk of accidents, improving safety, increasing road capacity and reducing traffic jams. Many research papers, for example Torok et al. (2008) and Sormani et al. (2006), pointed out that a significant reduction of traffic jams can be achieved through the use of vehicular ad-hoc networks (VANETs). Vehicles could serve as Traffic and Travel Information (TTI) collectors and transmit this information to other participants in the vehicular network Laborczi et al. (2006). The ITS applications could utilize this information to actively relieve traffic congestion. Practically, vehicles could detect traffic congestion automatically when the usual (historical) characteristics of traffic patterns drastically change, i.e. the number of neighboring vehicles is high and/or the average speed is too low. Then this information should be relayed, disseminated to the vehicles approaching the congested area; thus, they will have enough time to choose alternative routes.

Due to their inherent characteristics, viable communication is harder to support in ITS scenarios than in conventional wireless networks. Vehicles are usually moving much faster than traditional mobile nodes; moreover, a vehicular network might be very heterogeneous in terms of node density, becoming fragmented in many cases. Reliability is also compromised due to the usually high interference in urban scenarios. Thus, there is a need to reconsider the wireless ad hoc communication networking protocols, and to use new concepts that fit better the specificities of ITS applications.

Traffic and Travel Information (TTI) spreading in vehicular ad hoc networks is achieved by the means of a flooding mechanism. To overcome network fragmentation the vehicles usually maintain and carry a copy of the packets, which is disseminated along the road segments Zhao & Cao (2006), Burgess et al. (2006), Tian et al. (2004). The frequency of subsequent transmissions will control the quality of the TTI reports, in terms of delay and accuracy. If the frequency of TTI transmissions is high, the time necessary for the information to reach the outer bounds of the geographic area is lower. The accuracy of TTI also varies in function of the amount of communication involved in the travel information gathering and transmission. Frequent information exchange leads to a more accurate picture about the traffic situation, but also to superfluous dissemination. Superfluous forwarding can be reduced by using adaptivity in the flooding mechanisms. Adaptivity can be introduced by controlling the

frequency of information exchange (timely manner) or limiting the dissemination only to areas where the TTI is really necessary (spatial manner).
Besides the presentation of the most important spatial TTI dissemination protocols we also investigate the problem of determining the areas of interest of traffic jams. As we argue, the presented spatial dissemination protocols fail to properly define the places where the TTI is useful. These solutions are only effective when are employed with additional mechanisms, which provide context-aware information to calculate the areas of interest of specific traffic jams.

2. Literature review

This section presents protocols related to spatial adaptivity-based TTI dissemination, which can be achieved pro-actively, using a data-push model Sormani et al. (2006), Leontiadis & Mascolo (2007), or based on a data-pull model Dikaiakos et al. (2007), when the information is obtained on-demand. In the first case the data is usually disseminated from the traffic incidents towards the outer inbound road segments, while in the second case the data is pulled to the locations of interest on-demand. In both cases the question is how to control and limit the traffic information dissemination only to places where the respective information is useful.

2.1 Spatial adaptivity by using data-push protocols

2.1.1 Dissemination restricted through publish/subscribe

The possibility of restricting the TTI dissemination to certain areas is investigated in Leontiadis & Mascolo (2007). In their proposal the authors present a publish-subscribe method, as the members of the traffic will receive only messages of their interest. The solution works well with methods employing the data-push model, for example the one described in Sormani et al. (2006). The publish-subscribe process starts with a vehicle subscribing to a topic (e.g. traffic congestion information). When a vehicle publishes a message, the area of interest and validity time of the message is determined. Vehicles subscribed to the given topic will receive the message if they are within the area of interest and the message is still valid. The basic idea is to maintain the message in the notification area, so every vehicle approaching the area where the message was generated (for example a traffic accident) gets the notification and has a chance to consider its reaction to the event (e.g. taking an alternate route to its destination). This is achieved by generating a fixed number of replicas of the message, which means that only those vehicles will broadcast the message which have a replica of the message. This way the load of the communication network is reduced compared to the general flooding mechanism, where every node of the network retransmits the received message, resulting in a broadcast storm. If a vehicle carrying a replica of the message is leaving the notification area, then it hands over the replica to an other vehicle, preferably driving the opposite direction, to keep the message replica in the desired area.
There are two questions regarding the message replicas. How many replicas should be there, and who should carry them? Before the replica owner broadcasts the message, it poll its neighbouring vehicles regarding the topic of the message. There are three possible answers to this poll:

- Informed: The answering vehicle is already received a notification for the given topic (e.g. if the topic is parking spots, this vehicle already knows where are free parking spaces).
- Interested: This vehicle is subscribed to the topic.

- Not interested: The vehicle is not subscribed.

If there are interested vehicles the carrier broadcasts the message. Also if the carrier is leaving the designated area, it selects a new carrier heading for the notification area, with the most interested vehicles in its vicinity. The aim of this selection method is to get the message replica where the most uninformed vehicles come from. This mechanism results in the replicas converging to areas where the information is needed, and if there are two replicas in the same area, one of them will move to an other area where the message is needed.
The number of the replicas is determined adaptively. Every replica carrier keeps the result of the last k polls, and based on these statistics the following options are possible:

- If there was at least one uninformed subscriber in the last k polls, the replica is kept.
- If there were at least k uninformed subscribers then a new replica is generated and forwarded to a vehicle, determined by the new carrier selection mechanism.
- If there are no uninformed subscribers, the replica is marked for deletion. In order to avoid deleting replicas simultaneously, the replicas are merged and are deleted only if the carrier receives a broadcast from an other carrier.

This way the number of replicas are adapted to the demand for the message, and they are forwarded to areas with the most subscribers.
However, due to the carrier selection and TTI replication mechanisms, it is not always guaranteed that the information carriers will meet their subscribers. The chance that a replica survives insensitivity, and meets proper subscribers, depends on the estimate of the replica's necessity, which is represented by the number of last k polls. Thus, the successful outcome of the protocol highly depends on the topological context and the fine tuning of the system. For example, considering the simulation results presented in Figure 1 for a scenario where two intersections are interconnected through two one way roads (one with traffic jam), it turns out that the fraction of cars entering the jammed road depends highly on the frequency of TTI disseminations (Timer), respectively on the number of transmissions until a TTI remains alive (TTL). If the frequency is too high then the TTI message is not transported until the intersection where the vehicles must be informed, even considering higher values for TTL. This can be attributed to the fact that the TTI replication and propagation was determined based on the interest of other neighboring vehicles, and in this particular case all the vehicles are heading outwards the jammed area; thus, they are uninterested about this particular jam. In order to overcome such problems additional context information regarding the road infrastructure has to be taken in consideration.

2.1.2 Dissemination restricted through propagation functions

In Sormani et al. (2006) the authors investigate methods on how to propagate the traffic messages to areas where the respective information is useful. They outline a scenario, where an accident occurs on a highway. A message broadcast happens within one mile of the accident, telling the vehicles to slow down. A second message is delivered to key points of the highway, where drivers can take alternative routes, in this scenario these points are the highway exits. This method can be considered as a data-push model, where the message is disseminated even if the information wasn't requested by an entity. The main idea is the definition of a propagation function, which restricts the message propagation to areas where the message is important. For our example this represents the highway, there is no point in disseminating the message outside this area. This propagation function has minimum points at the target zones, and its value is increasing as the distance from the target zones increase.

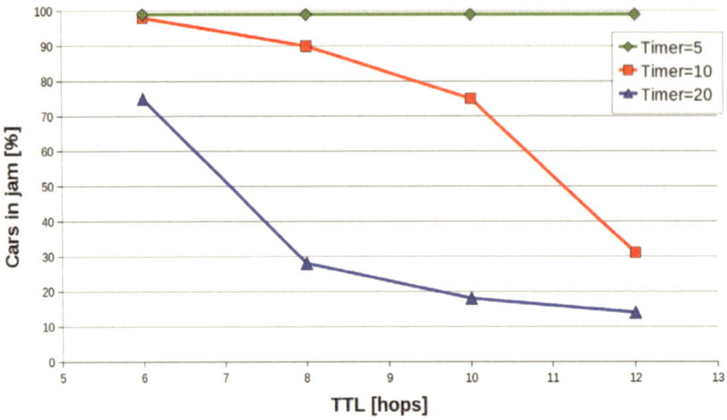

Fig. 1. Effect of TTI replication on alternative route selection

The message, originated the place of an event (e.g. a traffic accident), is always forwarded to a vehicle whose position results in the lowest value in the propagation function. This way the message will be routed towards the minimum of the propagation function, the target zone. The shape of this function is determined to follow the road network, where the vehicles can disseminate the message. The propagation function is either computed by the originator of the message, taking into account the current traffic situation and the road network in the vicinity, or it can be precomputed for important areas.

The authors consider some basic protocols to disseminate the traffic message in order to evaluate the effects of the propagation function. The most basic protocol is a modification of the flooding mechanism, where the received message is rebroadcasted only for the first time it has been seen and the value of the propagation function at the receiving vehicle is lower than at the sender of the message (One Zero Flooding, OZF). An other basic protocol is a further modification, taking into account the distance between the sender and the receiver (Distance Driven Probabilistic Diffusion, DDPD). This distance is used for probabilistic message forwarding, where the probability of forwarding is the distance between the vehicles divided by the communication range (approximately 200 meters for 801.11 capable devices). This way the surplus message retransmissions can be avoided. A more advanced protocol takes into account the shape of the propagation function (Function Driven Probabilistic Diffusion, FDPD). In this case the probability of forwarding is zero at the sender's position and is increasing as the value of propagation function decreases, and takes the value of one at the lowest value of the propagation function inside the communication radius of the sender of the message. This method yields to a more accurate routing, as a lower value of the propagation function is not enough, the algorithm tries to find the lowest possible value. The authors propose some store & forward variations of these algorithms, where after receiving a message the vehicle not only retransmits it immediately but carries it for some time and rebroadcasts the message periodically. The first store & forward variant (Function Driven Feedback-augmented Store & Forward Diffusion, FD-FSFD) is based on FDPD with the addition of a feedback augmented store & forward mechanism. The feedback augmention means, if the carrier receives a message from a vehicle whose position results in a lower value of the propagation function, then the further broadcasts are cancelled as the message already reached a lower point of the propagation function. The second store & forward algorithm (Direction-aware Function Driven Feedback-augmented Store & Forward

Diffusion, DFD-FSFD) is an extension of FD-FSFD by taking into account the direction of movement of the nodes. This means that only nodes moving towards lower points of the propagation function are used to carry the message. These methods are useful in sparse networks where the connection between clusters of vehicles is not guaranteed.
Unfortunately, there are no methods presented to calculate the propagation function, i.e., the locations where the information should be propagated. Therefore, this protocol is not ready to be applied for TTI dissemination in urban scenarios.

2.2 Spatial adaptivity by using data-pull protocols

In Dikaiakos et al. (2007) the authors outline an application-layer communication protocol (Vehicular Information Transport Protocol, VITP), which could be used in VANETs to disseminate location based information. Such location based information can be traffic information regarding road conditions (e.g. slippery road or congestion), or some kind of roadside service information (e.g. fuel prices at gas stations or menus of restaurants). These kinds of information are typically requested by someone; thus this method can be called as the data-pull model. The authors introduce the concept of virtual ad-hoc servers (VAHS), which means that an information request is processed by a number of peers at the target location of the request, and the result is sent back to the originator of the query. For example, if a vehicle wants to know the traffic condition on a road segment in its path, it sends a request to that road segment. When a vehicle in the target area receives the query, it attaches the requested information to the message, and retransmits the message, so other vehicles can contribute to the reply. The ones contributing to the reply constitute the virtual ad-hoc server. After a certain threshold is met, for example ten vehicles attached their velocity information to the message, the last vehicle generates the reply from the gathered data, and sends it back to the originator vehicle. This way the answer can be more accurate, than in the case where only one vehicle made the reply, or when separate replies were generated by multiple vehicles. The data-push method is also supported by the proposed protocol as it is favorable in some cases, for example in case of an accident. The vehicles couldn't be forced to constantly generate queries for accidents, instead the information is "pushed" to them. The described protocol is also capable of caching the information in some cases, so a reply could be made before the query reaches the target location, speeding up the process. The effect of caching is further elaborated in Dikaiakos et al. (2010), and it is shown that significant improvements can be achieved in both the data-pull and the data-push cases.

2.3 Aggregation scheme for roadside unit placement

The authors of Lochert et al. (2008) present a method for optimization of roadside unit placement in order to minimize the required bandwidth for traffic information dissemination. A domain specific aggregation scheme is presented, then a genetic algorithm is proposed to identify the most appropriate positions for the roadside units. The aggregation scheme is conceived in a hierarchical fashion: the farther away a region is, the coarser will be the information on its traffic situation. By using this scheme a vehicle traveling along the road network will obtain coarser and coarser approximations about the traffic situation, travel times will be summarized in regions that are farther and farther away. Thus, the aggregation scheme will allow to limit the bandwidth requirements for TTI dissemination, by reducing the network capacity necessary for information spreading. The aggregation scheme is based on the definition of special multi-level landmarks, which will cover the hierarchy of the road networks. The higher levels are constituted by highways or junctions of main roads, while the lower levels will include all higher level landmarks plus more and more intersections of smaller streets. Thus, cars can make investigations about the travel times between neighboring landmarks, which information will be disseminated in the surroundings of the respective road segment. A coarser picture, calculated from travel times between landmarks of higher levels, will be disseminated to a larger area, which leads to a summarized view of the travel times in the area. Roadside units are placed to form a backbone network, allowing them to exchange the TTI to be disseminated. In order to use a very limited number of roadside units the authors propose a toolchain for placement optimization. Since the identification of the optimal subset of roadside unit locations is a difficult optimization problem a genetic algorithm based approximation method is used to obtain a good result subset. The toolchain is composed from a network and traffic simulator (ns-2 and VISSIM), respectively from a closely interacting application simulator and the genetic algorithm. The application simulator is used to process the log file of the network-traffic simulator, perform the specific aggregation methods, decide about the dissemination of TTI beacons. At the level of the network simulator all the possible roadside unit locations are simulated, all of them transmit the periodic beacons. The non-existing roadside units are ignored at the level of the application simulator, the received beacons are not considered when its knowledge base is updated by the genetic algorithm. Thus, with the separation of movement and network issues from application behavior travel time savings are achieved by calculating the vectors of active roadside unit locations. These savings are used as a fitness metric, making the application-centric optimization through the genetic algorithm. The viability of the approach is confirmed through simulations by applying the proposed solution to a large-scale city scenario.

3. Spatially-aware congestion elimination (SPACE)

In this section the SPatially-Aware Congestion Elimination algorithm (SPACE) is designed. An algorithm is given to determine the locations, domain of interests, where a possible event (e.g. traffic jam) on a certain road influences the route choice of the driver. To illustrate the problem, a small example is presented, then we formulate it as a graph theoretical optimization problem. Both a heuristic and a linear programming optimization solution are provided. Thus, we give a well defined area (the Domain of Interest, DoI) where information about a specific traffic jam is useful.

3.1 Example

First let us consider an example of one way roads from left to right (Figure 2), which represents a subset of a larger road network.

We assume that a vehicle enters the network at node 1, its destination is at node 10. The vehicle has route decisions at nodes 2,3,4 and 5, respectively. It can take either Route A, Route B, Route C or Route D to reach its destination. Route A is shortest and fastest; consequently, the vehicle takes the middle route in the default case. If route A at road segment 6-7 is congested, this information has to be disseminated throughout the road network.

The *Domain of Interest (DoI)* is defined as the set of road segments, where the information about a traffic jam influences the route choice of the driver, i.e., the roads where the information should be disseminated. At these places, the vehicles are still able to change their routes, without a drastic deterioration in their travel time. However, if the vehicle leaves a critical junction, enters in the *zone of no return*, where is no possibility to avoid the traffic jam, or only with a major increase in the travel time. Our scope is to optimize the area of DoI in order to

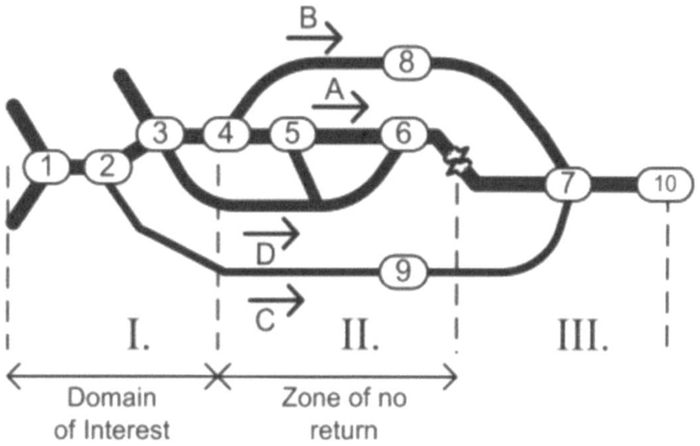

Fig. 2. Example road network

reduce the amount of TTI flooding and at the same time to achieve as low vehicle travel times as possible.

Traditional flooding methods disseminate this information towards any directions. However, in the example this information is only interesting at the decision point 4 (optimized DoI), since the second best choice is route B. It is useless to deliver the TTI further than junction 4, as vehicles are heading towards junction 4, anyway. There is no sense in providing this information to the whole DoI, like (1,2,3,8,9). However, if both routes A and B would be congested, this information should be provided to an earlier decision point (junction 2, segment 1-2), where both routes can be avoided by the by-pass route C. This means that the DoI can also present characteristics varying over the time.

3.2 Problem formulation

The road network is represented by a directed and weighted graph $G(\mathcal{V},\mathcal{E})$ with representation described in Speicys & Jensen (2008), using two types of edges $\mathcal{E}_r$ and $\mathcal{E}_t$ ($\mathcal{E}_r \cup \mathcal{E}_t = \mathcal{E}$, $\mathcal{E}_r \cap \mathcal{E}_t = \emptyset$). $\mathcal{E}_r$ is a directed edge representing a road between two intersections. One-way roads are represented with directed edges, while two-way roads with two opposite directed edges. The set of $\mathcal{E}_t$ represents the turning regulations, i.e., an edge from $n_1 \in \mathcal{V}$ to $n_2 \in \mathcal{V}$, where n_1 is the destination node of e_1 while n_2 is the origin node of e_2, is included in the graph if and only if a turn is allowed from e_1 to e_2. The weight of an edge represents the travel time on the corresponding road, or turning.

The event (traffic jam) is associated to a set of failed roads $\mathcal{E}_f$, which is a subset of the roads ($\mathcal{E}_f \subseteq \mathcal{E}$). We assume that the set $\mathcal{E}_f$ contains the core of the problem, where the actual speed decreases to a fraction of the normal speed.

We also assume that an estimated Origin-Destination (OD) function for the road network is known. The *OD* function $OD(n,m)$ represents the average amount of vehicles traveling from node $n \in \mathcal{V}$ to $m \in \mathcal{V}$. If the OD function is not known then it can be assumed that it has uniform distribution, i.e., $OD(n,m) = 1$ for each $n,m \in \mathcal{V}$.

The output of the algorithm is an Impact Vector $I_{\mathcal{E}_f}(e)$ that shows whether an event on edges of $\mathcal{E}_f$ has an impact on the route choice of vehicles travelling on edge e, and if yes, in what extent. The value of $I_{\mathcal{E}_f}(e)$ is zero if it has no impact on edge e, non-zero if it has an impact.

The value $I_{\mathcal{E}_f}(e)$ expresses the average amount of vehicles on edge e, whose route choice is impacted by the knowledge about an event on roads of $\mathcal{E}_f$.

3.3 SPACE algorithm

In this section the proposed SPACE Algorithm is described, which generates the Impact Vector I. The algorithm simulates the impact of an event (obstacle, traffic jam) on a set of edges of the graph. For each affected optimal path p of the graph it is assumed that the travel time on edges of $\mathcal{E}_f$ increases significantly. Then, the weights of the affected edges of $\mathcal{E}_f$ are increased, or these edges are excluded from the graph temporarily, and a new optimal path (the by-pass route) is calculated by running the shortest path algorithm in the temporary graph, resulting in path p^*. We define three parts of the optimal route p, and illustrate it on Figure 2, where p=(1,2,3,4,5,6,7,10), and the alternative route bypassing the edges with increased weights p^*=(1,2,3,4,8,7,10):

Part I. Set of edges common with p^*, before the disjoint part: e.g., road (1,2,3),

Part II. Set of edges not included in p^*, before the event: e.g., road (4,5,6),

Part III. Set of edges after the event: e.g., road (7,10).

In this case, an event on edges in $\mathcal{E}_f$ is important for the last X of edges in part I of p (in order to choose another route), and in the disjoint part before the event (part II.), in order to be informed about the obstacle (without the possibility of choosing the other route). These edges are called **relay** edges in the algorithm. For all these edges the impact vector I has to be increased with the amount of vehicles traveling on that route (or by 1, if an OD matrix is not available). The algorithm is summarized as follows:

Input: Directed Weighted Graph $\mathcal{G}$, OD matrix, Set of failed roads $\mathcal{E}_f$
Output: Impact Vector I
for all Pair of nodes $(n,m) \in OD$ **do**
  Calculate in $\mathcal{G}$ the optimal path p from n to m;
  Create a new temporary graph $\mathcal{G}_{\mathcal{E}_f}$: increase weights of edges of $\mathcal{E}_f$ significantly;
  Calculate by-pass route p^* from n to m in $\mathcal{G}_{\mathcal{E}_f}$;
  Calculate the set of **relay** edges: $\mathcal{E}_R \subseteq \mathcal{E}$;
  for all edge $e_r \in \mathcal{E}_R$ **do**
    Increase $I_{\mathcal{E}_f}(e)$ by $OD(n,m)$;
  end for
end for

3.4 SPACE_ILP algorithm

In this subsection the problem of finding the optimal Domain of Interest (DoI) is formulated as an Integer Linear Programming (ILP) problem, as presented in Torok et al. (2010). Although, solving an ILP by a solver has a long running time, we emphasize that this formulation has the following motivations: the formulation gives an exact definition of the TTI dissemination problem and it allows a precise analysis of the problem compared to heuristic algorithms.
First, let us define the normal route of the vehicles. For each edge (i,j), $i,j \in \mathcal{V}$, and origin and destination nodes $n,m \in \mathcal{V}$, we define the set of assignment variables, $\mathcal{X} = \{x_{ij}^{nm}\}$. The variable x_{ij}^{nm} takes value 1 if edge ij is used in the shortest path from n to m, and 0 otherwise.
The known flow conservation constraints for the default routes are as follows:
For each $j,n,m \in \mathcal{V}$ where $OD(n,m) > 0$:

$$\sum_{i \in \mathcal{V}} x_{ij}^{nm} - \sum_{k \in \mathcal{V}} x_{jk}^{nm} = \begin{cases} -1 & \text{if } j = n \\ 1 & \text{if } j = m \\ 0 & \text{otherwise} \end{cases} \quad (1)$$

Similarly, the by-pass route is defined for the vehicle. For each edge (i,j), $i,j \in \mathcal{V}$, and origin and destination nodes $n,m \in \mathcal{V}$, we define the set of assignment variables, $\mathcal{Y} = \{y_{ij}^{nm}\}$. The variable y_{ij} takes value 1 if edge ij is used in the by-pass route from n to m, otherwise 0. The flow conservation constraints for the by-pass routes are as follows:
For each $j,n,m \in \mathcal{V}$ where $OD(n,m) > 0$:

$$\sum_{i \in \mathcal{V}} y_{ij}^{nm} - \sum_{k \in \mathcal{V}} y_{jk}^{nm} = \begin{cases} -1 & \text{if } j = n \\ 1 & \text{if } j = m \\ 0 & \text{otherwise} \end{cases} \quad (2)$$

Furthermore, in the formulation, both the normal and the by-pass routes are to be split in several pieces. For this, five more assignment variables are defined: a_{ij}^{nm}, b_{ij}^{nm}, c_{ij}^{nm}, d_{ij}^{nm}, f_{ij}^{nm} (for each edge (i,j), $i,j \in \mathcal{V}$ and origin and destination nodes $n,m \in \mathcal{V}$) with following definitions:

- a_{ij}^{nm} is 1 if edge ij belongs to the common part of the normal and the by-pass route, 0 otherwise.
- b_{ij}^{nm} is 1 if edge ij belongs to the normal route after the fork of the normal route but before the traffic jam, 0 otherwise.
- c_{ij}^{nm} is 1 if edge ij belongs to the traffic jam ($(i,j) \in \mathcal{E}_f$), 0 otherwise.
- d_{ij}^{nm} is 1 if edge ij belongs to the normal route after the fork of the normal route but after the traffic jam, 0 otherwise.
- f_{ij}^{nm} is 1 if edge ij belongs to the by-pass route while not to the normal route, 0 otherwise.

For an example of these definitions see Figure 2 with origin=1, destination=10, optimal route (1,2,3,4,5,6,7,10) and by-pass route (1,2,3,4,8,7,10). $a_{ij}^{nm} = 1$ for roads (1,2),(2,3),(3,4) and (7,10). $b_{ij}^{nm} = 1$ for roads (4,5) and (5,6). $c_{ij}^{nm} = 1$ for road (6,7). $d_{ij}^{nm} = 0$ for all roads. $f_{ij}^{nm} = 1$ for roads (4,8) and (8,7).
The above definitions are ensured by the following equations:
For each $(i,j) \in \mathcal{E}$ and $n,m \in \mathcal{V}$ where $OD(n,m) > 0$:

$$a_{ij}^{nm} + b_{ij}^{nm} + c_{ij}^{nm} + d_{ij}^{nm} = x_{ij}^{nm} \quad (3)$$

$$x_{ij}^{nm} + f_{ij}^{nm} \leq 1 \quad (4)$$

$$a_{ij}^{nm} + f_{ij}^{nm} = y_{ij}^{nm} \quad (5)$$

Furthermore, the part of the default route after the jammed link (d) has to be distinguished form the part before the jam (b) with the following constraint:
For each $j,n,m \in \mathcal{V}$ where $OD(n,m) > 0$:

$$\sum_{i \in \mathcal{V}} d_{ij}^{nm} - \sum_{k \in \mathcal{V}} d_{jk}^{nm} = \begin{cases} -1 & \text{if } c_{ij}^{nm} = 1 \\ \geq 0 & \text{otherwise} \end{cases} \quad (6)$$

For each road (i,j) affected by the traffic jam $((i,j) \in \mathcal{E}_f)$ set c_{ij}^{nm} to 1 and y_{ij}^{nm} to 0, while for each other (not jammed) road $((i,j) \notin \mathcal{E}_f)$ set c_{ij}^{nm} to 0.
Next, the assignment variables for the propagation region are defined and we formulate the fact that vehicles does not by-pass the jam until they receive a message about it, i.e., the normal route and corresponding by-pass route are to be the same outside the propagation region. For each edge (pair of nodes) (i,j), $i,j \in \mathcal{V}$, we define the set of assignment variables, $\mathcal{R} = \{r_{ij}\}$. The variable r_{ij} takes value 1 if edge ij is included in the propagation region, otherwise 0.
In order to ensure a propagation region that reaches all places where normal and by-pass routes are to be forked, the following constraints are defined:
For each $n,m \in \mathcal{V}$ where $OD(n,m) > 0$:

$$r_{ij} \geq b_{ij}^{nm} \tag{7}$$

Finally, we define the objective by minimizing the weighted average of the length of all by-pass routes and the total length of the propagation region:

$$\min \sum_{(i,j)\in\mathcal{E}} \left(\alpha l'_{ij} \sum_{n,m\in\mathcal{V}} y_{ij}^{nm} + (1-\alpha) l''_{ij} r_{ij} \right) \tag{8}$$

l'_{ij} denotes the cost (length, travel time, etc.) of travelling on road ij while l''_{ij} denotes the cost (e.g., road length, communication cost) of propagating information on road ij. Parameter α $(0 \leq \alpha \leq 1)$ expresses the importance of minimizing the total length of all by-pass routes against the total propagation region.
In summary, for the ILP formulation we define constants: c_{ij}^{nm}, l'_{ij}, l''_{ij}; binary variables x_{ij}^{nm}, y_{ij}^{nm}, r_{ij}, a_{ij}^{nm}, b_{ij}^{nm}, d_{ij}^{nm}, f_{ij}^{nm}; objective: (8) and constraints: (1)-(7).

3.5 Query-based information gathering

As we presented above, in Dikaiakos et al. (2007) a query-based protocol is provided to achieve spatial adaptivity by the means of pull-based techniques. Unfortunately, there is no exact mechanism specified to calculate at what extent the queries, respectively the TTI reply/caching, should be propagated inside the road network. In the original paper the authors present simulations, where the queries are propagated only to a randomly selected value of 400-800 meters inside the road segments. This means that in certain situations the information will not be received in time to calculate the proper by-pass route. Therefore, additional mechanisms are necessary to determine the critical points until when the queries must be propagated, i.e., from where the TTI information has to be gathered. Considering a vehicle entering at junction 1 in the road graph of Figure 2, it is hard to decide when and how deep to inject the traffic information query in order to discover a possible traffic jam on route A. This problem is related to the calculation of the optimal DoI, and can only be solved by using additional information about the vehicle's context, i.e. the route graph traversed during the trip.
In Laborczi et al. (2010) an extension of the previous ILP formulation is given, with the aim to optimize the positions where vehicles along their routes should send query messages, in order to collect information about possible traffic jams. Thus, this formulation can be considered as an optimization of the query protocol presented in Dikaiakos et al. (2007). The numerical results of the formulation are presented in the following section.
The equations 1 - 6 from the previous section are the same also for the query-based information gathering.

In order to find an optimal point to send a query message, we have to define the following set of variables:

- s_{ij}^{nm} is 1 if edge ij belongs to the common part of the normal and the by-pass route and to the DoI as well.

With the following three constraints the properties of variables s_{ij}^{nm} are ensured:
For each $n,m \in \mathcal{V}$ where $OD(n,m) > 0$:

$$r_{ij} \geq b_{ij}^{nm} + s_{ij}^{nm} \tag{9}$$

$$s_{ij}^{nm} \leq a_{ij}^{nm} \tag{10}$$

Edges where s_{ij}^{nm} or b_{ij}^{nm} takes value 1 should be a coherent region, which ends at the traffic jam:

$$\sum_{i \in \mathcal{V}} (s_{ij}^{nm} + b_{ij}^{nm}) - \sum_{k \in \mathcal{V}} (s_{jk}^{nm} + b_{jk}^{nm}) = \begin{cases} 1 & \text{if } \exists k \; c_{jk}^{nm} = 1 \\ \leq 0 & \text{otherwise} \end{cases}$$

Finally, the propagation delay of the distributed messages has to be taken into account. It takes some time for the query message to reach the jam, and the reply message to reach the originator. The query message should reach the begin of jam, collect information and the reply message should reach the vehicle before it reaches the optimal decision point expressed by the following constraint:

$$\sum_{(i,j) \in \mathcal{V}} \left(s_{ij}^{nm} \frac{l_{ij}}{v_{\text{veh}}} - (2b_{ij}^{nm} + s_{ij}^{nm}) \frac{l_{ij}}{v_{\text{mess}}}\right) \geq 0, \tag{11}$$

Where l_{ij} is the length of the ij road segment, v_{veh} is the velocity of the vehicle, v_{mess} is the velocity of the message propagation.
Finally, we define the objective by minimizing the weighted average of the length of all by-pass routes and the total length of the DoI:

$$\min \sum_{(i,j) \in \mathcal{E}} \left(\alpha l'_{ij} \sum_{n,m \in \mathcal{V}} y_{ij}^{nm} + (1-\alpha) l''_{ij} r_{ij} \right) \tag{12}$$

where, l'_{ij} is the cost (length, travel time, etc.) of traveling on road (i,j) while l''_{ij} denotes the cost (e.g., road length, communication cost) of propagating information on road (i,j). Parameter α $(0 \leq \alpha \leq 1)$ expresses the importance of minimizing the total length of all by-pass routes against the total DoI.
In the next section, we use this formulation as follows. First, for each $n,m \in \mathcal{V}$ where $OD(n,m) > 0$ calculate shortest path and set variables x_{ij}^{nm} based on the result of the shortest path algorithm. Second, solve the following ILP problem: constants: x_{ij}^{nm}, c_{ij}^{nm}, l'_{ij}, l''_{ij}; binary variables: y_{ij}^{nm}, r_{ij}, a_{ij}^{nm}, b_{ij}^{nm}, d_{ij}^{nm}, f_{ij}^{nm}, objective: (12) and constraints: (2)-(6) and (9)-(11).

4. Numerical analysis of the SPACE and SPACE_ILP algorithms

In this section we present the evaluation of the proposed heuristic and linear programming algorithms. All the simulations were effectuated on the same section of a digital map of Budapest.

4.1 SPACE

The output of the SPACE algorithm is demonstrated on Figure 3 for two roads of the Budapest test network. A main road (bridge, solid line), and a side road (from down-town, dotted line) were considered for analysis. The x-axis represents the domain of interest, i.e., the sum of road lengths on which the information about the event is disseminated, while y-axis represent the impact factor. The information is sent to roads (e) of higher impact factors ($I_{\mathcal{E}_f}(e)$), while roads with minor impact factors can be neglected. First, we assume that there is a traffic jam (obstacle) on the main road (depicted with solid line). If the TTI information is flooded to the whole domain of interest, then it should be spread to 16,000 meters. Therefore, a threshold (e.g., TR = 4,000) should be set in order to avoid superfluous forwarding. In this way the information is carried to the majority of vehicles (just 1,000 from more than 15,000 do not receive the information in time), while the domain of interest is decreased to 4,000 meters, instead of 16,000. Second, we consider an obstacle on a side street (dashed line). As expected, the impact factor of such streets is less, i.e., if the threshold is set to 1,000 then the domain of interest is 500 meters.

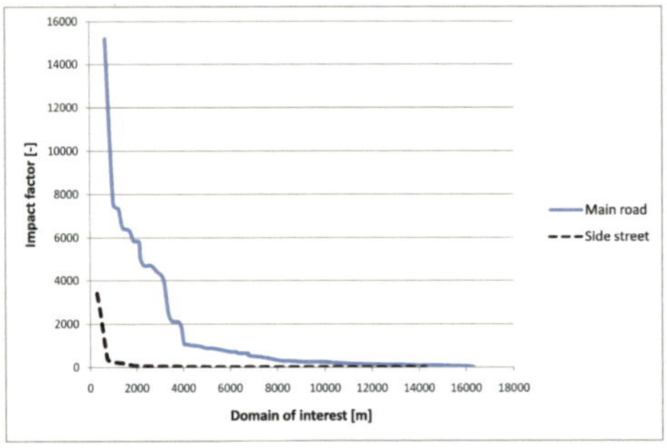

Fig. 3. Domains of Interest for different road types using the SPACE algorithm

4.2 SPACE_ILP

In order to have a better understanding, the results of the SPACE_ILP algorithm are presented below. A main road (bridge), and different side roads (from downtown area) were considered for analysis. The bridge graph represents averaged values for traffic demands initiated from both sides of the city, considering traffic jam on one of the bridge lanes. The downtown graph represents averaged values from different congested downtown roads (considering also the major roads leading to the bridge).

Figure 4 shows the Domain of Interest (DoI) depending on the parameter α (see Objective (8)). We recall that α ($0 \leq \alpha \leq 1$) expresses the importance of minimizing the total length of all vehicle by-pass routes against the importance of minimizing the propagation region (area of dissemination). The DoI is represented as the sum of the road segment lengths included in the propagation region.

It is obvious that for both graphs the DoI increases by increasing α. On the other hand, the figure shows that the two types of roads represent different dynamics considering their DoI. In case when the obstacle is on the bridge, the DoI increases steeply with the increase of α. This means that in order to reach all roads of the maximum DoI, higher efforts must be involved for TTI dissemination. However, after a limit ($\alpha \geq 0.6$) the DoI is not increasing significantly (only about 1 km). A crucial point for α is between 0.3 - 0.4, where the DoI increases significantly.

Considering congestion on downtown roads the situation is different. It can be seen, that the

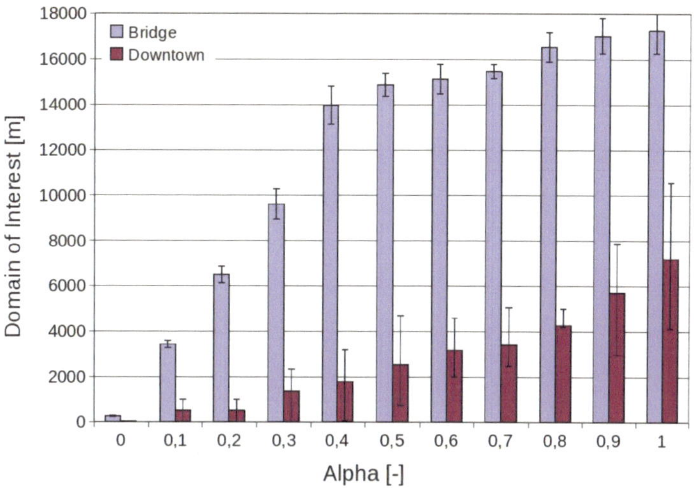

Fig. 4. Domains of interest in function of parameter α

variance of the DoI values is higher; however, the area of DoI for downtown scenarios is only a fraction of the values of the bridge scenarios.

These observations are also validated if we consider the length of alternative (by-pass) routes in function of α (Figure 5). As α increases, the length of by-pass routes will decrease, because more and more vehicles will be able to choose the ideal by-pass routes to avoid the congestion. For the bridge scenario the length of alternative routes decreases with about 30% if we disseminate TTI by employing $\alpha = 0.4$. In case of downtown congestions, we can observe that the length of alternative routes will not decrease significantly as we increase α, since the best by-pass routes are closer to the area of congestion. Thus, for downtown roads it is useless to disseminate the information further than the next couple of road segments (e.g. 200-300 meters), since the by-pass routes would not become shorter in any case.

Numerous analysis have been carried out that also show that the effect of α on the DoI and length of alternative routes is significant between 0.2 and 0.4 for most of the roads.

4.3 Query-based information gathering

In this section we present the numerical results of the generic query-based traffic information gathering protocol. The results were generated by creating a large amount of random source-destination route pairs on the road graph of Budapest. Optimal query locations were generated by solving the ILP formulation described in the previous section. A characteristic set of results, presenting interesting cases, were selected for presentation.

On Figure 6 the x-axis represents the distance of the source (n) of a vehicle from the traffic jam (while the destination (m) was fixed), while the y-axis represents the alteration in length of the respective metrics. The road length increase presents the difference between the original and the different by-pass routes ($\sum_{(i,j)\in\mathcal{E}} l'_{ij}(y^{nm}_{ij} - x^{nm}_{ij})$, for source n and destination m), while the query distance metric presents the distance from where the query is injected towards the point of interest ($\sum_{(i,j)\in\mathcal{E}}(l''_{ij}r_{ij})$, $l'_{ij} = l''_{ij} =$ length of road (i,j)).

In case when the jam was on a bridge (diagrams noted with (B)), the length of the original route was 3300 meters. When the distance was less than 200 meters from the traffic incident, the increase in the by-pass route length was nearly 1600 meters, an increase of around 50% of

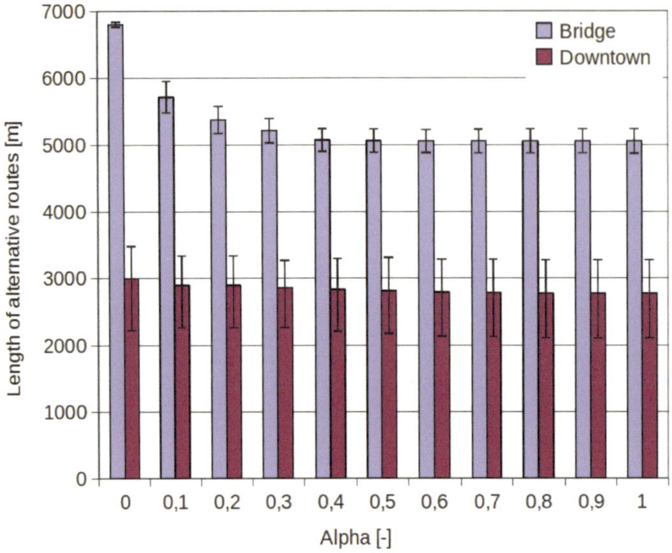

Fig. 5. Length of alternative routes in function of parameter α

the original route. As the source was generated further away from the traffic jam, the by-pass route length increase could be reduced significantly. The breakpoints in the graph are the points in the road network, where a new by-pass route could be taken. As we can see, the optimal query distance in this case is around 1000 meters. That is the point where the vehicles can take the shortest by-pass route according to the information contained in the returned TTI message.

On the diagrams noted with (M2), a major road in the city is displayed, where the length of the original route was 2500 meters. It can be seen that in case of short query distances the length of the by-pass routes could be as much as the double of the original route length. As the query distance is increased to 1000 meters, the by-pass route length decreases to around 600 meters, and with a query distance of 1200 meters, the route length increase is only 100 meters. This shows that finding the optimal query distance is really important, because the length of the by-pass route can be reduced significantly.

The diagrams noted with (M1) present a case when the traffic jam is on a main downtown road with plenty of nearby roads, which can be taken as by-pass routes. Thus, the query distance can be set to a small value, since the increase in by-pass route will become negligible.

4.4 Comparison of SPACE and SPACE_ILP

Until now we investigated the effect of traffic congestion on TTI dissemination separately studying the heuristic (SPACE) and the optimal (SPACE_ILP) Domain of Interest calculation algorithms. Considering the results from Section 4.1 and Section 4.2 we can affirm that the outcome of the algorithms present certain similarities. For example, from both approaches it turns out that the traffic jams can be classified in two major categories. One category is represented by traffic jams of main, crucial roads (e.g. bridge), with a large Domain of Interest and an increased length of the by-pass routes. The dissemination of TTI for such traffic incidents is extremely important, since the zone of no return of these traffic jams is also large. The second category of traffic jams is represented by downtown roads with small DoI

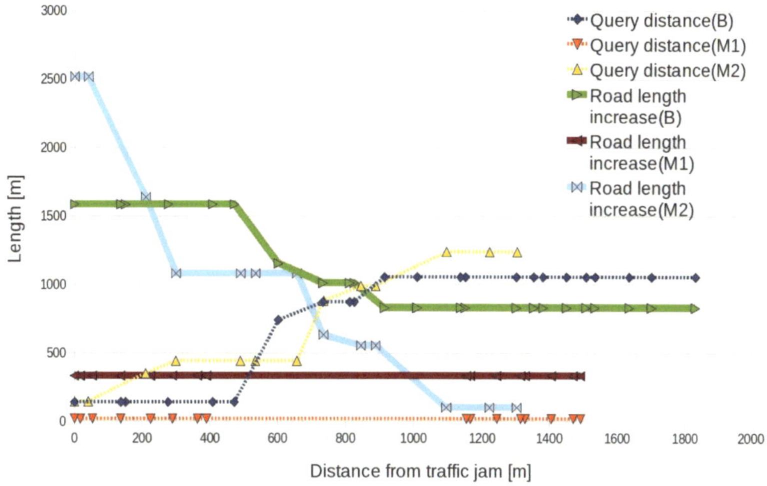

Fig. 6. Query results of different traffic jams

values. Such congestions can be avoided quite easily, since there is a large number of shorter by-pass routes around them.

Besides discovering resemblances of the algorithms' outcome it is also important to consider their potential benefits and differences. The main advantage of SPACE_ILP is the optimal size of DoI result sets for different parameters of the actual context (e.g. network load, reflected by parameter α). However, the SPACE_ILP algorithm can be employed only for the calculation of a limited number of DoI calculations, since its running time is relatively high; thus, it cannot be used to calculate the DoI for all the segments of a larger road network. That is where the SPACE algorithm comes into picture, because it presents much faster running times. Unfortunately, there is no method defined on how to select the most important road segments from the outcome of the SPACE algorithm. The impact factor metric gives us a good measure regarding the importance of different road segments, but it does not indicate a certain threshold , which would limit the DoI area for the respective result set. Therefore, the aim of this subsection is to provide a method, which provides a relationship, associates the results of the two DoI generator algorithms.

In order to find such a relationship we opted to compare and analyze the result sets of the two algorithms in a spatial manner. We designed and built an extension for our RUBeNS vehicular simulation environment Laborczi et al. (2006), which is able to load, store and analyze the DoI result sets of the algorithms. The extension is built in the PostgreSQL database management system and takes advantage of its geographic support, PostGIS and pgrouting. In this framework by using common map references we have an unified view of uploaded data, and through embedded functions we are able to define different spatial operations and methods for analyzing the uploaded result sets. For example, we can calculate the area or perimeter of DoIs, we can compare the DoI result sets regarding their spatial coverage and relationships, and we can design methods to intelligently reduce the area of SPACE DoIs.

For the reduction of SPACE DoI areas we implemented the following simple method. From the uploaded DoI sets of SPACE_ILP we selected certain cases, which represent characteristic DoIs (for large and small traffic jams). From these DoIs we get the optimized set of road segments, which in turn will be searched in the DoI set of the SPACE algorithm for the same traffic jam. Based on this we get a reduced set of road segments from the respective SPACE

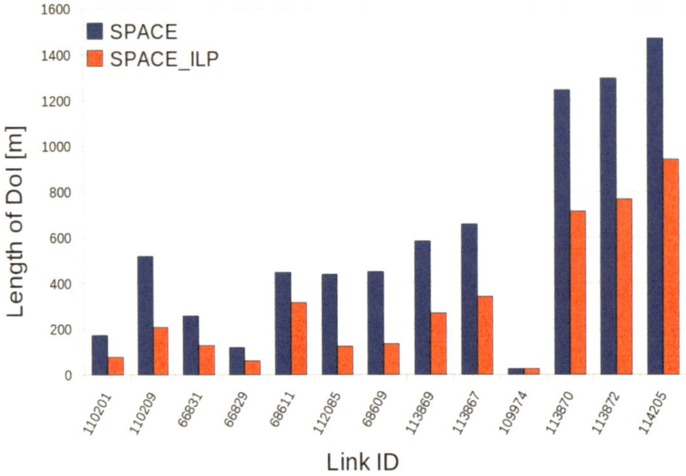

Fig. 7. Comparison of Domains of Interest of the SPACE and SPACE_ILP algorithms

DoI, for which the dissemination area is optimal. Then considering the impact factors and domain of interest lengths for this result set we calculate the derivative of the graph's slope. This will provide the threshold for limiting the area of other DoI result sets. For reducing the size of other SPACE DoI result sets we always select road segments until their graph's slope represents higher values than the calculated threshold. This would mean that these road segments are still important for TTI dissemination about the respective traffic incident. By using this calculation the size of all SPACE DoIs can be reduced; thus, a corresponding association between the two algorithms was identified.

The results of the method are presented on Figure 7, where we represent how the road segments with different relevance regarding the DoI are situated along a selected vehicle's route. A cross-bridge route was selected, where the length of DoI (considered only on this specific path) is represented in function of the link IDs along the path, the traffic jams were generated consecutively for the respective road segments. On the figure traffic jams with large influence (large DoI) can be observed between links with IDs 113870 and 114205 (critical part), where the DoI (along the route) of SPACE can reach even 1400 meters. This means that in case of a traffic jam situated along this critical part the TTI should be disseminated to a large part of the route, in order to avoid the traffic jam of the respective links with small by-pass routes. This critical part of the route contains also the bridge. For the rest of the route the congested links can be avoided easily, this is represented by small values for DoI.

It is important to observe that the results of DoI sets for SPACE and SPACE_ILP represent similar behavior, emphasizing the difference between the different kinds of traffic jams (with small, respectively large DoI values). The difference between the values of DoI length of the algorithm's output come from the fact that in the case of the SPACE algorithm we added a few more link segments (additional length increase), since we wanted to provide a larger zone for query and decision making during the trip along the respective route segment. The outcome of SPACE_ILP is a little bit to optimistic, since it does not take into account the delays of information propagation.

By applying the method for the whole set of uploaded SPACE DoIs (Figure 8) we can observe that size of DoIs (original, without reduction) shifts from the larger values towards much smaller ones (reduced DoIs). This can be attributed to the fact that only a few road segments

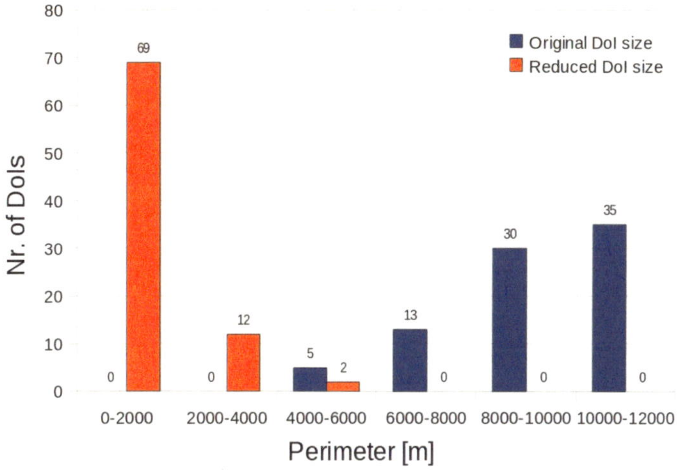

Fig. 8. Reshaping the Domains of Interest of the SPACE algorithm

are really important (large DoI areas), while the majority of traffic jams affect roads with small influence (from downtown areas). Similar results could be achieved by employing the SPACE_ILP algorithm for DoI calculations. Thus, the SPACE DoI reduction algorithm proves to be effective to determine the correct size of dissemination areas.

5. Conclusions

The current trends and problems of intelligent transportation systems have been presented in the beginning of this chapter. We presented a short overview of research on providing location-aware services in vehicular networks by disseminating information messages, for example regarding traffic congestion or fuel prices at a gas station, and maintaining these messages at key areas in the traffic network. This way the vehicles that traverse through these areas get informed about the content of the message and are able to alter their route accordingly. In the second part of the paper we presented SPACE, a heuristic algorithm to determine the domains of interest to a given event, for example a traffic jam, where the dissemination of the message is important. Following that we have given a linear programming formulation to determine the optimal domains of interest for a given traffic scenario. In the next section an extension of the linear programming formulation was described, to take the velocity of the vehicles and the message propagation delay into account. This allowed the extension of the DoI, so the vehicles could be notified before they reach the junction where the optimal alternative route starts. In the final section we evaluated the heuristic and the linear programming algorithms with different settings of the adjustable parameters using the RUBeNS simulation environment. It was shown, that the linear programming solution can be used to calibrate the heuristic algorithm, although SPACE does not presents the optimal solution like the linear ILP algorithm, but it runs significantly faster, thus it is more usable.

6. References

Burgess, J., et al., *MaxProp: Routing for Vehicle-Based Disruption-Tolerant Networks*, IEEE Infocom 2006, 2006 April, Barcelona, Spain.

Dikaiakos, M. D., et al. (2007), *Location-Aware Services over Vehicular Ad-Hoc Networks using Car-to-Car Communication*, IEEE Journal on Selected Areas in Communications, vol. 25, no. 8, October 2007.

Dikaiakos, M. D., et al. (2010), *On the Evaluation of Caching in Vehicular Information Systems*, 9th Hellenic Data Management Symposium; 2010 July 2.

Laborczi, P., et al., In: *Vehicle-to-Vehicle Traffic Information System with Cooperative Route Guidance*, 13th World Congress on Intelligent Transport Systems; 2006 October 8-12; London, United Kingdom.

Laborczi, P., Mezny B., Torok, A., Ruzsa, Z., *Query-based Information Gathering in Intelligent Transportation Systems*, International Symposium on Combinatorial Optimization, March 24-26, 2010, Hammamet, Tunisia

Leontiadis, I., Mascolo, C., *Opportunistic SpatioTemporal Dissemination System for Vehicular Networks*, The First International Workshop on Mobile Opportunistic Networking ACM/SIGMOBILE MobiOpp 2007; 2007 June 11; San Juan, Puerto Rico, USA.

Lochert C., et al., *Data Aggregation and Roadside Unit Placement for a VANET Traffic Information System*, ACM VANET'08, September 15, 2008, San Francisco, California, USA.

Sormani, D., et al., *Towards Lightweight Information Dissemination in Inter-Vehicular Networks*, The Third ACM International Workshop on Vehicular Ad Hoc Networks; 2006 September 29; Los Angeles, California, USA.

Speieys, L., Jensen, C. S., *Enabling Location-based Services Multi-Graph Representation of Transportation Networks*, *Journal GeoInformatica*, Publisher Springer Netherlands, ISSN 1384-6175

Tian, J., Han, L., Rothermel, K., *Spatially Aware Packet Routing for Mobile Ad Hoc Inter-Vehicle Radio Networks*, IEEE Intelligent Transportation Systems, 2004, 2004 October 12-15.

Torok, A., Laborczi, P., Gerhath, G., *Spatially Constrained Dissemination of Traffic Information in Vehicular Ad Hoc Networks*, IEEE Vehicular Technology Conference VTC2008-Fall, 21-24 Sept., 2008, Calgary, Canada.

Torok, A., Laborczi, P., Mezny, B., *Context-aware Traffic Information Flooding in Vehicular Ad Hoc Networks*, Tamkang Journal of Science and Engineering, Vol. 13, No. 1, 2010

Zhao, J., Cao, G., *VADD: Vehicle-Assisted Data Delivery*, Vehicular Ad Hoc Networks, IEEE Infocom 2006, 2006 April, Barcelona, Spain.

7

CARAVAN: Context-AwaRe Architecture for VANET

Sławomir Kukliński[1] and Grzegorz Wolny[2]
[1]*Orange Labs Poland, Warsaw University of Technology*
[2]*Orange Labs Poland, University of Warsaw*
Poland

1. Introduction

One of the new networking concepts which was born during the last decade is the communication between cars using short range wireless solutions, known as VANET (Vehicular Ad hoc NETwork). Despite the significant research efforts the VANET are still in the infancy and so far there are no implementations. In fact the VANET concept imposes a lot of serious problems, which so far have been resolved only partially. The most important problem is the intermittent connectivity of VANET nodes caused by high mobility of cars. The high dynamics of cars combined with usage of short range communications make the connectivity among the cars very unstable, so even the best effort service cannot be guaranteed. Of course, the communication quality differs in cities during traffic jams (where we may obtain dense and stable networks) from one in highways (where the VANET network is sparse and the topology is highly dynamic). Due to the variety of communication scenarios it is very hard to predict the generic transport characteristics of VANET networks. It is worth to mention that there have been created many routing protocols for VANET. Unfortunately, most of them are focused on specific scenarios only. Another VANET problem deals with the definition of services suitable for these networks. VANET can be used for safety applications (this is the primary goal), for driving support information services (information about parking places, points-of-interest, etc.) and, in some concepts, it can offer classic Internet services including high quality media streaming. Of course, the above mentioned VANET problem with highly unreliable communication will limit the service portfolio in the same way as it is observed in other networks. It has to be noted however, that in VANET the diversity of communication quality is enormous; in some situations high quality links of high throughput can be created (in case of parked cars) but at the other extreme there is highly intermittent communication which enables only the exchange of short messages between the nodes (the highway case). The next VANET specific problem is related to the identification of service recipients and the way in which the information is disseminated among them. It has to be noted that the client-server model, very popular in the Internet, has only limited or even no usage in VANET, because such networks are generally server-less. Using node address for service delivery also has limited usage, because it characterizes neither a node nor its service requirements. The classic unicast communication has very limited usage as well. The range controlled broadcast, multicast (if groups are identified) or anycast in some cases are better suited for, and in many (though not all) cases the geocasting is a solution of choice.

A very important problem concerns the incremental deployment of VANET and is linked to VANET business model. There is no doubt that the deployment of such networks will progress gradually, and it is expected that VANET nodes will be built in cars by car vendors. That way only new cars will be equipped with VANET nodes and it will take years before almost all cars will have them on board. Of course, such slow deployment creates several important issues; in the beginning the VANET nodes density will be extremely low resulting in sporadic communication only. On the other hand, in coming years, due to the technology progress, new and more advanced VANET concepts can be developed. The last problem calls for an open approach to VANET architecture that is able to accommodate new concepts on the component basis.

In this chapter we present a unified architectural VANET framework, called CARAVAN (Context-AwaRe Architecture for VANET), which is aimed to cope with all problems mentioned above. This framework is able to simultaneously handle different communication schemes (including disruption tolerant networking), to deal with rich service portfolio adapted to communication limitations, and to accommodate most of algorithmic concepts designed for VANET. The unified framework behavior and adaptation is driven by contexts (context-aware approach). Contexts are related to service/application requirements, communication ability, mobility vectors and the mutual space-time relations of cars/nodes. In the proposed approach the communication scheme is based on multiple protocols. The selection of the protocol is made accordingly to the quality and stability of links that typically depend on the distance between the nodes and on the mobility vectors. The heart of the proposed concept is the context engine which processes context coming from different layers of the architecture.

In Section 2 of this chapter a short introduction to VANET which will help in the understanding of the proposed solution is presented. Section 3 contains the related work and short review of main algorithmic concepts which are applicable to VANET. Section 4 is the main part of this chapter and it contains the CARAVAN concept description together with the description of its implementation and sample use case. Section 5 concludes the chapter.

2. Vehicular ad hoc networks

Vehicular ad hoc networks (VANETs) have recently become an important research area with contributions split between government and industrial consortia, as well as the academic community. Current efforts focus on the design of a new set of vehicular applications and services improving the comfort and security of the driving experience. The specification of this type of mobile networks enables many new possibilities, but from the other side it creates many non-trivial challenges. This section is a very short introduction to VANET. Much more detailed descriptions of VANETs can be found in (Olariu & Weigle, 2009) and (Moustafa & Zhang, 2009).

2.1 Characteristics

Vehicular ad hoc networks can be considered as a special case of mobile ad hoc networks (MANETs). However, there are several important factors, which make this type of networks specific and which allow to treat them as a separate category. Here are the fundamental VANET features:

- Very high dynamics of nodes resulting in fast topology changes. As the communication devices are installed inside vehicles, the network nodes are much more mobile and they move with much higher speeds. Vehicles are restricted to move using roads and to abide by the traffic rules, so some mobility patterns can be observed and some statistical mobility models for VANET have been designed (Härri et al., 2006).
- Information about the current position, movement direction, current velocity, city map and planned movement trajectory of VANET nodes is available, as more and more vehicles are equipped with GPS devices and navigation systems.
- VANETs impose lack of energy constraints, higher computational power and practically unlimited memory capacity, in comparison to some other ad hoc networks (especially to sensor networks).
- VANET networks are usually of very large size (case of traffic jams) but also may exist in a form of many small, neighboring networks with a high probability of splitting and joining.
- There is a big diversity of VANET services and applications, and one-to-one communication is less important than some intelligent broadcast (for example geocast) required by most safety related applications.

Besides the scenarios when vehicles communicate only with each other (sometimes called Vehicle to Vehicle (V2V) or Car to Car (C2C) communication) there exist two other scenarios which also distinguish VANETs from MANETs. Some research studies consider a case of communication between vehicles and fixed roadside equipment (Vehicle to Infrastructure (V2I) or Car to Infrastructure (C2I)). Such communication can be used for Internet access or in some new vehicular applications, e.g., vehicles exchanging data with a service station while being in repair. Last scenario is a hybrid case when vehicles can be treated as relays to increase the range of ground services.

2.2 Services and applications

VANET services and applications differ significantly from the classic ones known from MANETs. Possibly the most important group of services, which makes research studies on VANETs increasingly popular, are those related to driving safety. Safety applications include among others: current traffic reports dissemination, road obstacles warning (accidents, works and other unusual situations), driving maneuvers assistance (vehicles overtaking, lane changing) or traffic-aware trajectory planning.

Another group of applications which can gain popularity among road users are infotainment services. The most basic ones are related to advertising - distributing information about free hotels rooms, restaurants offers, discounts in stores, etc.. Gasoline prices, information about free parking spaces of the nearest service station can also be disseminated. Probably such services will have to be extended by some publish/subscribe mechanisms, so that driver is not spammed with a number of unwanted messages. Some infotainment services can be also useful for pedestrians - e.g., buses can estimate time of arrival at the bus stop using knowledge about traffic conditions and then distribute this information to the waiting passengers.

Some one-to-one services can also be considered, but probably they can be applied only in the limited range because of highly intermittent communication. Much more applications will be based on the intelligent context-aware data dissemination.

2.3 VANET related projects

Vehicular ad hoc networks are a hot topic nowadays, so many researchers are involved in national and international consortia leading great number of projects related to different challenges in this area. The best known consortia are Car 2 Car Communication Consortium (C2C-CC) (Baldessari et al., 2007) in Europe, Vehicle Safety Consortium (VSC), Collision Avoidance Metrics Partnership (CAMP), Vehicle Infrastructure Integration Consortium (VIIC) in United States and Advanced Safety Vehicle (ASV) in Japan. The Intelligent Car Initiative (Reding, 2006) is one of the biggest initiatives of the European Union which aim is to investigate the potential of information and communication technologies in improving life quality - also by development of intelligent vehicular systems in order to make cars smarter and safer. Here is a short list of selected projects related to VANETs:

- FleetNet (Franz et al., 2001) - a pioneer research project investigating the direct communication between cars,
- Network on Wheels (NoW) (Festag et al., 2008) - the successor of FleetNet which aimed at developing an open communication platform designed for safety, traffic efficiency and infotainment purposes,
- PReVENT (Schulze et al., 2005) - a R&D project on the use of different technologies to help the driver to avoid an accident with two main issues being investigated - wireless local danger warning and intersection safety,
- Co-operative Vehicle-Infrastructure Systems (CVIS) (Mietzner, 2007) - another R&D which is focused on providing methods for continuous V2I communication and cooperative services in order to increase road safety and traffic efficiency,
- SAFESPOT (Giulio, 2007) - a project designing Safety Margin Assistant which should help to detect in advance the dangerous situations and to make a driver more aware of the environment surrounding,
- SeVeCom (Leinmüller et al., 2006) - a project focusing on the security and privacy issues in vehicular communication.

Beside pure research oriented projects trying to resolve particular problems there is also whole big branch of consortia involved in regulation and standardization activities. Car-to-Car Communication Consortium composed of about 50 partners is working on the industry standard for vehicular communication using wireless LAN technology. An institution playing a major role is the European Telecommunications Standard Institute (ETSI) with a new technical committee for Intelligent Transport Systems (TC ITS). Different working groups are focusing on application requirements, architectural and cross-layer issues, transport and network, media and related issues, and security. CEN is the European Committee for standardization - it is a private and non-profit organization which works on such issues as electronic fee collection, dedicated short-range communication (DSRC) and identification of vehicles. Another standardization works are done by ISO and Internet Engineering Task Force (IETF).

A comprehensive review of the VANET related European consortia, projects and standardization activities can be found in (Le et al., 2009).

3. VANET supporting concepts – state of the art

The characteristics of the VANET listed in Section 2.1 call for a specific approach to VANET. It should consider proper routing in very dynamic network, content dissemination, service specific issues and finally a security and privacy. In fact some concepts developed for mobile ad hoc networks (MANET) that have been studied for a long time can be reused and adapted to VANET, however some VANET specific properties may require completely new approach. This chapter provides a short overview of the most popular concepts which were developed for VANET. This overview is very important in the context of the CARAVAN that has been designed in order to obtain the synergy by integration and dynamic selection of the already developed VANET algorithms.

3.1 Routing protocols

The detailed survey of routing protocols in the context of vehicular networks can be found in (Ros et al., 2009), which has been a main guide for creating this review.

The design of the routing protocols for VANET is especially challenging task due to the high mobility of nodes, large network size and the intermittent communication. The first attempts base on the usage of routing protocols developed for mobile ad hoc networks (MANET). These protocols can be divided into four groups - proactive, reactive, hybrid and geographic routing. In proactive routing, the paths between all pair of nodes are determined in advance, i.e., before they are needed. The most popular proactive protocol is Optimized Link State Routing (OLSR) (Clausen & Jacquet, 2003) in which the nodes discover network topology using beacon messages. Knowing the topology each node computes the shortest paths to all possible destinations and stores the next hops in its routing table. OLSR is a good approach for dense, small and relatively static networks, thus it can be applied to stable groups of VANET nodes. The reactive routing protocols try to find the path to the destination only when it is needed. Two most known protocols belonging to this category are Dynamic Source Routing (DSR) (Johnson et al., 2001) and Ad hoc On Demand Distance Vector (AODV) (Perkins & Royer, 1999). The first one finds a path to the destination by broadcasting route request messages into the network. Each node forwarding the request updates it by adding itself to the path. When the message achieves the destination it contains the complete path. The DSR protocol includes some optimization and the path maintenance procedures. Each packet sent from the source to the destination stores the path in its header - this is a big drawback in large networks, where routes are usually long. The AODV protocol uses quite similar approach but the paths are stored in the routing tables of the nodes instead in packets themselves. Each node builds such table whenever it is possible by storing next hop nodes on the paths to the destinations. The mechanism of the sequence numbers in route request packets guarantees loops freedom. Reactive protocols can handle well the dynamic topologies. Unfortunately, they are not well suited for large static networks, thus they can be used in limited ranges in city scenarios with increased nodes density, but not necessarily in case of traffic jams.

The main representative of hybrid routing protocols category is Zone Routing Protocol (ZRP) (Haas & Pearlman, 2001). All nodes have assigned their own zone including all neighbors that are at most k hops away, where k is a protocol parameter, which can vary depending on external conditions. Inside this zone the proactive Intra-Zone Routing Protocol (IARP) is used while to communicate with peripheral nodes the reactive Inter-Zone Routing Protocol (IERP) applies. This approach splits the network into interconnected zones, which provides increased scalability. Unfortunately, the criterion for the network split (i.e., the hop distance) has limited value in VANET.

The protocols belonging to the categories mentioned above can be used in vehicular networks, but they do not match well all possible VANET communication scenarios. The main drawbacks of these protocols are: scalability, incorrect assumption about full network connectivity and extensive use of flooding, that leads to high consumption of network resources and increases the contention to medium access as well as communication latency. Another important issue is that the paths established in VANET are usually valid only for short period of time, as a result of high mobility of nodes. These protocols usually do not take into account the VANET specific features, such as the access to information about the geographic position, city maps, node trajectories, mobility constraints and possibilities to predict the movement of nodes in the near future that is based on mobility patterns. However, there are also so called geographic routing protocols that use the information listed above.

The first geographic routing protocols were simple greedy algorithms. In Greedy Scheme (GS) (Finn, 1987) approach the forwarding node as a successor chooses a node assuring the greatest progress. In Compass Routing (CR) (Kranakis et al., 1999) protocol the idea is very similar but the choice is based on the smallest angle between the line to the destination and the line to the neighboring node. The Most Forward within *R* (MFR) (Takagi & Kleinrock, 1984) approach deals with loops problem which can occur in the previous concepts. One of the most important solutions in this category is Greedy-Face-Greedy (GFG) (Frey & Stojmenovic, 2006) protocol, which uses some simple geometrical properties of planar graphs. The nodes transform the network connections graph using some localized planarization algorithm and then they try to follow adjacent faces intersecting the line to the destination. These protocols form a good foundation for further work on geographic routing protocols optimized for vehicular networks.

Another group of geographic routing protocols can be characterized as source routing. The Geographic Source Routing (GSR) (Lochert et al., 2003) protocol tries to use a street map to find the shortest path with Dijkstra algorithm. The route is represented by a list of streets intersections. Then greedy forwarding is used to deliver packets between junctions on the list. The Spatial Aware Routing (SAR) (Reichardt et al., 2002) is a modification of GSR which deals with a local maximum problem (it appears when there is no node to which we can forward a message and make a progress). SAR introduces buffering of packets in nodes which are not able to forward immediately. Such nodes wait a predefined time for the suitable successor before dropping a packet. The next protocol named Anchor based Street and Traffic Aware Routing (A-STAR) (Seet et al., 2004) makes use of information about road traffic and tries to find a path consisting of road segments with the greatest possible traffic density. The Connectivity Aware Routing (CAR) (Naumov & Gross, 2007) protocol is based on similar idea of building a path using crossroads instead of nodes, but it achieves a goal without a map access. During the route finding phase the nodes which are close to the crossroads add their locations to the created path (so called anchor points). Moreover, such nodes create dynamic guards in the neighborhood of anchor points which are later used during packet forwarding.

The next set of geographic routing protocols is represented by Greedy Perimeter Coordination Routing (GPCR) (Lochert et al., 2005). As in CAR protocol there is no assumption on the street map data. The nodes which are near the junctions become coordinators and using beacon messages they build virtual topology graph with streets as edges and junctions as vertices. When such coordinator receives a message to forward, it makes a decision about the correct street to push it there.

Table 1 taken from (Ducourthial & Khaled, 2009) clearly shows that there is no single routing protocol that is suitable for all vehicular scenarios. Not all of them are mentioned in the presented short summary of the routing solutions. Reader will find a nice survey of the protocols in the book chapter mentioned above.

	Communication Kind				
	One-to-One		One-to-Many		One-to-All
Traffic Kind	Topology	Position	Geocast	Mobility	
Adapted to sparse networks		Epidemic, MDDV, VADD		Epidemic, MDDV, VADD	Epidemic, MDDV, VADD
General	AODV, DSR, OLSR	DREAM, GSR, MGF, MORA, MURU	DRG, GAMER, IVG, LBM, MGF	RBM, TRADE	DREAM
Adapted to dense networks	CBRP, HSR,	CAR, GPCR, GPSR	GeoGRID	LBF, OABS, ODAM, SB, SOTIS, UMB	

Table 1. Application-based taxonomy for routing protocols according to traffic density in VANET (Ducourthial & Khaled, 2009)

3.2 Location services

There is a silent assumption of many geographic routing protocols that nodes know the destination position. Depending on the applications, the requirements for location can vary considerably. Node position data and sometimes street topology obtained from GPS navigation system are usually sufficient. Other techniques for obtaining position of nodes include dead reckoning, which works well for short periods of GPS unavailability, cellular localization and relative distributed ad hoc localization. For many services information such as maximum range or direction of message propagation is enough. For the others - especially based on one-to-one communication - quite detailed knowledge is required. Some protocols (like GSR) find the destination node by flooding route request messages and only if this phase ends with success they can use their geographic properties. Other protocols use various independent location service mechanisms. For example in the CarTalk2000 project (Reichardt et al., 2002) nodes position is distributed only to nodes within a given number of hops. Researchers involved in FleetNet project proposed Grid Location Service (Li et al., 2000) using some nodes as "location servers", and Reactive Location Service (Käsemann et al., 2002), finding position of destination on demand. The V-Grid (Gerla et al., 2006) approach is based on two complementary location services - one in infrastructure network and the second in vehicular network. Node looking for destination position has to communicate with the nearest fixed infrastructure point providing location information.

3.3 Clustering

Nodes clustering algorithms are useful in order to identify "similar" or close in terms of the predefined clustering metric nodes and form their groups (clusters). Clustering in the routing enables partitioning of the whole network into smaller subnetworks, thus in some cases resolves the scalability problem of routing. In dynamic networks clustering helps to identify the regions of relatively stable topology. Having a partition of the network nodes different protocols can be used for the communication inside the clusters and outside of them (hierarchical routing). Examples of routing protocols that use clusters are Clustered OLSR (COLSR) (Ros & Ruiz, 2007) and Directional Propagation Protocol (DPP) (Little & Agarwal, 2005). There also exist pure clustering algorithms such as Modified Distributed and Mobility Adaptive Clustering (Modified DMAC) (Wolny, 2008) or Density Based Clustering (DBC) (Kukliński & Wolny, 2009), both of which use mobility patterns and nodes behavior prediction to form stable clusters. Another possible application of clustering technique is the automatic identification of user groups that can be interested in the same kind of services.

In conclusion, the clustering technique is a powerful mechanism and can have various applications in VANET.

3.4 Content dissemination

One of the biggest challenges in vehicular networks, besides the high mobility of the nodes causing constant topology changes, is the intermittent communication. In such environment it is extremely difficult to achieve a reliable content dissemination between the nodes. In VANET it is very common that the path between the source and the destination is not only unstable (has very short lifetime), but often it simply does not exist. Due to the high dynamics of nodes and the use of short range communication of VANET radio interfaces a permanent communication between the nodes cannot be guaranteed. A possible solution to this problem is the usage of some roadside fixed infrastructure or some additional communication channel, e.g., cellular networks. Such solutions have some obvious drawbacks like limited range (the first approach) and high costs (the second one). Another possible way of dealing with this problem is to use the Delay-Tolerant Networks (DTN) approach. DTNs are the main research topic of the Delay-Tolerant Networking Research Group (Fall & Farrell, 2002), which is focused on the application of DTN to satellite communications. DTN also has easily observable disadvantage, because it can be applied only for services which are not delay aware. Fortunately, in many potential VANET applications longer delays are perfectly acceptable - just to mention infotainment and traffic control services or even some safety ones, e.g., when the nodes have to spread information about some road obstacles.

The main idea of DTN is to aggregate messages into so called bundles. Bundles can be stored in nodes buffers when the immediate forwarding is not possible and forwarded later, when the communication is established again. This communication paradigm sometimes is described as store-carry-forward, which means that nodes have a possibility to place bundles in their local buffers and then carry them until a proper node, to which the bundle should be forwarded, is found. In case when no destination node is reached in a specific period of time, the bundle is discarded. It is clear that DTN forwarding decisions are more or less effective depending on the quality of information about network topology and mobility vector of nodes. DTN routing protocols can be split into deterministic and stochastic ones. Their common goal is to maximize delivery probability while minimizing the delay. Some deterministic DTN protocols assume that almost full knowledge about the network and its future topology evolution is given, sometimes even with the possibility to affect nodes behavior in order to optimize communication. Such assumption does not make sense in vehicular networks. A category of protocols which best suits the vehicular environment can be described as passive stochastic routing protocols. Epidemic Routing

(ER) (Vahdat & Becker, 2000) is an exemplary protocol belonging to this group. The idea is trivial - nodes carrying the bundles forward them whenever it is possible. This protocol works well in networks with large buffers, long interaction between nodes and low network load. In such a case the Epidemic Routing assures minimal delays and high success rates. It is the most popular benchmark for performance evaluation of newly designed algorithms. Another popular DTN routing protocol is called Spray and Wait (Spyropoulos et al., 2005) - this time the number of forwarded bundle copies is limited by a certain threshold. Moreover, there is also a Wait phase, during which nodes try to deliver bundle straight to the destination. If they do not succeed the new Spray phase begins. The interesting observation is that with increasing network density the lower copies threshold is needed for the same protocol performance. A slightly different solution is used in Probabilistic Routing Protocol using History of Encounters and Transitivity (PROPHET) (Lindgren et al., 2004), where nodes estimate probability of delivering message to each possible destination.

Research studies on DTN routing protocols for VANET resulted in a development of several new concepts. The Vehicle Assisted Data Delivery (VADD) (Zhao & Cao, 2008) uses knowledge about stree topology, mean traffic density, average and maximum speed on each each street in order to select a path with the smallest expected delivery delay - for example in case when there is no direct connection between source and destination the node will try to select streets with higher nodes speed and density so that vehicles carrying packets can do it faster. Motion Vector Scheme (MoVe) (Lebrun et al., 2005) is a solution which uses information about neighbors velocities to choose node which makes the biggest progress towards destination. Geographic Opportunistic Routing for Vehicular Networks (GeOpps) (Leontiadis & Mascolo, 2007) is a trajectory-based protocol which uses the vehicular mobility patterns properties as well as assumption that each node knows its complete trajectory from the navigation system.

A more detailed survey of DTN solutions for VANET can be found in (Shao et al., 2009). There is no doubt, that DTN is a viable content delivery solution which can not be ignored in VANET.

3.5 Context aware mechanisms

In the descriptions of VANET related concepts presented in the previous sections there is one common property of the majority of presented solutions - i.e., the use of the knowledge about the network, nodes environment and the mobility, in order to make optimized decisions. The collected information concerning the node itself as well as the network can be treated as node context. Using contexts led us to so called Context-Aware Networks paradigm. In case of VANET the Node Context may consists of, among others, node position, velocity vector, neighborhood information, street topology together with information such as vehicles density or speed limits, planned movement trajectory, communication capabilities, services in use and many more. All this context data can be used by the routing protocols to increase their performance. In VANET we may also use the context-awareness for efficient data dissemination. On the context basis we may use message addressing instead of node addressing; the message destination is described by the context, e.g., location or maximum distance from the source, not by a destination or identifier (e.g., the IP address). The message context can also include information about time validity of the message, priority or service requirements, e.g., whether it is delay tolerant or not.

One of the interesting approaches to data dissemination is called Conditional Transmissions technique (Ducourthial et al., 2007). Authors assume that most of the applications require in fact broadcast communication and the receiver can be described by some set of conditions. As the consequence to deal with highly dynamic environment the conditional addressing is considered instead of network addressing, the path maintaining instead of traditional unicast and the conditional transmissions instead of broadcast. Each application can use its own conditions (e.g., the geographic information, the time-related information, the trajectory related information, the node identity related information, any combination of the above or even more) to define destination nodes. Conditional transmission service has been implemented (it is called HOP) and in some simple scenarios it has proved to behave better than many existing routing protocols.

3.6 Security and privacy

Security and privacy issues - although it is a topic of great importance, especially as far as safety services are concerned - have not gained yet a big attention in VANET research community. Insecure safety services can lead to a counter effect. Gaps in privacy data protection can result in poor driver interest. Without going into details, there is a possible attack classification, which shows the challenges in designing security system for vehicular networks. It should be resistant to both internal and external attacks, where internal attacks are those by authenticated users and they can be the most dangerous ones. Another distinction is on intentional and unintentional attacks, with the second type caused usually by communication errors. There exist active (modification of network traffic) and passive (captured data used for later unauthorized use) attacks. We can also split attacks into independent and coordinated ones. The main security challenges for vehicular networks include real-time constraints, data consistency liability, low tolerance for errors, key distribution and high mobility of the nodes. Some security requirements which should be at least taken into account are: availability, message integrity, confidentiality, source authentication, mutual authentication, authorization and access control, non-repudiation and privacy protection. The outlined issues are only a short introduction into the problems which should be resolved before wider deployment of VANET.

A good introduction into a security related issues in VANETs together with a comprehensive list of references can be found in (Tchepnda et al., 2009).

4. CARAVAN

4.1 Motivation

This section presents CARAVAN, the unified VANET framework, which is able to accommodate most of the existing VANET mechanisms and use them in an optimal way. The first version of the concept was defined in (Kukliński et al., 2010). This framework is component based thus enables independent modification of every component functionality without the necessity to redesign other components or the overall architecture. In the proposed framework the usage of a specific mechanism is tuned individually to the node's environment and service requirements. The component based architecture enables easy deployment of new applications which can use well-defined, lower level services offered to the application platform.

There are several observations which led us to the development of the framework:

- The communication quality and reliability in VANET may take extremely different values that depend on node's specific situation. For example on the highways the combination of high mobility of nodes and the short range of radio coverage (50 - 300 metres) leads to the intermittent communication of low quality, but during traffic jams we may obtain stable links being able to handle HDTV services.
- There are car mobility models which can be used to predict car positions. Moreover, most cars are equipped with GPS or navigation systems, thus the information about car position, direction, speed and even about the travel destination is generally available to every node and can be disseminated to node neighbors. This information can be used for the proper selection of the communication scheme and services offered to a node.
- There is an easy way to determine the proximity of nodes or their communication ability. It can be done using periodic transmission of HELLO messages. That way it is possible to discover neighbors and nodes density. These HELLO messages and responses may contain the position of the node and the mobility vector. Subsequent analysis of this data can lead to the identification of the longevity and the quality of the possible communication links between the nodes and their potential belonging to groups (clusters), which can be formed. Such clusters may provide relatively stable intra-cluster communication. Thus the group membership can be used for communication purposes (selecting the communication scheme or protocol), but it is not limited to. From the service point of the view nodes proximity (group membership) has an important value - it is possible that group members can be interested in the same or similar set of services. So, the identification of the relative positions of nodes has an important impact on nodes communications abilities and on their potential interest in services. In such model every node can be treated as an isolated node, group membership candidate (during the group membership inclusion procedure) or a group member. For every node category a different communication and service scenario can and should be applied.
- There have been many routing protocols designed for VANET in order to resolve the problem of communication reliability. It can be improved by the specific mechanism of the routing protocols, applying clustering, or the multi-path routing. All the mentioned mechanisms can improve reliability, but still a lack of communications in case of sparse networks can be observed, and the intermittent communication still may occur in case of high speed moving cars. Thus, we cannot guarantee the existence of permanent communication. The disruption tolerant networking paradigm (DTN) which uses store-carry-forward mechanism seems to be a good solution for handling temporary lack of communication. The information about nodes mobility vectors and even the destination (GPS and navigation based) makes VANET a good candidate for efficient implementation of DTN. Additionally, DTN enabled cars which do not belong to any stable group of cars (cluster) can play an important role of mules, which can carry on the information between the groups, thus such an isolated node plays a positive role in the overall communication model. In conclusion, the communication capability of every node can be different for groups of nodes (clusters) and for isolated nodes. The communication protocol should take into account the individual node state. At present, there is no single approach which is able to handle all the mentioned cases. That observation has led to the conclusion that for every node a local environment (number of other nodes, topology stability, and group membership) should determine the protocol which is used for data exchange or content delivery.

- In a very conservative approach the number of VANET services is limited to driving safety applications only. These simple services usually transmit short local messages, which should be geocasted or broadcasted. In more advanced service scenario we may think about the inclusion of video services, voice services and all other, Internet-like services. The real-time services, like video or voice based, require higher communications QoS guarantees which in VANET networks are hard to fulfill in general. However, there are some cases, for example the one-hop communication in which the communication ability of VANET goes beyond the most demanding services. In the opposite case, the DTN example, no real-time services are possible at all. This observation leads to the well-known conclusion that the service offer is limited by network transport capabilities, but this conclusion in the mentioned case has more dramatic meaning that in the classic, wired networks - the variance of the network QoS is much, much bigger. So, before the services will be offered to the end users, their communication ability has to be checked first. It is obvious that these communications properties will change over time, in some cases pretty rapidly.
- Due to the distributed nature of VANET networks there is a lack of special nodes (servers) which can help in service offering. Because of that, the nodes do not have a list of the "preferred" addresses and the (IP) addresses of their neighbors have for them very limited usefulness - what is the reason to communicate with them? What is represented by an IP address? There is, of course, a set of messages which can be delivered to all nodes in a specific area, but such geocasting should not be used for all services. In some situations, the car driver can indicate which service he or she is looking for, but the mechanism of service selection by the end-user should be kept at the very minimum level - the end-user should not be attacked by new services, but he should be well informed only about these services on which he is really interested in. It means that the end-user should have a possibility to indicate which services are interested for him at the specific moment. In that context the publish-subscribe mechanism can be applied. The variety of possible services in terms of their QoS requirements and the dissemination range and type make the classical IP service not adequate for VANET.

All of the observations presented above have led to the conclusion that it is unrealistic to cover all the possible network configurations, communication issues and service scenarios by a single approach. The communications and services should be adapted according to nodes density, mobility, relative mobility, group membership and user preferences. In order to cope with all these problems the best solution is a rich set of well-defined tools that are appropriately selected accordingly to the environment status and/or to service preferences. In the proposed unified framework there are multiple sets of tools and the choice of the appropriate one depends on the set of node contexts. The overall behavior of the nodes is individually controlled by the Cross-context Processing Engine, which receives and sends context from different components of the architecture.

4.2 CARAVAN design

As it was outlined in section 3, a rich selection of algorithms has been developed for VANET in order to cope with different problems. Unfortunately, so far there is no single approach which enables to use them as components of a bigger system. The main idea of the proposed framework is to collect a set of algorithms (tools) that are useful in different VANET situations and for a specific application, nodes density and mobility select appropriate set of

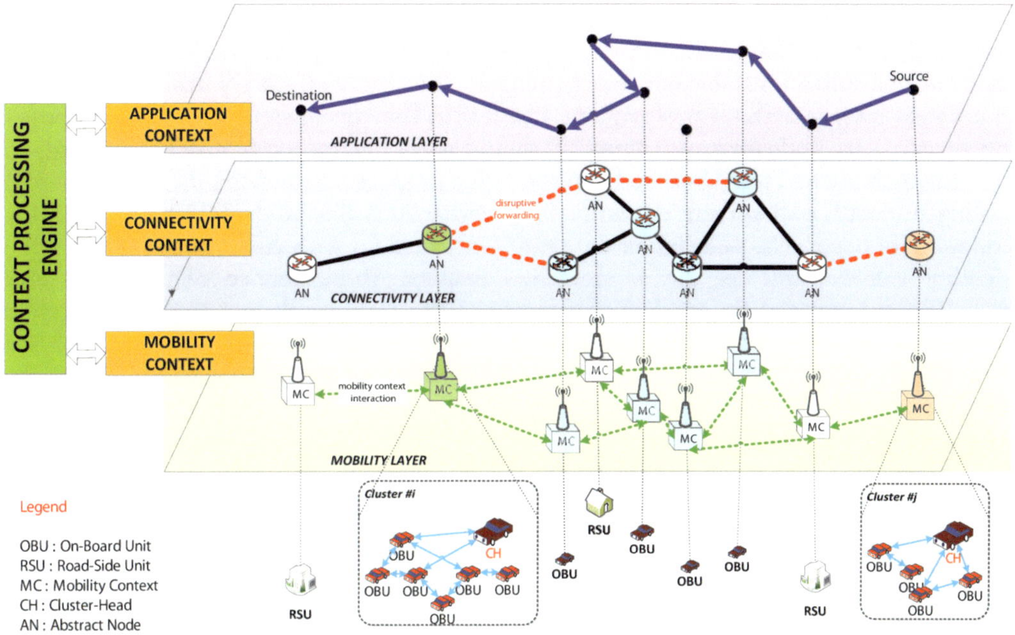

Fig. 1. Framework layers

them on a per node basis. Such individual and dynamic selection of tools provides obvious profits, but it imposes a new problem related to the criteria of algorithms selection and the way in which this process is implemented.

The discussion presented in the previous sections has shown that the communication ability of every node depends on the node mobility, the number of nodes in the neighborhood and their mobility vectors. The information about the node such as its current and averaged position, speed and direction and node track can be obtained from GPS. More information about the future node position and the final destination can be taken from the navigation system (if available and active). The GPS and navigation system provide the information about nodes position expressed in absolute coordinates. However, the information about the relative position of the nodes is very useful as well. Such information can be retrieved by processing the GPS data but can be also obtained directly when the neighboring nodes should respond to request sent over the radio channel (for example beacon messages). Using this mechanism we can determine in a very simple way the local density of nodes and using time averaging of responses we may find good candidates to create a cluster (Kukliński & Wolny, 2009). There is no doubt that for the estimation of the absolute and relative position of nodes, for creating clusters a plethora of algorithms exists, thus the proposed framework should be able to accommodate them. The information about the nodes mobility, their mutual communication relation and about nodes clusters is of great importance for routing as well as for services. This is the reason why in the framework we decided to introduce an independent component which offers to other elements of the framework the preprocessed information about nodes mobility, clusters etc. We named this component the Mobility Layer. The internal elements of the Mobility Layer are not fixed, however they should perform all the functions described above. In the proposed framework the context-aware approach is used. In-line with this philosophy the output of the Mobility Layer is the Mobility Context.

In Section 3.1 a short overview of different MANET and VANET routing protocols has been presented. Every of the described protocols has both advantages and deficiencies. Some of them are well suited for stable network topologies, other work efficiently in a sparse, but not in a dense network. These observations has led to the conclusion that in the proposed framework every node (or group of nodes) should select the routing protocol accordingly to the node mobility and neighborhood density. In the proposed framework the set of different routing protocols and content dissemination mechanisms (including DTN) composes the Connectivity Layer. The selection of the routing protocol for a specific node is based on the Mobility Context and on the service requirements. These service requirements and properties are exposed by another component of the framework that is the Application Layer. The Mobility and Application contexts have impact on the selection of the appropriate routing protocol; however they of course have no impact on the quality of the obtained connectivity. This connectivity is characterized by the Connectivity Context, which is exposed by the Connectivity Layer.

The Application Layer generates contexts that describe the applications and user requirements, but it also adapts the applications to the connectivity, quality and mobility information.

In the CARAVAN all the contexts of the Mobility Layer, the Connectivity Layer and the Application Layer are processed by the Context Processing Engine (CPE). The CPE is a heart of the proposed framework and it is responsible for the dynamic selection of the tools to the overall context that characterize node mobility, connectivity possibility and service requirements and restrictions. The details of implementation of CARAVAN are presented in the subsequent sections.

4.3 CARAVAN software architecture

The CARAVAN is composed of three functional layers, focused on mobility, connectivity and application. The internal behavior of layer components is controlled and described by a set of key parameters, represented as context information. This information is exchanged bi-directionally between the layers by applying cross-layer context adaptation. Transferring significant contexts in a unified format transversally between the layers facilitates the optimization of both important intra-layer operations, as well as the overall performance of an architecture based on this framework (e.g., selecting the best routing or forwarding scheme according to mobility information).

The entire framework is driven by context data exchange and decisions based on it, so node internal architecture can be defined around the idea of context exchange in a layered approach, by emphasizing the three key components - mobility, connectivity and application. Each component features mechanisms for processing context and feeds it to a cross-layer component which centralizes all of the context data (including that from the other components). The cross-layer component makes intelligent decisions and then feeds back key input context, influencing the behavior of the component. Such architecture can be applied to all types of entities, such as unclustered nodes, clusters and roadside infrastructure nodes. The architecture driven by context information exchange is based on:

- a layered functional structure centered on mobility, connectivity and application,
- a cross-layer transversal interaction, in order to optimize intra-layer and overall system performance,
- a relatively simple architecture, ideal for adding new functionality to improve intralayer operations.

This generic architecture for the Abstract Node (AN) being the basic entity inside the proposed framework is depicted in Figure 2. In order to enable incremental upgrades, an implementation calls for a modular design to be derived from the defined framework and applied to the AN.

4.4 Abstract node description

As mentioned in the previous section, we are applying a modular design for the AN. This design is based on a hierarchy of modules (see Figure 2), implementing specific functions related to mobility, connectivity and application.

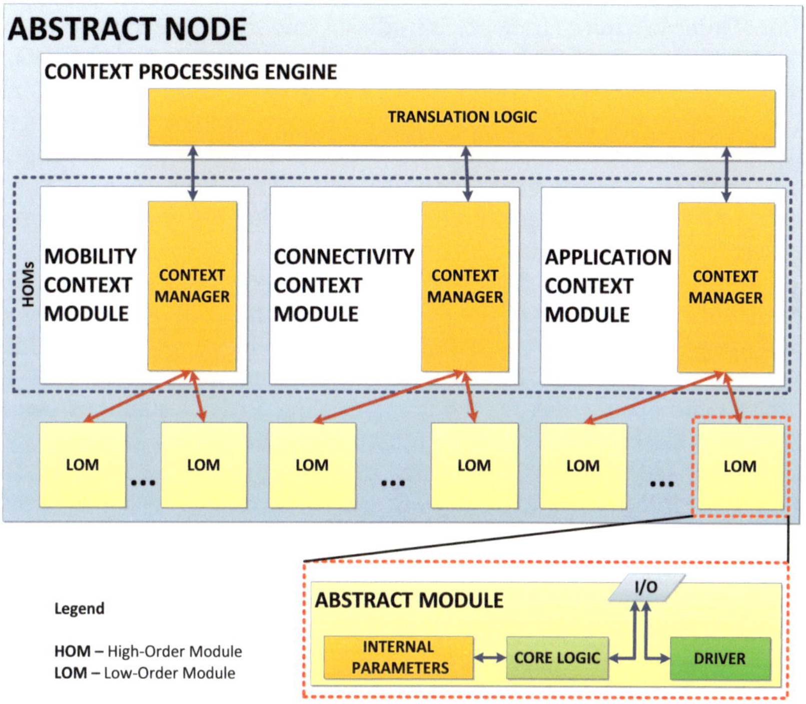

Fig. 2. Abstract Node and Abstract Module architecture

4.4.1 High-Order Modules

The proposed abstract node architecture has a hierarchical structure and consists of a Context Processing Engine (CPE) and three dedicated High-Order Modules (HOMs). HOMs, together with CPE, create a fixed core. Each of the HOMs, specifically the Mobility Context Module (MCM), Connectivity Context Module (CCM) and Application Context Module (ACM) correspond to a different layer of the framework and is functionally separated from the other modules. There is no direct transfer between them - the only way to communicate with each other is a bi-directional exchange of context information with the CPE using the same established interface in all three cases. The CPE performs a context adaptation and enables the cross-layer transfer of the relevant data in order to perform in the most suitable way. A dedicated Context Manager (CM) in each of the HOMs and a Translation Logic (TL) in CPE are directly responsible for data exchange between the top level modules. They all are also involved in the core logic of their parent modules. The typical communication looks as follows - the TL receives a context information from specific CMs, then it does some data processing, e.g., translation of the received data into some unified language, and afterwards it feeds the other modules with a newly obtained information according to their needs. Due to a single module gathering context information from all functional planes and its proper processing and distribution, it can be helpful in selecting some specific behaviors inside a particular HOM, e.g., choosing the routing/forwarding mechanism best suitable to a certain node mobility information.

4.4.2 Low-Order Modules

In addition to the above HOMs there exists the second tier of the modules hierarchy which are called Low-Order Modules. They are introduced into the framework to make it easily extensible by enabling a possibility of adding new mechanisms and algorithms, e.g., new routing schemes or new data dissemination mechanisms. Such approach allows the integration of the existing VANET concepts and leads to the diversity of choices in order to increase the overall performance of the system. LOMs are by design exchangeable user-defined modules which provide specific VANET algorithms. Each of the LOMs has to be attached to one of the HOMs depending on its destination for mobility, connectivity or application layer.

As it was already mentioned the fixed core of the architecture consisting of the CPE and three HOMs secures the integrity of the framework together with a functional separation between the defined layers. The role of the LOMs is to allow a flexible definition of new algorithms and their integration into the overall logic of the system. This makes the proposed architecture open - also for the many existing VANET solutions. An important fact is that all LOMs are built on a common internal definition of the generic module, called the Abstract Module (AM), which is presented in the bottom part of Figure 2. Due to this fact, all LOMs can be integrated into framework and handled in a very similar way.

The Abstract Module definition assumes the use of a simple interface to exchange data between LOM and the parent top level module. As the whole architecture is built around the idea of context-awareness, also in this case the exchanged data can be seen as some specific context encoded using some generic format. Depending on its role in the system the LOM can provide context information to the system or require such information. However, in most cases the LOM can do the both. The capabilities of each module together with its needs are registered in the system using a built-in Driver during a module initialization phase. If all the needs are fulfilled, which means the LOM can be fed with the required input context information; the module is ready to work. The received data are processed by the Core Logic and the proper output context information is provided as the result. The Core Logic implements the algorithm or mechanism for which the module is intended, e.g., a routing scheme or a scheduling mechanism. The processing part of the module can be constrained by a set of adjustable internal parameters.

The majority of developed LOMs implement functions related to one of the three HOMs corresponding to one of the three functional layers, although it is possible to define a LOM for some particular CPE functionality, such as scheduling of DTN bundles. Therefore the most typical connection will be between low and high order modules. LOMs are plugged in the proper Context Manager, so the role of the Context Manager is to register such LOM inside the TL of CPE and to manage all of the LOMs connected to it. This means the CM is actively involved in the HOM logic and the context processing is not focused on CPE, but rather it is distributed in the core of the framework with some of the decisions being shifted to the CMs.

4.5 High Order Modules description

The defined framework features three functional layers built on the importance of mobility, connectivity and application context information. As described in the previous section, each of these three layers has a corresponding High-Order Module in the module hierarchy defining the Abstract Node architecture.

4.5.1 Mobility Context Module

The Mobility Context Module is responsible for processing of nodes mobility data. Among its functionalities there is the network topology discovery. Whenever it is appropriate it can group the network nodes into a set of clusters, obtaining this way a topology composed of virtual entities called Abstract Nodes. When a clustering is not performed in the network, each Abstract Node will correspond to one real node. This HOM monitors a set of mobility parameters, such as geographical position and velocity vectors, as well as the neighboring nodes behavior and inter-nodes dependencies.

As the framework is thought to be mobility driven this module is of great importance. The collected and processed context mobility data can be used for many purposes. First of all some clustering algorithm can be fed with this data to optimize its operations. Clusters of nodes can be treated as virtual Abstract Nodes with their own mobility contexts defined for example in relation to distinguished real node being a cluster head.

The mobility context of the node itself and of the neighborhood can be used to make some movement prediction. Such information, distributed among the other modules using a Context Manager connected with a Translation Logic has a potentially great value for routing and forwarding schemes - especially those dealing with delay tolerant networking. An important advantage is that introducing MCM allows making all mobility context related data gathering and processing only once in a single dedicated place for many different protocols and mechanisms.

4.5.2 Connectivity Context Module

The Connectivity Context Module is presented in the Figure 3. The main goal of this module is providing connectivity between the virtual Abstract Nodes created in MCM. This means the module is responsible for routing and forwarding functionalities. The routing functionality uses a logic implemented in the Routing Manager (RM) which finds routes to particular network destinations using the best Routing Scheme selected from the available ones. The Context Manager managing all the connected LOMs participates in this selection process. The forwarding duties are performed by the Forwarding Manager (FM) which makes the decisions such as selecting the right forwarding scheme and choosing the next hop. The choices are depended on many factors, e.g., application context containing the information about QoS requirements. There is also some additional custody transfer mechanism implemented by the dedicated Custody Manager to support disruption-tolerant forwarding. Moreover, the CCM cares about its local routing table to be up-to-date. Another important group of the module duties is computing the performance metrics related to routing or even DTN forwarding and providing this data as a connectivity context to other modules.

4.5.3 Application Context Module

The Application Context Module implements functions similar to those included in the Application Layer in the OSI stack. It is closest to the end user of the system and it interacts

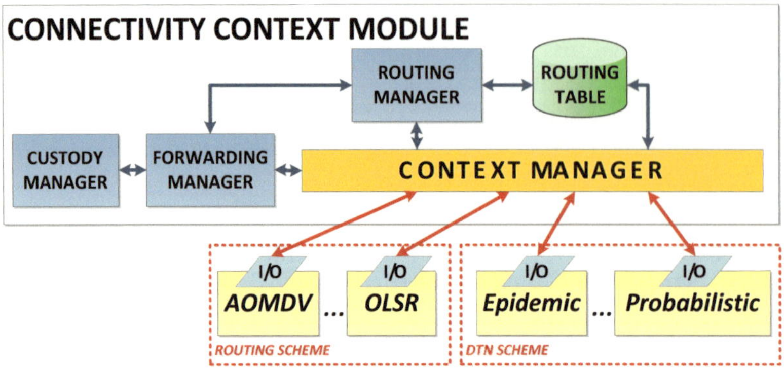

Fig. 3. Connectivity Context Module architecture

with some applications. New context based services can be defined and integrated in a similar way, by using LOMs.

One of key issues in implementing the proposed architecture is the addressing of nodes and services, especially if there is the requirement of enabling disruption-tolerant communication between the nodes. In the case of the highly dynamic vehicular environment, most applications involve a kind of controlled broadcast of information and there is little need for unicast applications.

In this case, assigning a constant address to the node is irrelevant. The address of the destination is not known and is not bound to the source, since the destination is constantly on the move. To address the groups of destination nodes characterized by high mobility, much more important is the context information related to location and neighborhood of the group, as well as its structure and interaction with other groups. In a dynamic network the services are context-addressable, hence the importance of context information exchange between modules.

4.5.4 Context Processing Engine

The Context Processing Engine which internal architecture is shown in Figure 4 is the most crucial part of the framework. It is a module where majority of system intelligence is hidden, which gathers and scatters important context information from and to the three HOMs and which manages a local Repository in which the context data is stored. The output data from the other modules is continuously monitored, filtered and processed, not to mention that sometimes future predictions are made in order to improve specific functions inside HOMs, e.g., the prediction related to the node mobility pattern can help to optimize routing and forwarding. CPE feeds the other modules with the context data according to their requirements reported in the initial registration phase. Hence, a registration is a moment when a set of rules and dependencies in relation to the already registered LOMs is created. These rules are then used to store and manage context data to meet the requirements of the newly attached LOM. Another area of responsibilities of CPE is scheduling of DTN bundles which is performed by a Scheduling Manager, while DTN bundles are queued and stored in the Repository.

The Translation Logic included in CPE adapts the received information and delivers it to the proper Context Module for being handled. The TL has some basic logic which makes use of Context Ontology (CO) engine. Use of ontology concept is necessary because an open architecture allows attaching many different LOMs which exchange data in many possibly

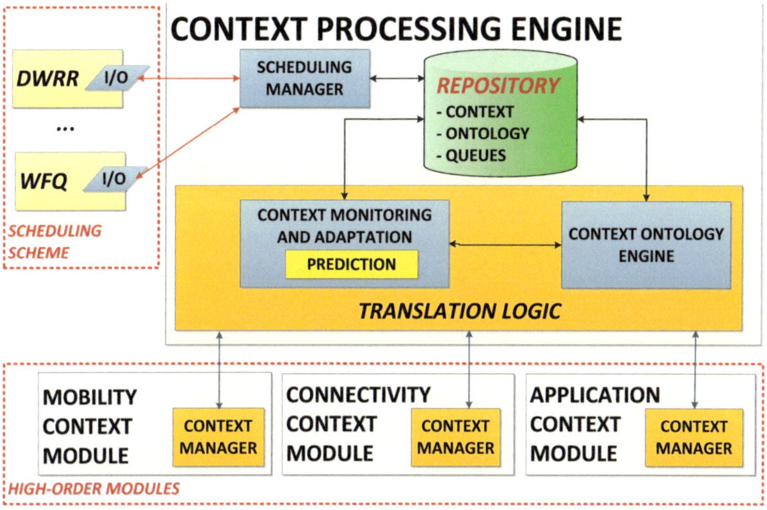

Fig. 4. Context Processing Engine architecture

different formats, so that translation into one formal context information representation is needed. The CO consists of a set of keywords, rules and structures for describing context data and all context relations using a common "language" which can be easily understandable and interpreted by the Context Managers. Usually the three representations are used. The CO approach allows for integration of new solutions into the framework which can be expressed using contexts, even if the offered capabilities were not available in the beginning stage of the designed system based on the framework.

Context in the presented framework is not just a set of external constraints on the system for a given instance. It is redefined to model every piece of information, be it internal control data, instance related data or external data. Building context information is done in accordance with the implemented ontology. Defining such a CO is in fact adding a new "template" to the architecture itself, which becomes context-aware, making it more robust.

Inside the CPE the Scheduling Manager is responsible for choosing the best Scheduling Scheme for DTN bundles before passing them to the CCM. There exists a possibility to integrate new scheduling schemes in the form of LOMs attached directly to the Scheduling Manager. The selection of a particular scheduling scheme, together with the context information monitoring and adaptation are part of a broader cognitive functionality inside the CPE, specifically the capability of the system to behave differently according to the given external context and learn from previous experiences.

To implement the architecture in a real network, other functions will need to extend the CPE logic, such as security functions related to data validation and user authorization, as well as convergence components to support inter-working with multiple communication stacks for different radio technologies. Although these functionalities are not yet handled, they are very important issues related to VANET and need to be solved in the future.

4.6 Interface description

A challenge in implementing the new system architecture is the design of the interfaces for data exchange between modules. Useful parameters and data are adapted to context information and passed between entities, to ensure compatibility and inter-working between them. The most important interfaces are described in Table 2.

All the listed interfaces are quite simple which allows for easier framework expansions by user developed modules. This simplicity can be assured due to context-aware design of CARAVAN.

Interface	Description
CM generic interface	Bi-directional generic interface for exchanging context information with Context Managers - both between a HOM and the CPE, specifically between a CM and the TL and between HOM and attached LOMs.
Bundle transfer interface	Aside from the standard CM-TL generic interface, there is also a second bi-directional interface between CPE and CCM, for sending and receiving DTN bundles. It is up to the CPE to provide the necessary adaptation of the Application Data Units (ADUs) to the bundles.
External I/O interface	The AN external I/O interface is responsible for physical communication with other devices in the network. This bi-directional interface connects to the CPE.
ADU transfer interface	Aside from the standard CM-TL generic interface, there is also a second bi-directional interface between the ACM and CPE, for sending and receiving ADUs (Application Data Units).

Table 2. Description of the interfaces

4.7 CARAVAN – a sample use case

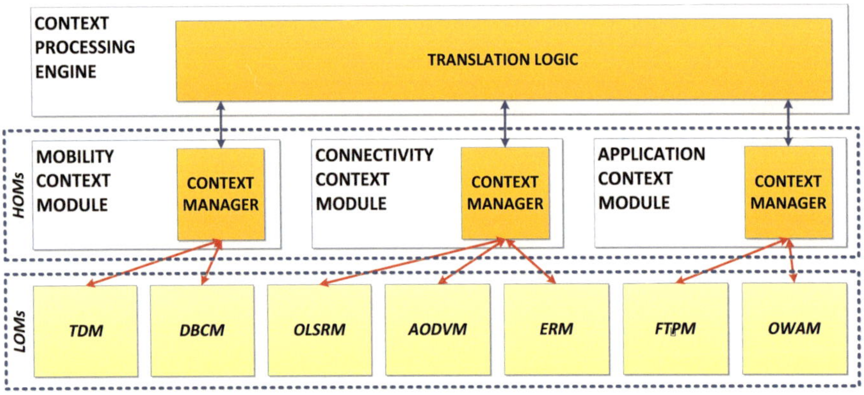

Fig. 5. Sample use case of the framework

To make the CARAVAN concept easier to understand a sample use case is presented in this section. As it is shown in Figure 5, the sample system built on the framework has following Low Order Modules attached:

- Topology Discovery Module (TDM) - the mobility layer module which uses GPS device and beacon messages to gather mobility context data of the node itself and from the neighboring nodes,
- Density Based Clustering Module (DBCM) - the mobility layer module implementing DBC clustering algorithm, responsible for assigning roles of cluster visitors, cluster

candidates and cluster members to neighboring nodes, for finding stable groups of nodes, for selecting clusterhead node being a cluster representative, and for choosing nodes being cluster border gateways,
- Optimized Link State Routing Module (OLSRM) - the connectivity layer module which makes sure that the routing tables for intra-cluster communication are up to date using OLSR routing protocol,
- Ad hoc On Demand Distance Vector Module (AODVM) - the connectivity layer module implementing AODV routing protocol for short range communication with nodes connected using the stable paths,
- Epidemic Routing Module (ERM) - the connectivity layer module which is used by context-aware services that can deal with longer delays,
- File Transfer Protocol Module (FTPM) - the application layer module allowing data transfer between node with a stable connection using FTP protocol,
- Obstacles Warning Assistant Module (OWAM) - the application layer module which warns other vehicles about road obstacles to increase driving safety.

All the Low Order Modules are connected using the generic CM interface for bi-directional exchange of the context data. Each of the modules has to be previously registered in the system using the built-in driver. For example the TDM will advertise itself that it is able to provide necessary nodes position and mobility data with no requirements for system input. The DBCM as the input needs some data generated by the TDM and as the output it offers information about nodes relations such as identification of stable clusters and about nodes which are bad candidates for cluster members, for example because they are moving faster than the group (however, this makes them potential candidates for passing data in DTN forwarding schemes). There can be observed dependence between DBCM and TDM. Due to the registration phase and the logic embedded in the top level modules, every such LOMs dependence can be tuned. The DBCM requires only at specified intervals and only some subset of data which TDM is able to provide. Hence, the TDM knows that there is no point to deliver neither more data nor to do it more frequently than it is needed. Of course, a demand for context data can vary in time as it depends on activity of different modules. The discussion on the relationship between the DBCM and TDM is also a good opportunity to clarify another introduced concept of Translation Logic inside CPE and ontologies. Let us consider the case when DBCM modules need velocity vectors of neighboring nodes for proper work and when TDM is able to provide information about direction and speed of nodes movement. Although it is not exactly the same, there exists a very simple one-to-one correspondence between these notions. Such rule can be easily encoded in the ontology and therefore the translation can be easily done in TL.

Similar dependencies occur between the other modules in the presented sample system - e.g., FTP data transfer can be applied only when a stable connection is detected, so it is possible inside the cluster (OLSR routing protocol is used) or in the traffic jam (AODV routing protocol is used). The OWAM module is designed to warn about obstacle the drivers which are moving towards it - so in this case the delay-tolerant forwarding can be applied combined with the context-aware (in a given direction) data dissemination. The best candidates for passing messages, that is nodes which are moving quickly in a right direction, can be selected using context information from TDM and DBCM.

It should be clear that the presented system can be easily extended by other modules implementing new applications or vehicular services, as well as new protocols to allow a selection of the most suitable solution depending on both external and internal circumstances in order to optimize the overall system performance.

5. Conclusions

In this chapter a new approach to VANET has been proposed. The main idea of the proposal is to integrate many VANET concepts into a common framework and use them on the dependency of the service requirements, connectivity properties and node mobility characteristics. In the proposed framework context-aware approach is used. Contexts are related to service/applications requirements, communications ability, mobility vectors of cars/nodes and the mutual space-time relations between them. The usage of contexts provides high level of adaptability and flexibility. In the CARAVAN we defined three layers: the Mobility Layer, the Connectivity Layer and the Application Layer. Such functional decomposition of the architecture provides ability to incremental modification of every layer via adding or modifying layer internal components without the necessity of the redesign of other components of the architecture. In fact the operations which are most influential on the system behavior are performed by the Cross-Context Processing Engine, i.e., the component that is responsible for the selection of appropriate tools for a specific, overall context. The presented software oriented view together with a sample use case give some clues how the CARAVAN can be implemented and deployed to make vehicular networks idea closer to the reality.

6. Acknowledgements

Authors would like to gratefully thank Zygmunt Wereszczyński from Orange Labs Poland for his invaluable help.

7. References

Baldessari, R., Bödekker, B., Deegener, M., Festag, A., Franz, W., Kellum, C., Kosch, T., Kovacs, A., Lenardi, M., Menig, C. et al. (2007). Car-2-car communication consortium-manifesto, *Car-2-Car Communication Consortium.*

Clausen, T. & Jacquet, P. (2003). RFC3626: Optimized Link State Routing Protocol (OLSR), *RFC Editor United States.*

Ducourthial, B. & Khaled, Y. (2009). Routing in Vehicular Networks: A User's Perspective, *in* H. Moustafa & Y. Zhang (eds), *Vehicular Networks: Techniques, Standards and Applications*, Auerbach Publications Boston, MA, USA, chapter 6, pp. 144–179.

Ducourthial, B., Khaled, Y. & Shawky, M. (2007). Conditional transmissions: Performance study of a new communication strategy in VANET, *IEEE Transactions on Vehicular Technology* 56(6 Part 1): 3348–3357.

Fall, K. & Farrell, S. (2002). Delay tolerant networking research group, Working group charter, Internet Research Task Force, URL: http://www. dtnrg. org.

Festag, A., Noecker, G., Strassberger, M., Lübke, A., Bochow, B., Torrent-Moreno, M., Schnaufer, S., Eigner, R., Catrinescu, C. & Kunisch, J. (2008). Now-network on wheels: Project objectives, technology and achievements, *Proceedings of 6th International Workshop on Intelligent Transportations (WIT), Hamburg, Germany.*

Finn, G. (1987). Routing and addressing problems in large metropolitan-scale internetworks, ISI Research Report ISU, *Technical report*, RR-87-180.

Franz, W., Eberhardt, R. & Luckenbach, T. (2001). Fleetnet-internet on the road, *Proc. 8th World Congress on Intelligent Transport Systems.*

Frey, H. & Stojmenovic, I. (2006). On delivery guarantees of face and combined greedy-face routing in ad hoc and sensor networks, *Proceedings of the 12th annual international conference on Mobile computing and networking*, ACM, pp. 390–401.

Gerla, M., Zhou, B., Lee, Y., Soldo, F., Lee, U. & Marfia, G. (2006). Vehicular grid communications: the role of the internet infrastructure, *Proceedings of the 2nd annual international workshop on Wireless internet*, ACM, p. 19.

Giulio, V. (2007). The SAFESPOT integrated project: an overview, *IEEE Intelligent Vehicles Symp*, p. 14.

Haas, Z. & Pearlman, M. (2001). ZRP: a hybrid framework for routing in ad hoc networks, *Ad hoc networking*, Addison-Wesley Longman Publishing Co., Inc., pp. 221–253.

Härri, J., Filali, F., Bonnet, C. & Fiore, M. (2006). VanetMobiSim: generating realistic mobility patterns for VANETs, *VANET '06: Proceedings of the 3rd international workshop on Vehicular ad hoc networks*, ACM, New York, NY, USA, pp. 96–97.

Johnson, D., Maltz, D., Broch, J. et al. (2001). DSR: The dynamic source routing protocol for multi-hop wireless ad hoc networks, *Ad hoc networking* 5: 139–172.

Käsemann, M., Füßler, H., Hartenstein, H. & Mauve, M. (2002). A reactive location service for mobile ad hoc networks, *Department of Computer Science, University of Mannheim, Tech. Rep. TR-02-014* .

Kranakis, E., Singh, H. & Urrutia, J. (1999). Compass routing on geometric networks, *Proc. 11th Canadian Conference on Computational Geometry*, pp. 51–54.

Kukliński, S., Matei, A. & Wolny, G. (2010). NGVN: A framework for Next Generation Vehicular Networks, *COMM2010: Proceedings of the 8th International Conference on Communications*, Bucharest, Romania, pp. 297–300.

Kukliński, S. & Wolny, G. (2009). Density Based Clustering algorithm for Vehicular Ad-Hoc Networks, *International Journal of Internet Protocol Technology* 4(3): 149–157.

Le, L., Festag, A., Baldessari, R. & Zhang, W. (2009). CAR-2-X Communication in Europe, *in* S. Olariu & M. C. Weigle (eds), *Vehicular Networks: From Theory to Practice*, Computer and Information Science Series, Chappman & Hall/CRC, chapter 10, pp. 1–36.

Lebrun, J., Chuah, C., Ghosal, D. & Zhang, M. (2005). Knowledge-based opportunistic forwarding in vehicular wireless ad hoc networks, *Proceedings of 61st IEEE Vehicular Technology Conference*, Vol. 4, pp. 2289–2293.

Leinmüller, T., Buttyan, L., Hubaux, J., Kargl, F., Kroh, R., Papadimitratos, P., Raya, M. & Schoch, E. (2006). SeVeCOM - secure vehicle communication, *Proceedings of IST Mobile Summit*.

Leontiadıs, I. & Mascolo, C. (2007). Geopps: Geographical opportunistic routing for vehicular networks, *IEEE International Symposium on a World of Wireless, Mobile and Multimedia Networks, 2007. WoWMoM 2007*, pp. 1–6.

Li, J., Jannotti, J., De Couto, D., Karger, D. & Morris, R. (2000). A scalable location service for geographic ad hoc routing, *Proceedings of the 6th annual international conference on Mobile computing and networking*, ACM, p. 130.

Lindgren, A., Doria, A. & Schelén, O. (2004). Probabilistic routing in intermittently connected networks, *Service Assurance with Partial and Intermittent Resources* pp. 239–254.

Little, T. & Agarwal, A. (2005). An information propagation scheme for VANETs, *2005 IEEE Intelligent Transportation Systems, 2005. Proceedings* pp. 155–160.

Lochert, C., Hartenstein, H., Tian, J., Fussler, H., Hermann, D. & Mauve, M. (2003). A routing strategy for vehicular ad hoc networks in city environments, *IEEE Intelligent Vehicles Symposium, 2003. Proceedings*, pp. 156–161.

Lochert, C., Mauve, M., F¨ ußler, H. & Hartenstein, H. (2005). Geographic routing in city scenarios, *ACM SIGMOBILE Mobile Computing and Communications Review* 9(1): 72.

Mietzner, R. (2007). CVIS-Cooperative vehicle-infrastructure systems [J], *COM Safety: Newsletter for European ITS Related Research Projects* 3(5).

Moustafa, H. & Zhang, Y. (2009). *Vehicular networks: techniques, standards, and applications*, Auerbach Publications Boston, MA, USA.

Naumov, V. & Gross, T. (2007). Connectivity-aware routing (car) in vehicular ad-hoc networks, *IEEE INFOCOM 2007. 26th IEEE International Conference on Computer Communications*, pp. 1919–1927.

Olariu, S. & Weigle, M. (2009). *Vehicular Networks: From Theory to Practice*, Computer and Information Science Series, Chapman & Hall/CRC.

Perkins, C. & Royer, E. (1999). Ad-hoc on-demand distance vector routing, *wmcsa*, Published by the IEEE Computer Society, pp. 90–100.

Reding, V. (2006). The Intelligent Car Initiative: raising awareness of ICT for Smarter, Safer and Cleaner vehicle, *Speech delivered at the Intelligent Car Launching Event, Brussels* 23.

Reichardt, D., Miglietta, M., Moretti, L., Morsink, P. & Schulz, W. (2002). CarTALK 2000: Safe and comfortable driving based upon inter-vehicle-communication, *IEEE Intelligent Vehicle Symposium, 2002*, pp. 545–550.

Ros, F. J., Ruiz, P. M., Sanchez, J. A. & Stojmenovic, I. (2009). Mobile Ad Hoc Routing in the Context of Vehicular Networks, *in* S. Olariu & M. C. Weigle (eds), *Vehicular Networks: From Theory to Practice*, Computer and Information Science Series, Chappman & Hall/CRC, chapter 9, pp. 1–48.

Ros, F. & Ruiz, P. (2007). Cluster-based OLSR extensions to reduce control overhead in mobile ad hoc networks, *Proceedings of the 2007 international conference on Wireless communications and mobile computing*, ACM, pp. 202–207.

Schulze, M., Nocker, G. & Bohm, K. (2005). PReVENT: A European program to improve active safety, *Proc. of 5th International Conference on Intelligent Transportation Systems Telecommunications, France.*

Seet, B., Liu, G., Lee, B., Foh, C., Wong, K. & Lee, K. (2004). A-STAR: A mobile ad hoc routing strategy for metropolis vehicular communications, *NETWORKING 2004, Networking Technologies, Services, and Protocols; Performance of Computer and Communication Networks; Mobile and Wireless Communications* pp. 989–999.

Shao, Y., Liu, C. & Wu, J. (2009). Delay-Tolerant Networks in VANETs, *in* S. Olariu & M. C. Weigle (eds), *Vehicular Networks: From Theory to Practice*, Computer and Information Science Series, Chappman & Hall/CRC, chapter 10, pp. 1-36.

Spyropoulos, T., Psounis, K. & Raghavendra, C. (2005). Spray and wait: an efficient routing scheme for intermittently connected mobile networks, *Proceedings of the 2005 ACM SIGCOMM workshop on Delay-tolerant networking*, ACM, p. 259.

Takagi, H. & Kleinrock, L. (1984). Optimal transmission ranges for randomly distributed packet radio terminals, *IEEE Transactions on Communications* 32(3): 246–257.

Tchepnda, C., Moustafa, H., Labiod, H. & Bourdon, G. (2009). Security in Vehicular Networks, *in* H. Moustafa & Y. Zhang (eds), *Vehicular Networks: Techniques, Standards and Applications*, Auerbach Publications Boston, MA, USA, chapter 12, pp. 331–353.

Vahdat, A. & Becker, D. (2000). Epidemic routing for partially connected ad hoc networks, *Technical report*, Citeseer.

Wolny, G. (2008). Modified DMAC Clustering Algorithm for VANETs, *3rd International Conference on Systems and Networks Communications, 2008. ICSNC'08*, pp. 268–273.

Zhao, J. & Cao, G. (2008). VADD: Vehicle-assisted data delivery in vehicular ad hoc networks, *IEEE Transactions on Vehicular Technology* 57(3): 1910–1922.

Part 2

Security and Caching in Ad Hoc Networks

8

Trust Establishment in Mobile Ad Hoc Networks: Key Management

Dawoud D.S.[1], Richard L. Gordon[2],
Ashraph Suliman[1] and Kasmir Raja S.V.[3]
[1]*National University of Rwanda*
[2]*University of KwaZulu Natal*
[3]*SRM University, Chennai,*
[1]*Rwanda*
[2]*South Africa*
[3]*India*

1. Introduction

Mobile ad hoc networks are complex wireless networks, which have little or no existing network infrastructure. These networks can be established in a spontaneous manner allowing organizations and network members to work together and communicate, without a fixed communication structure. The mobility, spontaneity and ad hoc nature of these networks makes them optimal solutions for disaster area communication and tactical military networks. Due to recent wireless technology advances, mobile devices are equipped with sufficient resources to realize implementation of these dynamic communication networks. However, for ad hoc networks to find a wide spread within both the military and commercial world, they must be secured against malicious attackers.

Mobile ad hoc networks have distinct characteristics, which make them very difficult to secure. Such characteristics include: the lack of network infrastructure; no pre-existing relationships; unreliable multi-hop communication channels; resource limitation; and node mobility. Users cannot rely on an outside central authority, like a trusted third party (TTP) or certificate authority (CA), to perform security and network tasks. The responsibility of networking and security is distributed among the network participants. Users have no prior relationship with each other and do not share a common encryption key. Therefore, only after the network has been formed, the users establish trust and networking links. The establishment of networking links is identified as being vulnerable to security attacks. Trust establishment should allow protection for the network layer and ensure that honest links are created.

The sporadic connectivity of the wireless links, inherent to mobile ad hoc networks, results in frequent link breakages. These characteristics introduce unique challenges to trust establishment. Both the routing and trust establishment protocols must be designed to handle the unreliable wireless communication channels: the dynamic topology changes and the distributive nature. The security solutions used for conventional wired networks cannot simply be applied to mobile ad hoc networks. More complex network management must be implemented to achieve trust establishment in mobile ad hoc networks.

Ad hoc network security research initially focused on secure routing protocols. All routing schemes however, neglect the crucial task of secure key management and assume pre-existence and pre-sharing of secret and/or private/public key pairs [Zhou & Haas, 1999]. This left key management considerations in the ad hoc network security field as an open research area. Security solutions which use cryptographic techniques rely on proper key management to establish trust. This chapter together with the next chapter focus upon key management which aids these cryptographic solutions.

Outlines of the Chapter

This chapter and the next chapter form one unit. The two chapters focus largely upon establishing trust in mobile ad hoc networks, and concentrate more specifically on secure key management on the network layer. Our research focuses upon providing a solution for the security issues found in mobile ad hoc networks.

The current chapter is organised in the following manner: Section-2 provides a theoretical background to mobile ad hoc networks and the security issues that are related to such networks. These networks and their characteristics are defined in terms of trust establishment. As the focus of this research is on the network layer, attacks specific to this layer are identified and explained.

Section-3 presents a survey of the existing key management solutions for mobile ad hoc networks. Discussions are based on: functionality; availability; security services; scalability; efficiency; and computational cost. A comparative summary is presented, which identifies the difference in the requirements and the application of each solution.

In the next chapter , Section-2, we continue the discussions given in Section-3 of this chapter by offering a survey of the existing secure routing protocols for mobile ad hoc networks. The two sections identify the problem that the two chapters are addressing. There exists secure routing mechanisms to address the unique characteristics of mobile ad hoc networks, however, these solutions assume that key management is addressed prior to network establishment. A novel, on-demand solution to the key management problem for mobile ad hoc networks will be introduced in next chapter. The implementation of the proposed model, simulation of the model, the results and there analysis are given in next chapter.

2. Mobile ad hoc networks

An ad hoc network is a network with no fixed infrastructure. It allows for users to enter and exit any time, while seamlessly maintaining communication between other nodes. Mobile Ad Hoc Networks (MANETs) are advanced wireless communication networks which operate in an ad hoc manner. The term ad hoc is defined as:

"Meaning "to this" in Latin, it refers to dealing with special situations as they occur rather than functions that are repeated on a regular basis." (The American Heritage Dictionary of the English Language, Fourth Edition. Houghton Mifflin Company, 2004)

This definition suggests that it is a network which is formed in a spontaneous manner so as to solve an immediate communication need between mobile nodes. Mobile ad hoc networks differ from existing wired networks because they do not rely on a fixed network infrastructure [Capkun et al., 2003] [Haas et al., 2002], such as base stations or mobile switching centres. Instead, network functionality (e.g., routing, mobility management, etc.) is adopted by the nodes themselves. When using a multi-hoping routing protocol, mobile nodes within each other's radio range communicate directly via wireless links. However the nodes that are far apart depend on the other nodes to relay the message in a multi-hop fashion. Figure 1 [Zhou & Hass, 1999] demonstrates these autonomous, multi-hop characteristics. Connection between nodes is made by means of other nodes within the network. In Figure 1, the circle represents wireless range of node A. In Figure 1, when node D appears within the range of node A, the topology changes to maintain the connection. Note that all network functions are performed by the nodes and no host or outside authority exists.

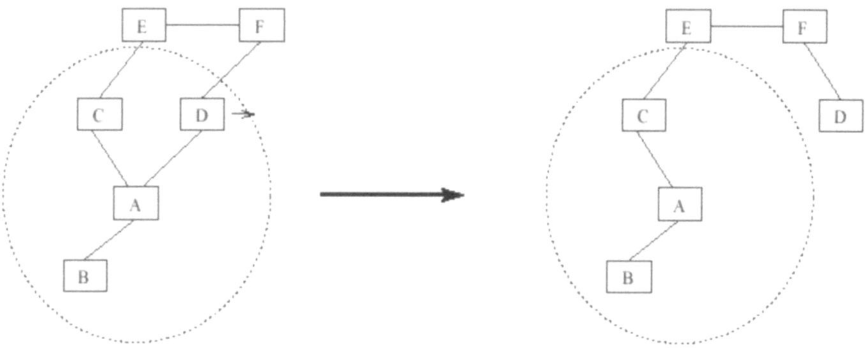

Fig. 1. Ad Hoc Network Topology

2.1 Application

Mobile ad hoc networks have become widely desired in military and commercial applications, due to the ever increasing development of mobile technology. The network's lack of infrastructure and independent nature allows for a robust network to be created within an unlikely networking environment.

a. Military Application

The first ad hoc networks were primarily deployed in the military domain in the early 1970's by the US Department of Defence, under the projects of DARPA and Packet Radio Network (PRnet) [Haas et al., 2002]. Ad hoc networks remain an important part of current and future military communication. They feature prominently in the following areas of military application: sensor networks; tactical networks; and positional systems.

Their application within the military field is based on the network's high mobility, survivability, and self-organized nature. This allows mobile military units to communicate effortlessly irrespective of the distance between each detachment. In a hostile environment, such as the battle field, an ad hoc network's distributive architecture eliminates the problem of a vulnerable network host or the loss of the network host. The modern battle field is characterized by highly mobile forces and the effect of a network which fails to maintain communication and high mobility is disastrous. An example of this can be seen in the experience of the Iraqi forces during the 1991 Gulf War. For this reason, soldiers would prefer mobile ad hoc networks, as opposed to existing local networks. Both invading and defending soldiers would avoid using the local operator, therefore ensuring communication stealth required for battle. Another illustration of the downfall of using an existing local network can be seen in Chechnya, where a general was killed by a missile which tracked the uplink signal of his portable phone. It is clear from these examples that mobile ad hoc networks provide stealth, mobility, and security in the battle field.

The military context is the most obvious application for mobile ad hoc networks. More recently in July 2008, DARPA invested $8.5 million in the Intrinsically Assurable Mobile Ad Hoc Network program (IAMANET) [Jameson, 2008]. This project aims to improve the integrity, availability, reliability, confidentiality, safety, non-repudiation of MANET communication and data in the future.

b. Commercial Application

Early application and developments were military focused. However, non-military applications have grown rapidly due to the availability and advances in mobile ad hoc research. The introduction of new standards such as IEEE 802.16e, IEEE 802.11g and IEEE 802.15.4, have significantly helped the deployment of wireless ad hoc network technology in the commercial domain [Haas et al., 2002]. In this sector the aforementioned networks are desirable due to their dynamic and self organized nature, which allows rapid network deployment. This is particularly useful in situations where infrastructure is damaged or does not exist, and where existing conventional networks are unaffordable or lack sufficient network coverage and need to be side-stepped. Some examples of these applications include: personal area networks; sensor networks; emergency networks; and vehicular communication

Personal area networks are created when a small number of nodes meet spontaneously to form a network for the purpose of teleconferencing, file sharing, or peer-to-peer communication. An example of this can be seen when attendees in a conference room share data using laptops or handheld devices.

Sensor networks are used to monitor data across an area. An example of these networks includes small sensor devices which are located in animals and other strategic locations that collectively monitor and analyze the environmental conditions. Sensor networks have also been developed, by the PermaSense Project, to monitor the permafrost found in the Swiss Alps [Talzi et al., 2007].

The application of this network to an emergency context often occurs in a hostile environment, similar to the military context. Natural or man-made disasters may result in the existing network infrastructure being unavailable or unreliable. Ad hoc emergency services could allow communication and sharing of video updates of specific locations, among relief workers and the command centre. An illustration can be seen in the event of the New York World Trade Centre disaster, on September 11, 2001. The majority of the phone base stations were knocked out in less than twenty minutes, after the attack. The remaining base stations were unable to operate because they could not work in ad hoc mode. The Wireless Emergency Rescue Team recommended afterwards that telecom operators provide ad hoc mode for their infrastructure in the event of emergency situations to enable co-operation between police, firemen and hospital networks [Karl & Rauscher, 2001]. Mobile ad hoc networks can allow for rapid network deployment in an emergency situation. Emergency networks can be set up in remote or hostile areas where there is no existing communication infrastructure, thereby assisting relief work and rescue missions.

A Vehicular ad hoc network provides communication between vehicles, roadside equipment and vehicles travelling in close proximity. Data is exchanged between nearby vehicles to provide traffic information and early warnings for accidents and road works. The purpose of Vehicular ad hoc networks is to provide a communication network of safety and information for users [Raya & Hubaux, 2005].

The benefits of ad hoc networks have realized new non-military communication opportunities for the public. Companies are starting to recognize the potential for commercial ad hoc network applications, and as a result laptops and handheld devices are being equipped with wireless functionalities. Businesses are offering products using ad hoc networking technology in areas of: law enforcement; intelligent transport systems; and community networking. These dynamic networks have still not reached their full potential, and it is clear that ad hoc technology has an imminent role to play in the development commercial technology of today and the future.

2.2 Ad hoc network challenges

An ad hoc network is a dynamic type of network which is both similar and very different to its parent fixed communication network. In the following we introduce the properties of an ad hoc network as a way of defining its shortcomings and to highlight its security challenges.

a. Dynamic Network Architecture

Ad hoc networks have no fixed or existing network infrastructure. The network architecture is continuously changing as the network evolves. There is no pre-existing or fixed architecture which handles all network tasks such as: routing security and network management. Instead, the network infrastructure is spontaneously set up in a distributive manner. Each participating node shares the network's responsibilities. Distribution of network functionality avoids single point attacks and allows for the network to survive under harsh network circumstances.

A fixed entity structure, such as a base station or central administration, is crucial for security mechanisms. A trusted third party member [William, 1999], which is expected in traditional networks, is similar to a fixed entity as both define security services; manage and distribute secret keying information (which allows secure communication of data through encryption and decryption techniques). Therefore the absence of such a control entity introduces new opportunities for security attacks on the network.

b. Self Organized Nature

Wireless ad hoc nodes cannot rely on an off-line trusted third party member. The security functions of the trusted third party member are distributed among the participating nodes. Each node takes responsibility for establishing and maintaining its own security and is, therefore, the centre of its own world and authority. A wireless ad hoc network is therefore referred to as a self organized network [Capkun et al, 2003].

c. No Prior relationships

In ad hoc networks, nodes can have no prior relationships with other nodes within the network. Prior acquaintance between nodes can be considered as pre-trust relationships between nodes. However, the ad hoc nature of these networks does not allow for these assumptions, as it cannot be assumed that secrets exist between the respective pair of nodes [Eschenauer & Gligor, 2002]. If nodes can join and leave the network at random without prior trust relationships with nodes, access control becomes a difficult task for the security mechanism.

d. Multi-hop communication channel

Wired networks include fixed nodes and fixed wired communication lines. Wireless ad hoc networks have mobile wireless nodes (often in the form of hand held devices) and, as suggested, their communication medium is wireless. This allows for greater network availability and easy network deployment. Each node's transmission range is limited and network communication is realized through multi-hop paths. Co-operation and trust along these paths is a crucial aspect of the security mechanism and ensures successful communication. The shared wireless communication medium means that any user can participate in the network. This creates access control problems for security mechanisms as adversaries are able eavesdrop on communication or launch active attacks to alter message data.

e. Mobility

Nodes are expected to be mobile within an ad hoc network, creating a dynamic and unpredictable network environment. In certain situations the nodes' mobility is not totally unsystematic and assumptions can be made in the form of mobility patterns [Capkun et al, 2006]. An example of these patterns is evident in a vehicular ad hoc network where vehicles move along fixed paths, or roads, at speeds which have a high probability of being within the local speed limit. However, nodes demonstrate random mobility within these predictions [Capkun et al, 2006].

Connectivity between nodes is sporadic. This is due to the shared, error-prone wireless medium and frequent route failures which caused by the unpredictable mobility of nodes [Van der Merwe & Dawoud, 2005]. Increased mobility can result in the multi-hop communication paths being broken and network services becoming unavailable. Security mechanisms must account for the weak connectivity and unavailability. Furthermore, due to mobility and sporadic connectivity, these mechanisms must also aim to be scalable with the changing network density.

f. Resource Limitations

Wireless nodes allow for the freedom of mobility and easy network establishment and deployment. Wireless nodes are often smaller hand-held devices that do not experience the same resource privileges of traditional wired nodes [Hass et al, 2002]. Mobile nodes are ideally low cost and small in size as to maximize node availability and mobility. In attempt to achieve these objectives wireless nodes have limited resource, specifically in the following areas:

- Battery life
- Communication range
- Bandwidth
- Computational capacity
- Memory resources

If mobility is to be attained, nodes must be battery powered. Battery powered nodes suffer from the consequences of power failures which break connectivity. They also run a high possibility of failing to be on-line the entire duration of the network. This could hinder network service availability. Cost and power restrictions limit the design features of wireless nodes. Power and transmission range are directly related, resulting in wireless devices having limited transmission ranges and bandwidths. Low powered, low cost CPU's are preferred, as this reduces the computational capacity and memory resources available for routing and security operations. As discussed above, network and security tasks are not performed by a central authority, but rather distributed among all the nodes. This creates a heavy burden upon the nodes to perform their own tasks as well as the network services. If the security mechanisms do not distribute the load fairly, adversaries can act in a selfish manner, forcing other nodes to perform extra tasks. In some instances malicious nodes will flood a single node with service requests in the aim of depleting its limited resources. A well designed security algorithms optimizes computational processing and operation to meet the limited resource requirements of these dynamic networks.

g. Physical Vulnerability

Another challenge in ad hoc networks is the physical vulnerability of nodes. In a mobile ad hoc network nodes are mobile and often small devices. This contributes to a higher probability of being capture or compromised when compared to traditional wired networks with stationary entities [Lidong &Zygmunt, 1999]. This means that wireless ad hoc networks are more prone to insider attacks and security mechanisms and must be designed with this in mind. An inside attacker could analyze the node to gain secret keying information or use the node to compromise other nodes. The same threats exist in wired formal networks. Although they may rely on a secure host to detect and recover compromised nodes. Sensitive security information may also be stored on that host, minimizing the consequences upon the network if a single node is captured. In an attempt to enhance security within hybrid ad hoc networks [Salem et al, 2005] a fixed architecture is combined with a volatile distributive architecture.

2.3 Security objectives and services

Securing mobile ad hoc networks requires certain services to be met. A security service is a made available by a protocol which ensures sufficient security for the system or the data transferred. The security objectives for mobile ad hoc networks are similar to that of fixed wired networks. The security objects are described in six categories, adapted from discussions in [Stalling, 2003]:

- Authentication
- Access Control
- Data Confidentiality
- Data Integrity
- Non-repudiation
- Availability Services

2.4 Attacks

Threats or attacks upon the network come from entities. They are known as adversaries. Mobile ad hoc networks inherit all the threats of wired and wireless networks. With these networks' unique characteristics, new security threats are also introduced [Zhou & Haas, 1999]. Before the development of security protocols, it is essential to study the attacks associated with these unique networks.

a. Attack characteristics

Attacks will be launched against either the vulnerable characteristics of a mobile ad hoc network or against its security mechanisms. Attacks against the security mechanism in all types of networks, including mobile ad hoc networks, include authentication and secret key sabotage. Mobile ad hoc networks have distinctive characteristics, as identified in Section 2.2. Attackers are expected to target these points of vulnerability, for example the multi-hop nature of communication routes. The attacks are classified by their different characters. The attacks, accordingly, are classified as follows:

Passive and Active Attacks

Security attacks can be classified by the terms active and passive [Stalling, 2002]. Passive attacks attempt to steal information from the network without altering the system resources. Examples of passive attacks include, eavesdropping attacks and traffic analysis attacks. It is difficult to detect passive attacks as they leave no traceable affect upon the system resources or network functionality. Although the results or the need for securing against these attacks may not be monitored or visibly present, it is still a priority to protect networks from these seemingly harmless attacks, particularly in a military context. Concerning this point, Bruce [Bruce, 2003] mentioned: "*If security is too successful, or perfect then the security expenditures are seen as wasteful because success is too invisible*". However, Schneier assures one that, despite the lack of visible results, the need to secure information still exists.

Active attacks attempt to modify system resources or network functionality. Examples of these attacks are message modification, message replay, impersonation and denial of service attacks.

Insider and Outsider Attacks

Malicious nodes are not authorized participants in the network, which launch outsider attacks. Impersonation, packet insertion, and denial of service are some examples of outsider attacks. In contrast to outsider attackers, inside attackers are more difficult to defend against. Inside attacks are launched from nodes which are authorized participants in the network. Insider attacks are common in pure mobile ad hoc network, where any user can freely join or exit. Security mechanism become vulnerable when participates are malicious and the confidentiality of keying information can be compromised. Thus, an advantage of the non-repudiation and authentication techniques, malicious insider nodes can be identified and excluded.

Layer Attacks

There are threats at each layer of the mobile ad hoc network communication protocol. The physical layer is vulnerable to passive and active attacks. The attacks found at the physical layer are as follows: eavesdropping; denial of service; and physical hardware alterations. Encrypting the communication links and using tamper-resistant hardware helps to protect the physical layer. However, at the data link layer adversaries can flood the communication links with unnecessary data to deplete network resources. Security mechanisms that provide authentication and non-repudiation can prevent this, as they allow invalid packets transfers to be identified. At the application layer messages are exchanged in an end-to-end manner using wireless multi-hop routes established by the network layer. The wireless multi-hop routes are invisible to the application layer. Conventional security techniques used for wired networks can be used to prevent expected attacks upon the application layer. The application layer is dependent upon the network layer to provide secure routes between the two communicating parties.

The network layer provides a critical service to the mobile ad hoc network, and the routing protocol. In the context of trust and security, the provision of secure routes is one of the most vital elements for trust establishment.

b. Attack Types

The different types of attacks are identified and described below. While there is a focus on the networking layer, attacks such as impersonation and denial of services can occur on any layer.

Wormhole attack

In a wormhole attack a compromised node receives packets at one place in the network. The attacker tunnels the packets to another destination (i.e. an external attacker) in the network, where the packets are resent back into the network [Qian & Li, 2007]. The tunnel created by the adversary is known as a wormhole. A wormhole allows adversaries to disturb the routing protocol, by intercepting routing messages and creating denial of service attacks. If the routing mechanism is not protected against such an attack mobile ad hoc routing protocols may fail to find valid routes.

Black hole attack

During route discovery a malicious node may falsely advertise itself as possessing the optimal route to the requested destination. The adversary, therefore, attracts all routing messages. The attacker then creates a black hole attack by dropping all routing packets, and disrupting the routing protocol and discovery phase.

Byzantine attack

During this type of attack a malicious node, or a group of malicious nodes, will launch attacks on the routing protocol. The aim is to direct routing packets to follow: non-optimal routes; routing loops; and selective dropping of packets [Awerbuch et al, 2002]. Byzantine behaviour is difficult to detect. A network could be operating with byzantine failures and be unaware of the attack on its routing mechanism.

Eavesdropping

An eavesdropping attack involves message or routing packet monitoring. It is a passive attack on the mobile ad hoc network. Eavesdropping attacks are performed by adversaries and can reveal confidential information about the network regarding: its topology; geographical locations; or optimal routes in the network. Attackers can use this information to launch other attacks at identified points of vulnerability. All networks are prone to passive eavesdropping attacks. It is the nature of wireless, mobile ad hoc networks that make them more vulnerable. In wireless networks adversaries do not need a physical wired communication link to monitor the routing packets. The wireless communication medium allows for any users, within range, to analyze the traffic. Attackers can also exploit the multi-hop nature of routes in mobile ad hoc networks. An adversary can position itself along a route path and forwarding the routing messages along the multi-hop path. This allows adversaries to also analyze every packet that is forwarded along the path. Eavesdropping is a common problem in networks and encryption techniques can protect routing protocols from these attacks.

Packet Replay

Like eavesdropping, replay is a passive attack where data is captured by monitoring adversaries. Old routing messages are then retransmitted to other nodes disturbing the routing process. Adversaries can, therefore, cause other node's routing tables to be updated with outdated information. Malicious attackers can also record authorized routing messages and replay them to gain unauthorized access to protected nodes.

Resource consumption attack

Mobile ad hoc nodes are restricted by their limited resources. Attackers exploit this by launching attacks that consume a node's resources hindering them from network participation. Resources targeted by attackers are: bandwidth, computational power and battery life.

Sleep deprivation attacks, are resource attacks which are, specifically aimed against mobile ad hoc node's battery power. Node's attempt to save power by going into a sleep mode, where a periodic scanning occurs and less battery power is used. Sleep deprivation attacks prevent nodes from going into sleep mode therefore draining the battery life and disabling the node itself. Attackers will flood a target node with redundant routing requests or routing packets to be processed, thereby keeping the node and its resources unnecessarily busy.

Packet replication is another type of resource attack where adversaries duplicate out of date packets and re-transmit them. This not only consumes battery life, bandwidth and computational power, but also disrupts the routing protocol.

Sleep deprivation attacks, flooding attacks and packet replication result in the depletion of precious resources. If this is not protected against, it will result in nodes and services becoming unavailable in the network.

Routing Table Poisoning

Malicious nodes will target the routing table in an attempt to sabotage the establishment of routes. One such attack is the routing table poisoning attack where malicious nodes send counterfeit routing updates or modify existing routing updates. This results in conflicting link information, unnecessary traffic congestion or denial of service.

Rushing attack

Mobile ad hoc networks that use on-demand routing protocols are vulnerable to rushing attacks [Hu et al, 2003a]. On-demand routing protocols, such as AODV [Perkins et al, 2003] and DSDV [Perkins & Bhagwat, 1994], use route request messages to discover the optimal route to a destination node. The network is flooded with route request messages. These messages are forwarded until the optimal route is found between the source and destination nodes. An adversary that receives a route request performs a rush attack by hurriedly flooding the network with that route request before other nodes, receiving the same route request, can respond. When other nodes receive the legitimate routing request, it is assumed to be a duplicate of the request which is distributed by the adversary, and the legitimate routing request is dropped. Therefore, the adversary will become part of the route that is discovered. This will result in an overall, insecure route.

Selfish attack

Misbehaving nodes will act in a greedy or selfish manner, resisting cooperating or participating in the network operations. This is a denial of service and the attack causes the nodes to refuse to make their resources available. Selfish nodes do not cooperate in network operations that do not benefit them. Rather they conserve their limited resources, such as battery life. Nodes may refuse to forward route request packets or turn off their devices when they are not transmitting data. The distributive architecture and multi-hop nature of mobile ad hoc networks means the network relies upon node cooperation [Molva & Michardi, 2003]. A security protocol should ensure fair distribution of network operation in order to provide reliable network services, and prevent node's resources becoming depleted because of selfish node attacks.

Impersonation

Impersonation attacks are also known as masquerading or spoofing attacks. The attacks occur when adversaries take the identity of an authorized node and breach the security of the network. Masquerading nodes are able to receive routing packets destined for other nodes. Mobile ad hoc networks can help protect against impersonation attacks by authenticating their routing messages.

Pure mobile ad hoc networks are more vulnerable as they have no access control. If there is no strong binding between the physical entity and the network identity, malicious nodes can adopt different identities. A severe attack which is prone to mobile ad hoc networks is the Sybil attack [Hashmi & Brooke, 2008] [Douceur, 2002]. A single adversary node launches a Sybil attack by adopting multiple identities and participating in the network with all identities at once. The result of such an attack gives the attacker a majority vote or considerable control in the network.

2.5 Security model

A security model for mobile ad hoc networks is illustrated, in general terms, in Figure 2. A message *M* is to be transmitted from the source *A*, across a network of nodes, to a destination node *B*. The two entities who are primary participants must collaborate for the transaction to occur. A routing protocol establishes a multi hop route between the primary

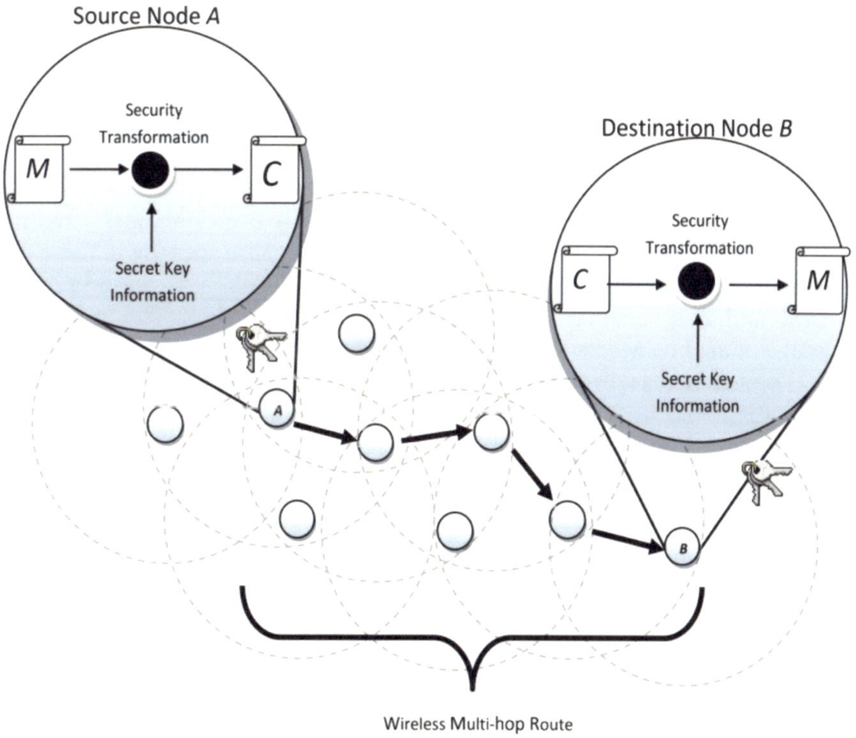

Fig. 2. General Security Model

participants. The multi-hop route will involve secondary participating nodes. Security is provided by two accompanying techniques: a security related transformation applied to the message (resulting in an encrypted message *C)* and secret keying information shared by the principal participants.

The general mobile ad hoc security model shows four basic tasks for a security mechanism:

1. The design of a security algorithm.
2. Generation of secret keying material used in conjunction with this security algorithm.
3. Distribution of secret keying material.
4. Protocol for the participants to follow which will achieve the required security services.

Tasks 1 and 2 deal with the cryptographic algorithm used to provide security services. It is widely recognized that existing cryptographic technology can provide sufficiently strong security mechanisms to ensure routing message confidentiality, authentication and integrity. The establishment of these security mechanisms is a dynamic problem in wireless ad hoc networks, as this network cannot adopt the same approaches of its wired predecessors. The focus of this chapter is upon tasks 3 and 4: the establishment of the security protocols in the mobile ad hoc environment.

3. Key management in mobile ad hoc networks

In Section-2 we discussed the different types of attacks upon wireless ad hoc networks. In this section the techniques used to prevent these malicious attacks and specifically key management techniques, will be looked at. Security solutions which use cryptographic techniques rely on proper key management to establish trust. This chapter focuses upon key management which aids these cryptographic solutions.

3.1 Description of key management

In any communication network, the cryptographic network security is dependent on proper key management. Mobile ad hoc networks vary significantly from standard wired networks. A specific efficient key management system is required to realize security in these networks. Key management is defined as a set of procedures employed to administrate the establishment and maintenance of secure key base relationship. The purposes of key management, as stated by Menezes et al [Menezes et al, 1996b], is to:

1. Initialize system users within a network.
2. Generate, distribute and install keying material.
3. Control the use of keying material within the network.
4. Update, revoke, destroy and maintain keying material.
5. Store, backup and recover keying material.

Key management systems are responsible for the secure distribution of keys to their intended destinations. Keys which are required to remain secret must be distributed in a way that ensures confidentiality, authenticity and integrity. For example, in symmetric key cryptography both, or all, the participants must receive the key securely. For asymmetric key cryptography, the key management system must ensure that private keys are kept secret and only delivered to the required, authorized participants. Public keys do not require confidentiality but, authentication and integrity is vital. The key management system must protect confidentiality and authenticity of the keys. This system must also prevent unauthorized use of keys, for example the use of keys which are out-dated and invalid.

Cryptographic algorithms can provide confidentiality, authentication and integrity. However, the primary goal of key management is to guarantee that the secret keying material is shared among the specific communicating participants securely. There are several methodologies of sharing the keying material. The main approaches are: key transport; key arbitration; key pre-distribution; and key agreement [Menezes et al, 1996b].

a. Key Transport

In a key transport system, one entity generates keys, or obtains keying material, and securely transports them to other entities in the network. The simplest key transport method is the key encrypting key method (KEK). This method assumes a prior shared key exists among the participating nodes. The prior shared key is used to encrypt new keys and transport them to all participating nodes. Prior shared keying relationships cannot be assumed in networks, especially in mobile ad hoc networks. If a public key infrastructure exists, then the new keys can be encrypted by the respective receiver's public key and transported without the existence of prior keying relationships. This approach assumes the existence of a trust third party (TTP) member which transports all the keying material. In pure mobile ad hoc networks a TTP member would not be available. Shamir's three-pass protocol [Shamir, 1979] is a key transport method, without prior shared keys.

b. Key Arbitration

A key arbitration system is a division of key transportation. In key arbitration a central arbitrator is assigned to create and distribute keys to all participants. The arbiter is often a wired node with no resource constraints. In mobile ad hoc networks nodes are wireless with resource constraints. The arbiter would be required to be online throughout the network communication and be accessible to every member in the network. This is difficult in mobile ad hoc networks because of the resource constraints such as: bandwidth; transmission range; and energy. A solution to these potential problems is a distributive system, where the arbiter is replicated at different nodes. Simple replication of the arbiter has severe resource expenses on certain nodes and creates multiple points of vulnerability in the network. If a single replicated arbiter is compromised the entire network can be at risk.

c. Key Pre-distribution

Keys are distributed to all participating member before the start of communication. Key pre-distribution requires prior knowledge of all participating nodes. Its implementation is simple and involves much less computation than other schemes. This method is suitable for mobile ad hoc sensor networks, as they have highly restrictive resource capabilities. The set of sensor nodes is also established before the network is deployed and data is tracked. Once the network is deployed there is no service which allows for new members to join or for keys to be changed. This method is extended by allowing sub-groups of communication to form in the network. Similarly, the decision is made prior to deployment, and not during communication.

d. Key Agreement

Key agreement is used to enable two participants to agree upon a secret key. In this way, keys are shared and establish a secure communication line over which a session can be run. Key agreement schemes are often based on asymmetric key cryptography and have high computational complexity, but little pre-configuration required. The most widely used key agreement scheme is the Diffie-Hellman key exchange [Steiner et al, 1996]. This is an asymmetric keying approach based on discrete logarithms.

3.2 Key management in mobile ad hoc networks

Ad hoc wireless networks have unique characteristics and challenges, which do not allow the simple replication of conventional key management methods that are used for wired networks. Mobile ad hoc network's lack of infrastructure poses the greatest threat to the establishment of a secure key management scheme. Fixed infrastructure such as: a trusted third party member; an administrative support or certificate authority; dedicated routers; or fixed reliable communication links, cannot be assumed in wireless ad hoc networks. Unique solutions are required for such unique networks. The focus of this chapter is around the investigation of the existing key management schemes for mobile ad hoc networks.

Key management schemes are investigated with regard to: functionality; scalability; availability; security services; efficiency; and computational cost. A key management solution, which is scalable, will effectively provide security services in a network which dynamically changes in size, as nodes join and leave the network. Availability is essential for a network whose topology is rapidly changing. Nodes should have easy access to authority members and keying services. A high priority is given to a key management solution that can successfully and efficiently provide crucial security services for the keying material. Such services include: key confidentiality; key authenticity; key integrity; and fresh key updates. These services are congruent with the security services described in Section 2.

Of the existing key management solutions, asymmetric cryptography is predominately used when managing trust via a public key infrastructure (PKI) of some sort. Existing PKI schemes utilize either the: hierarchical or web-of-trust model.

a. Hierarchical Trust models

The hierarchical trust models are more structured, as they use a PKI and a certificate authority as a source of trust. The certificate authority (CA) is a trusted entity used to verify; issue; and revoke certificates, therefore enabling successful public key cryptography. A key management service for public key cryptography would include the certificate authority service which has a public key, *K*, and private key, *k*. The CA's public key is distributed to all the nodes in the network. The nodes know that any certificate signed by the CA's private key may be trusted. Each node also has its own public/private key pair, which allows for nodal communication. The CA stores the public keys of all the network nodes and distributes the respective keys to the nodes that request to setup a secure communication with another node [William, 1999]. A fixed CA is not considered in this investigation, due to the limitations caused by no TTP.

The CA distributes trust in a hierarchical manner, as seen in Figure 3. A root CA issues certificates to delegated CA's or end users. The CA can issue certificates to user nodes or other CA nodes. The PKI X.509 framework is an example of such an infrastructure [Stalling, 2003].

The following types of hierarchal trust models have been investigated in the context of mobile ad hoc networks:

1. *Off-line trusted third party models:* use a trusted outside entity to achieve a large portion of the key management tasks.
2. *Partially distributed certificate authority models:* distribute the functionality of the CA to a small set of nodes.
3. *Fully distributed certificate authority models:* are self organized models, which are similar to the previous distribute to the CA. However, this model is across the entire network in a self organized manner.

4. *Cluster based model:* is a special kind of hierarchical trust in the form of group authentication, where *clustered groups* of nodes are treated as single trust entities and authenticated as a group.

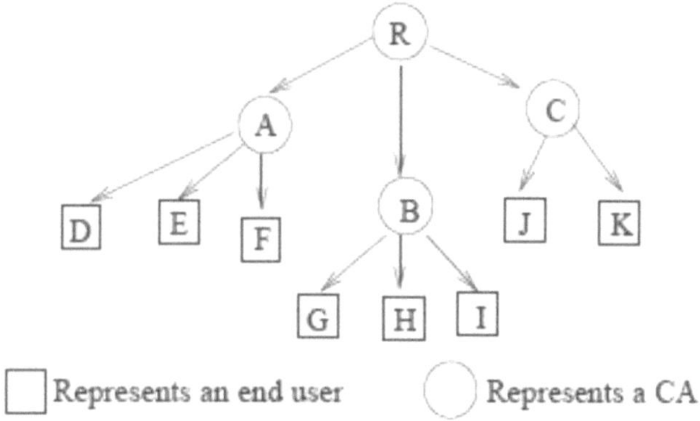

Fig. 3. Hierarchical trust

b. Web-of-Trust Models

The Pretty Good Privacy model (PGP) [Abdul-Rahman, 1997], also known as a "web-of-trust-model", enables nodes to act as independent certification authorities. There is no distinction between a CA and an end user node. Nodes provide individual trust opinions of other nodes, thereby creating a "web of trust", as illustrated in Figure 4. Each user node is the "centre of its own world" and is responsible for certificate management. The advantage of a PGP model is its dynamic, autonomous nature, which is seemingly ideal for application in decentralized environments such as ad hoc networks [Davis, 2004].

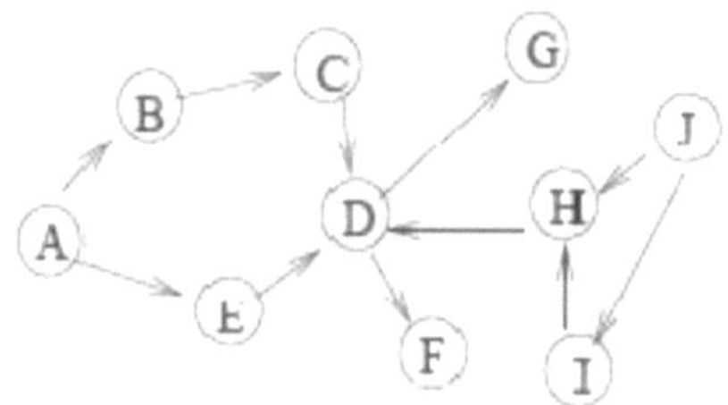

Fig. 4. PGP web-of-trust

Certificates are issued by the nodes themselves. However, a public certificate directory is required for their distribution. This directory is often located at an online, centralized, trusted third party entity. This makes the PGP model unsuitable for ad hoc network application. Steps are needed to be taken to localize such directories and realize certificate distribution. The autonomous nature of the "web-of-trust" model means that it is more susceptible to malicious attackers than to more structured networks. For example, if one entity is compromised a corrupt set of certificates is filtered throughout the network. The self issued certificate model is investigated as a foundation for PGP based solutions in mobile ad hoc networks. Figure 5 illustrates the key management solutions investigated in this chapter.

3.3 Off-line trusted third party models

A progress trust negotiation scheme was introduced by Verma [Verma et al, 2001]. It is a hierarchical trust model where authentication is preformed locally, but an off-line trusted third party performs trust management tasks like the issuing of certificates. The off-line trusted third party also manages the certificate revocation process. This scheme is extended through a localized trust management scheme proposed by Davis [Davis, 2004]. Davis attempts to localize Verma's solution. The only trust management task that is not implemented locally is the issuing of the certificates.

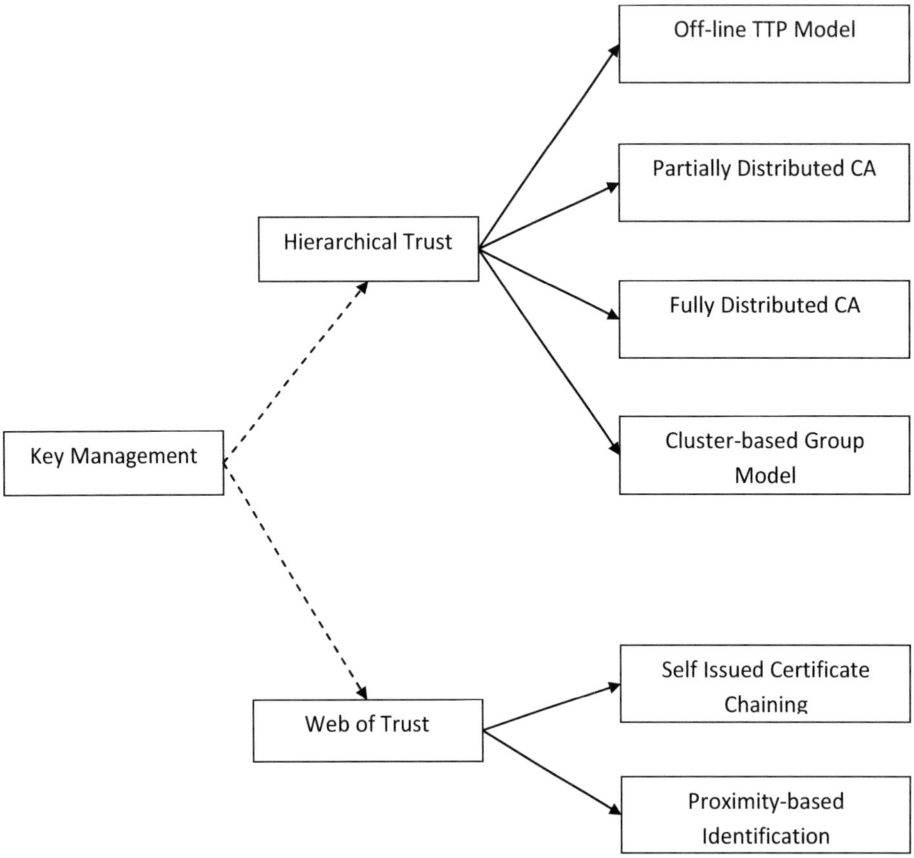

Fig. 5. Key Management Solutions

a. System Overview

Each node possesses its own private key and the trusted third party's public key. The maintenance of these keys is the responsibility of each node. Trust is established when the trustor provides the trustee with a certificate that has not expired, or has not been revoked and the trustee can verify it with the trusted third party's public key (possessed by the trustee). Furthermore, to realize certificate revocation, each node must possess two certificate tables: a status and profile table. The profile table, illustrated in Figure 6, describes the conduct or behaviour of each node. The status table describes the status of the certificate, i.e. revoked or valid. These two tables are maintained locally by the nodes themselves, with the purpose of maintaining consistent profiles.

Davis's scheme is a fully distributed scheme. It requires that a node broadcasts its certificates and its profile table to all the nodes in the network. It also requires that each node's profile table be kept updated, and distributed with synchronization of data content. The profile table contains information from which the user node may define if a certificate can be trusted or of it must be revoked. Node *i*'s profile table stores three pieces of data:

1. *Accusation info:* the identity of nodes that have accused node *i* of misbehaving.
2. *Peer n ID:* the identity of nodes that node *i* has accused, acting almost as a CRL (certificate revocation list).
3. *Certificate status:* a 1-bit flag indicating the revocation status of the certificate.

The fully distributed information in the profile tables should be consistent. If there is any inconsistency detected, an accusation is expected to be launched against the node in question. Inconsistent data can be defined as data which differs from the majority of data.

Fig. 6. Profile Table

The status table is then used to calculate the certificates status, i.e. revoked or not revoked. The node *i*'s status table stores and analysis the following factors: A_i (total number of accusations against node *i*); a_i (total number of accusations made by node *i*) ; *N* (expected maximum number of nodes in the network). These factors are used to calculate the weight of node *i*'s accusation and the weight of other nodes accusations against node *i*. A revocation quotient is then calculated, R_j, as a function of the sum of the weighted accusations. It is then compared to a network defined revocation threshold R_T. If $R_j > R_T$ then the node *i*'s certificate is revoked.

b. Analysis

This scheme uses a hierarchical trust model which relies upon an off-line trusted third party for aspects of key management. The off-line trust third party is to be resident as a trusted source if required. This scheme assumes the existence of a trusted off-line entity which initializes certificates, and securely distributes them amongst the network participants. This scheme is a pre-distributive key exchange model. It provides robust security; however, its implementation is more realistic within a hybrid infrastructure. A key management scheme with a hybrid infrastructure is a scheme which makes use of both wired and wireless architecture. A wired trusted off-line node performs all or a portion of the key management services to maximise security and efficiency. Hybrid infrastructures allow for greater security and a simple solution to the central problem of key distribution in mobile ad hoc networks.

Verma and Davis's solution does not specify that a wired node be the off-line authority for key pre-distribution. Nevertheless, a separate trusted entity capable of intense computation, high security and network distribution must exist for the success of Verma and Davis's model. Such assumptions cannot be made in pure mobile ad hoc networks. The hybrid nature of Davis's solution is displayed in Figure 7.

Verma localizes the task of authentication. Davis goes one step further by localizing the revocation module of the scheme by proactively maintaining accusation information in profile tables and locally, calculating revocation decisions. This scheme mitigates against malicious accusation exploits. This could result in a node being revoked based on single malicious offender's broadcast information. To solve this problem one must not treat all accusations equally, but rather use a sum of weighted accusations, which are calculated before the node is revoked. Davis's scheme succeeds in taking steps toward self-organization in ad hoc network trust establishment as it provides a protocol that enables revocation of certificates, without continual trusted third party involvement.

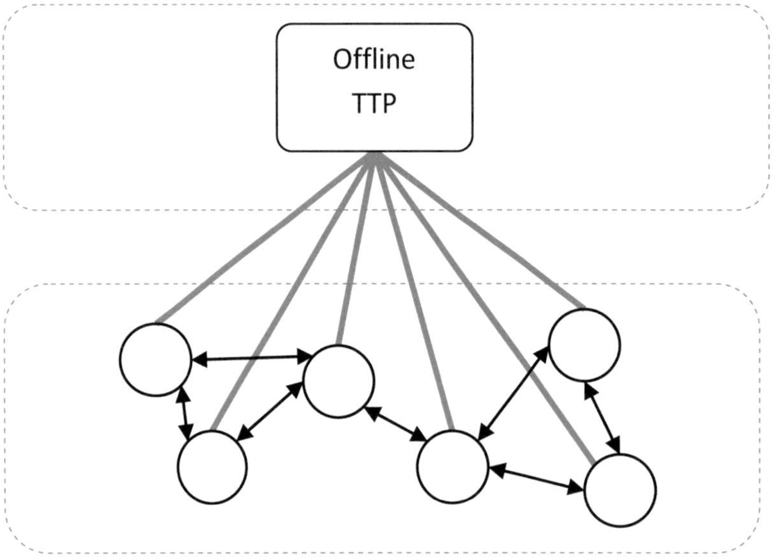

Fig. 7. Hybrid progressive trust negotiation scheme

3.4 Partially distributed certificate authority

The solution proposed by Zhou and Haas [Zhou & Hass, 1999] allows for the functionality of the certificate authority to be shared amongst a set of nodes in the network. This solution aims to create the illusion of an existing trusted third party. Zhou and Haas's proposal in 1999 was instrumental in the initial research of key management solutions for ad hoc networks. This approach has been extended to incorporate the heterogeneous nature of nodes in [Yi & Kravets, 2001].

a. System overview

The CA's public key, *K*, is known by all nodes (*m*) and the CA's private key, *k*, is divided and shared by *n* nodes where $n < m$. The distributed CA signs certificates by recreating the private key via a *t* threshold group signature method. Each CA node has a partial signature. The CA's signature is successfully created when *t* correct partial signatures are combined, at a combiner node. To prevent the distributed CA nodes from becoming compromised and the authentication becoming compromised, a preventive proactive scheme is implemented as to refresh the CA nodes. A simple partially distributed CA system is illustrated in Figure 8.

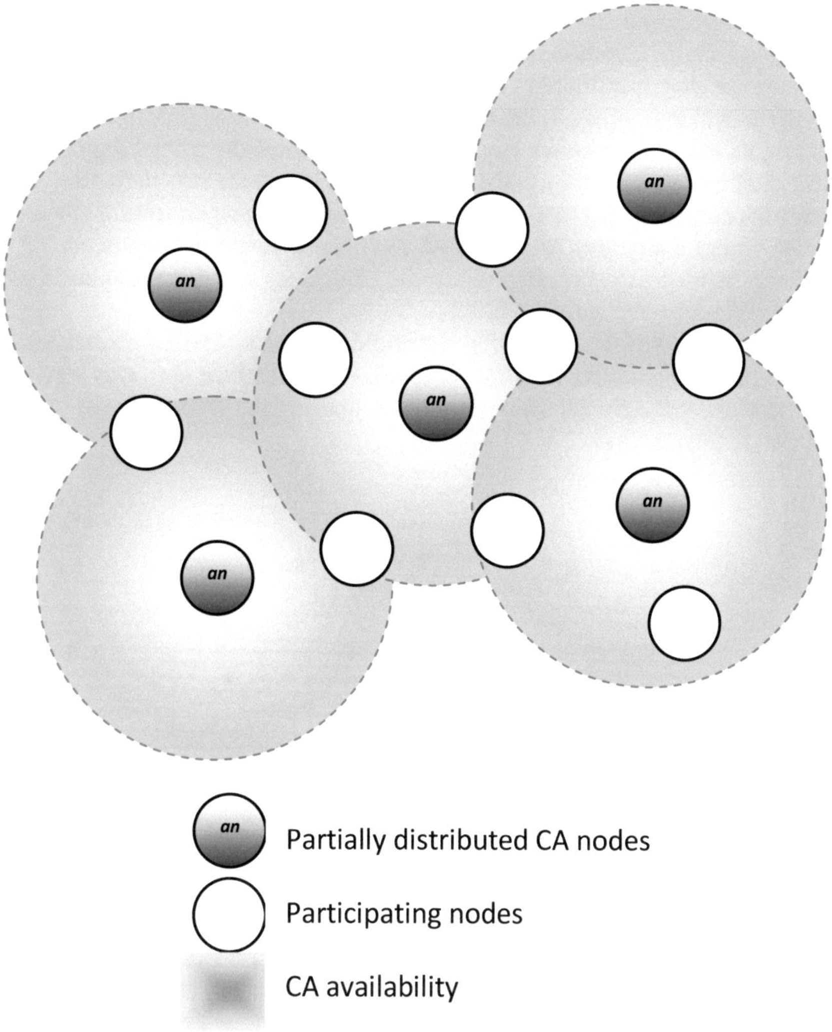

Fig. 8. Partially Distributed Certificate Authority

b. Threshold Scheme

Threshold cryptography is used to share the CA service between nodes. A threshold cryptography scheme allows the sharing of cryptographic functionality. A *(t-out-of-n)* threshold scheme allows n nodes to share the cryptographic capability. However, it requires t nodes, from the n node set, to successfully perform the CA's functionality jointly. Potential attackers need to corrupt t authority nodes, before being able to exploit the CA's functionality and analyze secret keying information. Therefore, a *(t-out-of-n)* threshold scheme tolerates t-1 compromised nodes, from the n node set [Aram et al, 2003].

When applying threshold cryptography to the shared CA problem, the CA service is shared by n nodes across the network called authority nodes. The private key k, crucial for digital signatures, is split into n parts $(k_1,k_2,k_3,\ldots,k_n)$ assigning each part to an authority node (*an*). Each authority node has its own public key, $K_{n,}$ and private key, $k_{n,}$ (as seen in Figure 9).It

stores the public keys of all the network nodes (including other authority nodes). Nodes wanting to set-up secure communication with node *i* need only request the public key of node *i* (K_i) from the closest authority node - therefore increasing the CA's availability. For the CA service to sign and verify a certificate, each authority node produces a partial digital signature using its respective private key, k_p, and then submit the partial digital signature to a combining node. Any node may act as a combiner in the ad hoc network. The partial digital signatures are combined at a combiner (*c*) to create the signature for the certificate, *t* correct partial digital signatures are required to create a successful signature. Therefore, protecting the network against corrupt authority nodes, up to *t-1* corrupt authority nodes may be tolerated [Lidong & Zygmunt, 1999].
For example, Figure 10 shows a (*2-out-of-3*) threshold scheme where the message *m* is signed by the CA, two partial signatures (*PS*) are accepted, while the third (an_2) was corrupted. The partial signatures meet the threshold requirements and the partial signatures are combined at *c* and applied to the message.

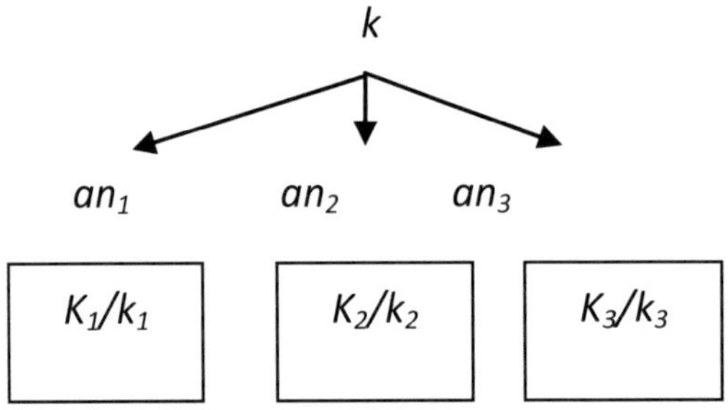

Fig. 9. (2-out-of-3) Threshold Key Management

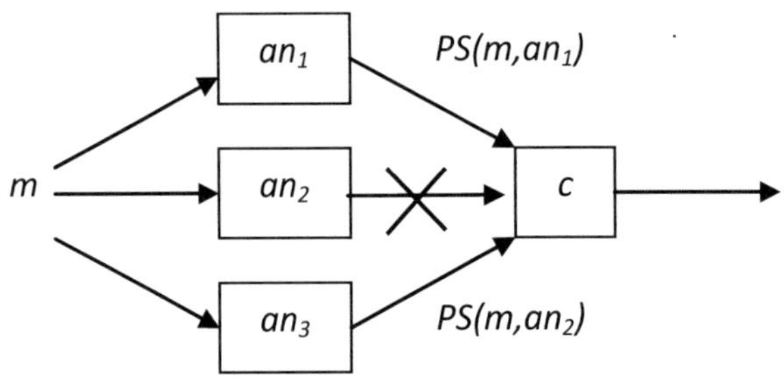

Fig. 10. (2-out-of-3) Threshold Signature

c. Proactive security

Threshold cryptography increases the availability and security of the network by de-centralizing the CA. Security is maintained with the assumption that all CA authority nodes cannot be simultaneously corrupt.
It is possible for a malicious attacker to compromise all the CA's authority nodes over time. An adversary of this type is then able to gain the CA's sensitive keying information. Proactive schemes [Van der Merwe & Dawoud, 2004] [Herzberg et al, 1997] [Frankel et al, 1997] [Jarecki, 1995] are implemented to avoid such adversaries.
A proactive threshold cryptography scheme uses share refreshing. This enables CA authority nodes to compute new key shares from old ones, without disclosing the CA's

public/private key. The new key shares make a new *(t-out-of-n)* sharing of the CA's public/private key pair. These are independent of the old pair [Herzberg et al, 1995].
Share refreshing relies on the following mathematical property:
If $(s_{11}, s_{21}, \ldots, s_{n1})$ is a *(t-out-of-n)* sharing of k_1 and $(s_{12}, s_{22}, \ldots, s_{n2})$ is a *(t-out-of-n)* sharing of k_2, then $(s_{11} + s_{12}, s_{21} + s_{22}, \ldots, s_{n1} + s_{n2})$ is a *(t-out-of-n)* sharing of $k_1 + k_2$. Therefore if k_2 is 0, then we get a new *(t-out-of-n)* sharing of k_1.
The share refreshing scheme is applied to a threshold CA. A threshold CA is a *(t-out-of-n)* system that shares the CA's private key k among n authority nodes $(an_1, \ldots, an_n)$ each with a share of the CA's private key. To generate a new *(t-out-of-n)* sharing $(an_1', \ldots, an_n')$ of k, each authority node an_i generates sub-shares $(an_{i1}, an_{i2}, \ldots, an_{in})$ a *(t-out-of-n)* sharing of 0, which represents the i'th column, as seen in Figure 11. Each sub-share an_{ij} is sent to the authority node an_j. When authority node an_j has received all sub-shares $(an_{1j}, an_{2j}, \ldots, an_{nj})$, which represents the jth row, seen in Figure 11, it then generates its new share an_1' by using the mathematical property described above.

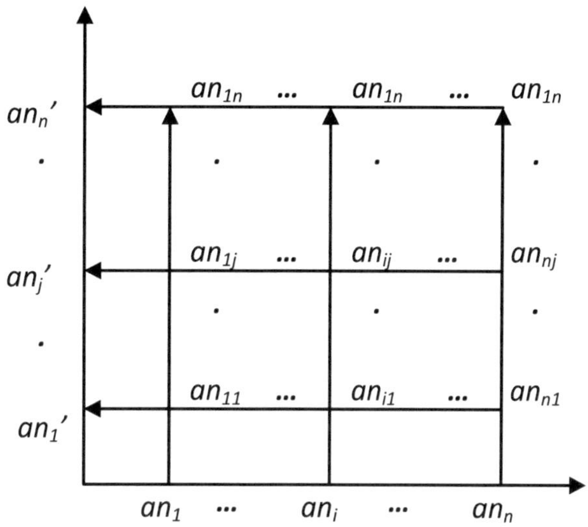

Fig. 11. (t-out-of-n) Share Refreshing

The communication of the sub-shares requires a secret redistribution protocol [Desmendt & Jajodia, 1997] [Chor et al, 1985] to ensure secure transmission. Note that share refreshing does not change the CA's private key pair. Share refreshing may occur periodically and be extended to occur upon events. These events can include the detection of compromised nodes or a change in network topology. Therefore, the key management service is able to transparently adapt itself to changes in the network and maintain secure communication.

d. Heterogeneous Extension

An extension to Zhou and Haas's scheme can be seen in the Mobile Certificate Authority (MOCA) scheme by Yi and Kravets [Yi & Kravets, 2003]. The MOCA scheme also uses threshold cryptography to implement a public key, which is a partially distributed certificate authority solution. The functionality of the certificate authority is distributed to n nodes, called MOCAs. The assumption is made that all nodes have heterogeneous visible qualities. These visible qualities act as initial trust evidence and are used when selecting the

MOCA nodes to distribute authority. Such visible evidence can include: computational power; physical security; or position. This evidence is based on a trust decision and authority distributed, accordingly. Similar to Zhou and Haas's scheme, nodes require *t+1* partial signatures from a set of *n* MOCAs to allow for certificate verification and trust relationship establishment, with a threshold of *t*. The MOCA scheme further builds on Zhou and Haas's solution by adding a revocation of certificates. Certificate revocation lists are stored at each MOCA. For certificates to be revoked, *t+1* MOCAs must sign a revocation certificate request with *t+1* partial signatures from the MOCAs. Once the partial signatures are gathered, the certificate revocation list is updated. Malicious nodes wanting to unnecessarily revoke another node's certificate can only do so with the approval of *t+1* trusted MOCAs, therefore ensuring the reputation of each node's certificate.

e. Analysis

This solution demonstrates some of the problems of an ad hoc network. Despite its obvious weaknesses, it is noted as one of the earliest key management solutions to ad hoc networks.

The partial distributive scheme proposed by Zhou and Haas requires that an off-line TTP member exists at the initialization phase in order to establish the distributive CA. The off-line TTP: generates the threshold private key; shares it among the appointed CA authority nodes; and distributes the CA's public key to all participating nodes in the network. All certificate related tasks including signatures, generation, distribution, refreshing and revocation, are performed by the participating nodes without the involvement of a TTP. The off-line TTP is not as involved in Verma [Verma et al, 2001] and Davis's [Davis, 2004] proposals. However, in spontaneous ad hoc networks such a trusted entity cannot be assumed at initialization.

The advantage of distributing the CA allows for the functionality of the CA to be distributed among the nodes. This avoids single point attacks and allows the computational overhead of the CA's services to be distributed. Although the CA is distributed, it still remains centralised between a few nodes.

The centralization of authority creates availability issues. The availability issues are sensitive as communicating nodes require communicating with *t* authority nodes before acquiring a signature. The CA's availability is dependent on the threshold parameters *t* and *n*. These parameters must be selected to provide a suitable trade-off between: availability; security; and cost of computation. The larger the threshold (*t*), the higher the security, but, the availability will pay the cost. The centralization of authority also results in a select group of nodes carrying the burden of security computations. This breaks the value of fair distribution in a network.

This solution requires that the CA authority nodes store all the certificates issued, which necessitates a costly synchronization mechanism. Furthermore, a share refreshing or proactive method is required. This is achieved by using a secret redistribution protocol [Desmendt & Jajodia, 1997]. With this in place, it is, therefore, certain that all the CA authority nodes are not compromised. The procedure of synchronization, updating and proactive refreshing is costly to resource constrained nodes.

Another potential problem is related to network participants addressing the CA authority nodes. A node requesting a service from the CA entity is required to contact *t* out of *n* nodes. The CA can then be given a multicast address and participating nodes can multicast their requests to the CA. The CA authority nodes can then unicast replies to the requesting participant. In ad hoc networks, which do not support multicasting, a participating node can broadcast its request. This approach is more common in mobile ad hoc networks, despite its potential of a large amount of network traffic.

Zhou and Haas's partially distributed certificate authority approach provides much of the groundwork for future solutions through the implementation of threshold cryptography in ad hoc networks.

3.5 Fully distributed certificate authority

The threshold scheme, investigated in [Luo & Lu, 2000] [Luo et al, 2002], uses ideas proposed by the partial distributive threshold scheme, found in [Lidong &Zygmunt, 1999]. *Luo* and *Lu* propose a scheme which embraces the distribution of the CA. In a network of *m* nodes, the network and security services are shared across *m* nodes. Therefore, a fully distributed system is realized, as seen in Figure 12. This scheme further differs from [Lidong &Zygmunt, 1999] in that there is no need to select specialized nodal authorities, as all nodes perform this role. Like the partial distributive scheme, the fully distributive scheme includes the use of share refreshing. This allows proactive security against significant nodes that are compromised. This scheme is designed for, and aimed at, long-term ad hoc networks which have the capacity to handle public key cryptography.

a. System overview

The Fully Distributive Certificate Authority scheme is a public key cryptography scheme. It takes the functionality of the certificate authority and distributes it across *m* nodes, where *m* is the total number of nodes in the network. This threshold scheme requires *k* or more nodes to act in collaboration to perform any operations of the CA. The CA's private key is divided and shared among all the participating nodes. This effectively enhances availability and allows nodes that are requesting the CA, to contact any *k* one-hop neighbour nodes. It is assumed that each node will have more than *k* one-hop neighbours [Luo & Lu, 2000]. Therefore, only one-hop certificate communication can occur. This allows for more reliable communication, in comparison with multi-hop communication. It is also easier to detect compromised nodes. Figure 12 illustrates the fully distributive network, where all nodes have a portion of authority in the form of a partial CA signature. Figure 12 shows a network with threshold $k=3$, where nodes *B*, *C* and *D* can find a coalition of partial CA nodes to form a group authentication CA signature. Node *A* is unable to find a sufficient coalition of nodes.

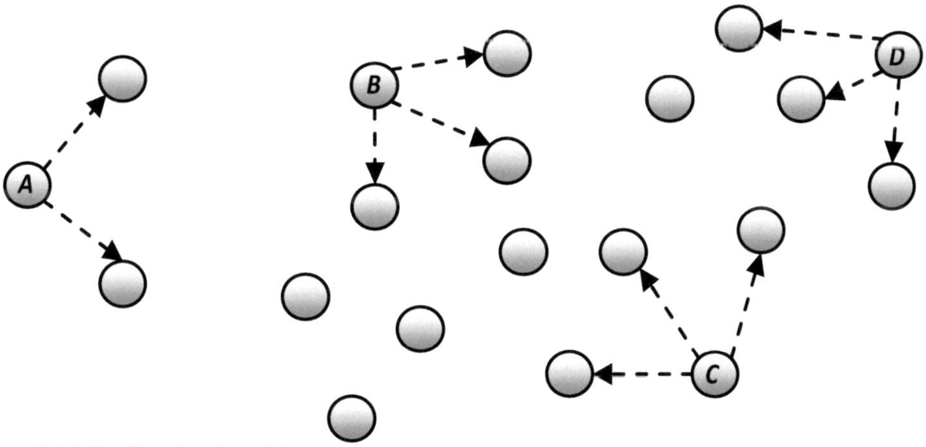

Fig. 12. Fully distributive CA system

b. Off-line Initialization

The initial phase of [Luo & Lu, 2000] [Luo et al, 2002] requires an off-line trusted third party (TTP) to establish the initial set of nodes. The off-line TTP will provide each node *i* with its own: certificate; the CA's public key; and a share of the CA's private key. A certificate is a binding between a nodes ID and its public key. The certificate is signed by CA's private key k_{CA} and can be verified by the CA's public key K_{CA} - which is made available to all the participating nodes. The off-line TTP initialises the threshold private key to the first *k* nodes by the following steps:

1. Generate the sharing polynomial $f(x) = a_0 + a_1x + ... + a_{k-1}x^{k-1}$ where $a_0 = k_{CA}$
2. Securely distribute node *i* identified by ID_i where $i \in k$ with its secret share $S_i = f(ID_i)$
3. Broadcast *k* public witnesses of the sharing polynomial's coefficients $\{h^{a_0}, \ldots, h^{a_{k-1}}\}$ and then the off-line TTP involvement is over.
4. Each node with ID_i that has received a secret share S_i verifies it by checking the sharing polynomial's coefficients such that $h^{S_i} = h^{a_0} \cdot (h^{a_0})^{ID_i} \cdot (h^{a_1})^{ID_i^2} \cdot \ldots \cdot (h^{a_{k-1}})^{ID_i^{k-1}}$.

After the initial establishment of the shared secret key amongst the first *k* nodes, the TTP is no longer responsible for the full distribution of the CA's private key. The off-line TTP maintains the responsibility of issuing new nodes with their initial certificates binding, and as a result impersonation attacks are prevented.

c. On-line Shared Initialization

New nodes entering the network need to be provided with their own share of the CA private key k_{CA} so that they can be part of the signing process. The participating nodes in the network perform this initialization process, without the interference of an off-line TTP. Shared initialization is modelled on Shamir's threshold secret sharing scheme [Shamir, 1979]. This scheme allows for a culmination of *t* nodes to initialize a joining node, with a share of the CA private key k_{CA}.

A node *i*, already initialized by the off-line authority, can generate a partial secret share $S_{p,i}$ for a joining node *p*. The combination of *k* partial secret shares results in node *p*'s secret share S_p. This is a partial share of the CA's private key.

$$S_p = \sum_{i=1}^{k} S_{p,i}$$

Node *i*'s secret share S_i can be derived from each partial secret share S_p, which is sent to node *p*. The joining node *p* must not be allowed to know the secret shares of other nodes, as this would breach confidentiality. The aim is to hide the actual partial secret shares $S_{p,I}$, while still transporting the combined secret share S_p to node *p*. A shuffling scheme is used to solve this problem. The shuffling scheme is illustrated in Figure 13. From Figure 13, nodes *i* and *j* wish to initialize node *p* with a secret share Sp. Nodes *i* and *j* agree upon a shuffling factor d_{ij}. The shuffling factor is combined with the partial secret shares $S_{p,i}$ and $S_{p,j}$. The sum of the shuffling factors is null. Therefore this allows for the secret share S_p to be calculated while hiding the secret shares of *i* and *j*. Figure 13 illustrates a system with a threshold of two nodes, to scale this to *k* nodes. Each pair of contributing nodes must decide on a shuffling factor resulting in *k(k-1)/2* shuffling factors which need to be distributed.

This key transport mechanism is described in the following steps:

1. Node *p* broadcast an initial request to a coalition of *k* neighbouring nodes.

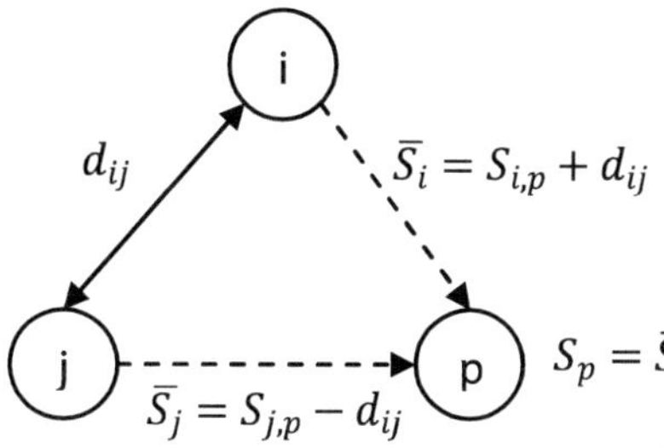

$$S_p = \bar{S_i} + \bar{S_j}$$
$$= S_{j,p} + d_{ij} + S_{j,p} - d_{ij}$$
$$= S_{i,p} + S_{j,p}$$

Fig. 13. Shuffling scheme of partial secret sharing

2. The coalition of nodes divides into i and j pairs and agree upon appropriate shuffling factors. An associated public witness $h^{d_{ij}}$ is generated and signed to identify any misbehaviour. The shuffling factor and the witnesses are sent to node p.
3. Node p routes all the shuffling factors and witnesses to the k coalition nodes.
4. Each coalition node j generates the partial secret share $S_{j,p}$ and shuffles it with the shuffling factors received by p such that $\overline{S_{J,p}} = S_{j,p} + \sum_{i=1}^{k} d_{ij}$ and sends $\overline{S_{J,p}}$ to p.
5. Node p verifies the shuffled share values $\overline{S_{J,p}}$ by checking the public witnesses that $h^{\overline{S_{J,p}}} = h^{S_p} \prod_{i=1}^{k} (h^{d_{ij}})$. If the verification is successful the shuffled share values are combines such that $S_p = \sum_{i=1}^{k} \overline{S_{p,\iota}}$.

After the joining node p has been issued with a part of the CA private key, it can perform the services of the CA in the network including certificate renewal and certificate revocation. System maintenance includes the initializing of joining nodes. System maintenance also encompasses the renewal of certificates, certificate revocation and proactive updating of the CA private key shares, therefore protecting against the CA's private key becoming compromised.

d. Share Updating

In a k threshold system, attacks can compromise k nodes over a period of time allow them to impersonate the CA and perform malicious communication attacks. A solution to this is secret share updates by the use of a proactive security method, similar to that used in partial distributed certificate authority methods.

The network will have an operation phase and an update phase where periodic updates will occur of the secret shares of the CA's private key will be updated. During the update phase all nodes participate in the updating procedure. Each node will have an equal probability of initiating the update phase, therefore fairly distributing the load. The secret share update phase following the following steps:

1. The node which is to initiate the update phase requests a coalition of k nodes and generates an update polynomial $f_{update}(x) = b_1 x + b_1 x^1 + \cdots + b_1 x^{k-1}$.
2. Each co-efficient of the polynomial is signed by the coalition CA and flooded through the network such that each node possesses the $f_{update}(x)$ polynomial.
3. Each node i generates its secret update share $\bar{S_\iota} = f_{update}(ID_i)$ and verifies it by a coalition of k nodes. Each node in the coalition returns a partial update to node i who

combines them to form its update share. This update share is added to the current share and a new updated share of the CA's private key is formed.

The share update procedure provides robust security against multi-point attacks but security comes at a high computational cost.

e. Certificate Renewal

Certificate issuing is assumed to be handled by the off-line TTP, which registers, initialises, and certifies new nodes joining the network. The issue of certificate renewal is performed by the distributed CA in the network. Each nodes certificate is only valid for a specified time period, after which they must renew the certificate before it expires. For successful certificate renewal in a k threshold fully distributive system, node i must request the renewal of certificate $Cert_i$ from a coalition of k nodes. One-hop neighbours are identified as more trust worthy coalition members. Each coalition node then generates a new partial signature and will send it to node i. Node i then act as a combiner (all nodes may act as combiners in the fully distributive certificate authority scheme) and combines the k partial signatures to produce the new certificate $\overline{Cert_i}$ [Luo &Lu, 2000]. In a similar manner, messages are signed by the coalition nodes and form a group signature as described in providing authenticity and security.

f. Certificate Revocation

Certificates can be revoked if nodes are found to be corrupt or compromised. This revocation service assumes that all nodes monitor their one-hop neighbour nodes and are capable of retaining their own certificate revocation list (CRL) [Luo & Lu, 2000]. When a user node identifies a neighbouring node is corrupt, it adds the node in question to its CRL and announces this to all neighbouring nodes. The neighbouring nodes in turn check if this announcement is from a reliable source, i.e. the source is not on the receivers CRL. If the source is reliable, the announced node is marked as suspect. If a threshold of k's reliable accusation is made against a single node then the node's certificate is revoked. This procedure allows for compromised nodes to be identified and explicitly quarantined from CA involvement, until such a time as they have become secure again. Implicit revocation is implemented by setting lifetimes for certificates t_{cert}. When the time has expired and the certificate has not been renewed it is implicitly revoked.

g. Analysis

This scheme is a hierarchical model. It is similar to the partially distributed certificate authority scheme. One can see that fully distributive networks possess similar weaknesses to partial distributive networks. Both schemes require prior knowledge and an off-line TTP for the initialization of certificates. The main advantages of the fully distributive scheme are its availability and implement revocation mechanism.

The fully distributive nature of the CA allows for high availability. It does require that each requesting node have k one-hop neighbours, which form a CA coalition. The localization of the coalition to the one-hop neighbours avoids transitive trust and reduces network traffic.

One can choose for the threshold parameter k to be larger, which will provide a higher level of security. This change requires an attacker to compromise a larger number of nodes in order to obtain the CA's private key. Increased security comes at the cost of availability. This scheme is non-scalable, as it lacks a mechanism that increases the threshold parameter k, dynamically, as the network density increases.

As the CA is distributed through the network its availability is greatly increased. However, an increase in availability of the CA requires a greater security and more focus upon the proactive share refreshing scheme. This scheme is a complex and computationally taxing maintenance protocol. It includes the share initialization and share update protocols. The trade-off between security and resources is an important issue in wireless ad hoc networks. The revocation mechanism allows for explicit and implicit revocation, while the assumption is made that all nodes are computationally capable of monitoring the behaviour of their one-hop neighbours. However, this assumption may not be true for certain ad hoc networks.

3.6 Cluster based model

This solution investigates the Secure Pebblenets [Basagni, 2001], which is a cluster or group based scheme. This solution uses symmetric key cryptography. It is a hierarchical distributive key management system. The focus of this scheme provides group authentication for user nodes, as well as message integrity and confidentiality. Group authentication is achieved by grouping nodes into clusters and treating them with blanket authentication. This solution is suited for planned, long-term distributed ad hoc networks. It is specifically aimed toward networks with low capacity nodes, which lack the resources to perform public key encryption.

a. System overview

This solution requires an initial infrastructure for setup. A secret group identity key k_G is set. This identity provides every node with authentication and integrity. Its key is kept constant for the duration of the network - unless an off-line authority re-initializes the network. k_G is used to generate further keys to provide message confidentiality [Basagni, 2001].

The life of the network is illustrated in Figure 14. The lifetime is divided into time slices, with three phases: the cluster generation phase; the operation phase; and the key update phase. Each time slice consists of these three phases. A network with low processing capacity nodes, authentication is complex and costly. Therefore authentication, confidentiality and integrity are provided for nodal groups or clusters. This maximizes efficiency and minimizes computational cost.

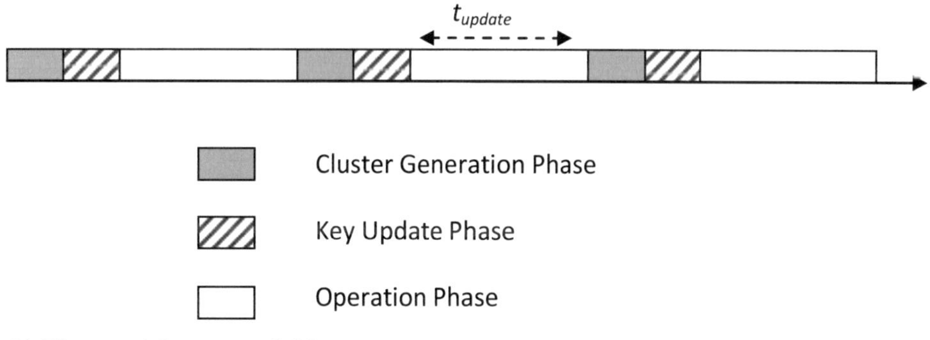

Fig. 14. Phases of the network lifetime

b. Cryptographic keying material

The network uses the following cryptographic keying material to provide message and group confidentiality and authentication:

1. Group identity key k_{GI} is shared prior to network establishment between all network nodes and is used to derive additional keys for security services.
2. Traffic encryption key k_{TEK} is used for symmetric data encryption and is updated during the network lifetime.
3. Cluster key k_C is used for cluster specific communication.
4. Backbone key k_B is used to encrypt communication between cluster heads.
5. Hello key k_H is used between neighbours in cluster generation phase.

The cluster key is generated by the cluster head. The k_{TEK} is randomly generated by the key manager, who is selected in the key update phase. The group identity key is used to derive the backbone and hello keys in the following manner:

$$k_B^0 = k_{GI}$$

$$k_H^i = h\left(k_B^{i-1}\right) = h^i(k_{GI})$$

$$k_B^i = h\left(k_H^{i-1}\right) = h^{i+1}(k_{GI})$$

where k^i represents the key in the i time slice and h^i represents a hash function to the order i. The three phases of operation use the described cryptographic keying material to provide cluster based security in a hierarchical manner.

c. Cluster Generation Phase

During the cluster generation phase, nodes decide to be either cluster heads or cluster members. This decision is based on a variable called weight [Basagni et al, 2001]. Node i's weight w_i is a representation of the node's current capacity status, which is made up of factors such as: battery power, and distance from other nodes etc. The cluster head will manage the group keying services for that cluster. The cluster heads then discover each other and establish a cluster head backbone, which is used to distribute updated traffic encryption key k_{TEK}.

The cluster generation phase follows the following three steps:

1. Nodes share their weights. Each node i calculates its weight w_i . It then broadcasts its *id* and w_i to its one-hop neighbours, and encrypts it with the hello key k_H . This provides confidentiality and, along with the group identity key, they provide authentication. The message is as follows.

$$E_{k_H}(w_i|id_i|E_{K_{GI}}(w_i|id_i))$$

2. After receiving the weighted messages from all its neighbours, node i will decide if it is a cluster head or cluster member. Once a role has been selected by node i it broadcasts its role to its neighbours in the following message.

$$E_{k_H}(w_i|id_i|role|E_{K_{GI}}(w_i|id_i|role))$$

   The *role* of node i is decided by its weight. The highest weighted node will broadcast a role of *ch*, cluster head, while other nodes will broadcast a role of id_j, where j is the identity of the cluster head that node i will belong to.
3. The cluster heads are then inter-connected. All cluster members inform their cluster head of any other cluster heads within a three hop radius. The network is effectively segmented and clusters are interconnected by a cluster head backbone, as illustrated in Figure 15.

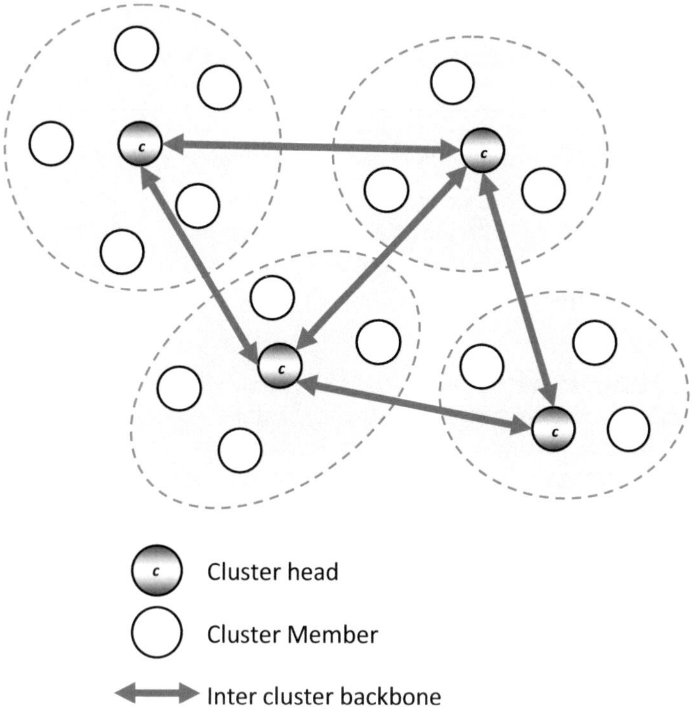

Fig. 15. Segmented network with cluster backbone

d. Operation Phase

During the operational phase, the nodes use the group identity key k_{GI} to authenticate nodes and provide message integrity. The traffic encryption key k_{TEK} is used to encrypt the application data and provide message confidentiality. These services are provided using the cryptographic functions of symmetric encryption algorithms and the one-way hash function [Basagni, 2001].

e. Key Update Phase

The traffic encryption key is updated periodically. This period is measured by an externally set parameter t_{update} (key update period). Updating occurs during the key update phase. Firstly, a key manager is selected from the pool of all the cluster heads. Selection is done by each cluster head, which checks if it is a potential key manager, by comparing its weight with the neighbouring cluster heads. Secondly, an exponential delay period, statistically averaged to Δ, is set aside, as to minimize the risk of multiple nodes becoming key managers [Basagni, 2001]. Thirdly, the cluster head with the highest weight value will arise as the selected key manager. The key managers purpose is to generate a new traffic encryption key k_{TEK} and then distribute this to all the cluster heads, effectively updating the traffic key (which provides message confidentiality). The new k_{TEK} is generated using a secure key generation algorithm. This new traffic key is distributed to the cluster heads securely using the backbone key k_B. The message sent to the cluster heads is:

$$E_{k_B}(w_c|id_c|\overline{k_{TEK}}|E_{K_{GI}}(w_c|id_c|\overline{k_{TEK}}))$$

Once the cluster heads have received the new traffic key this is distributed to the cluster members using the cluster key k_c , which is generated by the cluster head. The message sent to the cluster members is:

$$E_{k_C}(w_c|id_c|\overline{k_{TEK}}|E_{K_{GI}}(w_c|id_c|\overline{k_{TEK}}))$$

These three phases are repeated every network time-slice. The shorter this time-slice, the greater the security obtained. Similarly, this applies to the t_{update} period for the key update phase. However, in this case, it stands that the shorter the update period or time-slice, the more resources are required.

f. Analysis

This scheme is designed for large ad hoc networks, which are made up of nodes with limited processing power and storage capacity. Public key cryptography is unsuited for such a design, as this solution is realized through symmetric key cryptography. This solution requires a TTP to initialise the network nodes with the group identity key k_{GI} and set the parameters, such as the t_{update} time period.

The group identity key, which is distributed to all participating nodes, is required to remain secret throughout the lifetime of the network. In [Basagni, 2001] the authors of the Secure Pebblenets solution propose that nodes have tamper-resistant storage, which securely holds the group identity key. Standard network devices do not have such features and this limits its application for mobile ad hoc networks. If an attacker were to compromise the group identity key, all the nodes in the network would need to be re-initialized with a new group identity key, given by a TTP.

The clustering approach does benefit large ad hoc networks, as routing algorithms for long distances or large networks can become complex and expensive. Cluster based communication allows for packets travelling long distances to travel via the cluster backbone, until they reach their desired neighbourhood or cluster. From there the cluster head can transmit the packets more specifically. This approach reduces security computation and routing complexity in large networks.

A cluster head centralizes the authority in a network. In doing so, it provides a central point of attack for adversaries. Nodes within mobile ad hoc networks have unreliable characteristics because of their mobility and wireless sporadic connectivity. Selecting a reliable cluster head may become a problem in these dynamic networks. Nodes may also refuse to adopt the computational burden of being the cluster head. This is due to resource constraints inherent to mobile ad hoc networks.

Authentication is limited to groups to reduce computational requirements of nodes. It was found that if authentication was to be extended to the individual nodes, it would require the management of $n \times \frac{(n-1)}{2}$ symmetric keys [William, 1999]. Therefore, this solution is not feasible for peer-to-peer communication.

3.7 Proximity-based identification

Smetters et al [Smetters et al, 2002] proposed a solution called demonstrative identification. This solution allows nodes to establish initial trust relationships without prior knowledge or relationship and without the existence of an off-line TTP, which most key management systems assume. This solution uses close proximity channels to establish initial bootstrapping and provides a basis for more complex key establishment. Demonstrative identification approach is designed for spontaneous, small, localized short term ad hoc networks. An example of such a network can be seen in the gathering of people in a coffee shop, where each person wishes to establish temporary communication network, via their PDA's.

a. System Overview

Two nodes desiring to establish a secure communication link, initially engage across a location-limited channel. This channel is separate to the main communication channel, as displayed in Figure 16. Location-limited channels include: infrared; physical contact; and audio etc. Across the location-limited channel pre-authentication information is exchanged. For example, a user with a PDA who wants to communicate with a second user's PDA can use an infrared channel. They can direct the PDA's infrared device towards the second device and an exchange is made. The user can be assured that the pre-authentication information is from the chosen PDA, due to the nature and characteristics of infrared communication.

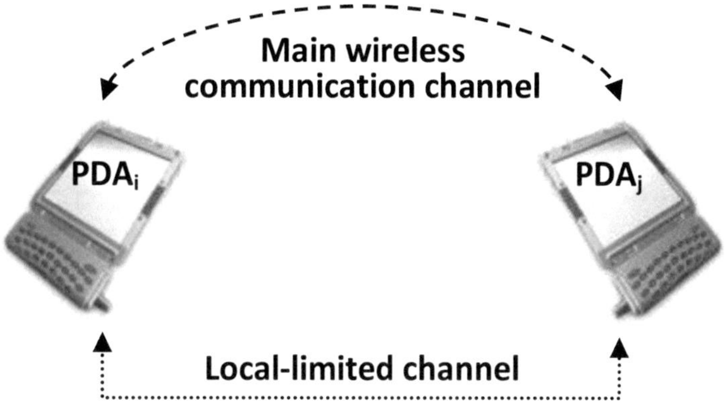

Fig. 16. Proximity based identification with location-limited channel

After the user has exchanged the pre-authentication information, a two-party (for example Diffie-Hellman) or group key exchange scheme can be implemented over the main communication channel. This is done in order to establish the keying material required for secure communication. A limited localized communication channel allows for communication without the existence of an off-line TTP or prior knowledge.

b. Two-Party Key Exchange

The key exchange between communication pair *i* and *j* is explained in the following steps:

1. Nodes *i* and *j* make close proximity contact with each other using a common location-limited channel.
2. Pre-authentication information is exchange across the common location-limited channel. Node *i* sends $h(K_i)$ to node *j* and *j* sends $h(K_j)$ to node *i*, where $h(K_j)$ is the irreversible one-way hash function of a node *j*'s public key.
3. Nodes *i* and *j* now exchange their public keys over the main channel such that *j* receives $\overline{K_i}$ and *i* receives $\overline{K_j}$. To avoid the impersonation attack which is common to mobile ad hoc networks, the public keys are then authenticated in step 4 using the pre-authentication information from step 2.

4. Authentication is checked using the one-way hash function h and verifies that $h(\overline{K}_i) = k(K_i)$ and $h(\overline{K}_j) = k(K_i)$.
5. Upon successful verification, any asymmetric key-exchange protocol can be implemented to allow for nodes i and j to share a secret key.

The two-party key-exchange described above is the basic formulae for demonstrative identification. This protocol can also be applied to heterogeneous nodes, where public key encryption is available to only one of the two communication members. This allows for nodes with limited complexity and computational capacity to participate in pair wise secret key exchange. The procedure for a two-party key exchange, where only one of the members (node i) is the public key competent, is described as follows:

1. Nodes i and j make contact on a location-limited channel, allowing i to send j, $h(K_i)$ and j to send i, $h(S_j)$, where $S_j$is a secret from j.
2. Node i sends j, $\overline{K}_i$over the main communication channel to realize authentication.
3. Node j authenticates node i's public key, K_i, by verifying that $h(\overline{K}_i) = h(K_i)$.
4. Upon successful authentication, node j sends $E_{K_i}(S_j)$ to i.
5. $E_{K_i}(\overline{S}_j)$ is decrypted at node i using K_i. $\overline{S}_j$ is then verified by checking that $h(S_j) = h(\overline{S}_j)$. Upon successful verification the two heterogeneous parties share a secret S_j , which can be used to establish secure communication keying material.

c. Analysis

This solution allows for a fully self-configured ad hoc network, as the initial trust establishment phase does not require the assistance of an off-line TTP. Users realize the initial trust relationship by localized communication. For example, a user with a PDA would point its PDA to another PDA to automatically exchange authentication information and establish a secure communication line.

This solution requires that nodes are equipped with location-limited communication devices. Examples of these devices are: infrared, audio or a wired link. This requirement limits the network participants to those possessing specific peripherals. The assumption is made that most portable wireless devices are equipped with some type of localized communication medium, such as infrared.

The location-limited pre-authentication exchange realizes demonstrative identification [Smetters et al, 2002]. It only allows key-exchange to occur in a localized manner, where nodes are in close proximity to each other. As a result, this solution is not suited to large networks, but it is best suited to small spontaneous networks. A solution presented by Capkun [Capkun et al, 2006] extends the self-issued certificate chaining approach as it implements a demonstrative identification approach in a PGP based network. Capkun's proposal uses location-limited communication to establish initial trust and relies upon mobility to distribute this trust in large networks. Such a proposal allows for demonstrative identification to be implemented in large to moderate networks.

More recently, the Amigo proximity-based authentication system proposed by Scannell et al [Scannell et al, 2009], uses shared radio environment evidences as proof of physical proximity to authenticate localized mobile communication nodes.

3.8 Self issued certificate chaining

A PGP-based security solution for ad hoc networks is proposed by Capkun and Hubaux [Capkun et al, 2003] [Hubaux et al, 2001]. This solution uses a certificate chaining approach.

It outlines a fully self-organized public key management system that allows users to: generate their public-private key pairs; issue certificates; and perform authentication, without the presence of an off-line trusted third party. Capkun and Hubaux focus on the key management and key distribution system. Without the need of prior relationships or an organizational TTP member, this solution is best suited to spontaneous ad hoc networks. However, due to its complex initialization phase it is not suited for small short-term networks.

a. System Overview

Public keys (*K*) and certificates are modelled as direct graphs *G(V,E)* where vertices, *V*, represent the public keys and the edges, *E*, represent a certificate between two vertices. The self-organized system proposed by Capkun and Hubaux [Capkun et al, 2003] [Hubaux et al, 2001] differs from PGP in that it relies on the users to store and distribute the certificates in a self-issued manner. Each user node carries a certificate memory, consisting of certificates limited to local neighbourhood. For a user to authenticate and certify another user's public key, a certificate chain is first found between the two users, by combining the users' certificate memory. Figure 17 illustrates a situation where node u and v request secure communication [Capkun et al, 2003]. Node u is required to verify the authenticity of the public key K_v for corresponding to node v. To do so nodes u and v combine their certificate memories to find a certificate chain or path between K_u and K_v , which is made up of valid public key certificates shared between the two communicating nodes.

The fully self-organized public key management system can be broken into four procedures of analysis, as follows:

- Public/private key creation
- Certificate exchange
- Authentication
- Certificate revocation
- Load sharing

During the initialization phase, the public-private keys are created and distributed with a certificate exchange procedure. Secure communication is realized and impersonation attacks are thwarted by the authentication of the available certificates. The certificate revocation protocol is outlined in order to maintain security and exclude malicious users. Optimization is implemented by a load sharing protocol that ensures fair distribution of the work load and prevents selfish nodes in a network.

Initialization phase is executed in a four step procedure which establishes trust in the network:

1. The user creates their own public/private key pair
2. The user issues public key certificates (vertices) based on the knowledge of the other public keys.
3. The user performs certificate exchange and collecting certificates, and creates a non-updated certificate repository.
4. The user constructs an updated certificate repository, modelled as a graph G_u. This is done by communicating with certificate graph neighbours or by a second method of applying the repository construction algorithm to the non-updated certificate repository.

After initialization is complete, authentication between two users can take place, through certificate chaining. Each step is explained in more detail below.

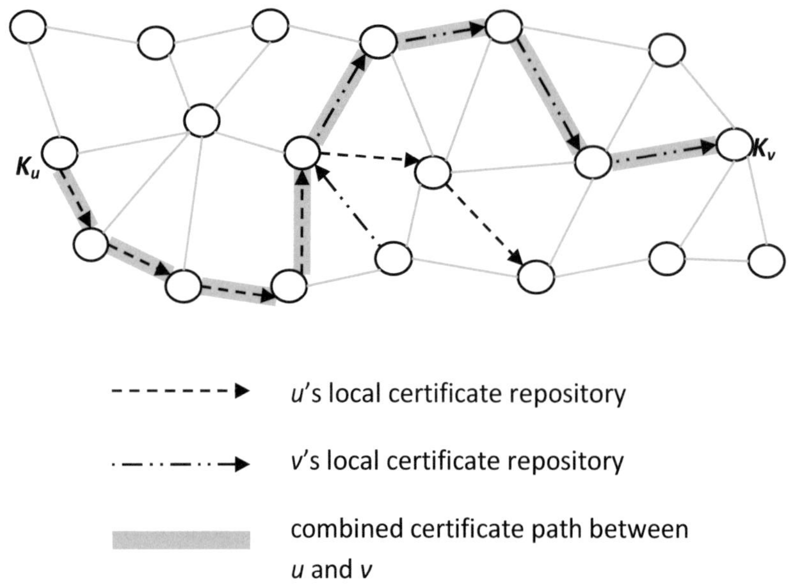

Fig. 17. A certificate chain or path between public keys K_u and K_v

b. Public/private Key Creation

Public and private keys for users are created locally. Public key certificates are issued by the user. If the user *u* believes that a public key K_v belongs to *v*, then the user *u* can issue a public key binding K_v to user *v*, by the signature of *u*. This certificate has an expiry time T_v. A periodic update may be issued which simply extends expiry time T_v. The reason for trust is not identified but assumed, for example through a physical side channel.

c. Certificate exchange

The certificate repositories are created automatically by exchanging certificates. A user *u* has two certificate repositories: an update certificate repository G_u and a non-updated certificate repository G_u^N. All certificates are stored twice, as when a certificate is issued, it is stored in both the certificate issuer *u* and certificate owner *v*'s repository. Therefore, initially each certificate repository has only the certificates it has issued and those that have been issued to it. Certificates are exchange periodically. Each node periodically polls its physical neighbour for certificates.

A certificate exchange is performed by the following procedure:

1. Node *u* broadcasts G_u and G_u^N to its physical neighbours. The broadcast contains only identities (hash values).
2. Neighbours reply with identities of their update repository G and non-update repository G^N.
3. Node *u* crosschecks the received sub-graphs and its sub-graphs for any additions.
4. Node *u* requests those certificates it does not hold.

After the initial convergence phase, all the certificates of the nodes are stored by all users. As a result, users' non-update repositories are created. After this phase the nodes exchange only new certificates at a rate of T_{CE}, which represents the time for a certificate to be exchanged throughout the network. Note that certificate expiration times are not considered thus far.

d. Construction of updated certificate repositories

The exchange of certificates provides an incomplete view of the graph and allows each node to create its own non-updated certificate repository. The updated repository G_u will consist of certificates which user *u* keeps updated. There are two approaches in this creation:

1. Apply algorithm *A* to G_u^N which results in G_u, and validity of each certificate is checked.
2. Communicate with certificate graph neighbours only.

The maximum degree algorithm is an algorithm *A* proposed by [Capkun et al, 2003] which is applied to the non-update repository G_u^N to create the update repository G_u in [Capkun et al, 2003] [Hubaux et al, 2001]. The algorithm selects a sub-graph that consists of two logically distinct paths: the out-bound path and the in-bound path, which are made up of outgoing edges and incoming edges, respectively. The selection of G_u's out-bound path is done in multiply rounds in the following manner [Capkun et al, 2003] [Hubaux et al, 2001]:

1. Each round runs from vertex K_{vert}, starting with vertex K_u.
2. User *u* requests the outgoing edge list of vertex K_{vert}. This is possible as every vertex stores this list locally.
3. An outgoing edge (with its terminating vertex *z*) is selected from the list in 2. Selection is based on the highest number of shortcuts of the terminating vertex *z*. Where a shortcut is defined as an edge, and removed, the shortest indirect path between the nodes, previously connected by that edge, becomes larger than two. User *u* can determine its number of shortcuts by gathering information about the outgoing and incoming edges of its adjacent users.
4. The selected vertex *z* is added to a set N_{out} of vertices selected, thus far. This is done to ensure that the selected out-bound paths are disjointed.
5. The round is finished and now the terminating vertex *z* becomes K_{vert} and a new round begins, starting from step 1.

The in-bound path selection is done in a similar way:

1. Each round runs from vertex K_{vert} , starting with vertex K_u.
2. User *u* requests the incoming edge list of vertex K_{vert}. Every vertex stores this list locally. Therefore, this step requires that each user be notified whenever another user issues a certificate to that user.
3. An incoming edge (with its originating vertex *y*) is selected from the list in 2. Selection is based on the highest number of shortcuts of the originating vertex *y*.
4. The selected vertex *y* is added to a set N_{in} of vertices selected so far, to ensure that the selected in-bound paths are disjointed.
5. The round is finish and now the originating vertex *y* becomes K_{vert} and a new round begins, starting from step 1.

The update repository is the union of the in-bound sub-graph and out-bound sub-graph. The pure method will operate on a single round. However, it is extended so the update repository consists of several vertex disjoint out-bound and vertex disjoint in-bound paths. The final sub-graph is star-like information.

e. Authentication

When initialization is complete, the user is prepared to perform authentication. Authentication is preformed between users *u* and *v* with public keys K_u and K_v respectively, as follows:

Firstly, user *u* and user *v* merge their update certificate repository (G_u and G_v) to find a certificate chain between *u* and *v*. User *u* then looks for a path in G_u and G_v. Validity and correctness checks are done to all certificates in the discovered path. Validity, checks that the certificates are not revoked. Correctness, checks the certificates contain the correct user-key bindings.

If no certificate chain is found, user u combines its two repositories of the updated and non-updated certificates to find a chain. User u searches for a path in G_u and G_u^N. If a chain is found, then u requests the updates of the expired certificates. Subsequently, the validity and correctness checks are made.

If there is still no certificate chain found between K_u and K_v then authentication is aborted. During authentication nodes that are one-hop physical neighbours (also known as helper nodes) are given precedence as to maximize performance. When a path is found, the certificates (edges) along this path are then used by user u to authenticate K_v.

f. Certificate revocation

Certificates are revoked when it is believed that the user-key binding is no longer valid. If a user believes his own private key is compromised then he can revoke his public key certificate binding. This is done in two ways, explicitly and implicitly:

1. Explicitly, a user u would revoke a certificate issued by u, by broadcasting a revoke statement broadcast to its G_u nodes. The certificate exchange scheme allows for this revoke to reach all other nodes at a time delay of T_{CE}.
2. Implicit revocation is based on the expiration of certificates. Certificates are valid for a given time T_v after which they must be updated.

This allows for comprised certificates and private keys, to be dealt with explicitly, and provides a higher level of confidence by implicitly maintaining validity.

The fully distributive nature of this scheme means every certificate is stored at each node allowing for nodes to cross-check conflict and detects inconsistent certificates.

To combat false certificate bindings the following two procedures are taken:

1. If a certificate is received which doesn't exist in G_u or G_u^N then it and the issuer are labelled *unspecified* until a period T_p where $T_p > T_{CE}$ where after if no conflicting certificates are received then it is marked *non-conflicting*. This does not prevent against Sybil attacks though.
2. If a certificate conflict is found where a user u has two certificate bindings (v,K_v) and (v,K'_v). Both certificates and the certificates that certified them are labelled as *conflicting*. To resolve such a conflict, validity of certificates is first checked with their issuers. If validity status remains true, then u will try to find chains of non-conflicting valid certificates to public keys K_v and K'_v. Confidence values are calculated based on the number and length of chains, and values compared to compute the correctness of the bindings. If no decision is made these bindings are labelled as *conflicting* and the node waits for more information to resolve the conflict.

In this case, a confidence algorithm is not identified but assumed. This conflict resolution mechanism can be further used: to evaluate trust in users; to issue correct certificates; and to detect malicious users.

g. Load Sharing

For an update to occur nodes contact the issuer of the certificates that they store. This approach is not efficient because one certificate issuer could be overloaded and unable to handle the computational work load. Simple load sharing is implemented which allows for relief. Each node u provides updates to up to s other nodes, where s is equal to size of u's updated repository. After which node u has provided s updates, it replies to update requests with a list of nodes that get updates directly from u. The requesting node then randomly selects a node from u's list and requests its update from that node.

h. Analysis

The self organized, self certificate issuing trust model is a web of trust type model inheriting PGP characteristics and applying them to an ad hoc network environment. In a similar way that PGP [Zhou & Hass, 1999] realizes trust, the certificate chaining approach is used to create chains of hierarchical trust between users. The main difference between PGP and the certificate chaining solution is that PGP stores certificates in a centralized manner, and this scheme decentralizes this procedure through local certificate repositories.

The main advantage of this scheme is that it is fully self-organized and does not require the presence of a TTP. Trust is established in a self-organised manner with self-certificate being issued by the nodes themselves. The initial phase requires nodes to interact and establish trust. Trust relationship can take time to establish. Therefore, in the early stages of the network, an initial time delay can be expected limiting the effectiveness of communication. For this reason, this network is not suited for short term mobile ad hoc network. An example of this shortcoming is illustrated in Figure 18, where node A wants to communicate with node B. At the early stage of the network only D and C have issued certificates and as a result no certificate chain exists between A and B. Only once the intermediate nodes have issued certificates will a certificate chain between A and B be possible.

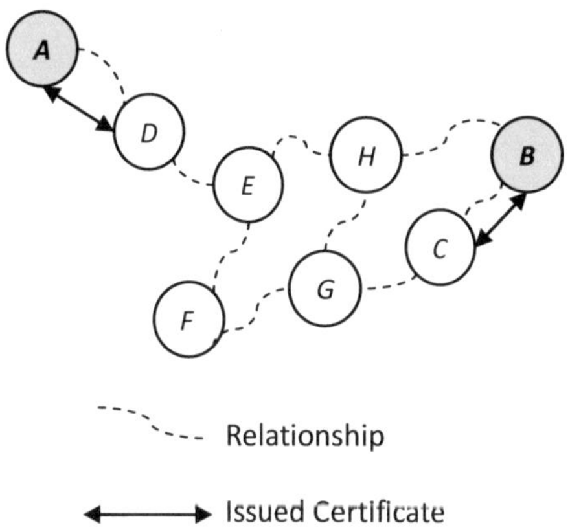

Fig. 18. Initial phase delay problem

The use of certificate chains is identified as vulnerable, because a chain of trust is 'only as strong as its weakest link'. A PGP hierarchical trust model is adopted that assumes transitive trust. This web-of-trust based approach allows for more flexibility than the other certificate approaches. However, a no central administration is present to enforce policy and trust assessment. Therefore, because of this lack of structure, it is more prone to attacks by malicious nodes. This solution is best suited to open mobile ad hoc networks, but may not be suited to applications where high degrees of security is required [Davis, 2004], like closed military mobile ad hoc networks.

This self-organized scheme is fully distributive which would result in a certificate updated to be computationally taxing. Certificate update repositories and load sharing relieve this expense. However, a better load balancing data management schemes can be introduced to further relieve the load [Hubaux et al, 2001].

The maximum degree algorithm *A* (or Shortcut Hunter Algorithm) is implemented to maximise effectiveness and optimise the update procedure. This proposal has been tested on PGP trust graphs. Nevertheless, an ad hoc network does not have the privilege of every node having public knowledge of all the certificates available. Step 3 of the maximum degree algorithm requires that an edge is selected from *vert* to *z* , where *z* is the vertex with the highest number of shortcuts. To determine *z* knowledge of the surrounding trust graph is required, which may not be available to all ad hoc network members.

One of the main disadvantages of a fully self-organized model is that nodes can adopt as many identities as they have resources, in order to support further steps which need to be taken to protect this solution from Sybil and impersonation type attacks [Capkun et al, 2003].

3.9 Discussion and summary

The solutions presented in this section give a summary of the work related to key management in mobile ad hoc networks. The solutions differ considerably in requirements, complexity and functionality. Each solution is suited for different types of ad hoc network environments. Criteria which these key management solutions can be grouped or differentiated included:

- Pre-configuration: *Planned vs Spontaneous*

This describes the pre-requisites and assumptions that are made for the nodes participating or joining the network. If an ad hoc network is planned then nodes can be assumed to have some pre-configured information, for example: initial shared secret; certificate; or authenticated identification. If the network is spontaneous then nodes have no prior security relationships or initial data assumptions. Pure ad hoc networks are more spontaneous allowing for nodes to join and leave the network without complex pre-configurations and assumptions made.

- Network Area: *Local vs Distributive*

This describes the area or space in which the key management scheme is operating. The physical topology of the network would result in more close proximity interaction or more multi-hop distributive interaction. A localized area is a network in which nodes come within a close proximity range of each other, such as in a classroom. A distributive area is a network where nodes are located some distance apart with little possibility of physical interaction. Certain key management schemes do not function in a distributive network area.

- Network Duration: *Short Term vs Long Term*

The duration of the network can dictate the initialization period of the key management scheme. For short term ad hoc networks, a group of nodes establish communication for a short time period and may never come into contact again. Short term ad hoc networks require speedy initialization and require communication to be available at the start of the network, without an initial period of weakened or delayed secure communication. Long term ad hoc networks consist of nodes that plan to be part of a network and in relationship with other nodes for a longer time period. Furthermore, nodes retain information and relationships with other nodes even when they leave the network. Long term ad hoc networks require more complex trust establishment.

- Off-line TTP Involvement

Ad hoc networks are characterized by their lack of infrastructure. Key management scheme often rely on an off-line trusted third party (TTP) for initialization and operational security. The extent of the off-line TTP involvement describes the self-organized nature of the network. Ideally, an ad hoc network has no off-line TTP involvement at the initialization or operational stages.

A summary of the presented key management solutions given in Table-1 with respects to the criteria discussed above. The off-line TTP model relies on an external TTP to establish and maintain security. This model is suited for networks which have available fixed infrastructure and will therefore have limited mobility. The partially and fully distributive CA solutions are similar using threshold cryptography, as they distribute the hierarchical trust of a certificate authority. They are suited to large planned ad hoc networks like military battlefield networks or disaster area networks. The Secure Pebblenet scheme is a cluster based model which is ideal for hierarchical group-oriented ad hoc networks where all nodes are distributed in a large network area and nodes have limited resources. An application of this cluster based approach is sensor networks.

The Self-Issued Certificate model or certificate chaining model uses a localized PGP web of trust approach. Its self-organized nature makes this solution most suited to spontaneous networks, such as peer-to-peer communication in a classroom or coffee shop. The proximity-based identification solution is suited to localized networks. Its greatest advantage is that it requires no prior knowledge to establish trust. The proximity-based identification method is, used in Capkun's mobility based approach, uses mobility of nodes to establish initial trust relationships across a large network.

This section shows that many of the solutions presented have issues which need to be resolved. Key management is an integral part of providing security and, as identified in Section-1, the routing layer is the focus of attack for adversaries. If these MANETS are to be recognized as secure, then mobile ad hoc network's security mechanism must strive to provide security on the routing and application layer.

	Pre-Configuration	*Network Duration*	*Network Area*	*Off-line TTP Involvement*
Off-line TTP Model	Planned	Long-term	Distributive	Full
Partially Distributed CA	Planned	Long-term	Distributive	Initialization
Fully Distributed CA	Planned	Long-term	Distributive	Initialization
Self Issued Certificates	Spontaneous	Long-term	Distributive	None
Cluster based Model	Planned	Long-term	Distributive	Initialization
Proximity-base Identification	Spontaneous	Short-term	Localized	None

Table 1. Summary of Key Management Solutions

4. References

[Abdul-Rahman, 1997] A. Abdul-Rahman, "The PGP trust model," *EDI-Forum: The Journal of Electronic Commerce,* vol. 10, pp. 27-31, 1997.

[Aram et al, 2003] K. Aram, K. Jonathan, and A. A. William, "Toward Secure Key Distribution in Truly Ad-Hoc Networks," in *Proceedings of the 2003 Symposium on Applications and the Internet Workshops (SAINT'03 Workshops)*: IEEE Computer Society, 2003.

[Awerbuch et al, 2002] B. Awerbuch, D. Holmer, C. Nita-Rotaru, and H. Rubens, "An on-demand secure routing protocol resilient to byzantine failures," in *Proceedings of the 1st ACM workshop on Wireless security* Atlanta, GA, USA: ACM, 2002.

[Basagni et al, 2001] S. Basagni, K. Herrin, D. Bruschi, and E. Rosti, "Secure pebblenets," in *Proceedings of the 2nd ACM international symposium on Mobile ad hoc networking \ & computing* Long Beach, CA, USA: ACM, 2001.

[Bruce, 2003] S. Bruce, *Beyond Fear: Thinking Sensibly about Security in an Uncertain World*: Springer-Verlag New York, Inc., 2003.

[Capkun et al., 2003] S. Capkun, L. Butty, and J.-P. Hubaux, "Self-Organized Public-Key Management for Mobile Ad Hoc Networks," *IEEE Transactions on Mobile Computing,* vol. 2, pp. 52-64, 2003.

[Capkun et al, 2006] S. Capkun, L. Buttyan, and J.-P. Hubaux, "Mobility Helps Peer-to-Peer Security," *IEEE Transactions on Mobile Computing,* vol. 5, pp. 43-51, 2006.

[Chor et al, 1985] B. Chor, S. Goldwasser, S. Micali, and B. Awerbuch, "Verifiable secret sharing and achieving simultaneity in the presence of faults (extended abstract)," *proc. 26th IEEE Annual Symposium on Foundations of Computer Science,* October, 21-23 1985.

[Davis, 2004] C. R. Davis, "A localized trust management scheme for ad hoc networks. ," *In: 3rd International Conference on Networking (ICN'04),* pp. 671–675, 2004.

[Desmendt & Jajodia, 1997] Y. Desmedt and S. Jajodia, "Redistributing Secret Shares to New Access Structures and Its Applications," Department of Information and Software Engineering, School of Information Technology and Engineering, George Mason University, Technical ReportJuly 1997.

[Douceur, 2002] J. R. Douceur, "The Sybil Attack," in *Revised Papers from the First International Workshop on Peer-to-Peer Systems*: Springer-Verlag, 2002.

[Eschenauer & Gligor, 2002] L. Eschenauer and V. D. Gligor, "A Key-Management Scheme for Distributed Sensor Networks," *proc. 9th ACM Conf. on Computer and Communication Security (ACM CCS'02),* November, 17-21 2002.

[Frankel et al, 1997] Y. Frankel, P. Gemmell, D. MacKenzie, and M. Yung, "Optimal resilience proactive public key cryptosystems," *proc. 38th Annual Symposium on Foundations of Computer Science (FOCS '97),* October, 19-22 1997.

[Haas et al, 2002] Haas J.D.Z., Liang B., P. Papadimitatos and S. Sajama, "Wireless ad hoc networks," in *Encyclopedia of Telecommunications* J. W. John Proakis, Ed., 2002.

[Hashmi & Brooke, 2008] Hashmi S. and J. Brooke, "Authentication Mechanisms for Mobile Ad-Hoc Networks and Resistance to Sybil Attack," in *Proceedings of the 2008 Second International Conference on Emerging Security Information, Systems and Technologies - Volume 00*: IEEE Computer Society, 2008.

[Herzberg et al, 1995] Herzberg A., S. Jaracki, H. Krawczyk, and M. Yung, "Proactive Secret Sharing Or: How to Cope With Perpetual Leakage," *proc. Advances in Cryptology - CRYPTO '95,* 1995.

[Herzberg et al, 1997] Herzberg A., M. Jakobsson, S. Jarecki, H. Krawczyk, and M. Yung, "Proactive Public Key and Signature Systems," *proc. 4th ACM Conf. on Computer and communications security,* April, 1-4 1997.

[Hu et al, 2003a] Hu Y.C., A. Perrig, and D. B. Johnson, "Rushing attacks and defense in wireless ad hoc network routing protocols," in *Proceedings of the 2nd ACM workshop on Wireless security* San Diego, CA, USA: ACM, 2003.

[Hubaux et al, 2001] Hubaux J.-P., L. Butty, and S. Capkun, "The quest for security in mobile ad hoc networks," in *Proceedings of the 2nd ACM international symposium on Mobile ad hoc networking \& computing* Long Beach, CA, USA: ACM, 2001.

[Jarecki, 1995] Jarecki S., "Proactive secret sharing and public key cryptosystems," Massachusetts Institute of Technology (MIT), 1995.

[Jameson, 2008] Jameson H., "Secure Military Networks: The war without weapons," 2008.

[Karl & Rauscher, 2001] Karl W.C., F. Rauscher, "Wireless Emergency Rescue Team (WRET) Final Report for the September 11, 2001 New York City World Trade Center Terrorist Attack," 2001.

[Lidong & Zygmunt, 1999] Lidong Z. and H. Zygmunt, "Securing Ad Hoc Networks," Cornell University1999.

[Luo & Lu, 2000] Luo H. and S. Lu, "Ubiquitous and robust authentication services for ad hoc wireless networks," Computer Science Department, University of California, Technical ReportOctober 2000.

[Luo et al, 2002] Luo H., P. Zerfos, J. Kong, S. Lu, and L. Zhang, "Self-securing Ad Hoc Wireless Networks," *proc. Seventh International Symposium on Computers and Communications (ISCC'02),* July 1-4 2002.

[Menezes et al, 1996b] Menezes A.J., S. A. Vanstone, and P. C. V. Oorschot, *Handbook of Applied Cryptography*: CRC Press, Inc., 1996.

[Molva & Michardi, 2003] Molva R. and P. Michiardi, "A Game Theoretical Approach to Evaluate Cooperation Enforcement Mechanisms in Mobile Ad hoc Networks (extended abstract)," in *proc. Modeling and Optimization in Mobile, Ad Hoc and Wireless Networks (WiOpt'03)*, 2003.

[Papadimitratos & Hass, 2003] Papadimitratos P. and Z. J. Haas, "Secure Link State Routing for Mobile Ad Hoc Networks," in *Proceedings of the 2003 Symposium on Applications and the Internet Workshops (SAINT'03 Workshops)*: IEEE Computer Society, 2003.

[Perkins & Bhagwat, 1994] Perkins C.E. and P. Bhagwat, "Highly dynamic Destination-Sequenced Distance-Vector routing (DSDV) for mobile computers," *SIGCOMM Comput. Commun. Rev.*, vol. 24, pp. 234-244, 1994.

[Perkins et al, 2003] Perkins C., E. Belding-Royer, and S. Das, *Ad hoc On-Demand Distance Vector (AODV) Routing*: RFC Editor, 2003.

[Qian & Li, 2007] Qian L., N. Song, and X. Li, "Detection of wormhole attacks in multi-path routed wireless ad hoc networks: a statistical analysis approach," *J. Netw. Comput. Appl.*, vol. 30, pp. 308-330, 2007.

[Raya & Hubaux, 2005] Raya M. and J. P. Hubaux, "The Security of Vehicular Ad Hoc Networks," in *proc. ACM Workshop on Security of Ad Hoc and Sensor Networks (SASN'05)*, 2005.

[Salem et al, 2005] Salem N.B., L. Buttyan, J.-P. Hubaux, and M. Jakobsson, "Node Cooperation in Hybrid Ad Hoc Networks," *IEEE Transactions on Mobile Computing,* 2005.

[Scannell et al, 2009] Scannell A., A. Varshavsky, A. LaMarca, and E. D. Lara, "Proximity-based authentication of mobile devices," *Int. J. Secur. Netw.,* vol. 4, pp. 4-16, 2009.

[Shamir, 1979] Shamir A., "How to share a secret," *Communications of the ACM,* vol. 22, pp. 612-613, 1979.

[Smetters et al, 2002] Smetters D.B., D. Balfanz, D. K. Smetters, P. Stewart, and H. C. Wong, "Talking To Strangers: Authentication in Ad-Hoc Wireless Networks," 2002.

[Stalling, 2002] Stallings W., *Cryptography and Network Security: Principles and Practice*: Pearson Education, 2002.

[Stalling, 2003] Stallings W., *Cryptography and Network Security: Principles and Practices*: Prentice Hall, 2003.

[Steiner et al, 1996] Steiner M., G. Tsudik, and M. Waidner, "Diffie-Hellman Key Distribution Extended to Groups," in *proc. Third ACM Conf. on Computer and Communication Security,* 1996.

[Talzi et al, 2007] Talzi I., A. Hasler, S. Gruber, and C. Tschudin, "PermaSense: investigating permafrost with a WSN in the Swiss Alps," in *Proceedings of the 4th workshop on Embedded networked sensors* Cork, Ireland: ACM, 2007.

[Tseng et al, 2003] Tseng C.Y., P. Balasubramanyam, C. Ko, R. Limprasittiporn, J. Rowe, and K. Levitt, "A specification-based intrusion detection system for AODV," in *Proceedings of the 1st ACM workshop on Security of ad hoc and sensor networks* Fairfax, Virginia: ACM, 2003.

[Van der Merwe &Dawoud, 2004] Van der Merwe J., D. Dawoud, and S. McDonald, "A Proactively Secure Threshold-multisignature Scheme based on Publicly Verifiable Distributed Key Generation and Publicly Verifiable Secret Redistribution," *IEEE Transactions on Parallel and Distributed Systems,* 2004.

[Van der Merwe &Dawoud, 2005] Van der Merwe J., D. Dawoud, and S. McDonald, "Fully Self-Organized Peer-to-Peer Key Management for Mobile Ad Hoc Networks," *proc. ACM Workshop on Wireless Security (WiSe'05),* September, 2 2005.

[Verma et al, 2001] Verma R., D. O'Mahony, and H. Tewari, "NTM- Progressive Trust Negotiation in Ad Hoc Networks," 2001.

[William, 1999] William S., *Cryptography and network security (2nd ed.): principles and practice*: Prentice-Hall, Inc., 1999.

[Yi & Kravets, 2001] Yi S. and R. Kravets, "Practical PKI for Ad Hoc Wireless Networks," Department of Computer Science, University of Illinois, Technical ReportAugust 2001.

[Yi & Kravets, 2003] Yi S. and R. Kravets, "MOCA: Mobile certificate authority for wireless ad hoc networks," in *proc. of the 2nd Annual PKI Research Workshop (PKI 2003),* 2003.

[Zhou & Hass, 1999] Zhou L. and Haas Z.J., "Securing Ad Hoc Networks," *IEEE Network: special issue on network security,* vol. 13, pp. 24-30, 1999.

9

Grouping-Enabled and Privacy-Enhancing Communications Schemes for VANETs

T.W. Chim[1], S.M. Yiu, Lucas C.K. Hui[1] and Victor O.K. Li[2]
[1]*Department of Computer Science, The University of Hong Kong*
[2]*Department of Electrical and Electronic Engineering, The University of Hong Kong*
Hong Kong, China

1. Introduction

A vehicular ad hoc network (VANET) is also known as a vehicular sensor network [Zhang, Lu, Lin, Ho & Shen (2008)] by which driving safety is enhanced through inter-vehicle communications or communications with roadside infrastructure. It is an important element of the Intelligent Transportation Systems (ITSs) [Wang et al. (2006)]. In a typical VANET, each vehicle is assumed to have an on-board unit (OBU) and there are road-side units (RSU) installed along the roads. A trusted authority (TA) and maybe some other application servers are installed in the backend. The OBUs and RSUs communicate using the Dedicated Short Range Communications (DSRC) protocol [Oh et al. (1999)] over the wireless channel while the RSUs, TA, and the application servers communicate using a secure fixed network (e.g. the Internet). Based on this infrastructure, vehicles can broadcast safety messages (e.g. road condition, traffic accident information), referred to as "ad hoc messages", to other nearby vehicles and RSU such that other vehicles may adjust their travelling routes and RSU may inform the traffic control center to adjust traffic lights for avoiding possible traffic congestion. Like other communication networks, security issues have to be well addressed. For example, the message from an OBU has to be integrity-checked and authenticated. Otherwise, an attacker can replace the safety message from a vehicle or even impersonate a vehicle to transmit a fake safety message. For example, an attacker may impersonate an ambulance to request other vehicles to give way to it or request nearby RSUs to change traffic lights to green. Besides, privacy is another important issue. A driver may not want others to know its driving routes by tracing messages sent by its OBU. Thus an anonymous communications protocol is needed. While being anonymous, a vehicle's real identity should be revealable by a trusted party when necessary. For example, the driver who sends out fake messages causing an accident should not be able to evade responsibility by using an anonymous identity.

In terms of integrity-checking and authentication, digital signature in conventional public key infrastructure (PKI) [Housley et al. (1999)] is a well accepted choice. However, requiring a vehicle to verify the signatures of other vehicles by itself using such schemes as in [Tsang & Smith (2008)] induces two problems as mentioned in [Zhang, Lin, Lu & Ho (2008)]. First, the computation power of an OBU is not adequate to handle all verifications in a short time, especially in places where the traffic density is high. Second, to verify a message from an unknown vehicle involves the transmission of a public key certificate which causes heavy message overhead. Therefore, the general approach is to let the nearby RSU help a vehicle verify the message of another. The volume of signatures to be verified can

be huge (each vehicle is expected to broadcast a safety message every few hundred ms [U.S. Department of Transportation (2006)]). An efficient method for verifying a batch of signatures within a short period of time is desirable.
Another motivation of our work is the observation that VANET can provide a platform for a group of known vehicles (e.g. police chasing a bank robber) to establish a secure communication channel (group communications). Since communication is through a wireless channel and is more vulnerable to attacks and member vehicles would leave the group and join the group again (e.g. at junctions) rather frequently, it is desirable to have an efficient frequent group key update procedure to accommodate dynamic membership in such a group communications scheme.
To conclude, besides security and privacy requirements, an ad hoc communications protocol for VANETs should have low message overhead and efficient message verification mechanism while having high success rate and low delay. The group communications protocol should be efficient for dynamic membership as well as frequent group key update. Existing solutions cannot satisfy all these requirements. Section 2 describes related work and their limitations.
In this chapter, we propose a Grouping-enabled and Privacy-enhancing communications Scheme (or GPS in short form) for VANETs. Our schemes can handle "ad hoc messages" (those sent by arbitrary vehicles) as well as allow vehicles that know one another in advance to form a group and send "group messages" securely among themselves. In summary, our schemes have the following features:
1) Our schemes are software-based and do not rely on any special hardware.
2) By establishing shared secrets with RSU and TA on the handshaking phase, a vehicle is allowed to use a different pseudo identity for each session (or message) to protect its privacy while the real identity is traceable only by TA.
3) We make use of the techniques of binary search in RSU message verification phase and bloom filter in notification messages to reduce the message overhead substantially and enhance the effectiveness of the verification phase.
4) Any vehicle can form a group with other vehicles after an initial handshaking phase with a nearby RSU and then authenticate and communicate with others (either to all members or to a dedicated member) securely without the intervention of RSU even after moving into the region of another RSU.
5) We support dynamic membership in a group. When a new member wants to join an existing group or an existing member wants to leave a group, there is no need to form a new group from scratch.
6) The group secure key can be updated periodically without any help from an RSU to increase the security level of the communication.
We provide a security analysis on our schemes and an analysis on the effectiveness of using bloom filter in the notification messages. Through analysis and simulation studies, we show that our schemes can reduce the message overhead and increase the verification success rate (will be formally defined in Section 8) by at least 45% while the additional overhead is insignificant when compared to the existing solutions.

2. State of the art

The problem of secure communications has been studied in other mobile ad hoc networks (e.g. [Kim et al. (2004)] and [Wong et al. (1998)]). However, VANET has its own characteristics which make the solutions for MANET not applicable. In brief, although mobile ad hoc network does not have a fixed topology, it may still be feasible to assume a rough topological structure such as cluster-based, tree-based or hierarchical, and nodes in a MANET move relatively slowly. On the other hand, these assumptions are no longer valid in VANETs. In a VANET, vehicles are moving at high speed and the topology changes rapidly. There are other issues that make the problem in VANETs unique. Interested readers can refer to [Hubaux et al. (2004)] and [Raya et al. (2006)] for more details.

Ad hoc communications in VANETs have been addressed in two recent work [Zhang, Lu, Lin, Ho & Shen (2008); Zhang, Lin, Lu & Ho (2008)]. In [Zhang, Lu, Lin, Ho & Shen (2008)], the IBV protocol was proposed for vehicle-to-RSU communications. The RSU can verify a large number of signatures as a batch using just three *pairing* operations (see Section 4 for what a pairing operation is). However, their work has some limitations. First, their protocol relies heavily on a tamper-proof hardware device, installed in each vehicle, which preloads the system-wide secret key. Once one of these devices is cracked, the whole system will be compromised. Second, a vehicle's real identity can be traced by anyone, thus the protocol does not satisfy the privacy requirement. Third, their protocol has a flaw such that a vehicle can use a fake identity to avoid being traced (anti-traceability attack) or even impersonate another vehicle (impersonation attack [1]). Fourth, in their batch verification scheme, if any of the signatures is erroneous, the whole batch will be dropped. This is inefficient because most signatures in the batch may actually be valid. Finally, the IBV protocol is not designed for vehicle-to-vehicle communications.

In a more recent work [Zhang, Lin, Lu & Ho (2008)], the RAISE protocol was proposed for vehicle-to-vehicle communications. The protocol is software-based. It allows a vehicle to verify the signature of another with the aid of a nearby RSU. However, no batch verification can be done and the RSU has to verify signatures one after another. On the other hand, to notify other vehicles whether a message from a certain vehicle is valid, a hash value of 128 bytes needs to be broadcasted. There can be tens or even up to thousands of signatures within a short period of time, thus the notification messages induce a heavy message overhead.

Group communications in VANETs, on the other hand, have been considered in three papers [Chim, Yiu, Hui, Jiang & Li (2009); Wasef & Shen (2008); Verma & Huang (2009)]. In [Chim, Yiu, Hui, Jiang & Li (2009)], a scheme is proposed to allow a set of vehicles to form a group with the help of an RSU such that subsequently, encrypted group messages can be broadcasted by any member to all other members without the intervention of RSU. While group messages can be authenticated to be sent by valid members, the scheme also satisfies the privacy-preserving property that the real identity of the sender cannot be linked to the messages (except by TA) to protect the route of the vehicles being traced by unauthorized parties. However, the scheme assumes that the vehicles in the group are static in the sense that no mechanism is provided if a new member wants to join an existing group or an existing member wants to leave the group. Also, wireless channel is more vulnerable to attacks, and it is important to have an efficient scheme to update the group key periodically without the help of a third party. So, to handle dynamic membership of a group and key update based on [Chim, Yiu, Hui, Jiang & Li (2009)], we need to go through the group formation scheme from scratch with the help of an RSU.

In [Wasef & Shen (2008)], the PPGCV protocol was proposed. In addition to a scheme for group formation, they provide a protocol to update the group key. However, the setting of their scheme is different from a typical VANET and the key update process relies heavily on a key server which holds the set of all keys distributed to the vehicles.

[1]Please refer to the Appendix or [Chim, Yiu, Hui & Li (2009)] for the attacks.

In [Verma & Huang (2009)], another group communications protocol, SeGCom, was proposed. However, its concern is totally different from ours and privacy is not considered. Also RSUs are assumed to be fully trusted but this is unrealistic in practice. Furthermore, a vehicle can only form a group with all vehicles that are in the same RSU's range. It cannot select vehicles to form a group with. In all three papers, the noise of the wireless channels was not adequately addressed. We found that the success rate for forming a group degrades substantially as the noise increases in the channel.

3. Problem statement

3.1 System model and assumptions

Recall that a vehicular network consists of on-board units (OBUs) installed on vehicles, road-side units (RSUs) along the roads, and a trusted authority (TA). We focus on the inter-vehicle communications over the wireless channel. We assume the following:

1. The TA is always online and trusted. RSUs and TA communicate through a secure fixed network. To avoid being a single point of failure or a bottleneck, redundant TAs which have identical functionalities and databases are installed.
2. RSUs have higher computation power than OBUs.
3. The RSU to Vehicle Communication (RVC) range is at least twice of the Inter-Vehicle Communication (IVC) range to ensure that if an RSU receives a message, all vehicles receiving the same message are in the feasible range to receive the notification from the RSU. Consider the following counter example. Assume that the RVC range and the IVC range are both r. Two vehicles V_1 and V_2 are r apart. The distance between V_1 and RSU is within r but that between V_2 and RSU is larger than r. If V_1 sends a message to V_2, V_2 has no way to verify it as it cannot receive the notification message from the RSU. However, this problem can be resolved if the RVC range is twice that of IVC.
4. There exists a conventional public key infrastructure (PKI) for initial handshaking. The public key of TA PK_{TA} is known by *everyone*. The public key of vehicle V_i PK_{V_i} is known by TA. Also any RSU R_k broadcasts its public key PK_{R_k} with hello messages periodically to vehicles that are travelling within RVC range of it. Thus PK_R is known by all vehicles nearby. The TA and RSU R_k keep their secret keys SK_{TA} and SK_{R_k} privately. There is no need for vehicles to know the public keys of other vehicles to avoid message overhead for exchanging certificates.
5. The real identity of any vehicle is only known by TA and itself but not by others.

3.2 Security requirements

We aim at designing schemes to satisfy the following security requirements:

1 *Message integrity and authentication*: A vehicle should be able to verify that a message is indeed sent and signed by another vehicle without being modified by anyone.

2 *Identity privacy preserving*: The real identity of a vehicle should be kept anonymous from other vehicles and a third party should not be able to reveal a vehicle's real identity by analysing multiple messages sent by it.

3 *Traceability*: Although a vehicle's real identity should be hidden from other vehicles, if necessary, TA should have the ability to obtain a vehicle's real identity and relate the message to the sender (for example, in case the real identity of the sender of a fake message causing an accident needs to be revealed).

4 *Confidentiality*: Group messages broadcasted to all members should not be decryptable by vehicles not in the group and a group message sent to a dedicated member should only be decryptable by that dedicated receiver, other vehicles (including other members) should not be able to decrypt the message.

4. Preliminaries

Our schemes are *pairing-based* and defined on two cyclic groups with a *bilinear mapping* [Menezes (1991)]. We briefly introduce what a bilinear map is and will discuss the basics on bloom filter which we apply in the RSU notification phase.

4.1 Elliptic curve cryptography (ECC)

Let $\mathbb{G}$ be a cyclic additive group with generator P and $\mathbb{G}_{\mathbb{T}}$ be a cyclic multiplicative group. Both groups $\mathbb{G}$ and $\mathbb{G}_{\mathbb{T}}$ have the same prime order q. The mapping $\hat{e}: \mathbb{G} \times \mathbb{G} \rightarrow \mathbb{G}_{\mathbb{T}}$ is called a *bilinear map* if it satisfies the following properties:

1. Bilinear: $\forall P,Q,R \in \mathbb{G}$ and $\forall a,b \in \mathbb{Z}$, $\hat{e}(Q,P+R) = \hat{e}(P+R,Q) = \hat{e}(P,Q) \cdot \hat{e}(R,Q)$. Also $\hat{e}(aP,bP) = \hat{e}(P,bP)^a = \hat{e}(aP,P)^b = \hat{e}(P,P)^{ab}$.
2. Non-degenerate: There exists $P,Q \in \mathbb{G}$ such that $\hat{e}(P,Q) \neq 1_{\mathbb{G}_{\mathbb{T}}}$.
3. Computable: There exists an efficient algorithm to compute $\hat{e}(P,Q)$ for any $P,Q \in \mathbb{G}$.

The bilinear map $\hat{e}$ can be constructed on elliptic curves. Each operation for computing $\hat{e}(P,Q)$ is a *pairing operation*. Pairing operation is the most expensive operation in this kind of cryptographic schemes. The fewer the number of pairing operations, the more efficient the scheme is. The groups $\mathbb{G}$ and $\mathbb{G}_{\mathbb{T}}$ are called bilinear groups. The security of our schemes relies on the fact that the discrete logarithm problem (DLP) on bilinear groups is computationally hard, i.e., given the point $Q = aP$, there exists no efficient algorithm to obtain a by given P and Q. The implication is that we can transfer Q in an open wireless channel without worrying that a (usually some secret) can be known by the attackers.

4.2 Bloom filter

A *bloom filter* is a method for representing a set $A = a_1,a_2,...,a_n$ of n elements to support membership queries. The idea is to allocate a vector v with m bits, initially all set to 0, and then choose k independent hash functions, h_1, h_2, ..., h_k, each with range 1, ..., m. For each element $a \in A$, the bits at the positions $h_1(a)$, $h_2(a)$, ..., $h_k(a)$ in v are set to 1 (A particular bit might be set to 1 multiple times). To answer if a value b is in A, we check the bits at positions $h_1(b)$, $h_2(b)$, ..., $h_k(b)$. If any of them is 0, then b is definitely not in the set A. Otherwise we conjecture that b is in the set although there is a certain probability that we are wrong (called a false positive). After inserting n keys into the vector with m bits with k hash functions, the probability that a particular bit is still 0 is $(1-\frac{1}{m})^{kn} \sim e^{-\frac{kn}{m}}$ assuming that on any input value, the hash functions pick each position with equal probability. Hence the probability of a false positive is $(1-(1-\frac{1}{m})^{kn})^k \sim (1-e^{-\frac{kn}{m}})^k$. Let $f(k) = (1-e^{-\frac{kn}{m}})^k$ and let $g(k) = lnf(k) = kln(1-e^{-\frac{kn}{m}})$. By finding $\frac{dg}{dk}$ and making $\frac{dg}{dk} = 0$, it can be shown that to minimize the probability of having false positives, k should be set to $\frac{m \ln 2}{n}$.

5. Our scheme for ad hoc communications

This section presents our scheme for ad hoc communications in details. There are some initial parameters to be generated by TA using the following steps. This needs to be done once

for the whole system unless the master key, or the real identity of a vehicle are believed to be compromised, or TA wants to update the parameters and the master key periodically to enhance the security level of the system.
(1) TA chooses $\mathbb{G}$ and $\mathbb{G}_{\mathbb{T}}$ that satisfy the bilinear map properties.
(2) TA randomly picks $s \in \mathbb{Z}_q$ as its master key and computes $P_{pub} = sP$ as its public key. The public parameters $\{\mathbb{G}, \mathbb{G}_{\mathbb{T}}, q, P, P_{pub}\}$ are publicly accessible by all RSUs and vehicles.
(3) TA assigns each vehicle V_i a real identity $RID_i \in \mathbb{G}$ and a password PWD_i. The drivers are informed about them during network deployment or during vehicle first registration.
Our scheme can be divided into the following modules:
(A) **Initial handshaking**: This module is executed when a vehicle meets a new RSU. The vehicle authenticates itself with TA via RSU. Note that TA is the only authorized party which knows the real identity of the vehicle, so TA will pass information to RSU to allow RSU to verify the vehicle's signature even if it uses pseudo identity to sign the message. Also, RSU will generate a shared secret with the vehicle. If this is the first time the vehicle authenticates itself with TA, TA will also pass its master key s and a shared secret to the vehicle. This only needs to be done once in the whole session. To increase the security level, s is not preloaded into any hardware on the vehicle as in [Zhang, Lu, Lin, Ho & Shen (2008)]. For the shared secret with RSU, a new secret is generated every time the vehicle moves into the region of another RSU.
(B) **Message signing**: When a vehicle wants to send out a message, it first creates a pseudo identity together with the signing key. This can be done *per message* to increase the difficulty of attackers attempting to trace its real identity. Then, it signs the message using the signing key of the pseudo identity.
(C) **Batch verification**: This module is used by the RSU to verify a set of messages using only *two* pairing operations in a batch mode. We also describe how to generate a notification broadcast message using bloom filter and how to handle the case in which there are some invalid signatures in the batch (recall that in [Zhang, Lu, Lin, Ho & Shen (2008)], once there is an invalid signature in the batch, the whole batch of signatures are assumed to be invalid and ignored).
(D) **Real identity tracking**: This module is used by TA to reveal the real identity of the sender of a given message.

5.1 Initial handshaking

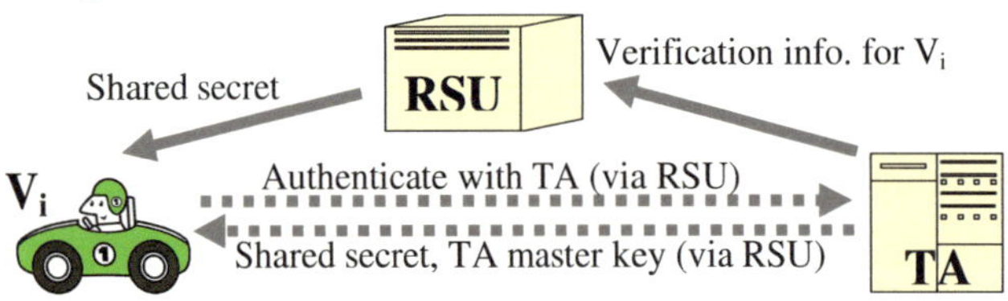

Fig. 1. Initial handshaking

We use the notations $CENC_Z(M)$, $CDEC_Z(M)$ and $CSIG_Z(M)$ to denote conventional encrypting, decrypting and signing, respectively, message M using the key Z. To enhance readability, we summarize the notations that will be used in this chapter in Table 1. The detailed processes in this module are as follows:

1. When a vehicle V_i meets the first RSU R_k, it encrypts its RID_i and PWD_i using TA's public key PK_{TA} and sends $ENC_{PK_{TA}}(RID_i, PWD_i, r)$ to the RSU which forwards it to TA. Here RID_i, PWD_i and r are concatenated in a pre-defined way and r is a random nonce. By including r, two similar blocks sent by the same vehicle cannot be related by an attacker.

Symbol	Meaning
$CENC_Z(M)$	Encrypting message M using the key Z
$CDEC_Z(M)$	Decrypting message M using the key Z
$CSIG_Z(M)$	Signing message M using the key Z
$H(.)$	MapToPoint hash function [Boneh et al. (2001)]
$h(.)$	One-way hash function such as SHA-1 [Eastlake & Jones (2001)]
r	random number
TA	Trusted authority
R_k	RSU number k
V_i	Vehicle number i
PK_{TA}	Public key of TA
SK_{TA}	Secret key of TA
PK_{R_k}	Public key of RSU R_k
SK_{R_k}	Secret key of RSU R_k
PK_{V_i}	Public key of vehicle V_i
s	TA's master key
$\{\mathbb{G}, \mathbb{G}_{\mathbb{T}}, q, P, P_{pub} = sP\}$	Public parameters
RID_i	Real identity of vehicle V_i
PWD_i	Password of vehicle V_i
t_i	Shared secret between TA and vehicle V_i
m_i	Shared secret between RSU R_k and vehicle V_i
M_i/M_r	Messages sent by vehicle / RSU
ID_i	Pseudo identity of vehicle V_i
VPK_i	Verification public key of vehicle V_i
$SK_i = (SK_{i1}, SK_{i2})$	Signing key of vehicle V_i
σ_i	ECC signature by vehicle V_i
LID_i	Local pseudo identity of vehicle V_i (for group communications)
GPK_i	Group public key of vehicle V_i (for group communications)
$LSK_i = (LSK_{i1}, LSK_{i2})$	Local signing key of vehicle V_i (for group communications)
ς_j	Local ECC signature by vehicle V_i (for group communications)
rr	Partial group secret key (for group communications)
β	Group secret key (for group communications)

Table 1. Notations used in this chapter

2. TA decrypts and verifies RID_i and PWD_i. If they are valid, it generates a shared secret t_i for V_i and computes V_i's ID Verification Public Key as $VPK_i = t_i \oplus RID_i$. TA then passes VPK_i to RSU to enable it to verify signatures from V_i even if V_i uses pseudo identity to sign the message. TA then stores the (RID_i, t_i) pair into its repository and forwards PK_{V_i}, VPK_i and $X = ENC_{PK_{V_i}}(s, VPK_i, SIG_{SK_{TA}}(s, VPK_i))$ to RSU, where PK_R and PK_{V_i} are conventional public keys of RSU and vehicle V_i, respectively. Note that to let V_i know that s and VPK_i are really sent by TA, TA includes its signature on s and VPK_i ($CSIG_{SK_{TA}}(s, VPK_i)$) into the encrypted text.
3. RSU chooses a random number m_i to be the shared secret between itself and vehicle V_i. It stores the (VPK_i, m_i) pair into its verification table for later usage. It then sends $Y = ENC_{PK_{V_i}}(m_i, CSIG_{SK_R}(m_i))$ and X to vehicle V_i. Again to let vehicle V_i know that m_i is really sent by RSU, RSU signs it.
4. Vehicle V_i decrypts Y to obtain m_i and verifies RSU's signature on it. Similarly, it decrypts X to obtain s and VPK_i and verifies TA's signature on them. It then computes its shared secret with TA using $t = VPK_i \oplus RID_i$.

This basically completes the initial handshaking phase. The following shows the procedure when vehicle V_i leaves the range of an RSU and enters the range of another. It includes a simpler authentication process with TA so that TA can pass the information to the new RSU for verifying V_i's signature and a new shared secret will be generated by this RSU.

5) V_i sends $ENC_{PK_{TA}}(RID_i)$ to TA via this new RSU. This time TA does not need to verify V_i's PWD anymore as it has already done that when V_i first starts up. Instead it directly generates a new t_i and a new VPK_i for V_i and sends VPK_i to the new RSU. TA then adds the new t_i into its repository. Next the new RSU chooses a random number m_i to be its shared secret with V_i. After storing (VPK_i, m_i) into its verification table, RSU sends $Y = ENC_{PK_{V_i}}(m_i, CSIG_{SK_R}(m_i))$ to V_i which then decrypts it using its conventional secret key. From now on, vehicle V_i starts to use the new shared secret with the new RSU for message signing.

5.2 Message signing

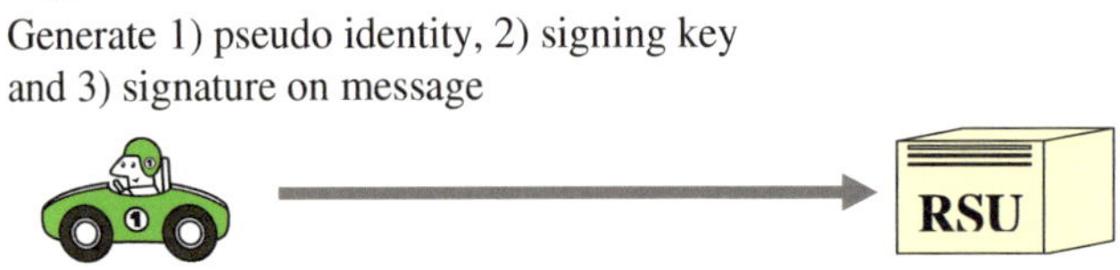

Fig. 2. Message signing

To sign a message, a vehicle generates a pseudo identity and the corresponding signing key. A different pseudo identity can be used for a different message.
To generate a pseudo identity, V_i first generates a random nonce r. Its pseudo identity ID_i contains two parts - ID_{i1} and ID_{i2} where $ID_{i1} = rP_{pub}$ and $ID_{i2} = VPK_i \oplus H(m_i ID_{i1})$. The corresponding signing key is $SK_i = (SK_{i1}, SK_{i2})$ where $SK_{i1} = sm_i ID_{i1}$ and $SK_{i2} = sH(ID_{i2})$. $H(.)$ is a MapToPoint hash function [Boneh et al. (2001)]. Then, to sign a message M_i, V_i computes the signature $\sigma_i = SK_{i1} + h(M_i)SK_{i2}$ where $h(.)$ is a one-way hash function such as SHA-1 [Eastlake & Jones (2001)]. Vehicle V_i then sends $< ID_i, M_i, \sigma_i >$ to others.

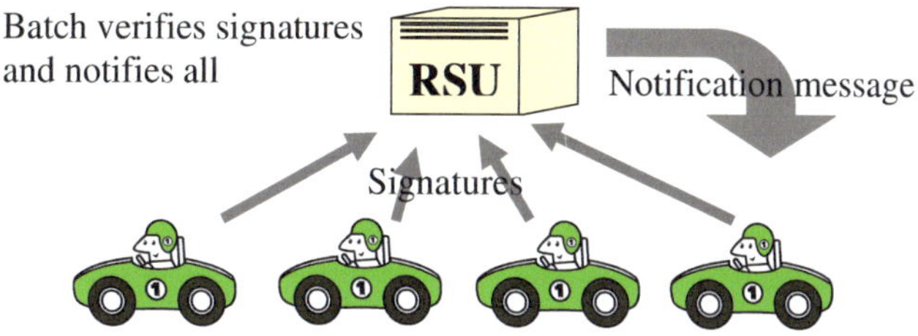

Fig. 3. Batch verification

5.3 Batch verification

This module allows an RSU to verify a batch of signatures using only two pairing operations based on the bilinear property of the bilinear map. We require an RSU to perform batch verification at a frequency higher than that with which a vehicle broadcasts safety messages so that a vehicle can verify the safety message of another before it broadcasts a more updated one. We first show the verification procedure. Then, we show how to make use of bloom filter to construct a notification message in order to reduce the message overhead. Lastly, we describe how to handle the case in which there are invalid signatures in the batch and how to extract valid ones from the batch instead of dropping the whole batch as in [Zhang, Lu, Lin, Ho & Shen (2008)].

Verification procedure. Assume that RSU wants to verify a batch of signatures σ_1, σ_2, ..., σ_n from vehicles V_1, V_2, ..., V_n on messages M_1, M_2, ..., M_n. With the shared secrets and the pseudo identities of the vehicles, RSU first determines their verification public keys $VPK_1, VPK_2, \ldots, VPK_n$ and shared secrets $m_1, m_2, \ldots, m_n$ by checking which of the stored (VPK_i, m_i) pairs satisfy $ID_{i2} = VPK_i \oplus H(m_i ID_{i1})$. It then verifies the signatures by checking if $\hat{e}(\sum_{i=1}^{n} \sigma_i, P) = \hat{e}(\sum_{i=1}^{n} m_i ID_{i1} + h(M_i)H(ID_{i2}), P_{pub})$ as

$$\begin{aligned}
&\hat{e}(\textstyle\sum_{i=1}^{n} \sigma_i, P)\\
&= \hat{e}(\textstyle\sum_{i=1}^{n} SK_{i1} + h(M_i)SK_{i2}, P)\\
&= \hat{e}(\textstyle\sum_{i=1}^{n} SK_{i1}, P)\hat{e}(\sum_{i=1}^{n} h(M_i)SK_{i2}, P)\\
&= \hat{e}(\textstyle\sum_{i=1}^{n} sm_i ID_{i1}, P)\hat{e}(\sum_{i=1}^{n} h(M_i)sH(ID_{i2}), P)\\
&= \hat{e}(\textstyle\sum_{i=1}^{n} m_i ID_{i1}, sP)\hat{e}(\sum_{i=1}^{n} h(M_i)H(ID_{i2}), sP)\\
&= \hat{e}(\textstyle\sum_{i=1}^{n} m_i ID_{i1}, P_{pub})\hat{e}(\sum_{i=1}^{n} h(M_i)H(ID_{i2}), P_{pub})\\
&= \hat{e}(\textstyle\sum_{i=1}^{n} m_i ID_{i1} + h(M_i)H(ID_{i2}), P_{pub}).
\end{aligned}$$

To avoid replay attack, an RSU stores the pseudo identities used by vehicles. If the pseudo identity in a vehicle's message matches any stored one, RSU rejects the message immediately. Note that if a vehicle does not know the shared secret with RSU, it cannot produce a valid signature. There may be a very small chance that the pseudo identities generated by two vehicles are the same. In that case, RSU will treat the signatures as invalid. The vehicles will sign again using a different pseudo identity.

Generating notification message. After RSU verifies vehicle V_i's signature σ_i, it notifies all vehicles within its RVC range the result. We first assume that all signatures are valid. For each valid message, we store a hash value $h(ID_i, M_i)$ of the message in the bloom filter (the hashing function is known to everyone) to minimize message overhead. However, as we have discussed in Section 4.2, there can be false positives in a bloom filter. To reduce this impact, we propose to use two bloom filters which contain opposite information: *Positive and Negative Filter*. The positive bloom filter stores the hash value of pseudo identities and messages of vehicles whose signatures are valid and the negative bloom filter stores the hash value of pseudo identities and messages of vehicles whose signatures are invalid.

If vehicle V_i wants to verify vehicle V_j's signature σ_j on message M_j, it first computes $h(ID_i, M_i)$ and then checks the positive filter and the negative filter as included in the RSU broadcast. There are four possible cases (see Table 2). For the first two cases, the resulting validity of σ_j can be confirmed. For the third case, V_j's hash appears in both filters. Then this must be a false positive in either filter, thus a re-confirmation procedure is needed. For the last case, V_j's hash does not appear in both filters. It means that RSU still has not yet verified σ_j and so V_i has to wait for RSU's next broadcasting message.

To facilitate re-confirmation, we require a vehicle to store the signatures of other vehicles which they are interested in upon receiving them for the first time for a short period. Also we require RSU to store the valid signatures that it has verified together with the sending vehicles' pseudo identities for at least one more batch verification period after that signature is lastly requested.

If Case 3 occurs, vehicle V_i re-sends σ_j to RSU. RSU searches for σ_j from those stored signatures. If σ_j can be found, RSU adds the hash of V_j into the positive filter. Otherwise, it adds it into the negative filter. All re-confirmation results can be embedded into a re-confirmation reply similar to a normal notification message. In practice, we can use one bit to distinguish whether the reply is a normal notification message or a re-confirmation reply.

Case	Positive Filter	Negative Filter	Validity of σ_j
1	True	False	Valid
2	False	True	Invalid
3	True	True	(Re-confirmation needed)
4	False	False	(Wait for next broadcast)

Table 2. Possible cases in bloom filters and their implications

There is still a chance that Case 3 occurs again. Our scheme allows the use of bloom filters for re-confirmation for K rounds. If after K rounds and Case 3 still occurs, RSU will send $h(ID_j, M_j)$ of V_j to vehicle V_i as a direct notification. To facilitate RSU to know what it should send in the re-confirmation reply, RSU stores the number of requests to each of its signature stored. See the next section for the performance of our schemes with different values of K.

Note that the size of each bloom filter m (i.e. the number of bits used) can be a variable in our schemes to save transmission overhead. To help the receiving vehicles to determine the size of the filters (so that they can adjust the range of hash functions accordingly), together with the valid and the invalid filters, RSU also transmits a value n to represent the total number of signatures in the batch (i.e. the number of values being added into any bloom filter cannot exceed n). To allow vehicles to confirm that a notification message is indeed sent by an RSU, RSU signs the bloom filters using its private key SK_R before broadcasting them.

Invalid signatures in the batch. A batch may contain tens and even up to thousands of signatures depending on the traffic density around RSU. In the IBV protocol, if any of the signatures inside the batch is invalid, the whole batch is dropped. This approach is inefficient in the sense that most of the signatures in the batch are actually valid and can be used. Thus in our schemes, we propose to adopt binary search in the verification process to extract those valid ones. Assume that the batch contains n signatures, we arrange them in a fixed order (say according to the senders' pseudo identities). If the batch verification fails, we first determine the mid-point as $mid = \lfloor \frac{1+n}{2} \rfloor$. Then we perform batch verification on the first half (the 1st to mid^{th} elements) and the second half (the $(mid+1)^{th}$ to n^{th} elements) separately. If any of the two batches causes a failure in the verification again, we repeat the same process on the invalid batch. If the pairing on any batch is valid, the RSU notifies all those signatures immediately.

The binary search stops if a batch contains only one signature or when a pre-defined level of binary search is reached. In Section 8, we evaluate the performance of our schemes using different number of levels in binary search and it is found that a full exploration may not be necessary in most cases.

5.4 Real identity tracking

To reveal the real identity of the sender of a message, TA is the only authorized party that can perform the tracing. Given vehicle V_i's pseudo identity ID_i and its shared secret with the connecting RSU m_i, TA can search through all the stored (RID_j, t_j) pairs from its repository. Vehicle V_i's real identity is the RID_j value from the entry that satisfies the expression $ID_{i2} \oplus t_j \oplus H(m_i ID_{i1}) = RID_j$ as $ID_{i2} \oplus t_j \oplus H(m_i ID_{i1}) = t_i \oplus RID_j \oplus H(m_i ID_{i1}) \oplus t_i \oplus H(m_i ID_{i1}) = RID_j$. No other party can obtain vehicle V_i's real identity since t_i is only known by TA and V_i itself.

6. Our scheme for group communications

This section presents our scheme for group communications in details. This scheme is based on the framework of our ad hoc communications scheme in Section 5. The scheme can be divided into the following modules:
(A) **Group formation**: This module is used when a set of vehicles want to form a group. A group partial secret key and a set of group public keys for group members will be generated by TA and forwarded by an RSU.
(B) **Secure one-to-many and one-to-one communications**: This module describes how a vehicle can send a message securely to all other members or to a dedicated member in the group.
(C) **New member joining**: This module is invoked when a new member wants to join an existing group.
(D) **Common group secret update**: This module shows how the common group secret can be updated without the help of RSU.
(E) **Member leaving**: This module is invoked when a member wants to leave a group.
(F) **Real identity tracking**: This module is used by TA to reveal the real identity of the sender of a given message.

6.1 Group formation

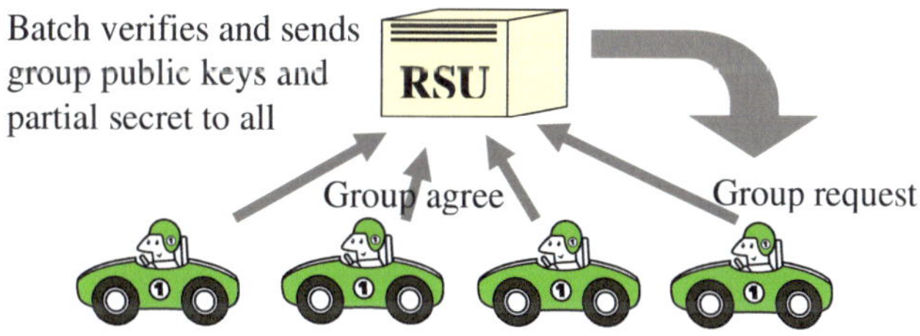

Fig. 4. Group Key Generation

When a group of vehicles want to form a group, each of them first creates a pseduo identity together with the signing key. This can be done per message to increase the difficulty of attackers to trace its real identity. Then, it signs a control message using the signing key of the pseudo identity. RSU verifies the set of control messages using only *two* pairing operations in a batch mode and distributes necessary information to vehicles in a group so that they can verify each others' messages without the aid of any RSU later on. A partial group secret key will also be generated. Details are as follow.

Assume that vehicles V_1, V_2, ..., V_n have already registered with the closest RSU and their shared secrets with RSU are $m_1, m_2, ..., m_n$ respectively. Also assume that these vehicles know pseudo identities of one another already.

Group request. Vehicle V_i generates the group request message $M_i = \{GPREQ, ID_1, ID_2, ..., ID_{i-1}, ID_{i+1}, ..., ID_n\}$ and its signature σ_i on M_i using the method in Section 5.2. Here ID_j for all $j \neq i$ is the pseudo identity lastly used by V_j as heard by V_i. V_i then broadcasts $< ID_i, M_i, \sigma_i >$ to RSU and others. Note that V_i can be anyone or the leader of the group.

Group agree. Any vehicle V_j receiving V_i's $GPREQ$ message checks whether its lastly used pseudo identity is included in the $GPREQ$ message. If yes, it composes the message $M_j = \{GPAGR, ID_j\}$ and its signature σ_j on it and sends $< ID_j, M_j, \sigma_j >$ to RSU. Note that vehicle V_j generates ID_j and σ_j using the method in Section 5.2 above.

Verification and key generation. At fixed intervals, RSU verifies the group request and group agree messages. Note that there is a method to batch verify a set of incoming signatures as we discussed in Section 5.3. For any vehicle V_x whose signature is found to be valid, it generates its group public key as $GPK_x = m_x P$ and stores it into the verification table. Besides, it also generates a random number rr which will be used to form the group secret key among the group of vehicles. Without loss of generality, assume the signatures from V_1, ..., V_x are valid, RSU broadcasts $M_r = \{ID_1, ID_2, ..., ID_x, GPK_1, GPK_2, ..., GPK_x, CENC_{m_1}(rr), CENC_{m_2}(rr), ..., CENC_{m_x}(rr)\}$ and its signature $CSIG_{CSK_R}(M_r)$ to the vehicles concerned.

In case the verification fails due to invalid signatures or vehicles inside the range have the same pseudo identity (although the chance is very small), RSU will stop the protocol and the group is required to repeat the protocol again for the sake of security.

Key reception and acknowledgement. Upon receiving RSU's broadcast, each vehicle V_i in the group acknowledges RSU by composing $M_i = \{KEYRECV\}$ and sending out the reply $< ID_i, M_i, \sigma_i >$. Note that ID_i and σ_i are generated in the same way as in the group request or group agree message. If after a timeout period (which is a system parameter), RSU still cannot receive the acknowledgement from V_i, it assumes that the message is corrupted on its way to V_i. More than one vehicle may not acknowledge. RSU then resends the previous broadcast to all these vehicles but this time, ID_j and $CENC_{m_j}(rr)$ of all vehicles V_j who have acknowledged do not need to be included in the broadcast anymore. In Section 8, we will show that acknowledgements are important in increasing the group formation success rate. Each vehicle in the group then stores all the group public keys and the decrypted rr values.

6.2 Secure one-to-many and one-to-one communications

In this sub-section, we describe how a vehicle can send a message securely to all other members or to a dedicated member in the group in detail. We also describe how a vehicle can sign a message so that another member can ensure the message is indeed sent by it. We consider the communication between two vehicles in a group as a *local* communication.

One-to-many communications. The vehicles in the group compute a common group secret as $\beta = s \times rr$ (note that RSU does not know how to compute β since s is only known by vehicles but not RSU) and they can communicate with each other securely using any symmetric key cryptographic algorithm such as DES [Brown et al. (1993)] from now on.

One-to-one communications. Based on the stored group public key GPK_j, when vehicle V_i wants to send a message M_i to another member V_j, it first generates a random number x. It then computes the ciphertext $C_i = \{xP, M_i + sxGPK_j\}$. To decrypt, it multiplies xP by sm_j and subtracts the result from the second part to obtain M_i as $M_i + sxGPK_j - xsm_jP = M_i + sxm_jP - xsm_jP = M_i$. We denote the encryption and decryption processes here as $C_i = EENC_{GPK_j}(M_i)$ and $M_i = EDEC_{m_j}(C_i)$, respectively.
Local message signing and verification. Next we look at the *local* pseudo identity generation, message signing and signature verification when group communications (either one-to-many or one-to-one) take place. To denote the *local* nature, we add the character L in front of the notations LID_i and LSK_i. When vehicle V_i wants to send a local message M_i, it first generates its *local* pseudo identity $LID_i = (LID_{i1}, LID_{i2})$ where $LID_{i1} = rP$ and $LID_{i2} = GPK_i + r\beta P$. r is again a random nonce here. Then it composes its *local* signing key $LSK_i = (LSK_{i1}, LSK_{i2})$ as $LSK_{i1} = sm_iLID_{i1}$ and $LSK_{i2} = sm_iH(LID_{i2})$ where $H(.)$ is a MapToPoint hash function as before. It signs the message M_i by computing the *local* signature $\varsigma_i = LSK_{i1} + h(M_i)LSK_{i2}$ where $h(.)$ is a one-way hash function (note that we use a different notation to differentiate it from a non-local signature). Finally, it sends to others $< LID_i, C_i, \varsigma_i >$ where C_i is the ciphertext corresponding to M_i.
Assume vehicle V_j wants to verify the signature of V_i on M_i. It first retrieves V_i's group public key GPK_i by computing $LID_{i2} - \beta LID_{i1}$ because $LID_{i2} - \beta LID_{i1} = GPK_i + r\beta P - \beta rP = GPK_i$. Then it decrypts C_i to obtain M_i and checks whether $\hat{e}(\varsigma_i, P) = \hat{e}(LID_{i1} + h(M_i)H(LID_{i2}), sGPK_i)$ as

$$
\begin{aligned}
&\hat{e}(\varsigma_i, P)\\
&= \hat{e}(LSK_{i1} + h(M_i)LSK_{i2}, P)\\
&= \hat{e}(LSK_{i1}, P)\hat{e}(h(M_i)LSK_{i2}, P)\\
&= \hat{e}(sm_iLID_{i1}, P)\hat{e}(h(M_i)sm_iH(LID_{i2}), P)\\
&= \hat{e}(LID_{i1}, sm_iP)\hat{e}(h(M_i)H(LID_{i2}), sm_iP)\\
&= \hat{e}(LID_{i1}, sGPK_i)\hat{e}(h(M_i)H(LID_{i2}), sGPK_i)\\
&= \hat{e}(LID_{i1} + h(M_i)H(LID_{i2}), sGPK_i).
\end{aligned}
$$

6.3 New member joining

Assume that vehicle V_k wants to join the group of V_i, namely, $(V_1, \ldots, V_n)$ which are in the range of the same RSU and the shared secret between V_k and RSU is m_k.
Group join. V_k first composes a group join message $M_k =< GPJOIN, ID_i >$ with its signature σ_k on it. It sends $< ID_k, M_k, \sigma_k >$ to the closest RSU. Note that ID_i is the pseudo identity lastly used by V_i as seen by V_k. It is not a local pseudo identity since V_k still has not joined V_i's group yet. Also V_k generates its pseudo identity ID_k and signature σ_k in the same manner as other vehicles when they send out their group request or group agree messages.
Group join agree. When V_i finds that its last used pseudo identity is inside V_k's group join message, V_i replies with a group join agree message $< ID_i, M_i, \sigma_i >$ where ID_i is generated as usual and $M_i = \{GPJOINAGR||ID_k||CENC_{CPK_R}(rr)\}$.
Verification by RSU and key generation. Upon receiving the group join and group join agree messages from V_k and V_i, respectively, RSU verifies them. RSU then generates V_k's group public key as $GPK_k = m_kP$. It broadcasts $M_r = \{ID_k||ID_i||GPK_k||GPK_i, CENC_{m_k}(rr)\}$ and its signature $CSIG_{CSK_R}(M_r)$ to V_k and V_i.
Key reception and acknowledgement. Upon receiving RSU's broadcast, V_j where $j \in \{i, k\}$ verifies RSU's signature and acknowledges it by composing $M_j = \{KEYRECV\}$ and sending out the reply $< ID_j, M_j, \sigma_j >$. Note that ID_j and σ_j are generated in the same way as in the

group join or group join agree message. If after a timeout period, RSU still cannot receive the acknowledgement from either V_k or V_i, it resends the previous broadcast to it. V_k then decrypts $CENC_{m_k}(rr)$ using its shared secret with RSU m_k and computes the common group secret as $\beta = s \times rr$.

Sharing of group public keys. Up to this moment, only V_i knows how to verify signatures by V_k. Thus V_i shares this piece of information with other members by composing the message $M_i = \{NEWMEMBER||GPK_k\}$ and broadcasting $< LID_i, M_i, \varsigma_i >$ to other members. Each member V_j verifies the signature of V_i and acknowledges V_i with the reply message $< LID_j, M_j, \varsigma_j >$ where LID_j is the local pseudo identity of V_j and $M_j = \{GPKRECV\}$. If after a timeout period, V_i still cannot receive the acknowledgement from any member, it resends the previous broadcast to it.

After all, one task is still missing. That is to inform V_k about how to verify other members' signatures. This task is again assigned to V_i. V_i composes the message $M_i = \{GPK_1, GPK_2, ..., GPK_{i-1}, GPK_{i+1}, ..., GPK_n\}$ and sends $< LID_i, M_i, \varsigma_i >$ to V_k. Upon receiving V_i's message, V_k acknowledges it like what other members do.

6.4 Common group secret update

Now we show how to update the common group secret β periodically without the help of RSU. Each member V_i can periodically request a key update by broadcasting the message $< LID_i, M_i, \varsigma_i >$ where LID_i is the local pseudo identity of V_i and $M_i = \{CGSUPDATE\}$. The requester V_i then generates a new random number rr_{new} and computes $\beta_{new} = rr_{new} \times s$. It sends to each other member V_j the message $< LID_i, M_i, \varsigma_i >$ where LID_i is the local pseudo identity of V_i and $M_i = \{NEWCGS, EENC_{GPK_j}(\beta_{new})\}$.

Each V_j acknowledges V_i with the reply message $< LID_j, M_j, \varsigma_j >$ where LID_j is the local pseudo identity of V_j and $M_j = \{NEWCGSRECV\}$. If after a timeout period, V_i still cannot receive the acknowledgement from V_j, it resends the previous message to it. Note that the acknowledgements here ensure that all members staying in the group can receive the new common group secret properly for ongoing one-to-many communications.

6.5 Member leaving

When a member V_k wants to leave a group, the group common secret should be updated so that V_k can no longer decrypt the group's ongoing communications. We can simply conduct a group key update protocol excluding V_k.

6.6 Real identity tracking

Again only TA can trace the real identity of the sender of a message. For vehicle V_i's group request or group agree message, TA can trace V_i's real identity using the routine in Section 5.4. For vehicle V_i's local message to other members, the connecting RSU first retrieves V_i's group public key GPK_i by computing $LID_{i2} - \beta LID_{i1}$ similar to what the receiver does. Then it looks up its verification table to retrieve V_i's verification public key VPK_i which was assigned by TA. By presenting VPK_i, TA can search through all the stored (RID_j, t_j, m_j) tuples from its repository. Vehicle V_i's real identity is the RID_j value from the entry that satisfies the expression $RID_j = t_j \oplus VPK_i$.

7. Analysis

7.1 Security analysis

We analyse our schemes with respect to the security requirements listed in Section 3.

Message integrity and authentication: For ad hoc messages, the signature σ_i on message M_i by vehicle V_i is composed of SK_{i1} and SK_{i2}. SK_{i1} is defined as sm_iID_{i1} where m_i is the shared secret between vehicle V_i and the RSU. Due to the difficulty of solving the discrete logarithm problem, there is no way for attackers to reveal m_i. Thus the attacker cannot forge a signature. Similarly, for group message, although all vehicles in the group know the group public key $GPK_i = m_iP$ of V_i, it is computationally hard to obtain m_i due to the same reason. Thus no other vehicle knows how to compose SK_{i1}. SK_{i2}, on the other hand, is defined as $sH(ID_{i2})$. Recall $ID_{i2} = VPK_i \oplus H(m_iID_{i1})$. Again, since no other vehicle knows m_i, only V_i can compute SK_{i2}. Therefore, no other vehicle can forge a valid signature by vehicle V_i. Note also that RSUs do not know the master secret s, and thus cannot forge a message either.

For local messages, the signature ς_i on message M_i by vehicle V_i is composed of LSK_{i1} and LSK_{i2}. LSK_{i1} is defined as sm_iLID_{i1}. Due to the difficulty of solving the discrete logarithm problem, there is no way for attackers to reveal m_i. LSK_{i2}, on the other hand, is defined as $sm_iH(LID_{i2})$. Again, since no other vehicle knows m_i, only V_i can compute LSK_{i2}. Therefore, no other vehicle can forge a valid signature by vehicle V_i. Again note that RSUs do not know the master secret s, and thus cannot forge a message either.

In practice, RSUs can be cracked easily and this is unavoidable. However, we can implement additional measures in our schemes to reduce the impact. For example, we can classify messages into different security levels. For critical messages, we can require them to be verified by TA instead of by RSUs. Or we can have another variation under which a message can only be trusted if it is verified by multiple consecutive RSUs. We believe with these measures, even if a few RSUs are cracked, the damage is limited.

Identity privacy preserving: We argue that if Decisional Diffie-Hellman (DDH) problem is hard, then the pseudo identity of a vehicle can preserve its real identity. The proof is as follows:

We consider a game between a challenger and an attacker. The challenger starts by giving a set of system parameters including P and P_{pub}. The attacker then freely chooses two verification public keys VPK_0 and VPK_1 and sends them to the challenger (these choices do not need to be random, the attacker can choose them in any way it desires). The challenger sets a bit $x = 0$ with probability $\frac{1}{2}$ and sets $x = 1$ with probability $\frac{1}{2}$. The challenger then sends the attacker the pseudo identity corresponding to VPK_b together with the group public key. The attacker tries to guess the value of x, and outputs its guess, x'. The attacker's advantage in the game is defined to be $Pr[x = x'] - \frac{1}{2}$. We say that our pseudo identity generation algorithm is semantically secure against a chosen plain text attack (CPA) if the attacker's advantage is negligible.

Next we assume that we have an algorithm A which runs in polynomial time and has a non-negligible advantage e as the attacker in the game described above. We will construct a DDH attacker B which has access to A and achieves a non-negligible advantage. B is given (P, aP, bP, T) as input. We let t denote the bit that B is trying to guess (i.e. $T = abP$ when $t = 0$ and is set randomly otherwise). B gives A $(P, P_{pub} = aP)$ as input. (Note that a now plays the role of s in our scheme.) A then chooses two verification public keys VPK_0 and VPK_1 which it has queried for the corresponding group public keys m_0P and m_1P before and sends them to B. B is playing the role of challenger here, so it sets a bit x randomly and generates the pseudo identity $ID = (ID_1, ID_2)$ where $ID_1 = raP$, $ID_2 = VPK_b \oplus H(rT)$ and r is a random nonce to send to A. B also sends A the group public key bP. (Note that b now plays the role of m_i in our scheme.) A sends B a bit x', which is its guess for x. B guesses that $t = 0$ if $x = x'$.

If $t = 0$ (so $T = abP$), then $ID_2 = VPK_b \oplus H(rabP) = VPK_b \oplus H(bID_1)$ is a valid pseudo identity. In this case, A will guess b correctly with probability $\frac{1}{2} + \epsilon$. Thus, $Pr[B\text{succeeds}|t =$

$0] = \frac{1}{2} + \epsilon$. If $t = 1$, we claim that $Pr[B\text{succeeds}|t = 1] = \frac{1}{2}$. To see why, we observe that when T is randomly chosen, $H(rT)$ cannot be cancelled by ID_1 and so there is no way to obtain VPK_b. Thus it reveals no information about x. In this sense, the value of x is hidden to A, so the probability that A will guess it is simply $\frac{1}{2}$. Hence, $Pr[B\text{succeeds}] = \frac{1}{2}(\frac{1}{2} + \epsilon) + \frac{1}{2}\frac{1}{2} = \frac{1}{2} + \epsilon/2$. Since ϵ is non-negligible, this shows that B violates the assumption that DDH is hard.
Furthermore, the random nonce r makes the pseudo identity of a vehicle different in different messages. Also since the verification public key VPK_i of a certain vehicle is different as seen by different RSUs, even if all RSUs collude, they have no way to trace a particular vehicle's travelling route.
Traceability: Section 5.4 shows that TA is able to trace a vehicle's real identity, thus traceability is satisfied.
Confidentiality: For one-to-one communications in a group, the message M_j to V_j is masked by the component $sxGPK_j$. To remove the mask, one has to present the first point xP, s and m_j. However, m_j is the shared secret between V_j and RSU. Also s is the master key of TA known by vehicles and TA only. Therefore, other than V_j, any other vehicles as well as RSU do not know how to obtain M_j.
For one-to-many communications in a group, any message is encrypted using the common group secret $\beta = s \times rr$. Since the partial secret rr is transmitted securely from RSU to each vehicle in the group, vehicles outside the group have no knowledge about it. In addition, since s is known by vehicles and TA only, RSU does not know how to decrypt the message either.

7.2 Analysis on bloom filter approach

This sub-section analyses our newly-proposed bloom filter approach in the verification notification phase. We first show that the probability of having false positives is very small if we set the parameters for the bloom filters appropriately, then we show that our message overhead is about 10 times lower than that under the RAISE protocol. Note that the IBV protocol does not have a notification phase, so we only compare ours with the RAISE protocol.
The probability of having a false positive in our bloom filter approach (i.e., Case 3 in Table 2) is equal to the probability that all k bits are set in one bloom filter while not all k bits are set in another bloom filter. Thus the probability of Case 3 is $Pr(Case3) = 2(1 - (1 - \frac{1}{m})^{kn})^k(1 - (1 - (1 - \frac{1}{m})^{kn})^k) \sim 2(1 - e^{-\frac{kn}{m}})^k(1 - (1 - e^{-\frac{kn}{m}})^k)$. Interestingly we find that the value of k that minimizes the false positive probability of a single bloom filter (i.e. $k = \frac{m \ln 2}{n}$) also minimizes Pr(Case 3) approximately (up to 5 decimal places) based on our empirical results. Hence we set the number of hash functions to $\frac{m \ln 2}{n}$ in our schemes and $Pr(Case3) \sim 2(0.6185^{\frac{m}{n}}(1 - 0.6185^{\frac{m}{n}}))$. It can be shown that when $\frac{m}{n} = 5$, $Pr(Case3)$ is about 0.16. When $\frac{m}{n} = 10$, $Pr(Case3)$ drops to 0.016 only. That is, if there are 100 signatures in a batch, on average only 1 to 2 signatures are affected by bloom filter false positive and need to be re-confirmed.
Now, we analyze the message overhead. Assume that there are n signatures in a batch. For the RAISE protocol, the HMAC() value sent by each vehicle is 16 bytes long while the H() value sent by the RSU in the notification phase is 16 bytes long per message. After that RSU signs the notification message using an ECDSA signature which is 56 bytes long. Together with a message header of 2 bytes, the total message overhead for verifying a batch of n signatures is $16n + 16n + 56 + 2 = 32n + 58$ bytes.
For our schemes, the ECC signature sent by each vehicle is 21 bytes long. In the notification phase, we use two bloom filters. To lower the false positive rate in any bloom filter, the total number of bits used in each bloom filter is set to 10 times the number of signatures in the batch

(i.e. $\frac{m}{n} = 10$). We have two bloom filters and so a total of $\frac{20n}{8} = 2.5n$ bytes are needed. We also use 2 bytes to represent the number of signatures in a batch. Together with a message header of 2 bytes, the total message overhead for verifying a batch of n signatures is $21n + 2.5n + 2 + 56 + 2 = 23.5n + 60$ bytes.

Note that when Case 3 occurs, additional message overhead is required for the re-confirmation procedures. If Case 3 only occurs in the first trial and does not occur in the second trial, the total message overhead for verifying a batch of n signatures becomes $23.5n + 60 + P(23.5n + 60) = (1 + P)(23.5n + 60)$ bytes where $P = Pr(Case3)$. Hence, if Case 3 occurs in all the first K trials and we switch to the hash approach after that, the total message overhead becomes $\sum_{i=1}^{k} P^i(23.5n + 60) + P^k(37n + 58)$ bytes. The component $P^k(37n + 58)$ represents the message overhead used for the hash approach after K trials. That is, 21 bytes for each ECC signature, 16 bytes for each H() value, 56 bytes for ECDSA signature and 2 bytes for message header. Since P is about 0.016, even if K is only 2, the overhead of our scheme is much lower than that of RAISE. And we found that as long as $K > 1$, the overhead is similar in different values of K since the probability of Case 3 is very low, so re-confirmation is quite unlikely.

In Fig. 5, we set the value of K to 1, 2, 3 and 5, respectively, and with $\frac{m}{n}$ set to 5 and 10. We can see that with all values of K and $\frac{m}{n}$, the message overhead by our schemes is far lower than that by the RAISE protocol due to the use of ECC signatures and bloom filters in the notification phase. For our schemes, when $\frac{m}{n} = 5$, the more the number of trials before switching to hash approach, the lower the message overhead. When $\frac{m}{n} = 10$, the lines with $K = 2$, $K = 3$ and $K = 5$ overlap. It means that with $\frac{m}{n} = 10$ and as long as $K > 1$, the probability of Case 3 is very low and so re-confirmation is quite unlikely.

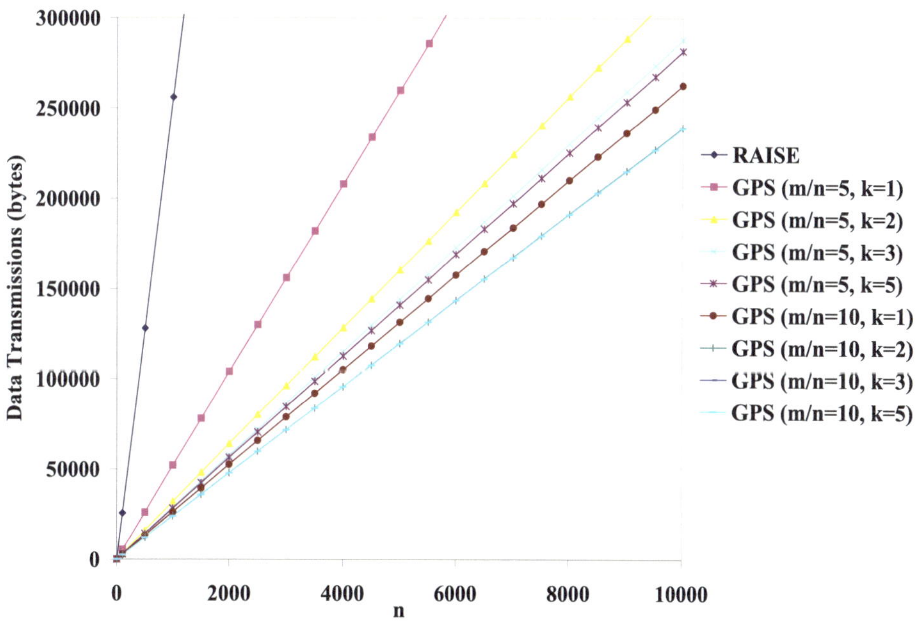

Fig. 5. Data transmission vs. number of signatures in the batch

8. Simulation results

In this section, we evaluate the network performance of our Grouping-enabled and Privacy-enhancing communications Scheme (GPS) in details. For ad hoc communications, we compare our scheme with the IBV protocol in terms of (1) the verification delay and (2) verification success rate through simulations. Note that IBV also uses a batch verification scheme, and so is much faster than the RAISE protocol. Thus, we compare the delay of our scheme with the IBV [Zhang, Lu, Lin, Ho & Shen (2008)] protocol. For success rate, we expect we will have a similar performance as RAISE as we will both identify all valid signatures even if there are invalid ones within the same batch. So, we compare our performance with the IBV protocol. We show that our scheme can verify more signatures while the additional delay required is insignificant. For group communications, we first compare our scheme with the RAISE protocol in terms of group message transmission delay. We expect we will have lower delay since group messages do not need to be verified by RSUs in our scheme. Next, we compare our GPS scheme with the SPECS protocol (group communications extension) [Chim, Yiu, Hui, Jiang & Li (2009)] in terms of (1) the group request delay and (2) group request success rate. We show that the success rate of forming a group using our scheme with the acknowledgement message is a lot higher than that of the SPECS scheme (with group communications extension) if the wireless channel is noisy. Then, we show that the delay introduced by retransmissions in our scheme is only marginal. Finally, we show the performance of our scheme in terms of key update average delay and member joining average delay as the number of vehicles in a group varies.

8.1 Simulation models

Some of the settings of our simulation are adopted from [Zhang, Lu, Lin, Ho & Shen (2008); Zhang, Lin, Lu & Ho (2008); Chim, Yiu, Hui, Jiang & Li (2009)]. We consider a highway of length 10 km and a number of RSUs are installed along it. The RVC and the IVC ranges are set to 600m and 300m, respectively. The bandwidth of the channel is 6 Mb/s and the average length of inter-vehicle message is 200 bytes. We compute the transmission time based on the bandwidth and the length of the message. The RSU performs batch verification every 300 ms and each pairing operation takes 4.5 ms [Scott (2007)]. We implement the simulation using C++.

For ad hoc communications, we assume that vehicles pass through an RSU at speeds varying from 50 km/h to 70 km/h. Our simulation runs for 1000 s. Inter-vehicle messages are sent every 500 ms from each vehicle. IEEE 802.11a is used to simulate the medium access control layer. (We simulate the IEEE 802.11a protocol by generating the time stamps for broadcasting messages of each vehicle. In case two stamps are identical, we randomly regenerate one of them.) We vary the total number of vehicles that have ever entered the RSU's RVC range during the simulation period from 200 to 1000 in steps of 200 to simulate the impact of different traffic densities. We also vary the inter-vehicle message signature error rate from 1% to 10% to study its impact on the performance of our scheme. For each configuration, we compute the average of 5 different random scenarios.

For group communications, the number of RSUs is a variable and these RSUs are evenly distributed along the given highway. Groups of vehicles are travelling on it at speeds varying from 50 km/h to 70 km/h. For each group, the number of vehicles is a variable and the vehicles are travelling on the road one after another. To simulate a noisy wireless channel (e.g. signal interference and signal blocking by obstacles or other signals), we use corruption rate. Since channel collision is also a kind of noise, we incorporate it into our corruption rate as well for simplicity. If the corruption rate is $cr\%$, a transmitted message is corrupted with probability $cr\%$. We vary the corruption rate from 5% to 50% in steps of 5% to investigate its impact on the group request success rate and group request average delay. For each configuration, we compute the average of 20 different random scenarios (since we find that the standard deviation of the results is larger than that in experiments for ad hoc communications). The simulation runs for 1 hour and for every minute, there will be a new group of vehicles. For each group, one group request is issued.

8.2 Simulation results for ad hoc communications

We fix the signature error rate (the percentage of invalid signatures) to 5% and vary the total number of vehicles that have entered the RSU's range throughout the simulation. We only consider batches that contain invalid signatures (Invalid batch). In [Zhang, Lin, Lu & Ho (2008)], the expression for verification success rate is defined. We extend its definition to handle invalid batch: $IBSR = \frac{1}{N}\sum_{i=1}^{N}\frac{M^i_{app}}{M^i_{mac}}$, where M^i_{app} is the total number of messages that are successfully verified by the RSU and are consumed by vehicle V_i in the application layer before vehicle V_i leaves RSU's RVC range, M^i_{mac} is the total number of messages received by both vehicle V_i and RSU in the medium access control layer from other vehicles. For our scheme, we can have different levels of binary search as mentioned in Section 5. We use the notation GPS(BSx) to denote our scheme with x levels of binary search.

The verification success rates for the scheme are shown in Fig. 6. Note that the success rates for IBV and GPS(BS0) are 0% as both will drop the whole invalid batch. From our simulation, we found that even if we only have 1 level of binary search, the success rate of GPS(BS1) is already raised to about 45%. If we increase the number of levels to 4, the success rate can be raised to more than 90%.

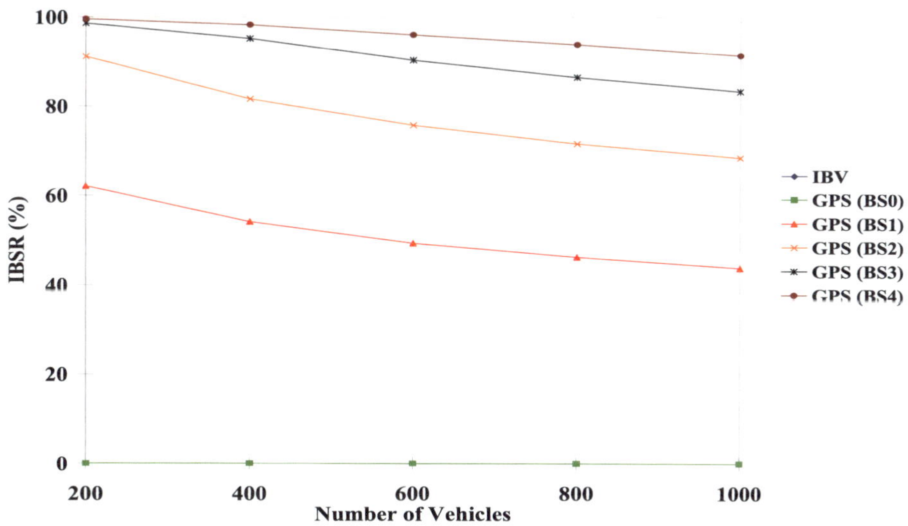

Fig. 6. Invalid batch success rate vs. number of vehicles

While the results are quite obvious, next we will show that the delay incurred by binary search procedure is minimal. Fig. 7 shows the delay performance. We define the average delay suffered by vehicles as $MD = \frac{1}{N}\sum_{i=1}^{N}\frac{1}{M}\sum_{m=1}^{M}(T^m_{verf} - T^m_{recv})$, where M is the number of messages received by vehicle V_i, T^m_{verf} is the time that vehicle V_i receives the verification notification message of message m from RSU and T^{m}_{recv} is the time that vehicle V_i receives message m from its neighboring vehicle. From Fig. 7, we can see that the delay under the IBV protocol and our scheme are very close to each other. For our scheme, as expected, with higher levels of binary search, longer delay is induced because more pairing operations are involved. However, even in the worst case (i.e. using 4 levels of binary search), our scheme only consumes an additional 10 ms which is roughly equivalent to the delay caused by 2 pairing operations. This is due to two main reasons. Not all cases require 4 levels of binary search and the time for each pairing operation is comparatively smaller than the transmission delay, so we can afford to do more pairing operations. One more interesting point to note is that without binary search, our scheme consumes 5 less ms than the IBV protocol. The reason is that our scheme requires 2 pairing operations only while the IBV protocol requires 3 as mentioned in Section 5.

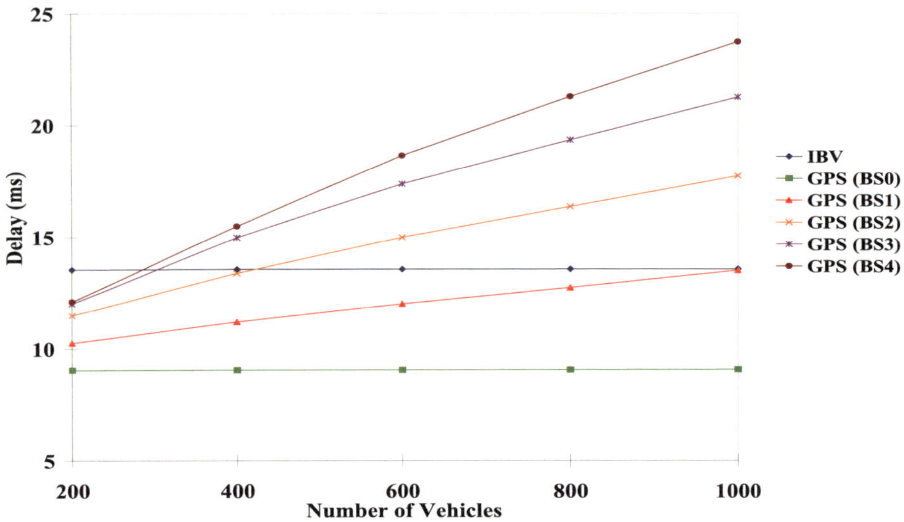

Fig. 7. Delay vs. number of vehicles

In the second set of experiments, we fix the number of vehicles that have entered RSU's RVC range during the simulation period to 300 and vary the signature error rate from 1% to 10% to investigate its impact on the invalid batch success rate and the message delay. We only consider batches that contain invalid signatures. Fig. 8 shows the results. The IBV and GPS(BS0) cases are not interesting as they drop all invalid batches. And it is also quite obvious that as the level of binary search increases, the success rate increases. The interesting point is that as the error rate increases from 1% to 10%, our scheme only degrades less than 10%.
The corresponding delay performance is shown in Fig. 9. As discussed earlier, GPS(BS0) gives a lower delay than the IBV protocol due to the saving of one pairing operation. As the error rate increases, more batches contain invalid signatures. Additional pairing operations are required to locate valid signatures. This increases the average delay. But the gap between our scheme and the IBV protocol is only about 10ms even when the error rate is 10%.

8.3 Simulation results for group communications

In the first set of experiments, we put aside the impact of interference and obstacles by setting the corruption rate to 0%. We vary the number of RSUs along the highway from 2 to 10. These RSUs are then evenly distributed along the highway. We define the group message

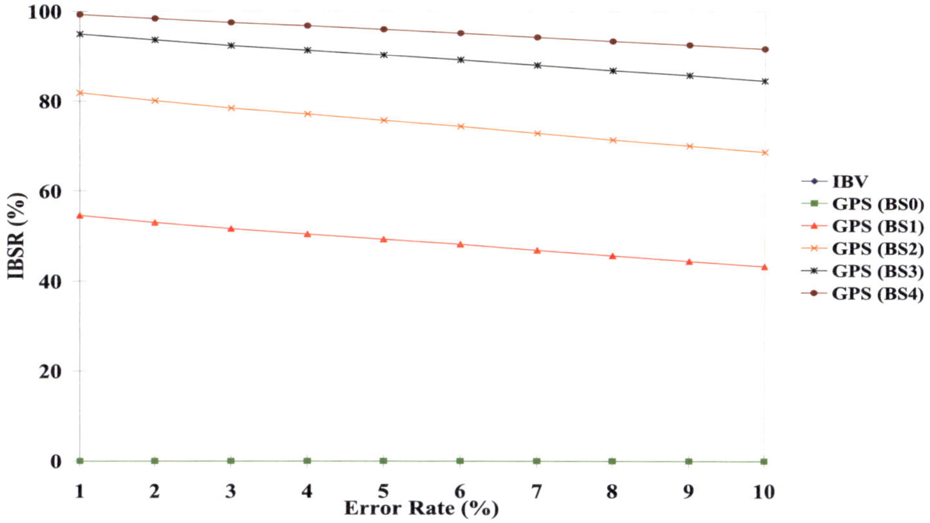

Fig. 8. Invalid batch success rate vs. error rate

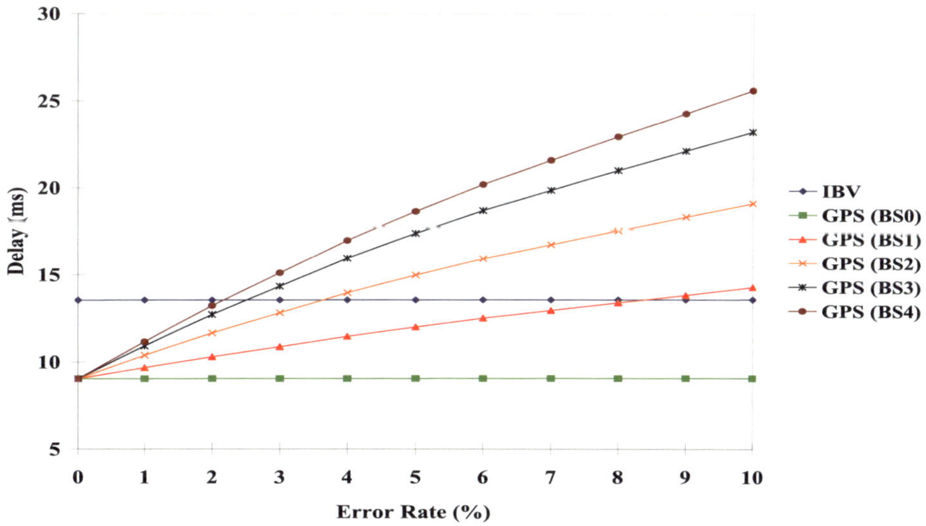

Fig. 9. Delay vs. error rate

transmission delay as the period from when a vehicle sends a message to when another vehicle in the same group verifies the message properly. We investigate the average group transmission delay under the RAISE protocol and our scheme as more RSUs are installed along the highway. From Fig. 10, we can see that under the RAISE protocol, the average group message transmission delay increases as RSUs become less dense along the highway. However, under our scheme, the average group message transmission delay remains constant (and near zero) no matter how dense the RSUs are. It is because under the RAISE protocol, all messages need to be verified by RSUs. When a vehicle wants to send a message to another group member, it first waits for a nearby RSU ahead of its journey. Then it waits for its batch verification period to expire. However, in our scheme, an initial RSU has already passed the necessary verification information (group public keys) to all vehicles in the group and so they know how to verify the signatures of each other without further support from RSUs. Thus the above two waiting periods are no longer needed.

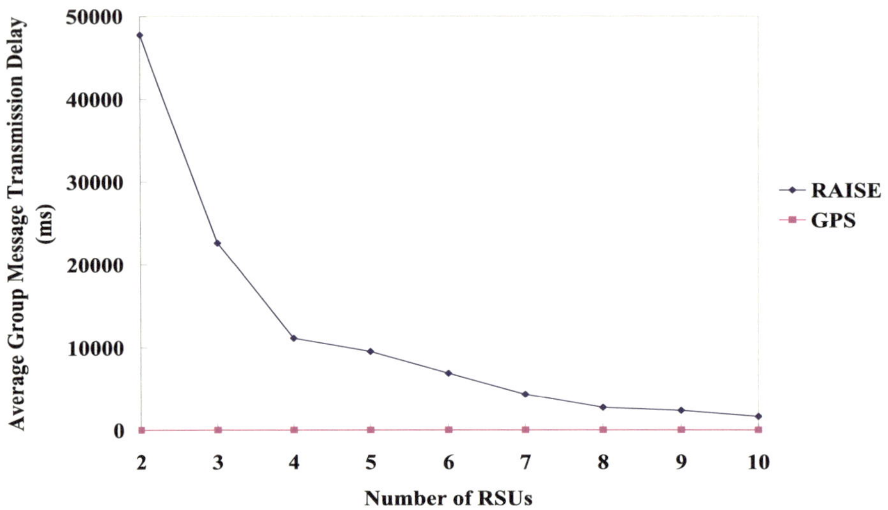

Fig. 10. Average group message transmission delay vs. number of RSUs

In the second set of experiments, we vary the corruption rate from 5% (low interference environment) to 50% (high interference environment) and study its impact on the group request success rate and the corresponding average delay. For each corruption rate, we further try different initial (before any new member joins) group sizes (n = 5, 10, 15 and 20). A group request is considered to be successful if all members in the group receive necessary information (i.e. group public keys of all others and the group partial secret) from the initial RSU for group communications. The group request success rate is defined as the number of successful requests divided by the total number of group requests made. The group request delay is defined a bit differently under SPECS and our GPS scheme. For the SPECS protocol, it is defined as the period from when any vehicle in the group first sends out a group request message to when all vehicles in the group receive necessary information for group communications. For our scheme, it is defined as the period from when any vehicle in the group first sends out a group request message to when RSU received and verified the acknowledgement messages from all vehicles in the group. The group request average delay is just the average of the group request delay among all successful group requests.
From Fig. 11, we can see that by requiring vehicles to acknowledge RSU, our scheme always

gives 100% group request success rate no matter how many vehicles are in the group. For SPECS, without any acknowledgement mechanism, the group request success rate drops gradually as the corruption rate increases from 5% to 50%. In particular, the more vehicles in the group, the lower the group request success rate. It is because with more vehicles in a group, the probability that all vehicles receive a message properly becomes lower.

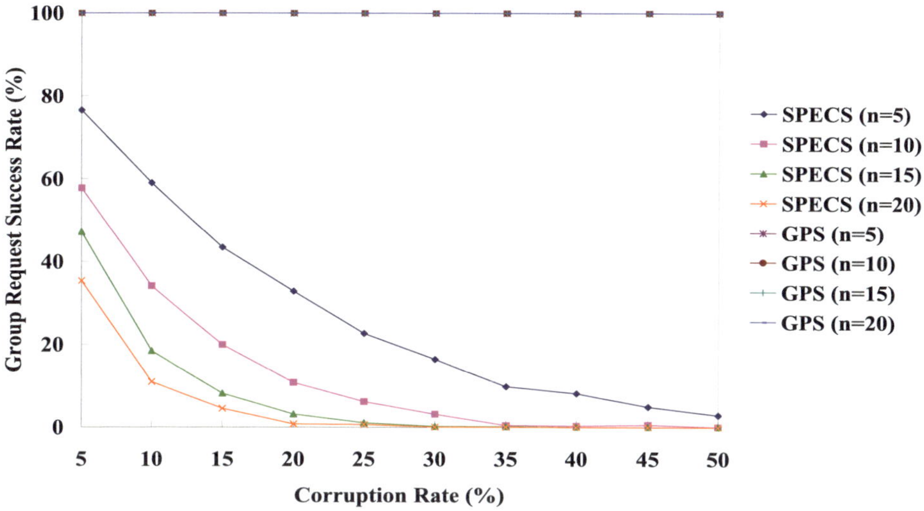

Fig. 11. Group request success rate vs. corruption rate

Fig. 12 shows the corresponding delay performance. We can see that larger groups are subjected to higher group request delay. This makes sense since with more vehicles in a group, the probability that all vehicles receive a message properly becomes lower. As a result, to ensure that all vehicles receive the message, more re-transmissions are needed. Thus higher delay is caused.

Next we focus on the case with 10 vehicles in a group to investigate the performance difference between SPECS and our scheme. From Fig. 13, we can see that our scheme gives a bit higher delay as the corruption rate increases due to increased number of re-transmissions. However, the increase is just marginal (less than 2 ms).

In the third set of experiments, we again vary the corruption rate from 5% to 50% and study its impact on the key update average delay under our scheme. A key update is considered to be successful if all vehicles in the group obtain the new group common secret. The key update delay is defined as the period from when any vehicle in the group sends out a key update request message to when it receives and verifies acknowledgement messages from all other vehicles in the group. The key update average delay is defined as the average of the key update delay among all successful key updates.

Fig. 14 shows that as more vehicles are involved in the group, higher key update delay is required. It is because with more vehicles in a group, the probability that all vehicles receive a message properly becomes lower. As a result, to ensure that all vehicles receive the message, more re-transmissions are needed. Thus higher delay is caused.

In the last set of experiments, we also vary the corruption rate from 5% to 50% but this time, we study its impact on the member joining average delay under our scheme. A member joining event is considered to be successful if all old members in the group obtain the new

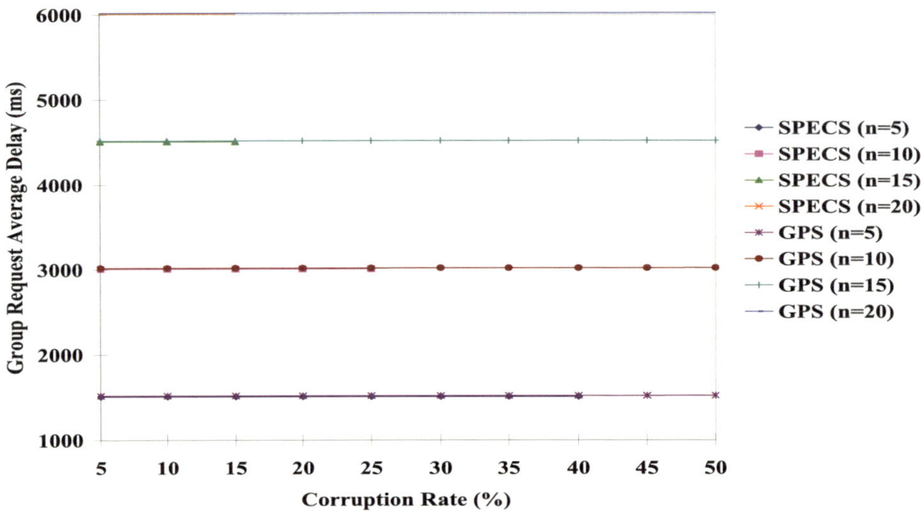

Fig. 12. Group request average delay vs. corruption rate

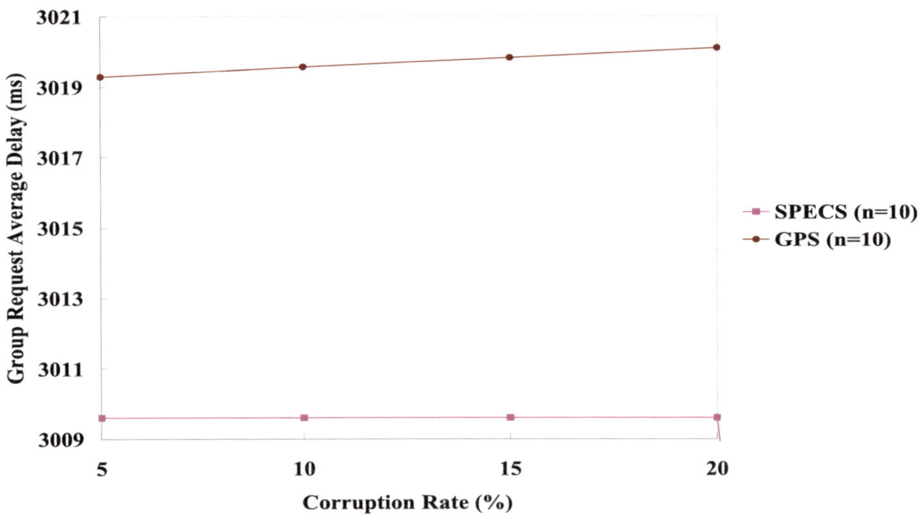

Fig. 13. Group request average delay vs. corruption rate

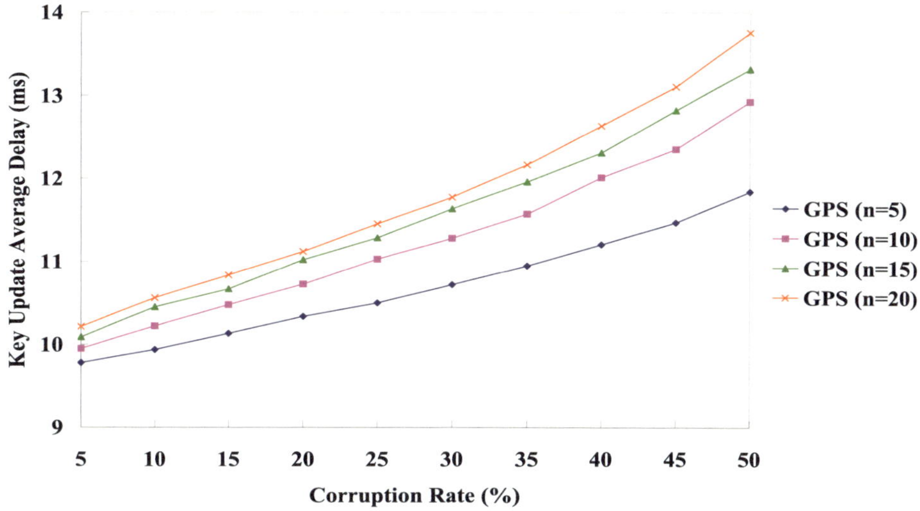

Fig. 14. Key update average delay vs. corruption rate

member's group public key while at the same time, the new member obtains all old members' group public keys and the group common secret for future communications. The member joining delay is defined as the period from when the new member sends out a group join request message to when the old member it contacts receives and verifies acknowledgement messages from all members. The member joining average delay is defined as the average of the member joining delay among all successful member joining events.

In Fig. 15, we study two initial (before any new member joins) group sizes ($n = 5$ and 10), and two new member set sizes, ($j = 5$ and 10) under our scheme. All configurations show slightly increasing trends as the corruption rate increases because of the same reason above. Also the delay performance under all configurations are actually close to each other and the difference is just about 10 ms.

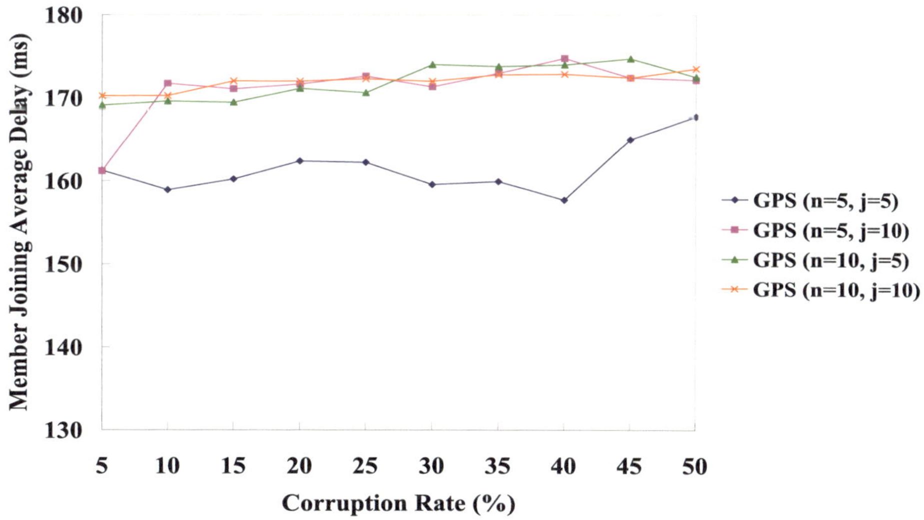

Fig. 15. Member joining average delay vs. corruption rate

9. Conclusions

In this chapter, we discussed the Grouping-enabled and Privacy-enhancing communications Schemes (GPS) for VANETs to handle ad hoc messages and group messages for inter-vehicle communications. For ad hoc messages, we follow the approach of letting RSU aid the signature verification process. We show that our schemes satisfy the security and privacy requirements. In terms of effectiveness, we show that our solution gives lower message overhead and at least 45 % higher success rate than previous work. For group messages, we proposed a set of modules for VANETs which allows dynamic membership and periodic update of the group key. RSU is needed only for group formation and member joining. In addition, we provide an add-on function for a group member to send a secure message to all other members or to a dedicated member. Again we show that our schemes satisfy all the security requirements. By simulation, we verify that our schemes are effective and the delay introduced by re-transmitting lost messages is negligible.

Note that in the early stage of VANET deployment, we may not have RSUs installed in all road sections. However, our protocol can be completed within the coverage of one RSU, and so can still be applied. Individual vehicles just cannot communicate on those sections of roads without RSUs, however, vehicles in the same group can still communicate without RSU. For future work, we will extend our group communications schemes to allow group merging and group splitting. We will also consider other secure applications in VANETs.

10. References

Boneh, D., Lynn, B. & Shacham, H. (2001). Short Signatures from the Weil Pairing, *Proceedings of Asiacrypt '01*, pp. 514 – 532.

Brown, R. H., Good, M. L. & Prabhakar, A. (1993). Data Encryption Standard (DES), *Federal Information Processing Standards (FIPS) Publication 46-2* .

Chim, T., Yiu, S., Hui, L. C., Jiang, Z. L. & Li, V. O. (2009). SPECS: Secure and Privacy Enhancing Communications Schemes for VANETs, *ADHOCNETS '09*.

Chim, T., Yiu, S., Hui, L. C. & Li, V. O. (2009). Security and Privacy Issues for Inter-vehicle Communications in VANETs, *IEEE Proceedings of the SECON '09 (Poster Session)*.

Eastlake, D. & Jones, P. (2001). US Secure Hash Algorithm 1 (SHA1), *IETF RFC3174* .

Housley, R., Ford, W., Polk, W. & Solo, D. (1999). Internet X.509 Public Key Infrastructure Certificate and CRL Profile, *IETF RFC2459* .

Hubaux, J. P., Capkun, S. & Lui, J. (2004). The Security and Privacy of Smart Vehicles, *IEEE Security and Privacy Magazine, 2(3)* pp. 49 – 55.

Kim, Y., Perrig, A. & Tsudik, G. (2004). Tree-based group key agreement, *ACM Transactions on Information Systems Security, 7(1)* pp. 60 – 96.

Menezes, A. (1991). An Introduction to Pairing-Based Cryptography, *1991 Mathematics Subject Classification, Primary 94A60*.

Oh, H., Yae, C., Ahn, D. & Cho, H. (1999). 5.8 GHz DSRC Packet Communication System for ITS Services, *IEEE Proceedings of the VTC '99*, pp. 2223 – 2227.

Raya, M., Papadimitratos, P. & Hubaux, J. P. (2006). Securing Vehicular Communications, *IEEE Wireless Communications Magazine, Special Issue on Inter-Vehicular Communications* pp. 8 – 15.

Scott, M. (2007). Efficient implementation of cryptographic pairings, Available online: http://ecrypt-ss07.rhul.ac.uk/Slides/Thursday/mscott-samos07.pdf.

Tsang, P. P. & Smith, S. W. (2008). PPAA: Peer-to-Peer Anonymous Authentication, *Proceedings*

of ACNS '08, pp. 55 – 74.
U.S. Department of Transportation, N. H. T. S. A. (2006). Vehicle Safety Communications Project Report.
Verma, M. & Huang, D. (2009). SeGCom: Secure Group Communication in VANETs, *IEEE Proceedings of the CCNC '09*, pp. 1 – 5.
Wang, F., Zeng, D. & Yang, L. (2006). Smart Cars on Smart Roads: an IEEE Intelligent Transportation Systems Society Update, *IEEE Pervasive Computing, Vol. 5, No. 4* pp. 68 – 69.
Wasef, A. & Shen, X. (2008). PPGCV: Privacy Preserving Group Communications Protocol for Vehicular Ad Hoc Networks, *IEEE Proceedings of the ICC '08*, pp. 1458 – 1463.
Wong, C. K., Gouda, M. & Lam, S. S. (1998). Secure group communications using key graphs, *IEEE Proceedings of the SIGCOMM '98*, pp. 68 – 79.
Zhang, C., Lin, X., Lu, R. & Ho, P. H. (2008). RAISE: An Efficient RSU-aided Message Authentication Scheme in Vehicular Communication Networks, *IEEE Proceedings of the ICC '08*, pp. 1451 – 1457.
Zhang, C., Lu, R., Lin, X., Ho, P. H. & Shen, X. (2008). An Efficient Identity-based Batch Verification Scheme for Vehicular Sensor Networks, *IEEE Proceedings of the INFOCOM '08*, pp. 816 – 824.

11. Appendix - attacks to IBV protocol

In this section, we first describe the IBV protocol. Then, we describe in details three security problems of the protocol - privacy violation, anti-traceability attack, and impersonation attack.

11.1 The IBV protocol

Before network deployment, TA sets up the parameters using the following steps:

1. TA chooses $\mathbb{G}$ and $\mathbb{G}_{\mathbb{T}}$ that satisfy the bilinear map properties.
2. TA randomly picks $s_1, s_2 \in \mathbb{Z}_q$ as its master keys. These two master keys are preloaded into each vehicle's tamper-proof hardware device.
3. TA then computes $P_{pub_1} = s_1 P$ and $P_{pub_2} = s_2 P$ as its public keys. The parameters $\{\mathbb{G}, \mathbb{G}_{\mathbb{T}}, q, P, P_{pub_1}, P_{pub_2}\}$ are then preloaded into all RSUs and OBUs.
4. TA also assigns each vehicle V_i a real identity $RID_i \in \mathbb{G}$ and a password PWD_i. The drivers are informed about them during network deployment or during vehicle first registration.

When a vehicle V_i starts up, its RID_i and PWD_i are input by the driver into a tamper-proof device. If they are valid, the tamper-proof device starts its role in generating pseudo identities, secret keys and message signing. Vehicle V_i's pseudo identity is generated as $ID_i = (ID_{i1}, ID_{i2})$ where $ID_{i1} = rP$ and $ID_{i2} = RID_i \oplus H(rP_{pub1})$ where r is a per-session random nonce. Its secret key is then generated as $SK_i = (SK_{i1}, SK_{i2})$ where $SK_{i1} = s_1 ID_{i1}$ and $SK_{i2} = s_2 H(ID_{i1}||ID_{i2})$. Here $H(.)$ is a MapToPoint hash function as in our schemes. When vehicle V_i wants to send the message M_i, it generates the signature $\sigma_i = SK_{i1} + h(M_i)SK_{i2}$ where $h(.)$ is a one-way hash function such as SHA-1. V_i then broadcasts ID_i, M_i and σ_i to the RSU.

RSU verifies the signature σ_i by checking whether $\hat{e}(\sigma_i, P) = \hat{e}(ID_{i1}, P_{pub1})\hat{e}(h(M_i)H(ID_{i1}||ID_{i2}), P_{pub2})$ as

$$\hat{e}(\sigma_i, P)$$
$$= \hat{e}(SK_{i1} + h(M_i)SK_{i2}, P)$$

$= \hat{e}(SK_{i1},P)\hat{e}(h(M_i)SK_{i2},P)$
$= \hat{e}(s_1 ID_{i1},P)\hat{e}(h(M_i)s_2 H(ID_{i1}||ID_{i2}),P)$
$= \hat{e}(ID_{i1},s_1 P)\hat{e}(h(M_i)H(ID_{i1}||ID_{i2}),s_2 P)$
$= \hat{e}(ID_{i1},P_{pub1})\hat{e}(h(M_i)H(ID_{i1}||ID_{i2}),P_{pub2})$
Having the pseudo identity ID_i of vehicle V_i, TA can trace its real identity by using the *TA RID Tracing Routine*: $ID_{i2} \oplus H(s_1 ID_{i1}) = RID_i \oplus H(rP_{pub_1}) \oplus H(s_1 rP) = RID_i$.

11.2 Privacy violation

Any vehicle can obtain $ID_i = (ID_{i1},ID_{i2})$ from V_i's transmissions. Also s_1 is preloaded into each vehicle's tamper-proof device during network deployment. Thus any vehicle can obtain V_i's RID_i by following the *TA RID Tracing Routine*.

11.3 Anti-traceability attack

We describe how a vehicle can make TA unable to trace its real identity from its message sent under the IBV protocol. We denote this kind of attack as an anti-traceability attack.
Assume that in a certain session, the attacking vehicle V_a generates its pseudo identity as $ID_a = (ID_{a1},ID_{a2})$ where $ID_{a1} = rP$ and $ID_{a2} = GARBAGE \oplus H(aP_{pub1})$ where $GARBAGE \in \mathbb{G}$ and r is again a per-session random nonce. V_a then proceeds to generate its secret keys $SK_a = (SK_{a1},SK_{a2})$ where $SK_{a1} = s_1 ID_{a1}$ and $SK_{a2} = s_2 H(ID_{a1}||ID_{a2})$, signs the message M_a by generating the signature $\sigma_a = SK_{a1} + h(M_a)SK_{a2}$ and sends out ID_a, M_a and σ_a to the RSU. Note that RSU can verify the message successfully because $\hat{e}(\sigma_a,P) = \hat{e}(ID_{a1},P_{pub1})\hat{e}(h(M_a)H(ID_{a1}||ID_{a2}),P_{pub2})$. Assume that at a later time, V_a's message M_a causes an accident on the road. RSU forwards V_a's pseudo identity to TA so as to reveal V_a's real identity. However, upon computing $ID_{a2} \oplus H(s_1 ID_{a1}) = GARBAGE \oplus H(rP_{pub_1}) \oplus H(s_1 rP) = GARBAGE$, TA finds that $GARBAGE$ does not match any record at TA. V_a can thus evade its responsibility of causing the accident.

11.4 Impersonation attack

We describe how a vehicle can send messages on behalf of another under the IBV protocol. We denote this kind of attack as an impersonation attack.
Assume that at a certain instance, vehicle V_i with real identity RID_i generates its pseudo identity $ID_i = (ID_{i1},ID_{i2})$, secret keys SK_i and signs message M_i by generating the signature σ_i as usual. While V_i is transmitting, an attacker V_a records ID_i. Later, V_a generates the message M_a. It generates its pseudo identity as $ID_a = (ID_{a1},ID_{a2}) = ID_i = (ID_{i1},ID_{i2})$ and its secret keys as $SK_a = (SK_{a1},SK_{a2})$ where $SK_{a1} = s_1 ID_{a1} = s_1 ID_{i1}$ and $SK_{a2} = s_2 H(ID_{a1}||ID_{a2}) = s_2 H(ID_{i1}||ID_{i2})$. It then signs the message M_a by generating the signature $\sigma_a = SK_{a1} + h(M_a)SK_{a2}$ and sends out ID_a, M_a and σ_a to the RSU.
Similar to the anti-traceability attack, upon receiving V_a's message, RSU can verify it successfully because $\hat{e}(\sigma_a,P) = \hat{e}(ID_{i1},P_{pub1})\hat{e}(h(M_a)H(ID_{i1}||ID_{i2}),P_{pub2})$. Assume at a later time, V_a's message M_a causes an accident on the road. RSU forwards V_a's pseudo identity ID_a as shown in its message to TA so as to reveal its real identity. After computing $ID_{a2} \oplus H(s_1 ID_{a1}) = ID_{i2} \oplus H(s_1 ID_{i1}) = RID_i \oplus H(rP_{pub_1}) \oplus H(s_1 rP) = RID_i$, both RSU and TA think that M_a is being sent by V_i because V_i's instead of V_a's identity is traced. Thus V_a can evade and pass its responsibility of causing the accident to V_i.

10

APALLS: A Secure MANET Routing Protocol

Sivakumar Kulasekaran and Mahalingam Ramkumar
Mississippi State University
USA

1. Introduction

In infra-structured networks dedicated routers relay packets between network hosts. In contrast, in mobile ad hoc networks (MANET) nodes are simultaneously network hosts and routers. Due to their reduced dependence on communication infrastructure MANETs have useful applications in scenarios where it may be impractical to set up expensive communication infrastructure, and in scenarios involving failure of communication infrastructure.

MANET routing protocols are rules to be followed by every mobile node to cooperatively discover optimal paths and route packets between end-points (source and destination nodes). In this chapter we restrict ourselves to the dynamic source routing (DSR) protocol (Johnson & Maltz, 1996). In DSR paths between end-points are established by flooding route-request (RREQ) packets originating from the source, in response to which route-response (RREP) packets are sent along the reverse path. The original DSR protocol (Johnson & Maltz, 1996) implicitly assumes that all nodes will abide by the rules. The presence of nodes that do not adhere to the rules, either deliberately, or due to malfunctioning, can have a deleterious effect on the MANET subnet. Secure MANET routing protocols strive to improve the reliability of the routing process under the presence of non cooperative nodes.

1.1 Securing DSR

Non confirming nodes can inflict a wide variety of passive and active attacks on DSR. Two basic tools employed to thwart attacks are i) mandating cryptographic authentication; and ii) monitoring neighbors to estimate their trustworthiness. A mechanism for cryptographic authentication is also a prerequisite for monitoring. Bootstrapping cryptographic authentication mechanisms mandates an authority trusted by all nodes.

1.1.1 Role of the trusted authority

The trusted authority (TA) provides secrets to nodes, thereby conferring upon them the eligibility to take part in ad hoc subnets. The TA may also revoke the privileges of some nodes, by disseminating revocation lists. The *network size* N is the total number of nodes afforded the privilege of participating in MANET subnets. Any random subset of the N eligible nodes, say N_S nodes (which may accidentally be in the same geographical region), can form a connected ad hoc subnet.

For retaining the most compelling advantage of MANETs (their reduced infrastructural support) it is desirable that the TA is *off-line*. In other words, subnets disconnected from the TA, or any other form of infrastructural support, should still be able to carry out their tasks.

In (Sanzgiri et al., 2002) such MANET networks, where the infrastructural support is required only for predistribution of keys, are referred to as *managed open* networks.

1.1.2 Resisting attacks

Active attacks involve modifying routing packets in ways that violate the prescribed protocol. For resisting active attackers secure routing protocols will require features to *detect* inconsistencies in routing information resulting from active attacks, *identify* perpetrators responsible for such inconsistencies, and have mechanisms in place to limit the role of such nodes. Effective strategies are also required to *deter* active attacks in the first place. For example, the ability to obtain non repudiable proof of active attacks can be an effective deterrent. Such proofs can be submitted to the TA and lead to revocation of such nodes.
Passive attackers may attempt to eavesdrop on messages exchanged between end-points or take selective part in routing. Submitting non-repudiable proof may not be possible[1] for passive attacks involving selective participation. Nevertheless, effective strategies are required to promote self-less participation.
Several secure routing protocols (Abusalah et al., 2008) have been proposed in the literature. Some focus on cryptographic authentication (Hu et al., 2005)-(Capkun and Hubaux., 2003), taking into consideration the resource constraints inherent to battery operated mobile devices. Some have suggested strategies like listening in the promiscuous mode (Marti et al., 2000)-(Marshall et .al, 2003), unambiguously identifying misbehaving nodes (Burmester et al., 2003)- (Awebuch et al., 2002), assigning trust metrics to nodes, and employing such metrics to advantageously influence the routing process.

1.2 Contributions

The contribution of this chapter is APALLS, a secure routing protocol based on DSR. Some of the shortcomings of current protocols that are addressed by APALLS are as follows:

1. Non-repudiable authentication, while necessary, is *not* sufficient for obtaining non-repudiable proof of misbehavior. While the importance of non repudiable authentication is well understood (Sanzgiri et al., 2002),(Sun et al., 2007), investigation of issues in obtaining non repudiable proof of misbehavior (in the context of adhering to a MANET routing protocol) has not received attention thus far. To our knowledge, APALLS is the first protocol which addresses this issue.
2. Protocols that focus on monitoring strategies (Marti et al., 2000)-(Marshall et .al, 2003) have predominantly ignored the relevance of cryptographic authentication. Likewise, protocols that focus on cryptographic authentication ignore strategies for monitoring. Comprehensive routing protocols should productively employ both.
3. While many key distribution strategies have been proposed for ad hoc networks, they have not considered realistic network models. The choice of schemes in APALLS are driven by realistic models for large scale MANET deployments.

APALLS[2] borrows some features from Ariadne (Hu et al., 2005), a popular secure extension of DSR. Ariadne does not prescribe strategies for monitoring, and consequently, does not possess mechanisms to address passive attacks involving selective participation. The cryptographic authentication strategies in Ariadne permit the source or destination to detect inconsistencies

[1] While a packet signed by a node B can be used by A to demonstrate that B violated the protocol, it is arguably infeasible for a node A prove to some third party that B did *not* forward a RREQ/RREP.

[2] Ariadne with Pairwise Authentication and Link Layer Signatures.

in path advertisements in (RREQ or RREP packets) resulting from active attacks. However, Ariadne does not attempt to *identify* the reason for the observed inconsistency, namely, the perpetrator responsible for the inconsistency. An implication of the fact that perpetrators remain unidentified is that an attacker faces *no risk* in carrying out (several types of) active attacks. Such attacks, which can reduce the efficacy of the path discovery mechanism, can be carried out repeatedly due to lack of deterrents.
APALLS provides a severe deterrent for active attacks, the threat of revocation from the network, as non repudiable proof can be submitted to the off-line TA. APALLS also recognizes that the process of revocation, involving submission of verifiable proof to the TA (when ever the node submitting the proof has access to the TA), followed by network wide dissemination of revocation lists by the TA, will not yield immediate relief from the active attacker(s) in ad hoc subnets. For this purpose, APALLS includes features to route around nodes suspected of misbehaving. APALLS also includes elements to keep an effective watch on neighbors to address passive attacks involving selective participation. Nevertheless, due to careful choice of cryptographic primitives, APALLS incorporates all these features without placing unreasonable demands on the capabilities of nodes.
The rest of this chapter is organized as follows. In Section 2 we provide an overview of DSR and Ariadne. In Section 3 we outline a generic managed-open MANET model consisting of an off-line trusted authority (TA). The threat model and the intended goals of APALLS are enumerated in Section 3.1. This is followed by a description of mechanisms for bootstrapping trust relationships between nodes of the network.
Section 4 provides the description of the APALLS protocol. The security analysis of APALLS, including rationale for the choices made, are described in Section 5. A discussion of related work can be found in Sections 6.1 and 6.2. Conclusions are offered in Section 6.3.
The following notations are used in this chapter

1. $A, B, C \cdots$ (upper case alphabets) represent unique identities of nodes;
2. $\Sigma_A = \langle M \rangle_A$ - digital signature of A for a message M.
3. $h()$ - a cryptographic second pre-image resistant hash function (not necessarily colission resistant);
4. $h(M, K)$ - hashed message authentication code (HMAC) for a message M based on a key K.
5. $C = K[P]$ encryption of a plain-text[3] P using a key K and some block cipher like AES/DES;
6. $P = K^{-1}[C]$ decryption of cipher-text C using key K.
7. $K_A^0 \cdots K_A^{L-1}$ - hash-chain of length L with commitment K_A^L, where $K_A^i = h(K_A^{i-1}), 1 \leq i \leq L$.

2. Background

Dynamic source routing (DSR) is an on-demand protocol where a node S desiring to find a path to a node T broadcasts a route-request (RREQ) packet indicating the source S, sequence number q, destination T, and a hop-limit n_h. RREQ packets are flooded. In general, every node in the same connected subnet will receive one RREQ (with the same source and sequence number) from each neighbor, but will rebroadcast only one (usually the first RREQ received). Every node rebroadcasting an RREQ inserts its identity.

[3]If the plain-text P is larger than the block size it is assumed that some appropriate mode of operation like cipher-block-chaining (CBC) is used, and the initial value is prepended to the cipher-text C.

For an RREQ from a source S, intended for a destination T, traversing a path $(A,B,C,\ldots)$

$$\begin{aligned} \mathcal{Q}^S_{(q,S)} &= \mathcal{Q}_{(q,S)} = [S,q,T,n_h] \\ \mathcal{Q}^A_{(q,S)} &= [\mathcal{Q}^S_{(q,S)},(A)] \\ \mathcal{Q}^B_{(q,S)} &= [\mathcal{Q}^S_{(q,S)},(A,B)] = [\mathcal{Q}^A_{(q,S)},(B)] \\ \mathcal{Q}^C_{(q,S)} &= [\mathcal{Q}^S_{(q,S)},(A,B,C)] = [\mathcal{Q}^B_{(q,S)},(C)], \\ &\vdots \end{aligned} \tag{1}$$

where the $\mathcal{Q}^X_{(q,S)}$ represents all RREQ fields relayed by a node X, and received by the destination.

When the RREQ packet reaches the destination[4], it contains the fields $\mathcal{Q}_{(q,S)}$ specified by the source, and the list of all nodes in the path traversed by the RREQ. The destination sends a route-response (RREP) packet along the reverse path. The source and destination may in general discover multiple paths as the destination can receive one RREQ from every neighbor.

2.1 Attacks on DSR

Attacks on DSR can be broadly classified into passive, semi-active, and active attacks. Passive attackers may perform eavesdropping on application data packets exchanged between nodes, or take selective part in the network. Semi-active attackers may act as invisible relays to create misrepresentations of the network topology. Such an attacker positioned between two nodes A and B (who are out of each other's range) may simply echo packets broadcast by A such that B can receive such packets. However by echoing RREQ packets and not echoing RREP packets, the attacker can cause the route discovery to fail. Interconnected invisible relays that are physically well separated can create *worm-holes* (Hu et al., 2001) to misrepresent the subnet topology. By periodically turning on and off such worm-holes they can create rapid changes to the perceived topology, thereby inflicting significant bandwidth overhead.

Active attackers intentionally modify routing packets in violation of the protocol - for example, by inserting nonexistent nodes in the path, or deleting nodes that are actually in the path, or modifying the values inserted in the RREQ by the source and/or other upstream nodes. DSR is also susceptible to rushing attacks (Hu et al., 2003) which result from the ability of an attacker to forward RREQ packets with the intent of either creating sub-optimal routes, or even causing complete failure of the route establishment process. As each node forwards only one RREQ packet, a rushed bad packet can *preempt* other good packets.

Two basic tools employed to thwart attacks attacks are i) mandating cryptographic authentication; and ii) monitoring neighbors to estimate their trustworthiness. As an important pre-requisite for monitoring is the need to know *who* is being monitored[5], cryptographic authentication is necessary for the ability to monitor.

2.2 Cryptographic authentication

Facilitating cryptographic authentication schemes requires a trusted authority (TA) who conveys secrets to every node and/or public values associated with every node to every

[4]While in the original DSR even intermediate nodes with the knowledge of a path to the destination can invoke an RREP, in most secure DSR extensions only the destination is allowed to do so.

[5]A node A observing the behavior of a neighbor claiming-to-be-B should be able to ascertain that the observed node is indeed B before A can make any inferences about the B's trustworthiness.

node in the network. Such values provided by the TA are necessary for computing verifiable *cryptographic authentication tokens*. As packets without verifiable cryptographic authentication will be ignored, the TA confers the eligibility for nodes to take part in MANET subnets.
Cryptographic authentication can be classified into mutual (or one-to-one) authentication and broadcast (one-to-many) authentication. A secret K_{AB} known only to A and B can be used by A and B for computing and verifying pairwise authentication tokens. For a mesage M the token is typically a hashed message authentication code (HMAC) $T_{AB}(M) = h(M, K_{AB})$. Schemes for pairwise authentication thus rely on key distribution schemes that facilitate pairwise secrets between all node-pairs.
An one-to-many authentication token $T_A(M) = (R_A, M)$ for a message M can be *created* only by a node with access to a secret key R_A. The token can be verified using a public counterpart U_A of the secret R_A. The verification function $f_{ver}(M, T_{AM}, U_A)$ provides a binary output (pass or fail). If the verification test passes, the verifier can conclude that only an entity with the private counterpart of U_A could have computed the token $T_A(M)$. In addition, if it is possible for the verifier to establish that the public value U_A is indeed associated with a node A, the verifier can conclude that the token was created by A. Facilitating such schemes calls for the TA to distribute authentic public values associated with every node to every node.

2.2.1 MANET layers

In the context of DSR, "application layer" cryptographic mechanisms are required for end-points to protect the privacy and integrity of data exchanged between them. "Link layer" mechanisms are required for authentication of neighboring nodes. "Network layer" mechanisms are required for authentication of intermediate nodes that take part in relaying packets between end-points.
How end-points choose to protect application data should be left to the end-points themselves; ideally, such strategies should *not* not be under the control of the TA. However, strategies for link layer and network layer authentication *should* be promulgated by the TA.

2.3 Ariadne

In Ariadne the end-points share an application layer secret. The authentication strategies facilitated by the TA include i) a sequence-number hash chain to limit the number of RREQs that can be sent by any node; ii) TESLA broadcast authentication (Perrig et al., 2001) for authentication of intermediate nodes by an end-point; and iii) a network-wide secret at the link layer to deny access to external nodes.
In the sequence-number hash chain of a node A, $\mathcal{S}_A = \{R_A^0, R_A^1, \ldots, R_A^{n_r}\}$, $R_A^i = h(R_A^{i-1})$ for $1 \leq i \leq n_r$, and the commitment $R_A^{n_r}$ is made known to every node by the TA. Along with an RREQ with sequence number q initiated by A the value $R_A^{n_r - q}$ is released from its sequence-number hash chain.
In the TESLA hash chain of A, $\mathcal{H}_{t_0,\Delta}^A = \{K_A^0, K_A^1, \ldots, K_A^{L-1}, K_L^A\}$ with commitment K_L^A, Δ is a time-interval (for example, 1 second), and t_0 is an absolute value of time (for example, Dec 1, 2009, 0200:00, GMT). The parameters associated with A's chain (commitment K_L^A, Δ and t_0) are made known to all nodes by the TA. This TESLA chain can be used by A only for a limited segment of time (between t_0 and $t_0 + (L-1)\Delta$).
A is expected to keep the key K_A^{L-i} in its chain private at least till time $t_i = t_0 + i\Delta$. This key can be used for authenticating a value M by appending a HMAC $h(M, K_A^{L-i})$, provided the HMAC reaches potential verifiers *before* time t_i. Once K_A^{L-i} is made public (*after* time t_i) the

verifier can i) verify the TESLA HMAC; and (if consistent) ii) repeatedly hash K_A^{L-i} (i times) and verify that the result is indeed K_L^A. The verifier can then conclude that the HMAC was computed by A (as only A had access to the value K_A^{L-i} at time t_i).

2.3.1 RREQ and RREP

The steps involved in RREQ propagation from a source S to a destination T through a path $(A,B,C,\ldots)$ are depicted in Table 1 (left column). $\bar{K}_{ST}$ is the application layer shared by the end-points. The RREQ fields $Q^S_{(q,S)}$ specified by the source includes a time t_i before which the destination should receive the RREQ, and a value $R_S^{n_r-q}$ from S's sequence number chain. Intermediate nodes and the destination will process the RREQ only if i) (S,q) is fresh (the node had not previously seen an RREQ from S with a sequence number q or higher); ii) hashing $R_S^{n_r-q}$ q-times yields the commitment $R_S^{n_r}$ of S; and iii) the RREQ is received before time t_i.
Apart from the fields $Q^S_{(q,S)}$ which are carried forward all the way to the destination, the source S broadcasts a value β_S which is intended only for neighbors. The value β_S is chosen such that the destination T (which shares the secret $\bar{K}_{ST}$ with the source) can also compute β_S.
A node A downstream of S appends two values before rebroadcasting the RREQ as $Q^A_{(q,S)} = [Q^S_{(q,S)},(A,M_A)]$, where M_A is a TESLA HMAC computed using a value from A's TESLA chain which will remain A's secret till time t_i. In addition, a per-hop hash value $\beta_A = h(\beta_S,A)$ is also broadcast by A. Similarly, a node B downstream of A broadcasts $Q^B_{(q,S)} = [Q^A_{(q,S)},(B,M_B)]$ and $\beta_B = h(\beta_A,B)$. Unlike TESLA HMACs, the per-hop hashes are not carried forward (they are intended only for neighbors).
The destination T computes β_S (as computed by the source) and verifies that the per-hop hash submitted by the last node in the path is consistent with the list of nodes in the path. The per-hop hash is intended to prevent node deletion attacks. Without this, a node C can simply remove the values (B,M_B) in the RREQ, and thus illegally delete B from the path. With the per-hop hash, to trick the destination into accepting a path $(A,C,\ldots)$ as valid, C will need access to the per-hop hash β_A which is privy only to neighbors of A (and C is not one). The destination will invoke an RREP only if the per-hop hash is consistent. The RREP includes all fields of the RREQ packet and is authenticated to the source using a HMAC (based on secret $\bar{K}_{ST}$).
After time t_i, the destination relays the RREP along the reverse path. Every node releases the value from the TESLA chain used for computing the TESLA HMACs during the forward path. At the end of the reverse path the source i) verifies the TESLA HMACs, and that ii) the TESLA keys are consistent with the time t_i and their respective commitments. Mandating authenication prevents illegal node insertions. In Ariadne node deletion attacks can be detected by the destination at the end of the forward path; node insertion attacks can be detected at the end of the reverse path, by the RREQ source.

3. APALLS: application model and goals

As an application of MANETs consider a (fictitious) network operator, Tingular, who desires to provide subscribers with a wide variety of services with minimal investment in infrastructure. Tingular subscribers can find other Tingular subscribers within range and form multi-hop MANET subnets. Nodes (Tingular subscribers) in a connected subnet can exchange messages with each other. Apart from communicating with other nodes in the subnet, any node with

wide-area connectivity (for example, access to the Internet) can act as a gateway and extend Internet access to all other nodes in the subnet.
The total number of Tingular subscribers at any time t, say $N(t)$, can be of the order of several millions (and varies with time t as new subscribers may join, and some may leave the Tingular network). Any small subset of current subscribers who happen to be in a geographical region like a mall, or an airport, can come-together to form a Tingular subnet (a small subnet may have just two nodes). Several Tingular subnets may operate simultaneously at different locations. While some subnets may have one or more nodes with wide-area network access, some may be completely isolated from the rest of the world. Tingular subnets will thus be able to function even during periods of natural disasters, when other communication infrastructure fail.
The main tasks to be performed by Tingular for *managing* the network are

1. to specify rules (the secure routing protocol) to be followed (a one-time process);
2. induct subscribers into the Tingular network, by providing them with secrets, possibly in exchange for a small subscription fee (performed once for every node inducted into the Tingular network); and
3. (periodically) revoke subscribers who violate rules, or do not renew their subscriptions.

Inducting a subscriber can be as simple as providing a SIM card (subscriber identity module) to the subscriber, preloaded with the secrets necessary for the node to communicate with other Tingular subscribers, and thus take part in any Tingular subnet. A "Tingular node" can be any WiFi enabled general-purpose computer (hand-held, laptop or desktop) which can house the SIM card, and can run Tingular software (which dictates the rules to be followed by Tingular nodes). Revoking nodes can be through periodic dissemination of revocation lists (posted in Tingular's website). Due to the modest investment required, Tingular can afford to provide useful services for a low subscription fee.

3.1 Threat model and goals

Some Tingular nodes in a subnet may engage in active attacks. Some such attacks may result from mere malfunctioning of nodes. Some may be perpetrated by Tingular nodes under the control of attackers. One goal of APALLS is to obtain non repudiable proof of active attacks. Such proofs can be submitted to the TA at a convenient time (as access to the TA may not be available from some ad hoc subnets), perhaps leading to revocation of such nodes from the Tingular network.
The revocation process will involve i) submission of proof to the TA, ii) verification of such proofs by the TA, and iii) dissemination of revocation lists by the TA. While the threat of revocation can be an effective deterrent, the process of revocation may not provide immediate relief from attackers in the subnet. For this purpose, APALLS will include explicit features to improve the chance of finding paths free of suspected active attackers.
Passive attacks include acts like not forwarding RREQ packets, not accepting RREP or data packets, etc. Selfish subscribers may not desire to take part in the subnet unless they require to communicate with another node. There may also exist other external attackers (who may not be Tingular nodes) performing semi-active attacks. While obtaining non repudiable proof of misbehavior is not possible for passive and semi-active attacks, APALLS aims to have tangible measures for promoting self-less behavior, and "living with" semi-active attackers.

3.2 Key distribution for APALLS

The operator of the Tingular network employs a web-server, with a certified trustworthy computer (for example, a cryptographic co-processor) at the back end. The trustworthy module is the TA. Tingular also possesses a facility for securely preloading SIM cards with secrets provided by the TA.

The TA i) generates an asymmetric key pair (R_{TA}, U_{TA}); and ii) chooses a master secret μ. Potential subscribers create a Tingular user account. On payment of the subscription fee subscribers receive a Tingular SIM card with some secrets.

The identity A assigned to a subscriber is of the form $A = [Q_A \parallel A']$ where Q_A is a serial number, and A' may be the user name. The node A is also issued an asymmetric key pair (R_A, U_A), and a certificate $C_A = E(R_{TA}, A \parallel U_A \parallel T_e)$, where T_e is time of expiry of C_A. The certificate can be decrypted any any node with access to the public key U_{TA} as $[A \parallel U_A \parallel T_e] = D(U_{TA}, C_A)$. The specifics of the functions $E()$ and $D()$ depend on the type of asymmetric primitive employed.

The subscriber sequence number is issued in a chronological order. The sequence number for the first subscriber is 1. A node A with (say) $Q_A = 1{,}000{,}000$ is the millionth subscriber. The values included in the SIM card provided to A are

1. a secret $K_A = h(\mu, A)$;
2. the private key R_A;
3. the public key of the TA, U_{TA}; and
4. the certificate C_A.

In addition, A receives $Q_A - 1$ public values of the form

$$P_{AX} = h(K_A, X) \oplus h(K_X, A), \forall X = [Q_X \parallel X'] | Q_X < Q_A,$$

where a specific $X = [Q_X \parallel X']$ represents the identity assigned to a subscriber inducted *before* A (or $Q_X < Q_A$). The public values could be downloaded from Tingular web site or mailed by Tingular (in a flash card / optical disc) over the postal network. The millionth subscriber A will receive 999,999 such public values.

The shared secret K_{AB} between two subscribers $A = Q_A \parallel A'$ and $B = Q_B \parallel B'$ is computed as follows. If $Q_A < Q_B$ then B has access to the public value P_{AB} (and A does not). The secret K_{AB} is computed as

$$K_{AB} = \begin{cases} h(K_A, B) & \text{by } A \\ h(K_B, A) \oplus P_{AB} & \text{by } B \end{cases} \quad (2)$$

The scheme described above for facilitating pairwise secrets is based on the modified Leighton-Micali scheme (MLS) (Ramkumar , 2008). If the public values (and consequently, the pairwise secrets) are 80 bits long, the ten-millionth subscriber will require about 100 MB of storage (perhaps in a flash card plugged into the hand-held computer). The maximum network size is not a hard limit. Newer subscribers (with larger sequence numbers) will just need more storage for public values.

If unlimited network sizes are desired, scalable KPSs are viable options (see Appendix 8.1).

4. The APALLS protocol

Any subscriber, after receiving the secrets and the required public values, can take part in any ad hoc subnet created by any subset of Tingular subscribers.

4.1 Joining a subnet

A subscriber A sends a probe $[A,c]$ (where c is a randomly chosen challenge) to determine other subscribers within range. A node B in the neighborhood responds with $[B,A,K_{BA}(c)]$. In general, the node A may receive one response from every neighbor within the reliable delivery neighborhood[6] (RDN) of A. Let us assume that A receives responses from 3 nodes X,Y and Z in its RDN. Node A then chooses a random one-hop group secret G_A and broadcasts individual encryptions of the secret as $[X,Y,Z,K_{AX}(G_A),K_{AY}(G_A),K_{AZ}(G_A)]$ to its neighbors. In response, A's neighbors X,Y,Z are expected to send their respective one-hop group secrets (G_X,G_Y,G_Z) to A.

Every packet sent/relayed by a node A is encrypted with the secret G_A. In other words, every node enforces a *private logical neighborhood* (PLN) (Sivakumar and Ramkumar, 2008). The PLN of a node A is (in general) a subset of nodes in the RDN of A. Node A explicitly invites some nodes into its PLN by providing a secret G_A individually to each node. If a node A, with nodes X, Y and Z in its PLN, suspects X of i) violating the protocol, or ii) acting in a selfish manner, or iii) suspect a semi-active attacker between A and X, then A can simply cut-off X from its PLN by providing a new secret to Y and Z (and withholding the secret from X). In a scenario where A has provided its PLN secret G_A to node X, but X did not send its PLN secret G_X to node A (or X did not accept A into its PLN), node A will eject X from its PLN during the next broadcast by A - which is encrypted using a new PLN secret $G_A{}'$, and the secret $G_A{}'$ conveyed only to nodes that A desires to retain in its PLN (and possibly other nodes that A desires to induct into its PLN). Thus, the PLN secret of A is updated whenever a node leaves the PLN (or is ejected by A from its PLN).

Every transmission by A is monitored by all nodes in the PLN of A, and checked for consistency with the protocol. Nodes that are observed to violate rules face the risk of being ejected from the PLN of the observer. Every node also maintains a revocation list periodically disseminated by the network operator. The revocation list can be a list of sequence numbers, signed using the TA's private key. Nodes are expected check their revocation lists before accepting a node into their PLNs. Nodes also broadcast their public key certificates to their PLN neighbors. The certificates are decrypted, and the authenticated public key of neighbors are cached.

4.2 RREQ propagation in APALLS

Table I depicts the sequence of operations involved in RREQ propagation from a source S to a destination T, through a path $(A,B,C,\ldots)$. To facilitate comparison with Ariadne the sequence of operations performed in Ariadne are also shown in the left column of Table I.

4.2.1 RREQ source S

The RREQ $Q^S_{(q,S)}$ from source S specifies a maximum hop-count n_h. The RREQ can (optionally) include a list of nodes (BL) black-listed by the RREQ source. The RREQ $Q^S_{(q,S)}$ includes the digital signature Σ_S of the source. In addition, the broadcast by S includes a per-hop hash β_S (as in Ariadne) intended only for nodes in the PLN; In Table I values which are not carried forward (and intended only for nodes in the PLN) are shown enclosed in flowered braces.

[6] The RDN of A includes all nodes with which A has confirmed that bi-directional links exist.

The PLN secret is used for encrypting all transmissions to neighbors. Every transmission by a node X is prepended[7] with the identity X of the node (in the clear) to enable the receiver to determine that the secret G_X should be used for decrypting the packet.

4.2.2 Intermediate nodes

Intermediate nodes verify the signature of the RREQ source. The public key certificate C_S required for verification of the signature Σ_S of the source can be included in the first RREQ sent by S in the subnet. Certificates are cached by other nodes in the subnet. If a node reviving the RREQ from S does not have access to C_S, it requests the upstream node to provide the certificate. Nodes will not rebroadcast the RREQ until they gain access to the public key of the source. If an intermediate node is included in the list BL, it simply ignores then RREQ. If an intermediate node receives an RREQ forwarded by a node in the list, the RREQ is ignored.
A neighbor A downstream of S decrypts the broadcast from S (using the PLN secret G_S provided by S) and verifies the signature Σ_S. Every intermediate node appends three values that are carried forward all the way to the destination. The fields $\mathcal{Q}^A_{(q,S)}$ broadcast by A include the triple (A, M_A, ν^S_A). M_A is a HMAC for the destination T computed using the secret K_{AT}. The value $\nu^S_A = K_{AT}[\beta_S]$ (which is not present in Ariadne) is the "encrypted upstream per-hop hash" (Sivakumar and Ramkumar, 2008).
Intermediate nodes also append 4 values which are intended only for nodes in the PLN. The 4 values broadcast by a node C are i) the per-hop hash β_C (as in Ariadne); ii) hash of the signature of the previous hop B - or $\sigma_B = h(\Sigma_B)$; iii) a value v_B, which is a one-way function of signatures appended by all nodes upstream of B; and iv) the signature Σ_C, computed over the values $\mathcal{Q}^C_{(q,S)}$, $v_c = h(v_B, \sigma_B)$, and β_C. For the first node in the path (A) the values σ, and v are not required. For notational consistency values that are not required are indicated as NULL. For nodes like B that are in the PLN of the A, the value v_A is NULL (as there is no intermediate node upstream of A).
In both Ariadne and APALLS (and more generally, any DSR-based protocol) a node with k neighbors will receive k RREQs (one from each neighbor) with the same source and sequence number. In Ariadne an intermediate nodes needs to store only one RREQ (corresponding to which an RREQ was relayed by the node), for a small duration, within which it is reasonable to expect an RREP. In APALLS an intermediate node with k neighbors will need to verify $k+1$ signatures for every RREQ (with the same source and sequence number): i) the signature of the RREQ source and ii) the signature appended by k neighbors. All k RREQs (along with the 4 one hop values) are cached for a small duration. Each node broadcasts one signed RREQ. While the signature of the RREQ source is verified by all nodes, signatures of intermediate nodes are verified *only* by PLN neighbors.

4.3 Route response

In Ariadne if any inconsistency is detected the destination simply drops the RREQ. In APALLS the destination takes some proactive steps to isolate the problem.
The destination computes $\beta_S = h(\mathcal{Q}_{(q,S)}, K_{ST})$ in the same manner computed by the RREQ source. The destination then proceeds to check the *self-consistency* and *consistency* of every node in the path. The values appended by an intermediate node C, viz., M_C and ν^B_C are deemed self-consistent by the destination T if

$$M_C = h(\mathcal{Q}^B_{(q,S)}, (C, \nu^B_C), h(K^{-1}_{CT}[\nu^B_C], C), K_{CT}). \quad (3)$$

[7]This is not shown in Table 1 to reduce notational complexity.

At Node S	
$\mathcal{Q}_{(q,S)} = [S,q,T,t_i]$	$\mathcal{Q}_{(q,S)} = [S,t,T,n_h,\langle BL\rangle]$
$\beta_S = h(\mathcal{Q}^S_{(q,S)},\bar{K}_{ST})$	$\beta_S = h(\mathcal{Q}_{(q,S)},K_{ST}), u = h(\beta_S)$
	$\Sigma_S = \langle \mathcal{Q}_{(q,S)},u\rangle_S$
$\mathcal{Q}^S_{(q,S)} = [\mathcal{Q}_{(q,S)},R_S^{n_r-q}]$	$\mathcal{Q}^S_{(q,S)} = [\mathcal{Q}_{(q,S)},u,\Sigma_S]$
$S \rightarrow * K_U[\mathcal{Q}^S_{(q,S)},\{\beta_S\}]$	$G_S[\mathcal{Q}^S_{(q,S)},\{\beta_S\}]$
At Node A	
$\beta_A = h(\beta_S,A)$	$\beta_A = h(\beta_S,A)$
	$\nu^S_A = K_{AT}(\beta_S)$
$M_A = h(\mathcal{Q}^S_{(q,S)},\beta_A,K_A^{L-i})$	$M_A = h(\mathcal{Q}^S_{(q,S)},\nu^S_A,\beta_A,K_{AT})$
$\mathcal{Q}^A_{(q,S)} = [\mathcal{Q}^S_{(q,S)},(A,M_A)]$	$\mathcal{Q}^A_{(q,S)} = [\mathcal{Q}^S_{(q,S)},(A,\nu^S_A,M_A)]$
	$\Sigma_A = \langle \mathcal{Q}^A_{(q,S)},\beta_A\rangle_A$
$A \rightarrow * K_U[\mathcal{Q}^A_{(q,S)},\{\beta_A\}]$	$G_A[\mathcal{Q}^A_{(q,S)},\{\beta_A,\Sigma_A,NULL,NULL\}]$
At Node B	
$\beta_B = h(\beta_A,B)$	$\beta_B = h(\beta_A,B)$
	$\nu^A_B = K_{BT}(\beta_A)$
$M_B = h(\mathcal{Q}^A_{(q,S)},\beta_B,K_B^{L-i})$	$M_B = h(\mathcal{Q}^A_{(q,S)},\nu^A_B,\beta_B,K_{BT})$
$\mathcal{Q}^B_{(q,S)} = [\mathcal{Q}^A_{(q,S)},(B,M_B)]$	$\mathcal{Q}^B_{(q,S)} = [\mathcal{Q}^A_{(q,S)},(B,\nu^A_B,M_B)]$
	$\sigma_A = h(\Sigma_A), \upsilon_B = h(\sigma_A,NULL)$
	$\Sigma_B = \langle \upsilon_B,\mathcal{Q}^B_{(q,S)},\beta_B\rangle_B$
$B \rightarrow * K_U[\mathcal{Q}^B_{(q,S)},\{\beta_B\}]$	$G_B[\mathcal{Q}^B_{(q,S)},\{\beta_B,\Sigma_B,\sigma_A,NULL\}]$
At Node C	
$\beta_C = h(\beta_B,C)$	$\beta_C = h(\beta_B,C)$
	$\nu^B_C = K_{CT}(\beta_B)$
$M_C = h(\mathcal{Q}^B_{(q,S)},\beta_C,K_C^{L-i})$	$M_C = h(\mathcal{Q}^B_{(q,S)},\nu^B_C,\beta_C,K_{CT})$
$\mathcal{Q}^C_{(q,S)} = [\mathcal{Q}^B_{(q,S)},(C,M_C)]$	$\mathcal{Q}^C_{(q,S)} = [\mathcal{Q}^B_{(q,S)},(C,\nu^B_C,M_C)]$
	$\sigma_B = h(\Sigma_B), \upsilon_C = h(\sigma_B,\upsilon_B)$
	$\Sigma_C = \langle \upsilon_C,\mathcal{Q}^C_{(q,S)},\beta_C\rangle_C$
$C \rightarrow * K_U[\mathcal{Q}^C_{(q,S)},\{\beta_C\}]$	$G_C[\mathcal{Q}^C_{(q,S)},\{\beta_C,\Sigma_C,\sigma_B,\upsilon_B\}]$
At Node D	
$\beta_D = h(\beta_C,D)$	$\beta_D = h(\beta_C,D)$
	$\nu^C_D = K_{CT}(\beta_C)$
$M_D = h(\mathcal{Q}^C_{(q,S)},\beta_D,K_D^{L-i})$	$M_D = h(\mathcal{Q}^C_{(q,S)},\nu^C_D,\beta_D,K_{DT})$
$\mathcal{Q}^D_{(q,S)} = [\mathcal{Q}^C_{(q,S)},(D,M_D)]$	$\mathcal{Q}^D_{(q,S)} = [\mathcal{Q}^C_{(q,S)},(D,\nu^C_D,M_D)]$
	$\sigma_c = h(\Sigma_C), \upsilon_D = h(\sigma_c,\upsilon_C)$
	$\Sigma_D = \langle \upsilon_D,\mathcal{Q}^D_{(q,S)},\beta_D\rangle_D$
$D \rightarrow * K_U[\mathcal{Q}^D_{(q,S)},\{\beta_D\}]$	$G_C[\mathcal{Q}^D_{(q,S)},\{\beta_D,\Sigma_D,\sigma_C,\upsilon_C\}]$
$\vdots$	

Table 1. RREQ (From Source S to Destination T, Sequence Number q) Propagation in Ariadne (left) and APALLS (right) over a path $(A,B,C,D,\ldots)$.

A self-consistent node C, with an upstream node B is deemed *consistent* only if the per-hop hash C claims to have received from B, viz, $h(K_{CT}^{-1}[\nu_C^B])$, matches what B claims to have broadcast, viz., $h(K_{BT}^{-1}[\nu_B^A],B)$. Verifying the consistency of C is possible only if B is found to be self-consistent, or

$$M_B = h(\mathcal{Q}_{(q,S)}^A, (B,\nu_B^A), h(K_{BT}^{-1}[\nu_B^A],B), K_{BT}). \tag{4}$$

Corresponding to every path through which the destination T receives an RREQ with *no* inconsistencies, the destination invokes an RREP of type $SUCC$. For an RREQ received through a path (A,B,W,G,H,P), the RREP is of the form

$$\mathcal{P}_q = [(u,\beta_S),\{SUCC:(A,B,W,G,H,P)\}] \tag{5}$$

The RREP is authenticated to the source using a HMAC based on the secret K_{ST}. The value u is a convenient index to the RREQ. That the RREP includes β_S (the preimage of u) informs the intermediate nodes that the destination did indeed invoke an RREP.
Some paths may contain one or more inconsistent nodes. As an example, consider a scenario where an RREQ indicates a path (A,B,C,D,E,F), and the destination determines that while D, E and F are consistent, C's consistency cannot be verified because B was not self-consistent. The RREP sent by T is then

$$\mathcal{P}_q = [(u,\beta_S),\{FAIL:((B\lambda C),(D,E,F))\}] \tag{6}$$

The RREP is authenticated individually (using individual HMACs) for verification by the consistent nodes D,E,F and the last self-consistent node C. The special code λ indicates that an inconsistency was detected in the RREQ. Nodes that are identified as consistent (D, E and F) consider this RREP as an instruction to drop subsequent RREQs from S to T *if* they include C or B in the path (for example, if the source S sends a second RREQ after the first one times out). This is also considered by nodes D and C as a request to store RREQs they had received from their respective upstream nodes (C and B) in non volatile storage, for submission to the TA at the earliest opportunity (for example, when the node has access to the Internet).

5. Rationale and security analysis

Some significant differences between Ariadne and APALLS are i) use of pairwise secrets instead of TESLA; ii) mandating the RREQ source to append a digital signature; iii) enforcing a PLN; iv) use of an additional upstream per-hop hash; and v) a digital signature appended by every node broadcasting an RREQ, intended only for one-hop neighbors.

5.1 Choice of cryptographic authentication strategies

While Hu et al (Hu et al., 2005) preferred TESLA, Ariadne was also designed to support pairwise secrets (Ariadne-PS) or digital signatures (Ariadne-DS) instead of TESLA. In Ariadne-DS every intermediate node appends a digital signature (instead of an HMAC), which is carried forward all the way till the destination. The obvious disadvantage of using digital signatures is the overhead. This is exacerbated in Ariadne due to the fact that signatures and public key certificates appended by every node in the RREQ path have to be carried over all the way to the destination. Another disadvantage is the increased susceptibility to trivial denial of service (DoS) attacks due to the high verification complexity. An attacker can send random "signed packets" which can be recognized as bogus only after the expensive verification process.

When pairwise secrets are used the HMACs by intermediate nodes are computed as in APALLS. End-points can also readily use a secret K_{ST} for authenticating RREQ/RREP instead of relying on an out-of-network mechanism for establishing an application layer secret $\bar{K}_{ST}$. No additional values need to be released during the reverse path; the destination can detect both deletion and insertion attacks. Furthermore, issues related to the delay sensitivity of TESLA based authentication (Sivakumar and Ramkumar, 2008) can also be avoided.

Another advantage of the fact that an out-of-network mechanism is not required to establish pairwise secrets between end-points is that it opens up mechanisms for salvaging broken routes, or re-routing RREPs. In a scenario where a node D in the path between S and T detects that the path is broken, if D does not share readily a secret with S or T (in Ariadne-TESLA and Ariadne-DS), it cannot raise an RREQ to S or T to find alternate paths. In Ariadne-PS, D can do so.

The primary differences in rationale for the choice of key distribution strategies in Ariadne and APALLS stem from the differences in the assumed network model.

5.1.1 Pairwise secrets instead of TESLA

The reason Hu et al (Hu et al., 2005) preferred TESLA over pairwise secrets was that "broadcasting N commitments is easier than distributing $\binom{N}{2}$ secrets." This is indeed true if a trusted entity is available in every subnet to broadcast commitments. Broadcasting $N_s = 200$ (where N_s is the number of nodes in a specific subnet) TESLA commitments within a subnet is indeed preferable to unicasting $\binom{N_s}{2} = 19,000$ values (or unicasting 199 unique values to every node).

However, keeping in mind that one of the most compelling advantages of MANET is the ability to operate without infrastructural assistance (like a trusted entity in every subnet), for large scale dynamic MANET networks ($N(t)$ of the order of several millions) distributing TESLA commitments can be more expensive. The only practical approach may be for each node to possess a certified asymmetric key pair which is used for authenticating TESLA commitments. In Ariadne, where the TESLA HMACs can be verified only by the destination, an intermediate node X will need to release (during the RREP), a value from its TESLA chain, and in addition, i) the TESLA commitment; ii) a digital signature of X for the commitment; and iii) a certificate (issued by the TA) for the public key X.

5.1.2 RREQ signatures

In Ariadne the choice of the strategy for controlling RREQ floods is also based on the implicit assumption that a trusted entity is available in every subnet. In the absence of such an entity in each subnet, an attacker can collect sequence-number hash chain values made public in one subnet to flood RREQs in other subnets. Thus, for the assumed network model, sequence-number chains are not useful.

Furthermore, while the RREQ hash chain value provides implicit authentication to two fields (source and sequence number), it does not protect other fields like the destination. Thus, if an active attacker forwards the RREQ after changing the destination field, such an RREQ will be propagated by downstream nodes. To authenticate all RREQ fields specified by the source Hu et al suggested the use of chained one-time signature (OTS) schemes (Hu et al., 2003). In chained OTS schemes, in a chain $C_0, C_1, \ldots C_L$ the values (C_i, C_{i+1}) are keys pairs of an OTS scheme which can be used to sign the RREQ. The signer can use C_i to compute a signature which can be instantaneously verified by any node with access to C_{i+1}. The next RREQ from the source will make C_i public and use C_{i-1} to sign the message. This approach also relies on

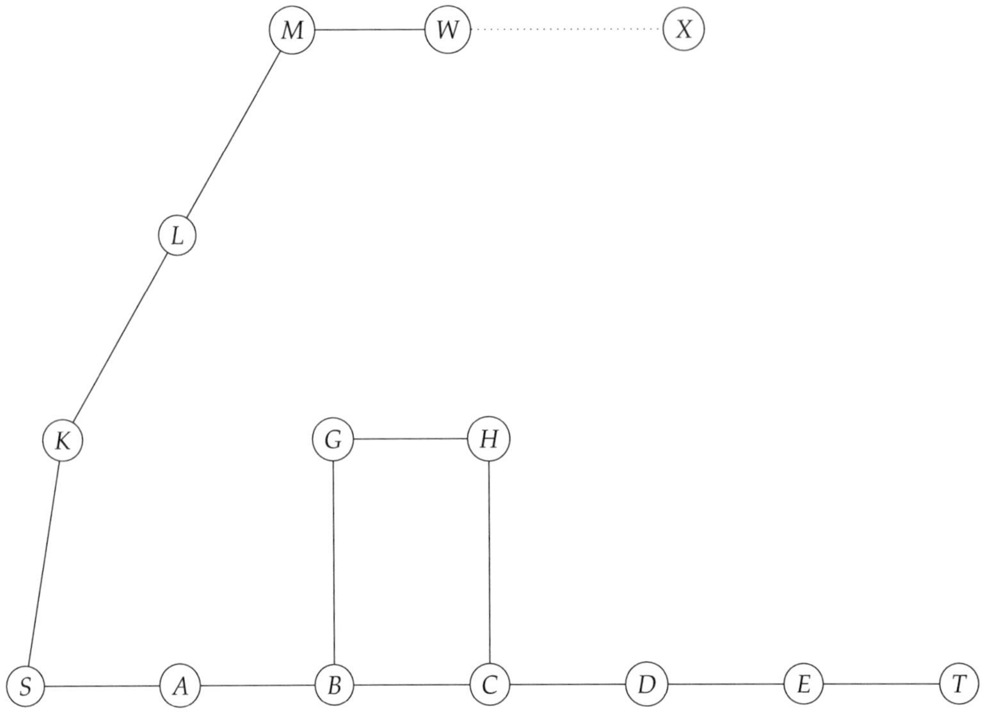

Fig. 1. Network Topology Used for Illustrations.

the same assumption (that the network is the subnet) as a value made public in one subnet can be used in another subnet.

Mandating digital signatures by RREQ source can prevent such attacks. Furthermore, as RREQs have to be authenticated, if a node S floods too many RREQs other nodes will simply drop RREQs from S in the future.

5.2 Rationale for PLN

Perhaps the most important implication of the fact that inexpensive schemes for computing pairwise secrets exist (MLS requires only one hash computation) is that it becomes practical to enforce PLNs, and derive many tangible benefits out of this ability. More specifically, enforcing the PLN is intended to achieve the following goals: i) eliminate trivial risk-free active attacks; ii) validate the assumption behind the security of the per-hop hashing strategy; iii) avoid one-way links and "links" created by semi-active attackers; iv) deter selfish behavior.

5.2.1 Trivial risk-free attacks

Consider a representative topology in Figure 1, for a specific scenario where a malicious node C receives a RREQ originating from a node S (intended for the destination T) through the path (A,B).

In Ariadne the shared network-wide secret K_U does not prevent nodes from impersonating other nodes for purposes of fooling their neighbors. For example, C, claiming to be "node C'," can relay the RREQ indicating a path (A,B,C'). All that D (which receives the packet from "node C'") can verify is that the C' has access to the network-wide shared secret K_U. While paths that include C' will be dropped by the source (in Ariadne-PS) or the destination (in Ariadne-TESLA), such RREQs can preempt good RREQs. Furthermore, C does not face any risk of being identified.

Note that Ariadne-DS is not susceptible to such attacks as the signature appended by C can be verified by its neighbors before the RREQ is forwarded. APALLS employs two independent forms of link-layer authentication: one based on the one-hop PLN secret, and one based on digital signatures. The PLN authentication also serves as a protection against DoS attacks that could be launched due to the higher computational overhead required for verifying digital signatures. RREQ packets relayed by a node C are first decrypted using PLN secret G_C. Only if the resulting packet is a valid[8] RREQ packet will the receiver proceed to verify the signature Σ_C.

5.2.2 Protecting per hop hash

Many secure routing protocols (including Ariadne) simply assume bidirectional links. Often the justification provided for this is that "the handshake used in colission avoidance protocols ensures bidirectional links" (Kim & Tsudik., 2005). Unfortunately, this is not sufficient to prevent a node C which *can* overhear packets sent by a node B from *pretending* that it cannot (Sivakumar and Ramkumar, 2006).

When B relays a RREQ from S indicating a path (A, B) and a per-hop hash value β_B, C can wait for the RREQ to be relayed along another path, say (A, B, G, H). Now, with access to β_B (as C *can* hear B), C has the ability to remove its immediate upstream neighbor H (or G and H) from the path. Imposing a PLN is *necessary* to validate the assumption behind the security of the per-hop hashing technique - that nodes that are not neighbors cannot gain access to the per-hop hash. If C pretends to be out of the range of B, B will not include C in its PLN (thus, C will not gain access to the per-hop hash broadcast by B).

5.2.3 Semi active attacks

If a PLN is enforced a semi active attacker between B and C has to rebroadcast packets from B and the response from C when they induct each other into their PLNs. B and C may have reasons to suspect a semi active attacker when they hear an echo of their own transmissions, or if an abnormally large delay is observed in handshakes between B and C (Hu et al., 2003). Under such suspicions they will simply not induct the other node into their PLNs.

5.2.4 PLN as a deterrent

Without the ability to impose a PLN all that a node B can do to reduce the participation of a bad neighbor C is to ignore packets from C. However, B cannot prevent C from forwarding packets sent by B. The ability to cut-off neighbors in the physical neighborhood from the PLN facilitates DoS-free countermeasures to reduce the ill-effects of malicious nodes. Nodes cut off by all neighbors are effectively cut off from the subnet. Mandating a PLN can also deter selfish selective participation. A node C will have to be *inducted* into the PLNs of its neighbors before C can actually monitor traffic. Once inducted, C is pressured to participate self-lessly due to the fact that it is under constant observation by its neighbors, who may cut C off if they sense selfish participation.

[8]Valid RREQ packets may start with a magic number. In addition, for a packet from C, the last entry in the path field should be C.

5.3 Identifying and revoking active attackers

Enforcing a PLN does *not* address *all* risk-free attacks. As an illustration, consider a scenario where C receives a RREQ (from S to T) along a path (A, B), and relays the RREQ indicating a path (Q, R, C) instead, where Q and R are fictitious nodes inserted by C. Nodes downstream of C have no reason to suspect that Q and R do not exist, and B does not have access to β_S and hence $\beta_Q = h(\beta_S, Q)$ or $\beta_R = h(\beta_Q, R)$ to verify that the value β_C is indeed inconsistent.

Note that *if* C had instead advertised a path (A, R, C) with a random β_C, B (which has access to β_A) can determine that $\beta_C \neq h(h(\beta_A, R), C)$. Similarly, if C had modified any of the fields specified by the "real" upstream nodes (A and B), then B can recognize such attempts. Thus, while there are some *blatant* active attacks which can be easily be recognized by neighbors, some subtler attacks can not.

Assume that the destination receives the tainted RREQ indicating a path (Q, R, C, D, E, F, G) and a per-hop hash β_G. Assume that the actual reason for the inconsistency in the RREQ was that C had preformed an active attack. In Ariadne all that the destination can detect at this point is that "the per-hop hash β_G is inconsistent." In Ariadne-DS and Ariadne-PS T can also conclude with certainty that node G exists (as T can verify the HMAC / signature of G). However, T cannot verify the authentication appended by F as T does not access to the value β_F (which had gone into the computation of the authentication appended by F). Thus T cannot even determine if the node F actually exists in the path.

If T desires to determine *who is responsible* for perpetrating this attack, it can come to several likely conclusions: like i) G is a malicious node and every other node in the path has been maliciously inserted by G; or ii) G is a good node, but F may have maliciously inserted nodes (A, B, C, D, E) in the path; or iii) both G and F are good nodes and the node E may have inserted nodes (A, B, C, D) in the path; and so on.

In Ariadne-DS the destination can then demand all intermediate nodes (A, B, C, D, E, F, G) to produce the per-hop hash they had received from *their* upstream neighbor, which is simultaneously consistent with the signature of the upstream node (which was already included in the RREQ sent to the destnation). Now node D can produce a value β_C consistent with the signature Σ_C, and $\beta_D = h(\beta_C, D)$ consistent with D's signature Σ_C. Likewise, all nodes that had *not* violated the protocol can also do so.

However, the attacker C cannot produce a value β_R consistent with the "signature Σ_R." The obvious recourse for C is to not respond to this demand (C could just power off or leave the subnet). Now, as it is not possible to compute the value β_R (which according to C, was sent by R) from $\beta_C = h(\beta_R, C)$, one cannot deduce that Σ_R is indeed inconsistent with β_R. If C has to be convicted based on its inability to provide an "affirmative defense" (providing β_R consistent with Σ_R) it is indeed possible that an innocent D, which had suddenly crashed (and thus loses the value β_C) can also suffer the same fate.

5.3.1 Proof of active attacks in APALLS

The encrypted upstream per-hop hash ν in APALLS serves two purposes. Firstly, it makes it possible for the destination to narrow down active attackers. For example, if in the path (A, B, C, D, E, F, G) the destination is able to determine that nodes (D, E, F, G) were consistent, and C, while self-consistent, cannot be verified to be consistent (as B is self-inconsistent), the destination can narrow down the active attacker to B or C. Secondly, when used in conjunction with one-hop signatures, it facilitates unambiguous identification of active attackers, and avoids the need for nodes to provide affirmative defense.

In other words, even without carrying over all signatures (thereby saving bandwidth overhead for signatures and public-key certificates) APALLS can provide non repudiable proof of active attacks. Irrespective of the nature of the active attack, a signed packet from the attacker (stored temporarily by a neighbor, and submitted to the TA at a convenient time) can be used for this purpose.

Note that the values broadcast by C is effectively a non repudiable statement to the effect "the fields $\mathcal{Q}^B_{q,S}$, $\beta_B = K^{-1}_{CT}[\nu^B_C]$, and v_B, were broadcast by B, and verified by me (C) to be consistent with the signature of B (Σ_B), the preimage of σ_B."

When the values stored by D (the contents of the RREQ broadcast by C) are submitted to the TA, the TA takes the following steps:

1. Verify that Σ_C is consistent with $\mathcal{Q}^C_{q,S}$, β_C, σ_B and v_B;
2. Check if B is a valid node in the network; if not, C is an active attacker (C had inserted a nonexistent node in the path);
3. If B is a valid node, compute the signature $\Sigma_B{}'$ for the values $\mathcal{Q}^B_{q,S}$ and $\beta_B = K^{-1}_{CT}[\nu^B_C]$ and v_B (which according to C, were broadcast by B), and
4. Verify if $h(\Sigma_S{}') = \sigma_B$. If so, B is an active attacker (as B advertised self-inconsistent values (B, M_B, ν_B)). If not, C is the active attacker (as C had accepted a packet with an invalid signature).

If the TA has access to the private keys of all nodes the TA can simply compute $\Sigma_B{}'$. If private keys are *not* escrowed by the TA, the TA will need to request B to produce a verifiable signature $\Sigma_S{}'$ for the values $\mathcal{Q}^B_{q,S}$ and $\beta_B = K_{CT}[\nu^B_C]$ and v_B. Thus even in scenarios where the private keys are not escrowed by the TA, unlike Ariadne-DS, nodes will only need access to their private key to avoid being penalized (revoked[9]) accidentally.

A compelling advantage of escrowing private keys by the TA is that the verification of proof of attacks can be performed immediately. This is especially useful in scenarios where access to the TA is available (for example, if at least one node in the subnet has Internet access), as the revocation message (signed by the TA) can be immediately distributed within the subnet.

5.4 Routing around attackers

In scenarios where access to the TA does not exist, nodes in the subnet will have to "live with" active attackers for some (indefinite) duration. APALLS includes two strategies for improving the ability to route around nodes suspected of active attacks. The first is by using black-lists specified by the RREQ source. The second is by employing RREPs with a *FAIL* code.

The list of nodes in S's black list can include nodes which were possibly S's neighbor at some time in the past, and observed by S to violate the protocol, or engage in selfish behavior. The list can also include nodes which have been recognized as active attackers when S was a destination node in some RREQ. That a node X is black-listed by a node S is not interpreted by other nodes to mean that "X is malicious." All this means is that the source S desires to avoid X in paths where S is an end-point. Thus, the black-list of S will only influence routing of RREQ packets in which S is the source or the destination.

The second strategy is intended to improve the success of the second RREQ that may be sent by the source S after the first RREQ times out. For instance, in a scenario where the *FAIL* RREP indicates $[(B\lambda C), (D, E, F, G)$, during the second RREQ the nodes (D, E, F, G) will drop

[9]Any node which claims to not have access to its private key should be revoked in any case.

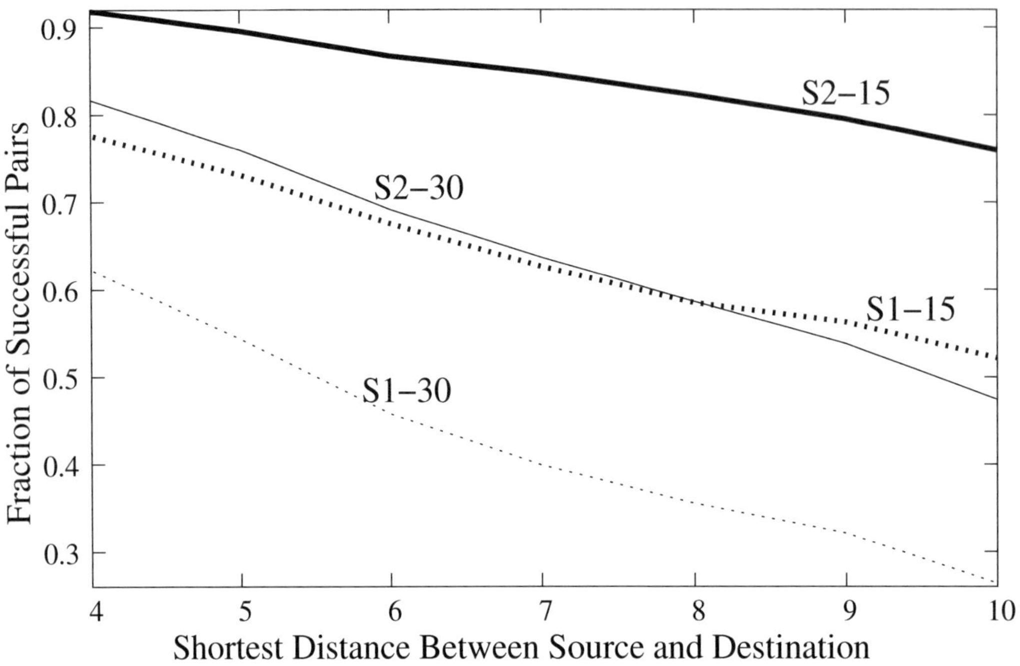

Fig. 2. Simulation results depicting the utility of the ability to narrow down the perpetrator.

RREQs that include C or B. Note that without this measure (and if the subnet topology has not changed) the second RREQ may suffer the same fate as the first RREQ.
To evaluate the benefit of this strategy simulations were performed with random realization of subnets with $N_s = 200$ nodes with uniformly distributed x and y coordinates in a square region with unit edges. The range of the nodes was chosen as 0.1 units (each node had 5 neighbors on an average). Of the $N_s = 200$ nodes, b randomly chosen nodes were labeled malicious. RREQ propagation was simulated between every pair of nodes.
Three different realizations of the network were simulated with different sets and numbers of "bad" nodes. The simulation results are depicted as fraction of node-pairs that succeed in discovering a path free of bad nodes (y-axis) vs the shortest number of hops between the pair (between which RREQ propagation was simulated) as the x-axis. RREQ propagation was simulated for over 400,000 pairs separated by hop lengths between 4 and 10. Path discovery between a pair is assumed to succeed if at least one of the established paths is free of b malicious nodes. In Figure 2 plots labelled S1 depict the success rates of first RREQs. Simulation results are shown for $b = 15$ (S1-15) and $b = 30$ (S1-30).
The plots labelled S2 indicate fraction of successful node pairs after the second RREQ (either the first or the second RREQ attempt succeeds). As can be seen from the simulation results the success rate after the second RREQ for the scenario with 30 bad nodes (S2-30) is comparable to the success of the first RREQ with just 15 bad nodes (S1-15). It is important to note that in the absence of this strategy, the second RREQ has only as much chance of succeeding as the first. Thus, in this particular instance, it can be argued that the additional upstream per-hop hash helps in realizing a two-fold improvement in resistance to malicious nodes in the subnet.

5.5 RREP authentication

In Ariadne the authentication appended by the destination for the RREP (which is verifiable only by the RREQ source) is indistinguishable from a random number for all intermediate nodes that relay the RREP. This can be exploited by attackers to send spurious RREPs over long (fictitious) paths to cause unnecessary bandwidth overhead for other nodes in the subnet. Nodes specified in the path will simply forward the RREP along the path specified. This attack is particularly dangerous in Ariadne as every intermediate node will need to release a TESLA key *and* a certificate for a commitment.

Consider a scenario where an RREQ from a source S to some destination T indicates t_i (as the upper limit before which the destination T should receive the RREQ). Assume that such an RREQ through a path (K, L, M) is heard by an attacker W. Just by overhearing any RREP packet in response to *any* RREQ (not necessarily a response for the RREQ from S) *after* time t_i, it is possible for the attacker W to harvest a preimage K_X^i corresponding to time t_i of some node X. The node X may even be many hops away from W. A malicious W can now send a fictitious RREP indicating a path (K, L, M, X) to M with a random HMAC by "destination T". All that nodes (K, L, M) can verify is that K_X^i is indeed i^{th}pre-image of K_X^0. Obviously this serves very little purpose without the ability to recognize the authentication appended by the RREP destination (which conveys the crucial information that the HMACs were received *before* time t_i). Effectively, *any* node can send such spurious RREP packets in response to any RREQ packet, impersonating some other node which may be several hops away.

In APALLS the destination includes a value β_S in the RREP which was until then known only to the source and destination. Thus, even while supercilious RREPs can be sent by nodes (which will be detected by the source as inconsistent), such RREPs can be raised only by nodes which had actually seen an RREP from the destination. Furthermore, such an attack is not worthwhile for any attacker as the RREP overhead is small in any case in APALLS.

6. Related work and conclusions

Several authors have investigated strategies for securing DSR, and mechanisms for cryptographic authentication.

6.1 Other secure DSR protocols

Papadimitros (Papadimitratos and Haas., 2002) et al propose a secure routing protocol (SRP) where only the source and destination share a secret. Marshall et al (Marshall et .al, 2003) argued that SRP cannot avoid malicious behavior by intermediate nodes during the route establishment phase, as long as the (malicious) behavior is consistent in the forward and reverse path. They also suggest techniques to mitigate issues in SRP by employing promiscuous mode of operation (Marti et al., 2000).

Kim et al (Kim & Tsudik., 2005) (SRDP) propose a general protocol for securing route discovery in DSR, where the primary deviation from Ariadne is that they strive to reduce the bandwidth overheads by aggregating the authentication appended by intermediate nodes (for Ariadne-PS and Ariadne-DS where the destination can verify authentication appended by intermediate nodes). The disadvantage of aggregating authentication is that the destination cannot verify *which* node was responsible for the inconsistency. As Ariadne does not strive to do that in any case, aggregating authentication can reduce RREQ overhead for Ariadne. However, aggregating HMACs can not be done for APALLS as it would not permit detection of self-consistency of nodes.

APALLS is an extension of an earlier work (also by the authors of this chapter) (Sivakumar and Ramkumar, 2008) which sought to improve the resiliency of Ariadne-PS. The improvements suggested in (Sivakumar and Ramkumar, 2008) include i) use of the upstream per-hop hash to narrow down active attackers; and ii) enforcing a PLN. The modifications in APALLS compared to (Sivakumar and Ramkumar, 2008) are: i) the use of one-hop digital signatures for non-repudiation; ii) mandating digital signature by the RREQ source; and iii) a modified strategy for authenticating RREPs.

6.2 Key distribution

Several key distribution schemes have been proposed in the literature for ad hoc networks. Zhou et al (Zhou and Haas., 1999) propose a key management service with distributed CA, using threshold cryptography to distribute shares of the CAs private key to several nodes. Capkun et al (Capkun and Hubaux., 2003) propose a strategy for "building secure routing from an incomplete set of security associations" (BISS), in which a combination of predistribution of keys (which facilitates only an incomplete set of pairwise secrets) and public key primitives are used. The motivation for BISS seems to be that schemes for establishing pairwise secrets between a fraction of nodes is more practical than schemes that permit *every* pair nodes to establish a secret.

Zhang et al (Zhang et al., 2005) propose the use of identity based encryption an signature (IBE / IBS) schemes for ad hoc networks. IBS schemes can reduce the bandwidth overhead for signatures as i) public keys and public key certificates are not required; and ii) the signatures are also generally smaller than (say) RSA signatures. This advantage is not compelling in APALLS as signatures are not carried forward. Unlike RSA signatures where we can reduce signature verification complexity by choosing small public exponents, IBS schemes do not have practical strategies to reduce verification complexity. High verification complexity can lead to simple DoS attacks. However, in APALLS, this is not a disadvantage as the low complexity PLN-based authentication (which is verified before signatures are verified) can prevent such DoS attacks. Thus, both the advantages and disadvantages of IBS schemes are less relevant in APALLS.

6.3 Conclusions

We have outlined a comprehensive secure routing protocol, APALLS, based on DSR. To the extent of our knowledge, APALLS is the first secure routing protocol which is designed to provide non repudiable proof of active attacks.

Non-repudiable authentication is necessary, but not sufficient to provide non repudiable proof of active attacks. In general, any active attack involves violation of the prescribed protocol. The protocol prescribes the steps that a node (say) C should take in response to a packet sent from a neighbor (say) B. For example, in distance vector based protocols, if a node B announces a hop-length of 5 to a node S, the neighbor C downstream of B is expected to announce a hop-length 6.

In a scenario where C advertises a hop-length 7, proving that C did (or did not) violate the protocol requires several *contextual* information like (for example) i) if B was indeed a neighbor of C at that time; ii) the hop count advertised by B at that time ; iii) if C did indeed process the information advertised by B (the packet broadcast by B did not suffer colission), etc.. Thus, even while some ad hoc routing protocols like ARAN (Sanzgiri et al., 2002) and Ariadne-DS employ non repudiable authentication, they do not address the issue of *how* a packet sent from a node can be used for proving an active attack. As pointed out in this chapter, even

while Ariadne-DS carries forward all signatures, it still has practical issues in providing non repudiable proof.
One of the motivations for APALLS stem from the fact that the main advantage of MANET based networks is their ability to operate without any infrastructural support. Ideally, while we would desire to eliminate even an off-line TA, this is simply not possible to do so as an authority is required to i) specify the rules (the protocol) that should be followed by every node; and ii) to boot-strap cryptographic associations between nodes.
While APALLS borrows some features from Ariadne, the major differences between Ariadne and APALLS stem from the network model. Several elements in Ariadne like i) the preference of TESLA over pairwise secrets; ii) the choice of the strategy to suppress RREQ floods; and iii) ignoring the risk of supercilious RREPs (RREP bandwidth can be high if a TA is not available in the subnet) assume the presence of a TA in every subnet. While APALLS can take advantage of access to TA (when at least one node in the subnet has access to the Internet) for quickly disseminating revocation lists, APALLS can operate effectively even in subnets that may be completely isolated from the rest of the world.
The choice of cryptographic authentication schemes in APALLS are also driven by the need to keep the overhead low. Storage is an inexpensive resource for mobile devices; any mobile device can easily afford several GBs of pluggable storage. However computational and bandwidth overheads are expensive for battery operated devices. This renders key predistribution schemes for pairwise secrets (which impose low computational and bandwidth overhead) well suited even for dynamic large scale networks.
That digital signatures appended by intermediate nodes are verified only by neighbors renders just about any scheme well suited for this purpose. More specifically, it also opens up the feasibility of non repudiable one-time signature (OTS) schemes[10] which do not require asymmetric primitives. That only neighbors need to verify the signature renders the scheme proposed by Merkle et al (Merkle, 1987) for constructing infinite OTS trees substantially more efficient. That OTS schemes require only block-cipher/ hash operations implies that even very low complexity SIM cards can perform the operations required for this purpose. Such low complexity SIM cards which need to perform only symmetric cipher operations can be realized at lower cost.
Some of the ongoing work of the authors include i) investigation of the suitability of OTS schemes; and ii) use of one-hop signatures for providing non repudiable proof of active attacks for other MANET routing protocols like AODV (Perkins et al., 2002), TORA (Park and Corson, 1997) and OLSR (Jacquet, 2001).

7. References

Johnson, P., Maltz, D. (1996). Dynamic source routing in ad hoc wireless networks, *Mobile Computing, Kluwer Publishing Company,*, ch. 5, pp. 153-181.

Sanzgiri, K., Dahill, B., Levine, N., Shields, C., Belding-Royer, E.M. (2002). A Secure Routing Protocol for Ad Hoc Networks, *Proceedings of the 2002 IEEE International Conference on Network Protocols (ICNP),* November 2002.

Abusalah, L., Khokhar, A., Guizani,M. (2008). A Survey of Secure Mobile Ad Hoc Routing Protocols, *IEEE Communications Surveys and Tutorials,* 10(4), 2008.

[10]This does *not* include chained OTS schemes which cannot be used for non repudiation as private keys are revealed eventually.

Hu, Y.C., Perrig, A., Johnson. D.B. (2005). Ariadne: A Secure On-Demand Routing Protocol for Ad Hoc Networks, *Journal of Wireless Networks,*11, pp 11–28, 2005.

Kim, J., G. Tsudik. (2005). SRDP: Securing Route Discovery in DSR, *IEEE Mobiquitous'05,* July 2005.

Zhang, Y., Liu, W., Lou, W., Fang, Y., Kwon, Y. (2005). AC-PKI: anonymous and certificate less public key infrastructure for mobile ad hoc networks, *IEEE International Conference on Communications (ICC'05),* Seoul, Korea, May 2005.

Zhou, L., Haas, Z. (1999). Securing Ad Hoc Networks, *IEEE Network,* 13(6), pp 24-30, 1999.

Capkun, S., Hubaux, J-P. (2003). BISS: Building Secure Routing out of an Incomplete Set of Security Associations, *In Proceedings of the Wireless Security Workshop (WISE) 2003,* San Diego, September 2003.

Marti, S., Giuli, T J., Kevin Lai., Mary Baker. (2000). Mitigating routing misbehavior in mobile ad hoc networks, *Proceedings of the 6th Annual International Conference on Mobile Computing and Networking,* Boston,2000.

Marshall, J.,Thakur,V., Yasinsac.A. (2003). Identifying flaws in the secure routing protocol, *Proceedings of the 2003 IEEE International Performance, Computing, and Communications Conference,* 2003.

Burmester, M., Van Le, T., Weir. M. (2003). Tracing Byzantine Faults in Ad Hoc Networks, *Proceedings of Communication, Network, and Information Security (CNIS),* NY, Dec 2003.

Awerbuch, B., Holmer, D., Nita-Rotaru, C., Rubens, H. (2002). An On-Demand Secure Routing Protocol Resilient to Byzantine Failures, *ACM Workshop onWireless Security (WiSe–02),* September 2002.

Sun, J., Zhang, C., Fang, Y. (2007). An id-based framework achieving privacy and non-repudiation in vehicular ad hoc networks, *MILCOM,* 2007.

Hu, Y.C., Perrig, A., Johnson. D.B. (2001). Packet Leashes: A Defense against Wormhole Attacks in Wireless Ad Hoc Networks, *Rice University Department of Computer Science Technical Report TR01-384,* Dec 2001.

Hu, Y.C., Perrig, A., Johnson. D.B. (2003). Rushing Attacks in Wireless Ad Hoc Network Routing Protocols, *WiSe 2003,* San Diego, CA, September 2003.

Perrig, A., Canetti, R., Song, D., Tygar, D. (2001). Efficient and Secure Source Authentication for Multicast, *In Network and Distributed System Security Symposium,* NDSS '01, Feb. 2001.

Ramkumar, M. (2008). On the Scalability of a Nonscalable Key Distribution Scheme, *IEEE SPAWN 2008,* Newport Beach, CA, June 2008.

Sivakumar, K. A., Ramkumar, M. (2008). Improving the Resilience of Ariadne, *IEEE SPAWN 2008,* Newport Beach, CA, June 2008.

Sivakumar, K A., Ramkumar, M. (2009). Private Logical Neighborhoods for Wireless Ad Hoc Networks, *5-th ACM International Symposium on QoS and Security for Wireless and Mobile Networks (Q2SWinet),* Canary Islands, Spain, October 2009.

Hu, Y.C., Perrig, A., Johnson. D.B. (2005). Efficient Security Mechanisms for Routing Protocols, *Symposium on Networks and Distributed Systems Security (NDSS),* 2003.

Sivakumar, K A., Ramkumar, M. (2006). On the Effect of Oneway Links on Route Discovery in DSR,*Proceedings of the IEEE International Conference on Computing, Communication and Networks, ICCCN-2006,* Arlington, VA, October 2006.

Papadimitratos, P., Haas, Z.J. (2002). Secure Routing for Mobile Ad Hoc Networks, *Proceedings of the SCS Communication Networks and Distributed Systems Modeling and Simulation Conference(CNDS 2002),* San Antonio, Texas,2002.

Merkle, R.C. (1987). A digital Signature based on Conventional Encryption Function, *Conference on the Theory and Applications of Cryptographic Techniques on Advances in Cryptology, Lecture Notes In Computer Science;* 293, pp 369 – 378, 1987.

Perkins, C., Royer,E., Das.S. (2002). Ad hoc On-demand Distance Vector (AODV) Routing, *Internet Draft, draft-ietf-manet-aodv-11.txt, Aug 2002. The 6th World Multi-Conference on Systemics, Cybernetics and Informatics (SCI 2002),* 2002.

Park, V.D ., Corson, M.S. (1997). A Highly Adaptive Distributed Routing Algorithm for Mobile Wireless Networks, *Proceedings of IEEE INFOCOM,* Kobe, Japan, 1997.

Jacquet, P., Mühlethaler., Clausen, T., Laouiti, A., Qayyum, A., Viennot,L. (2001). Optimized link state routing protocol for ad hoc networks,*Proceedings of the 5th IEEE Multi Topic Conference (INMIC 2001),* 2001.

Ramkumar, M. (2009). On the Complexity of Probabilistic Key Predistribution Schemes, *to be presented in the Embedded Systems and Communications Security Workshop (ESCS 2009),* Niagara, NY, September 2009.

8. Appendix

8.1 A scalable key predistribution scheme

Unlike MLS, scalable KPSs are susceptible to collusions. For an (n,p)-secure KPS, an attacker with access to secrets of n nodes can compute a fraction p of all possible pairwise secrets. As long as p is low enough (say 2^{-64}) it is computationally infeasible for an attacker to even identity *which* pairwise secrets can be compromised by using the pool of secrets accumulated from n nodes.

8.2 A scalable key predistribution scheme

In the subset keys and identity tickets (SKIT) scheme (Ramkumar, 2009) defined two parameters m and M, the KDC chooses mM secrets, say, $K_{i,j}, 1 \leq i \leq m, 1 \leq j \leq M$ (which can be derived from a single master secret μ as $K_{i,j} = h(\mu, i, j)$).

The KDC chooses a public pseudo random function (PRF) $f()$ which generates a $m \log_2 M$ pseudo-random bits. For a node with identity A the output of the PRF $f(A)$ is interpreted as $m \log_2 M$-bits values, $a_i, 1 \leq i \leq m, 0 \leq a_i \leq M-1 \forall i$. Corresponding to the m indices, A is issued m secrets $K_{i,a_i}, 1 \leq i \leq m$. Node A is also issued mM identity tickets $I_{i,j} = h(K_{i,j}, A), 1 \leq i < m, 1 \leq j \leq M$. Identity tickets are conceptually similar to HMACs; however, while HMACs are not intended to be secrets, identity tickets provided to A are intended only for A.

Two nodes A and B can compute $2m$ common tickets. Computing any pairwise secret (say when A requires to compute K_{AB}) will require generating $m \log_2 M$ pseudo random bits to determine the indices of the m secrets assigned to B, followed by computation of m hashes. Every node requires storage for mM certificates.

An attacker with access to secrets of n nodes $O_1 \cdots O_n$ can compute K_{AB} if the m secrets of each of the n nodes include $K_{i,a_i}, 1 \leq i \leq m$ *and* $K_{i,b_i}, 1 \leq i \leq m$. The probability of such an event is

$$p(n) \approx (1 - \mathrm{e}^{-\mathrm{n}/\mathrm{M}})^{2\mathrm{m}}. \quad (7)$$

For $m = 32$ and $M = 2^{16}$, $p(45{,}000) < 2^{-64}$, and $p(84400) \approx 2^{-30}$. For $m = 32$ and $M = 2^{16} \times 5$, $p(225{,}000) < 2^{-64}$, and $p(422{,}000) \approx 2^{-30}$.

If each node can afford 100 MB storage we can choose $m = 32$ and $M = 2^{16} \times 5$ to realize a scheme for which $p(225{,}000) \approx 2^{-64}$ and $p(422{,}000) \approx 2^{-30}$. Only the storage complexity is increased. The computational overhead, which is influenced by the value $m = 32$ remains the same. Due to the low computational overheads, the computations can be easily performed inside the modest SIM cards to further alleviate the issue of exposure of secrets from a large number of nodes. An attacker desiring the exploit the collusion susceptibility of SKIT will have to successfully tamper with and expose secrets from several hundred thousand SIM cards.

11

Meta-heuristic Techniques and Swarm Intelligence in Mobile Ad Hoc Networks

Floriano De Rango and Annalisa Socievole
DEIS Department, University of Calabria
Rende (Cs),
Italy

1. Introduction

The infrastructure-less and the dynamic nature of mobile ad hoc networks (MANETs) demands new set of networking strategies to be implemented in order to provide efficient end-to-end communication. MANETs employ the traditional TCP/IP structure to provide end-to-end communication between nodes. However, due to their mobility and the limited resource in wireless networks, each layer in the TCP/IP model requires redefinition or modifications to work efficiently in MANETs. One interesting research area in MANETs is routing. Routing is a challenging task and has received huge attention from researches. Due to the adaptive and dynamic nature of these networks, the *Swarm Intelligence* approach is considered a successful design paradigm to solve the routing problem. Swarm intelligence is a relatively new approach to problem solving that takes inspiration from the social behaviours of insects and of other animals. In particular, the collective behaviour of ants have inspired a number of methods and techniques among which the most studied and the most successful is the general purpose optimization technique known as *Ant Colony Optimization* (ACO) meta-heuristic. ACO takes inspiration from the foraging behaviour of some ant species. These ants deposit a chemical substance called *pheromone* on the ground in order to mark some favourable path that should be followed by other members of the colony. This behaviour has led to development of many different ant based routing protocols for MANETs. In this chapter, a description of swarm intelligence approach and ACO meta-heuristic is given, an overview of a wide range of ant based routing protocols in the literature is proposed and finally other applications related to ACO in MANETs and new directions are discussed.

2. The swarm intelligence approach

Swarm Intelligence (Bonabeau et. al, 1999) is a property of natural and artificial systems involving multiple individuals interacting with each other and the environment to solve complex problems exhibiting a collective intelligent behaviour. Examples of systems studied by swarm intelligence are colonies of ants and termites, schools of fish, flocks of birds, herds of land animals. Some human artifacts also fall into the domain of swarm intelligence, notably some multi-robot systems, and also certain computer programs written to solve optimization and data analysis problems.

Swarm intelligence has a multidisciplinary character. It is usual to divide swarm intelligence research into two areas according to the nature of the systems under analysis: in *natural* swarm intelligence research biological systems are studied while in *artificial* swarm intelligence human artifacts are studied. A different classification of swarm intelligence research can be given based on the goals that are pursued: it is possible to identify a *scientific* and an *engineering* stream. The goal of the scientific stream is to model swarm intelligence systems in order to understand the mechanisms allowing a system to behave in a coordinated way as a result of local individual-individual and individual-environment interactions. On the other hand, the goal of the engineering stream is to employ the biological behaviours in order to design systems able to solve problems of practical relevance.
The typical swarm intelligence system has the following properties:

- it is composed of many individuals;
- the individuals are either all identical or belong to a few typologies;
- the interactions among the individuals are based on simple behavioural rules that make use of local information exchanged directly or via the environment;
- the overall behaviour of the system results from the interactions of individuals with each other and with their environment.

The characterizing property of a swarm intelligence system (Tarasewich & MecMullen, 2002) is its capability to act in a coordinated way without the presence of a coordinator. In nature there are many examples of swarms performing some collective behaviour without any individual controlling the group. Wasps build nests with a highly complex internal structure that is well beyond the cognitive capabilities of a single wasp. Termites build nests whose dimensions can reach many meters of diameter and height. When compared to a single termite, which can measure as little as a few millimetres, these nests are huge. Schools of fish and flocks of birds are other examples of highly coordinated groups. Scientists have shown that these elegant behaviours can be understood as the result of a self-organized process where there is no leader and each individual bases its movement decisions solely on locally available information: the distance, the perceived speed, and the direction of movement of neighbours.
The most interesting swarm-level behaviours belongs to ants. What is fascinating is that ants are able to discover the shortest path to a food source and to share that information with another ants through *stigmergy* (Deneubourg et al., 1990; Dorigo et al., 1999). Stigmergy is a form of indirect communication used by ants in nature to coordinate their problem-solving activities. Ants realize stigmergetic communication by depositing on the ground a chemical substance called *pheromone* that induces changes in the environment which can be sensed by other ants. From the observation of real ant colonies, ant algorithms were inspired and applied to many different optimization problems.
The main advantages of the swarm intelligence approach compared with a classical approach are the following:

- flexibility: the group can quickly adapt to a changing environment;
- robustness: even when one ore more individuals fails, the group can still perform its tasks;
- self organisation: the group needs relatively little supervision or top down control.

These properties make swarm intelligence a successful design paradigm.

2.1 Ant foraging behaviour

The observation of ant's behaviour inspired the implementation of different optimization algorithms (Bonabeau et al., 2000). An ant colony is able to find the shortest path between the nest and a food source using simple local decisions. Ants use a signalling communication system based on the deposition of pheromone over the path it follows, marking a trail. Pheromone is a hormone produced by ants that establishes a sort of indirect communication among them.

An ant foraging for food lay down pheromone over its route. When this ant finds a food source, it returns to the nest reinforcing its trail. Other ants in the proximities are attracted by this substance and have greater probability to start following this trail and thereby laying more pheromone on it. This process works as a positive feedback loop system because the higher the intensity of the pheromone over a trail, the higher the probability of an ant start travelling through it. The following example (see Fig. 1) will show how this process leads the colony to optimize a route:

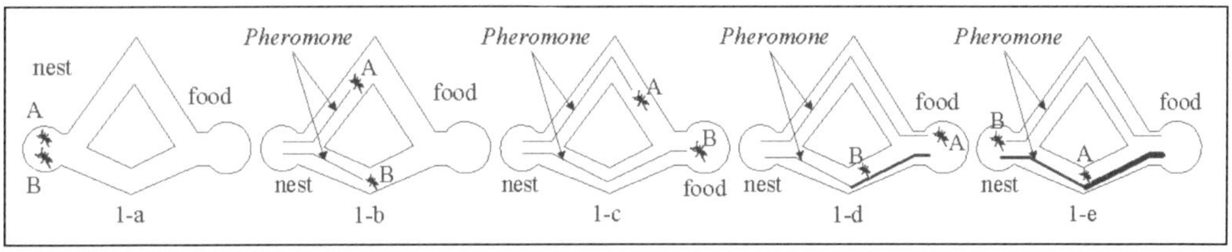

Fig. 1. Two ants exploring the shortest path

Suppose two ants, called A and B, were randomly searching for food when they found two different routes between the nest and the source. Since the route chosen by ant B is shorter, first ant B will reach food. Going back to the nest, ant B will choose the same path laying more pheromone over it. When ant A will also find the food, it will choose the path with the higher pheromone concentration to reach the nest. So, ant A will follow the same B's path to the nest. As the process continues, the pheromone concentration on this trail will increase while the longest route will be discarded because of the pheromone evaporation process.

When more paths are available from the nest to a food source, a colony of ants may be able to exploit the pheromone trails left by the individual ants to discover the shortest path from the nest to the food source and back.

2.2 ACO meta-heuristic

The ant colony foraging behaviour has attracted a lot of attention in combinatorial optimization problems, and has been reverse-engineered in the context of *Ant Colony Optimization* (ACO) meta-heuristic (Deneubourg et al., 1990). A *meta-heuristics* is a set of algorithmic concepts that can be used to define heuristic methods applicable to a wide set of different problems. In other words, a meta-heuristic is a general purpose algorithmic framework that can be applied to different optimization problems with relatively few modifications. Examples of meta-heuristics include simulated annealing (Cern'y, 1985), tabu search (Glover & Laguna, 1997), iterated local search (Lourenço et al., 2002), evolutionary computation (Dorigo et al. 2006), and ant colony optimization (Dorigo et al. 1996; Dorigo et al., 1999; Dorigo & Stützle, 2004).

In ACO, a number of artificial ants build solutions to an optimization problem and exchange information on the quality of these solutions via a communication scheme that is reminiscent of the one adopted by real ants.

The computational resources are allocated to a set of relatively simple agents (artificial ants) that communicate indirectly by stigmergy. Artificial ants have been enriched with some capabilities which do not find a natural correspondence in order to make them more effective and efficient. In particular, the use of a colony of cooperating individuals, an (artificial) pheromone trail for local stigmergetic communication, a sequence of local moves to find shortest paths, and a stochastic decision policy using local information and are stemmed from real ants. The other features which do not find their counterpart in real ants are the following:

- artificial ants live in a *discrete world* and their moves consist of transitions between discrete states;
- artificial ants have an *internal state* containing the memory of the ant past actions;
- artificial ants deposit an amount of pheromone which is a function of the *quality of the solution* found;
- artificial ants timing in pheromone laying is problem dependent and often does not reflect real ants behaviour;
- to improve overall system efficiency, ACO algorithms can be enriched with *extra capabilities* like lookahead, local optimization, backtracking, and so on, that cannot be found in real ants.

In ACO algorithms a finite size colony of artificial ants with the above described characteristics collectively searches for good quality solutions to the optimization problem under consideration. The complexity of each ant is such that even a single ant is able to find a (probably poor quality) solution. High quality solutions are only found as the emergent result of the global cooperation among all the agents of the colony concurrently building different solutions.

The model of a combinatorial optimization problem is used to define the pheromone model of ACO. A pheromone value is associated with each possible *solution component* and the set of all possible solution components is denoted by C. In ACO, an artificial ant builds a solution by traversing a fully connected *construction graph* $G_c(V,E)$, where V is a set of vertices and E is a set of edges. This graph can be obtained from the set of solution components C in two ways: components may be represented either by vertices or by edges. Artificial ants move from vertex to vertex along the edges of the graph, incrementally building a *partial solution*. Additionally, ants deposit a certain amount of pheromone on the components; that is, either on the vertices or on the edges that they traverse. The amount Δ_τ of pheromone deposited may depend on the quality of the solution found. Subsequent ants use the pheromone information as a guide toward promising regions of the search space. The ACO meta-heuristic algorithms is the following:

Set parameters, initialize pheromone trails
SCHEDULE_ACTIVITIES
   ConstructAntSolutions
   ApplyLocalSearch {optional}
   UpdatePheromones
END_SCHEDULE_ACTIVITIES

After initialization, the meta-heuristic iterates over three phases: at each iteration, a number of solutions are constructed by the ants; these solutions are then improved through a local search (this step is optional), and finally the pheromone is updated.

The interest of the scientific community in ACO meta-heuristic has risen sharply. Different ACO algorithms have been proposed in the literature (Dorigo et al. 1996; Dorigo et al., 1999; Dorigo & Stützle, 2004). Although ACO has been applied in many combinatorial optimization problems this chapter focuses on surveying ACO approaches in networks routing and load-balancing. In the following sections the most relevant ACO algorithms for routing and load balancing problems will be analyzed.

2.3 Approaches to mitigate stagnation

A major weakness of ACO algorithms is the *stagnation* in which all ants are taking the same position. Stagnation occurs when a network reaches its convergence (or equilibrium state) (Sim & Sun, 2003); an optimal path p_0 is chosen by all ants and this recursively increases an ant's preference for p_0. This may lead to the congestion of p_0 and to a dramatic reduction of the probability of selecting other paths. These two consequences are undesirable for a dynamic network since p_0, becoming congested, may become nonoptimal and disconnected due to network failure. Moreover, other nonoptimal paths may become optimal due to changes in network topology, and new or better paths may be discovered.

To alleviate the stagnation problem of ACO algorithms, different approaches have been proposed (Dorigo & Stützle, 2004) and can be categorized as follows:

- pheromone control;
- pheromone-heuristic control;
- privileged pheromone laying.

Pheromone control adopts several approaches to reduce the influences from past experience and encourages the exploration of new paths or paths that were previously nonoptimal: *evaporation, aging, limiting* and *smoothing pheromone.*

The approach called evaporation is typically used in conjunction with ACO in order to reduce the effect of past experience. Evaporation prevents pheromone concentration in optimal paths from being excessively high and preventing ants from exploring other (new or better) alternatives. In each iteration, the pheromone values $\tau_{i,j}$ in all edges *(i,j)* are discounted by an *evaporation factor* called *p.*

Additionally, past experience can also be reduced by controlling the amount of pheromone deposited for each ant according to its age. This approach is known as *aging*. In aging, an ant deposits lesser and lesser pheromone as it moves from a node to another one. Aging is based on the rationale that "old" ants are less successful in locating optimal paths since they may have taken longer time to reach their destinations. Both aging and evaporation prefer recent encouraging discoveries of new paths that were previously nonoptimal.

Limiting pheromone mitigate stagnation by limiting the amount of pheromone in every path. By placing an upper bound τ_{max} on the amount of pheromone for every edge *(i,j)*, the preference for optimal paths over nonoptimal paths is reduced. A variant of such an approach is *pheromone smoothing,* in which the amount of pheromone along an edge is reinforced as follows:

$$\tau_{i,j}(t') = \tau_{i,j}(t) + \delta \cdot (\tau_{max} - \tau_{i,j}(t)) \tag{1}$$

where δ is a constant between 0 and 1. It can be noticed that as $\tau_{i,j} \rightarrow \tau_{max}$, a smaller amount of pheromone is reinforced along an edge *(i,j)* .While evaporation adopts a uniform discount rate for every path, pheromone smoothing places a relatively greater reduction in the reinforcement of pheromone concentration on the optimal path(s). Consequently, pheromone smoothing seems to be more effective in preventing the generation of dominant paths.

Pheromone-heuristic control configures ants so that they do not solely rely on sensing pheromone for their routing preferences. This can be accomplished by configuring the probability function $P_{i,j}$ for an ant to choose an edge *(i,j)* using a combination of both pheromone concentration $\tau_{i,j}$ and heuristic function $\eta_{i,j}$. $\eta_{i,j}$ is function of the cost of edge which may include factors such as queue length, distance, and delay . $P_{i,j}$ at time t is given as follows:

$$P_{i,j}(t) = \frac{[\tau_{i,j}(t)]^{a} \cdot [\eta_{i,j}]^{\beta}}{\sum [\tau_{i,j}(t)]^{a} \cdot [\eta_{i,j}]^{\beta}} \tag{2}$$

where a and β represent the respective adjustable weights of $\tau_{i,j}$ and $\eta_{i,j}$. The routing preferences of ants can be altered by selecting different values of a and β. If $a > \beta$, ants choose paths with more optimistic heuristic values.
By adopting the policy of *privileged pheromone laying*, a selected subset of ants to have the privilege to deposit extra or more pheromone on the best paths (in terms of trip time and length). This approach reduces the probability of ants reinforcing stagnant paths that are nonoptimal or congested.

3. ACO routing algorithms

ACO routing algorithms (Dorigo et al., 1999) are a subset ACO algorithms which model the behaviour of insect swarms to solve the routing problem.
ACO routing algorithms show a number of interesting properties compared to traditional routing algorithms. Firs of all, they are adaptive by means of continuous path sampling and probabilistic ant forwarding which leads an interrupted exploration of the routing capabilities. Moreover, they are robust because routing information is the result of the repeated sampling of paths. The use of sampling implies that routing information is based on direct measurements of the real network situation, which enhances its reliability.
In the following subsections, the main ACO algorithms solving the routing problem will be discussed. In order to illustrate the differences between them clearly, the example of the *travelling salesman problem* will be analyzed.
In the TSP (Dorigo & Gambardella, 1997) a set of locations (e.g. cities) and the distances between them are given. The problem consists of searching a closed tour of minimal length that visits each city once and only once. To apply ACO to the TSP, the graph is defined by associating the set of cities with the set of vertices of the construction graph. Since in the TSP it is possible to move from any given city to any other city, the construction graph is fully connected and the number of vertices is equal to the number of cities. The lengths of the edges between the vertices are proportional to the distances between the cities represented by these vertices and pheromone values and heuristic values are associated with the edges of the graph. Pheromone values are modified at runtime and represent the cumulated experience of the ant colony, while heuristic values are problem dependent values that, in the case of the TSP, are set to be the inverse of the lengths of the edges. The ants construct the solutions as follows. Each ant starts from a randomly selected city (vertex of the construction graph) and at each construction step it moves along the edges of the graph, keeping a memory of its path. In subsequent steps ant chooses among the edges that do not lead to vertices that it has already visited. A solution will be constructed once an ant has visited all the vertices of the graph. At each construction step, an ant probabilistically

chooses the edge to follow among those that lead to yet unvisited vertices. The probabilistic rule is biased by pheromone values and heuristic information: the higher the pheromone and the heuristic value associated to an edge, the higher the probability an ant will choose that particular edge. Once all the ants have completed their tour, the pheromone on the edges is updated. Each of the pheromone values is initially decreased by a certain percentage. Each edge then receives an amount of additional pheromone proportional to the quality of the solutions to which it belongs (there is one solution per ant). This procedure is repeatedly applied until a termination criterion is satisfied.

3.1 AS: Ant System

Ant system (AS) (Fenet & Hassas, 1998) was the first ACO algorithm to be proposed in the literature. The pheromone values are updated by *all* the ants that have completed the tour. Solution components, denoted with $c_{i,j}$, are the edges of the graph, and the pheromone update for $\tau_{i,j}$, that is, for the pheromone associated to the edge joining cities i and j, is performed as follows:

$$\tau_{i,j} \leftarrow (1-\rho)\cdot\tau_{i,j} + \sum_{k=1}^{m} \Delta\tau_{i,j}^{k} \tag{3}$$

Where $\rho \in (0,1]$ is the evaporation rate, m is the number of ants, and $\Delta\tau_{i,j}^{k}$ is the quantity of pheromone laid on edge (i,j) by the k-th ant:

$$\Delta\tau_{i,j}^{k} = \begin{cases} \dfrac{1}{L_k} & \text{if } k-th\ ant\ travels\ on\ edge(i,j) \\ 0 & otherwise \end{cases} \tag{4}$$

where L_k is the tour length of the k -th ant.
In order to construct the solutions, the ants traverse the construction graph and make a probabilistic decision at each vertex. The transition probability of the k-th ant moving from city i to city j is given by:

$$P(c_{i,j} \mid s_k^p) = \begin{cases} \dfrac{\tau_{i,j}^{a}\cdot\eta_{i,j}^{\beta}}{\sum_{c_{i,j}\in N(s_k^p)} \tau_{i,j}^{a}\cdot\eta_{i,j}^{\beta}} & if\ j \in N(s_k^p) \\ 0 & otherwise \end{cases} \tag{5}$$

where $N(s_k^p)$ is the set of components that do not belong yet to the partial solution s_k^p of ant k, and parameters a and β control the relative importance of the pheromone versus the heuristic information $\eta_{i,j}=1/d_{i,j}$, where $d_{i,j}$ is the length of component $c_{i,j}$.

3.2 Ant Colony System

The Ant Colony System algorithm (Dorigo & Gambardella, 1997) was proposed as an improvement over the original AS algorithm. The first relevant difference between ACS and AS is the decision rule used by the ants during the construction process. Ants in ACS use the so-called *pseudorandom proportional* rule: the probability for an ant to move from city i to city j depends on a random variable q uniformly distributed over [0,1], and a parameter q_0; if $q \leq$

q_0, then, among the feasible components, the component $\tau_{i,j}\eta^{\beta}_{i,j}$ that maximizes the product is chosen; otherwise, the same equation as in AS is used. This rather greedy rule, which favours exploitation of the pheromone information, is counterbalanced by the introduction of a diversifying component: the *local pheromone update* (Ducatelle et al., 2005). The local pheromone update is performed by all ants after each construction step. Each ant applies it only to the last edge traversed:

$$\tau_{i,j} = (1-\varphi)\cdot\tau_{i,j} + \varphi\cdot\tau_0 \tag{6}$$

where $\varphi \in (0,1]$ is the pheromone decay coefficient, and τ_0 is the initial value of the pheromone. The interesting goal of the local update is to diversify the search performed by subsequent ants during one iteration. In fact, decreasing the pheromone concentration on the edges as they are traversed during one iteration encourages subsequent ants to choose other edges and hence to produce different solutions. This also prevents that several ants produce identical solutions during one iteration. Additionally, because of the local pheromone update in ACS, the minimum values of the pheromone are limited.
As in AS, also in ACS at the end of the construction process a pheromone an *offline* pheromone update is performed. This update is performed only by the best ant and only edges visited by the best ant are updated, according to the equation:

$$\tau_{i,j} \leftarrow (1-\rho)\cdot\tau_{i,j} + \varphi\cdot\Delta\tau^{best}_{i,j} \tag{7}$$

where $\Delta\tau^{best}_{i,j} = 1/L_{best}$ if the best ant used edge *(i,j)* in its tour, $\Delta\tau^{best}_{i,j} = 0$ otherwise. L_{best} can be set to either the length of the best tour found in the current iteration (L_{ib}) or the best solution found since the start of the algorithm (L_{bs}).

3.3 MMAS: MAX-MIN Ant System

MAX-MIN ant system (MMAS) algorithm (Stützle & Hoos, 1998) is another improvement of the original AS algorithm. Unlike AS, only the best ant adds pheromone trails, and the minimum and maximum values of the pheromone are explicitly limited (in AS and ACS these values are limited implicitly as a result of the algorithm working rather than a value set explicitly by the algorithm designer).
The pheromone update equation (applied, as in AS, to all the edges) is the following:

$$\tau_{i,j} \leftarrow (1-\rho)\cdot\tau_{i,j} + \Delta\tau^{best}_{i,j} \tag{8}$$

where $\Delta\tau^{best}_{i,j} = 1/L_{best}$ if the best ant used edge *(i,j)* in its tour, $\Delta\tau^{best}_{i,j} = 0$ otherwise. As in ACS, L_{best} can be set (subject to the algorithm designer decision) to either the length of the best tour found in the current iteration (L_{ib}) or the best solution found since the start of the algorithm (L_{bs}), or to a combination of both.
The pheromone values are constrained between a max value τ_{max} and a minimum value τ_{min} by verifying, after they have been updated by the ants, that all pheromone values are within the imposed limits: $\tau_{i,j}$ is set to τ_{max} if $\tau_{i,j} > \tau_{max}$ and to τ_{min} if $\tau_{i,j} < \tau_{max}$. The minimum value $\tau_{i,j} < \tau_{min}$ is most often experimentally chosen (however, a theory about how to define its value analytically has been developed). The maximum value τ_{max} may be calculated analytically using the optimum ant tour length value. For the TSP, $\tau_{max} = 1/(\varphi\cdot L^{*})$, where L^{*} is the

length of the optimal tour. If L^* is not known, it can be approximated by L_{bs}. It is important to underline that the value of the trails is set to τ_{max}, and that the algorithm is restarted when no improvement can be observed for a given number of iterations (Stützle, 1999).

3. ACO routing algorithms for MANETs

A mobile ad-hoc network (MANET) is a set of mobile nodes which communicate over radio. These networks have an important advantage, they do not require any existing infrastructure or central administration. Therefore, mobile ad-hoc networks are suitable for temporary communication links.

Due to the limited transmission range of wireless interfaces, usually communication has to be relayed via intermediate nodes. Thus, in mobile multi-hop ad-hoc networks each node also has to be a router. To find a route between different endpoints is a major problem in mobile multi-hop ad-hoc networks. Many different approaches to handle this problem were proposed in literature (Buruhanudeen et al., 2007), but so far no routing algorithm has been suitable for all situations.

Analyzing some important features of mobile ad-hoc networks, the following considerations explain why ant algorithms could perform well in these networks:

- Dynamic topology: this property is responsible for the unfulfilling performances of many classical routing algorithms in mobile ad-hoc networks. The ant algorithms are based on autonomous agent systems imitating individual ants. This allows a high adaptation to the current topology of the network.
- Local information: in contrast to other routing approaches, the ant algorithms make use of local information; no routing tables or other similar information have to be transmitted to other nodes of the network.
- Link quality: it is possible to integrate the connection/link quality into the computation of the pheromone concentration, especially into the evaporation process. This will improve the decision process with respect to the link quality.
- Support for multi-path: each node has a routing table with entries for all its neighbours. Adding the information about the pheromone concentration, the decision rule for selection of the next node could be based on the pheromone concentration at the current node.

In this section, an overview of the main ant based routing algorithms proposed explicitly for MANETs will be presented.

Ad hoc Networking with Swarm Intelligence (ANSI). ANSI is a reactive routing protocol (Rajagopalan & Shen, 2005) which defines two kinds of mobile agents called *forward reactive ants* and *backward reactive ants*. The routing tables in ANSI contain an entry for each reachable node and next best hop while the ant decision tables store the pheromone values. In ANSI, the forward reactive ants are generated only when a node has to transmit data to another node. The forward reactive ants are broadcast while the backward reactive ants retrace the path of forward reactive ants and update the pheromone values at the nodes. The data packets choose the next hop deterministically i.e., the hop which contains the largest pheromone value is chosen as the next hop.

Ant-colony-based Routing Algorithm (ARA). ARA is another reactive routing protocol (Günes & Spaniel, 2003) for MANETs. The routing table entries in ARA contain pheromone values for the choice of a neighbour as the next hop for each destination. The pheromone values in the routing tables decay with time and the nodes enter in a sleep mode if the pheromone in the routing table has reached a lower threshold. As in ANSI, route discovery in ARA is performed by two kind of mobile agents: forward ants and backward ants. During route discovery, the forward and backward ant packets characterized by unique sequence numbers to prevent duplicate packets, are flooded through the network by the source and destination nodes, respectively. The forward and backward ants update the pheromone tables at the nodes along the path for the source and destination nodes respectively. At the end of the route discovery process for a particular destination, the source node does not generate new mobile agents for the destination instead the route maintenance is performed by the data packets.

Probabilistic Emergent Routing Algorithm (PERA). Also in PERA (Baras & Mehta, 2003) route discovery is performed by forward and backward ants. These ant agents create and adjust probability distribution at each node for the node's neighbours. The probability related to a neighbour reflects the relative likelihood of that neighbour forwarding and eventually delivering the packet. Each forward node contains the IP address of its source node, the IP address of the destination node, a sequence number, a hop count field and a dynamically growing stack The stack contains the information about the nodes traversed by the forward ant and the times at which the nodes have been traversed. When a node does not have a record of a route to a destination, it creates a forward ant and the node pushes its own IP address on to the stack of the forward ant as well as the time at which the ant is created. Henceforth, the node keeps sending forward ants periodically to the destination for as long as a route is required. When a forward node reaches the destination, the destination node creates a new backward ant. It uses the information contained in the forward ant on the reverse path to modify the probability distribution at each node and update routing tables to reflect the current status of the network. Since the forward ant is broadcast at the source and intermediate nodes, each forward ant will cause the broadcast of multiple forward ants, several of which may find different paths to the destination, generating multiple backward ants.

POSition based ANT colony routing algorithm (POSANT). POSANT is a reactive routing algorithm (Kamali & Opatrny, 2008) based on ant colony optimization and location of nodes. This protocol is able to find optimum or nearly optimum routes when a given network contains nodes with different transmission ranges. Each node is assumed to be aware of its position, the position of its neighbours and the position of the destination node. A route in POSANT is searched only when there is a collection of data packets that are to sent from a source node to a destination node. Sending the data packets will start after a route from source to destination is established. Before that, only forward and backward ants are being exchanged. In order to minimize the time that POSANT spends to find a route while keeping the number of generated ants as small as possible, information about the position of nodes is used as a heuristic value. Neighbours in POSANT are partitioned into three zones in dependence of the position. The use of location information as a heuristic parameter results in a significant decrease of the time required to establish routes from a source to a destination. Moreover, having a short route establishment time, POSANT reduces greatly the number of control messages. POSANT has also a higher delivery rate with a shorter average packet delay than other position based routing algorithms.

Ant Routing Algorithm for Mobile Ad hoc networks (ARAMA). ARAMA (Hossein & Saadawi, 2003) is a proactive routing algorithm. As in other ACO algorithms for MANETs, the forward ant has to collect path information. However, in ARAMA, the forward ant takes into account not only the hop count factor but also the links local heuristic along the route such as the node's battery power and queue delay. ARAMA defines a value called *grade*, calculated by each backward ant, which is a function of the path information stored in the forward ant. At each node, the backward ant updates the pheromone amount of the node's routing table, using the *grade* value. The protocol uses the same *grade* to update pheromone value of all links. In ARAMA the route discovery and maintenance overheads are reduced by controlling the forward ant's generation rate.

HOPNET. This is a hybrid ant colony optimization routing protocol (Wanga et al., 2008) based on ants hopping from one zone to the next. HOPNET is highly scalable for large networks compared to other hybrid protocols. The HOPNET algorithm consists of the local proactive route discovery within a node's neighbourhood and reactive communication between the neighbourhoods. The network is divided into zones which are the node's local neighbourhood. A routing zone consists of the nodes and all other nodes within the specified radius length measured in hops. A node may be within multiple overlapping zones and zones could vary in size. The nodes can be categorized as interior and boundary (or peripheral) nodes with respect to the central node. Each node has two routing tables: *Intrazone Routing Table* (IntraRT) and *Interzone Routing Table* (InterRT). The IntraRT is proactively maintained so that a node can obtain a path to any node within its zone quickly. This is done by periodically sending out forward ants to sample path within its zone and determine any topology changes. Once a forward ant reaches a destination, a corresponding backward ant is sent back along the path discovered. The InterRT stores the path to a node beyond its zone. This source routing table is setup on demand as routes outside a zone is required. The peripheral nodes of the zone are used to find routes between zones. For small number of nodes, due to the constant movement of border nodes, new routes have to be determined continuously resulting in more delay than other hybrid routing protocols.

Distributed Ant Routing (DAR). In DAR (Rosati et al. 2008) routes are created on-demand, in order to have a low routing signalling load. Forward ants collect information only about the identities of the crossed nodes and move towards the destination choosing the next hop only on a pheromone basis. The amount of pheromone deposited by backward ants on each crossed link is constant. In DAR, in each node the routing tables are stochastic: next hop is selected according to weighted probabilities, calculated on the basis of the pheromone trails left by ants. When a node receives a datagram with destination *d*, if the routing entry for *d* is available, then the datagram is forwarded. Otherwise, the datagram is buffered and forward ants are sent out at constant rate r_{ae} (ant emission rate) in order to search a path to *d*. The forward ant goes to each node according to the probabilities for the next hop in the routing table at the current node. Thus, the forwarding of the forward ant is *probabilistic* and allows exploration of paths available in the network. Datagrams are routed *deterministically* based on the maximum probability at each intermediate node from the source node to the destination node. This process creates a complete global route by using local information. The simplicity of the protocol could be helpful in achieving seamless routing in networks constituted by heterogeneous elements.

Ant-based Distributed Route Algorithm (ADRA). In ADRA (Zheng et al., 2008) ants move across the network between randomly chosen pairs of nodes. Along the path, ants deposit simulated pheromones as a function of their hop distance from their source node, the quality of the link, the congestion encountered on their journey, the current pheromones the nodes possess and the velocity at which the nodes move. The node also ages the link by pheromones evaporating. An ant selects its path at each intermediate node according to the distribution of simulated pheromones at each node. In order to accelerate the convergence rate of the congestion problem and the shortcut problem, the parameters are given with different weight values to update the probability routing table. The ADRA system exhibits many attractive features of distributed control.

Ant-based Energy Aware Disjoint Multipath Routing Algorithm (AEADMRA). Earlier research has proposed several unipath routing protocols for MANETs. However, due to the dynamic topology of these networks, the single path is easily broken leading to a new route discovery process and an increase in both delay and control overhead. AEADMRA (Wu et al., 2007) was proposed to alleviate these problems. This algorithm is based on swarm intelligence and especially on the ant colony based meta-heuristic. AEADMRA has been designed to enable path accumulation in route request/reply packets and discover multiple energy aware routing paths with a low routing overhead.

ImProved Ant Colony Optimization algorithm for mobile ad hoc NETworks (PACONET). PACONET is a reactive routing protocol (Osagie et al., 2008) which also uses two kinds of agents: forward ant (FANT) and backward ant (BANT). The FANT explores the paths of the network in a restricted broadcast manner in search of routes from a source to a destination. The BANT establishes the path information acquired by the FANT. These agents create a bias at each node for its neighbours by leaving a pheromone amount from its source. Data packets are stochastically transmitted towards nodes with higher pheromone concentration along the path to the destination. FANTs also travel towards nodes of higher concentration but only if there is no unvisited neighbour node in the routing table. The rows of the routing table represent the neighbours of a node and the columns represent all the nodes in the network. Each pair (row, column) in the routing table has two values: a binary value indicating if the node has been visited and the pheromone concentration. All possible paths are explored to find the best path towards the destination. The node with the highest pheromone is chosen as the next hop after the FANT has determined that it has not visited the node before.

AntHocNet. This is a hybrid routing protocol (Caro et al. 2004) consisting of both reactive and proactive components. Nodes do not maintain routes to all possible destinations at all the times and generate mobile agents only at the beginning of a data session. The mobile agents search for multiple paths to the destination and these paths are set up in the form of pheromone tables indicating their respective quality. During the course of the data session, the paths are continuously monitored and improved in a proactive manner.

4. ACO techniques in load balancing

Routing problem in MANET is very challenging and difficult due to the mobility of nodes. Ant colony optimization is an efficient optimization technique used to find the optimum shortest route in the ad-hoc network. However, other problems has to be addressed in order to obtain full efficiency. Network congestion is one of these problems and is present when load is not perfectly balanced. In this case the simple implementation of ant behaviour is not sufficient and some adjustments have to be applied. Load-balancing becomes one of the important issues since the network performance such as network throughput and end-to-end delay can be improved if the loads are well balanced. In the following subsections some ACO algorithms for load balancing, improving efficiency and stability of classical ACO algorithms, will be described.

4.1 ABC: Ant Based Control

Ant based control system (ABC) (Schoonderwoerd et al., 1996) was designed to solve the load-balancing problem. Each row in the pheromone table represents the routing preference for each destination, and each column represents the probability of choosing a neighbour as the next hop. Along the paths, incoming ants update the entries in the pheromone table of a node. In order to mitigate stagnation, three approaches are adopted:

- aging;
- delaying;
- noise.

Aging is designed to discourage ants from following the trails of an ant that has travelled a longer path to some destination. In contrast to evaporation, aging may induce an ant to select a nonoptimal link, if the path from a node to its destination is very long. Used in conjunction with aging, delaying is designed to reduce the flow rates of ants from a congested node to its neighbours. By slowing down the ants originating from a congested node, the amount of pheromone they deposit reduced with time because of the aging process. Noise approach enables ants to choose a path randomly not taking into account the influence of the pheromone table. Thus, ants can explore new and better routes, particularly in dynamic networks.

In one of the ramifications of the ABC system (Guérin, 1997), *smart ants* are adopted to enhance performance. While in classic ABC an ant updates only the entry corresponding to the source node in the pheromone table of each node it passes, smart ants update all the entries in the pheromone table at each node. By performing more pheromone updates at every intermediate node, smart ants are more complex but fewer smart ants are needed in order to achieve the same routing purpose.

In another ramification of the ABC system (Subramanian et al., 1997), two kinds of ants are proposed: *regular ant* and *uniform ant*. Regular ant uses the accumulated cost of a path to determine the amount of pheromone to deposit. A regular ant that travels a higher cost path to a destination node deposits lesser pheromone. Unlike regular ants, uniform ants choose their next nodes in a random way. Moreover, while regular ants use the accumulated cost in the direction from source to destination, uniform ants use the accumulated cost in the reverse direction to establish the amount of pheromone to deposit.

4.2 Ant-Net

Ant-Net algorithm (Caro & Dorigo, 1997) was originally designed for routing in packet-switched networks. Unlike traditional routing algorithms which focused on shortest path routing, AntNet aims to optimize the performance of the entire network. In AntNet, forward ants are launched at regular intervals from a source node N_s to a destination node N_d to discover a feasible low-cost path. Backward ants travel from N_d to N_s to update pheromone tables at each intermediate node. From N_s to N_d, a forward ant selects the next hop node N_i using a random scheme that take into consideration of both the probability of choosing N_i, called P_{id} and a heuristic correction factor I_{ni}. While I_{ni} depends on the queue length at N_i, P_{id} is a selection probability which can be viewed as a pheromone concentration that can be reinforced by other ants.

As a forward ant travels from source node to destination node, it collects statistics such as the local data traffic condition on each intermediate node and the trip time to reach N_i. When a forward ant arrives at destination, a backward ant will be activated. This ant updates the probabilistic pheromone table at each intermediate node N_i and the estimated trip time for the path N_s - N_i. Backward ants update the selection probability by determining the *goodness* of the trip times of forward ants, and the amount of reinforcement using a *squash* function.

The *goodness* of the trip time is a relative measure determined comparing the current trip time to the current statistical estimates and the confidence interval of the best trip time. The squash function is a nonlinear function that is more sensitive in rewarding solutions with higher goodness values.

This algorithm (called Ant-Net-CL) alleviates the problem of stagnation. However, using both forward and backward ants generally doubles the routing overhead.

In another version of Ant-Net, called Ant-Net-CL (Caro & Dorigo, 1998) forward ants travel from a source to a destination in high priority queues, and backward ants estimate the trip time (by size of queuing data, links' bandwidth and delay), update local traffic statistics, and determine and deposit the amount of probability to reinforce. Since backward ants determine the amount of reinforcement using real time statistics, the routing information is comparatively more accurate and up-to-date.

Another ramification of AntNet (Baran & Sosa, 2000) is characterized by the five following distinguishing features from AntNet:

1. intelligent initialization of AntNet;
2. intelligent pheromone updates after link or node failures;
3. use of noise to mitigate stagnation;
4. deterministic rather than probabilistic selection of a node;
5. restricting the number of ants inside a network.

The first feature was included to regulate the exploration ants in the initial stage. The original entries in a routing table consist of a uniform distribution of probabilities which may not reflect the states of the network. Taking into consideration the a-priori knowledge of the network, ants in this work are configured to select neighbouring nodes with a higher initial probability. While AntNet did not consider situations of link failures, this version suggests that in case of link failures, the corresponding probability of a link that fails will be set to zero and will be distributed evenly among the remaining neighbouring nodes. The third feature deals with noise, where some ants select paths uniformly without considering the effect of pheromone concentration. The fourth feature uses a deterministic approach for the selection of the next hop. However, this approach may lead to a possible infinite looping. The fifth feature suggests to fix an upper bound in number of ants inside a network. Although restricting the number of ants may reduce routing overhead and possible congestion, it also places a restriction on the frequency of launching ants which may lead to possible reduction in the adaptiveness of the routing algorithm.

4.3 ASGA (Ant System with Genetic Algorithm) and SynthECA (Synthetic Ecology of chemical Agents)

Ant system with genetic algorithm (ASGA) was designed to solve problems of point-to-point, point to multipoint and cycle (multipath) routing in circuit-switched networks (White et al. 1998). In ASGA explorer ants are used to update pheromone tables. Although similar to AntNet, explorers travel in a round trip, but unlike backward ants in AntNet, explorers deposit the same amount of pheromones in their return trips. In addition, evaporation agents and pheromone heuristic control were used to mitigate stagnation. The genetic algorithm was added to increase the adaptivity of ants. For instance, if the best path is congested, it increases the likelihood of ants to find an alternative path. However, unlike the ABC system, ASGA was not designed to solve the load-balancing problem in circuit-switched networks.

Subsequently, in order to solve this problem, ASGA was generalized to a framework called *Synthetic ecology of chemical agents* (*SynthECA*) (White, 2000). SynthECA was also designed to solve other problems such as fault location detection in circuit-switched networks. Although SynthECA was not designed with any specific type of ants, all ants in SynthECA are characterized with a combination of the following:

- emitters;
- receptors;
- chemistry;
- migration decision function;
- memory.

Emitters are used to generate different types of *chemical* pheromone. Pheromones are represented by strings such as "1100" or "10#1." While each type of pheromone corresponds to a *genotype*, each string corresponds to a *chromosome* in GA. Pheromone is generated by an emitter decision function (EDF). As in GA, the operations of crossover and mutation are applied in the EDF to evolve the pheromone types. With various pheromone types and pheromone reactions, ants can be designed to send and sense more types of signals in their stigmergic communication.

In order to sense local pheromone changes generated by emitters, a *receptor* is used. Using receptor detection function (RDF), a receptor senses different types of pheromone. By configuring ants with different EDFs and RDFs, more sophisticated pheromone manipulation techniques such as privileged pheromone laying and pheromone heuristic control can be realized.

Chemistry is a set of rules (inspired by GA) that specifies pheromone reactions. In SynthECA, ants use pheromone reactions to send out control information to other ants. In the set of rules, five types of pheromone reactions are specified as follows:

1. X→"nothing:" this is similar to evaporation;
2. X+Y→Y: this is applied when two ants are competing for a path and only one ant will prevail;
3. X+Y→Z: this rule is used to report the status of network resources (e.g., poor connection quality);
4. X+Y→X+Z: this rule, in computational terms, represents a conditional construct. A pheromone type Y is transformed into another type of pheromone Z in the presence of a specific type of pheromone X;
5. X+Y→W+Z: this rule allows two ants X and Y to jointly communicate both inhibitory (e.g., W) and excitatory (e.g., Z) messages to other ants.

While a *migration decision function* is a set of rules determining the next hop of an ant, pheromones (i.e., labels and concentrations) and the state of an ant are stored in the ant's *memory*.

Using a combination of the above five components, several types of ants such as *route finding agent* (RFA), *connection monitoring agent* (CMA) and *fault detection agent* (FDA) can be configured to solve different networking problems. RFAs include *explorers, allocators* and *deallocators*. An explorer is used to find a path from a source to a destination and is configured with an emitter for a single type of pheromone and three receptors for sensing

pheromone, measuring link costs and detecting quality of links. Using a probability function, an explorer chooses a path taking into account the pheromone and the cost of the path. Travelling from source to destination, explorer records all the nodes it passed. When it reaches destination, it returns to via the same path and deposits pheromone along the way, which may influence the pheromone concentration of other types. Explorers are also programmed to also take into consideration the quality/reliability of the link. While an allocator is used to obtain link resources, a deallocator release resources previously acquired by an allocator.
CMA's are activated if the quality of service changes. A CMA evaluates the quality of a link using local traffic statistics and it deposits a special type of pheromone (called *q-chemical*) to indicate the quality of the associated link. CMAs use q-chemical to indirectly communicate the quality of links to FDAs while they circulate the network for diagnostics purposes.

4.4 MACO: Multiple Ant Colony Optimization

In MACO (Sim & Sun, 2003), more than one colony of ants are used to search for optimal paths, and each colony of ants deposits a different type of pheromone represented by a different colour. Although ants in each colony respond to pheromone from its own colony, MACO is augmented with a *repulsion* mechanism preventing ants from different colonies to choose the same optimal path. In order to establish connections between two gateways, two groups of mobile agents (e.g., MAG1 and MAG2), acting as routing packets, construct, manipulate and consult their own routing tables. In MACO, each group of mobile agents corresponds to a colony of ants, and the routing table of each group corresponds to a pheromone table of each colony. Even though MAG1 and MAG2 may have their own routing preferences, they also take into consideration the routing preferences of the other group. While the routing preferences of ants are recorded in their pheromone tables, the routing preferences of mobile agents are stored in their routing tables. In constructing its routing table, MAG1 (respectively, MAG2) consults the routing table of MAG2 (respectively, MAG1) in order to avoid routing packets to those paths that are highly preferred by the other group. This increases the chance of distributing data traffic. By adopting the MACO approach, it may be possible to reduce the likelihood that all mobile agents establish connections using *only* the optimal path. The advantage of using MACO is that it is more likely to establish connections through multiple paths to help balance the load but does not increase the routing overhead.

5. Applications and new directions

The works surveyed in the previous sections addressed the application of swarm intelligence and in particular ACO algorithms to solve the routing problem and/or load balancing in MANETs. However, ACO algorithms have been applied to solve different kinds of problems in MANETs. Reduction of power consumption is one of these important issues in ad hoc wireless networks. Mobile nodes are powered by battery and an efficient utilization of battery energy is very important. When a node exhausts its available energy, it ceases to work and the lack of mobile nodes can result in network partitioning. In recent years, some improvement in ACO routing algorithms were proposed in order to reduce the communication load related to energy spent with communications (De Rango & Tropea, 2009; Zyiadi et al., 2009; Li & Shi, 2009). In (De Rango & Tropea, 2009) has been proposed a novel routing algorithm able to satisfy multiple metrics for a multi-objective optimization like end-to-end delay, load balancing and energy savings. Innovation factor of this proposal has been the pheromone updating policy and the joint metric used.

The demand for quality of services (QoS) in MANETs suggested also the development of QoS routing strategies computing paths that are suitable for different type of traffic generated by various applications while maximizing the utilizations of network resources. The main problem to be solved by QoS routing algorithm is the Multi-Constraint Path problem. Instead of using a shortest path algorithm based on statically configured metrics, as in traditional routing protocols, the algorithm must select several alternative paths that are able to satisfy a set of constraints regarding, for instance, end-to-end delay bounds and bandwidth requirements. Several approaches (Shokrani & Jabbehdari, 2009; Liu et al., 2007; Liu et al. 2008, Zhang & Li, 2008) have been proposed to address the complexity of multi-constrained path computation problem using ACO approach.

Another interesting issue in MANETs, in which has been employed the behavioural principle present in ant colonies, is address management. In ad-hoc networks, address management is a particularly tough challenge, because of their dynamically changing topology, and the sort of events that occur in their environment. In (Pachon & Madrid, 2009) has been proposed a solution to this problem, involving the self-organization and emergency principles governing the behaviour of ant colonies.

All the works presented in this chapter show how swarm intelligence and in particular the behaviour of ant colonies have inspired a number of successful methods and techniques to solve different problems in MANETs. However, the potential of this distributed intelligent technique is more clear when applied to other dynamic network scenarios. In (De Rango et al., 2008), for example, has been showed how minimum hop count and load balancing metrics based on ant behaviour over a HAPs mesh can lead to a better management of the system resources and an increase in the number of calls admitted by the system. This is only one example of the wide range of applications covered by swarm intelligence techniques which underlines the importance of this approach in communication networks.

6. References

Baran, B. and Sosa, R. (2000). A new approach for AntNet routing, *Proc. 9th Int. Conf. Computer Communications Networks*, 303-308, 0-7803-6494-5, Las Vegas, USA

Baras, J. S. & Mehta, H. (2003). A Probabilistic Emergent Routing Algorithm for Mobile Ad Hoc Networks, *Proceedings of the Conference on Modeling and Optimization in Mobile, Ad Hoc and Wireless Networks (WiOpt '03)*, Sophia-Antipolis, France, (March 2003)

Bonabeau, E.; Dorigo, M. & Theraulaz, G. (1999). *Swarm Intelligence. From Natural to Artificial Systems*, Oxford University Press, 0-19-513159-2, Oxford

Bonabeau, E.; Dorigo, M. & Theraulaz, G. (2000). Inspiration for optimization from social insect behaviour, *Nature*, Vol. 406, No. 6, (July 2000) 39-42, 0028-0836

Buruhanudeen, S.; Othman, M., Othman, M. & Mohd Ali, B. (2007). Existing MANET routing protocols and metrics used towards the efficiency and reliability - an overview, *Proceedings of the 14th Int. Conference on Telecommunications and 8th Malaysia. Int. Conference on Communications (ICT - MICC 2007)*, 231-236, Penang, Malaysia, (May 2007)

Caro, G. D. & Dorigo, M. (1998). AntNet: Distributed stigmergetic control for communications networks. *J. Artif. Intell. Res.*, Vol. 9, No. 3, 317-365

Caro, G. D. & Dorigo, M. (1997). AntNet: A Mobile Agents Approach to Adaptive Routing, In : *IRIDIA – Technical Report Series*, IRIDIA, 97-12, Universitè Libre de Bruxelles.

Černý, V. (1985). A thermo dynamical approach to the traveling salesman problem, *Journal of Optimization Theory and Applications*, Vol. 45, No. 1, (January 1985) 41-51

Deneubourg, J. L.; Aron, S. Goss, S. & Pasteels, J. M. (1990). The self-organizing exploratory pattern of the argentine ant, *Journal of Insect Behavior*, Vol. 3, No. 2, 159-168

De Rango, F.; Tropea, M.; Provato, A.; Santamaria, A.F. & Marano, S. (2008). Minimum Hop Count and Load Balancing Metrics Based on Ant Behavior over HAP Mesh, *Global Telecommunications Conference, 2008. IEEE GLOBECOM 2008. IEEE* , 1-6, 1930-529X, New Orleans, (Nov.-Dec. 2008)

De Rango, F. & Tropea, M. (2009). Energy saving and load balancing in wireless ad hoc networks through ant-based routing, *International Symposium on Performance Evaluation of Computer & Telecommunication Systems, SPECTS 2009.*, Vol. 41, (July 2009), 978-1-4244-4165-5

Di Caro, G.; Ducatelle, F. & Gambardella, L. M. (2004). AntHocNet: an Ant-Based Hybrid Routing Algorithm for Mobile Ad Hoc Networks, In: *IDSIA-25-04-2004 Technical Report*, IDSIA/USI-SUPSI, 1-12, Dalle Molle Institute for Artificial Intelligence, Switzerland

Dorigo, M.; Maniezzo, V. & Colomi, A. (1996). The Ant System: Optimization by a colony of cooperating agents, *IEEE Transactions on Systems, Man and Cybernetics - Part B*, Vol. 26, No. 1, 29-41

Dorigo, M. & Gambardella, L. M. (1997). Ant Colony System: A Cooperative Learning Approach to the Traveling Salesman Problem, *IEEE Transactions on Evolutionary Computation*, Vol. 1, No. 1, (April 1997) 53-66, 1089-778X(97)03303-1

Dorigo, M.; Caro, G. D. & Gambardella, L. M. (1999). Ant algorithms for discrete optimization, *Artificial Life*, Vol. 5, No. 2, (April 1999) 137-172, 1064-5462

Dorigo, M. & Stützle, T. (2004). *Ant Colony Optimization*, MIT Press, 0-262-04219-3, Cambridge, MA, UK

Dorigo, M.; Birattari, M. & Stützle, T. (2006). Ant Colony Optimization. Artificial Ants as a Computational Intelligence Technique, In: *IRIDIA - Technical Report Series* , IRIDIA, 1-12, 1781-3794, Université Libre de Bruxelles

Ducatelle, F.; Di Caro, G. & Gambardella, L. M. (2005). Using ant agents to combine reactive and proactive strategies for routing in mobile ad hoc networks, *International Journal of Computational Intelligence and Applications*, Vol. 5, No. 2, (June 2005) 169-184

Fenet, S. & Hassas, S (1998). An ant system for multiple criteria balancing, *Proceedings of 1st International Workshop on Ants Systems*, Brussels, (September 1998)

Glover, F. & Laguna, M. (1997). *Tabu Search*, Kluwer Academic Publishers, 079239965X, Norwell, MA, USA

Guérin, S. (1997). Optimization Multi-Agents en Environment Dynamique: Application au Routage Dans les Réseaux de Telecommunications, In: *DEA*, Univ. Rennes I, Ecole Nat. Supér. Télécommun. Bretagne, France

Günes, M. & Spaniel, O. (2003). Ant-Routing-Algorithm for mobile multi-hop ad-hoc networks, *Proceedings of Int. Conference on Network Control and Engineering for QoS (Net-Con 2003)*, 120-138, Muscat, Oman, (October 2003)

Hossein, O. & Saadawi, T. (2003). Ant routing algorithm for mobile ad hoc networks (ARAMA), *Proceedings of the 22nd IEEE International Performance, Computing, and Communications Conference*, 281-290, Phoenix, USA, (April 2003)

Kamali, S. & Opatrny, J. (2008). A Position Based Ant Colony Routing Algorithm for Mobile Ad-hoc Networks, *Journal of Networks*, Vol. 3, No. 4, (April 2008) 31-41

Li, Z. & Shi, H. (2008). A Data-Aggregation Algorithm Based on Adaptive Ant Colony System in Wireless Sensor Networks, *Congress on Image and Signal Processing2008, CISP '08*. Vol. 4, 449-453, 978-0-7695-3119-9, , Sanya, China, (May 2008)

Liu, M.; Sun, Y.; Liu, R. & Huang, X. (2007). An Improved Ant Colony QoS Routing Algorithm Applied to Mobile Ad Hoc Networks, *International Conference on Wireless Communications, Networking and Mobile Computing, 2007. WiCom 2007*, 1641-1644, 978-1-4244-1311-9, Shanghai, (September 2007)

Liu, Y.; Zhang, H.; Ni, Q.; Zhou, Z. & Zhu G. (2008). An Effective Ant-Colony Based Routing Algorithm for Mobile Ad-Hoc Network. *Circuits and Systems for Communications, 2008. ICCSC 2008. 4th IEEE International Conference on* , 100-103, 978-1-4244-1707-0, Shanghai, China, (May 2008)

Lourenço, H. R.; Martin, O. & Stützle, T. (2002). Iterated local search, In: *Handbook of Metaheuristics of International Series in Operations Research Management Science*, Glover, F., & Kochenberger, G., (Eds.), 321-353, Kluwer Academic Publishers, Norwell, MA, USA

Osagie, E.; Thulasiraman, P. & Thulasiram, R. K. (2008). PACONET: imProved Ant Colony Optimization routing algorithm for mobile ad hoc NETworks, *Proceedings of 22nd International Conference on Advanced Information Networking and Applications*, 204-211, Okinawa, Japan, (March 2008)

Pachon, A. & Madrid, J.M. (2009). Application of an ant colony metaphor for network address management in MANETs. *Communications, 2009. LATINCOM '09. Conference on IEEE Latin-American*, 1-6, 978-1-4244-4387-1, Medellin, (September 2009)

Rajagopalan, S. & Shen, C. (2005). ANSI: a unicast routing protocol for mobile ad hoc networks using swarm intelligence, *Proceedings of the International Conference on Artificial Intelligence*, 24-27, Las Vegas, USA, (June 2005)

Rosati, L.; Berioli, M. & Reali, G. (2008). On ant routing algorithms in ad hoc networks with critical connectivity, *Ad Hoc Networks*, Vol. 6, No. 6, (August 2008) 827-859

Schoonderwoerd, R.; Holland, O., Bruten, J. & Rothkrantz, L. (1996). Ants for Load Balancing in Telecommunication Networks, In: *HPL-96-35 Technical Report*, Hewlett Packard Laboratory, Bristol, UK

Sim, K. M. & Sun, W. H. (2003). Ant Colony Optimization for Routing and Load-Balancing: Survey and New Directions, *IEEE Transactions on Systems, Man and Cybernetics - Part A: System and Human*, Vol. 33, No. 5, 560-572

Shokrani, H. & Jabbehdari, S. (2009). A novel ant-based QoS routing for mobile ad hoc networks, *Proceedings of the First international Conference on Ubiquitous and Future Networks*, 79-82, 978-1-4244-4215-7, Hong Kong, (June 2009)

Stützle, T. & Hoos, H. (1998). The MAX-MIN Ant System and Local Search for Combinatorial Optimization Problems: Towards Adaptive Tools for Combinatorial Global Optimization, In: *Meta-Heuristics: Advances and Trends in Local Search*

Paradigms for Optimization, Vos, S.; Martello, S., Osman, I. H. & Roucairol, C., (Eds.), 313-329, Kluwer Academic Publishers, Norwell, MA, USA

Stützle, T. (1999). Local Search Algorithms for Combinatorial Problems: Analysis, Improvements, and New Applications, In: *DISKI*, Infix Publishers, Vol. 220, Sankt Augustin, Germany

Subramanian, D.; Druschel, P. & Chen, J. (1997). Ants and reinforcement learning: A case study in routing in dynamic networks, *Proc. Int. Joint Conf. Artificial Intelligence*, 832–838, 1-555860-480-4, Palo Alto, USA

Tarasewich, P. & McMullen, P. R. (2002). Swarm Intelligence: Powers in numbers, *Communications of the ACM*, Vol. 45, No. 8, (August 2002) 62-67, 0001-0782

Wang, J.; Osagie, E., Thulasiraman, P. & Thulasiram, R. K. (2009). HOPNET: A hybrid ant colony optimization routing algorithm for mobile ad hoc network, *Ad Hoc Networks*, Vol. 7, No. 4, (June 2009) 690-705

White, T.; Pagurek, B. & Oppacher, F. (1998). ASGA: Improving the ant system by integration with genetic algorithms, *Proceedings of 3rd Genetic Programming Conf.*, 610–617, (July 1998)

White, T. (2000). SynthECA: A Society of Synthetic Chemical Agents. *Ph.D.dissertation*, Carleton University, Northfield, MN

Wu, Z.; Song; H., Jiang, S. & Xu, X. (2007). Ant-based Energy Aware Disjoint Multipath Routing Algorithm in MANETs, *Proceedings of International Conference on Multimedia and Ubiquitous Engineering*, 674-679, Vol. 1, No. 1, Seoul, Korea, (April 2007)

Zhang, Y. & Li, Z. (2009). HCRS: A Routing Scheme for Ad Hoc Networks as a QoS Guarantee Primitive, *Wireless Communications, Networking and Mobile Computing, 2009. WiCom '09. 5th International Conference on* , 1-4, 978-1-4244-3692-7, Beijing, (September 2009)

Zheng, X.; Guo, W. & Liu, R. (2004). An ant-based distributed routing algorithm for ad-hoc networks (ADRA), *Proceedings of International Conference on Communications, Circuits and Systems (ICCCAS 2004)*, 412-417, Vol. 1, No. 1, Chengdu, China, (June 2004)

Ziyadi, M.; Yasami, K. & Abolhassani, B. (2009). Adaptive Clustering for Energy Efficient Wireless Sensor Networks Based on Ant Colony Optimization, *Communication Networks and Services Research Conference*, CNSR '09, Seventh Annual, 330-334, 978-1-4244-4155-6, May 2009, Moncton, NB

12

Impact of the Mobility Model on a Cooperative Caching Scheme for Mobile Ad Hoc Networks

F.J. Gonzalez-Cañete and E. Casilari
University of Malaga
Spain

1. Introduction

In the last few years more and more mobile devices have been developed and they are now part of our lives. Cellular phones, PDAs and laptops are daily used in mobile environments such as airports or train stations. For many applications these terminals require access to data networks or the Internet. In order to share information among them the mobile nodes have to cooperate to forward and route the traffic from one node to another forming the so-called Mobile Ad Hoc Networks (MANETs). Due to the mobility of the nodes the mobile nodes can enter or leave the coverage area of other nodes forcing to recalculate the routes in order to make possible the packet forwarding. In addition, wireless networks have a more limited bandwidth and a greater error probability than the wired medium, since the radio medium is shared and prone to interferences and packet collisions. Moreover, the mobile devices have to be portable and hence the processing and battery capabilities also have important restrictions.

Due to the limitations of the MANETs and the mobile devices, some caching mechanisms can be implemented in order to reduce the traffic in the network. Reducing the traffic along the network, and hence the number of forwarded packets, also reduces the battery consumption.

Let us suppose a MANET with conectivity to an external network such as the Internet as depicted in Figure 1. The mobile nodes cooperate to route packets to the Access Routers that provide access to the external networks. As all the traffic in the MANET is routed to the Access Routers these devices can turn into a bottleneck. In addition, the Access Routers can become temporally inaccessible because they can be out of the coverage area of any mobile node in the network because of the nodes' mobility. This situation causes temporal disconnections to the external networks. The caching mechanisms reduce the impact of the temporal disconnection to the Access Routers as the mobile nodes can cooperate to serve the documents they have previously cached to the rest of the nodes. On the other hand, since the mobile nodes also have the capability of serving information the bottleneck produced in the Access Routers is reduced as the traffic does not reaches them.

As the mobility model influences the behaviour of the nodes in the MANET and the cooperative caching mechanisms depend on the connectivity among the mobile nodes this paper evaluates and compares the performance of a caching scheme under different mobility models. Consequently, in this paper we propose a cooperative caching mechanism and evaluate its performance using two different mobility models.

The rest of this paper is organized as follows. Section 2 comments the related work about the mobility models and the caching architectures. Section 3 describes the caching architecture evaluated in this paper. Section 4 illustrates the simulation model. Section 5 describes the performance evaluation of the proposed caching scheme using two different mobility patterns. Finally, Section 6 outlines the main conclusions of this work and proposes future research directions in this topic.

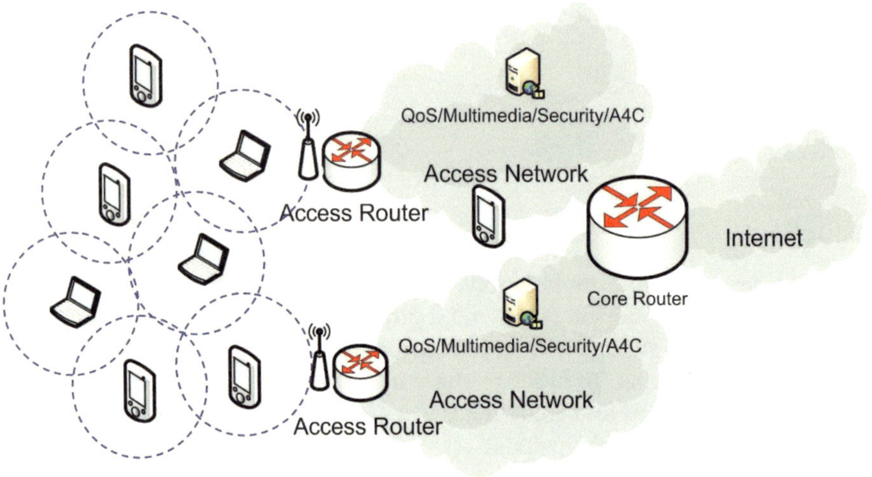

Fig. 1. MANET with Internet conectivity

2. Related work

In order to reproduce the mobility behaviour of mobile nodes in an ad hoc wireless network some mobility models have been proposed in the last few years. These mobility models can be categorized into three groups:

- Unrestricted random models: the next mobile node destination waypoint is decided randomly according to some heuristics depending on the mobility model. The most used models are: the RWP (Random Way Point) (Broch et al., 1998) mobility model simply selects a random destination in the simulation area; the RD (Random Direction) (Royer et al., 2001) mobility model selects a direction that is followed by the node until the simulation boundary area is reached and then another direction is selected; the Markovian Way Point (MWP) (Hyytia et al., 2006a) and Gauss-Markov (GM) (Liang et al., 1999) mobility models select the next destination using Markovian probabilities among waypoints.
- Geographic-based models: the next mobile node destination is decided according to some geographical constraint. In this category we can mention: the Obstacle Model (OM) (Jardosh et al., 2003) which defines a set of obstacles in the simulation area that must be avoided; the FreeWay and ManhattanGrid (Bai et al., 2003) mobility models limit the nodes' mobility to predefined ways within the simulation area.
- Group mobility models: the nodes' mobility tries to imitate typical human collective movements. The RPGM (Reference Point Group Mobility) (Hong et al., 1999) includes the possibility of having dynamic groups of mobile nodes with a leader that decides the next target that the entire group must reach. The DartMouth model (Kim et al., 2006)

chooses the destination of the node movements according to real data sets of human behaviour in a simulation area; the Clustered Mobility Model (CMM) (Lim et al., 2006) divides the simulation area into clusters so that the mobile nodes are assigned to the clusters. The nodes move or not between clusters depending on the number of the mobile nodes in the clusters; ORBIT (Ghosh et al., 2007) randomly defines a set of clusters and the mobile nodes are assigned to some of them moving only between the assigned clusters; SLAW (Self-similar Least Action Walk) (Lee et al., 2009) mobility model represents social contexts present among people sharing common interests using fractal waypoints and heavy-tail flights on top of the waypoints.

As it can be observed, each mobility model tries to reproduce some mobile characteristics although none of them is able to be general enough to be considered the most reliable mobility model under any circumstances. Specifically, mobility models such as Freeway and Manhattan Grid are suitable for vehicular networks since ORBIT or SLAW are adapted for human mobility.

On the other hand, some cooperative caching architectures have been proposed in order to reduce the traffic among the mobile nodes in a wireless network and hence the power consumption. The caching procedures aim to reduce the number of requests sent to the network as some of them can be resolved by the caches implemented in the mobile nodes. Moreover, the cooperation among mobile nodes using caching techniques also reduces the traffic in the data servers or the routers to external networks because the requests are replied on its way to the servers.

The cooperative caching strategies can be divided into four categories:

- Broadcast-based: the mobile nodes broadcast the requests in order to find a mobile node to reply with the requested document. The data server is a static node and hence it can also reply to the request.
- Information-based: the mobile nodes interchange or store information about where the documents are located in the network.
- Role-based: Each mobile node has a function in the network which can be organised in clusters. Depending on the architecture some mobile nodes are selected as information coordinators, clients, etc.
- Directed requests: The requests are directly sent to the server and it is expected to be replied in their way.

MOBEYE (Dodero & Gianuzzi, 2006) is a broadcast based caching scheme that proposes implementing a cache with the LRU (Least Recently Used) replacement policy into each mobile node. When a mobile node needs a document (and it does not have a valid copy in its local cache) it broadcasts a request message. If a mobile node receives the request message and it has a valid copy in its local cache, the mobile node replies using an *ack* (acknowledge) message to the requester. Finally the document is requested to the first mobile node that acknowledges the request.

SimpleSearch (Lim et al., 2006) is another broadcast based caching scheme very similar to MOBEYE. If a mobile node needs a document that is not stored in its local cache, a broadcast request message is sent a limited number of hops away. When a mobile node with a document copy is found it replies with an *ack* message that stores the path between the node with the document and the requester. Finally, a *confirm* message is sent by the requester to the node with the document following the inverse path. Three replacement policies were proposed to use with this scheme:

- TDS_D (Time and Distance Sensitive - Distance) - The first criteria to evict documents from the local cache is the distance in hops to the server node. Thus, nearest copies are evicted first.
- TDS_T (Time and Distance Sensitive - Time) - The documents with the highest time from the last access are evicted first.
- TDS_N - Distance and frequency are pondered in order to choose a document to be removed.

SimpleSearch also defines an admission control that avoids to store in the local cache those documents that are served from less than a certain number of hops away from the requester. In that way very popular documents are avoided to be stored in all the caches.

DGA (Distributed Greedy Algorithm) (Tang et al., 2008) is an information based scheme that implements for all the network nodes a table informing about the location of the documents in the network. The nodes store which is the closest and the second closest node where the documents are stored. In addition the mobile nodes send *AddCache* and *DeleteCache* broadcast messages in order to inform the rest of the nodes about the insertion and deletion of documents in the local cache so that they can update their information tables. When a mobile node requests a document it first checks if there is a valid copy in its local cache. If not, it checks if the corresponding table includes the possible document locations. If so, the document is requested to the node stored in the table. If this fails, the document is requested to the data server.

Similarly to DGA the GroupCaching scheme (Ting & Chang, 2007) proposes the mobile nodes to implement a local cache and a group table that stores information about the documents stored in the nodes located only one hop away. Every second the nodes send information to their neighbours informing about its local cache changes in order to maintain the group table updated. On the other hand Hello messages are used to know if a node leaves the group.

COACS (Cooperative and Adaptive Caching System) (Artail et al., 2007) is a role-based scheme that obliges the mobile nodes to adopt one of two roles: QD nodes caches the requests and the CN nodes caches the documents. The QD nodes maintain a distributed table about where the documents are located. In that way, if a QD node receives a request and it does not know where to find the document, the request is forwarded to the closest QD. The documents are stored in the CN only if they are served by the data server. In that case, the CN informs the nearest QD about this fact.

Another role based schema was proposed in (Denko, 2007), where the mobile nodes create clusters with a cluster head node (CH) (responsible of the communication among clusters), a data source node (DS) (that stores the data about where the documents are located), caching agents (CA) (that implements a local cache) and mobile hosts (MH). When a node needs a document it is requested to its neighbours, to the CA, the DS and the CH respectively. If the document is not found in any of them it is requested to another cluster using the CH.

The above mentioned cooperative caching schemes have been evaluated to measure their performance using only the Random Waypoint mobility model. As the employed mobility model influences the behaviour of the nodes in the MANET and hence their connectivity we consider necessary to evaluate the caching schemes not only using a unique mobility model, but also using at least one more model in order to compare the obtained performance results.

3. Caching scheme proposed

The caching scheme proposed follows the same request-reply model mentioned in the related work. There is one or more static data server in the MANET that stores the universe of documents and the rest of the devices are mobile nodes that periodically request documents to the data servers. When a node requests a document it waits for the reply a certain amount of time. If the document is not received during this time the node will request it again.

3.1 Local caching

Firstly, all the mobile nodes implement a local cache that stores the received documents. Therefore, if the mobile node have to request a document it first searches in its local cache for a valid copy of the document. If the document is found the request is avoided and hence there is no traffic generated in the network and the server load is also diminished.

The local cache has some parameters that have to be taken into account: the replacement policy, the cache size and the document's expiration.

The replacement policy defines which documents have to be evicted of the local cache in order to make room for a new one. The replacement policy objective is to select for eviction those documents that have the least probability to be requested again in the near future. Unfortunately this is not trivial because it depends on the traffic characteristics. As the traffic in real MANETs is not as well known and studied as the Internet traffic only a few replacement policies have been proposed such as the classic LRU; TDS_D, TDS_T and TDS_N proposed for SimpleSearch; and SXO (Size x Order) proposed in (Yin & Cao, 2006). We adopt the LRU replacement policy because of its simplicity.

The cache size is another important parameter that has to be considered because the bigger the cache is, more documents it will store and the probability to find a previously requested document increases. Due to the fact that the mobile devices may have some restricted storage capabilities the cache sizes of actual equipment will not be too large.

Finally, all the documents in the network have an associated expiration time or TTL (Time To Live) that defines when the information contained in the document is considered obsolete and hence the document has to be requested again if needed. On the other hand, the obsolete documents stored in the local cache can be deleted because they are not valid.

3.2 Interception caching

The functionalities of the mobile nodes can be expanded if they are enabled to perform as a proxy for the other nodes. Since the mobile nodes have to forward the requests to the data servers, they can also check the requested document and search for a valid copy in its local cache. If found, the mobile node replies with the document to the requester node instead of forwarding the message to the server. Using this capability the latency perceived by the user is reduced because the document is served by a closer node in the route to the server. Similarly, the network traffic and server load are also decreased because the request is replied before it reaches the data server.

This operation is illustrated in Figure 1. In the ad hoc network snapshot shown in the figure, *DS* represents the data server (or a node that accesses to an external network which provides the documents), nodes *1, 2, 3, 4, 5* and *6* are mobile nodes and the lines between them symbolise the existing routes. In this situation if node *2* requests the document *A* to DS the request will pass through node *1* and will reach *DS* that will reply to node *2* through node *1*. As the document *A* is received in node *2* it will be cached. If now node *3* requests the same document *A*, the request will reach node *2* that will search for the document *A* in its local cache. As it has a valid copy of *A*, node *2* will reply to node *3* with document *A* and will not forward the request to the *DS*. Using the request interception the number of hops and messages has been reduced from 6 (3-2-1-DS-1-2-3) in the case of no interception to 2 (3-2-3). As the number of hops is reduced the latency perceived by node *3* is also reduced. In addition, the server load is also reduced as the request does not reach the *DS*. This obviously is achieved at the cost of a higher processing load in the MANET nodes, as they are obliged to analyse all the document requests passing thorough them.

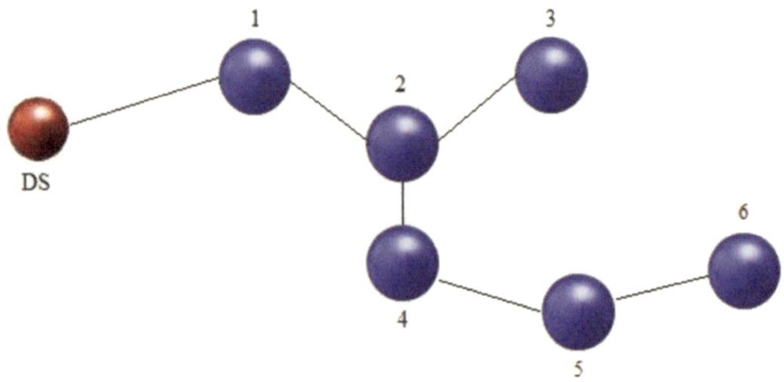

Fig. 1. Example of ad hoc network

3.3 Redirection caching

As the mobile nodes in the MANET have to forward requests and replies from other nodes and the data servers, they can use this information to learn where and how far (in number of hops) the documents are stored in the MANET. Using the information collected from the traffic the forwarding mobile nodes can redirect the requests to other node that is known to have the requested document and that is located closer than the original destination of the request.

Let us suppose that node *3* in Figure 1 requests the document *A* to *DS*. The request will pass through nodes *2* and *1* to DS and they will annotate that node *3* has requested document *A* and it will be available there in the near future at one and two hops away respectively. When *DS* replies with the document *A* to node *3* through nodes *1* and *2* they annotate the document's Time To Live (TTL) in order to know the expiration time. Nodes *1* and *2* do not store information about the reply because the document was served by the *DS*. If node *4* then requests the document *A* to DS the request will reach node *2*, which will realize that *DS* is located two hops away and node *3* is one hop away and both have a copy of the document *A*. As node *3* is nearer than *DS*, node *2* will redirect the request to node *3* instead of forwarding it to *DS*. When node *3* receives the request it replies to node *4* with the copy of the document *A* stored in its local cache. Using the redirection feature the number of hops and messages have been reduced from 6 (4-2-1-DS-1-2-4), in the case of no redirection, to 4 (4-2-3-2-4). As the number of hops is reduced the latency perceived by node *4* is also reduced. The server load is also reduced as the request is served by node *3*.

In the previous example the information stored was relative to the requester because the first reply was performed by the data server. Let us suppose that node *6* requests the document *B* to *DS* and node *2* has a copy of *B* in its local cache. The request will pass through nodes *5*, *4* and *2*, which will annotate that node *6* will probably store the document *B* in its local cache in the local cache. As node *2* receives the request it will reply to node *6* intercepting the request. The reply will pass through nodes *4* and *5* to node *6* that will annotate that the document *B* is stored in node *2* and they will also update the TTL information of the request. Under this situation if node *4* requests the document *B* it will realize that *DS*, node *2* and node *6* are located 3, 1 and 2 hops away respectively and hence node *4* will request the document directly to node *2*, which is the closer node that is known to have a valid copy of the document *B*.

We have to remark that if the TTL of the information about the document location is not set, the redirection is not allowed. This constraint prevents from redirecting a request to a node that has not received a certain requested document as the TTL is obtained from the reply and not from the request.

Unfortunately, although the TTL assigned to the redirection information prevents from redirecting requests to nodes that have an obsolete copy of the document, this mechanism does not avoid the requests redirection to a node that has evicted the document because of the replacement policy. To cope with this situation we propose that the node that receives a redirected request and it has not a valid copy of the document in its local cache sends a special error message to the requester in order to send the request again. This message will pass through the redirecting node that will update the information about the incorrect redirection.

Let us suppose that after the situation described previously in the Figure 1, node *6* deletes the document *B* from its local cache and then node *5* requests the document *B*. Node *5* has stored that nodes *6* and *2* have the document *B* and they are located at 2 and 1 hops away respectively. As node *6* is closer the request will be redirected to node *6*. When node *6* receives the request it realises that there is not a valid copy of the document *B* in its local cache and replies with a redirection error message to node *5* that deletes the information about the location of the document *B* in node *6*. Then node *5* will proceed to request the document to the node *2*. The redirection errors generate more traffic in the network as well as the latency perceived by the requester node because the number of hops also increases.

Aiming at reducing the number of redirection errors produced by the eviction of documents in the local caches we propose to set as validity time for the redirection information the minimum between the document TTL and the mean time the documents are stored in the local cache. This value is easily calculated by each node considering the amount of time since the document has been stored and the instant in which it is evicted from the local cache.

Figure 2 lists the pseudo-code for the redirection mechanism.

4. Simulation model

We have evaluated by means of simulations the performance of the caching scheme described in the previous section. In order to evaluate the mobility model influences we compare the performance results obtained using the Random Waypoint and the Manhattan Grid mobility models. The simulations are based on the network simulator NS-2.33 which is a popular simulator for the researchers on ad hoc networking (Kurkowski et al., 2005). The BonnMotion (Aschenbruck et al., 2010) and the *setdest* mobility generators were used to create the mobility scenarios for the Manhattan Grid and Random Waypoint models respectively.

```
for each message (msg) to be sent or forwarded

msg.method - Request (GET) or response (RESP)
msg.id - Document identification
msg.hops - Number of hops from the source node
msg.TTL - Document's TTL

switch (msg.method)
 case GET:
         redirectNode = lookRedirectionCache(msg.id)
         if (exists(redirectNode) and distanceHops(redirectNode) < distanceHops(server(msg.id))
            redirectMessageTo (redirectNode)
         else
            forwardMessage

         savePassingByInformation(GET, msg.id, msg.hops)
         break
 case RESP:
         updateTTL(msg.id, msg.TTL)

         if (msg does not come from a server)
            savePassingByInformation(RESP, msg.id, msg.hops, msg.TTL)

         break
end
```

Fig. 2. Pseudo-code for the redirection caching mechanism

Table 1 summarises the main simulation parameters. We will assume a default scenario with 50 mobile nodes distributed in a square area of 1000x1000 meters. The scenarios with 25, 75 and 100 mobile nodes have also been evaluated in order to study the influence of the density of nodes in the network. There are two fixed servers (*DS*) located at the coordinates (x,y)=(0,500) and (x,y)=(1000,500) respectively. There are 1000 documents (identified by a number) with a size of 1000 bytes equally distributed between the two servers. Thus, documents with an odd identification number will be stored in one server and the documents with an even identification number will be stored in the other server. All the documents have an associated TTL modeled as an exponential distribution with mean of 2000 seconds. Additionally, we have also tested a mean TTL time of 250, 500, 1000 and infinite (the documents do not expire) in order to study the influence of the document expiration time.

The mobile nodes request documents to the servers following a Zipf-like traffic pattern distribution with a default slope of 0.8 although the 0.4, 0.6 and 1.0 slopes have also been tested aiming at studying the influence of the Zipf slope in the caching scheme proposed. The Zipf-like distribution has been chosen as a traffic pattern because it has been demonstrated to properly characterize the popularity of the Web documents in the Internet (Adamic & Huberman, 2002). The Zipf law asserts that the probability *P(i)* for the *i-th* most popular document to be requested is inversely proportional to its popularity ranking as shown in the Equation 1.

$$P(i) = \beta / i^{\alpha} \quad \text{with } \alpha \text{ close to } 1 \qquad (1)$$

Parameter		Default	Tested values
Simulation area (meters)		1000x1000	
Nodes		50	25-50-75-100
Servers		2	
Documents		1000	
Document size (bytes)		1000	
Timeout (s)		3	
TTL (s)		2000	250-500-1000-2000-∞
Mean time between requests (s)		25	5-10-25-50
Traffic pattern (Zipf slope)		0.8	0.4-0.6-0.8-1.0
Replacement policy		LRU	
Cache size (number of documents)		35	5-10-35-50
Simulation time (s)		20000	
Warm-up time (s)		4000	
MAC protocol		802.11 b	
Radio propagation model		Two Ray Ground	
Coverage radio (meters)		250	
Ad hoc routing protocol		AODV	
Mobility pattern	Random WayPoint	Min. and max. speed: 1m/s Pause time: 0s	Min. and max. speed: 1-3-5 m/s Pause time: 0s
	ManhattanGrid	Min. and max. speed: 1m/s Pause time: 0s 8 blocks	Min. and max. speed: 1-3-5 m/s Pause time: 0s 6, 8 and 10 blocks

Table 1. Simulation parameters

The parameter a is the slope of the log/log representation of the number of references to the documents as a function of its popularity rank (i) while β is the displacement of the function. Each time a mobile node requests a document it will wait for a timeout to receive the reply. If the document is not received during this time it will be requested again. Once the requested document has been received the node will wait during a certain amount of time modelled by an exponential distribution with a mean of 25 seconds before proceeding to a new request. Waiting times of 5, 10 and 50 seconds have also been tested. Using this wide range of mean time between requests we can explore the influence request looad.

The LRU replacement policy has been chosen for the caches with a default storage space of 35 documents. Cache sizes with a capacity of 5, 10, and 50 documents have also been simulated aiming at testing the influence of the cache size.

The simulation time has been set to 20000 seconds. 20% of this time (4000 seconds) has been used to warm-up the caches and avoid cold start influences. Consequently the statistics collected from the simulations are those corresponding to the time after the warm-up.

The 802.1b MAC protocol with the Two Ray Ground propagation model and a coverage radio of 250 meters were used. The popular AODV (Perkins et al., 2003) (Ad hoc On Demand Vector) protocol was selected as the MANET routing protocol.

The default speed of the nodes is 1 m/s. No pause time is considered between consecutive movements. Speeds of 2 and 5 m/s have also been tested in order to study the speed influence in the caching mechanism.
For the Manhattan Grid mobility model 8x8 blocks have been chosen as a default scenario. In addition, scenarios with 4x4, 6x6 and 10x10 blocks have been also simulated since these scenarios will allow us to evaluate the influence of the connectivity. Figure 3 illustrates the scenario with the Manhattan Grid mobility model with 8x8 blocks. The mobile nodes (represented by small circles) move along the grid using the lanes defined by the blocks. The two servers A and B (represented as big circles in the figure) are located in the middle of the left and right sides of the scenario.

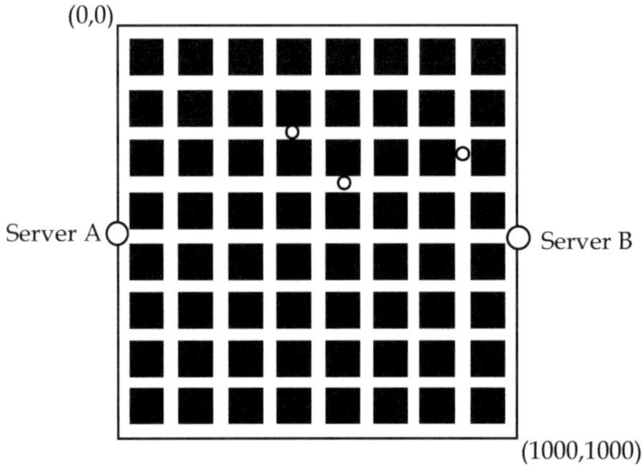

Fig. 3. Example scenario using the Manhattan Grid with 8x8 blocks

5. Performance evaluation

The goal is to evaluate the performance of a MANET with the proposed caching scheme taking into consideration the speed and density of nodes, the traffic load (mean time between requests), the mean document expiration time (TTL), the traffic pattern (Zipf slope) and the cache size. For all these analysis, the network performance is studied using both the Random Way Point and the Manhattan Grid mobility models.
For the study of the influence of the density and speed of the nodes every simulation scenario has been executed five times using the same TTL for each document, mean time between requests and request distribution but using a different starting point within the simulation area and a different mobility pattern for each mobile node. The simulation of the rest of scenarios have been executed five times using the same TTL for each document, time between requests and mobility pattern for each node but using a different request distribution. The performance evaluation presented is the mean of the results obtained for the five simulations. Again, the presented results are the mean of the measurements obtained for the five simulations.
As performance metrics we use the following measurements:

- Traffic - The amount of traffic that each mobile node in the network has to process because the node generates the packets or because the packets have to be forwarded. This measurement includes not only the traffic corresponding to document requests and replies but also the overhead introduced by the routing protocol.

- Hops – Defined as the number of nodes that a document has to traverse to be served. It includes the request from the requester to the node that serves the document and back again to the requester node.
- Delay - Defined as the time elapsed between a document request and the reception of the corresponding reply.
- Percentage of timeouts - Defined as the proportion of requests that must be retransmitted again because the reply does not reach the destination before the timeout is reached.
- Local hit ratio – It is the ratio between the number of documents served by the local cache and the total number of documents requested by each node. The higher the local cache hit ratio, the lower the traffic injected in the network is.
- Remote hit ratio – It is the ratio between the number of documents served by a node that is not a server (because of an interception or a redirection) and the total number of documents requested by each node. As the remote hit ratio increases, the server load decreases because more requests are served by the mobile nodes instead of the servers.

5.1 Effect of the network load

Figure 4 represents the mean traffic processed by the nodes (a), the mean delay (b), the mean number of hops (c), the percentage of timeouts (d) and the cache hits (e) as a function of the mean time between requests.

Figure 4.a shows that the traffic generated in the scenario using RWP is greater than that using MG. This is caused by the AODV broadcast messages employed to create the routes between the mobile nodes (Saad & Zukarnain, 2009). As the RWP mobility model tends to concentrate the mobile nodes in the centre of the simulation area (Hyytia et al. 2006b), more nodes receive the broadcasted RREQ (Route Request) messages.

In Figure 4.b we can observe that as the periodicity of document requests increases, the delay is also augmented. As the time between requests increases, the number of documents expired in the nodes' local caches is also increased and the documents in the local caches are less updated. This can be observed in Figure 4.e where the cache hits decreases as the network load decreases. Therefore, the reduction of the cache hits increases the delay as less requests are served by the local or remote caches. On the other hand, the delay perceived by the RWP (Random Way Point) mobility model is slightly smaller than the Manhattan Grid using 6x6 (MG6) and 8x8 (MG8) blocks but greater than the 10x10 (MG10) blocks. This behaviour is due to the fact that the connectivity is improved as the number of blocks increases because the nodes can communicate with more nodes located at adjacent lanes as long as the distance between lanes is shorter.

In addition, the route TTL configured in AODV is ten seconds and hence the network with a mean time between requests less or equal to this time will take advantage of the already created routes while greater time between requests will have to create the routes again. However, Figure 4.c shows that under RWP nodes need less hops to obtain the documents than under MG although the difference declines as the number of blocks increases. This can be explained as before, the probability to find a shorter route with RWP is higher because the nodes move freely along the simulation area so that they are not restricted to move along the lanes defined by the blocks. Finally, Figure 4.d shows that the number of timeouts is diminished as the network traffic decreases (the mean time between requests increases) until 25 seconds between requests but for 50 seconds between requests the number of

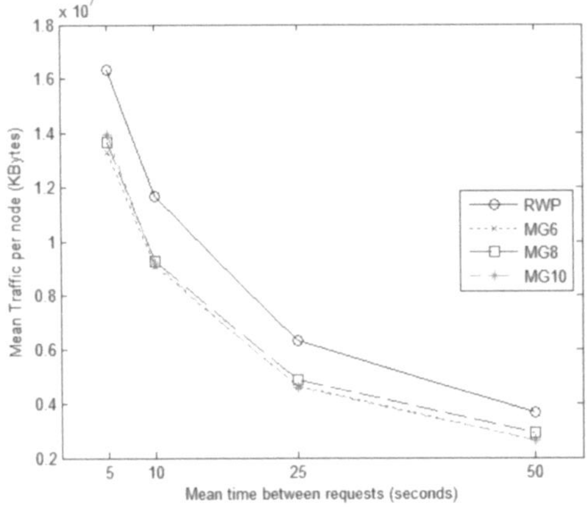

(a)

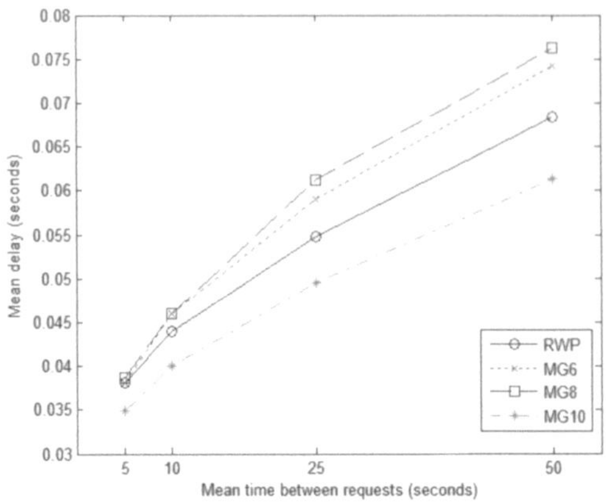

(b)

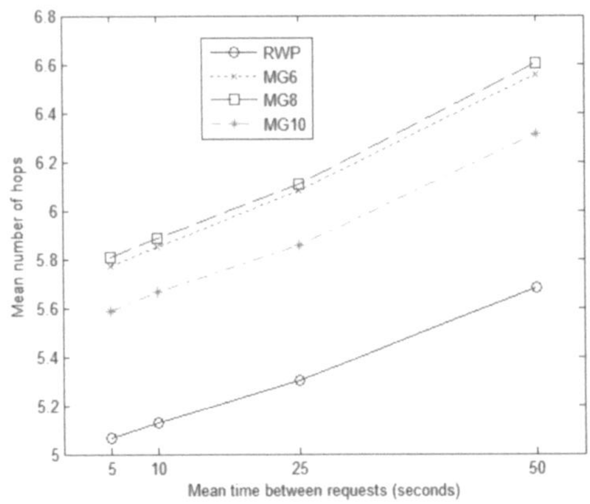

(c)

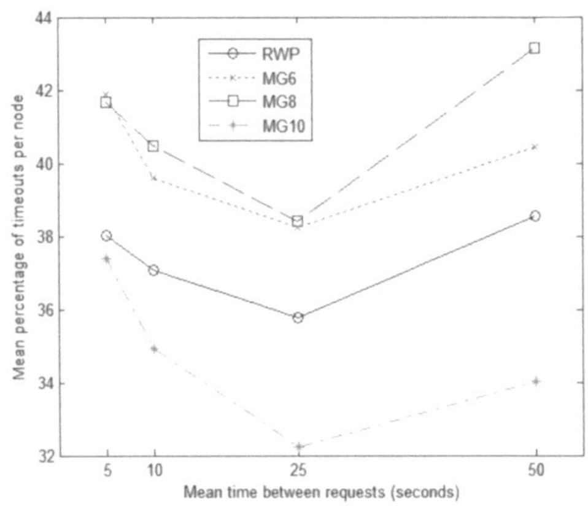

(d)

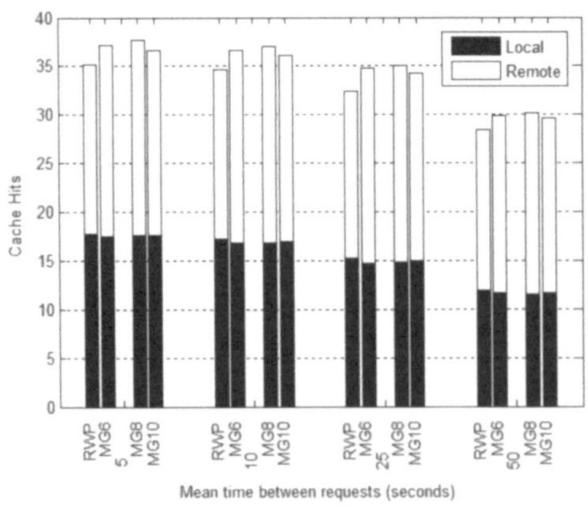

(e)

Fig. 4. Mean traffic (a), mean delay (b), mean hops (c), percentage of timeouts (d) and cache hits (e) as a function of the mean time between requests

timeouts is increased. This can be explained similarly as in the case of the delay. Obviously, when the document TTL expires, the effectiveness of the local and remote caching mechanisms decreases and hence the probability to have to request the documents to the data servers increases. As the data servers could remain unavailable due to the nodes' mobility the probability of timeouts is also increased.

5.2 Effects of the TTL

Figure 5 shows the mean traffic (a), the mean delay (b), the mean number of hops (c), the percentage of timeouts (d) and the percentage of cache hit (e) as a function of the mean documents' TTL.

The TTL defines the time that the documents are stored in the local caches. We have tested the situations from a low mean TTL (the documents expire after a short interval and they are deleted from the caches very soon) to an infinite TTL (the documents never expire). As the TTL increases the percentage of cache hits is also increased from about 10% to 35% as shown in Figure 5.e and then more requests are served by the local caches. This fact causes the progressive reduction of the traffic generated in the network (Figure 5.a), the delay perceived by the nodes (Figure 5.b), the mean number of hops (Figure 5.c) and the percentage of timeouts (Figure 5.d).

Figure 5.a shows that the traffic generated under RWP mobility model is also greater than with MG as in the studies presented in section 5.1.

Finally the figures show a similar behaviour as the presented in section 5.1, the mean delay and the mean number of timeouts is higher using MG6 and MG8 than RWP while MG10 obtains the lowest delay values. However, the RWP obtains a better performance in terms of the number of hops as it is able to find shorter routes.

5.3 Effects of the traffic pattern

Figure 6 shows the mean traffic (a), the mean delay (b), the mean number of hops (c), the percentage of timeouts (d) and the percentage of cache hit (e) as a function of the Zipf parameter α.

As the Zipf parameter is closer to one the probability to request again a popular document is higher. This fact drastically enhances the number of local hits as shown in Figure 6.e where the local hit ratio evolves from about 3% to 30% for a equal to 0.4 and 1.0 respectively.

The remote hit ratio is also slightly increased as the parameter a is closer to 1.0. The higher cache hits obtained as a is increased causes the reduction of the generated traffic (Figure 6.a), the delay perceived by the nodes (Figure 6.b), the number of hops needed to obtain the documents (Figure 6.c) and the number of timeouts (Figure 6.d).

The mobility models follow the same behaviour as the previous studies. Under RWP, the network performance obtains intermediate results between MG6, MG8 and the best results obtained by MG10 for the mean delay and mean percentage of timeouts. On the other hand RWP mobility generates more traffic than MG although it requires a lower number of hops to obtain the documents.

5.4 Effects of the cache size

Figure 7 depicts the mean traffic (a), the mean delay (b), the mean number of hops (c), the percentage of timeouts (d) and the percentage of cache hit (e) as a function of cache size.

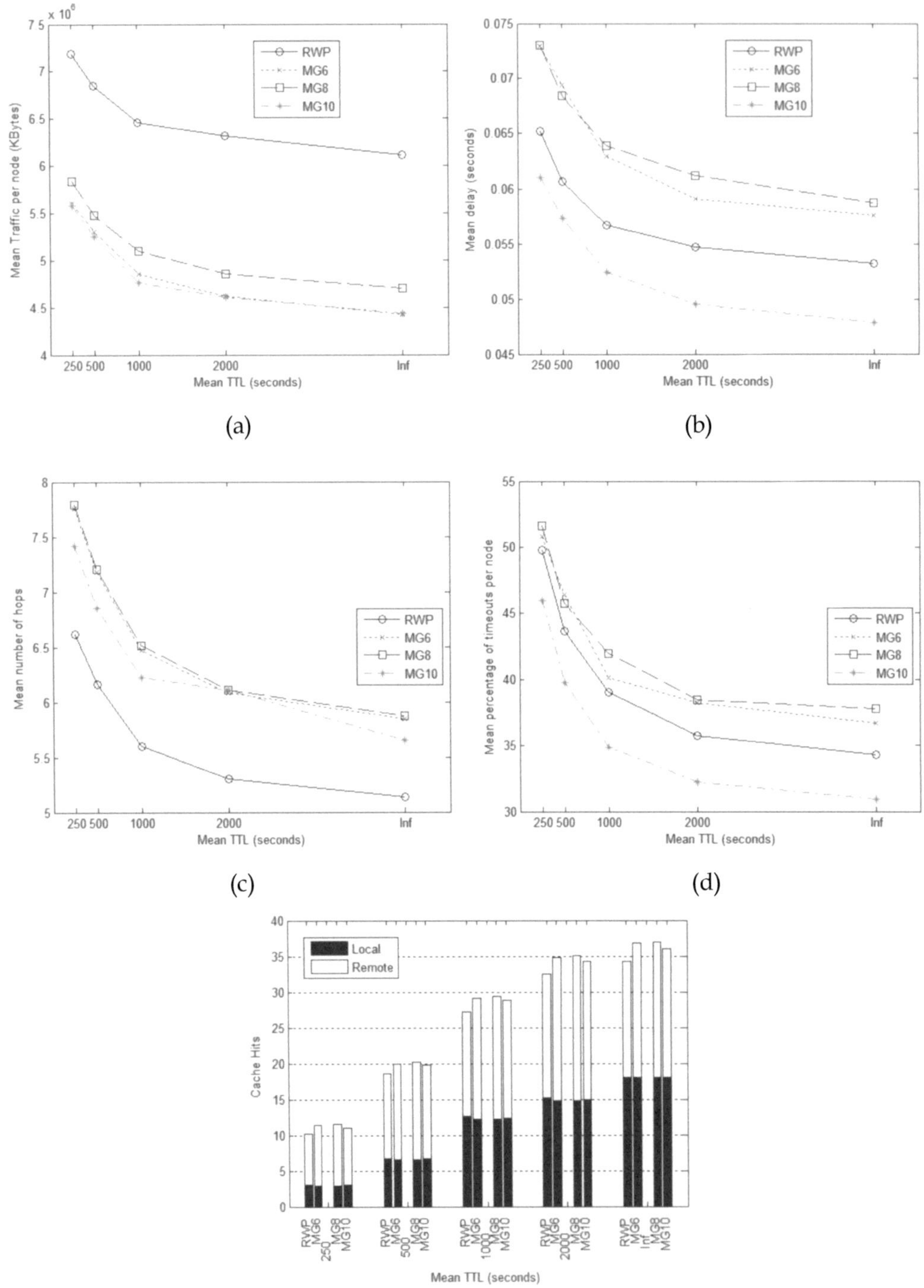

Fig. 5. Mean traffic (a), mean delay (b), mean hops (c), percentage of timeouts (d) and cache hits (e) as a function of the mean document's TTL

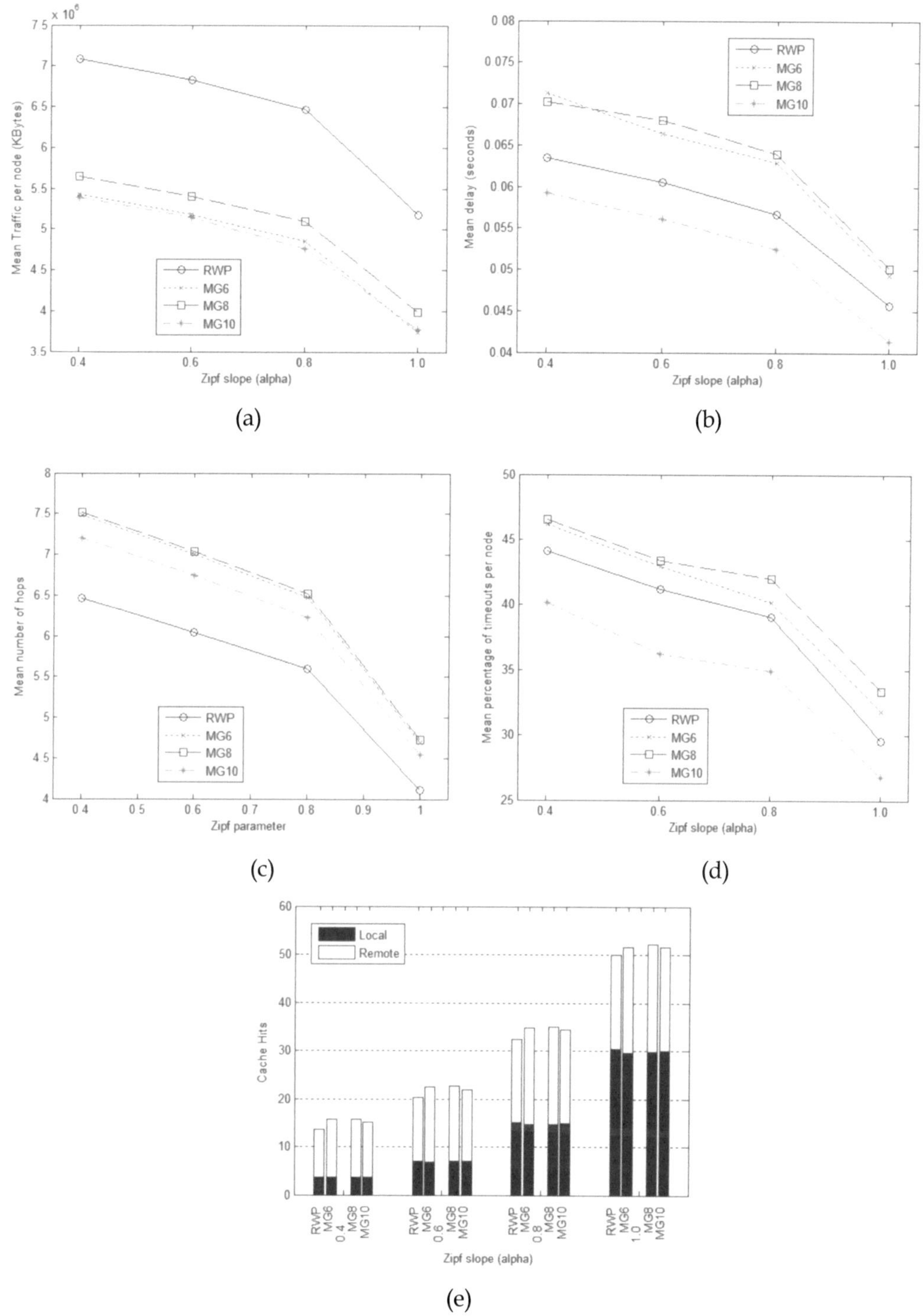

Fig. 6. Mean traffic (a), delay (b) and hops (c), percentage of timeouts (d) and cache hits (e) as a function of the Zipf slope α

The cache size determines the number of documents that fit in the local cache. As more documents are stored in the nodes' local cache the probability of a local or remote cache is increased as shown in Figure 7.e. In this figure we can observe that the cache hit ratio increases from about 18% for the smaller cache (10 documents) to about 36% for the larger cache (50 documents). As the hit ratio increases the amount of documents that have to be requested to the servers is decreased and the number of requests served for the mobile nodes is increased. As a consequence the traffic in the network is reduced (Figure 7.a) as well as the mean delay (Figure 7.b), the mean number of hops (Figure 7.c) and the mean number of timeouts (Figure 7.d).

The RWP mobility generates more traffic than MG for all the cache sizes although it obtains the better performance if we consider the mean number of hops. For the rest of the metrics (delay and percentage of timeouts) the RWP mobility model achieves a better performance than MG6 and MG8 but worse than MG10.

5.5 Effects of the density of nodes

Figure 8 illustrates the mean traffic (a), the mean delay (b), the mean number of hops (c), the percentage of timeouts (d) and the percentage of cache hit (e) as a function of the number of mobile nodes in the network.

As the node density increases the probability to find a route between the requester node and the server is also increased. So, the mean percentage of timeouts is reduced drastically (from 80~90% to 25%) as shown in Figure 8.d. For the lowest tested density of nodes (25 nodes) the RWP performs better than the MG because it obtains a better cache hit ratio (Figure 8.e). For node density greater than 25 nodes the difference in percentage of timeouts between the mobility models is reduced and all the scenarios obtain similar results for a network with 100 nodes.

Similarly, RWP obtains a lower mean delay than MG for low density networks as depicted in Figure 8.b. while for higher densities the mean delays are very similar. This fact is produced by the higher cache hit obtained by RWP. On the other hand, the RWP mobility model, as in the previous studies, obtains a lower mean number of hops (Figure 8.c) at the cost of injecting more traffic in the network (Figure 8.a).

5.6 Effects of the nodes' speed

Figure 9 shows the mean traffic (a), the mean delay (b), the mean number of hops (c), the percentage of timeouts (d) and the percentage of cache hit (e) as a function of the node's speed.

Figure 9.e shows that the cache performance does not depend on the nodes' speed as the performance results are the same for the considered values of the speed.

As the nodes' velocity increases the routes created between them are broken more frequently. Thus, the routes to the servers have to be created again. Consequently, the perceived delay augments as the nodes' speed increases as shown in Figure 9.b. Due to the same reason, the percentage of timeouts is also increased as the nodes' speed increases (Figure 9.d). On the other hand, RWP needs less hops to obtain the documents than MG as showed in the previous sections (Figure 9.c) while the required traffic is higher (Figure 9.a).

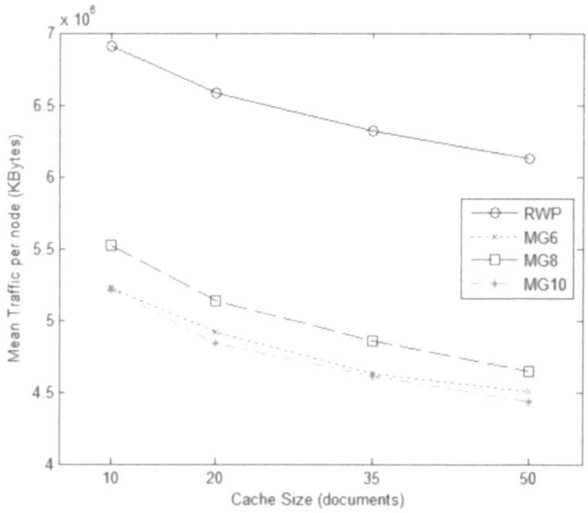

(a)

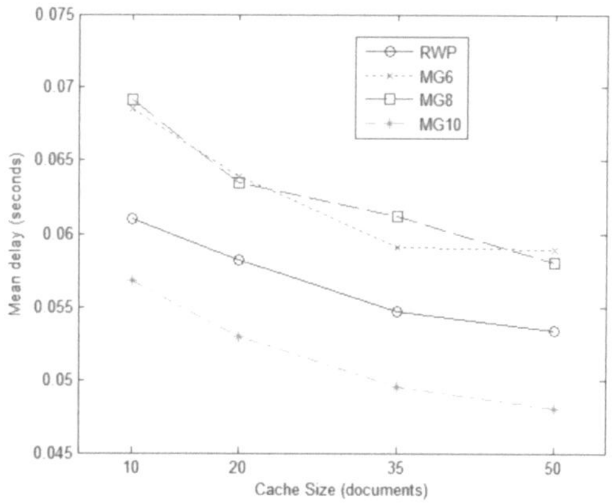

(b)

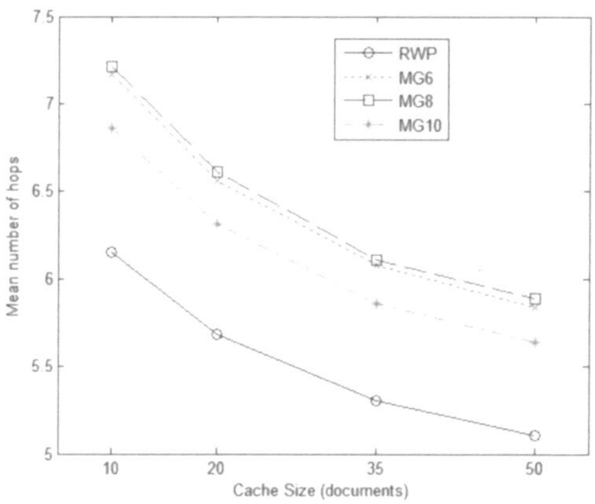

(c)

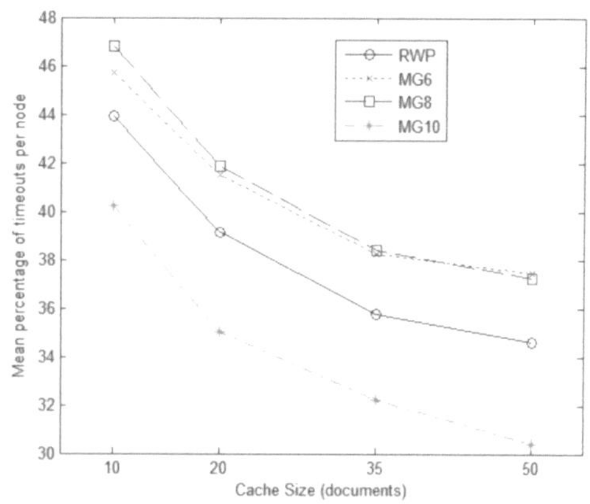

(d)

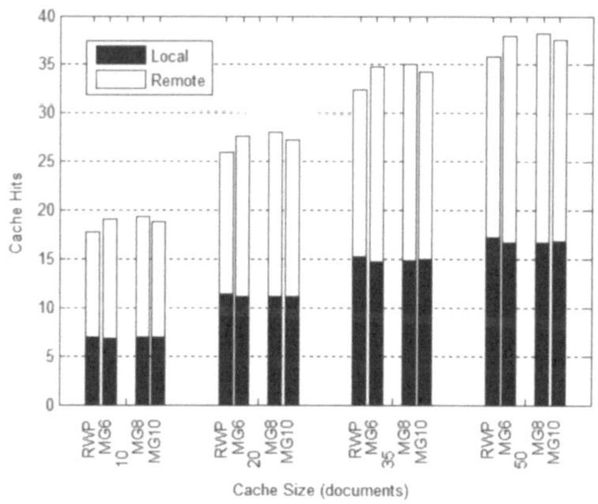

(e)

Fig. 7. Mean traffic, delay and hops, percentage of timeouts and cache hits as a function of the cache size

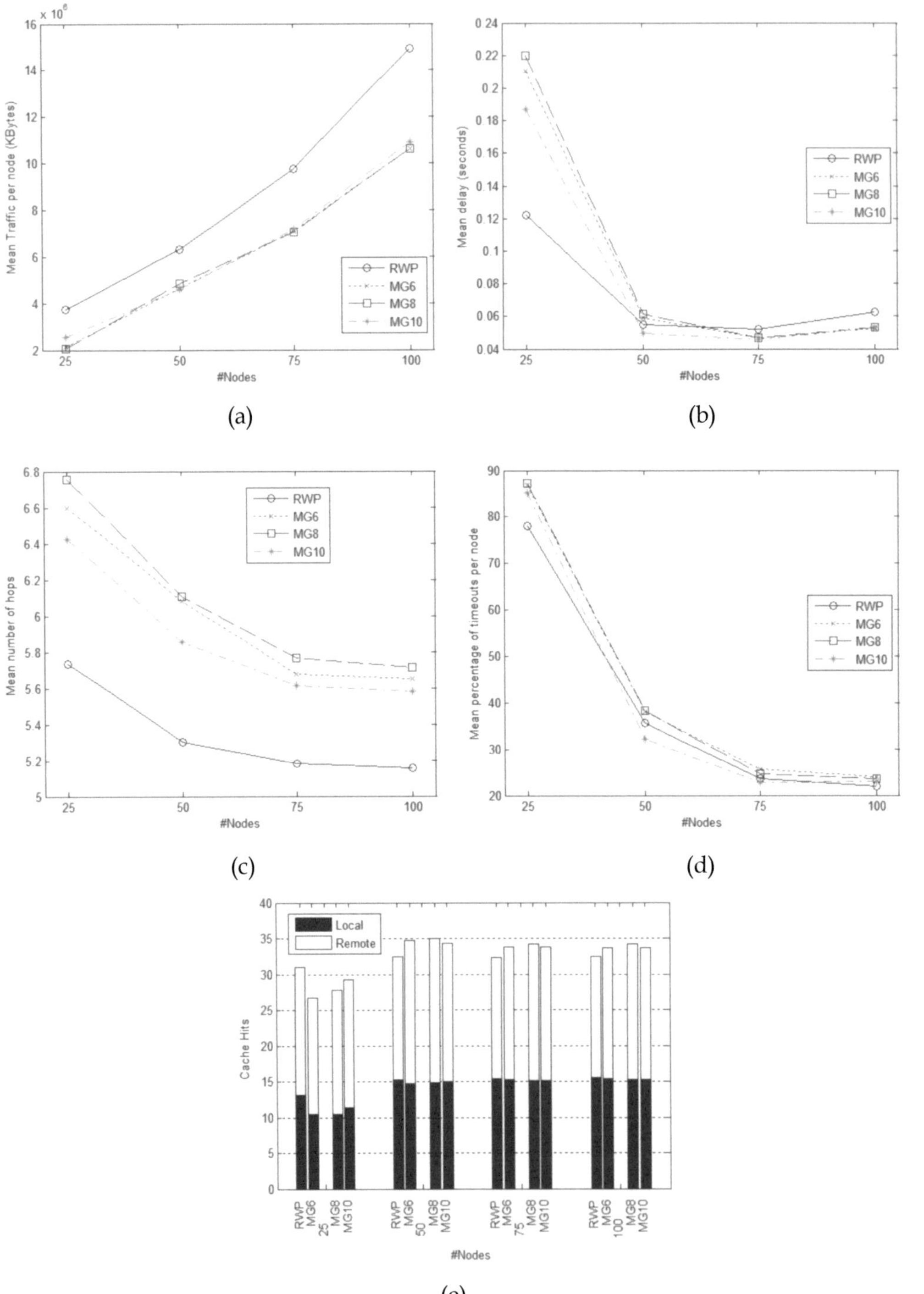

Fig. 8. Mean traffic (a), delay (b) and hops (c), percentage of timeouts (d) and cache hits (e) as a function of the number of nodes

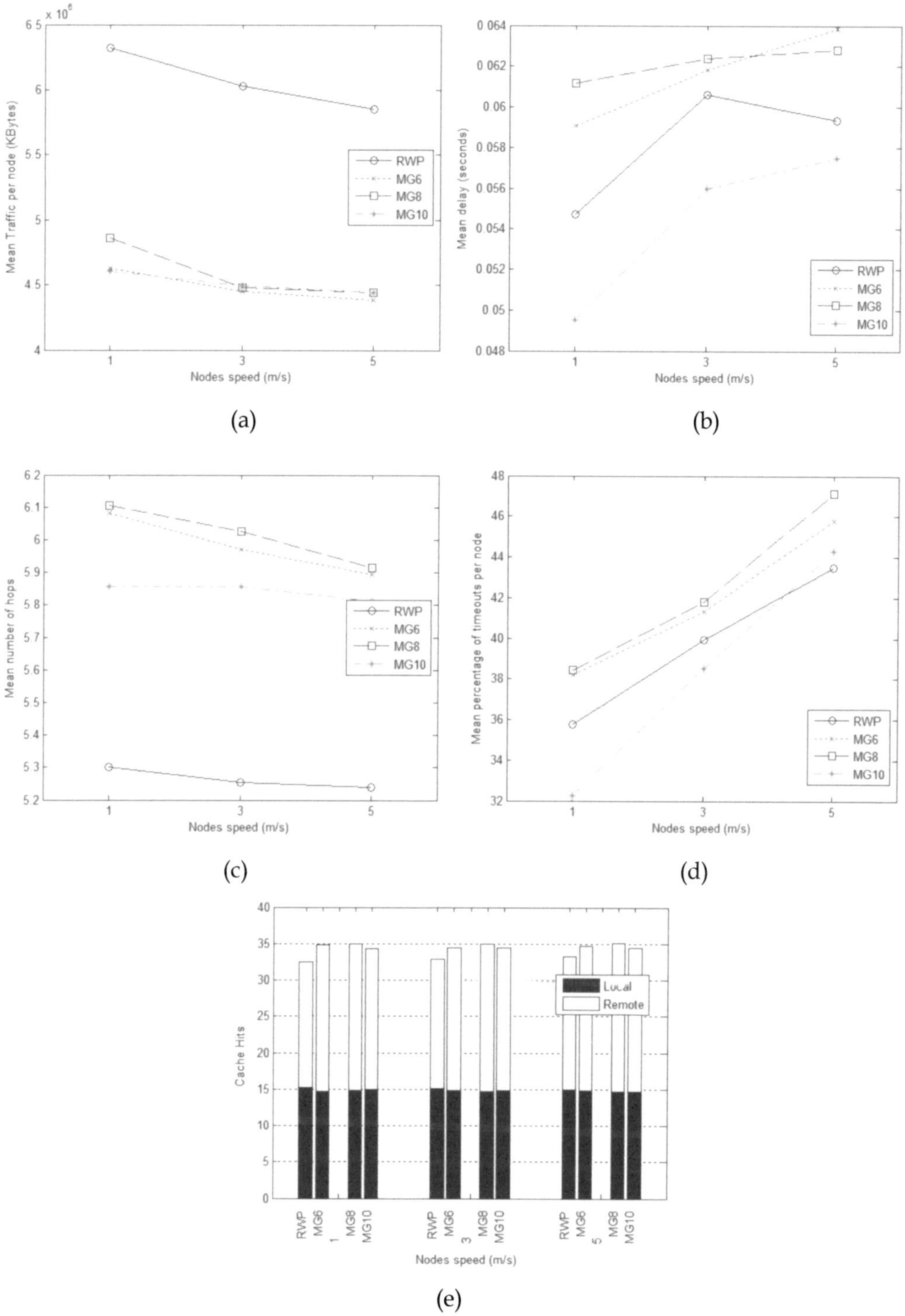

Fig. 9. Mean traffic (a), delay (b) and hops (c), percentage of timeouts (d) and cache hits (e) as a function of the nodes' speed

6. Conclusions

In this paper we have presented a caching scheme for Mobile Ad Hoc Networks that implements a local cache in each mobile node of the network. The mobile nodes have the capability of intercepting and responding the requests that they have to forward to the data server if they find a copy of the requested document in its local cache. On the other hand, the mobile nodes also implement a cache of document location in order to redirect the received requests to another mobile node that is known to be closer than the original destination of the request. This redirection cache is filled using the information obtained from the requests and replies that the nodes have to forward.

We have evaluated the performance of the proposed caching scheme through simulations using the mean generated traffic, the delay, the number of hops, the percentage of timeouts and the percentage of cache hits as performance metrics. We have compared the proposed caching scheme using the popular Random Way Point and the Manhattan Grid mobility models. The Manhattan Grid model has been evaluated using different topographical configurations (6x6, 8x8 and 10x10 blocks). In addition, we have evaluated the effect of several factors such as the mean time between requests, the documents' TTL, the request pattern (Zipf slope), the cache size, the nodes' density and the nodes' speed.

As main conclusions we can assert that the traffic generated using the RWP mobility model is greater than the traffic generated by the MG for all the parameters evaluated. Similarly the mean number of hops used by RWP is lower than that used by MG for all the performed simulations. If we consider the mean delay, the RWP mobility model performs better than MG when the distance between parallel lanes reduce the node connectivity (6x6 and 8x8 blocks) but worse than MG with a higher proximity of the lanes (10x10 blocks). The same results are obtained if the mean percentage of timeouts is taken into consideration. The cache performance is similar for all the studied parameters except for a low nodes' density where the network using the RWP mobility model obtains a better performance.

As the mobility model defines how the mobile nodes behaves in the network and the cooperating caching schemes depends on the behaviour of the mobile nodes, we can conclude that the mobility model used to evaluate a caching scheme clearly influences the obtained performance results of the network.

As a future research direction we suggest to evaluate the proposed caching scheme using more mobility models as those presented in section 2. On the other hand, the presented caching scheme has to be compared with other caching schemes in order to evaluate its effectiveness.

7. Acknowledgement

We would like to thank Adela Isabel Fernández Anta for revising the syntax and grammar of this paper. This work was partially supported by the public Project TEC2009-13763-C02-01.

8. References

Adamic, L.A.; & Huberman, B.A. (2002). Zipf's law and the Internet, *Glottometrics*, Vol. 3, (2002) 143-150, 1617-8351

Artail, H.; Safa, H., Mershad, K., Abou-Atme, Z. & Sulieman, N. (2008). COACS: A Cooperative and Adaptive Caching Systems for MANETs, *IEEE Transactions on Mobile Computing*, Vol. 7, No. 8 (August 2007) 961-977, 1536-1233

Aschenbruck, N.; Ernst, R., Gerhards-Padilla, E. & Schwamborn, M. (2010). BonnMotion - A Mobility Scenario Generation and Analysis Tool, *Proceedings of the 3rd International Conference on Simulation Tools and Techniques (SIMUTOOLS 2010)*, 78-963-9799-87-5, Torremolinos (Spain), March 2010

Bai, F.; Sadagopan, N. & Helmy, A. (2003). Important: a framework to systematically analyze the impact of mobility on performance of routing protocols for adhoc networks. *Proceedings of the 22th Annual Joint Conference of the IEEE Computer and Communications Societies (INFOCOM)*, pp. 383-403, 0-7803-7753-2, San Francisco (USA), April 2003, IEEE

Broch, J.; Maltz, D.A., Johnson, D.B., Hu, Y.C. & Jetcheva, J. (1998). A performance comparison of multi-hop wireless ad hoc network routing protocols, *Proceedings of the Fourth Annual ACM/IEEE International Conference on Mobile Computing and Networking (Mobicom98)*, pp. 85-97, 1-58113-035-X, Texas (USA), October 2008, ACM, New York

Denko, M.K. (2007). Cooperative Data Caching and Prefetching in Wireless Ad Hoc Networks, *International Journal of Business Data Communications and Networking*, Vol 3, No. 1, (January 2007), 1548-0631

Dodero, G.; and Gianuzzi, V. (2006). Saving Energy and Reducing Latency in MANET File Access, *Proceedings of the 26th International Conference on Distributed Computing Systems Workshops (ICDCSW'06)*, 0-7695-2541-5, Lisboa (Portugal), July 2006, IEEE

Ghosh, J.; Philipb, S.J. & Qiao, C. (2007). Sociological orbit aware location approximation and routing (SOLAR) in MANET. *Ad Hoc Networks*, Vol. 5, No. 2, (March 2007) 189–209, 1570-8705

Hong, X.; Gerla, M., Pei, G. & Chiang, C. (1999). A Group Mobility Model for Ad Hoc Wireless Networks, *Proceedings of the ACM International Workshop on Modelling and Simulation of Wireless and Mobile Systems (MSWiM)*, pp. 53-60, Seattle (USA), August 1999, ACM

Hyytia, E.; Lassila, P. & Virtamo, J. (2006a). A markovian waypoint mobility model with application to hotspot modeling. *Proceedings of the IEEE International Conference on Communications (ICC 2006)*, Istanbul (Turkey), June 2006, IEEE

Hyytia, E.; Lassila, P. & Virtamo, J. (2006b). Spatial Node Distribution of the Random Waypoint Mobility Model with Applications. *IEEE Transactions on Mobile Computing*, Vol. 5, No. 6, (June 2006) 680-694, 1536-1233

Jardosh, A.; Belding-Royer, E., Almeroth, K. & Suri, S. (2003). Towards realistic mobility models for mobile ad hoc networks, *Proceedings of 9th International Conference on Mobile Computing and Networking (MobiCom)*, pp. 217–229, 1-58113-753-2, San Diego (USA), September 2003, ACM

Kim, M.; Kotz, D. & Kim, S. (2006). Extracting a mobility model from real user traces, *Proceedings of the 25th Annual Joint Conference of the IEEE Computer and Communications Societies*, 1-4244-0222-0, Barcelona (Spain), April 2006, IEEE

Kurkowski, S.; Camp, T. & Colagrosso, M. (2005). MANET Simulation Studies: The Incredibles, ACM's Mobile Computing and Communications Review, Vol. 9, No. 4, (October 2005) 50-61, 1559-1662

Lee, K.; Hong, S., Kim, S.J., Rhee, I. & Chong, S. (2009). SLAW: A Mobility Model for Human Walks, *Proceedings of the 28th Annual Joint Conference of the IEEE Computer and Communications Societies (INFOCOM)*, Rio de Janeiro (Brazil), April 2009, IEEE

Liang, B.; and Haas, Z.J., (1999). Predictive Distance-Based Mobility Management for PCS Networks, *Proceedings of the 18th Annual Joint Conference of the IEEE Computer and Communications Societies (INFOCOM)*, pp. 1377-1384, 0-7803-5417-6, New York (USA), April 1999, IEEE

Lim, S.; Yu, C. & Das. C.R. (2006). Clustered mobility model for scale-free wireless networks, *Proceedings of the 31st IEEE Conference on Local Computer Networks (LCN 2006)*, Tampa (USA), November 2006, IEEE

Lim, S.; Lee, W.C., Cao, G. & Das, C.R. (2006). A novel caching scheme for improving Internet-based mobile ad hoc networks performance, *Ad Hoc Networks*, Vol. 4, No. 2, (March 2006) 225-239, 1570-8705

Perkins, C. E.; Belding-Royer, E. M. & Das, S. (2003). Ad Hoc On Demand Distance Vector (AODV) Routing. IETF RFC 3561

Royer, E.M.; Melliar-Smith, P.M. and Moser, L.E. (2001). An Analysis of the Optimum Node Density for Ad hoc Mobile Networks, *Proceedings of the IEEE International Conference on Communications (ICC01)*, pp. 857-861, Helsinki (Finland), June 2001, IEEE

Saad, M.I.; & Zukarnain, Z.A. (2009). Performance Analysis of Random-Based Mobility Models in MANET Routing Protocol, *European Journal of Scientific Research*, Vol. 32, No. 4 (2009) 444-454, 1450-216X

Tang, B.; Gupta, H. & Das, S.R. (2008). Benefit-Based Data Caching in Ad Hoc Networks, *IEEE Transactions on Mobile Computing*, Vol 7, No. 3, (March 2008) 289-304, 1536-1233

Ting, Y.; & Chang, Y. (2007). A Novel Cooperative Caching Scheme for Wireless Ad Hoc Networks: GroupCaching, *Proceedings of the International Conference on Networking, Architecture and Storage (NAS 2007)*, pp. 62-68, 0-7695-2927-5, Guilin (China), July 2007, IEEE

Yin, L.; & Cao, G., Supporting Cooperative Caching in Ad Hoc Networks, *IEEE Transaction on Mobile Computing*, Vol. 5, No. 1, (January 2006) 77- 89, 1536-1233

Part 3

Applications of Ad Hoc Networks

13

Ad Hoc Networks for Cooperative Mobile Positioning

Francescantonio Della Rosa[1], Helena Leppäkoski[1], Ata-ul Ghalib[1], Leyla Ghazanfari[1], Oscar Garcia[1], Simone Frattasi[2] and Jari Nurmi[1]

[1]Tampere University of Technology (TUT), Finland.
[2]Aalborg University (AAU), Denmark

1. Introduction

Wireless ad-hoc networks have received huge attention during recent years due to the potential applications in different fields such as emergency, disaster relief, battle-fields, automotive, social networks and entertainment. They are rapidly deployable, self-organizing, and require no fixed infrastructure for communications. (Huang et al., 2008)
At the same time, localization in wireless networks is becoming a hot topic for society, industry and research. The needs of location information has driven companies to build mobile handsets with embedded GPS receivers (which is nowadays the most popular mass market solution for positioning), causing huge increase in costs, size, battery consumption, and a long time for a full market penetration (Sayed et al., 2005). However, it is also known that the GPS is not always the most suitable solution for localization. In adverse environments, such as outdoor urban canyons and indoor, it is not an easy task to obtain location information, due to the signal blocking, multipath conditions and the infeasibility to have a continuous tracking of at least four satellites (Mayorga et al., 2007).
The Fourth generation (4G) communication systems also stimulate the need of providing alternative ubiquitous localization solutions, regardless the environment (i.e., outdoors and indoors), which should overcome, or at least complement, the drawbacks of GPS-based and GPS-free systems (Della Rosa, 2007). Traditional alternative technologies make use of time difference of arrival (TDOA) measurements from the serving cellular system where the Base Stations (BSs) are considered as fixed reference points (Sayed et al., 2005).
Different type of measurements, such as received signal strength (RSS), are widely used in local area scenarios, where Wi-Fi Hot Spots deployed in big cities allow user terminals to predict their locations by means of known fixed positions (Sayed et al., 2005).
Unfortunately, when localization is performed in indoor environments the accuracy is highly dependent on the wireless channel conditions since several error sources cause huge signal fluctuations detected at terminal level, severely decreasing the final location estimation accuracy (Della Rosa et al., 2010).
Recently, in alternative to traditional methods, a new branch of positioning techniques has been developed: the *Cooperative Mobile Positioning* (Figueiras & Frattasi, 2010), which makes use of hybrid schemes and exploits the benefits in terms of accuracy of short-range measurements provided by the ad-hoc networks (Della Rosa, 2007).

In this chapter we will explain the basics of Cooperative Mobile Positioning and demonstrate the applicability of the technique in real cases, demonstrating that the exploitation of the most reliable RSS measurements detected in the ad-hoc links represent a valid and complementary approach to traditional non-cooperative methods, and that the *hybrid network model* adopted is the most natural environment in which cooperation among terminals is established and best exploited without additional hardware components (Figueiras & Frattasi, 2010) (Della Rosa et al., 2010).

2. Mobile positioning

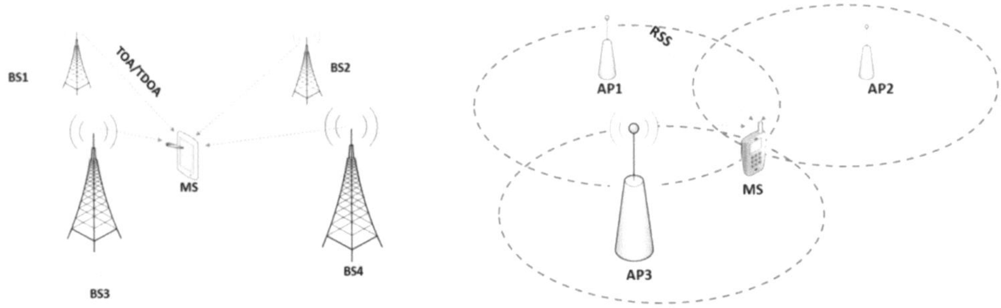

(a) Outdoor Scenario (b) Indoor Scenario

Fig. 1. Conventional Positioning

Several different radio navigation methods, based on different measurements and operation principles, have been used in practical positioning applications (Sayed et al., 2005; Syrjarinne, 2001). The positioning techniques can be categorized as mobile based and network based methods. In mobile based methods, the mobile station (MS) measures parameters from signals it receives from BS and uses the measurements to determine its position. In network based methods the base stations measure parameters from signals coming from the mobile, and the position calculation is performed in a positioning server connected to the network. The following measurements and positioning techniques can be used for positioning a MS in communication systems:

2.1 Angle Of Arrival (AOA)

AOA utilizes multi-array antennas which are used to estimate the angle of the line of arrival of the signal. The position of the MS can be located at the intersection of the lines if more than one AOA measurement is performed, as shown in Fig.2.a. This positioning method is called triangulation. Antenna arrays capable for AOA measurements are large in size, and thus more suitable to be measured by BS rather than MS. Therefore AOA lends itself easily to network-based positioning. The AOA is considered mainly as outdoor positioning method using BSs of cellular networks (Sayed et al., 2005), but results on AOA positioning in WLAN infrastructure have also been reported (Wong et al., 2008). Reflections and non-line-of-sight (NLOS) conditions distort the direction of arrival of the signals, deteriorating the accuracy of AOA positioning.

2.2 Time Of Arrival (TOA)

TOA information from the MS to a station with known coordinates (navigation satellites, BSs of wireless communication networks, etc.) or vice versa can be estimated if both entities

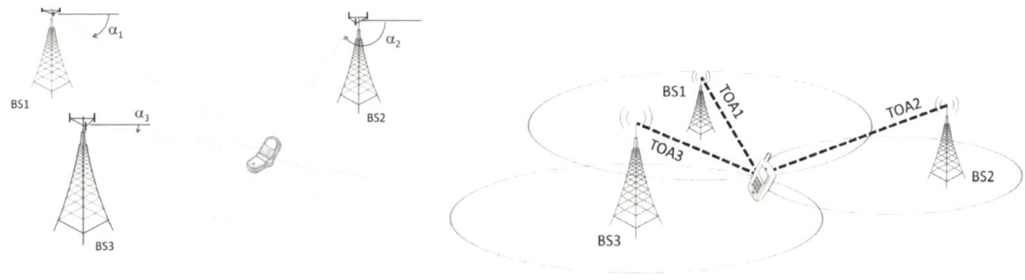

(a) Angle of Arrival (AOA) (b) Time of Arrival (TOA)

Fig. 2. Angle of arrival and Time of Arrival based positioning

are precisely synchronized in time. The distance between MS and BS can be obtained from TOA, since electromagnetic waves propagate at constant speed of light. To estimate the position of the MS, TOAs to at least three stations in different locations are required for trilateration; all the BSs have also be perfectly synchronized to each other (Rappaport et al., 1996). In trilateration the position estimate is the intersection of circles with radii determined from TOA measurements and centers at the known BS coordinates, as shown in Fig.2.b. Reflections and non-line-of-sight (NLOS) conditions distort the TOA of the signals.

2.3 Time Difference Of Arrival (TDOA)

TDOA is based on estimating the difference in the arrival times of the signals coming from two different transmitters to the receiver. Geometrically a particular TDOA value defines a hyperbola between the two receivers on which the MS may be located. As seen in the (Fig.3.), the position of the MS can be estimated at the intersection of the hyperbolas if more than one TDOA measurement is performed (Misra et al., 2006). One of the benefits of this technique is that it does not require knowledge of the absolute time of the transmission, i.e., the receiver time does not need to be synchronized with the transmitter, but the transmitters need to be synchronized with each other.

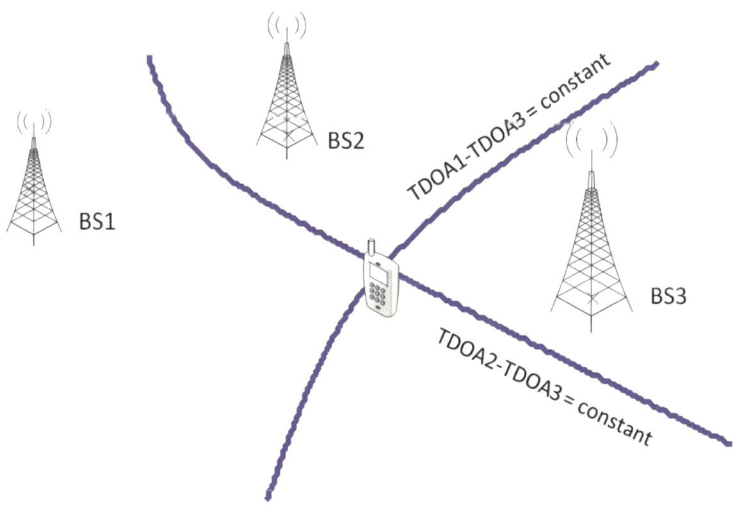

Fig. 3. Time Difference of Arrival (TDOA)

2.4 Received Signal Strength (RSS)

In RSS based positioning, the MS location is estimated using models that relate the strength of the received radio signal either to the distance between the MS and the signal emitter or to the MS location directly. Typically at least some parameters in the applied models are determined experimentally to adapt the model to the application environment. RSS based positioning methods can be divided into three main categories: cell identifier based, pathloss-based, and fingerprinting. For consumer market positioning applications, the RSS observables are considered to be more easily available than AOA or TOA, as the RSS can be passively listened from the access points (APs) of the infrastructure WLAN, without adding any extra load to the network. According to IEEE 802.11 standard, the infrastructure APs periodically transmit beacon frames, which contain information for network identification, broadcasting network capabilities, and for other control and management purposes (Wallbaum et al., 2005). The MS can sweep from channel to channel and record information from any beacon it receives. This process is performed regularly to determine the AP with the best link quality. This allows the MS to determine the cell identifiers and signal strengths of all APs visible for the MS. In many mobile devices, such as mobile phones, PDAs and laptop computers, this information is easily available through Application Programming Interfaces (API) of standard WLAN services.

2.4.1 Cell ID based positioning

In cell identifier method, the MS scans the available WLAN channels. As the position estimate it reports the position of the AP from which it received the strongest signal (Fig.4). In cell identifier method, a MS needs prior information about the locations of APs and their unique Media Access Control (MAC) addresses. Therefore, the system set-up for positioning is relatively easy. Granularity of the position estimate is determined by the distances between MS and AP. Because of the coarse granularity of the estimate and noise introduced by the environment, this method is applicable in scenarios where rather coarse accuracy is sufficient.

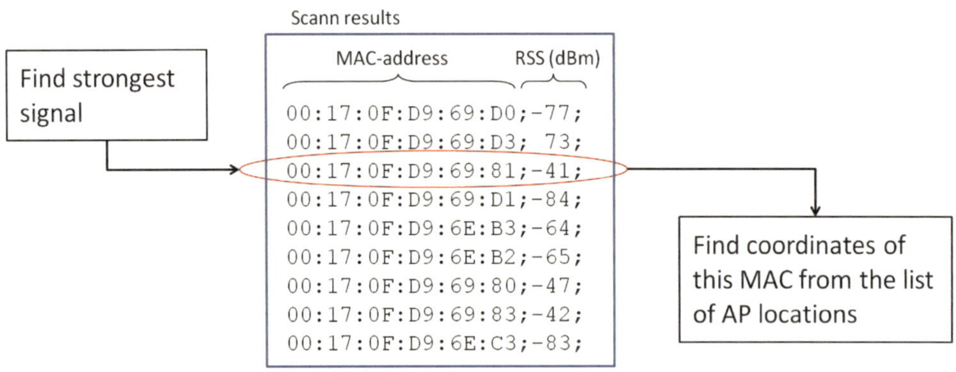

Fig. 4. Cell ID positioning based on RSS measurements

2.4.2 Fingerprinting

Fingerprinting approaches are based on experimental models that relate the measured RSS values directly to the measurement position. These models are generated from off-line collected data from several locations that sufficiently cover the area where positioning is needed. The principle of fingerprinting based positioning is illustrated in Fig.5. For each location, from the off-line collected data a typical signal pattern is extracted and saved to the fingerprint database with the coordinates of the location (Fig.5.a). In positioning phase, the current set of RSS measurements from the APs in the coverage area are compared to the patterns stored in database. The coordinate estimate is obtained from the database entry whose stored signal pattern has the closest match with the measured signal vector (Fig.5.b). Compared to other RSS based methods, fingerprinting algorithms are considered to be more robust against signal propagation errors such as multipath or attenuations generated by walls and other structures; fingerprinting actually make use of these location dependent error characteristics of radio signals. In estimation phase, new measurement vectors are related with the information stored in fingerprint database. A known disadvantage in fingerprinting approaches is the fact that the collection of the data for fingerprint database is laborious and time consuming (Wallbaum et al., 2005; Bahl et al., 2000).

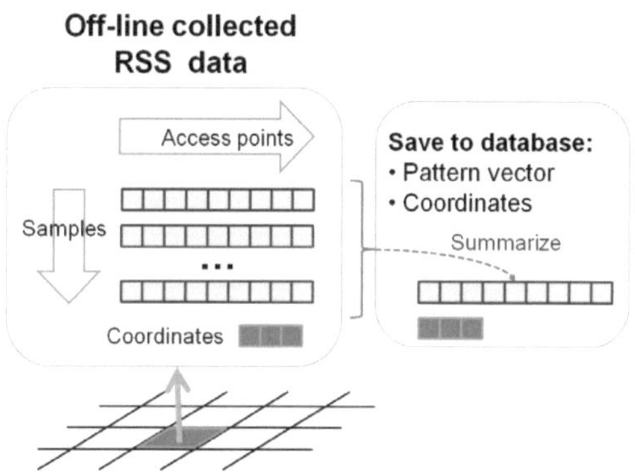

(a) Generating fingerprint database

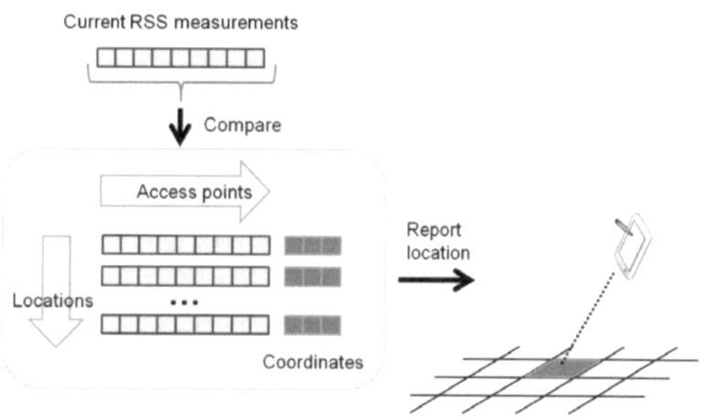

(b) Positioning using fingerprints

Fig. 5. Fingerprinting

2.4.3 Pathloss-Based positioning

Pathloss models of radio signals are used to translate RSS measurements to distances between the MS and APs. After the distances are estimated from RSS measurements, trilateration methods are used to estimate the position of the MS (Fig.6). To obtain a unique solution, the MS needs from measure RSS to at least three distinct APs. As in cell ID based methods, the MS needs prior information about the MAC addresses and locations of APs, which is easily acquired, at least when compared with fingerprint databases. In indoor environments, multipath and attenuation caused by walls, other structures, and even people complicate the modeling of signal propagation. Because of this, the positioning errors in pathloss-based positioning are typically larger than in fingerprinting (Bahl et al., 2000). On the other hand, methods that utilize path-loss models to estimate distances are needed for example if signal properties of ad-hoc WLAN connections between two MSs need to be used for positioning, because dynamic information about moving AP locations is difficult if not impossible, to be incorporated in fingerprint databases. Because of the low system set up cost of pathloss-based positioning, and its better suitability for incorporating measurements from ad-hoc connections, we concentrate on pathloss-based positioning in this research.

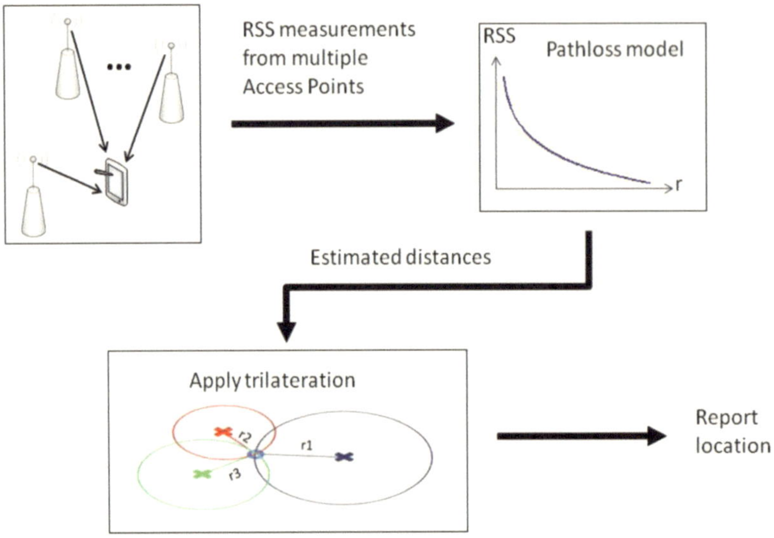

Fig. 6. Pathloss based positioning

3. Cooperative mobile positioning

Cooperative Mobile Positioning is a recent research topic for wireless communication systems (Figueiras & Frattasi, 2010) (Sand et al., 2008) concerning the development of innovative techniques and positioning schemes to enhance location accuracy in adverse scenarios, where conventional methods are not able to offer desired levels of accuracy. In such context, heterogeneous technologies and mobile terminals coexist and cooperate with the objective of helping each others for enhancing accuracy of their estimated positions. This can be accomplished by sharing link information with peer nodes connected in ad-hoc mode and exploiting their spatial diversity with advanced positioning algorithms (Sand et al., 2008).

Raising up as a new branch of positioning techniques, the fundamental idea is simple: making use of the short-range mobile-to-mobile measurements connected in ad-hoc mode, where usually unreliable long-range measurements coming from the deployed fixed reference points are provided (Frattasi & Monti, 2007) (Della Rosa, 2007). In this scenario ad-hoc connections play the dual important role of serving as medium for collecting the RSS information and exchanging data among neighboring nodes.

The exploitation of spatial proximity estimated within a group of neighboring mobiles has the strong potential to enhance the location estimation accuracy (Mayorga et al., 2007) and it can be easily applied in case of (i) outdoor environments, by merging the measurements from the cellular links and ad-hoc networks; (ii) indoor environments, by replacing the cellular and ad-hoc segments with wireless local area network (WLAN) communications in infrastructure and ad-hoc mode, respectively; and (iii) GPS-equipped mobiles, where the location estimation can be enhanced in areas where the stand-alone GPS might not be sufficient (Mayorga et al., 2007). Sharing radio signals that are just enough to ensure network connectivity among mobiles, the ad-hoc network model achieves better performances over the stand-alone cellular one (Della Rosa et al., June 2007) (Mayorga et al., 2007).

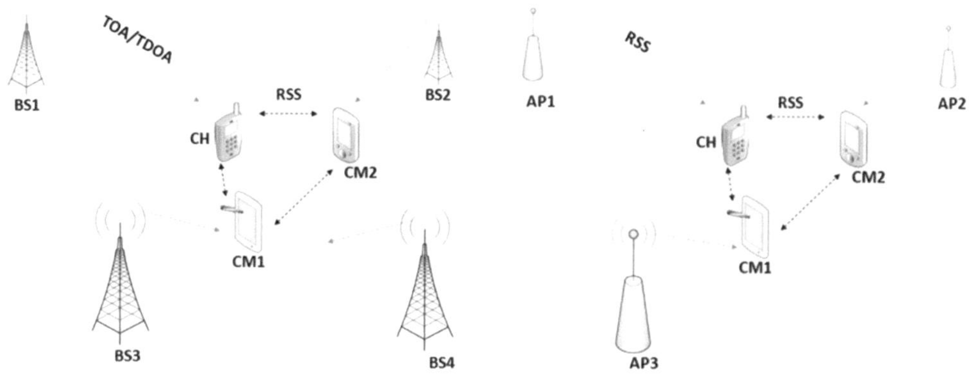

(a) Outdoor Scenario (b) Indoor Scenario

Fig. 7. Cooperative Positioning

3.1 Data-fusion and cooperative filtering

The use of data-fusion and positioning algorithms is fundamental to combine heterogeneous long- and short-range measurements and estimate the final location of MSs. Efficient methods and mathematical models able to deal with error sources are needed in wireless positioning. The most promising approaches proposed in (Figueiras & Frattasi, 2010) (Frattasi, 2007) (Della Rosa et al., June 2007) make use of Least Squares (LS) methods and Bayesian filters. LS methods allow the estimation of the position by minimizing the error between detected and expected measurements, by making use of Non-Linear-least-Squares

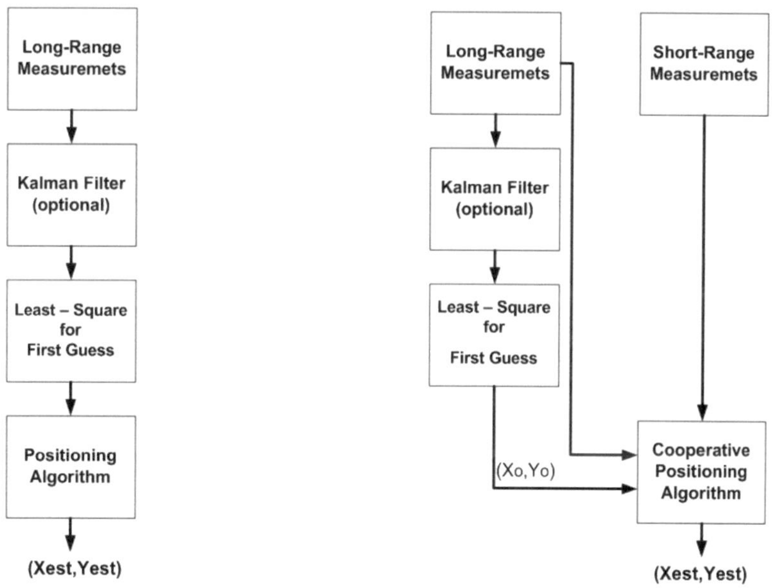

(a) Conventional (b) Cooperative

Fig. 8. Block Scheme for conventional and cooperative positioning (Della Rosa et al., June 2007) (Della Rosa et al., 2010).

(NLLS) and Weighted- Non-Linear-least-Squares (W-NLLS) (Frattasi, 2007), where the objective function to be minimized represents the main engine for processing the hybrid measurements (Della Rosa, 2007). Bayesian filters are a valid alternative to the previous ones. However, the non-linear characteristics between measurements and positions make the common Kalman Filter (KF) not applicable for solving this problem. Better results come from Extended Kalman Filters (EKF), widely used for both positioning and tracking by linearizing the models and applying then the classical KF to the linearized system (Figueiras & Frattasi, 2010) (Sand et al., 2008). In the examples proposed in this chapter, we will show results achieved by using an EKF (Della Rosa, 2007) in simulations and a NLLS algorithm (Della Rosa et al., 2010) in the experimental activity.

3.2 Ad-hoc networks and Measurements-Sharing protocol

Exploiting the ad-hoc connectivity in wireless communications has the advantage of not depending on fixed infrastructures. A central BS (or AP) is not needed at all and the overall serving area is self-defined by the area where the nodes (MSs) are deployed (Hekmat, 2006). Ad-hoc networks can be formed fast, just when they are needed and for the specific needs of each user. When used in a mesh fashion all the nodes are aware of the others within the coverage range.

An interesting ability of the ad-hoc networks is also that it is self-configuring. If one of the nodes linking the others accidentally fails, the overall network adapts the other nodes to the new configuration and rebuilds by itself.

With the benefits offered by the ad-hoc networks, the cooperative mobile positioning can be handled in terms of scalability, self-configuration, re-configuration and flexibility (Breed, 2007). In every-day life, peer-to-peer connections and data exchange are more and more common among users, becoming one of the most efficient methods for exchanging inter-systems data. During the years, several applications have been proposed for ad-hoc networks but only recently it has also been recognized as a complementary technology for enhancing location accuracy of mobile terminals (Figueiras & Frattasi, 2010) (Della Rosa, 2007). However, using ad-hoc networks in localization is still not fully independent on fixed infrastructures. On the other hand, the localization information obtained and shared by ad-hoc networks can complement and improve the infrastructure based localization, especially in cases when relation between the localization result and fixed reference systems (coordinates, geographic location) is desired.

Localization of mobile nodes is always a difficult task, due to the radio signal that is both environment and hardware dependent. A common situation is that the receiving mobiles are in NLOS (Fig. 9(a))with respect to the transmitting BS or AP, meanwhile measurements coming from ad-hoc neighbors are much more reliable. Fig. 9(b) show the RSS measured in a typical indoor environment as depicted in Fig. 17 where more fluctuating measurements (in black) are detected at AP-MS link if compared with the short-range ones (in red and blue) detected between MS-MS connected in ad-hoc mode.

Recognizing the beneficial impact of the ad-hoc links, the cooperative technique proposed (Frattasi, 2007) (Figueiras & Frattasi, 2010) (Della Rosa, 2007) can be applied as follows. Assuming a cluster of MSs connected in ad-hoc mode in a mesh configuration, a MS can be nominated as the cluster-head and its neighbors as cluster-members; (i) all the mobiles perform long-range measurements (TOA/TDOA or RSS)from the available BSs/APs links broadcasting the signals; (ii) the cluster-head looks for potential cooperative peers in the ad-hoc coverage area; (iii) it sends its cooperation-requests with an ack/nack procedure;

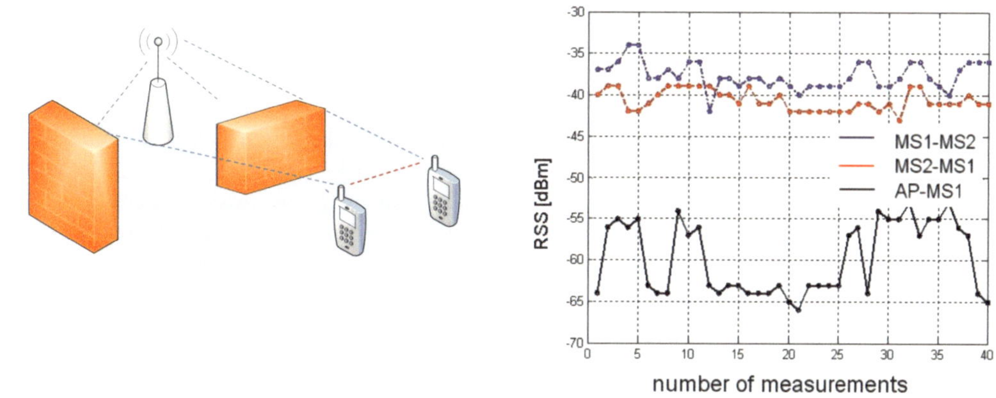

(a) Long-range and short-range links. (b) Long-range and short-range RSS measurements.

Fig. 9. Long- and short-range RSS measurements in indoor environments.

(iv) if cluster-members accept the request, the connected mobiles measure the RSS of their ad-hoc links; (v) cluster-members measure the RSS from the available APs and send the recorded data sets to the cluster-head; (vi) after receiving all the needed information, the position of each member is obtained by a cooperative data-fusion method implemented in each mobile by appropriately combining and weighting the long- and short-range raw measurements using the chosen algorithm (EKF/NLLS/W-NLLS)(Della Rosa, 2007) (Frattasi, 2007) (Mayorga et al., 2007)[10]. A potential protocol for outdoor and indoor environments (using TDOA and RSS for long-range measurements respectively) is shown in Fig. 10(a) and Fig. 10(b).

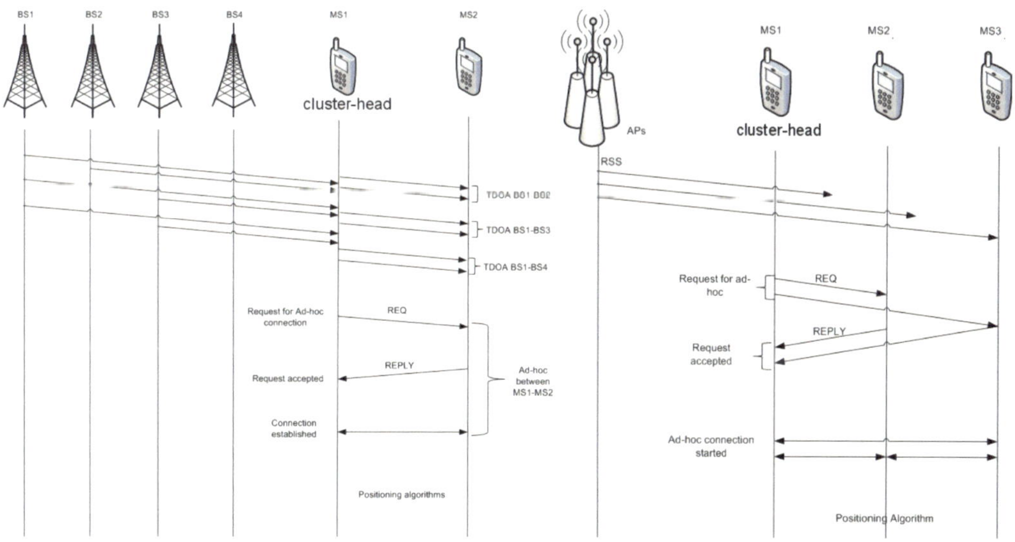

(a) Outdoor (b) Indoor

Fig. 10. Protocol for measurements and data exchange.

4. Results

This section analyzes the results, where computer simulations and experiments have been performed by developing proof of concepts for different scenarios: (i) a hybrid cellular/ad-hoc framework implemented in Matlab (Della Rosa, 2007) (Mayorga et al., 2007) and (ii)a small-scale experiment using real devices in a WLAN/ad-hoc network (Della Rosa et al., June 2007) (Della Rosa et al., 2010).
While the cellular/ad-hoc scenario is a simulated hybrid MobileWiMAX/WLAN system, the WLAN/Ad-hoc framework proves the feasibility of the cooperative techniques for heterogeneous MSs with different embedded wireless cards. In the latter it is also shown that the cooperation can be used to avoid long time-consuming calibration phases of different mobiles when performing RSS-to-distance conversions for AP-MS and MS-MS links(Della Rosa et al., 2010).

4.1 Outdoor: Cellular/Ad-hoc:

The system architecture of the simulator is shown in Fig. 11. While the cellular system is simulated according to the IEEE 802.16e standard (Mayorga et al., 2007), the ad-hoc links between MSs are modeled according to the IEEE 802.11a PHY (Mayorga et al., 2007). The scenario reproduces four synchronized BSs, with maximum synchronization error of 1ms among them. The cell radius is r = 3 km, and two MSs placed at distance of 20m from each others. MSs are assumed to be connected to the serving BS, (e.g. BS1). A mobility model simulates users moving with constant velocity of 3 km/h along parallel straight lines.
Typically (Della Rosa, 2007) 20 meters are enough for establishing ad-hoc connections; specially when the devices are in LOS, as in our simulated environment.
The full chain of blocks (cellular environment, mobility models, positioning estimators) is depicted in Fig. 11 where the physical layer (PHY) of the IEEE 802.16e standard is Orthogonal Frequency Division Multiplexing (OFDM) modulation. While in free-space the traveling time of the radio signal is only dependent upon the distance BS-MS, in real situations it is strongly delayed by channel impairments, having a direct impact on the TDOA values estimated at the receiver. For this reasons a channel model has been simulated according to (Della Rosa, 2007) (Mayorga et al., 2007).

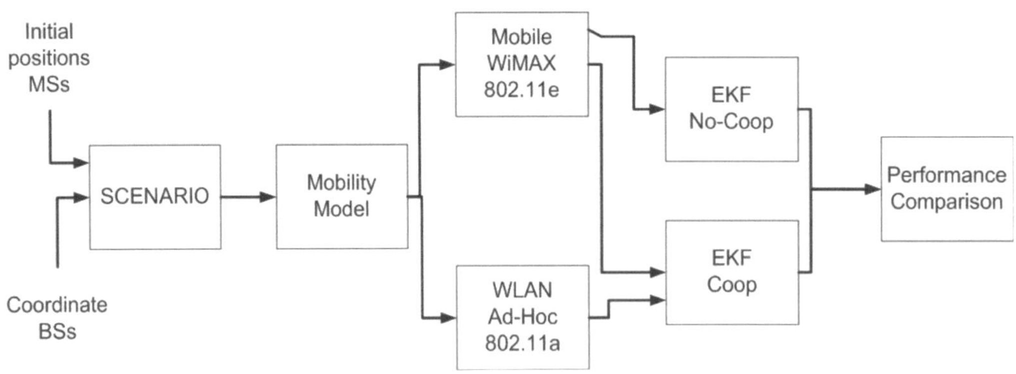

Fig. 11. Simulator Blocks (Della Rosa, 2007).

TDOA measurements are calculated at terminal level for each MS by performing cross-correlations of the signals arriving from the deployed BSs with respect to the reference one.

Also the IEEE 802.11a PHY is based on OFDM modulation (for more details the reader can refer to (Della Rosa, 2007) and (Mayorga et al., 2007)). But, differently from the AP-MS links, the MS-MS links measure RSS values, meaning that the implementation of a path loss model with small scale fading effects for a LOS scenario is also required.

Finally an EKF is used as data-fusion algorithm and positioning filter according to (Figueiras & Frattasi, 2010) (Della Rosa, 2007).

TDOA measurements are generated according to the 802.16e standard and combined with the RSS measurements within the ad-hoc network in the cooperative case. In non-cooperative case only TDOA measurements are considered.

Fig. 12 describes the simulated and estimated path of the users moving in parallel where the estimated positions for MS1 and MS2, respectively, with and without cooperation are shown. The average Root-Mean-Squared-Error (RMSE) is evaluated through the estimated path and the resulting Cumulative Distribution Function (CDF) of the RMSE describes the improvements by using only two cooperative MSs in the simulated environment (Fig. 13). It is worth mentioning that the proposed example requires the handsets to be equipped both with WiMAX and Wi-Fi modules. The resulting performances achieved show that cooperation reduces the average RMSE with respect to conventional stand-alone positioning methods (Figueiras & Frattasi, 2010) (Della Rosa, 2007).

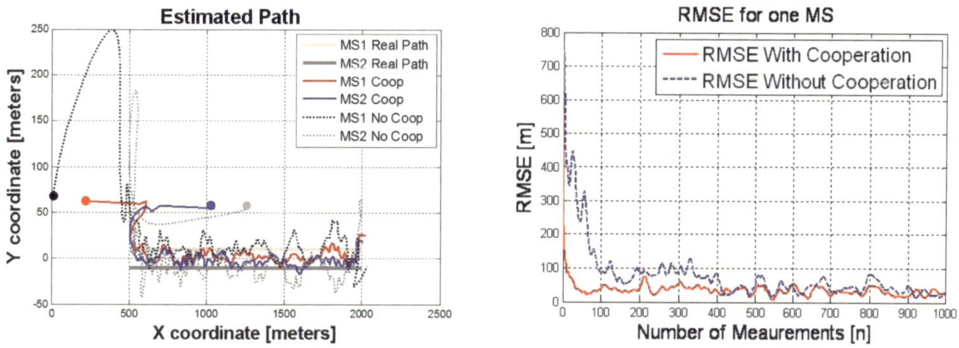

(a) Estimated Path (b) Example of RMSE improvements for one MS.

Fig. 12. Estimated Path and RMSE with and without cooperation.

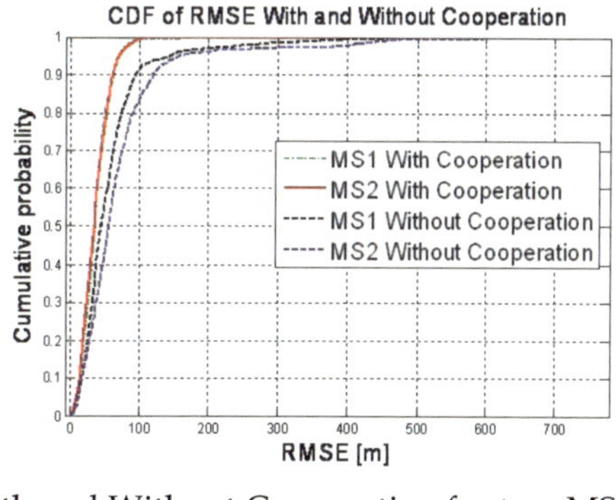

Fig. 13. CDF of RMSE With and Without Cooperation for two MSs.

4.2 Indoor: WLAN /Ad-hoc:

In this section a proof of the applicability of the cooperative techniques is shown with a real-life small scale experiment, performed in an indoor scenario as in (Della Rosa et al., 2010) (Della Rosa et al., June 2007), where the long-range measurements are represented by RSS from APs-MSs and the short-range measurements are the RSS measured at MS-MS ad-hoc links. Having precise enough measurements is an important first step for wireless positioning. However the behavior of data collected in real environments differ from the simulated ones since unpredictable errors appear quite often, causing huge fluctuations in RSS and consequently degradation of the final position estimation accuracy. Hence it is not straightforward to understand the distance-dependent behavior of the wireless signals propagating in the air. Converting power measurements for estimating the distance among APs-MSs and MSs-MSs is a crucial and time consuming activity since several parameters affect the accuracy of the measurements. Multipath, shadowing, presence of humans and objects, signal blocking, overlapping channels, walls, noise and sensitivity of the wireless cards embedded in the MSs (Della Rosa et al., 2010) introduce several complications when developing positioning applications aiming to locate heterogeneous mobiles, especially because different vendors use different chipsets with different Radio Frequency (RF) characteristics.
Experiments (Della Rosa et al., 2010) show that different laptops placed at the same distance from APs record RSS values which differ several dBm from each others due to the different embedded wireless cards (Fig. 14).

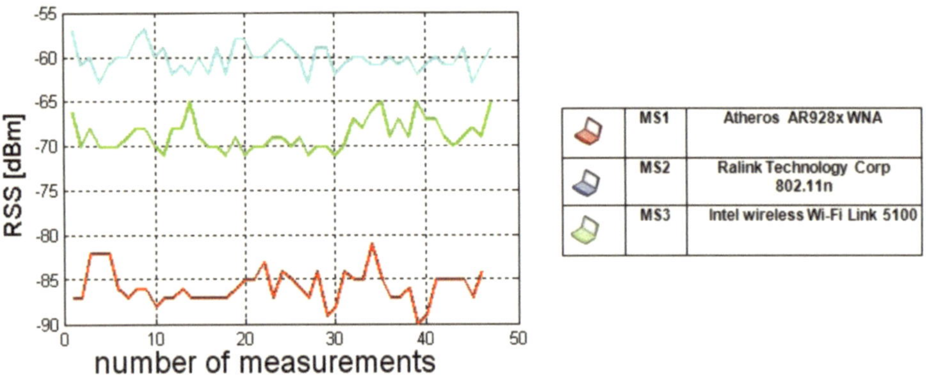

Fig. 14. RSS of laptops placed at same distance from AP, with different embedded wireless cards.

Theoretical path-loss models provided in literature are not accurate enough to reach high localization accuracy performances and exhaustive device calibrations are needed to find precise models for each mobile in use. Even after calibration, the obtained model is usually useful only for the calibrated one (Della Rosa et al., 2010).
What if we would like to develop robust and more scalable positioning applications? Every mobile (every wireless card) should be accurately re-calibrated. The cooperative technique helps in the aforementioned problem by exploiting ad-hoc connections and spatial constrains allowing the on-the-fly calibration of peer heterogeneous mobiles with different embedded wireless cards. We can imagine the situation described in Fig. 15.
One MS, (MS1) is calibrated according to the accurate procedure depicted in Fig. 15(a) (and discussed in (Della Rosa et al., June 2007) (Della Rosa et al., 2010)) and another MS, the non-

calibrated (MS2) enters the coverage area of the ad-hoc network. MS1 and MS2 are placed at distances d1 and d2, respectively, from AP1 as shown in Fig. 15(b), and recording the RSS from AP1. MS2 sends the recorded RSSs to MS1 via ad-hoc connection. MS1, after having measured also the RSS of the ad-hoc connection with MS2, estimates the distance between the MSs; it is assumed that the MS2 transmits also info about its transmission power. MS1 estimates the distance d1 from AP1 and the distance d3 from MS2. The distance d2 should not exceed the radius of d3 estimated by MS1. At this point MS1 calculates a correction parameter for MS2, to allow MS2 to apply the path-loss model of MS1. After receiving the correction parameter, MS2 can finally estimate the distance from AP1.

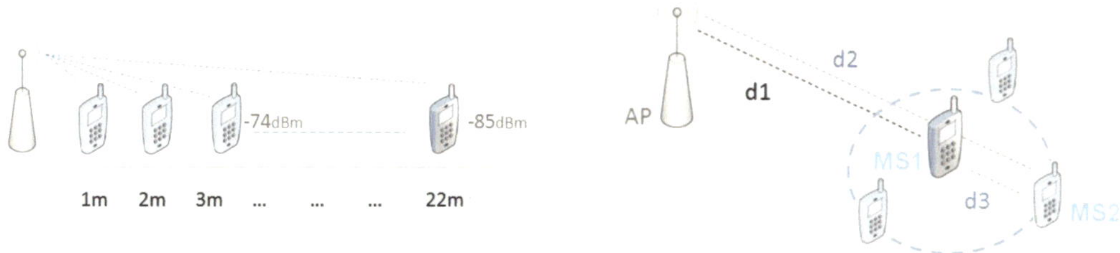

(a) Conventional (b) Cooperative

Fig. 15. Conventional and Cooperative calibration for multiple devices.

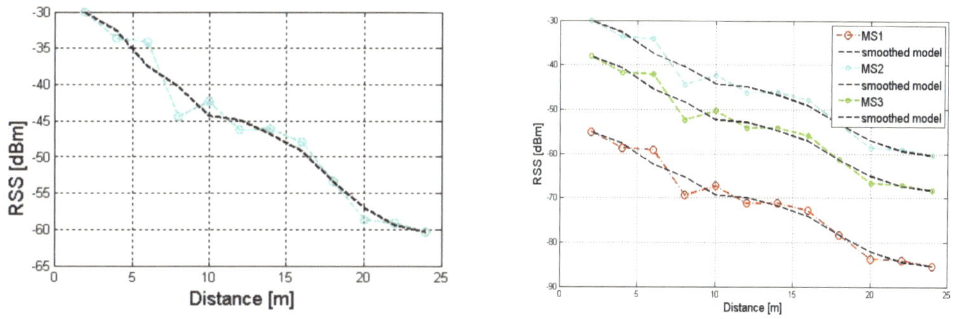

(a) Experimental pathloss for one MS (b) Pathloss models with correction factor.

Fig. 16. Pathloss model for different mobiles.

It is worth mentioning that the closer the MSs are in the cooperative calibration phase, the better accuracy in calibration can be achieved. This process is performed iteratively; it is more precise if the two mobiles are static during the calibration procedure. Once the MSs are calibrated, the cooperative mobile positioning technique can be applied using the protocol proposed in Fig. 10(b).

The experiment took place at the 3rd floor of Tietotalo building, Tampere University of Technology, Department of Computer Systems, Finland. A typical office area with dimensions of 50x50 square meters was used as testing environment, where several objects, rooms, walls and furniture are deployed inside the area, causing severe signal obstructions between APs and MSs as expected.

Four APs Cisco 802.11 a/b/g and three laptops with their own embedded wireless cards were used. A C++ application has been developed for measuring and recording real-time RSS from each AP and also the RSS from ad-hoc links among MSs. All the measurements were logged into text files and processed with Matlab scripts in both calibration and positioning phase. A Cooperative-NLLS algorithm was performed according to (Figueiras & Frattasi, 2010) (Della Rosa et al., 2010) (Frattasi, 2007) and results were compared with the non cooperative approach (Mayorga et al., 2007).

Fig. 17 shows the averages of the estimated positions for the three MSs with cooperation (circles with border) and without cooperation (circles without border). Laptops icons represent the real positions of the mobiles. It is demonstrated as in such adverse environments, the ad-hoc network has a beneficial impact in positioning accuracy for all the devices. Moreover, as the number of cooperative users increases, also the positioning accuracy gets improved (Figueiras & Frattasi, 2010).

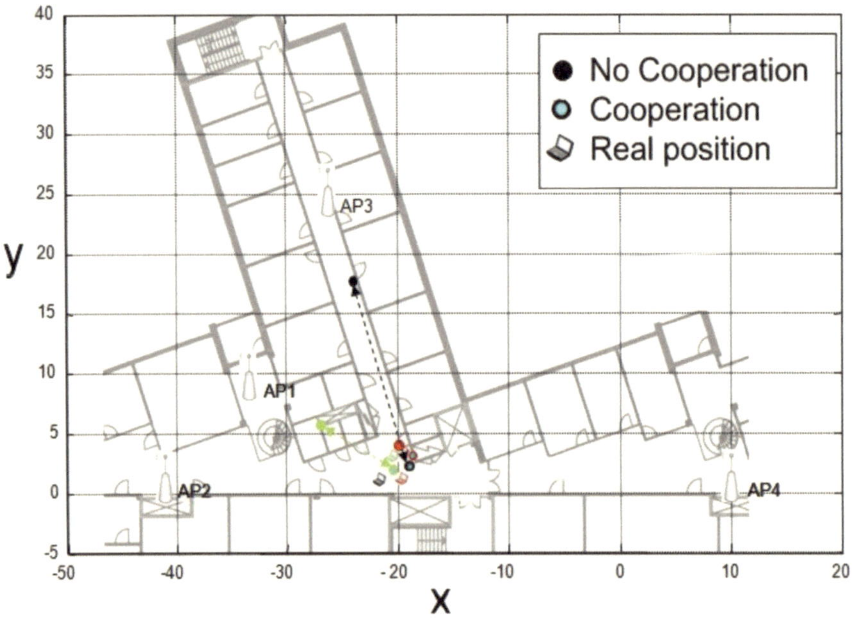

Fig. 17. Estimated Positions.

The achieved performances are highlighted in Fig. 18 by showing the CDF of the RMSE of the three mobiles.

5. Conclusion

In this chapter we have described the basics of Cooperative Mobile Positioning and the exploitation of ad-hoc networks in adverse positioning environments. Our test results from simulations and real life experiments show that, thanks to the short-range measurements available from ad-hoc links, the positioning accuracy is improved when compared to the accuracy of the non-cooperative approach. The ad-hoc link measurements present lower absolute errors than measurements in long-range cellular links; they are more stable and contain less signal fluctuations.

Although we have provided examples on Mobile WiMAX and WiFi technology, the cooperative technique can be adapted and exploited by replacing one or both technologies with different and newer ones.

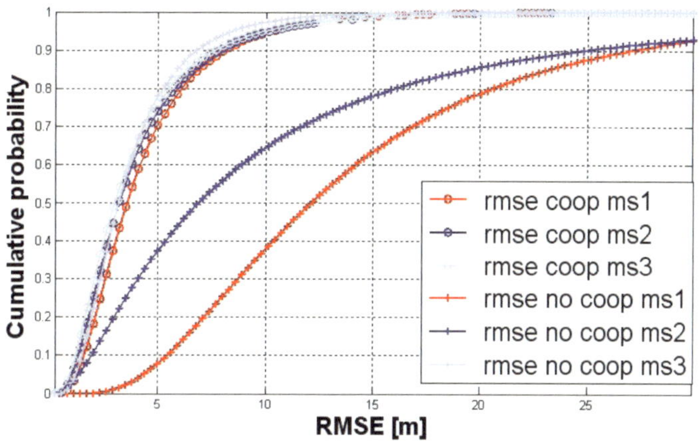

Fig. 18. Cumulative Distribution Function of the RMSE.

6. References

Bahl, P. & Padmanabhan, V.N. (2000). Radar: An in-building RF-based user location and tracking system, *IEEE INFOCOM 2000 Conference on Computer Communications*, vol. 2, pp. 775-785, Tel Aviv , March 2000, IEEE.

Breed, G. Wireless Ad Hoc Networks:Basic Concepts. *Summit Technical Media, LLC.*.

Della Rosa, F. (2007). Cooperative Mobile Positioning and Tracking in Hybrid Mobile WiMAX/WLAN. *M.Sc. Thesis*, Aalborg University (AAU), Denmark, June, 2007.

Della Rosa, F., Paakki, T., Leppäkoski, H., Nurmi, J., (2010). A Cooperative Framework for Path Loss Calibration and Indoor Mobile Positioning, Proceedings of *7thWorkshop on Positioning, Navigation and Communication 2010 (WPNC'10)*,Dresden, Germany, March 2010.

Della Rosa, F. Wardana, S.A., Flores Mayorga, C.L., Simone, G., Raynal, M.C.N., Figueiras, J., Frattasi, S. (2007)Experimental Activity on Cooperative Mobile Positioning in Indoor Environments. Proceedings of *2nd IEEE Workshop on Advanced Experimental Activities on Wireless Networks and Systems (EXPONWIRELESS)*, ,Helsinki, Finland, June, 2007.

Figueiras, J., Frattasi, S., (2010). Mobile Positioning and Tracking: From Conventional to Cooperative Techniques. (1st Edition),Wiley, ISBN 978-0470694510.

Frattasi S.(2007). Link layer techniques enabling cooperation in fourth generation (4g) wireless networks, *Ph.D. Thesis*, Aalborg University AAU, Denmark,(September, 2007).

Frattasi S., Monti M. (2007). Ad-Coop Positioning System (ACPS): positioning for cooperative users in hybrid cellular ad-hoc networks. *EUROPEAN TRANSACTIONS ON TELECOMMUNICATIONS*. Wiley InterScience. 2007.

Huang E., Hu W., Crowcroft J., Wassell I. (20xx). Towards Commercial Mobile Ad Hoc Network Applications: A Radio Dispatch System, *x*Urbana-Champaign, Illinois, USA.

Hekmat R. (2006). Ad-hoc Networks: Fundamental Properties and Network Topologies. *EUROPEAN TRANSACTIONS ON TELECOMMUNICATIONS*.ISBN: 978-1-4020-5165-4.

Mayorga, C.L.F, Della Rosa, F., Wardana, S.A., Simone, G., Raynal, M.C.N., Figueiras, J., Frattasi, S. (2007). Cooperative Positioning Techniques for Mobile Localization in 4G Cellular Networks. *IEEE International Conference on Pervasive Services (ICPS)*, Instanbul, Turkey, July, 2007.

Misra, P. & Enge, P. (2006). *Global Positioning System; Signals, Measurements, and Performance. 2nd ed.*, Ganga-Jamuna Press, ISBN 0-9709544-1-7, Lincoln, MA.

Rappaport, T.S., Reed, J.H. &Woerner, B.D. (1996). Position Location UsingWireless Communications on Highways of the Future. *IEEE Communications Magazine*, Vol. 34, No. 10, (Oct. 1996) page numbers (33-41).

Sand, S., Mensing, C., Ma, Y., Tafazolli, R., Yin, X., Figueiras, J., Nielsen, J., Fleury, B.H. (2005). Hybrid Data Fusion and Cooperative Schemes for Wireless Positioning. *Vehicular Technology Conference, 2008. VTC 2008-Fall. IEEE 68th*. Calgary, 2008.

Sayed, A.H., Tarighat, A., Khajehnouri, N. (2005). Network-Based Wireless Location: Challenges Faced in Developing Techniques for Accurate Wireless Location Information. *IEEE Signal Processing Magazine*, Vol. 22, No. 4, (July, 2005) page numbers (24-40).

Syrjrinne, J. (2001). *Studies on Modern Techniques for Personal Positioning*, PhD thesis, Tampere University of Technology.

Wallbaum, M. & Diepolder, S. (2005). Benchmarking wireless LAN location systems, *Second IEEE International Workshop on Mobile Commerce and Services WMCS'05,*, pp. 42-51, Munich , July 2005, IEEE.

Wong, C., Klukas, R. & Messier, G.G. (2008). Using WLAN Infrastructure for Angle-of-Arrival Indoor User Location, *68th Semi-Annual IEEE Vehicular Technology Conference, 2008. VTC 2008-Fall*, pp. 1-5, Calgary, BC, Sept. 2008, IEEE.

14

Ad-hoc Networks As an Enabler of Brain Spectroscopy

Salah Sharieh
Mcmaster University
Canada

1. Introduction

The purpose of this chapter is to show the feasibility of using ad-hoc networks as an enabler of brain spectroscopy. Ad-hoc networks have many applications. The application which this chapter explains provides full mobility in everyday environment using a near-infrared light sensor designed to monitor brain function in humans. Multiple wireless networks employing several different protocols are used for data carriage and provide new freedom to conduct tests in real environment outside a lab. An Ad-hoc network (Bluetooth) is one of the wireless networks used to support the application. The value of this application is to measure the changes in the concentration of oxyhemoglobin (HbO2) and deoxyhemoglobin (Hb) in tissues in the real-life environment. This might lead to better understanding of tissue pathologies. This type of application was not available before.

A fully mobile functional brain spectroscopy system has been developed to allow the possibility of testing subjects to be monitored in their real environment. To test this hypothesis, communication software was developed to allow for the collection of physiological data from a mobile near-infrared sensor via a mobile telephone that has a Bluetooth support. The developed application is used to track the changes in the concentrations of HbO2 and Hb during various activities and send the data to a computer at a remote monitoring site.

The specific aims of this application have been to build a fully mobile system to monitor the concentrations of HbO2 and Hb in near real time, to monitor the concentrations of HbO2 and Hb during smoking, as well as to analyze the gathered data, and to try to understand the correlation between HbO2 and Hb during smoking. Performance and data accuracy were the key for this application to provide the sought value.

Java portability allows the developed application to run on a wide range of operating systems and devices. Java Standard Edition (J2SE) was used for server code; Java Micro Edition (J2ME) was used to run code in the phone; C language was used to build the Bluetooth code and the protocol in the sensor; and Eclipse was used as the integrated development environments (IDE) to build and debug the application.

Java has native network support. It is possible to create applications to support different kinds of networks and protocols. Java has native libraries that support wired and wireless communications. It supports Bluetooth, WiFi, and more. Several popular network protocols and standards are also supported. By default, Java libraries support Transmission Control Protocol (TCP), User Datagram Protocol (UDP), and binary stream communications.

In this application a reliable network is required. To meet part of reliability requirements, TCP protocol was fond to be the best supported protocol in the mobile device used in this system. TCP protocol is a reliable protocol used in communication when a reliable connection is required (Comer, 1997). It allows two hosts to communicate and exchange data streams and guarantees the data delivery (Stevens, 1994). Data packets are delivered in the same order they were sent. In contrast, UDP does not provide guaranteed delivery and does not guarantee packet ordering (Comer, 2007). Selecting which protocol to choose for a particular application mainly depends on the application requirements. These protocols have proven their value and made their way into Bluetooth and GSM networks. Bluetooth networks support both TCP and UDP communications (Bray & Sturman, 2002). Applications running on the Bluetooth networks can use any of these protocols to send and receive data. The most common way to send TCP and UDP packets over Bluetooth is using Bluetooth Radio Frequency Communications (RFCOMM) (Ganguli, 2002). RFCOMM is a transport protocol that provides RS-232 serial port emulation. Bluetooth Serial Port Profile (SPP) is based on this protocol (Huang, 2007; Bluetooth Core Specifications Version 2.1. 2007).

GSM networks are similar to Bluetooth networks and wired local area networks. They support TCP and UDP communication protocols (Delord et al., 1998; Eberspächer et al., 2001; Chakravorty et al., 2003). Since wireless networks support the same communication protocol as wired local area networks, applications running on wireless networks can communicate and exchange data with the applications running on wired local area networks.

Application level protocols are created to support specific applications. These protocols can run on top of either TCP or UDP protocols. KREIOS protocol and LayerPro protocol in this application are examples of such protocols. It contributes to the overall reliability of the application. KREIOS is a packet-oriented protocol created to support data exchange between the sensor used in this application and any other application running in another device (Arquatis GmbH, 2007). LayerPro is a protocol created in this research based on KERIOS protocol to allow global communication between the sensor, the PDA, and the server over Bluetooth and GSM networks.

The wireless sensor used in this application implemented KREIOS protocol, which was created by (Muhlemann, 2006) and implemented by Arquatis GmbH, Rieden Switzerland (Arquatis GmbH, 2007) in the wireless sensor. The KREIOS is a packet-oriented protocol between two devices: one acts as a master and the second one acts as a slave; both communicate through a request and response transaction. In this application, the master is the PDA and the slave is the sensor.

Several methods have been devised for imaging the human brain, in particular Electroencephalography (EEG), Computed Tomography (CT), Magnetic Resonance Imaging (MRI), Functional Magnetic Resonance Imaging (fMRI), Positron Emission Tomography (PET), Single Photon Emission Computed Tomography (SPECT), Near-infrared Spectroscopy (NIRS), and Diffuse Optical Tomography (DOT). These methods vary in their strengths (Strangman et al., 2002). In recent years, researchers have started using NIRS and DOT, either alone or in combination with other methods, to image brain functions. The non-invasive nature of the NIRS is appealing to researchers to measure changes in HbO2 and Hb during brain function activities (Izzetoglu et al., 2003).

Functional Optical Brain Spectroscopy using Near-infrared Light (fNIRS) has been introduced as a new method to conduct functional brain analysis. fNIRS is a method that uses the reflection of infrared light to observe changes in the concentration of HbO2 and Hb in the blood, and can provide a similar result to fMRI (Villringer & Chance, 1997). fNIRS takes advantage of the absorption and scattering of near-infrared light to provide information about brain activities (Gratton et al., 1997). For a long time, it was thought that it was only possible to collect information from the superficial layers of tissue (e.g., microscopy) due to light scattering. However, about 25 years ago, it was discovered that functional information could be obtained from brain tissue using light shone at the scalp and detected from the scalp (Jobsis, 1977). This discovery motivated the development of diffuse optics as a method for brain monitoring. This method has different names: Near-infrared Spectroscopy (NIRS), Diffuse Optical Tomography, and/or Near-infrared Imaging (NIRI). Today, several types of NIRS devices have been built to image brain functions. These devices differ in their capabilities, designs, and costs (Strangman et al., 2002; Bozkurt et al., 2005).

The NIRS devices can be classified into three main types: Continuous Wave Spectroscopy (CWS), Time-resolved Spectroscopy (TRS), and Frequency Domain Spectroscopy (FDS). The CWS device consists of a continuous light source, which transmits light waves with constant amplitude, and a detector that locates the attenuated incident light after it passes through the tissues. The TRS device transmits short incidents of light pulses into tissues and measures the light after it passes through the tissues. On the other hand, the FDS device transmits a sinusoidally modulated light wave into the tissue (Strangman et al., 2002).

Each of these types of NIRS devices has limitations and strengths (Hong et al., 1998). CWS has the advantage of low cost; however, with CWS it is difficult to distinguish contributions of absorption and scattering to light attenuation. FDS, on the other hand, is known for its good spatial resolution, penetration depth, and accurate separation of absorption and scattering effects. Nevertheless, FDS is significantly more expensive than CWS. As for TRS, although theoretically, it can provide a better spatial resolution than FDS, it has a lower signal-to-noise ratio. Since TRS requires short pulsed lasers and photon counting detection, it is the most expensive type of the NIRS instrumentation. Despite the advancements in NIRS technology, NIRS still has limitations, such as the short path length and the artifacts' movements during measurements.

Absorption and scattering are the main physical processes affecting the transmission of light photons in tissues. Light photon absorption and scattering causes the light intensity to decrease. Both absorption and scattering are wavelength dependent. The amount of absorbed light photons is also impacted by the concentration of blood HbO2 and Hb in tissues which vary in time, reflecting physiological changes in tissues' optical properties (Villringer & Chance, 1997).

When light photons travel through tissues, they are scattered several times before finally reaching the receiver. Scattering increases light optical path length, causing photons to spend more time in tissues which in turn affects the tissues' absorption characteristics.

Despite the fact that both absorption and scattering play a major role in light transmission, scattering is more dominant than absorption. When light travels through tissues and blood, photon absorption leads to a loss of energy to tissues and blood chromophores, or induces either fluorescence (or delayed fluorescence), or phosphorescence. The main substances of biological tissues that contribute to light photon absorption in the near-infrared light are water, fat, and hemoglobin. While water and fat remain fairly constant over a short period of time, the concentrations of oxygenated and deoxygenated hemoglobin change according to the function and metabolism of the tissues. Thus, the corresponding changes in absorption can provide clinically useful physiological information (Villringer & Chance, 1997).

Near-infrared light, in the range of 700-900 nm, can travel relatively deep into body tissues. It is also worth mentioning that such light can easily travel through soft tissues and bones, such as those of neonates and infants. Therefore, it is suitable to use near-infrared devices to monitor brain activities or other oxygen-dependent organs in this category of humans (Germon et al., 1998).

NIRS relies on a simple principle: light in the range of near-infrared light emitted on the organ of interest passes through the different layers above the organ. When it passes through the tissues, light photons go through physical interactions, such as scattering and absorption that leads to a loss of energy in the emitted light. When the remaining light exits the organ, it is measured by a detector.

In neuroimaging applications, the light is injected through the scalp, so the photons pass through several layers of tissue surrounding the brain, such as the scalp, skull, Cerebrospinal Fluid (CSF) and meninges. Then, the NIR light reaches the brain and the blood vessels, and backscattered light gets detected by a set of detectors. The light in this case follows the so-called banana-shaped path due to scattering effects caused by the tissues. Due to the fact that water and lipids are relatively transparent to near-infrared light and the optical properties of the layers surrounding the brain and blood are fixed within a given period of time, it was found that light is mainly absorbed by oxygenated and deoxygenated hemoglobin. Here, it must be noted that the scattering of the near-infrared light in the human tissues is much larger than its absorption, while absorption of this kind of light is much larger in the blood. This leads to the belief that the optical properties of the blood, which in fact change based on the amount of oxygen in the blood, can play a vital role in determining the amount of backscattered light from the brain. The amount of blood volume and blood oxygen concentration can be an indicator of hemodynamic activities that are related to brain functions. Analyzing the amount of backscattered light during the oxygenation and deoxygenation process of the blood flow in the brain can lead to a better understanding of the brain function (Benni et al., 1995).

NIRS measures the optical properties of HbO2 and Hb in near-infrared light. The effects of the changes in concentration levels of HbO2 and Hb in the blood stream on light absorption can be described by the Beer-Lambert's Law. A Modified Beer-Lambert Law can be used to predict the amount of blood chromophoers (HbO2 and Hb) in tissues (Bozkurt et al., 2005).

2. Brain spectroscopy

Functional brain imaging using fMRI and Positron Imaging Tomography (PET) have increased our understanding of the neural circuits that support cognitive and emotional processes (Cabeza & Nyberg, 2000; Davidson & Sutton, 1995). However, these methods are expensive, uncomfortable, and might have side effects such as exposure to radioactive materials (with PET) or loud noises (with fMRI) (Hong et al., 1998; Chance et al.,1993). Such disadvantages make these imaging methods inappropriate for many uses that require the monitoring of brain activities under daily, real-life conditions.

Functional Optical Brain Spectroscopy Using Near-infrared Light (fNIRS) is another method to conduct functional brain analysis. fNIRS is a non-invasive method that uses infrared light reflection to gather changes in the concentration of HbO2 and Hb in the blood(Jobsis, 1977). The main advantages of fNIRS are: ability to measure concentration of chemical substance; device's low cost; device's low power requirements; non-invasiveness nature, and device's portability. Low cost and portability have made it possible to use fNIRS to monitor patients in their homes for an extended period of time, allowing health care providers to monitor slowly developing diseases in patients. The non-invasive nature of fNIRS has also made it possible to perform as many tests as needed without worrying about side effects (Boas et al., 2002).

Blood carries oxygen and nutrients to tissues. Also, it carries carbon dioxide and other products of metabolism away from tissues, so the body can eliminate them. Red blood cells contain hemoglobin, which is the main oxygen transporter. When the red blood cells pass through the lungs, they collect oxygen where it becomes bound with the hemoglobin. Furthermore, red blood cells release carbon dioxide to the lungs. Blood vessels form a comprehensive network inside the body where they deliver blood to different tissues and organs. Arteries, arterioles, and capillaries deliver oxygenated blood to tissues whereas veins and venules collect deoxygenated blood from them (Boas et al., 2002).

The human brain is protected by several layers. These layers provide a safe and secure environment for the brain. Near-infrared light, used to measure changes in the blood oxygenation, has to pass through all the protective layers: scalp, periosteum, skull, and the meninges (Fig. 1). The meninges contain three layers: dura mater, arachnoid mater, and pia mater (Porth, 2005).

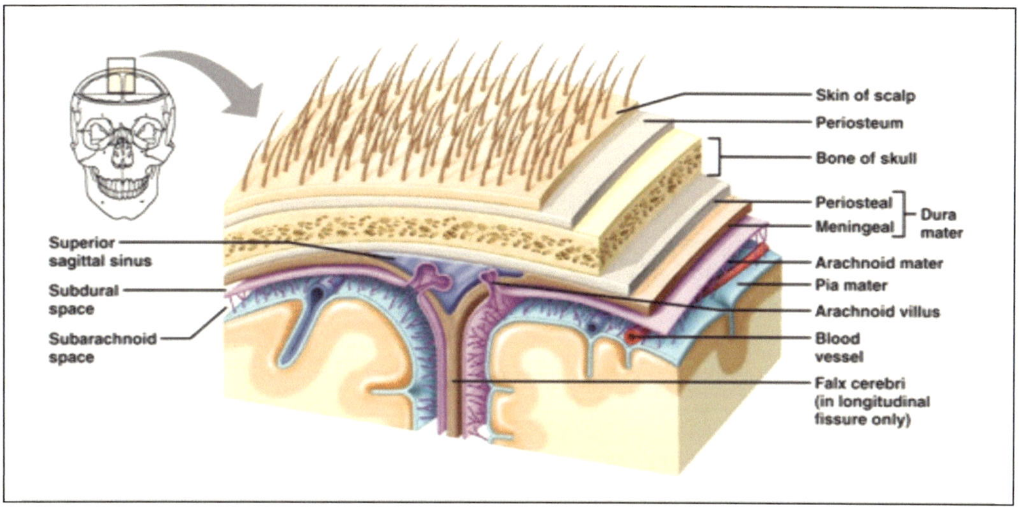

Fig. 1. The Brain's Protective Layers [18]

3. System design

The system developed for this application consists of three main hardware components. The first component is a Bluetooth wireless sensor (built by Arquatis GmbH, Rieden, Switzerland), which is the data acquisition device (Muhlemann, 2006; Muhlemann et al., 2006). The second component is a PDA and is the main controller for the measurement process and the data communication bridge between the sensor and the central computer. The third component is a central computer (Server, or Host Computer, or PC) that stores the data for later analysis. See Fig. 2 for a full display of the system's architecture.

Two different ranges of communication are used in the developed system. First, the communication between the sensor and the PDA is carried over a Bluetooth network. The signal range between the PDA and the sensor is approximately 10 m (short range). This short range is enough to perform bedside monitoring without the need to carry the PDA. The other range of communication occurs between the PDA and the central computer and is carried over the GSM network (wide range). The range of the GSM is very wide indeed since the system employs the mobile phone network with a roaming feature. Technically, it is possible to monitor a test subject wearing the sensor in any part of the world as long as they are within the range of a GSM network with roaming capabilities.

The PDA and the sensor are light weight devices that make it possible to carry them easily. The sensor has a set of programs developed in the C language required to enable the data acquisition and data transmission. The PDA runs the Java ME program that performs the data transmission between the sensor and the host PC. The host PC works as a server and a database server as well. Additionally, the PC is configured with a public IP address to make it accessible through the Internet and to the GSM network. The communication between the PDA and the sensor is bidirectional and the communication between the PDA and the PC is unidirectional - from the PDA to the server.

The combination of these communication technologies allowed the creation of a fully mobile system for Functional Optical Brain Spectroscopy using Near-infrared Light (fNIRS) technology extending the range and the mobility of an existing solution (78; 79; Muhlemann et al., 2006; Trajkovic, 2006).

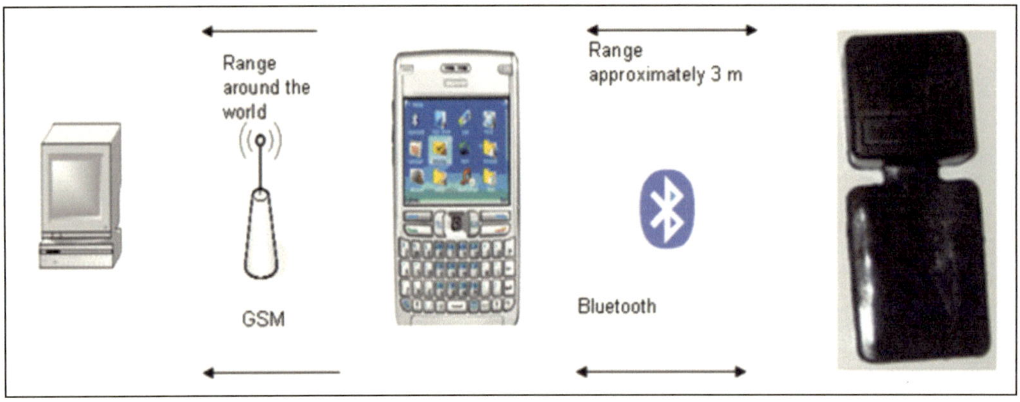

Fig. 2. System Architecture

4. Protocols and algorithms

Initially the system used HTTP protocol as a data encapsulation protocol. HTTP protocol is designed to be a request-response protocol to transmit text based data. This makes it not suitable for binary transmission without adding performance overhead.

This application required continuous fast binary data upload. After reviewing existing upload protocols and approaches, we came to conclusion that a new protocol is needed to be created. Performance and native binary upload were key requirements for the protocol. Based on the requirements the protocol was designed and extended KERIOS protocol. The protocol achieved the requirement through minimizing the control data and the number of the overall transactions. Moreover the protocol packet was designed to hold binary data which reduced the data representation overhead.

The protocol (see Fig. 3) for this application was created to encapsulate only the acquired data and send it to the server; it is based on KERIOS protocol. The extension was necessary to ensure data integrity and improve KERIOS protocol parsing. LayerPro carries only the KERIOS data packet and adds 3 extra bytes as a sequence number. The sequence number ensures that the packets are continuous and no packet loss will occur during transmission. Moreover LayerPro has a fixed length; it has 35 bytes, while KERIOS has a variable length. These modifications made LayerPro packet parsing easier and faster on the server. This protocol is stateless and supports limited transactions, of which it allows three: open, close, and send.

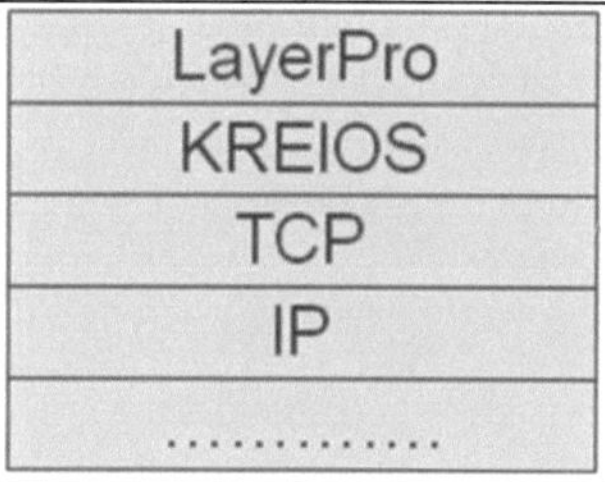

Fig. 3. Protocol Stack

LayerPro protocol has two parts: head and tail (see Fig. 4). The head contains 3 bytes representing the transmission sequence number and 1 byte describing the packet type (Data or Control). There are four possible values for the packet type field: 0-data; 1-open; 2; send; 3-close. The tail contains the actual binary data. In this protocol, the fixed length is used to determine the end of the packet.

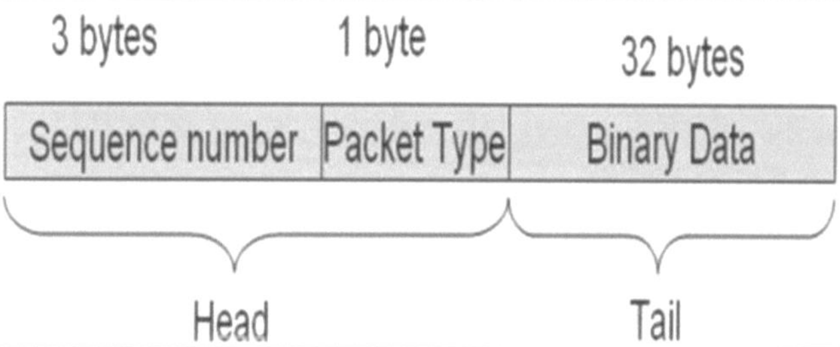

Fig. 4. LayerPro Packet Format

To start the data streaming, the source system sends an open transaction packet. This transaction packet indicates to the destination system (server) the beginning of a transmission. The sequence number value in the packet head is "00 00 00"; the packet type field contains the open command, and no data in the packet tail. The open transaction packet is followed by a send transaction packet that contains the acquired data from the source (sensor) in the tail, the send command in the packet type and a sequence number in the sequence number field. The close transaction packet indicates to the destination (server) the end of transmission. The packet type field has a close command; the sequence number value in the packet head in this transaction is the last data sequence number with no data in the packet tail. Fig 5 demonstrates these transactions and flow between the source system (PDA) and the destination system (server).

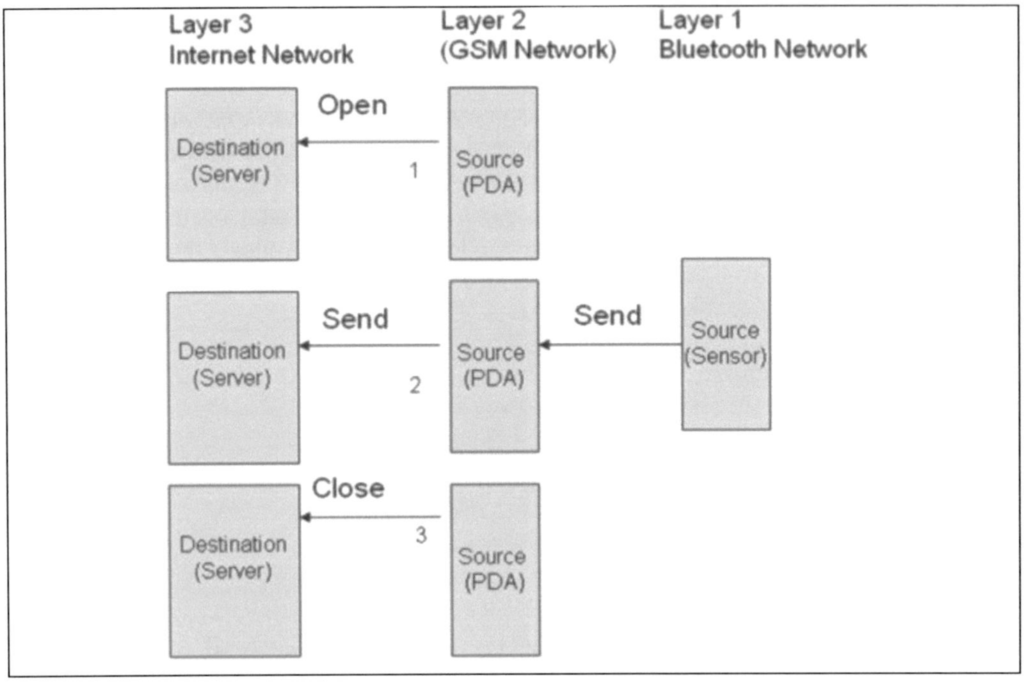

Fig. 5. Layer Pro Transactions

5. Network integration

To validate the protocol's basic functionality more than 100 tests were performed. They were designed to monitor brain functions during smoking outside the lab environment to collect changes in oxygenation concentration levels in the brain during breath holding and finally to measure the changes in oxygenation concentration levels in dogs' brains when presented with their favorite toys.

The tests were focused on performance, data integrity, availability and the effectiveness of the developed protocol. The system worked in all cases, but different amounts of delay were experienced in the data transmission. The delays vary between 1 to 5 seconds. The delay is impacted by the networks' speed during the time when the experiments were performed.

The protocol design allowed the sending of one packet at a time. This approach reduced the overall packet size which made it possible to send the data with a very short delay (1 second) most of the time. Whereas the packet size was very small (36 bytes) due to the protocol design, the network bandwidth requirements became very small. Therefore, the system required only a few resources to transmit the data to the server which made it possible to transmit the data without data lost despite unpredictable changes in the networks load.

To compare LayerPro performance versus HTTP protocol performance, two version of the system were implemented. The first version implemented LayerPro protocol and the second version implemented HTTP protocol. The results demonstrate that LayerPro protocol provides better near-real-time binary data transmission than the HTTP protocol. Table 1 shows a sample result compares LayerPro protocol and the HTTP protocol.

Layer Pro Average Delay	HTTP Average Delay
1 Second	8 Second
1 Second	5 Second
1 Second	5 Second
1 Second	7 Second
1 Second	5 Second
1 Second	5 Second
1 Second	5 Second

Table 1. LayerPro Protocol versus HTTP Protocol

The system was tested in two different locations to ensure that the protocol can support true mobility. The test subject was wearing the sensor and carrying the PDA while he was moving around between two cities (Toronto: big city has 5 million people and Markham: small city has 0.5 million people). The tests were performed over several days and different times. The combination of location, date and time were necessary to investigate the effect of the mobile network and the internet load on the quality of the transmitted data during low usage and peak usage of the heterogonous networks. Moreover the location, time and date combination were used to validate how well the protocol can handle the communication during different networks load.

Fig. 6 shows a direct comparison between LayerPro and HTTP protocol. From the figure we can see that LayerPro protocol provides better near-real-time binary data transmission than the HTTP protocol. The figure also shows that the network load effect is minimal on LayerPro protocol.

In biomedical applications data integrity is very important. Even one packet dropping sometimes means losing valuable information. Tests also showed that all data packet were streaming correctly and in a timely manner.

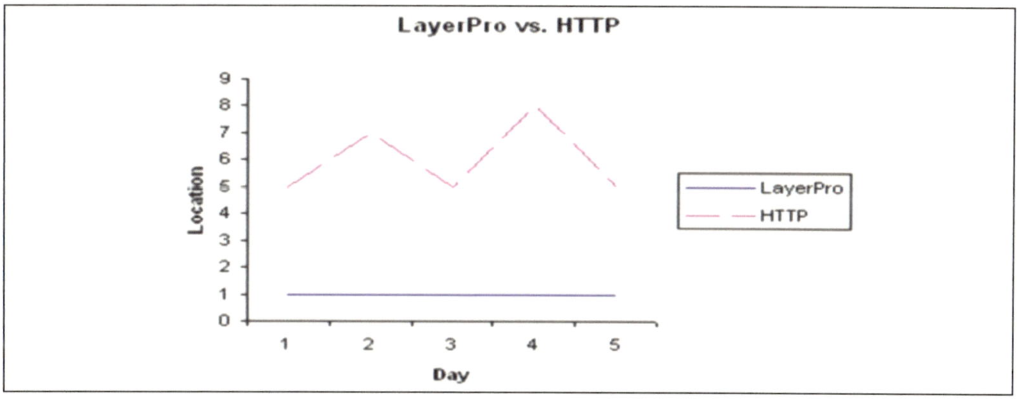

Fig. 6. Averafe delay LayerPro vs. HTTP

6. Ad-hoc networks embedding in medical sensors

The application software architecture (Fig. 7) has three major layers: a data acquisition layer (DAL), a control layer (CL), and a data storage layer (DSL). The DAL software component in the sensor controls data acquisition and packet transmission. It is composed of a set of programs that implements the data communication protocol, the RFCOMM Bluetooth protocol, and the sensor's low-level controls. The second layer (CL) resides on the PDA and acts as the central control unit for the application. The majority of the system components reside in this layer. The third layer (DSL) is mainly used to accept connections from the PDA and stores the received data packets in the server for later analysis. The PDA creates a persistent connection with the sensor and with the PC during the duration of the measurements. The system is designed to support a wide range of measurements and acquisition activities. Several types of tests can be performed using the system without the need to modify the programs. Most of the components are designed to be configuration-driven. The system architecture provides high interoperability between heterogeneous hardware and software.

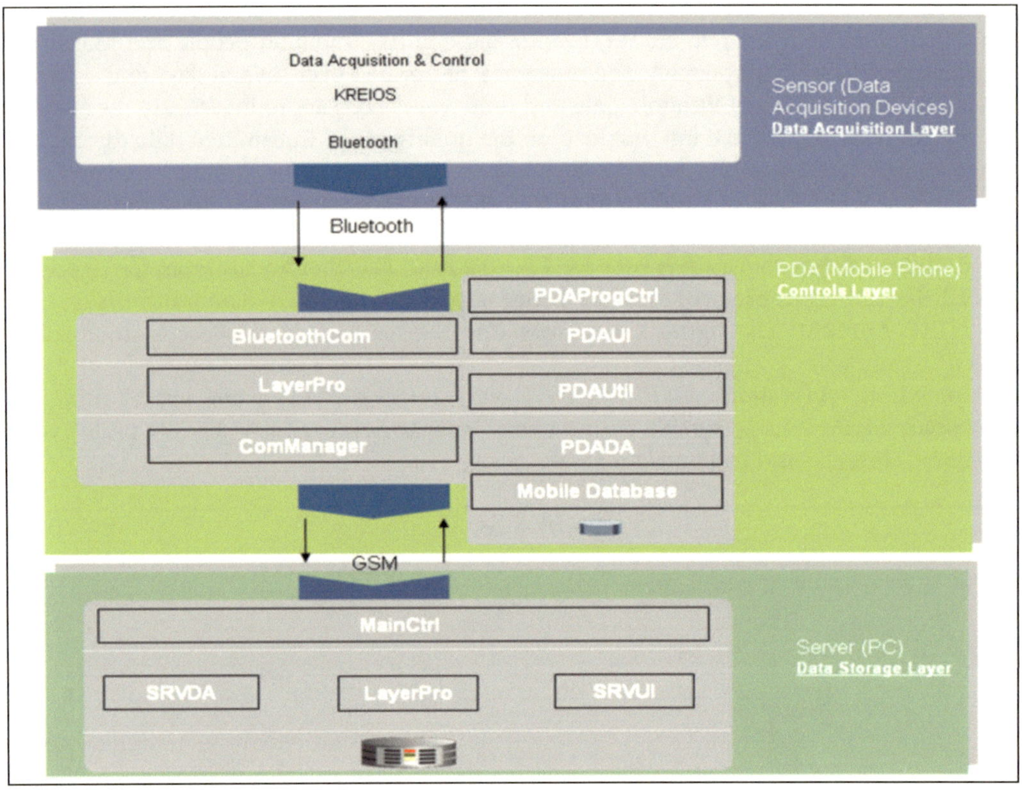

Fig. 7. The Application Software Architecture

All user interactions (Fig. 8) in the system are initiated by the User Interface Component (PDAUI) that is controlled by the program control component (PDAProgCtrl). Program control calls the LayerPro component to create command and data packets. All commands are encapsulated by a LayerPro packet before they are sent to and received from the sensor; this is performed by the LayerPro component. A LayerPro packet is sent and received over the air using the Bluetooth communication component. When data is collected from the sensor it is sent to the server using the communication manager component (ComManager); then a local copy of the packet will be saved to the mobile local file system using the Mobile Database Access Component (PDADA).

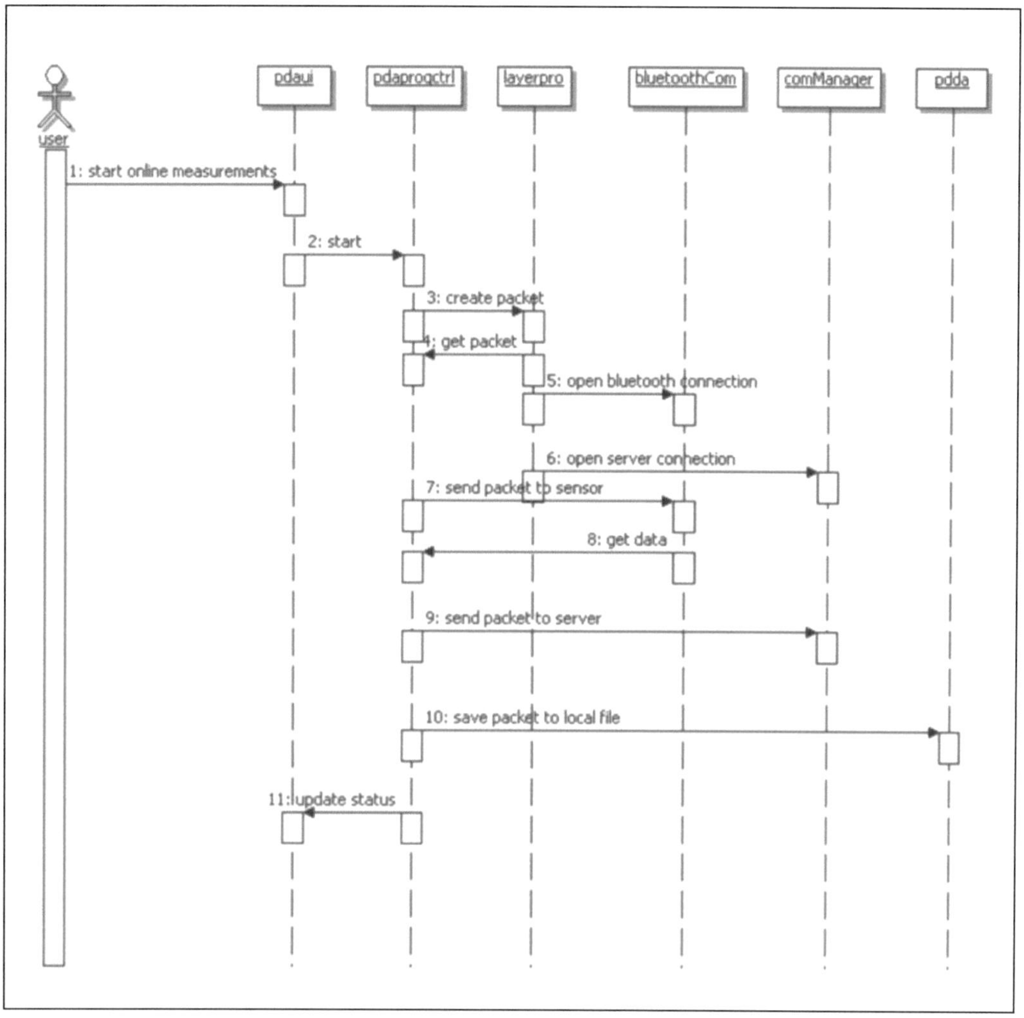

Fig. 8. The application overall interactions

7. Case studies summary

The system was designed to support a wide range of measurement activities. We wished to ensure that a variety of data be available for testing. In order to achieve this goal we performed tests on both humans and trained dogs with tests being conducted both inside and outside a lab environment. HbO2 and Hb changes in brain and tissue were collected for both species in different circumstances. In total, three major types of biomedical experiments were conducted using our system.

The first experiment was a breath holding experiment. The test was used as a validation experiment in order to ensure that our system worked correctly and could collect biological data.

The second experiment was related to smoking and was conducted entirely outside the lab. This experiment was performed to understand the effect of smoking on the brain in a real

environment away from the distractions and unrealities of a rigid laboratory environment—an environment where smoking actually takes place.
The third experiment was conducted to monitor a trained canine's brain activities. The experiment was conducted in order to determine if it was possible to monitor the brain activity of animals. We found the second and third tests to be particularly compelling. Smoking is an addictive behavior that occurs in the real world. In order to understand the factors that cause this behavior more accurately, we believed that any measurement must occur in the true circumstances of the activity. The third experiment involving trained canines was motivated by both the need for data outside the human realm and because we believed it could be possible to determine elements of mental activity within working animals—specifically canines--that directly relate to the activity that the animal is about to engage in. This is significant because it implies a certain level of predictability. Whether this is actually feasible is beyond the scope of this application; however, Helton et. al. have run a similar test in a lab environment without the benefit of our system (Helton et al., 2007) However, if testing is ever to be done in a real world setting, there must be a mechanism for allowing it.

8. Case studies 1

To validate that the system was functioning as expected, a breath holding experiment was performed on humans. The result was compared with a lab method (Zhang et al., 2005). Test subjects were asked to rest for 20 seconds, then to hold their breath for 20 seconds, and thereafter exhale and breathe normally for 20 seconds. The trial for each test subject lasted for 120 seconds. The rest duration between trials for each test subject was approximately 2 days.
We performed 15 breath holding trials. We asked three different test subjects (two males and one female) to hold their breath. The first test subject was a 23-year-old healthy female, non-smoker; the second test subject was a 46-year-old healthy male, non-smoker; and the third test subject was a 36-year-old healthy male smoker. During the lab trials, the test subjects were asked to wear the sensor on their forehead near the hair line and lay down on their backs on the test bed; they were asked not to move and not to speak. Instructions to inhale and exhale were communicated to them by the person running the trials. In the outside trials, the test subjects were asked to wear the sensor on their foreheads and sit on a chair in the open and they were asked not to move or speak while performing the breath holding trial.
After analyzing the collected data using our system, we can see that each breath holding trial had a measured impact on the HbO2 and Hb concentration. The result was compared to a result obtained from a similar experiment using fMRI (Zhang et al., 2005). This experiment proved that the system can provide results similar to the ones previously obtained by other test methods (Zhang et al., 2005). Clearly the system worked as expected.
Fig. 9 shows an example of data obtained during a breath holding trial. The graph shows that HbO2 increases during the breath holding. The arrow indicates when the increase happens due to breath holding. The brain compensates for the lack of oxygen by increasing the blood flow (Zhang et al., 2005). Then the HbO2 level goes down after breathing was resumed.

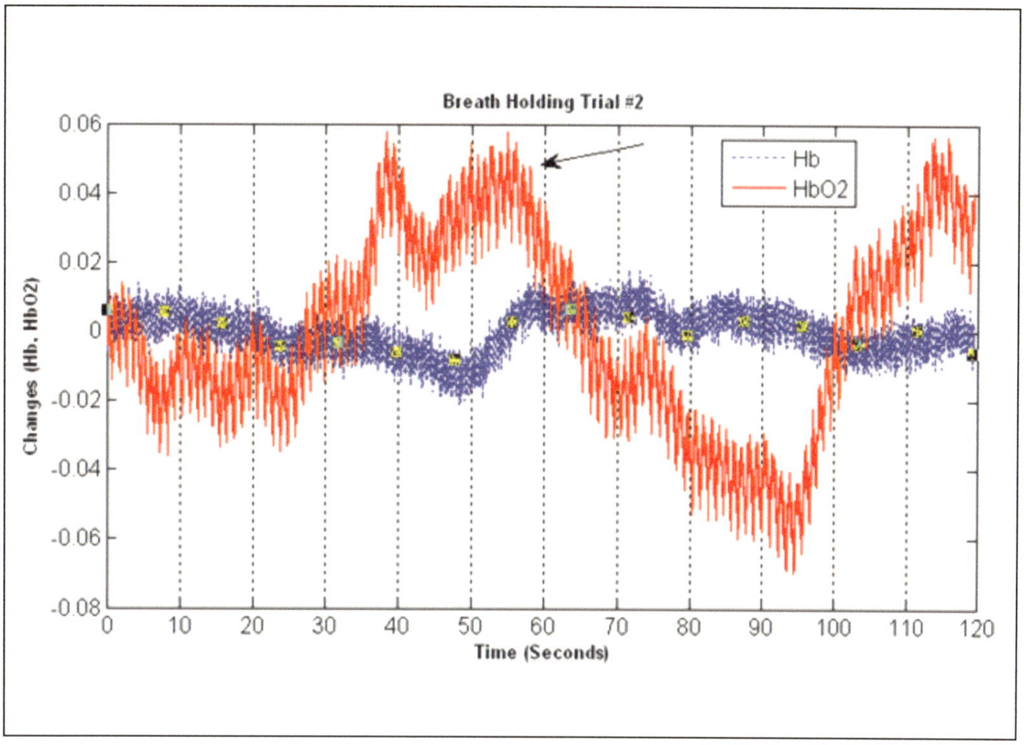

Fig. 9. Sample breath holding trails result

9. Case studies 2

There is an agreement among scientists that cigarette smoking causes lung cancer, heart diseases, and other serious illnesses (Carmines, 2002; Giessing et al., 2006). Almost five million Canadians smoke 15 times or more per day (Flight, 2007; Health Canada, 2007). The chemical substances, including nicotine, found in cigarettes Hoffmann et al., 2001; Baker et al., 2004; Rodgman et al., 2000; Frederick et al., 2007)entering the human body during smoking can cause several physiological changes. Few studies have applied fMRI to detect the oxygen level changes in the human brain under the effect of direct nicotine administration. The results have proven that nicotine can impact the level of oxygen in the hemoglobin in the brain Giessing et al., 2006; Siafaka et al., 2007). It is important to emphasize, however, that all these studies have tested the impact of the nicotine on the oxygen level in the brain using direct nicotine administration rather than actual smoking.

To understand the real effect that cigarettes (nicotine and other chemicals) have on the brain, as opposed to direct administration of nicotine, smoke testing must be performed in a natural way rather than in a controlled environment. One contribution of the developed system is to address this need. In fact, there are pragmatic health and safety reasons why this method is superior to in-lab testing. Because test subjects can be tested independently of the environment, no collateral damage from smoking need be accidentally inflicted on auxiliary participants in the test. Thus this method is safer, does not require special lab modifications and is as effective as other methods.

In total, six smoking trials were conducted. The experiment's purpose was to examine the relationship between smoking and HbO2 and Hb changes in the brain. Five healthy human males and one healthy human female participated in the experiment. The test subjects' ages ranged from 30 to 40 years old. All the test subjects were active smokers for a period of more than 2 years. During the trials, the test subjects were asked to wear the sensor on their foreheads and sit on a chair in the open where they were asked not to move more than they had to in order to smoke and not to speak. The sensor was fixed with a bandage on the test subject's head to improve the sensor's stability on the head and minimize the effect of the test subject's movement during the smoking process. Instructions as to when to smoke were communicated to the subjects by the person running the trials. Each trial lasted for 15 minutes, which included a five-minute baseline, five minutes of smoking, and a five-minute recovery after smoking.

Baseline data was recorded for 5 minutes before the test subject started smoking. The test subject was asked to smoke for 5 minutes. The test subject inhaled every 20 seconds for the duration of the test. After the 5-minute smoking period, the test subject was asked to keep wearing the sensor for another 5 minutes. The data collection continued during the 5-minute waiting period after the smoking was complete. The recovery period allowed us to capture any delayed after-effect changes that occurred due to smoking.

When we analyzed the data, we observed HbO2 and Hb changes during the baseline, the smoking, and the recovery periods. Fig. 10 illustrates the results from a smoking experiment. The graph shows that during the baseline duration, changes in HbO2 and HHb reflected normal physiological states. Sharper changes in HbO2 and Hb were appeared during smoking. These changes were similar to changes that occur during functional brain activities. Usually, such changes occurred due to the increase in the blood flow (Toronov et al., 2001).

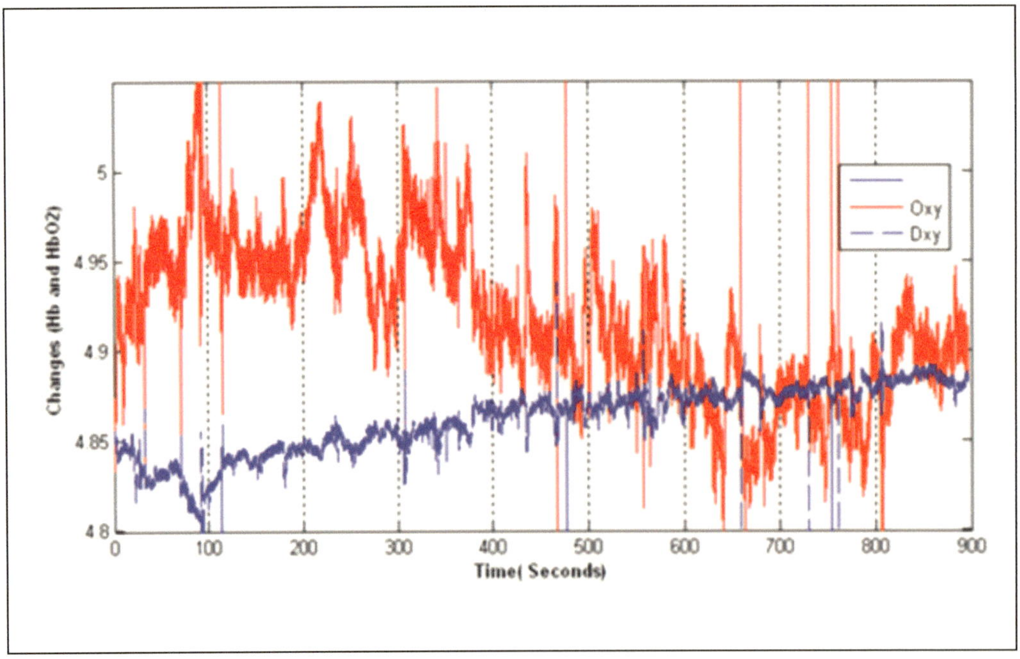

Fig. 10. Sample smoking trial result

10. References

Lima, P.; Bonarini, A. & Mataric, M. (2004). *Name of Book in Italics,* Publisher, ISBN, Place of Publication

Li, B.; Xu, Y. & Choi, J. (1996). Title of conference paper, *Proceedings of xxx xxx,* pp. 14-17, ISBN, conference location, month and year, Publisher, City

Siegwart, R. (2001). Name of paper. *Name of Journal in Italics,* Vol., No., (month and year of the edition) page numbers (first-last), ISSN

Arai, T. & Kragic, D. (1999). Name of paper, In: *Name of Book in Italics,* Name(s) of Editor(s), (Ed.), page numbers (first-last), Publisher, ISBN, Place of publication

Arai, T. & Kragic, D. (1999). Name of paper, In: *Name of Book in Italics,* Name(s) of Editor(s), (Ed.), page numbers (first-last), Publisher, ISBN, Place of publication

[67] Comer, D. (1997). In Stevens D. L. (Ed.), Internetworking with TCP (Windows sockets version. ed.). Prentice Hall, Upper Saddle River, N.J.

[68] Stevens, W. R. (1994-). Addison-Wesley Pub. Co., TCP. Reading, Mass.

[69] Comer, D. (2007). *The internet book: Everything You Need To Know about Computer Networking and how the Internet Works,* Pearson Prentice Hall, 0132335530, Upper Saddle River, NJ.

[70] Bray, J. & Sturman, C. F. (2002). *Bluetooth: Connect Without Cables,* Prentice Hall, 0130661066, Upper Saddle River, NJ., U.S.A.

[71] Ganguli, M. (2002). *Getting started with Bluetooth,* Premier Press, 1931841837, Cincinnati, Ohio, U.S.A.

[72] Huang, A. S. (2007). In Rudolph L. (Ed.), Bluetooth essentials for programmers. New York, NY: Cambridge University Press.Muller, N.J. (2001).

[73] Bluetooth Core Specifications Version 2.1. 2007. Available at <http://www.bluetooth.com/NR/rdonlyres/F8E8276A-3898-4EC6-B7DA-E5535258B056/6545/Core_V21__EDR.zip>.

[74] Delord, X.; Perret, S. & Duda, A. (1998). Efficient Mobile Access to the WWW over GSM, *Proceedings of the 8th ACM SIGOPS European Workshop on Support For Composing Distributed Applications,* pp. 1-6, Sintra, Portugal, September 1998, ACM, New York, U.S.A.

[75]. Eberspächer, J.; Vögel, H.J. & Bettstetter, C. (2001). *GSM: Switching, Services and Protocols,* John Wiley & Sons, 047149903X, Toronto, Canada.

[76] Chakravorty, R.; Clark, A. & Pratt, I. (2003). GPRSWeb: Optimizing the Web for GPRS Links, *Proceedings of the 1st International Conference on Mobile Systems, Applications and Services,* pp.317-30, San Francisco, California, U.S.A, May 2003, ACM, New York, USA.

[78] Arquatis GmbH, Rieden, Switzerland. Developed in cooperation with the Biomedical Optics Research Laboratory at the Clinic of Neonatology(2007) <http://www.arquatis.com>: product name nScan W1200.

[79] Muhlemann, T.L. (2006). New Wireless Probes for Near In-frared Spectroscopy. Master Thesis, Swiss Federal Institute of Technology, Zurich.

[80] Muhlemann, T.L.; Haensse, D. and Wolf, M. (2006). A Wireless Near-Infrared Imaging Device. The 34th Annual Meeting of the International Society on Oxygen Transfer

to Tissue. (Louisville, Kentucky, August 12-17), available at <http://louisville.edu/conference/isott06/WebProgram.pdf>.

[81] Trajkovic, I. (2006). Examination of the Brain with Light: Integration of a Wireless Sensor into a Graphical User Interface based on Java. Semester Thesis, Swiss Federal Institute of Technology, Zurich.

[82] Strangman, G.; Boas, D.A. & Sutton, J.P.(2002). Non-Invasive Neuroimaging Using Near-infrared Light. *Biological psychiatry*, 52, 7, (October 2002) 679-693, 0006-3223.

[83] Izzetoglu, K.; Yurtsever, G.; Bozkurt, A.& and Bunce, S. (2003). Functional Brain Monitoring via NIR Based Optical Spectroscopy, *Proceedings of the 29th IEEE Annual Conference-Bioengineering Conference*, pp.335-336, 0-7803-7767-2, Newark, N.J., USA, March 2003).

[84] Villringer, A. & Chance, B. (1997). Non-invasive optical spectroscopy and imaging of human brain function. *Trends in Neuroscience*, 20, 10, (October 1997), 435-42, 0166-2236.

[85] Gratton, E.; Fantini, S.; Franceschini, M.A.; Gratton, G. & Fabiani, M. (1997). Measurements of scattering and absorption changes in muscle and brain. *Philosophical Transactions: Biological Sciences*, 352,1354, (June 1997), 727-735, 0962-8436.

[20] Jobsis, F.F. (1977). Noninvasive infrared monitoring of cerebral and myocardial oxygen sufficiency and circulatory parameters. *Science*, 198, 4323, (December 1977)1264-1267, 0036-8075.

[86] Bozkurt, A.; Rosen, A.; Rosen, H. & Onaral, B. (2005). A portable near infrared spectroscopy system for bedside monitoring of newborn brain. Biomedical Engineering Online, 4, 1, (April 2005),
< http://www.biomedical-engineering-online.com/content/4/1/29 > 1475-925X.

[87] Hong, L.; Worden, K.; Li, C.; Murray, T.; Ovetsky, Y.; Pidikiti, D. & Thomas, R. (1998). A novel method for fast imaging of brain function, non-invasively, with light. *Optics Express*, 2, 10, (May 1998) 411-423, 1094-4087.

[88] Germon, T.J.; Evans, P.D.; Manara, A.R.; Barnett, N.J.; Wall, P.& Nelson, R.J. (1998). Sensitivity of near infrared spectroscopy to cerebral and extra-cerebral oxygenation changes is determined by emitter-detector separation. *Journal of Clinical Monitoring and Computing*, 14, 5, (July 1998) 353-360, 1387-1307.

[89] Benni, P.B.; Bo , C.; Amory, D. & Li, J.K. (1995). A novel near-infrared spectroscopy (NIRS) system for measuring regional oxygen saturation., Proceedings of the 1995 IEEE 21st Annual Northeast, pp. 105-107, 0-7803-2692-X, May 1995, IEEE.

[16] Cabeza, R. & Nyberg, L. (2000). Imaging cognition II: an empirical review of 275 PET and fMRI Studies. *Journal of Cognitive Neuroscience*, 12,.1 (January 2000), 1-47, 0898-929X.

[17] Davidson, R. J. & Sutton, S. K. (1995). Affective neuroscience: the emergence of a discipline. *Current Opinion in Neurobiology*, 5, 2 (April 1995), 217-224, 0959-4388.

[19] Chance, B.; Zhuang, Z.; UnAh, C.; Alter, C.; & Lipton, L. (1993). Cognition-activated Low-frequency Modulation of Light Absorption in Human Brain, *Proceedings of the National Academy of Sciences*, 3770-3774,, 90, 8, (April 1993),.

[21] Boas, D.A.; Franceschini, M.A.; Dunn, A.K. & Strangman, G. (2002). Non-invasive imaging of cerebral activation with diffuse optical tomography, In: In *Vivo Optical Imaging of Brain Function*, R. Frostig, (ed.), 193-221, CRC Press, 0849323894, Boca Raton, Florida.

[94] Porth, C. (2005). *Pathophysiology: Concepts of Altered Health States*, Lippincott Williams & Wilkins, 0781749883, Philadelphia.

[97] Helton, W.S.; Hollander, T.D.; Warm,J.S.; Tripp, L.D.; Parsons, K.; Matthews, G.; Dember, W.N.; Parasuraman, R.; & Hancock, P.A.(2007). The abbreviated vigilance task and cerebral hemodyamics. *J. of Clinical and Experimental Neuropsychology*, 29, 5, (July 2007) 545-552, 1380-3395.

[98] Zhang, X.; Toronov, V.; Webb, A. (2005). Methodology development for simultaneous diffuse optical tomography and magnetic resonance imaging in functional human brain mapping, *Proceedings of SPIE*, 453-463, San Jose, CA, USA, April 2005, International Society for Optical Engineering, Bellingham, WA, USA.

[99] Carmines, E.L. (2002). Evaluation of the potential effects of ingredients added to cigarettes. Part 1: Cigarette design, testing approach, and review of results. *Food and Chemical Toxicology*, 40,1,(January 2002), 77-91, 0278-6915.

[100] Giessing, C.; Thiel, CM.; Rösler., F & Fink GR. (2006). The modulatory effects of nicotine on parietal cortex activity in a cued target detection task depend on cue reliability. *Neuroscience* 137,3, (February 2006) 853-864, 0306-4522.

[101] Flight, J. (2007). Canadian Addiction Survey (CAS): A national survey of Canadians' use of alcohol and other drugs: substance use by youths. Health Canada, Ottawa.

[102] Health Canada (2007). Canadian Tobacco Use Monitoring Survey. Health Canada, Ottawa.

[103] Hoffmann, D.; Hoffmann, I.; & El-Bayoumy, K. (2001). The less harmful cigarette: A controversial issue. A tribute to Ernst L. Wynder. *Chemical Research in Toxicology* 14,7, (July 2001) 767- 790, 0893-228X.

[104] Baker, R.R.; Massey, E.D. & Smith, G. (2004). An overview of the effects of tobacco ingredients on smoke chemistry and toxicity. *Food and Chemical Toxicology*, 42 Suppl: 1, (March 2004) 53-83, 0278-6915.

[105] Rodgman, A.; Smith, C.J. & Perfetti, T.A. (2000). The composition of cigarette smoke: a retrospective, with emphasis on polycyclic components. *Human and Experimental Toxicology*, 19,10,(October 2000) 573-595, 0960-3271.

[106] Frederick, B.; Lindsey, KP.; Nickerson, LD.; Ryan, ET. & Lukas SE. (2007). An MR-compatible device for delivering smoked marijuana during functional imaging. *Pharmacology, Biochemistry, and Behavior*, 87,1,(May 2007) 81-89, 0091-3057.

[107] Siafaka, A.; Angelopoulos, E.; Kritikos, K.; Poriazi, M.; Basios, N.; Gerovasili, V.; Andreou, A.; Roussos, C.& Nanas, S.(2007). Acute effects of smoking on skeletal muscle microcirculation monitored by near-infrared spectroscopy. *Chest Journal*, 131,5,(May 2007) 1479-1485, 0012-3692.

[108] Toronov, V.; Webb, A.; Choi, JH.; Wolf, M.; Michalos, A.; Gratton, E.& Hueber, D.. (2001). Investigation of human brain hemodynamics by simultaneous near-infrared spectroscopy and functional magnetic resonance imaging. *Medical Physics*, 28, 4, (April 2001) 521-527, 0094-2405.

15

MANET Mining: Mining Association Rules

Ahmad Jabas
Department of Computer Science and Engineering, Osmania University
India

1. Introduction

The growing advances in mobile devices, processing power, display and storage capabilities, together with competitive market has enabled *information technology* to be more affordable and available to almost everybody around the world. Moreover, with the advent of wireless communications and mobile computing, another type of wireless communications, called Mobile Ad hoc NETworks (*MANETs*), came into existence.

The operation of *MANET* does not depend on pre-existence infrastructure or base stations, since there is no central node in the network and nodes collaboratively share all the network activities. The simplicity of *MANET* deployment comes with a cost of complexity of the algorithms in different layers. In addition, the absence of the infrastructure induces new challenges to wireless networks in the fields of routing, security, power conservation, quality of service, and so on.

For better perception of the new concepts in this chapter, a summary of the necessary background in Data Mining (*DM*) is given, and particularly more emphasis and in depth explanation is given on *association rule mining technique*, an area upon which the new concepts of this chapter revolves.

DM or Knowledge Discovery in Databases (*KDD*) is defined as "The nontrivial extraction of implicit, previously unknown, and potentially useful information from data" (Frawley et al., 1992). *DM* is the process of finding hidden relationships in data sets and summarizing these patterns in models. These patterns can be utilized to understand the whole data sets. In simplified terms, *DM* is a technology that allows an applicant to discover knowledge, which is hidden in large data sets, by applying various algorithms (Hofmann, 2003).

This chapter shows how *DM* approaches are applied to *MANET*, in that the traffic of *MANET* is mined in a simple way called "*MANET* Mining using Association Rule Techniques". *MANET* Mining enables the establishment of the fact that there are still some hidden relationships (patterns) amongst routing nodes, even though nodes are independent of each other. These relationships may be used to provide useful information to different *MANET* protocols in different layers. Precisely, *MANET* Mining, discovers hidden patterns (meta-data) in the third layer to be used as common tokens (keys) in the application layer in a bid to address one challenging security problem in *MANET*, namely, key distribution. This is the first time this approach has been used to solve key distribution problem in *MANET*.

Interestingly, security in *MANET* has been paid a lot of attention over the past few years. One of the most challenging security issue in *MANET* is key management where there is no on-line access to trusted authorities. Key management is the central part of any secure communication, and is the weak point of system security and protocol design. Most

cryptographic systems rely on the underlining secure, robust, and efficient key management system.
Key management scheme is the prerequisite for all security primitives and thus, it is the basis for secure *MANETs*. However, the performance of existing key distribution schemes developed so far is undesirable in the terms of efficiency and scalability. Besides, these schemes revolve around Third Trusted Party (*TTP*) and therefore, compromising this *TTP* means disclosing all the issued keys. Surprisingly, the fully distributed and self-organized key distribution schemes without *TTP* are still not robust to changing topology or intermittent links commonly encountered in *MANETs* (Chan, 2004).
Section 2 gives an overview of *association rule* and its application to social networks. Section 3 explains how data mining approaches are applied in *MANET* and introduces a new distributed algorithm, *MANET* Mining. Section 4 provides a detailed explanation of applying Association Rule Techniques to *MANET* traffic. Section 5 shows how Association Rule Mining Techniques are used on *MANET* traffic with a Step Threshold. Section 6 shows an important application of *MANET* Mining to key distribution. Section 7 concludes the chapter and draws some future research directions.

2. Data Mining: An overview of association rule mining technique

2.1 Association rule

This section presents a methodology known as association rule mining, useful for discovering interesting relationships hidden in huge data sets. Association rules have received lots of attention in *DM* due to their many applications in marketing, advertising, inventory control, and many other areas (Simovici & Djeraba, 2008). Association Rules can be derived using supervised and unsupervised processes (Joe, 2009).
Let $A = \{l_1, l_2, l_3, l_4, ..., l_m\}$ be a set of items. Let T be a set of transactions on a database. A transaction t is said to support an item l_i, if l_i is present in t. Moreover, t is said to support a subset of items $X \subseteq A$, if t supports each item l in X (Pujari, 2001).
$X \subseteq A$ is said to have a Support s in T, denoted by s(X), if s percent of transactions in T support X.
A subset X is said to be a Frequent Set (FS) in T with respect to σ (where σ is a user-specified minimum Support), if

$$s(X) \geq \sigma$$

FS is called Maximal Frequent Set (MFS) if no supper set of this set is FS. The following are important properties of MFS:

- Downward Closure: Any subset of FS is FS.
- Upward Closure: Any supper set of an infrequent set is an infrequent set.

Moreover, the set of all Maximal Frequent Sets ($MFSs$) is called maximum frequent set.
For a given database, an association rule is an expression of the form:

$$X \Longrightarrow Y$$

where X and Y are subsets of A. The intuitive meaning of such a rule is that a transaction of the database which contains X tends to contain Y.
Some used measures of rule interestingness are:

1. Confidence (τ): The association rule $X \Longrightarrow Y$ holds with confidence τ if τ% of transactions in T that supports X also supports Y.

2. Support (σ): The association rule $X \Longrightarrow Y$ has Support σ in the transaction set T if σ% of transactions in T support $X \cup Y$.

Association rules have another synonym, *market basket*. *Market basket* analysis (Association Rule Mining) is a research technique for retailers that is used to discover customer purchasing patterns (Post, 2005). In direct marketing, *DM* has been used extensively to identify potential customers for a new product (target selection) (Javaheri, 2007). Accordingly, accumulated data is analyzed to know the behavior of the customers.
The supermarket may be interested in identifying associations between item sets; for example, it may be interested to know how many of the customers who bought bread and cheese also bought butter (Simovici & Djeraba, 2008). Furthermore, nowadays a `market basket` is applied to e-commerce rather than supermarkets. For example, whenever customers shop an item online, they might read a recommendation after that "Customers who bought this item also bought ..." or "Buy these two items together and save ...".
Binary format can be used to represent *market basket*, each row is a transaction and each column is an attribute (item). An item is represented as a binary variable, if the item is present the value of the variable is one, otherwise its value is zero.
The problem of mining association rules can be decomposed into two subproblems (Agrawal & Shafer, 1996):

1. Find all set of items (itemsets) whose support is greater than the user-specified minimum Support (σ). Itemsets with minimum Support are called frequent sets (itemsets).
2. Use the frequent itemsets to generate the desired rules. The general idea is that if, say for example, $ABCD$ and AB are frequent itemsets, then we can determine if the rule $AB \Rightarrow CD$ holds by computing the ratio:

$$confidence = \frac{Support(\{ABCD\})}{Support(\{AB\})} \geq \tau$$

   Note that this rule has minimum support because ABCD is frequent.

Because of the multiplicity and variety of Association Rules Mining (*ARM*) techniques, Apriori algorithm is chosen and applied in this section as a de facto algorithm for mining association rules.

2.2 Apriori algorithm

The problem of deriving association rules from data was first formulated by Agrawal, Imielinski and Swami in 1993 and is called the *market-basket* problem (Agrawal et al., 1993). They introduced in their work the Apriori algorithm, which is the most commonly used association rule discovery algorithm that utilizes the frequent sets. This algorithm make use of the downward closure property. Algorithm 1 shows the pseudo-code of Apriori algorithm (Agrawal & Shafer, 1996; Yao et al., 2003).
One of the advantages of the method is that before reading the database at every level, it graciously prunes many of the sets which are unlikely to be frequent sets. Apriori algorithm has become a reference algorithm, and has been improved in several ways in terms of time complexity, the number of scans of the database, size of transaction, threshold and so forth.
Since association rules are derived from $MFSs$, the terms MFS and association rules are used interchangeably.

Algorithm 1 Apriori

```
Initialize: k := 1, C_1 = all the 1-itemsets;
read the traffic bit-matrix to count the Support of C_1 to determine L_1
while L_{k-1} ≠ φ do
    C_k = gen-candidate-itemsets with the given L_{k-1}
    prune(C_k)
end while
L_1 := {frequent 1-itmesets};
k := 2; //k represents the pass number
for all rows ∈ bit-matrix do
    increment the count of all candidates in C_k that are contained in r;
    L_k := All candidates in C_k with minimum Support;
    k := k + 1
end for
Answer L := ⋃_k L_k;
```

Association Rule	Confidence
{ budget resolution = no, MX-missile = no, aid to El Salvador = yes} ⟶ {Republican}	91.0%
{ budget resolution = yes, MX-missile = yes, aid to El Salvador = no} ⟶ {Democrat}	97.5%
{ crime = yes, right-to-sue = yes, physician fee freeze = yes} ⟶ {Republican}	93.5%
{ crime = no, right-to-sue = no, physician fee freeze = no} ⟶ {Democrat}	100%

Table 1. Association rules extracted from the 1984 US Congressional Voting Records.

2.3 Application of association rule mining technique to social networks

Market basket is used not only in supermarkets but also in social networks. For example, one of the hidden knowledge in social networks is mining criminal relationship (Fard & Ester, 2009).

Tan (Tan et al., 2006), gave a simple and clear example of a social network in small community and applied association analysis to United States congressional voting records. The data-set is maintained in University of California Irvine (*UCI*) machine learning repository and includes votes for each of the U.S. house of representatives congressmen on the 16 key votes, 1984 (Asuncion & Newman, 2007). Figures 1(a) and 1(b) show random and voting data respectively. Even though both figures look random, there are still underlying relationships/patterns in the data and these relationships can be revealed through *DM* techniques, for example, Apriori algorithm. As a result, table 1 shows some of the relationships/outcome obtained by applying Apriori algorithm on the voting data set. Notably, at confidence of 91%, the first association rule is derived, which says that most of the members who voted yes for "aid" to "El Salvador" and no for "budget resolution" and "*MX* missile" are Republicans; while at 97.5% another association rule is derived which says that those who voted no for "aid" to "El Salvador" and yes for "budget resolution" and "*MX* missile" are Democrats. Of course, by changing the confidence level new rules can be found.

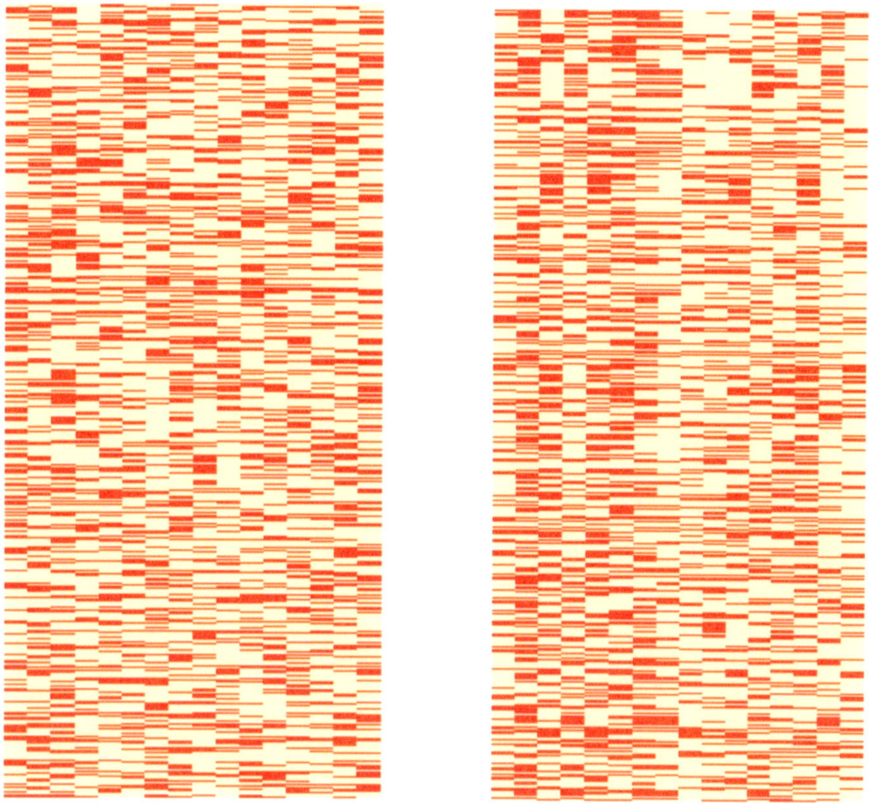

(a) Random data. (b) Voting data.

Fig. 1. Congress voting records

3. MANET mining

3.1 Introduction

The powerful methods for discovering knowledge from data go beyond the boundaries of traditional statistics, machine learning and database querying, to be applied in the field of *MANET*.

Section 3.2 compares *DM* and *MANET* Mining, while the rest of the subsections study how to mine traffic in a network called *MANET*, present a general distributed algorithm named *MANET* Mining with major concentration on *ARMs*, and explain the mathematical analysis of *MANET* itemsets.

3.2 Comparison of MANET and data mining: Visions of convergence

Data mining, in reference to transactions is similar, to a large extent, to mining packets in *MANET* in the following aspects (Jabas et al., 2008c):

1. Each transaction in data mining is a set of items (attributes). In case of *MANET*, the nodes are the attributes and the transaction is the transmission of one packet.
2. Data mining is applicable to a database with a large number of transactions. *MANET* mining is applicable to a traffic with a large number of packets.

3. The purpose of mining association rules in a database is to discover all rules that have Support and Confidence (predictability) greater than or equal to the user-specified minimum Support and minimum Confidence. In case of $MANET$, the rules represent the most likely patterns among the cooperating/routing nodes.
4. Each Frequent Set FS in data mining is equivalent to the common nodes of different paths in $MANET$.

3.3 MANET mining algorithm

$MANET$ consists of a finite collection of computational entities (nodes) communicating by means of messages (packets). Accordingly, $MANET$ communication algorithms are distributed by nature.

A distributed algorithm or protocol for given entities is a set of rules that specify the functionality of each entity. The collective but autonomous execution of those rules, possibly without any supervision or synchronization, must enable the entities to perform the desired task to solve the problem (Santoro, 2007). Algorithm 2 and its subprograms (algorithms 3 and 4) represent the pseudo codes of the new general distributed $MANET$ Mining protocol (algorithm).

The algorithm shows that mining techniques are applied with a Threshold to the mining procedure. Depending on the address information in a given packet, a node can be a source node, rely node or destination node. The bit-matrix is constructed from the bit-vectors of routed packets, which can be carrying data or acknowledgement.

For a $MANET$ with n nodes, the header of each packet should have a bit-vector of length n bit. Each bit represents inrormation about the participation of a node in the routing process of a packet. The default values of these vector's bits are zeros. Only the traversed node assigns a value of one to its entry in the packet's bit-vector, that means, the vector's values corresponding to untraversed nodes remain unchanged/unaltered at the default value of zero.

The algorithm does not overload the network by introducing new packets. Nodes do not add new traffic to $MANET$, but just capture the passing packets and assign a value of one to their corresponding entry in the packet's bit-vector (header). Since one bit is enough to represent one node, few bytes in the packet's header are required to represent the network. Each node is capable of extracting a sample of the traffic in $MANET$ to construct its own bit-matrix without any request from any other node.

The final status ϕ in the algorithm indicates that the algorithm works continuously to build the bit-matrix preparing it for mining at anytime later, i.e., the algorithm is proactive and the nodes do not reach a final status.

Spontaneously, whenever the threshold value is attained, the mining procedure (algorithm) is invoked. Despite the many types of thresholds, namely, time stamp, size of bit-matrix, number of differences between successive entries, columns, and so forth, two thresholds are studied in this chapter. The first is timestamp threshold, which is utilized in section 4 while the second, called the Step Threshold, and defined as the number of differences between two successive entries in the bit-matrix, is utilized in section 5.

3.4 Mathematical modelling and analysis of MANET itemsets

Hegland introduced a formal mathematical model to describe itemsets and associations in DM domain (Hegland, 2005). This section explains how Hegland's mathematical model can be applied in $MANET$.

Algorithm 2 MANET Mining

```
• Status Values: S = {SOURCE, RELAY, DESTINATION};
  S_INIT = {SOURCE, RELAY, DESTINATION};
  S_TERM = φ.
• Restrictions:
  R = {Connectivity}.
SOURCE
          Spontaneously
                    BEGIN
                              contribute to the bit-matrix;
                              ROUTE
                    END
          Spontaneously
                    BEGIN
                              MINE
                    END
          receiving (Acknowledgement)
                    BEGIN
                              contribute to the bit-mtrix;
                    END
RELAY
          Spontaneously
                    BEGIN
                              MINE
                    END
          Receiving (Data-Packet)
                    BEGIN
                              contribute to the bit-matrix;
                              ROUTE;
                    END
          Receiving (Acknowledgement)
                    BEGIN
                              contribute to the bit-matrix;
                              ROUTE;
                    END
DESTINATION
          Spontaneously
                    BEGIN
                              MINE
                    END
          Receiving (Data-Packet)
                    BEGIN
                              contribute to the bit-matrix;
                              ROUTE
                    END
```

Algorithm 3 Procedure MINE

BEGIN
if the user-specified threshold is met **then**
    apply mining algorithms (Apriori) to the bit-matrix;
end if
END

	N_0	N_1	N_2	...	N_j	...	N_{n-1}
$itemset_0$	$a_{0\,0}$	$a_{0\,1}$	$a_{0\,2}$	...	$a_{0\,j}$	...	$a_{0\,n-1}$
$itemset_1$	$a_{1\,0}$	$a_{1\,1}$	$a_{1\,2}$	...	$a_{1\,j}$	...	$a_{1\,n-1}$
$itemset_2$	$a_{2\,0}$	$a_{2\,1}$	$a_{2\,2}$	...	$a_{2\,j}$	...	$a_{2\,n-1}$
...	...	...	...	...	...	...	...
$itemset_i$	$a_{i\,0}$	$a_{i\,1}$	$a_{i\,2}$	...	$a_{i\,j}$	...	$a_{i\,n-1}$
...	...	...	...	...	...	...	...
$itemset_{m-1}$	$a_{m-1\,0}$	$a_{m-1\,1}$	$a_{m-1\,2}$	...	$a_{m-1\,j}$	...	$a_{m-1\,n-1}$

Table 2. The bit-matrix.

Consider a $MANET$ with n nodes, whose source node is N_s and destination node is N_d. Nodes are enumerated from N_0 to N_{n-1}. $MANET$ is applying some routing protocol, where each delivered packet carries an itemset. Itemsets (bit-vectors) are sets of strings of n binary numbers, where $a \in A := \{0,1\}^n$. In table 2, the value of the item j is set to one in the corresponding itemset iff the jth node contributed to the process of routing the corresponding packet, otherwise, the item's value remains at the default value of zero. The set of itemsets (bit vectors) forms a bit-matrix, where the jth column represents the node N_j and the ith row represents the ith itemset.
The nodes involved in routing are chosen randomly. Thus, the corresponding itemsets and bit-matrix $A \in \{0,1\}^{m,n}$are random, where m is the number of itemsets. The elements $a_{i\,j}$ are binary random variables.
Assume the probability distribution function $p :\rightarrow [0,1]$, where:

$$\sum_{a \in A} p(a) = 1$$

and $A = \{0,1\}^n$. The probability with distribution p is denoted by P and has:

$$P(A) = \sum_{a \in A} p(a)$$

Algorithm 4 Procedure ROUTE

BEGIN
set the node's bit-vector value to one
use the given routing algorithm
END

The data can be represented as an empirical distribution (Cumulative Distribution Function (CDF)) with:

$$P_{emp}(a) = \frac{1}{n}\sum_{i=1}^{n}\delta(a - a^{(i)})$$

where $\delta(a)$ is the indicator function; $\delta(0) = 1$ and $\delta(a) = 0$ if $a \neq 0$. For simplicity, the empty *market basket* is denoted by 0 instead of (0, ..., 0).
Mining of frequent itemsets means to find itemsets (bit-vectors) that occur frequently in the traffic. Accordingly, the itemsets are partially ordered with respect to the inclusion. In other words, $a \leq b$ if the set with representation a is a subset of the set with representation b or $a = b$. Now, the Support of an itemset a can be defined with the partial order as follows:

$$s(a) = P(c \mid a \leq c)$$

$s(a)$ is also called anticummulative distribution function of the probability P. The Support is function $s : A \to [0,1]$ and $s(0) = 1$. The Support is antimonotone, i.e., if $a \leq c$ then $p(a) \geq p(b)$. Previous equation can be reformulated in terms of $p(a)$ as:

$$s(a) = \sum_{c \geq a} p(c)$$

This is a linear system of equations which can be solved recursively using $s(e) = p(e)$, where $e = (1, ..., 1)$ is the maximum item set and:

$$p(a) = s(a) - \sum_{c \geq a} p(c)$$

That means the Support function $s(a)$ provides an alternative description of the probability measure P, which is equivalent to p.

3.5 The random itemsets of MANET nodes

Assume that the nodes that contribute to routing are chosen randomly, i.e., the items (bits) in a are chosen independently with probability p_0. This corresponds to routing protocol that randomly chooses nodes as packets move along the way from the source to the destination. Then, the distribution is:

$$p(a) = p_0^{|a|}(1 - p_0)^{n - |a|}$$

where $|a|$ is the number of bits (contributing nodes) in the itemset a and n is the total number of nodes in $MANET$. As any $c \geq a$ has at least all the bit sets which are sets in 'a' extracted for the Support

$$s(a) = p_0^{|a|}$$

and the frequent itemsets are those with at most the following number of items:

$$|a| \leq log(\sigma_0)/log(p_0)$$

This relation finds itemsets that have a specific Support and and the probability of choosing a node.
Assume that the items are chosen independently with different probabilities p_j, then the probability of choosing an itemset is:

$$p(a) = \prod_{j=1}^{n} p_j^{a_j}(1 - p_j)^{1 - a_j}$$

and

$$s(a) = \prod_{j=1}^{n} p_j^{a_j}$$

Using Zif's law:

$$p_j = \frac{\alpha}{j}$$

where α is a constant. This means that itemsets with few popular itemsets are most likely.

4. MANET mining: Mining temporal association rules (TARs)

4.1 Introduction

MANET is an autonomous system of mobile routers (and associated hosts) connected by wireless links, the union of which forms an arbitrary graph. The ability to specify the topology of a *MANET* could also prove to be crucial if limitations in the scalability and total capacity of *MANETs* are found to exist (Rashmi, 2009; Robinson, 2007).

This section proves that even though the topology is changing rapidly, i.e., the graph is not constant, there are still hidden relationships among the nodes. The association rule techniques are responsible for revealing these relationships, and seek to identify what items go together (Olson & Delen, 2008). The Correlation Ratio (*CR*), defined in the next subsection, measures the strength of the relationships.

Section 4.2 demonstrates how to apply the association rule technique periodically, every Δ second as a Threshold, to the bit-matrix constructed from *MANET* traffic. Accordingly, these rules are denoted in this chapter by Temporal Association Rules (*TARs*). Section 4.3 is concerned with simulation with different parameters and the results show that there are still some relationships among the nodes even though the topology is changing.

4.2 Demonstration on a simple MANET

This section shows the application of association rule techniques to *MANET* and explains the use of Apriori algorithm to mine *MANET* traffic (Jabas et al., 2008b). Consider *MANET* scenarios in Fig. 2 each representing a 15 node *MANET*, in which node N_1 is the source and node N_2 is the destination. For simplicity, a small number of packets is considered, i.e., not more than five packets in each of the scenarios. The initial path at time t_0, depicted in Fig. 2(a), is $<N_1, N_3, N_5, N_6, N_9, N_{14}, N_{13}, N_2>$ and the source sends 3 packets, after which the topology changes, in the sense that node N_{14} leaves and node N_8 enters the route leading to the second scenario in Fig. 2(b), in which the source sends 5 packets through the path $<N_1, N_3, N_5, N_6, N_9, N_8, N_{13}, N_2>$ at time t_1. The third scenario, in Fig 2(c), is derived from the second scenario as a result of change in topology, i.e., node N_9 leaves while node N_{14} and node N_{15} enter. 4 packets are sent through the new path $<N_1, N_3, N_5, N_6, N_{15}, N_8, N_{14}, N_{13}, N_2>$. Finally, the fourth scenario is derived from the third scenario by some change in topology as shown in Fig. 2(d) such that node N_5 leaves and node N_4 joins. 4 packets are sent through the path $<N_1, N_3, N_4, N_6, N_{15}, N_8, N_{14}, N_{13}, N_2>$.

The bit-matrix in both the source and destination corresponding to this transmission of packets is shown in table 3. Apriori algorithm is applied to the bit-matrix with support $\sigma = 70\ \%$. Shown below are the stepwise application of Apriori algorithm on the bit-matrix:

i:=1

$C_1 = \{\{N_1\}, \{N_2\}, \{N_3\}, \{N_4\}, \{N_5\}, \{N_6\}, \{N_7\}, \{N_8\}, \{N_9\}, \{N_{10}\}, \{N_{11}\}, \{N_{12}\}, \{N_{13}\}, \{N_{14}\}, \{N_{15}\}\}$

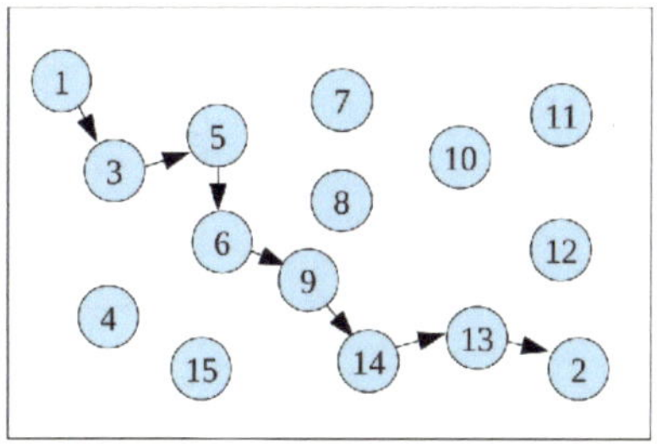

(a) *MANET* at time t_0.

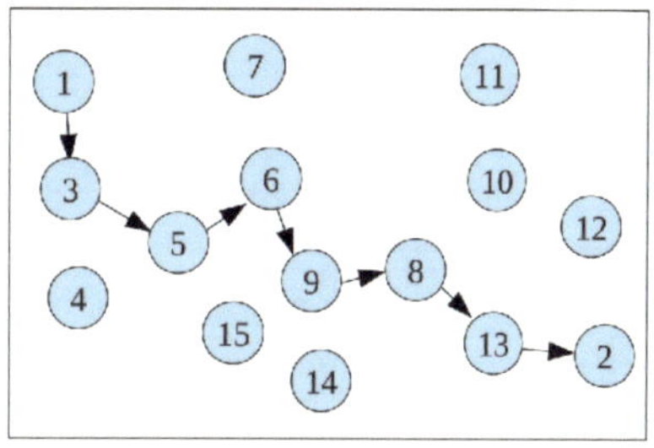

(b) *MANET* at time t_1.

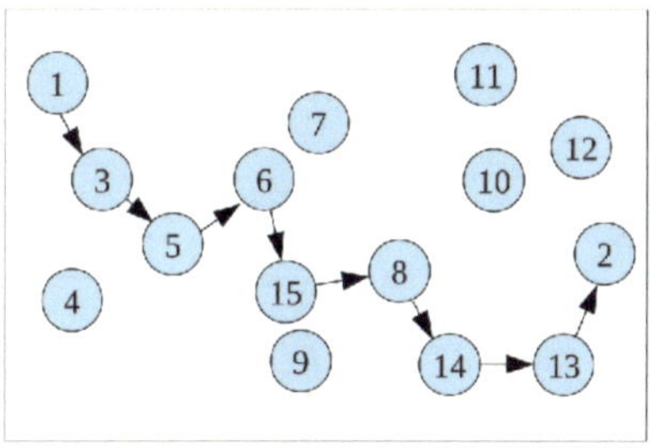

(c) *MANET* at time t_2.

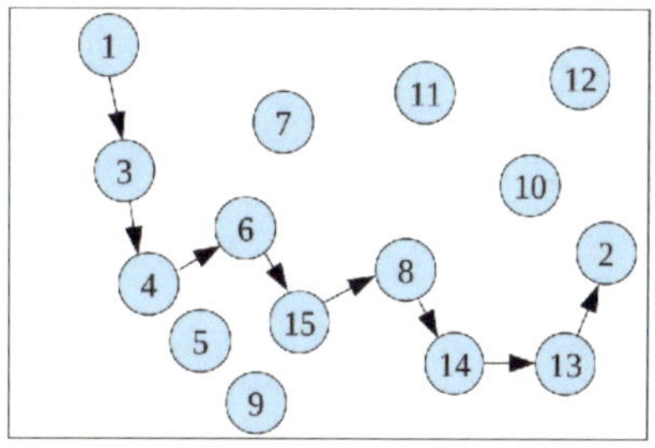

(d) *MANET* at time t_3.

Fig. 2. Simple *MANET* demonstration of TARs.

After pruning C_1 the set of frequent sets with one element is:
$L_1 = \{\{N_1\},\{N_2\},\{N_3\},\{N_5\},\{N_6\},\{N_8\},\{N_{13}\}\}$
i:=2
$C_2 = \{\{N_1,N_2\},\{N_1,N_3\},\{N_1,N_5\},\{N_1,N_6\},\{N_1,N_8\},\{N_1,N_{13}\},\{N_2,N_3\},\{N_2,N_5\},$
$\{N_2,N_6\},\{N_2,N_8\},\{N_2,N_{13}\},\{N_3,N_5\},\{N_3,N_6\},\{N_3,N_8\},\{N_3,N_{13}\},\{N_5,N_6\},\{N_5,N_8\},$
$\{N_5,N_{13}\},\{N_6,N_8\},\{N_6,N_{13}\},\{N_8,N_{13}\}\}$
After pruning C_2, the set of frequent sets with two elements is:
$L_2 = \{\{N_1,N_2\},\{N_1,N_3\},\{N_1,N_5\},\{N_1,N_6\},\{N_1,N_8\},\{N_1,N_{13}\},\{N_2,N_3\},\{N_2,N_5\},$
$\{N_2,N_6\},\{N_2,N_8\},\{N_2,N_{13}\},\{N_3,N_5\},\{N_3,N_6\},\{N_3,N_8\},\{N_3,N_{13}\},\{N_5,N_6\},\{N_5,N_{13}\},$
$\{N_6,N_8\},\{N_6,N_{13}\},\{N_8,N_{13}\}\}$
i:=3
$C_3 = \{\{N_1,N_2,N_3\}, \{N_1,N_2,N_5\}, \{...\}, ...\}$
After pruning C_3, the set of frequent sets with three elements is:
$L_3 = \{\{N_1,N_2,N_3\}, \{N_1,N_2,N_5\}, \{...\}, ...\}$
i:=4
$C_4 = \{\{N_1,N_2,N_3,N_5\}, \{...\}...\}$
After pruning C_4, the set of frequent sets with four elements is:
$L_4 = \{\{N_1,N_2,N_3,N_5\}, \{...\}, ...\}$
i:=5
$C_5 = \{...\}$
After pruning C_5, the set of frequent sets with five elements is:
$L_5 = \{...\}$
i:=6
$C_6 = \{ ...\}$
After pruning C_6, the set of frequent sets with six elements is:

No.	N_1	N_2	N_3	N_4	N_5	N_6	N_7	N_8	N_9	N_{10}	N_{11}	N_{12}	N_{13}	N_{14}	N_{15}
01	1	1	1	0	1	1	0	0	1	0	0	0	1	1	0
02	1	1	1	0	1	1	0	0	1	0	0	0	1	1	0
03	1	1	1	0	1	1	0	0	1	0	0	0	1	1	0
04	1	1	1	0	1	1	0	1	1	0	0	0	1	0	0
05	1	1	1	0	1	1	0	1	1	0	0	0	1	0	0
06	1	1	1	0	1	1	0	1	1	0	0	0	1	0	0
07	1	1	1	0	1	1	0	1	1	0	0	0	1	0	0
08	1	1	1	0	1	1	0	1	1	0	0	0	1	0	0
09	1	1	1	0	1	1	0	1	0	0	0	0	1	1	1
10	1	1	1	0	1	1	0	1	0	0	0	0	1	1	1
11	1	1	1	0	1	1	0	1	0	0	0	0	1	1	1
12	1	1	1	0	1	1	0	1	0	0	0	0	1	1	1
13	1	1	1	1	0	1	0	1	0	0	0	0	1	1	1
14	1	1	1	1	0	1	0	1	0	0	0	0	1	1	1
15	1	1	1	1	0	1	0	1	0	0	0	0	1	1	1
16	1	1	1	1	0	1	0	1	0	0	0	0	1	1	1

Table 3. The bit-matrix both in the source and the destination nodes

$L_6 = \{\{N_1, N_2, N_3, N_5, N_6, N_{13}\}, \{N_1, N_2, N_3, N_6, N_8, N_{13}\}\}$
i:=7
The maximum frequent set is:
$MFSs = L_6 = \{\{N_1, N_2, N_3, N_5, N_6, N_{13}\}, \{N_1, N_2, N_3, N_6, N_8, N_{13}\}\}$
The set of all frequent sets is:
$L = L_1 \cup L_2 \cup L_3 \cup L_4 \cup L_5 \cup L_6$
The maximum frequent set (MFS) gives the relationships (patterns) among the nodes for a set of entries regardless of the total number of nodes in these entries.
A new metric which defines the strength of relationship among nodes in the pattern is shown below:

$$Correlation\ Ratio\ (CR) = \frac{Average\ Size(MFSs)}{Length(route)}$$

Where:
Average Size ($MFSs$): refers to the average number of the nodes in the pattern extracted from the bit-matrix for a specific period of time as a Threshold.
Length ($route$): refers to the total number of nodes in the bit-matrix that are routing for the same period of time.
Higher CR indicates better correlation among nodes. CR may approach 1 for wireless static topology, which means that the entries in the bit-matrix are participating in the routing process.
By mining the whole bit-matrix shown in Table 3 with Support $\sigma = 70$ %, $MFS = \{\{N_1, N_2, N_3, N_5, N_6, N_{13}\}, \{N_1, N_2, N_3, N_6, N_8, N_{13}\}\}$ is obtained. Notably, the average number of nodes in $MFSs$ is 6 and the number of nodes involved in routing (active nodes) in this bit-matrix is 11, i.e., columns that contain at least a one value (unshaded columns in Table 3), therefore,

$$CR\ (with\ \sigma = 70\ \%) = \frac{6}{11}$$

Parameter	Value
Number of the nodes	100
Routing protocol	*AODV*, *DSDV*, *DSR*
Mobility model	Random way point
Pause time	1 *s*
Radio transmission range	250 *m*
Channel capacity	1 *mbps*
Data flow	*CBR*, *FTP*
Data packet size	512 bytes
Node placement	random
Terrain area	$1500 \times 1500\ m^2$
Simulation time	120 *s* for *CBR* with *AODV*, *DSDV* and *DSR*, 200 *s* for *TCP* with *AODV* and *DSR* 400 *s* for *TCP* with *DSDV*
Propagation model	Two Ray Ground
Node Mobility Speed (NMS)	5, 10, ..., 50 m/s with *AODV* and *DSR* 1, 2, ..., 10 m/s with *DSDV*

Table 4. Simulation parameters.

In this example, for simplicity's sake, few nodes and less network traffic is considered. Practically, more nodes may communicate for longer time resulting into a huge number of heterogeneous packets (traffic). The mining of such traffic for a long period of time results into a low *CR*. This is because, after some time, the topology completely changes with reference to the initial topology. In other words, the number of common nodes decreases with time and consequently, the *CR* decreases. As a prerequisite, the rate of mining Δ should be in consonance with the rate of changing of topology.

4.3 Simulation and results

This section shows how Apriori algorithm is applied to extract *MFSs* from different types of *MANET* traffic. The simulation is performed by *NS2* (ns2, 2009; Greis, 2007; Fall, 2007). Parameters used in the simulator are summarized in Table 4. Hundred nodes are distributed randomly in the simulation area of $1500 \times 1500\ m^2$ and with a 250 *m* transmission range for each node. The Propagation model of the signal is "Two Ray Ground". The channel capacity is 1 *mbps*. The random mobility mode of the nodes is generated by the *CMU's* node-movement utility "setdest" with different Node Mobility Speeds (*NMS*) within the range of 5-50 m/s. The nodes do not move through out the simulation time, i.e., they stop according to a constant pause time parameter which lasts for one second. The packet size is 512 bytes.

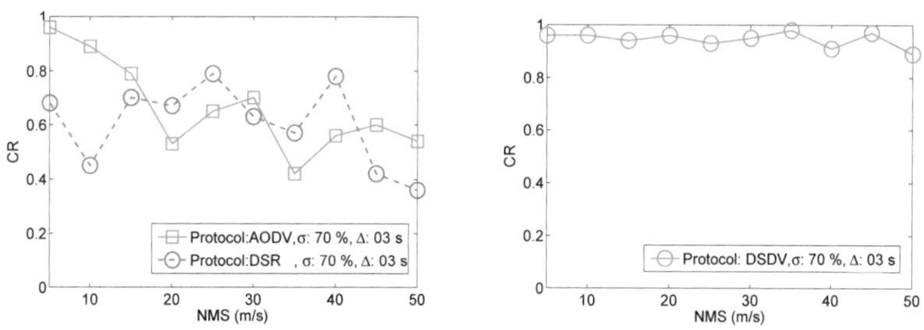

(a) Data traffic. (b) Data traffic.

Fig. 3. The variation of CR with NMS for connection-less traffic.

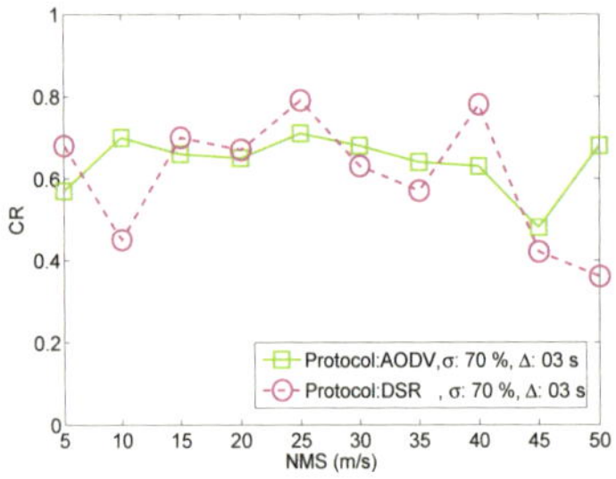

(a) Data traffic.

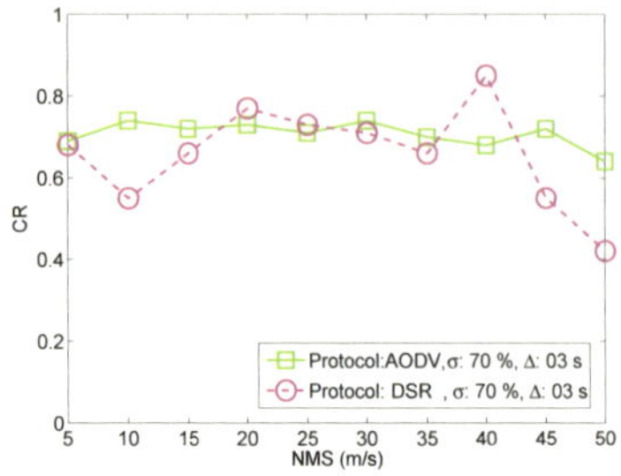

(b) Acknowledgement traffic.

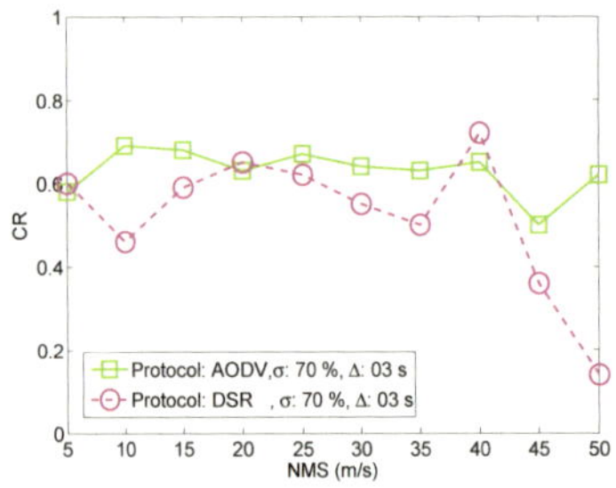

(c) Both acknowledgement and data traffic.

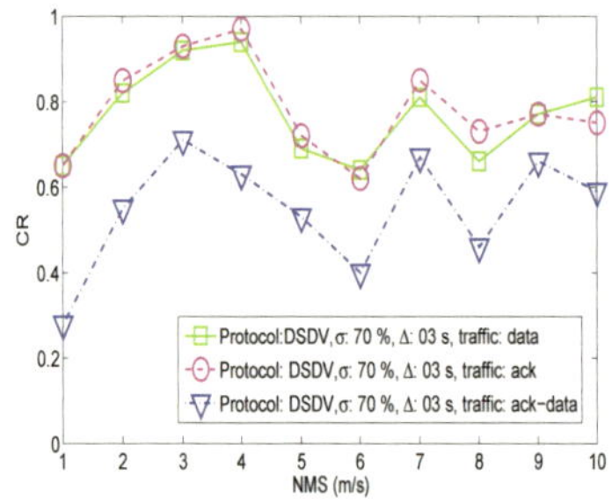

(d) All types of traffic.

Fig. 4. The effect of increasing NMS on CR for connection oriented traffic.

Three standard routing protocols are used in the simulation to deliver the packets from the source to the destination. Two of them are reactive: Ad hoc On-demand Distance Vector (*AODV*) and Dynamic Source Routing (*DSR*); and one is proactive, Destination Sequenced Distance Vector (*DSDV*).

With reactive routing protocol the simulation time of the second application lasts for 200 *s*, while with proactive protocol, namely *DSDV*, the simulation lasts for 400 *s*. The *NMS* varies discretely by one unit between 1 m/s and 10 m/s with *DSDV*, while for the other two protocols the *NMS* varies discretely in five unit step between 5 m/s and 50 m/s.

The *CR* metric defined in previous subsection is applied to different scenarios generated in this subsection to evaluate the strength of the hidden relationships (*MFS* patterns) among the nodes.

Two applications are chosen. The first, the Constant Bit Rate (*CBR*) application is simulated along with the connection-less transmission protocol/User Datagram Protocol (*UDP*) for 120 *s*. Figures 3(a) and 3(b) show how *CR* varies with *NMS* for *AODV*, *DSR* and *DSDV* protocols with connection-less traffic. The second, the File Transfer Protocol (*FTP*) is simulated along with connection-oriented Transport Control Protocol (*TCP*).

Figures 4(a), 4(b) and 4(c) show the behavior of *CR* as *NMS* varies for *AODV* and *DSR* for data, acknowledgment and both types of packets. Similarly, Fig. 4(d) analyzes the *CR* for the connection-oriented traffic of *DSDV* routing protocol for the three types of traffic packets.

It is evident from the above *CR*/*NMS* graphs that there are relatively good relationships (*TARs*) among the nodes, and therefore, *MANET* traffic is a raw material for mining and for revealing these relationships. Despite the high *NMS* of nodes, up to 50 m/s for *AODV* and *DSR* and 10 m/s for *DSDV*, there are still good relationships among the routing nodes.

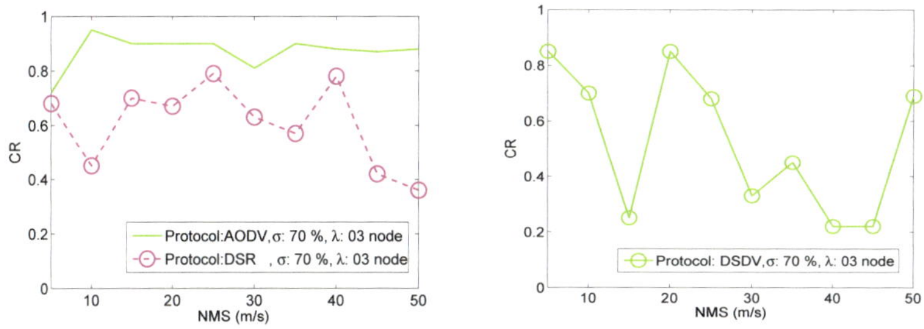

(a) Data traffic. (b) Data traffic.

Fig. 5. The variation of CR with NMS for connection-less traffic.

These relationships can be interpreted as topologies representing the routing nodes for a small period of time.

5. MANET mining: Mining step association rules (SARs)

5.1 Introduction

Section 4 explains how Temporal Association Rules (*TARs*) are mined from the *MANET* traffic, extracted and mined at a rate of Δ. In this section, a new threshold is imposed on the mining process to monitor the difference (change) between successive entries in the bit-matrix. This difference is denoted by Step and the mined rules are denoted, in this chapter, by Step Association Rules (*SARs*). The length of the Maximal Frequent Set (*MFS*) depends on the

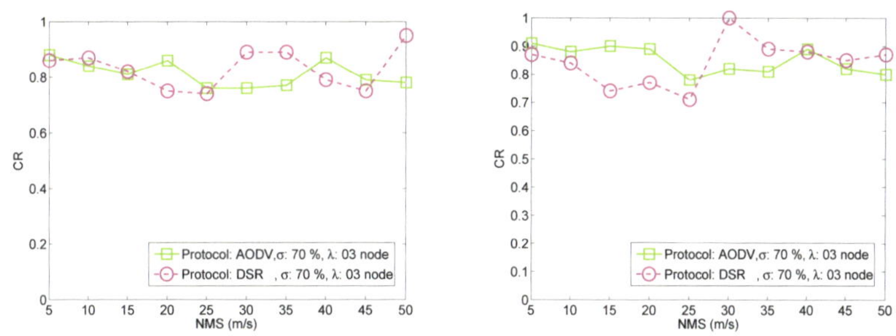

(a) Data traffic. (b) Acknowledgement traffic.

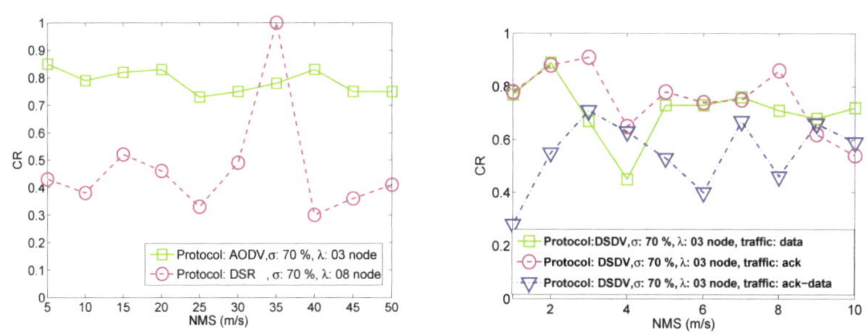

(c) Both acknowledgement and data traffic. (d) All types of traffic.

Fig. 6. The variation of CR with NMS for connection oriented traffic.

similarity among the entries (bit-vectors) in the bit-matrix. Similar bit-vectors leads to longer $MFSs$ and higher CR. In fact, longer $MFSs$ mean more packets are carried in the same routes. $TARs$ do not support sudden changes in the $MANET$ topology (routes), while $SARs$ support and detect the modification in the topology.
An explanation of how to mine $SARs$ is shown in section 5.2, while the application of SAR techniques to $MANET$ traffic and illustration of the results of imposing the Step Threshold λ on the mining procedures is demonstrated in section 5.3.

5.2 Demonstration using a simple MANET

This section explains how SAR mining is executed in $MANET$ (Jabas et al., 2008a). Whenever there is a change in route followed by the packets a corresponding change is reflected in the bit-matrix; for example, in reference to table 3, there are two bit changes between the third entry and the fourth because of change in route/topology .
This demonstration involves three steps. First, the bit-matrix is mined without imposing a Step Threshold. Second, it is splitted into two bit-matrices according to the Step Threshold $\lambda > 2$ nodes. Finally, the new bit-matrices are mined separately.
Following these steps, the bit-matrix in table 3 is used to obtain the results. First, the bit-matrix is mined with Support $\sigma = 70\ \%$, and the resultant MFS ($SARs$) obtained are:
$L_6 = MFSs = \{\{N_1, N_2, N_3, N_5, N_6, N_{13}\}, \{N_1, N_2, N_3, N_6, N_8, N_{13}\}\}$.
The average number of nodes in $MFSs$ is 6 and the number of nodes involved in routing in this bit-matrix is 11 (i.e. columns that contain at least one "1"value). $CorrelationRatio\ (with\ \sigma = 70\ \%) = \frac{6}{11}$.
Second, Step Threshold $\lambda > 2$ is applied. Accordingly, the first 8 entries saved in the first sub-bit-matrix and the next remaining 8 entries are saved in the second sub-bit-matrix. By mining these sub-bit-matrices with the same Support as in the first step, the following $MFSs$ and their corresponding CRs are obtained as shown below:
$MFSs_1 = L_7 = \{\{N_1, N_2, N_3, N_5, N_6, N_9, N_{13}\}\}$
$MFSs_2 = L_8 = \{\{N_1, N_2, N_3, N_6, N_8, N_{13}, N_{14}, N_{15}\}\}$
$CR_1(\sigma = 70\ \%) = \frac{7}{9}$
$CR_2(\sigma = 70\ \%) = \frac{8}{10}$
Note that CR_1 and CR_2 are higher than CR obtained in the first step. This means that application of Step Threshold leads to stronger CR.

5.3 Simulation and results

This section applies the same simulation scenarios and parameters used in section 4 to show "Step Threshold" effects. Simulation parameters are summarized in Table 4.
Again, the two transport protocols, connection-less and connection-oriented, along with proactive and reactive routing protocols are studied. The same metric, CR, defined in 4.2 is utilized to evaluate the strength of the relationship. The two parameters, the Step Threshold λ and the Support s, with their effect on $MANET$ with different routing protocol are studied in this section.
Figures 5(a) and 5(b) show the effect of increasing NMS on CR with Step Threshold. We conclude that there are relatively strong relationship among the nodes in $AODV$ and DSR protocols as depicted in figure 5(a). However, in 5(b), CR is fluctuating indicating that relationship is not reliable for $DSDV$. Figures 6(a), 6(b), 6(c) and 6(d) show the effect of increasing NMS on CR with Step Threshold, and therefore, it is possible to mine the $MANET$ traffic.

Generally, step threshold improves CR or the strength of the relationships among the nodes.

6. Case Study: Key management in MANET

6.1 Introduction

Like in many distributed systems, security in $MANET$ widely depends on the use of a secure key management mechanisms. Most of the cryptographic mechanism in $MANET$ depends on a key management infrastructure (Ghoreishi & Analoui, 2009).

A key is a piece of input information for cryptographic algorithm. Key disclosure is tantamount to revealing the encrypted information itself, therefore, there is need for Key Encryption Key (*KEK*) algorithm applied at local host level (Fumy & Landrock, 1993). In view of this, specific key management systems have been developed to fit the characteristics of $MANET$.

Two principles for ad hoc key management are identified (Yi & Kravets, 2004):

1. The first principle is the node participation principle which states that "A key management framework for ad hoc network should rely on a large number of nodes for availability purposes, but on small group of nodes for security purposes ".
2. The second principle, requires the services of TTP principle and states that "A key management framework should use a TTP to improve the quality of authentication of the framework ".

Luo presented Ubiquitous and Robust access control Solution (*URSA*) for *MANETs* (Luo et al., 2003). $URSA$ implements ticket certification services through multiple-node consensus and fully localized instantiation, and uses tickets to identify and grant network access to cooperative nodes. The merits of this protocol are high efficiency, secure local communication and system availability. The demerits include extremely large threshold compare to the network degree and requirement for off-line configuration before accessing the network.

Seung, (Yi & Kravets, 2003), proposed a new key management framework, MObile Certificate Authority (*MOCA*) for *MANETs*, where the certificate service is distributed to n *MOCA* nodes that are most secure. *MOCA* scheme uses threshold cryptography, which is an application of secret sharing. The concept of secret sharing is that it is mathematically possible to divide up a secret to n pieces in such way that anybody who requires the full secret can collect any k pieces out of those n *MOCA* to reconstruct the full secret. k becomes the threshold needed to reconstruct the secret.

Later, Seung presented a composite key management scheme that uses virtual CA and certificate chaining simultaneously in a single ad hoc network (Yi & Kravets, 2004), thereby, combining the central trust with the fully distributed trust models.

Sections 4 and 5 explain how ARM techniques can be applied to $MANET$ traffic to extract hidden patterns. This section shows an important application of these patterns in the security field. A new method for key distribution in $MANET$ has been developed to mine the traffic moving in $MANET$ in a distributed manner to obtain *MFSs* that are used as tokens (keys) (Jabas et al., 2010).

Section 6.2 explains new security framework that relies on *MFSs*. Section 6.3 explains the key revocation and renewal issues for the new framework. Section 6.4 gives the mathematical model of the new framework and analyzes its effectiveness in $MANET$. Section 6.5 discusses the experimental analysis of the the new framework. Section 6.6 lists the features that make the new scheme outstanding.

6.2 Key distribution through traffic mining (KDTM)

A group of nodes, the routing nodes, contribute to the bit-vector of the routed packet, which in turn is used to build the bit-matrices in both the source and the destination nodes. Both end nodes apply Apriori algorithm to these bit-matrices, and interestingly, the same $MFSs$ (patterns) are obtained at only these two nodes.

For a $MANET$ with n nodes, a bit-vector of size n is required for each packet routed. Before the source sends a packet to the destination, the entries of the bit-vector are initialized to zero and the bit-vector is attached to the header of the packet. As the packet passes through the routing nodes, the corresponding entries in the bit-vector are set to one. When the packet reaches the destination node the bit-vector is extracted from the packet and attached to the acknowledgement packet that is sent back to the source node, and also this extracted bit-vector is stored in the node's bit-matrix. The size of the bit-matrix, i.e., the number of entries, is a function of packets received and extracted.

Now, the source and destination nodes mine the bit-matrices to extract their $MFSs$, and both nodes must obtain the same result, i.e., the same $MFSs$, used as common tokens between the source and the destination.

The advantage of MFS mining algorithms is that they are tolerant, in the sense that if the bit-matrix is changed slightly, intentionally or by accident, either in the source or in the destination or in both, the same $MFSs$ are obtained in the source and destination nodes. At the same time, if any node other than the source or destination tries to build and mine its own bit-matrix, different $MFSs$ are obtained. This difference in $MFSs$ is brought about by the fact that mining is applied to different bit-matrices.

In addition to the end-nodes, any routing node may also extract the bit-vectors of passing packets for an interval of time and may cache them in its bit-matrix. This means that the routing node does not need any request or permission from any other node to perform these activities. $KDTM$ does not increase the network traffic by inducing control packets as is the case with other key distribution schemes.

A summary of $KDTM$ steps are shown below:

- Nodes synchronization: Nodes should be at least loosely synchronized. There are several distributed algorithms for synchronization (Lamport, 1987; Simons et al., 2006)
- The sender node attaches a bit-vector to each sent packet.
- The receiver deattaches the bit-vector of each received packet and firstly, caches it to its bit-matrix; secondly, attaches it to the corresponding acknowledgement to be sent to the source/sender.
- The sender deattaches the bit-vector of each received acknowledgement and add it to the corresponding bit-matrix of the sending node.
- Later, the two parties, the sender/the receiver, trigger the mining algorithm to mine bit-matrices using either Mining Rate or Step as a threshold. The resultant MFS obtained, is used as the secret key. Full details of the steps can be obtained from algorithm 2.

Blackhole attacks do not affect the new $KDTM$. Blackholes either dump the data packets on their way to the destination or dumps the acknowledgement on their way to their source. In the first scenario, the bit-matrices at the end nodes are not affected and so is the MFS (key) obtained from them. In the second scenario, because the source does not receive the ACK the source retransmit the packet. At the destination, since the retransmitted packet has the same id as the previous packet a swap is performed in the corresponding entries in the bit-matrix.

This swap is equivalent to dropping one of the two similar bit-vectors in the bit-matrix. Since there is utterly no difference between the source and the destination matrices the same *MFSs* (key) are obtained.
Wormhole attacks do not affect *KDTM*. Wormhole leaks routed packets at a node to the outside world. Still the *MFSs* built from the leaked packets is not the same as that of the end nodes because not all traffic from the source to the destination pass through the same route.
KDTM is immune to Man-in-Middle (*MIM*) attacks. In passive *MIM* attack, the malicious node just builds and mines its bit-matrix, however, the resultant *MFS* obtained is different from that of the end nodes. Still in this situation, the *MFS* obtained at the end nodes is not affected because *MIM* does not alter the bit-vector of both passing data packets and passing *ACK*. Two scenarios are observed in active *MIM* attacks. The first scenario, the malicious node forges/modifies the bit-vector of the passing data packets. This means the same alteration is reflected in both the bit-matrices of the end nodes. In the second scenario, the *MIM* alters the bit-vector of the passing *ACK*. This means the same change is induced in the source bit-matrix but not in the destination bit-matrix. The difference induced between source and destination bit-matrices is insufficient, because a small number of *ACK* pass through the same route; and therefore, the same *MFS* obtained at the end nodes.
Notably, active *MIM* can be identified through checking of bit-vector by routing nodes before sending it to the next node; and if any node discover that its bit (or the bits of her neighbors who have not received the packet) is changed, then this node should send a warning message to the other nodes in the *MANET* that there is an active *MIM* in the network.
Simulation results show that *KDTM* is tolerant, in that adding/deleting bit-vectors randomly to/from bit matrix up to 30 % does not change the resultant *MFS*. Further more, *KDTM* allows concatenating several *MFSs* or keys in a bid to develop a stronger key.
KDTM may applies Nitin's watch dog and Pathrater concepts to eliminate malicious nodes in the transmission range of the end nodes so that the extracted key is not compromised (Kyasanur & Vaidya, 2003).
KDTM is a new cross layer key distribution scheme, which extracts *MFS* from network layer to be used in other layers, for instance, the application layer.

6.3 Key revocation

Key disclosure is very frequent in *MANET*. There is no guarantee that the route between the communicating nodes is free of malicious nodes.
In contrast to using static long-term keys, dynamic short-term cryptographic keys can be used to minimize the availability of ciphertext, encrypted with the same key, and therefore, making it difficult to compromise the key (Menezes et al., 1996). Accordingly, key renewal is compulsory to reduce the amount of disclosed packets in case the key is compromised. In the new method, key renewal, not affected by any other factor and is very simple because the key is mined as long as there is traffic, may be done at any time.
Key can be changed periodically between the two communicating nodes. The parameters such as Support σ, Mining Rate Δ and step threshold λ may be changed to mislead the *MIM*. This is somehow similar to frequency hopping in wireless communication used for security purpose.
The next two sections analyze mathematically and experimentally the new framework.

$$
\begin{array}{ll}
n=0 & C(0,0) \\
n=1 & C(1,0)\quad C(1,1) \\
n=2 & C(2,0)\quad C(2,1)\quad C(2,2) \\
n=3 & C(3,0)\quad C(3,1)\quad C(3,2)\quad C(3,2) \\
\ldots & \ldots\quad \ldots\quad \ldots\quad \ldots\quad \ldots \\
n=n & C(n,0)\quad C(n,1)\quad \ldots\quad \ldots\quad C(n,i-1)\quad C(n,i)
\end{array}
$$

Fig. 7. Pascal triangle

6.4 Mathematical analysis of the new framework

One of the main features of Apriori algorithm is tolerance, in the sense that arbitrarily adding some rows (bit-vectors) with random values to the data set (bit-matrix) does not affect the end result (outcome), and therefore, the same *MFS* is obtained. Further more, deleting some rows (bit-vectors) randomly from a data set (bit-matrix), does not change the output of the algorithm. At the same time, it is very difficult to guess the output of the algorithm without acquiring the whole bit-matrix.

The algorithm can be applied on three different types of traffic. The first type is the data traffic. The algorithm extracts the *MFSs* from the bit-matrix of bit-vectors of data packets. The second type is the acknowledgement traffic and the third type is a mixture of data and acknowledgement packets.

Consider a *MANET* with a set of *n* nodes. The output of Apriori algorithm is *MFSs* in an increasing order and without repetition. The number of ways to form *MFS* of length *i* is:

$$C(n,i) \tag{1}$$

i \ n	100	150	200	250	300	350	400	450	500	550	600	...
03	2^{18}	2^{20}	2^{21}	2^{22}	2^{23}	2^{23}	2^{24}	2^{24}	2^{25}	2^{25}	2^{25}	...
04	2^{23}	2^{25}	2^{27}	2^{28}	2^{29}	2^{30}	2^{31}	2^{31}	2^{32}	2^{32}	2^{32}	...
05	2^{27}	2^{30}	2^{32}	2^{33}	2^{35}	2^{36}	2^{37}	2^{38}	2^{38}	2^{39}	2^{39}	...
06	2^{31}	2^{35}	2^{37}	2^{39}	2^{40}	2^{42}	2^{43}	2^{44}	2^{45}	2^{46}	2^{46}	...
07	2^{35}	2^{39}	2^{42}	2^{44}	2^{46}	2^{47}	2^{49}	2^{50}	2^{51}	2^{52}	2^{52}	...
08	2^{39}	2^{43}	2^{47}	2^{49}	2^{51}	2^{53}	2^{54}	2^{56}	2^{57}	2^{58}	2^{58}	...
09	2^{42}	2^{47}	2^{51}	2^{54}	2^{56}	2^{58}	2^{60}	2^{61}	2^{63}	2^{64}	2^{64}	...
10	2^{46}	2^{51}	2^{55}	2^{59}	2^{61}	2^{63}	2^{65}	2^{67}	2^{68}	2^{70}	2^{70}	...
11	2^{49}	2^{55}	2^{60}	2^{63}	2^{66}	2^{68}	2^{71}	2^{72}	2^{74}	2^{76}	2^{76}	...
12	2^{52}	2^{59}	2^{64}	2^{68}	2^{71}	2^{73}	2^{76}	2^{78}	2^{79}	2^{81}	2^{82}	...
13	2^{55}	2^{63}	2^{68}	2^{72}	2^{75}	2^{78}	2^{81}	2^{83}	2^{85}	2^{87}	2^{87}	...
14	2^{58}	2^{73}	2^{72}	2^{76}	2^{80}	2^{83}	2^{85}	2^{88}	2^{90}	2^{92}	2^{93}	...
15	2^{61}	2^{76}	2^{76}	2^{80}	2^{84}	2^{87}	2^{90}	2^{93}	2^{95}	2^{97}	2^{98}	...
16	...	...	...	...	...	...	...	...	...	...	...	...

Higher Security (→, increasing n) / Higher Security (↓, increasing i)

Table 5. A combinatoric relationship (C(n, i)) between n and i, where $n \equiv$ number of nodes and $i \equiv$ length of MFS.

Accordingly, all the possible ways to form an MFS of variable length i is:

$$C(n,2) + C(n,3) + ... + C(n,i) + ... + C(n,n-1) + C(n,n) \quad (where\ 2 \leq i \leq n) \tag{2}$$

See figure 7, the sum of the nth row of Pascal triangle is given by (Mott et al., 1992):

$$C(n,0) + C(n,1) + C(n,2) + \ldots + C(n,i) + ... + C(n,n-1) + C(n,n) = 2^n \tag{3}$$

From 2 and 3, the total number of ways is:

$$C(n,2) + \ldots + C(n,i) + ... + C(n,n-1) + C(n,n) = 2^n - (n+1) \tag{4}$$

If $i = 2$, then the source and the destination are neighbors, that means no intermediate nodes. If $i = n$ then the topology is chained.
Equation 4 assumes that the MFS may contain any number of nodes not exceeding n. In fact, this may be correct in one case only, a chain network topology. For example, queue of soldiers following their commander.
The number of routing nodes related to several factors, namely the routing protocol, sending/receiving range, and so on.

6.5 Experimental analysis of the new framework

In this section, the length of $MFSs$ that are used as tokens (keys), is measured experimentally. The $NS2$ simulator is utilized to generate different scenarios. Same parameters that are used in sections 4and 5, and listed in table 4, are used in this section except for the density of nodes. In reference to the density of nodes in $MANET$, Royer (Royer et al., 2001) shows that the optimum number of neighbors, for 0 m/s mobility or stationary nodes, is around seven or eight per node. This number differs only slightly from what Kleinrock proved for a stationary network (Kleinrock & Silvester, 1978). The density of nodes in wireless network is given by:

$$Density\ (8\ for\ optimal) = n * (\pi * R^2)/(X * Y)$$

where R is the radio transmission range of the node; X and Y are the dimensions of the terrain area, whose area is defined by product $X * Y$.
Tables 5 and 6 show that the bigger the size of MFS, the safer or more secure is the key obtained. In reference to table 6, the evaluation of average size of MFS eliminates short distances, i.e., distances less than five nodes for $AODV$ and DSR protocols.
For example, the average length of the key (MFS) is $i = 15$, which corresponds to the strength of the key of $C(300,15) = 2^{84}$, using the following parameters for simulation: NMS = 10 m/s; mining rate Δ=5 s; number of nodes = 300; terrain area = 2700×2700 m^2; Support σ=40 %; routing protocol is DSR; and data traffic.

Speed (m/s)	Time (s)	Nodes (#)	Area (m × m)	Protocol	The Average Size of MFS (token) Support = 40% Data	Ack	Ack-Data	Support = 50% Data	Ack	Ack-Data	Support = 60% Data	Ack	Ack-Data	Support = 70 % Data	Ack	Ack-Data
10	02	200	2200 X 2200	AODV	8	8	8	7	8	7	7	7	7	6	6	6
				DSR	11	11	11	10	10	10	10	7	7	10	6	6
		250	2400 X 2400	AODV	7	8	8	7	7	6	6	6	5	5	5	5
				DSR	12	12	12	12	12	10	10	11	8	10	10	6
	05	300	2700 X 2700	AODV	7	7	7	6	7	7	6	6	6	5	5	5
				DSR	15	15	13	14	13	13	14	13	12	12	11	12
		350	3000 X 3000	AODV	7	7	8	7	6	7	6	6	6	5	5	5
				DSR	11	12	11	11	10	9	10	9	8	9	6	8
		400	3200 X 3200	AODV	13	12	13	13	13	12	13	12	11	12	12	11
				DSR	15	15	15	15	15	15	15	15	15	15	15	14
		450	3500 X 3500	AODV	14	14	13	13	13	13	13	13	13	12	12	11
				DSR	17	15	14	17	14	13	16	14	13	14	13	11
01	02	100	1500 X 1500	DSDV	5	5	5	5	5	5	5	5	4	5	5	3
02					6	5	6	6	5	6	6	5	3	6	5	3
03					4	4	4	4	4	4	4	4	3	4	4	2
04					5	5	5	5	5	4	5	5	3	5	4	2
05					5	4	5	4	4	4	4	4	3	4	4	2

Table 6. The average length of MFS.

6.6 Outstanding features of the new Scheme

Several features make the new scheme more effective, more flexible, more tolerant and more secure than the present key distribution schemes in *MANET*. These features include:

- Robustness: The protocol is flexible and works in all circumstances, In other words, the absence of any number of nodes in the network topology at any time does not affect the the new protocol. All nodes in other schemes, such as schemes proposed by (Becker et al., 1998; Burmester & Desmedt, 1994; Kim et al., 2001), should be online before the key establishment process is completed (Chan, 2004).
- Transparency: The new scheme is transparent and works in all scalable routing protocols.
- Packet Size Independence: The new security protocol is independent of the packet size and type. In other words, it operates on all types of traffics, such as data, acknowledgement and control.
- Key Revocation and Renewal: The key can be renewed or removed any time even before its expiry time. These activities reinforce the security of the key.
- Overhead at Intermediate Nodes: The new scheme has low overhead on intermediate nodes, achieved through eliminating cryptographical checking of packets at intermediate nodes. The present schemes which use public key cryptography have high overhead on intermediate nodes.
- Scalability: The new scheme allows the number of nodes to be adjusted. Notably, the bigger the number of nodes in the network the bigger the number of ways to choose *MFSs* and the higher the security.
- Time and Space Complexities: Experimental results of the new protocol show that the time-complexity of the protocol for *MANETs* is of second order. These complexities depend

directly on the number of node (*MANET* size), the distance (in terms of number of nodes) between the communicating nodes, and the speed of *ARM* algorithms used. The space complexity is Sizeof(bit-vector) * Numberof (bit-vectors), where bit-vectors is equivalent to the number of contributing packets.

- Message Complexity: The new scheme has a message complexity of zero for all routing protocols. For source routing protocols such as *DRS* , which need not attach the bit-vector at all because each data packet has its route; still the message complexity is zero. Even for other protocols the complexity is zero because the bit-vector is attached to packets, and therefore, no security-dedicated packets are sent.
- Fault Tolerance: The failure of a number of nodes does not affect the new protocol because the same bit-entries are dropped from all bit-vectors.
- Adjustability: The new scheme is adjustable. For instance, Apriori is tunable through the Support parameter of *MFS*, size of bit-matrix and bit-vector extraction time. It is not necessary to attache bit-vector to each packet.

7. Conclusion and future research directions

KDTM, a cross layer scheme, shows that *MANET* traffic in the third layer is raw material that can be mined and utilized in other layers. In addition, the scheme shows how to collect dynamic data from complex and chaotic *MANET* with large population of mobile nodes and convert it into knowledge. The algorithm mines the *MFS* patterns through *ARM* technique employing two methods *TAR* and *SAR* mining.
The new concepts generated by *KDTM* and this chapter as a whole can be extended in several ways. Described below are some of the possible enhancements and extensions:

- Security Enhancement: *MANET* mining techniques can be used in identifying malfunctioning or blackholes or compromised nodes in *MANETs* through analyzing the *MFSs*. Such nodes, if identified by a number of other nodes in *MANET*, are discarded/excluded from the list of trusted nodes.
- Maximizing the Network Life Span: Energy conservation is of paramount importance in *MANET*, therefore, uniform energy consumption of nodes increases considerably the lifetime of the network. *MFS* can be used to identify active and dormant nodes. Dormant nodes in *MANET* increase the workload on active nodes and thereby decreasing their lifespan. It is therefore evident that decreasing the number of dormant nodes translates into increasing the life span of the *MANET*. Accordingly, *MFSs* may be considered as a life span metric.
- Load Balancing: Heavily-loaded nodes may become a bottleneck that lowers the network performances through congestion and longer time delays. *MFSs* can be used as an indicator to avoid over utilized nodes and select energy rich nodes for routing.
- Activity Based Clustering: Similar to other clustering metrics, like power, distance and mobility, among others, node activity levels can be considered as a metric for cluster formation. Nodes belonging to one *MFS* (pattern) are most likely connected and can be used as a cluster. Another metric for clustering is the Support parameter, i.e., the higher the Support level the higher the relationship among the routing nodes.
- Routing and Multicasting: Nodes belonging to one *MFS* are most likely connected. Accordingly, delivery or sending of packets is guaranteed amongst nodes in the same *MFS*.

- Applying Different Association Rules Mining Types: This chapter applies positive association rules mining techniques that mine binary attributes and considers that the utilities of the itemsets are equal. The frequency of an itemset may not be a sufficient indicator of interest. Non-boolean fuzzy association rule mining such as weighted/utility association rules, may find and measure all the itemsets whose utility values are beyond a user specified threshold that suggest different decisions. For example, in battlefield a commander can give higher weight/utility to his higher rank commanders and less weight to soldiers in order to find the hidden relationships (rules) amongst them. These rules may give an idea about soldiers who are in touch with each other, with commanders, and so on.
- Wireless Sensor Networks (WSN) has the inherent characteristics of *MANETs*, and therefore, the aforementioned benefits of using *MFS* in *MANETs* may also be applicable in *WSN*.

8. References

Agrawal, R., Imielinski, T. & Swami, A. (1993). Mining association rules between sets of items in large databases, *Proceeding of the 1993 ACM SIGMOD International Conference on Management of Data*, ACM, New York, NY, USA, Washington, D.C., United States, pp. 207–216.

Agrawal, R. & Shafer, J. C. (1996). Parallel mining of association rules, *IEEE Transactions on Knowledge and Data Engineering* 8(6): 962–969.

Asuncion, A. & Newman, D. J. (2007). UCI machine learning repository. URL: *http://www.ics.uci.edu/~mlearn/MLRepository.html*

Becker, K., Wille, U. & Wille, U. (1998). Communication complexity of group key distribution, *Proceedings of the 5th ACM conference on Computer and communications security*, ACM New York, NY, USA, San Francisco, California, United States, pp. 1–6.

Burmester, M. & Desmedt, Y. (1994). Vol. 950/1995 of *Lecture Notes in Computer Science*, Springer Berlin, Heidelberg, chapter A Secure and Efficient Conference Key Distribution System, p. 275.

Chan, A. C. F. (2004). Distributed symmetric key management for mobile ad hoc networks, *Proceeding of IEEE INFOCOM 2004. Twenty-third Annual Joint Conference of the IEEE Computer and Communications Societies*, Vol. 4, IEEE Press Piscataway, NJ, USA, Hong Kong, pp. 2414–2424.

Fall, K. (2007). *The NS Manual*, The VINT Project, University of California.

Fard, A. M. & Ester, M. (2009). Collaborative mining in multiple social networks data for criminal group discovery, *International Conference on Computational Science and Engineering*, IEEE CS Digital Library, Vancouver, Canada, pp. 582–587.

Frawley, W. J., Piatetsky, G. & Matheus, C. J. (1992). Knowledge discovery in databases: An overview, *AI Magazine* 13(3): 57–70.

Fumy, W. & Landrock, P. (1993). Principles of key management, *IEEE Journal on Selected Areas in Communications* 11(5): 785–793.

Ghoreishi, S. M. & Analoui, M. (2009). Design a secure composite key-management scheme in ad-hoc networks using localization, *International Journal of Computer Science and Network Security* 9(9): 35–49.

Greis, M. (2007). *Tutorial for the Network Simulator NS2*, http://www.isi.edu/nsnam/ns/tutorial/.

Hegland, M. (2005). Wspc/lecture notes series: The apriori algorithm - tutorial, *Technical report*, Australian National University, CMA, John Dedman Building, Canberra ACT

0200, Australia.

Hofmann, M. (2003). *The development of a generic data mining life cycle (dmlc)*, Master's thesis, MSc. in Computing Science , Dublin Institute of Technology, Duplin, USA.

Jabas, A., Abdulal, W. & Ramachandram, S. (2010). An efficient and high scalable key distribution scheme for mobile ad hoc network through mining traffic meta-data patterns, *Fifth IEEE International Conference on Network and System Security (IEEE NSS'10)*, IEEE CS Digital Library, Melbourne, Australia.

Jabas, A., Garimella, R. M. & Ramachandram, S. (2008a). Manet mining: Mining step association rules, *Fifth IEEE International Conference on Mobile Ad-hoc and Sensor Systems (IEEE MASS'08)*, IEEE CS Digital Library, Atlanta, Goergia, USA, pp. 589–594.

Jabas, A., Garimella, R. M. & Ramachandram, S. (2008b). Manet mining: Mining temporal association rules, *Third International Workshop on Intelligent Systems Techniques for Ad hoc and Wireless Sensor Networks (IEEE IST-AWSN 2008), Sydney, Australia*, IEEE CS Digital Library, Sydney, Australia, pp. 765–770.

Jabas, A., Garimella, R. M. & Ramachandram, S. (2008c). Proposing an enhanced mobile ad hoc network framework to the open source simulator ns2, *Mosharaka International Conferences on Communications, Computers and Applications (IEEE MIC-CCA'08)*, IEEE CS Digital Library, Amman, Jordan, pp. 14–19.

Javaheri, S. H. (2007). *Response modeling in direct marketing, a data mining based approach for target selection*, Master's thesis, Continuation Courses, Marketing and e-commerce, Department of Business Administration and Social Sciences, Division of Industrial marketing and e-commerce.

Joe, B. (2009). Do association rules represent supervised or unsupervised learning, *Technical report*. http://wardselitelimo.com/2009/07/02/.

Kim, Y., Perrig, A., & Tsudik, G. (2001). Communication-efficient group key agreement, *In 17th International Information Security Conference (IFIP SEC01)*, Kluwer Academic Publishers Norwell, MA, USA, Paris, France, pp. 229–244.

Kleinrock, L. & Silvester, J. (1978). Optimum transmission radii for packet radio networks or why six is a magic number, *Proceedings of the IEEE National Telecommunications Conference*, IEEE CS Digital Library, Birmingham, Alabama, p. 4.3.14.3.5.

Kyasanur, P. & Vaidya, N. H. (2003). Detection and handling of mac layer misbehavior in wireless networks, *International Conference on Dependable Systems and Networks (DSN'03)*, IEEE CS Digital Library, San Francisco, California, pp. 173–182.

Lamport, L. (1987). Synchronizing time servers, *Technical report*, Digital Equipment Corporation. Systems Research Center.

Luo, H., Kong, J., Zerfos, P., Lu, S. & Zhang, L. (2003). Ursa: Ubiquitous and robust access control for mobile ad-hoc networks, *58th IEEE Vehiclular Technology Conference VTC'03*, Vol. 3, IEEE Press Piscataway, NJ, USA, Orlando, Florida, USA, pp. 2137–2141.

Menezes, A., Oorschoot, P. V. & Vanstone, S. (1996). *Handbook of Applied Cryptography*, CRC Press, San Antonio, Texas.

Mott, J. L., Kandel, A. & Baker, T. P. (1992). *Discrete Mathematics for Computer Scientists and Mathematicians*, Reston Publishing Company, Inc.

ns2 (2009). The network simulator (ns2), Information Sciences Institute.
URL: *http://nsnam.isi.edu/nsnam/index.php/Main-Page*

Olson, D. L. & Delen, D. (2008). *Advanced Data Mining Techniques*, Springer, Verlag Berlin

Heidelberg.
Post, G. V. (2005). *Database Management Systems: Designing And Building Business Applications*, McGraw-Hill, Irwin.
Pujari, A. K. (2001). *Data Mining Techniques*, Universities Press, 3-6-747/1/A and 3-6-754/1, Himayatnagar, Hyderabad 500 029, Andhra Pradesh, India.
Rashmi (2009). Manet (mobile adhoc network), http://www.saching.com/Article/MANET--Mobile-Adhoc-NETwork–/334 [Access time: 20 Oct., 2009].
Robinson, J. A. (2007). Connecting the edge: Mobile ad-hoc networks (manets) for network centric warfare, *Technical report*, AIR UNIV MAXWELL AFB, Maxwell-Gunter Air Force Base Montgomery, Alabama, USA.
Royer, E. M., Melliar-Smith, P. M. & Mosery, L. E. (2001). An analysis of the optimum node density for ad hoc mobile networks, *IEEE International Conference on Communications, ICC*, Vol. 3, IEEE CS Digital Library, Helsinki, Finland, pp. 857–861.
Santoro, N. (2007). *Design and Analysis of Distributed Algorithms*, John and Wiley and Sons, Inc. Hoboken, New Jersey, Hoboken, New Jersey.
Simons, B., Welch, J. L. & Lynch, N. (2006). *Fault-tolerant distributed computing*, Vol. 448/1990 of *Lecture Notes in Computer Science*, Springer, Berlin / Heidelberg, chapter An overview of clock synchronization, pp. 84–96.
Simovici, D. A. & Djeraba, C. (2008). *Mathematical Tools for Data Mining, Set Theory, Partial Orders, Combinatorics*, Springer-Verlag Limited, Uk, London.
Tan, P.-N., Steinbach, M. & Kumar, V. (2006). *Introduction to Data Mining*, Addison-Wesley.
Yao, J., Li, X. & Jia, L. (2003). A new method based on ltb algorithm to mine frequent itemsets, *International Conference on Machine Learning and Cybernetics*, IEEE CS Digital Library, Xian, China, pp. 71–75.
Yi, S. & Kravets, R. (2003). Moca: Mobile certificate authority for wireless ad hoc netwroks, *Proc. of the 2nd Annual PKI Research Workshop (PKI)*, National Institute of Standards and Technology, Gaithersburg, USA.
Yi, S. & Kravets, R. (2004). Composite key management for ad hoc networks, *The First Annual International Conference on Mobile and Ubiquitous Systems: Networking and Services. MobiQuitous'04*, IEEE CS Digital, Boston, USA, pp. 52–61.

16

Wired/Wireless Compound Networking

Juan Antonio Cordero[1], Emmanuel Baccelli[1], Philippe Jacquet[1] and Thomas Clausen[2]
[1]*INRIA Saclay*
[2]*École Polytechnique*
France

1. Introduction

Routing, and more precisely routing within an Autonomous System (AS), is the most basic and still outstanding wireless ad hoc networking challenge. As the properties of ad hoc networks are *a priori* unpredictable and may change dynamically during the lifetime of the network, no assumptions can be made in general concerning topology, link reliability, routers positions, capabilities, and other such aspects. Routing protocols operating within an AS – *i.e.* interior gateway protocols (IGP) – must enable each router to acquire and maintain the information necessary to forward packets towards an arbitrary destination in the routing domain. Currently, the dominant IGP technology is link state routing, as acknowledged by reports of Cisco Systems, Inc. such as Halabi (2000).

Routing protocols that were designed for wired, static environments do not perform well in ad hoc networks: even for small networks, as Henderson *et al.* (2003) points out, control traffic explodes in a wireless, dynamic context. Many efforts have been deployed over the last decade, aiming at providing routing protocols suitable for ad hoc networks. In such context, information acquisition and maintenance has to be provided by distributed mechanisms, since neither hierarchy nor centralized authority can be assumed to exist. Moreover, the typical bandwidth scarcity experienced in wireless ad hoc networks calls for mechanisms that are extremely efficient in terms of communication channel utilization. In the realm of link-state routing two main strategies have been explored: (i) the design of ad hoc specific routing protocols; and (ii) the reuse and adaptation of existing generic routing protocols so that they can handle ad hoc conditions. The first strategy has mainly led to the emergence of the Optimized Link State Routing protocol, OLSR, standardized as RFC 3626 (2003). The second approach has led to protocol extensions such as RFC 5449 (2009), which enable the operation of Open Shortest Path First (OSPF) on ad hoc networks.

This chapter focuses on scenarios where the AS consists in *compound networks*: networks gathering both potentially mobile ad hoc routers, and fixed wired routers. Such scenarios may become frequent in a near future where wireless ad hoc and sensor networks play an increasing role in pervasive computing. Obviously, it is possible to employ multiple routing protocols within a compound network (*e.g.* one for wireless ad hoc parts of the network, and another for the wired parts of the network). However, a single routing protocol makes more economical sense for the industry, and furthermore avoids the potential sub-optimality of having to route through mandatory gateways between different routing domains. Thus a single protocol is desired to route in compound networks, and (ii) is deemed the best strategy

to do so. The main reason for this is, that (ii) takes advantage of wide-spread, generic protocols which on one hand already provide very elaborate modules for various categories of wired networks, and on the other hand can easily accommodate a new module for efficient operation on ad hoc networks.
This chapter thus explores techniques that enable efficient link state routing on compound networks. These techniques rely on the selection and maintenance of a subset of links in the network (i.e. an *overlay*) along which the different operations of link-state routing can be performed more efficiently. The following provides a formal analysis of such techniques, a qualitative evaluation of their specific properties and example applications of such techniques with a standard routing protocol.

1.1 Terminology

In this chapter, the following notation is used:

- The 1-hop and 2-hop (bidirectional) neighborhoods of a router x are denoted by $N(x)$ and $N_2(x)$, respectively.
- The usual notation of graph theory is assumed: $G = (V, E)$ stands for a (connected) network graph, in which the set of vertices is $V = V(G)$ and the set of edges is $E = E(G)$. Overlay subgraphs are denoted accordingly, as subsets of G.
- Given two vertices (routers) $x, y \in V$, $dist(x, y)$ is the cost of the optimal path between x and y. Similarly, given two vertices $x, y \in V$ reachable in 2 hops, it will be denoted by $dist_2(x, y)$ the cost of the optimal path between x and y in 2 hops or less (*local shortest path*). For two neighbors x and y, $m(x, y) = m(\overline{xy})$ denotes the cost of the direct link from x to y.

1.2 Chapter outline

The chapter is organized as follows. Section 2 describes the key operations providing link-state routing. Section 3 elaborates on the constraints that ad hoc networking imposes on link-state routing, with a specific focus on compound networks. Section 4 introduces to the notion of overlay for performing these key operations, analyzes the properties of several overlay-based techniques and discusses their advantages and drawbacks of their use in the context of a concrete routing protocol. Section 5 applies and evaluates the performance of such techniques as ad hoc OSPF extensions. Finally, section 6 concludes this chapter.

2. Communication aspects in link-state routing

This section provides a structural high-level description of the operations of link-state routing. Section 2.1 presents a short summary of link-state routing. Sections 2.2, 2.3 and 2.4 describe in more detail the main tasks associated to such operation: neighbor discovery, network topology dissemination and route selection for data traffic, respectively.

2.1 Link-state routing overview

Link-state routing requires that every router learns and maintains a view of the network topology that is sufficiently accurate to compute valid routes to every possible destination. This, typically (as for OSPF or IS-IS[1]), in form of shortest paths w.r.t. the metrics used. Such shortest paths are computed among the available (advertised) set of links by means

[1]Intermediate-System-to-Intermediate-System, specified in ISO 8473 (2002).

of well-known algorithms such as Dijkstra (1959), and will provide effectively optimal routes when the view of the topology is up to date.

These objectives require that every router in the network performs two operations, other than the shortest path computation: first, take efficient flooding decisions for the forwarding of topology information messages; and second, describe accurately its links in order to advertise them to the rest of the network. Three tasks emerge thus as necessary for the performance of link-state routing operation:

1. participation in the flooding of topology information (both of self-originated messages and of messages from other routers),
2. selection of links to advertise to enable shortest route construction and,
3. discovery and maintenance of the neighborhood, as a pre-requisite for the two previous tasks.

2.2 Neighbor discovery and maintenance

The discovery and maintenance of neighbors is a prerequisite for performing efficient link-state routing. Without neighborhood knowledge, link-state routing can only be deployed by means of pure flooding, which has been proven by Ni *et al.* (1999) to be dramatically inefficient when dealing with ad hoc networks (the *broadcast storm* problem); or with counter-based or similar approaches, which have severe performance limitations, as shown in Tseng *et al.* (2003). The most widespread and basic mechanism for neighbor sensing consists of the periodic transmission of Hello packets by every router in the network (Hello protocol). Exchange of such Hello packets enable routers to learn their neighborhoods and establish bidirectional communication, if possible, with neighbors within its coverage range. Aside from this use, Hello exchange may be useful for acquiring additional information about the neighbors (geographic position, remaining battery power, willingness to accept responsibilities in communication), the links to them (link quality measures) or the neighbors of such neighbors (2-hop neighborhood acquisition).

2.3 Topology information dissemination

Consistency of the distributed LSDB and correctness of routing decisions require that every router maintains an updated view of the network topology. When a router detects a relevant change in its neighborhood, it needs to advertise it by flooding a topology update message, so that any other router can modify accordingly its link-state database and, if necessary, recalculate optimal routes.

In ideal conditions [2], such mechanism would be sufficient for keeping identical LSDBs in every router in the network. Since these conditions are not found in wireless ad hoc scenarios, additional mechanisms might be considered:

- **Reliable flooding of topology messages.** Reception of such messages is acknowledged by the receiver, or retransmitted by the sender/forwarder in the absence of such acknowledgment, in a hop by hop fashion. Reliable flooding is provided by the main wired routing protocols (OSPF, IS-IS), but its cost in mobile ad hoc networks discourages its use in MANET-specific solutions such as OLSR.
- **Periodic re-flooding of messages.** After a certain interval, even if no changes have been registered in the neighborhood, the routers reflood to the network an advertisement

[2]That is, static, always-connected networks in stationary state with error-free links.

containing the current state of the links between themselves and their neighbors. The length of the interval is typically related to the mobility pattern of the network: the faster nodes in the network move, the shorter the interval between consecutive topology messages from the same source needs to be.

- **Point-to-point link-state database synchronization.** A link between two routers is said to be *synchronized* when the routers have completed a synchronization process of their respective LSDB. This involves the exchange of the database contents and the installation of the most updated topology information in each of them. This mechanism is implemented in the major wired routing protocols (OSPF, IS-IS), but the conditions in which such synchronization is performed are not completely adapted to mobile ad hoc operation. Therefore, the mechanism as-is is not considered in specific protocols such as OLSR, and its use is widely restricted, for instance, in the different OSPF MANET extensions.

These mechanisms handle different issues concerning topology dissemination. Reliable transmission permits overcoming phenomena such as wireless channel failures or collisions. Periodic re-flooding and point-to-point synchronization provide up-to-date topology information to routers appearing in the network after some of the disseminated messages were flooded across the network. Periodic reflooding by itself enables every router to acquire the latest topology information (maybe with a non-negligible delay, depending on the re-flooding interval). In contrast, full synchronization is not capable on its own to assure database convergence from all routers in link-state routing[3]. Point-to-point synchronization is, at best, a complementary mechanism to periodic re-flooding that allows a router that has not received all the topology updates to get within a shorter delay the last topology information from an updated neighbor.

Synchronization techniques implicitly introduce the concept of a *synchronized overlay*. A router is *included* into the synchronized overlay if it is aware of the last topology update messages that were flooded across the network, and, correspondingly, it is *removed* from the overlay when it does not receive one of more topology information messages. In that context, the periodic re-flooding of topology messages permits including every reachable router into the network within a maximum delay equal to the interval between two consecutive refloods. Point-to-point LSDB synchronization between a router and a synchronized neighbor permits, in turn, including routers immediately into the overlay (by means of the database exchange process), *i.e.*, to restore or establish for the first time the router's synchronism with the rest of the network.

In wired networks, the synchronized overlay is expected to grow monotonically until it contains all routers – then the network is said to converge. Router removals from the synchronized overlay are rare events mostly caused by physical link disconnections or router shut-downs. In ad hoc networks, the nature of the synchronized overlay is far more unstable. Alternative inclusion and removal events may thus occur due to router mobility or wireless link quality variations, preventing the network to converge in the usual sense.

2.4 Route selection for directed communication

The final goal of any routing protocol is that every router is able to route traffic to any other router (and any destination provided by such router) in the network. For a link-state routing

[3]This is different, for instance, in proactive distance-vector routing, in which the network is expected to converge through *repeated* database synchronization processes. In the considered link-state context, synchronization occurs *once* in a link lifetime, which is not sufficient for assuring convergence.

protocol, such ability is provided by disseminating the topology updates of all routers across the network. Such dissemination permits every router to construct and maintain updated routing tables, as Figure 1 describes schematically.

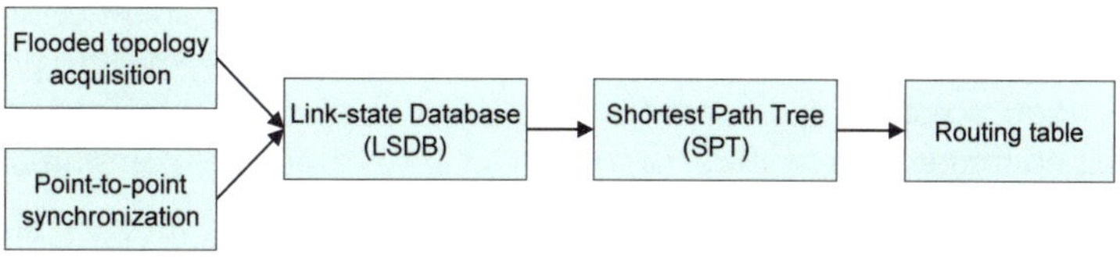

Fig. 1. Construction of the routing table for a link-state routing protocol.

The tree of the optimal routes to every destination (Shortest Path Tree) is then computed by means of well-known minimum paths algorithms. Typically, link-state routing protocols (OSPF, IS-IS, OLSR) use Dijkstra (1959), while distance-vector protocols (RIP[4], EIGRP[5]) rely on Bellman-Ford [Bellman (1958); Ford & Fulkerson (1962)]. These algorithms operate over a graph in which vertices correspond to routers in the network and edges mostly correspond to links advertised by the received topology update messages[6]. The routing table is thus extracted from the next hop, according to the Shortest Path Tree, to every possible destination. In general, the reconstructed link-state database should bring every router exactly the same perspective of the network topology, which would require that all links are advertised. In practice, the set of links that a router advertises to the rest of the network can be restricted as far as it does not prevent the shortest path algorithm to select network-wise optimal routes.

3. Link-state routing with ad hoc constraints

This section exposes the main challenges for link-state routing in ad hoc networks. These are mainly related to (i) the efficient dissemination of topology information across the network, in presence of lossy channels and dynamic topologies as is typical in these networks, and (ii) the ability of the network to acknowledge and react quickly to topology changes. Section 3.1 presents the most relevant implications of the ad hoc nature in the performance of link-state routing, while section 3.2 focuses on the specific case of compound networks integrated by wired and wireless groups of routers.

3.1 General issues of ad hoc link-state routing

Wireless ad hoc networking presents a certain number of unique communication conditions that link-state routing needs to accommodate:

- **Unreliability of wireless links.** Wireless links are inherently unreliable: channel failures and collisions are more frequent than in wired links. Wireless link quality can be also highly dynamic. Both circumstances make necessary continuous monitoring of the state and characteristics of links.

[4]Routing Information Protocol, specified in RFC 1058 (RIPv1), RFC 1723 and RFC 2453 (RIPv2) and RFC 2080 (RIPng, designed for IPv6).

[5]Enhanced Interior Gateway Protocol, Cisco proprietary routing protocol that improves Cisco's previous IGRP.

[6]Not necessarily all edges have been acquired by means of topology update messages. Section 4 explores some techniques in which some additional edges, not advertised in such messages, might be included as well.

- **Semibroadcast nature of wireless multi-hop communication.** Wireless communication entails shared bandwidth among not only the routers participating in the communication, but also those within the radio range of the transmitting routers. This reduces drastically the available bandwidth for a router, since it is affected by the channel utilization of its neighbors. Applications may take advantage of such bandwidth sharing phenomenon by privileging, when possible, multicast transmissions in place of a unicast (point-to-point) approach that no longer corresponds to the physical conditions of communication.
- **Asymmetry and non-transitivity of links.** Semibroadcast communication also implies that the set of nodes receiving a transmission if not (necessarily) the whole network. Moreover, the set of nodes receiving a transmission may be different for two routers, even when such routers are neighbors. This means that wireless links in a multi-hop ad hoc network cannot be expected to be transitive: the fact that a router x can directly communicate with routers y and z does not imply that routers y and z can also communicate directly ($x \leftrightarrow y, y \leftrightarrow z \not\Rightarrow x \leftrightarrow z$). Asymmetric links (*i.e.*, links in which a router can hear the other's transmissions, but not the other way around) are also possible due to specific channel conditions or different router capabilities.
- **Topology acquisition and maintenance.** Neither hierarchy nor specific routers relationships can be *a priori* assumed in an ad hoc network. Dynamic configuration of hierarchical schemes becomes unfeasible due to difficulties on electing top-level routers (related to non-transitivity of links) and cost of performing hierarchy recompositions (caused by node failures, node mobility or channel quality variations). Distributed approaches are thus encouraged in place of hierarchical ones. Moreover, unreliability of wireless links makes necessary to complement topology dissemination with a periodic and frequent reflooding of topology messages that ensures that nodes acquires the last updates with a relatively short delay.

3.2 Dissemination in compound networks

In addition to wireless ad hoc routers, compound networks also contain wired static components, for which the typical link lifetime is much higher than for standard ad hoc communications. The coexistence of wired and wireless ad hoc components poses some additional constraints to those presented in the previous section 3.1. Frequent flooding updates from the wired components lead to inefficient use of the available bandwidth, as the information about wired links carried by consecutive messages would be unchanged. Low update frequencies (with intervals in the order of wired networks) may however be insufficient to accommodate communication failures in the wireless and/or mobile components of the network.

Link synchronization between selected pairs of neighboring routers (in addition to topology changes flooding and periodic topology reflooding) helps to alleviate this issue. Point-to-point link synchronization enables highly dynamic routers to acquire updated topology information from wired links even long time after its origination, without requiring frequent refloods of the same link-state description by the corresponding wired (stable) source.

Consider Figure 2, where fixed routers (1 and 2) can handle changes in their wired (stable) links by transmitting topology updates at relatively low rate (with the time interval between updates in the order of minutes). Mobile routers (such as 5, 6 and 7) and, more in general, routers maintaining wireless links (also the hybrid routers 3 and 4) should use significantly lower time intervals (in the order of seconds, depending on their mobility pattern). If, for any reason, a mobile router (such as 5, 6 or 7) did not receive a topology update from a wired one as router 1, it will be unable to update its LSDB until the next flooding from the wired router, failing at computing valid routes that involve that router in the meanwhile.

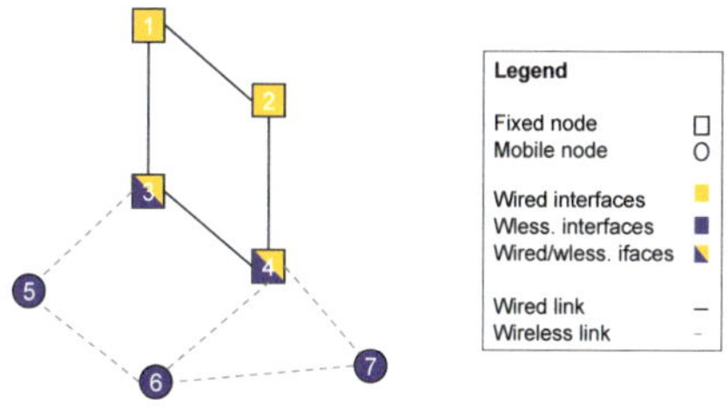

Fig. 2. Example of compound (wired/wireless) network.

The inclusion of a LSDB synchronization mechanism addresses the coexistence of wired and wireless components without having to reflood unnecessary topology updates from wired routers nor compromising the accuracy of network topology view of ad hoc (mobile) routers. This, at the expense of an additional dissemination mechanism (in addition to regular flooding of topology changes and periodic topology reflooding) and the corresponding additional complexity in the flooding operation.

4. Overlay techniques for compound networks

This section proposes and analyzes various techniques for performing link-state routing in ad hoc compound networks. Section 4.1 introduces the notion of overlay and reformulates the main operations of link-state routing in terms of overlays. Subsequent sections 4.2, 4.3 and 4.4 describe three overlay-based techniques (Multi-Point Relays, Synchronized Link Overlay and Smart Peering, respectively) and analyze their most relevant properties, both from theoretical and experimental (simulation-based) perspectives.

4.1 The notion of overlay

The three main operations of link-state routing in ad hoc networks can be reduced to overlay definition problems. Intuitively, an *overlay* of an ad hoc network is a restricted subset of routers and links of the network in which a certain operation is performed. More formally, the overlay of a network graph $G = (V, E)$ corresponds to a subgraph $S \subseteq G$ containing a subset of vertices $V(S) \subseteq V(G) = V$ and a subset of links $E(S) \subseteq E(G) = E$ of the underlying network graph G. In an ad hoc network, link-state routing operations are performed locally (independently by every router in the network) and thus, the corresponding overlays are built in a distributed fashion and may change dynamically during the network lifetime. Three different types of overlays can be identified, one for each of the following operations:

- **Topology update flooding.** The flooding overlay has to be dense (in the mathematical sense) in every of its connected components – meaning that, in case the overlay is not connected, each of its *pieces* is at distance ≤ 1 (number of hops) of every router in the network. This condition guarantees that a topology update generated in any of such components reaches all routers. Due to the impact of any additional router in the flooding overlay (an additional transmission, and the corresponding utilization of the channel of all its neighbors for every topology update generated in the network), the size of such overlay should be minimized.
- **Point-to-point synchronization.** The synchronized overlay contains links between those routers having exchanged their LSDBs. Formally, such overlay needs to form a spanning

connected subgraph of the general network graph[7], in order to facilitate the distribution of the LSDB over the whole network. The number of LSDB synchronization processes induced by a synchronized overlay is related to the overlay density (the number of links in the overlay), and also depends on the lifetime of the synchronized links (given that synchronization is performed once during the existence of the link). Therefore, minimization of overhead caused by LSDB synchronization requires a low density overlay with stable links.

- **Topology selection.** In wired deployments, all links are typically advertised to ensure that all routers in the network have an identical view of the network topology. In wireless ad hoc networks, this condition is often relaxed, and every router is only expected to acquire a consistent topological view of the network accurate enough to perform correct route computation. Hence, selection of advertised links trades-off the size of the topology update messages and the accuracy of the topological view of the network in all routers. A topology selection rule must, however, produce a connected and spanning subgraph (otherwise there would be non-reachable destinations) and whose set of edges contains all network-wide shortest paths – otherwise the computation would be asymptotically suboptimal[8].

Table 1 summarizes the requirements of each operation to the corresponding overlay.

	Graph / Overlay	Topology requirements	Minimization targets
Full Network	$G = (V, E)$	Connected	-
Flooding	$G_F = (V_F \subseteq V, E_F \subseteq E)$	Dense for every conn. cp.	Number of links
Link-State DB Synchronization	$G_S = (V, E_S \subseteq E)$	Connected and spanning	Number of links & link change rate
Advertised Links (topology selection)	$G_R = (V, E_R \subseteq E)$	Connected and spanning Includes sh.-paths of G	Link change rate

Table 1. Summary of overlay requirements.

4.2 Multi-point relays – MPR

Multi-Point Relaying (MPR) is primarily a technique for efficient flooding. It reduces the number of required transmissions for flooding a message to every 2-hop neighbor of the source by allowing a restricted subset of 1-hop neighbors (*multi-point relays* of the source) to forward it. Figure 3 illustrates that a clever election of 1-hop neighbors as relays can achieve the same coverage as allowing every 1-hop neighbor to transmit (pure flooding, see Fig. 3.a) while reducing significantly the number of redundant transmissions.

The subset of selected relays must satisfy the condition of full 2-hop coverage:

MPR coverage criterion Every 2-hop neighbor of the computing router must be reachable by (at least) one of the selected multi-point relays.

Therefore, an MPR set of a router x can be formally defined as follows:

$$R(x) \subseteq N(x) \text{ is an MPR set of } x \Longleftrightarrow \forall z \in N_2(x), \exists y \in R(x) : z \in N(y) \quad (1)$$

---

[7] *I.e.,* has to include every vertex (router) in the network.

[8] In real conditions, the computation may be suboptimal due to stale topology information, transmission failures and such. *Asymptotic suboptimality* implies that even in ideal conditions (message transmission delay $\longrightarrow 0$, collision probability $\longrightarrow 0$, channel failure probability $\longrightarrow 0$) the computation would be suboptimal.

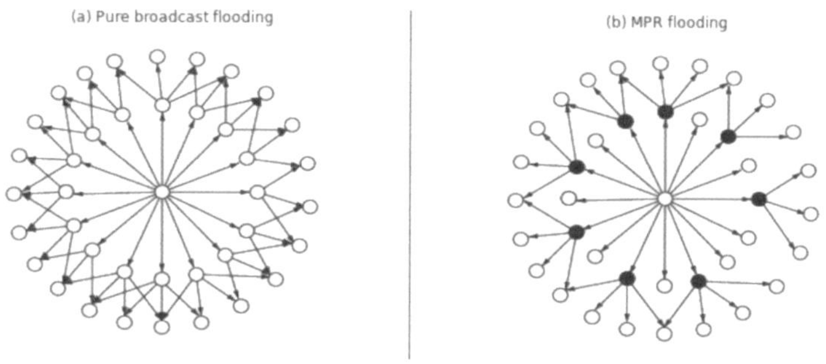

Fig. 3. **(a)** Pure flooding *vs.* **(b)** flooding based on the Multi-Point Relays (MPR) principle. *Solid dots in (b) represent multi-point relays.*

Different heuristics can be used for selecting multi-point relays, all valid as long as they satisfy the MPR coverage criterion. This chapter uses the heuristic in Figure 4, presented and analyzed in Qayyum *et al.* (2002).

$$\begin{cases} MPR(x) = \{\varnothing\} \\ MPR(x) \longleftarrow \{y_{excl} \in N(x) : y_{excl} \text{ provides exclusive coverage to one or more 2-hop neighbor(s) of } x\} \\ while(\exists \text{ uncovered 2-hop neighbors of } x), \\ MPR(x) \longleftarrow y \in N(x) : y \text{ covers the maximum \# of uncovered 2-hop neighbors of } x \end{cases}$$

Fig. 4. Summary of the MPR heuristic.

This heuristic assumes that the source is aware of its 2-hop neighbors. Acquisition of the 2-hop neighborhood is thus required. Dependence on 2-hop neighbors has yet another side effect on the MPR properties: given that an MPR selection may become obsolete due to a change in the 2-hop neighborhood of the computing source, stability of the MPR set is not only affected by conditions in MPR links[9], but also by the MPR recalculations due to changes within the 2-hop neighbors or they way in which they are connected to the 1-hop neighbors of the source (see Figure 5). Such sensitiveness of the MPR set of a router to variations in its 2-hop neighborhood has further implications for the MPR overlay that will be further detailed in section 5.

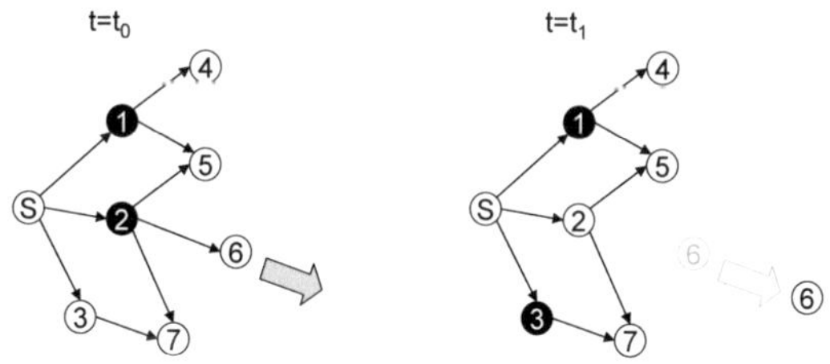

Fig. 5. MPR recalculation due to changes in the 2-hop neighborhood. *Solid dots represent relays of router S.*

4.2.1 MPR as a flooding overlay principle

MPR flooding introduces a directed overlay for every flooded message, by allowing a router to forward such message if and only if the following two conditions are satisfied:

[9]An *MPR link* is a link connecting a router to one of its multi-point relays.

1. the message comes from a MPR selector (that is, a neighbor that has selected that router as multi-point relay), and
2. it is the first time the message is received in that router.

Note that condition (2) ensures that the flooding process terminates in a finite number of steps. The (re)transmission of a message by a router triggers a number of retransmissions for which an upper bound is the number of multi-point relays (MPRs) of such router (see Fig. 12), and the process iterates recursively. The number of retransmissions triggered by a single transmission is close to the size of the MPR set in the first steps of MPR flooding. As the flooding advances over the network, an increasing part of the MPR links of the transmitting routers have already received the message and thus do not forward it again (condition (2)), until the message reaches routers for which every neighbor has received a copy, and the flooding terminates.
The flooding overlay formed by the MPR links of every router in an ad hoc network does not need to be connected. *Lemma 1* shows that each of its connected components (in case there are several) are dense in the network. For proofs of the results presented in this section, as well as for examples of disconnected MPR overlays, see Cordero (2010).

Lemma 1 *Let $G = (V, E)$ be a network connected graph, and $H \subseteq G$ the subgraph of G containing the links from every vertex in the graph to all its MPRs. Then, every connected component of H is dense over G.*

Note that this lemma addresses an asymptotic topological property of the overlay generated by condition (1), depending only on the ad hoc network topology. Condition (2) is not contradictory with this property by its own nature, since it removes from the overlay those links which produce no additional coverage. Thus, the conclusion is valid also for the overlay resulting from conditions (1) and (2).

4.2.2 MPR as a synchronized overlay

Multi-Point Relays can also be used for synchronization purposes. A link between two neighbors becomes synchronized if any of its endpoints has selected the other as multi-point relay. The overlay derived from this contains the same links as those described by condition (1) of section 4.2.1. Unlike the flooding overlay, the MPR synchronized overlay is undirected. This is due to the symmetric nature of the LSDB synchronization operation (see section 2.3), and leads to a denser overlay (that is, with more links per router) than the MPR flooding one, as it can be observed in Figure 12.
A synchronized overlay needs to be asymptotically connected[10]. This is not necessarily the case for an overlay containing MPR links of all routers in the network, as it was pointed out in section 4.2.1. *Lemma 2* provides a sufficient condition for *connecting* the MPR overlay.

Lemma 2 *Let $G = (V, E)$ be a network connected graph, and $H \subseteq G$ the subgraph of G consisting of:*

1. *$H_1 \subseteq G$: For every vertex $x \in V$, the edges from x to the neighbor vertices selected by x as MPRs.*
2. *$H_2 \subseteq G$: For a certain $s \in V$, the edges from s to every neighbor of s.*

Then, H is connected.

[10] An overlay defined over a network is *asymptotically connected* if its definition ensures connection in conditions of instantaneous transmission (delay $\longrightarrow 0$), error-free and collision-free links (probability of error/collision $\longrightarrow 0$). Note that an overlay may be asymptotically connected, but not connected in practice due to stale information stored in routers, loss of messages and such.

Under these conditions, the MPR-based overlay G_S defined in (2) is asymptotically connected. Despite fulfilling which topological condition, Multi-Point Relaying does not fill well in the requirements for a synchronized overlay, as they were defined in section 4.1. The link density (average number of links per node) of the MPR synchronized overlay, even without considering any additional router s, is significantly higher than the MPR flooding overlay (see Figure 12, below). The reduction with respect to the full network overlay (bidirectional links) is less than a 60%, even for dense networks. In following sections there are presented techniques able to minimize in a higher degree the synchronized overlay.

$$\begin{cases} V(G_S) = V(G) \\ E(G_S) = \{\overline{xy} \in E(G) : x \in MPR(y) \vee y \in MPR(x) \vee (x \equiv s) \vee (y \equiv s)\} \end{cases} \tag{2}$$

In addition to the high overlay density, the MPR synchronized overlay also presents a high overlay link change rate. Changes in the 1-hop of 2-hop neighborhood of a router may cause changes in the MPR set of such router (see Figure 5). This turns useless part of the synchronized links (those connecting with neighbors that are no longer MPRs) and increases the amount of synchronizations to perform (to newly elected MPRs), thus increasing the overhead dedicated to maintain the synchronized overlay.

The *persistent* MPR synchronized overlay overcomes partially these issues. This overlay includes, for each router, existing links to all neighbors that were elected as MPR by this router, even if they were later removed from the MPR set. The persistent mechanism produces significantly larger synchronized overlays (see Figure 13), but these persistent overlays are more stable than the non-persistent ones. Section 5 empirically evaluates the impact of the persistent mechanism in the size and stability of the MPR synchronized overlay (see Figs. 13 and 14).

4.2.3 MPR as a topology selection rule – Path MPR

Section 4.1 points out that the main requirement for an overlay of advertised links (topology selection overlay) is that it is a spanning subgraphs that contains the network-wide shortest paths to all destinations.

Computation of shortest paths involves a metric, that is, a link cost function which gives sense to the notion of *shortest*. But the MPR mechanism is defined in terms of coverage requirements, rather than cost minimization objectives. It becomes thus necessary to translate the cost-based optimality considerations in terms of optimal coverage, in order to reuse and extend MPR as efficient topology selection mechanism.

This section elaborates on the *Path MPR* mechanism, based on the previously stated conditions. Figure 6 displays the input/output block diagram of such approach.

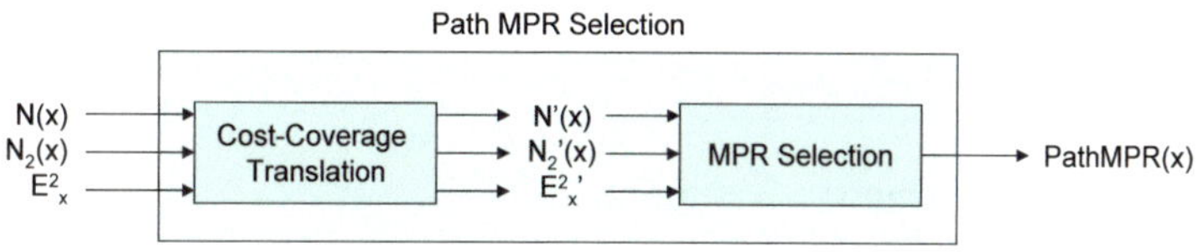

Fig. 6. Block diagram for an MPR-based topology selection algorithm. $E_x^2 \subset E(G)$ *stands for the set of edges connecting vertices within* $x \cup N(x) \cup N_2(x)$.

The cost-coverage translation block (see Fig. 6) extracts the subgraph of (local) shortest paths from the 2-hop and 1-hop neighbors of x to x. Vertices of this subgraph include x, $N'(x)$ and $N_2'(x)$, while the edges are represented by $(E_x^2)'$. $N'(x)$ extracts from $N(x)$ those neighbors for which the direct link from x is also the optimal (shortest) one; and correspondingly, $N'_2(x)$ extracts from $N_2(x)$ those neighbors for which the optimal path from x has 2 hops. Finally, $(E^2_x)'$ contains those edges (links) of E^2_x that participate in at least one shortest path from a 1-hop or 2-hop neighbor of x to x. The formal definition of the translation block's output is as follows:

$$\begin{cases} N'(x) = & \{n \in N(x) | m(x,n) = dist_2(x,n)\} \subseteq N(x) \\ N'_2(x) = & \{n \in N(x) \cup N_2(x) | n \notin N'(x), \exists m \in N'(x) : m(n,m) + m(m,x) = dist_2(n,x)\} \subseteq N(x) \cup N_2(x) \\ (E^2_x)' = & \{\overline{nm} \in E(G) : n \in N'(x), m \in N'_2(x), m(x,n) + m(n,m) = dist_2(x,m)\} \cup \\ & \cup \{\overline{xn} \in E(G) : n \in N'(x)\} \subseteq E^2_x \end{cases}$$

From these definitions, it is immediate that the Path MPR mechanism, as defined in Figure 6, returns a set of relays that provide (local) shortest paths from every 2-hop neighbor of x to x: if a path $p_{zy} = \{\overline{zy}, \overline{yx}\}$ is not optimal, with $y \in N'(x)$ and $z \in N'_2(x)$, then $\overline{yz}$ will not belong to $E(S'_x)$. That ensures that this extension of MPR is able to select the local (2 hops) shortest paths to the computing router x, given that every 2-hop neighbor of x is included in $N'_2(x)$.
A topology selection mechanism based on the advertisement by each router of the Path MPR set, as it has been defined, induces a network-wide overlay that contains, for every router x, the 1-hop neighbors of x that provide shortest paths (in a 2 hop scope) from 2-hop neighbors of x to x. The requirements for topology selection overlays identified in section 4.1 included however:

- Overlay connection.
- Preservation of network-wide (and not only local) shortest paths.

Connection of an MPR overlay can be achieved (*Lemma* 2) by adding to the overlay all the links maintained by a single arbitrary router. *Lemma* 3 shows that the overlay that results of adding such additional router (the computing router itself, for Path MPR) contains network-wide shortest paths from every destination of the network to the computing router:

Lemma 3 *Let $G = (V,E)$ be a connected network graph, an edge metrics function $cost(e \in E(G))$, a router $s \in V(G)$ and a subgraph $G'_s = (V, E'_s)$ including:*

1. *the edges connecting s to its 1-hop neighbors, and*
2. *for every router x of the network, the edges from x to those 1-hop neighbors of x providing local shortest paths from every 2-hop neighbor of x to x.*

Then, the Dijkstra algorithm computed on a source router s over G'_s selects the shortest paths in G from the source to every possible destination.

Note that, as other improvements are possible (such as including not only $N(x)$ but also $N_2(x)$), the previous lemma states a sufficient condition for the asymptotic correctness of an MPR-based topology selection overlay.

4.3 The Synchronized Link Overlay-Triangular – SLO-T

The Synchronized Link Overlay (SLO) is an overlay-based technique inspired by the Relative Neighborhood Graph (RNG), first presented in Toussaint (1980). Given a set of points S in a plane, the relative neighbor graph of S is the graph that results from considering links between points in S, except those connecting points for which there are points *closer*[11] to them than the

[11]Even though RNG was originally defined for Euclidean distances (so the notion of close has to be understood under such distance), it can be easily generalized to other metrics.

routers themselves to each other. Included links thus connect pairs of points $\{u,v\}$ for which the intersection of circles centered on u and v, with radius the distance from u to v, contains no other points of S (see Figure 8, the intersection corresponds to the dotted region). More formally, the relative neighbor graph of S is defined as follows:

$$RNG(S) = \{\overline{xy}, x,y \in S : \nexists z \in S : dist(x,z), dist(z,y) < dist(x,y)\}$$

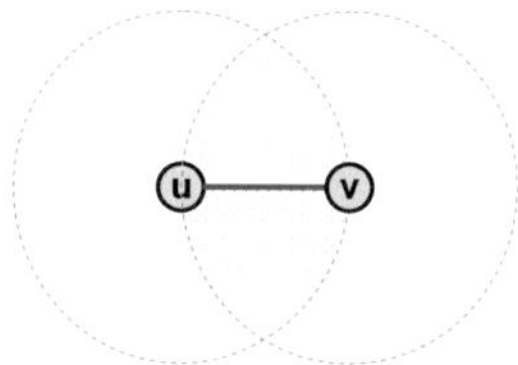

Fig. 7. Link $\overline{uv}$ belongs to $RNG(S)$ if the dotted region does not contain any other point of S.

where *dist* represents the standard, Euclidean distance in the plane. A similar principle is used in SLO. A link of a network graph $G = (V,E)$ is not synchronized under this rule if there is a chain of common neighbors to both endpoints of the link such that the links in the chain are cheaper (w.r.t. the metrics) than the considered link.

$$\overline{ab}) \notin SLO(G) \Longleftrightarrow \exists c_1, c_2, c_3, ..., c_n : \begin{cases} \forall i \leq n, c_i \in N(a) \cap N(b) \\ m(a,b) > max\{m(a,c_1), m(c_1,c_2), ..., m(c_n,b)\} \end{cases}$$

This section elaborates on a simplified version of the SLO, the Synchronized Link Overlay Triangular (SLO-T). This version restricts the chain of intermediate common neighbors $\{c_1, c_2, ..., c_n\}$ to a single neighbor. It consists of synchronizing a link between two neighbor routers u and v if and only if it does not exist any router w that is common neighbor of u and v and is closer or at the same distance to u and v than they are to each other. Note that this simplification generalizes RNG for arbitrary metrics m. In case of link cost equality (*i.e.*, $m(\overline{uw}) = m(\overline{wv}) = m(\overline{uv})$, m being the metric function), the tie is broken by excluding from synchronization the link connecting the routers with lowest ids.
Different metrics lead to different SLO-T rules. Two variations are considered in this section: the unit link cost (associated to the SLOT-U rule), and the distance-based cost (associated to the SLOT-D rule). Note that the tie breaking applies for the former (as all the link costs are equal to 1), while the main rule is implemented for the latter. Both variations are formally defined as follows:

$$\begin{cases} SLOT_U(G) = \{\overline{xy} \in E(G) : (\nexists z \in V(G), z \in N(x) \cap N(y) : id_z > max\{id_x, id_y\})\} \\ SLOT_D(G) = \{\overline{xy} \in E(G) : (\nexists z \in V(G), z \in N(x) \cap N(y) : m(x,y) \geq max\{m(x,z), m(z,y)\})\} \end{cases}$$

SLOT-U can be implemented more easily since it does not require any particular mechanism to monitor and measure the link cost: all the available links are treated equally, with the same uniform metric. For SLOT-D, in contrast, it is needed a mechanism to estimate the distance between two neighbor routers, something that can be achieved by location-based means (such as GPS).

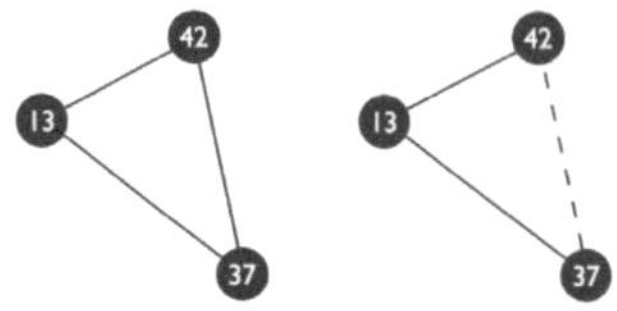

Fig. 8. The SLOT triangular elimination under unit link cost. *The link connecting routers with the highest ids, $\overline{42,37}$ in the picture, is excluded.*

SLOT inherits the properties required for a synchronized overlay (connection and spanning subgraph) from the Relative Neighbor (RNG). For any set of points S of the plane, Toussaint (1980) shows that $RNG(S)$ contains the Minimum Spanning Tree (MST) of G. Hence, SLOT contains it also and, in particular, is connected and a spanning subgraph of G.
Link synchronization and flooding operations require low density overlays that contain the most stable links, as mentioned in section 4.1. The two following sections elaborate on the overlay density and link stability for SLOT-U and SLOT-D from two different perspectives: theoretical analysis on mobile conditions and simulations of static scenarios. Proofs for the results presented in the remaining of the section are detailed in Baccelli *et al.* (2010). Theoretical analysis assumes a unit disk graph model, in which routers are distributed uniformly (approximated by Poisson distribution) over a large enough scenario (area $A \longrightarrow \infty$, not considering border effects) and move following isotropic random walks with an average speed s.

4.3.1 Overlay density

An overlay containing the full network has $M_{full} = \pi\nu$ links per router in average under the unit disk graph model, where ν is the network density. *Theorems 1* and *2* show how the overlay density is reduced when using SLOT with unit cost and distance-based cost, respectively.

Theorem 1 *The average number of SLOT-U links per router satisfies, as a function of the network density ν,*

$$M_u(\nu) = \int_{\frac{\pi}{3}}^{\frac{\pi}{2}} d\theta \frac{8\pi}{\nu(A(\theta))^2} sin(2\theta)(\nu A(\theta) + e^{-\nu A(\theta)} - 1)$$

and tends when network density $\nu \longrightarrow \infty$, to

$$M_u = \int_{\frac{\pi}{3}}^{\frac{\pi}{2}} d\theta \frac{8\pi sin(2\theta}{2\theta - sin(2\theta)} + O\left(\frac{1}{\nu}\right) \approx 3.604$$

Theorem 2 *The average number of SLOT-D links per router satisfies, as a function of the network density ν,*

$$M_d(\nu) = \int_0^1 dr 2\pi\nu r e^{-r^2 A(\frac{\pi}{3})}$$

and tends when network density $\nu \longrightarrow \infty$, to

$$M_d = \frac{\pi}{2\frac{\pi}{3} - \frac{\sqrt{3}}{2}} + O(\nu e^{-\nu(\frac{2\pi}{3} - \frac{\sqrt{3}}{2})}) \approx 2.558$$

where $A(\theta) = 2\theta - sin(2\theta)$. Figure 9.a indicates the evolution of SLOT-U and SLOT-D overlay densities depending on the network density ν. It can be observed that the density reduction, while being relevant for both SLO-T variations, is more significant for the distance-based cost: in this case, routers have more information about the network topology and can thus perform a more accurate synchronized links selection.

Theorems 1 and *2* shows that SLOT overlay (both the unit cost and distance-based cost variations) densities are upper-bounded by finite limits (V_u and V_d) which do not depend on the network density. This is a outstanding advantage of SLOT-like solutions with respect to other overlays for which the size (number of links) grows with the full network density, mainly for very dense networks.
Figure 12 (see below) confirms the previous theoretical analysis with an experiment that measures the average number of synchronized links in static uniformly distributed networks over a finite square scenario, for different network densities. Distance-based costs are implemented by means of a discrete function $m_d(\overline{xy}) = \lceil \frac{K}{r} d(x,y) \rceil \in \mathbb{N}$ ($d(x,y)$ measuring the Euclidean distance between x and y), that quantizes the link length into a number between 1 and K.
It can be observed that SLOT overlays are in general less dense than the MPR overlays studied in section 4.2, in particular with very dense networks. For low densities, however, SLOT-U produces overlays with a very similar asymptotic density to the directed MPR flooding (directed) overlay.

4.3.2 Link stability

Let $\Delta(s)$ be the average relative speed between two routers. Then, the link rate change under the unit disk graph, for an isotropic random walk router mobility, corresponds to $V_{full} = 2\Delta(s)\nu$. *Theorems 3* and *4* show that links belonging to SLOT variations have a significantly lower change rate. Figure 9.b illustrates such stability for a moderate mobility scenario (constant router speed $s = 5m/s$).

Theorem 3 *The average number of SLOT-U links per router satisfies, as a function of the network density* ν,

$$V_u(s,\nu) = \Delta(s) \int_{\frac{\pi}{3}}^{\frac{\pi}{2}} d\theta \frac{32\theta sin(2\theta)}{\nu(A(\theta)^3)} (A(\theta)\nu - 2 + e^{-\nu A(\theta)}(2 + \nu A(\theta))) \tag{3}$$

where $\Delta(s)$ *is the average relative speed between routers. For constant speed* ($\Delta(s) = \frac{4}{\pi}s$)*, equation (3) becomes*

$$V_u(s,\nu) = \frac{128s}{\pi} \int_{\frac{\pi}{3}}^{\frac{\pi}{2}} d\theta \frac{\theta sin(2\theta)}{(2\theta - sin(2\theta))^2} \approx 4.146s + O(\frac{4s}{\pi\nu})$$

Theorem 4 *The average number of SLOT-D links per router satisfies, as a function of the network density* ν,

$$V_d(s,\nu) = \frac{4}{3}\Delta(s) \int_0^1 2\pi\nu^2 r^2 e^{-r^2 \nu A(\frac{\pi}{3})} \tag{4}$$

where $\Delta(s)$ *is the average relative speed between routers. For constant speed* ($\Delta(s) = \frac{4}{\pi}s$)*, equation (4) becomes*

$$V_d(s,\nu) \approx 3.471s\sqrt{\nu}$$

Note that SLOT-U presents a higher stability than SLOT-D, which is caused by the sensitivity of the latter variation to router position (and thus distance to the other link endpoint) changes. Changes in the link cost may lead to new SLOT-D elections, while the unit cost (SLOT-U) ensures that there will be no changes in the synchronization decisions as long as there are no new routers forcing new triangular eliminations (see Fig. 8).

4.3.3 Link characterization depending on distance

The two considered variations of SLOT (SLOT-U and SLOT-D) assume different behaviors with respect to the distance of the links selected for synchronization. Intuitively, the longer

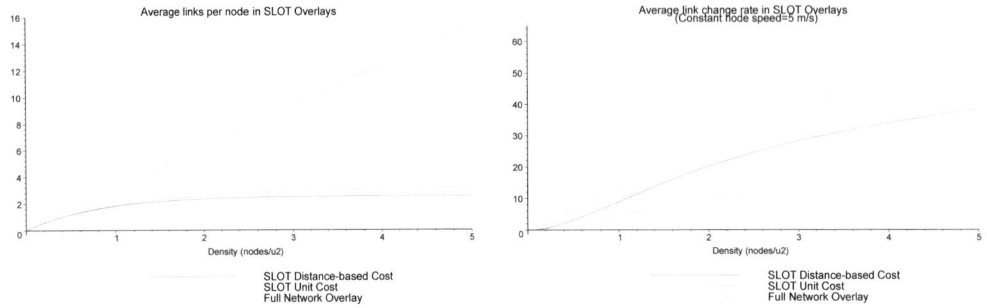

Fig. 9. **(a)** Average SLOT overlay density (links per router), and **(b)** average SLOT links change, for constant speed $s = 5m/s$.

the link is, the less likely is that there is a common neighbor to both endpoints whose identity is higher than those of the involved routers (and thus excludes the link from the synchronized overlay, according to the tie breaking rule of SLOT-D). On the contrary, the more far two neighbor routers are, the easier is that a common neighbor is closer to both endpoints – thus, the more likely is that SLOT-D discards such link.
This intuition can be formalized as follows. Let us denote the synchronization relationship by the symbol $\sim$. Then, the probability that a link $x \longleftrightarrow y$ is synchronized under the SLOT-U rule is:

$$P(x \sim y)_U = \left(\frac{2}{3}\right)^{n_{x,y}} \tag{5}$$

where $n_{x,y}$ is the number of common neighbors of x and y.
In consequence, the probability that a link between two routers x and y at distance $d < r$ is selected as part of the synchronized link can be defined as:

$$\begin{aligned} P(x \sim y|m(\overline{xy}) = d)_U &= \sum_{k=0}^{\infty} P(n_{x,y} = k) P(x \sim y|m(\overline{xy}) = d, n_{x,y} = k) = \sum_{k=0}^{\infty} \left(\frac{2}{3}\right)^k e^{-\nu A_r(d)} \frac{(\nu A_r(d))^{k+2}}{(k+2)!} = \\ &= e^{-\nu A_r(d)} \sum_{k=0}^{\infty} \left(\frac{2}{3}\right)^k \frac{(\nu A_r(d))^{k+2}}{(k+2)!} = \frac{3}{2} e^{-\nu A_r(d)} \left(\frac{3}{2} e^{\frac{2}{3}\nu A_r(d)} - \frac{3}{2} - \nu A_r(d)\right) \end{aligned} \tag{6}$$

where ν is the router density in the network and $A_r(d)$ is the intersection area between two circles of radius r at a distance d:

$$A_r(d) = 4 \int_{\frac{d}{2}}^{r} \sqrt{r^2 - x^2} dx \tag{7}$$

Figure 10 indicates the probability that a link is selected for synchronization, depending on its length.
The same argument applies for the distance-based cost of SLOT-D: a link between routers at distance d is selected for synchronization if there are no routers which are closer to any of the link endpoints that both endpoints to each other. If the link cost corresponds exactly to its length, this condition leads to:

$$P(x \sim y|m(\overline{xy}) = d)_D = 1 - e^{-\nu d^2 (2\frac{\pi}{3} - sin(2\frac{\pi}{3}))} \tag{8}$$

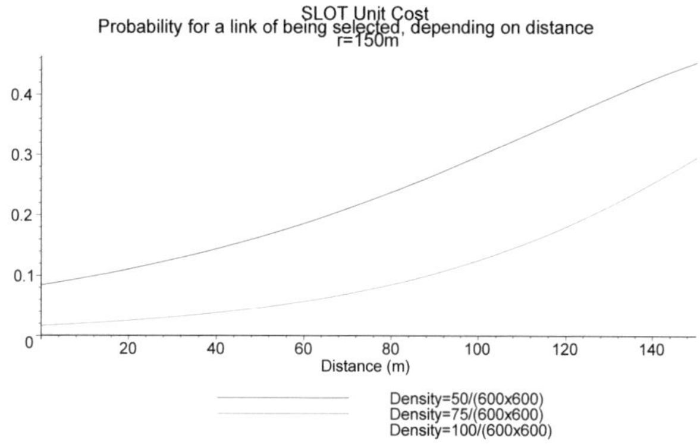

Fig. 10. Probability for a link of being selected, under SLOT (unit cost), for different network densities ν.

With a the more realistic model of link cost (*e.g.*, $cost = \lceil K\frac{d}{r} \rceil$), (8) becomes

$$P(x \sim y | m(\overline{xy}) = d)_D \quad = \quad 1 - e^{-\nu \lceil \frac{K}{r} d \rceil^2 (2\frac{\pi}{3} - sin(2\frac{\pi}{3}))} \tag{9}$$

where K stands for the number of discrete values for the distance-based quantized cost.

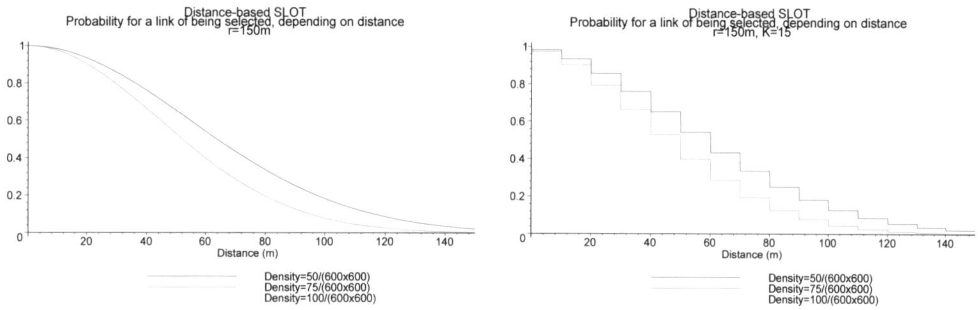

Fig. 11. Probability for a link of being selected, under SLOT based on distance, for different network densities ν.

The upper quantization of the link cost reduces the probability of selecting a router for synchronization. This is consistent with the effect observed in Figures 9.a and 12, in which the theoretical number of links per router achieved by SLOT-D (with an ideal link cost equal to the length) was significantly higher than the average number of links per router obtained in the static simulations (performed with a quantized link cost, $K = 10$).

4.4 The Smart Peering rule – SP

The Smart Peering rule was presented in Roy (2005) as a mechanism for link-state database synchronization and flooding in ad hoc networks ruled by OSPF. Under this rule, a router x synchronizes its link-state database with a bidirectional neighbor y if and only if:

- There are not *enough* available paths from x to y within the synchronized overlay (consisting on links selected through the Smart Peering rule).

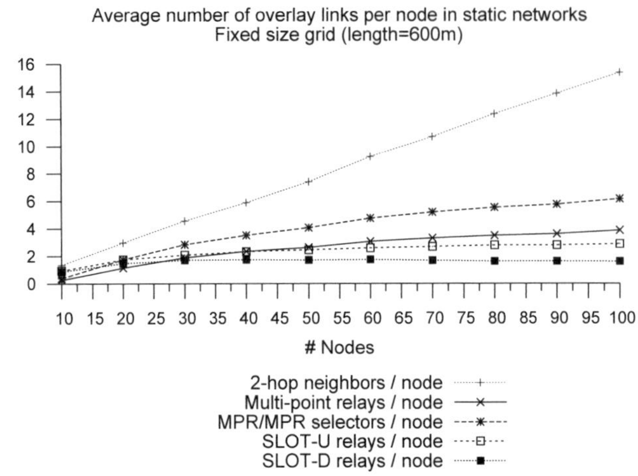

Fig. 12. Density of MPR and SLOT overlays in static networks. *The SLOT-D simulations are computed with the quantized cost:* $cost(\overline{xy}) = \lceil K\frac{d(\overline{xy})}{r}\rceil$, *with* $K = 10$ *and* $d(\overline{xy})$ *being the Euclidean distance between* x *and* y.

- The link between x and y provides a *significantly* cheaper path from x to y than those already present in the synchronized overlay.

The precise meaning of *enough* and *significantly* defines the different possible variations of Smart Peering. This section considers the most basic version: a neighbor is synchronized if and only if there are no paths within the synchronized overlay to it. In all variations a link synchronization is triggered when any of the involved routers (not necessarily both) decides to allow such operation.

Performing SP-based decisions requires that every router determines whether a synchronized path between itself and the router candidate to synchronization exists. When using a link-state routing protocol such as OSPF, such verification can be done by means of the Shortest Path Tree (SPT), if synchronized links are included and can be identified within the SPT. In this sense, Smart Peering needs to rely on a topology selection mechanism that advertises the synchronized links of the routers in the network.

From an asymptotic perspective, the overlay induced by the Smart Peering rule fulfills the topological requirements for a synchronized and a flooding overlay: *Lemma 4* shows that Smart Peering decisions lead to an asymptotically connected overlay. By construction, this overlay includes every router belonging to a connected ad hoc network, which trivially implies the density of the Smart Peering overlay.

Lemma 4 *Using the Smart Peering rule, every pair of routers* (x,y) *of a connected network are connected through at least one SP-synchronized path.*

Proof: Let d be the minimum distance in bidirectional hops from x to y ($d < \infty$).

- $d = 1$: if x and y are not already connected via an SP-path, the two routers will synchronize their link-state databases, by definition of Smart Peering.
- $d \Rightarrow d+1$. Let us consider the set of bidirectional neighbors of x, $N(x)$. There exists at least one $z \in N(A)$ for which $d(z,y) = (d+1) - 1 = d$, and is thus SP-connected to y (induction hypothesis). Calling $\overline{xz}$ the SP-route between x and z (which exists as proved for the case $d = 1$), $\overline{zy}$ the SP-route between z and y, it is clear that the route $\overline{xz} \cup \overline{zy}$ is an SP-route between x and y, and that concludes the proof. □

Unlike the techniques presented in previous sections, Smart Peering decisions are not taken under local (neighborhood) considerations. The overlay produced by the Smart Peering rule for a given ad hoc network cannot thus be deduced from the relations between routers. Rather, it may be significantly affected by aspects such as the order of appearance of the routers in the network, the trajectory of routers or the mobility patterns of the network. This latter is probably one of the most interesting features of the Smart Peering rule for mobile ad hoc networks.

For a static and stable network with error-free links, in which synchronization decisions are taken independently and concurrently, the overlay induced by Smart Peering is roughly equivalent to the full network overlay. When a router first appears in a network, and is discovered by its neighbors, none of them has any trace of it in the link-state database maintained locally. In consequence, all of them initiate synchronization processes with the new router (the argument is also valid from the point of view of such new router).

In wireless ad hoc networks, the characteristics of the Smart Peering relay are more unpredictable. For mobile scenarios, the SP rule filters the less stable links, those between routers with high relative speed. Once the first LSDB synchronization of a router has been completed and advertised to the whole network, no other router will accept a new synchronization with it as long as the trace of the first one remains. Highly mobile routers will therefore have difficulties to establish synchronized links after the completion of the first one, while routers presenting a lower relative speed to their neighbors will have more chances to keep up their synchronized links by means of their initial performed synchronization. This behavior is confirmed empirically (via simulations) in section 5.4.

5. Application: OSPF extensions for ad hoc operation

This section addresses an experimental evaluation of the overlay techniques presented in section 4, implemented as extensions to the modules of flooding, LSDB synchronization and topology selection of OSPF. These extensions are tested in ad hoc networks, both static and mobile, but would coexist with classic OSPF in compound wired/wireless networks. Section 5.1 indicates the parameter set and the implementations used for the simulations. Section 5.2 describes briefly the main elements of OSPF, and section 5.3 presents each of its analyzed ad hoc extensions. Sections 5.4, 5.5, 5.6 and 5.7 discuss the performance of such extensions and their corresponding techniques in the different link-state routing modules.

5.1 Simulation parameters

Implementations of the extensions are publicly available[12], and were simulated with the Georgia Tech Network Simulator (GTNetS, see Riley (2003) for reference). Unless otherwise specified, the set of simulation parameters corresponds to the set described in Baccelli *et al.* (2009), except for the following aspects:

- Node mobility: constant node speeds, $0\frac{m}{s}$ (static scenario) and $5\frac{m}{s}$ (mobile scenario).
- Pause time: $0 sec$.
- Time interval between periodic topology refloods (*LSRefreshInterval*): $20 sec$.

5.2 Overview of OSPF

The Open Shortest Path First protocol (OSPF) is, together with IS-IS, one of the most widespread protocols for link-state routing within an Autonomous System [Halabi (2000)].

[12] INRIA OSPF Extensions for MANET Code: `www.emmanuelbaccelli.org/ospf`

Although it supports a hierachical 2-level structure based on *areas*, this section focuses on a single area scheme[13].

Routers in OSPF maintain an identical Link-State Database (LSDB) and thus share the exact same view of the network topology. The routes are extracted from the Shortest Path Tree (SPT), which is computed over the LSDB by means of the Dijkstra algorithm. Network topology information is disseminated through topology update messages called *Link State Advertisements* (LSA). These LSAs are flooded in a reliable manner (that is, implying hop-by-hop acknowledgements in case of successful transmission and retransmission in case of failure), both periodically and following a topology change event.

In OSPF, the flooding operation depends on the type of interface performing the transmission. For Non-Broadcast Multiple Access (NBMA) interfaces, the flooding is centralized by a *Designated Router* (DR), which is elected by all routers in the link. Such DR synchronizes its LSDB with those of all its neighbors (in OSPF terminology, synchronized links are denominated *adjacencies*), and it is responsible of diffusing topology updates (LSAs) originated in the link to all its synchronized (adjacent) neighbors. Point-to-point synchronization is performed by exchanging *Database Description* (DBD) packets and requesting the most updated LSAs that are missing in a reliable fashion. LSAs originated by the routers (Router-LSAs) advertise the adjacent links of the originator.

Routers annouce their presence in the network and learn the presence of its neighbor through the periodical exchange of *Hello* messages. Typically, such messages advertise the source identity and the list of neighbhors. That allows every router in the network to keep track of its 2-hop neighborhood, as well as to establish symmetric (*bidirectional*) communication with its 1-hop neighbors. Flooding, synchronization and routing decisions are performed over the available bidirectional links in the network.

These properties define implicitly an OSPF link model: adjacencies are selected among the set of bidirectional links; and the Shortest Path Tree is computed over the adjacent overlay. The fact that flooding is performed over adjacent links implies that control traffic (LSAs and database exchange packets) only flows through synchronized links, while data traffic is sent via shortest paths. These two principles constitute the core of the OSPF routing philosophy.

5.3 OSPF-based configurations

This section examines different configurations of OSPF that explore other principles for data and control traffic forwarding than those from classic OSPF. They combine the overlay techniques presented in section 4 to optimize the performance in ad hoc networks of the link-state routing modules (operations). Table 2 summarizes the architecture of such configurations.

	MPR-OSPF	MPR+SP	OR/SP	SLOT-OSPF
Flooding	MPR	MPR	MPR(SP)	MPR
Synchronization	MPR+synch	SP	SP	SLOT-U
Routing	Path MPR	Path MPR	SP	Path MPR

Table 2. OSPF MANET Configurations.

MPR-OSPF and OR/SP are standard extensions for OSPF MANET (RFC 5449 (2009) and RFC 5820 (2010)). The considered configuration of Overlapping Relays (OR) performs flooding, synchronization and topology selection based exclusively in Smart Peering (SP), without considering additional links (denominated *unsynchronized adjacencies* in RFC 5820) in any of

[13]For a more detailed description of OSPF and its area-based architecture, see Moy (1998b).

these operations. SLOT-OSPF is a variation of MPR-OSPF which uses SLOT-U for LSDB synchronization, while keeping MPR as a flooding and topology selection overlay (Path MPR). Finally, MPR+SP incorporates Smart Peering as synchronization criterion in the framework of MPR-OSPF, also keeping MPR for flooding and topology selection.

5.4 Overlay properties validation

The simulations validate the main properties of the overlays described in section 4. Figure 13.a displays the average density of the considered overlay techniques, for a static and moderately mobile scenario (constant router speed, 5 m/s). In the static deployment, it can be observed that the Smart Peering rule produces the most dense overlay, showing a linear increase with respect to the full network density. Its size is only comparable to the one achieved by the MPR synchronized overlay for low density networks, but it has to be pointed out that a significant part of such overlay consists of *persistent* sychronized links which are not costly in terms of database exchange. Therefore, the network overlay reduction performed by SP is very low when routers do not move and thus the corresponding router traces cannot be used to reject (a part of the) new synchronizations – this effect disappears in mobile scenarios (see Figure 13.b). The MPR flooding overlay achieves a significantly lower density than the sychronized MPR overlay, and the Synchronized Link Overlay Triangular (SLOT) for unit link costs remains below the theoretical upper bound shown in *Theorem 1*.

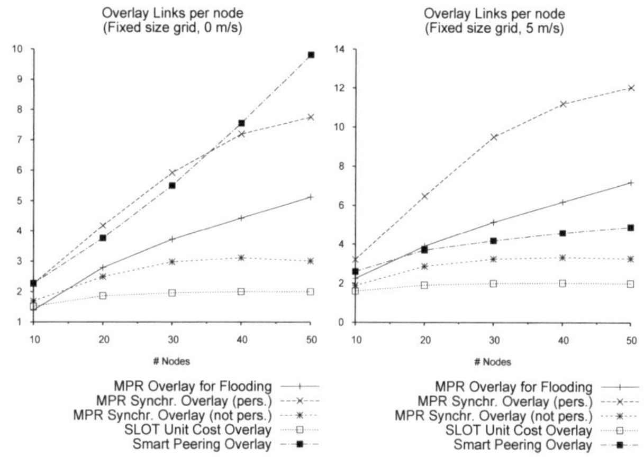

Fig. 13. Overlay link density for **(a)** static scenarios and **(b)** mobile scenarios (5 m/s).

5.5 Link synchronization – adjacencies

The average lifetime of the synchronized links (adjacencies) for each configuration is displayed in Figure 14.a. It confirms that adjacencies selected through the Smart Peering rule (both in configurations MPR+SP and Overlapping Relays) are more stable than those selected by MPR-OSPF. The Smart Peering ability to choose the most stable links for synchronization is also visible in Figure 14.b, where the adjacent set of SP configurations remains stable for a significant range of link quality[14] values. On the contrary, MPR-OSPF keeps increasing the number of adjacencies as α grows (the channel becomes more reliable).

Adjacency stability of MPR+SP and Overlapping Relays present a significant difference, although both configurations rely on the same technique (Smart Peering) for selecting synchronized links. This gap is caused by the neighbor keep-alive mechanism. In OSPF,

[14]For more details on the link quality model and the parameter $\alpha \in [0,1]$, see Henderson *et al.* (2005).

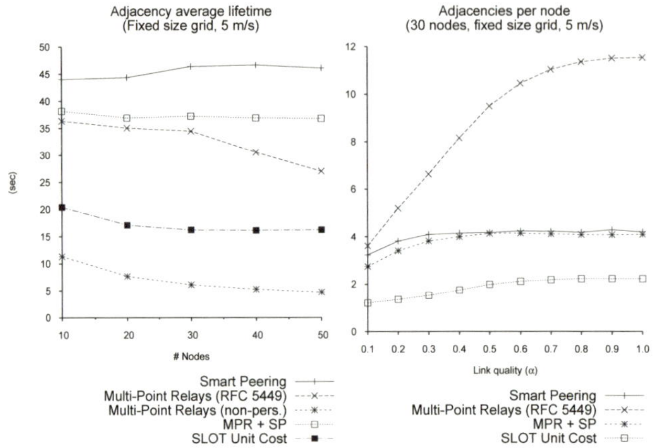

Fig. 14. **(a)** Average adjacency lifetime, depending on the network density, and **(b)** Average number of adjacencies depending of link quality (30 routers, 5 m/s).

a router declares a neighbor *dead* if it has not received a Hello packet from it during a *DeadInterval* period. In the simulated moderately lossy channel ($\alpha = 0.5$), the probability of losing a packet is related to the length of such packet. Since Hello packets of MPR+SP are significantly longer than those of Overlapping Relays (see Figure 15.a), the loss of packets in the former configuration is more relevant, causing the breakup of more links (and, in particular, adjacencies) than for the latter. The impact of such keep-alive mechanism has been measured in Figure 15.b, which shows the variation of the adjacency lifetime achieved for the same MPR+SP configuration when Hellos are the only packets assuming a keep-alive role, and when other packets (Link State Updates) are used as well for the same purpose.

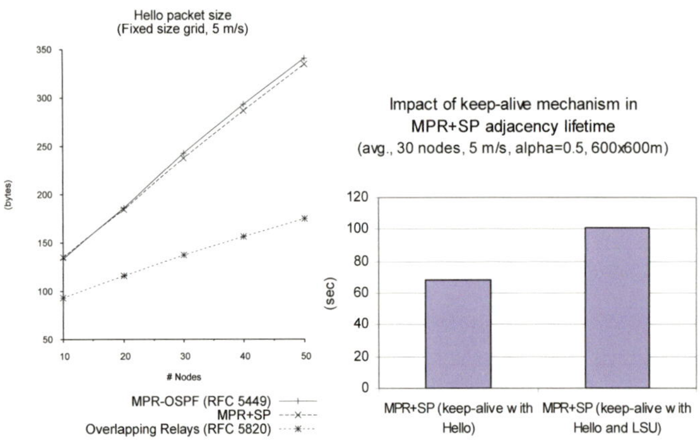

Fig. 15. **(a)** Average size of Hello packets, **(b)** Impact of keep-alive configuration in adjacency lifetime (fixed grid of length $600m$, with speed $5m/s$).

5.6 Route selection and routing quality

From the considered overlay techniques, only Multi-Point Relaying can be used as a basis for an efficient (optimal) topology selection mechanism. In section 4 it was shown that the Path MPR algorithm, which adapts the MPR to the requirements of a topology selection overlay, generates an overlay that contains the network-wide shortest paths. The impact of such property is shown in Figure 16.a, which compares the average path length for data traffic of configurations using Path MPR, with the path length achieved by a configuration that uses Smart Peering, which does not advertise in general optimal routes – the Overlapping Relays configuration, without *unsynchronized adjacencies*. Suboptimal routing of data traffic may lead to a significant waste of bandwidth dedicated to forward data packets through non-optimal routes – see Figure 16.b for data delivery ratio: the configuration not providing shortest paths performs significantly worse than the others.

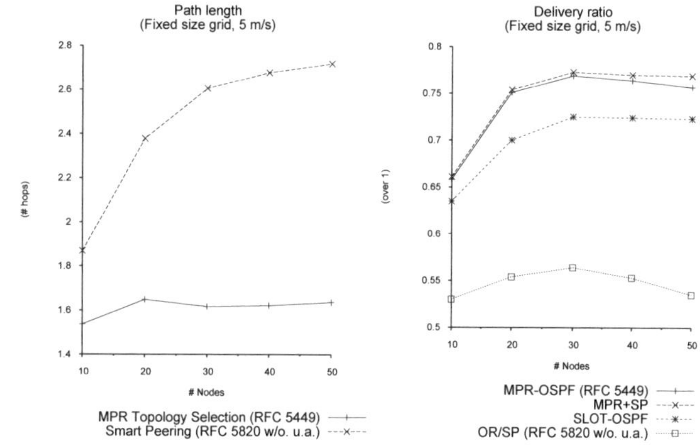

Fig. 16. **(a)** Path length, **(b)** Data delivery ratio, for $5m/s$.

5.7 Flooding and control traffic overhead

MPR and SP are used for reliable flooding in the configurations analyzed in this section. MPR-OSPF relies on (directed) MPR links to flood control traffic, expecting acknowledgement from MPR synchronized (persistent) links. In the Overlapping Relays configuration, the flooding operation is performed in two rounds. The first one involves the MPRs computed among the Smart Peering links, called *active Overlapping Relays*. Absent this acknowledgement, other SP-synchronized neighbors may retransmit the pending packet until it is acknowledged.

Let us first discuss the implications of the MPR election over the SP-synchronized overlay. When compared with the selection of Multi-Point Relays among the bidirectional neighbors of a source, Overlapping Relays presents a lower amount of MPRs per router and a significantly higher stability of such relays, as shown in Figure 17.a and 17.b.

The drawbacks of this approach are however significant. In first term, computing MPRs over a restricted overlay weakens the main advantage of using Multi-Point Relays for flooding, which is the ability of reaching all the 2-hop neighbors of the source while avoiding redundant transmissions. Since the neighborhood topology in which MPR operates is distorted by the Smart Peering selection rule, the set of reachable 2-hop neighbors becomes also affected and the quality of the flooding operation becomes damaged, as Figure 17.c shows.

In second term, MPR selection over the Smart Peering overlay makes the MPR computation nearly irrelevant. If the probability of relaying an MPR flood is close to $\frac{M_r}{M}$ (with M_r being the average number of relays per router and M the average number of bidirectional neighbors), the situation in sparse networks (such as the Smart Peering overlay) is close to $M_r = M$, meaning that almost every SP-synchronized neighbor will become a multi-point relay, thus making wasteful the relay selection process.

The control traffic mobilized by every configuration is displayed in Figure 18, together with the overall traffic. Such overall control traffic has two main components: the traffic dedicated to adjacency-forming processes, which depends on the synchronized overlay, and the reliable

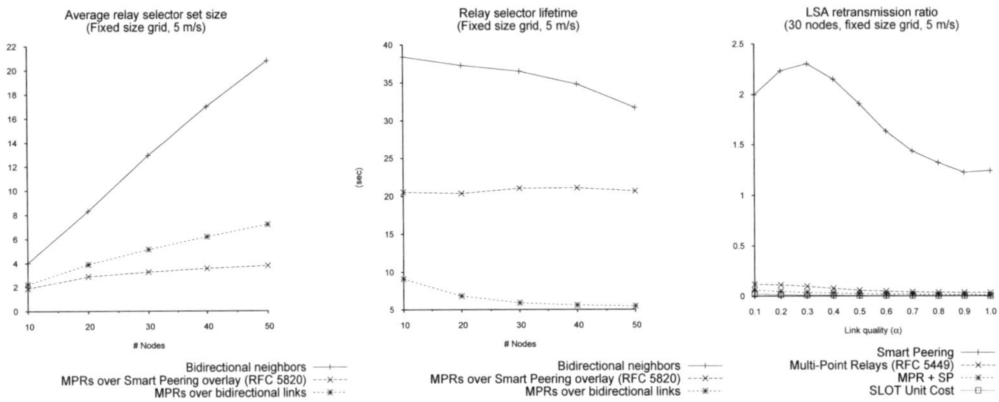

Fig. 17. **(a)** Average size of the MPR set and **(b)** average relay lifetime ($5m/s$). **(c)** LSA retransmission, depending on the link quality (30 routers, $5m/s$. *The LSA retransmission ratio is the number of backup LSA retransmissions over the number of primary LSA transmissions.*

flooding traffic. The figure shows that the analyzed configurations can be grouped in three categories: the one not using MPR at all (Overlapping Relays, with relies on Smart Peering), those using MPR only as a flooding rule (SLOT-OSPF and MPR+SP), and the configuration using MPR both for flooding and synchronization purposes. The results indicate that, while MPR flooding has in general a better performance (in terms of overhead) than other flooding overlays, the use of MPR as a synchronization rule has significant shortcomings in terms of overhead – a conclusion that is consistent with section 4.2. Therefore, configurations exploring less dense overlays for synchronization, while keeping MPR as the reliable flooding overlay, present more balanced trade-offs between flooding quality and control traffic overhead.

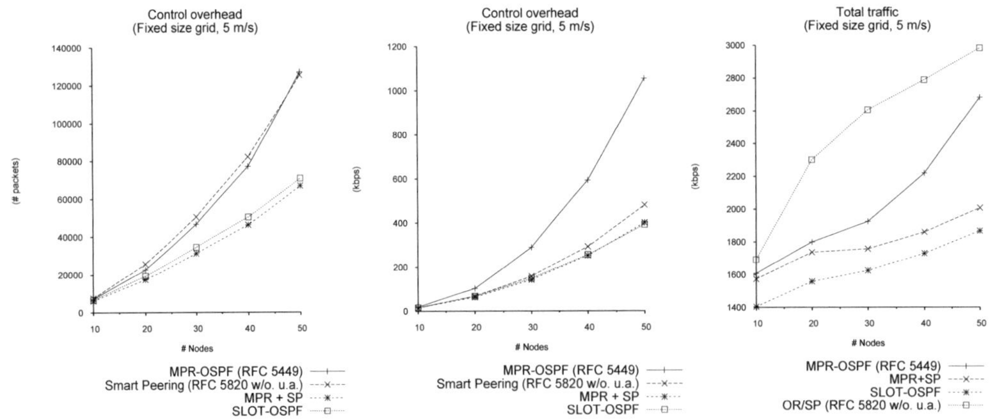

Fig. 18. **(a)** Control traffic overhead, in number of packets, **(b)** in kbps, and **(b)** Total (data+control) traffic, in kbps ($5m/s$).

6. Conclusion

This chapter has investigated the subject of compound networks, *i.e.* networks comprising of fixed wired routers as well as wireless mobile ad hoc routers. A single protocol is desired to provide routing over such networks in order to avoid sub-optimality due to paths through

gateways between incompatible protocols, and lack of efficient traffic engineering. This chapter has thus reviewed various mechanisms that enable OSPF to fulfill this task. OSPF is indeed a designated candidate for this job, as it is both a popular routing solution for wired IP networks, and quite similar to OLSR, the most deployed MANET routing protocol, also based on a link state algorithm.
This chapter analyzed the specific mechanisms that enable efficient link state routing in compound networks, focusing in particular on different overlay techniques. These techniques were compared via simulations their performance when applied to OSPF. The resulting analysis may be useful in order to design appropriate routing protocols for compound networks.

7. References

Adijh, C.; Baccelli, E.; Clausen, T.; Jacquet, P. & Rodolakis G. (2004). Fish Eye OLSR Scaling Properties. *IEEE Journal on Communications and Networks (JCN)*, Special Issue on Mobile Ad Hoc Wireless Networks, December 2004, pp. 343-351.

Baccelli, E. (2006). *Routing and Mobility in Large Heterogeneous Packet Networks*. PhD Thesis, École Polytechnique, Paris (France).

Baccelli, E.; Jacquet, P.; Nguyen, D. & Clausen, T. (2009). *OSPF Multipoint Relays (MPR) Extension for Ad Hoc Networks*, RFC 5449, IETF, February 2009.

Baccelli, E.; Cordero, J. A. & Jacquet, P. (2009). Multi-Hop Relaying Techniques with OSPF on Ad Hoc Networks, *Proceedings of the 4th IEEE International Conference on Sensor Networks and Communications*, pp. 53-62, IEEE ComSoc, Porto (Portugal), Sept. 2009.

Baccelli, E.; Cordero, J. A. & Jacquet, P. (2010). Using RNG for Reliable Database Synchronization in MANETs, *Proceedings of the 5th IEEE SECON Workshop on Wireless Mesh Networks (WiMESH)*, pp. 13-18, IEEE ComSoc, Boston, MA (United States), June 2010.

Bellman, R. (1958). On a Routing Problem, In: *Quarterly of Applied Mathematics*, No. 16, pp. 87-90.

Clausen, T.; Jacquet, P. (2003). *Optimized Link State Routing Protocol (OLSR)*, RFC 3626, IETF, October 2003.

Coltun, R.; Ferguson, D.; Moy, J. (2008). *OSPF for IPv6*, RFC 5340, IETF, July 2008.

Cordero, J. A. (2010). MPR-based Pruning Techniques for Shortest Path Tree Computation, *Proceedings of the 18th International Conference on Software Telecommunications and Computer Networks*, IEEE ComSoc, Split (Croatia), September 2010. *(to appear)*

Dijkstra, E. W. (1959). A Note on Two Problems in Connection with Graphs, In: *Numerische Mathematik*, No. 1, pp. 269-271.

Ford, L. R. Jr. & Fulkerson, D. R. (1962). *Flows in Networks*, Princeton University Press.

Halabi, S. (2000). *Internet Routing Architectures*, Cisco Press.

Henderson, T.; Spagnolo, P. & Kim, J. H. (2003). A Wireless Interface Type for OSPF, *Proceedings of the IEEE Military Communications Conference (MILCOM)*, pp. 137-145, IEEE ComSoc, Boston, MA (United States), October 2003.

Henderson, T.; Spagnolo, P.; Pei, G. (2005). *Evaluation of OSPF MANET Extensions*, Technical Report, D950-10897-1, Boeing, July 2005.

International Organization for Standardization (2002). *Intermediate System to Intermediate System intra-domain routeing information exchange protocol*, International Standard, Ref. Number ISO/IEC 10589:2002(E), ISO Secretariat, Geneva (Switzerland), 2002.

Jacquet, P. (2006). Control of Mobile Ad hoc Networks, *Proceedings of the IEEE Information*

Theory Workshop, pp. 97-101, IEEE, Punta del Este (Uruguay), March 2006.
Moy, J. (1998a). *OSPF Version 2*, Request For Comments 2328, IETF, April 1998.
Moy, J. (1998b). *OSPF: Anatomy of an Internet Routing Protocol*, Addison-Wesley.
Ni, S.Y.; Tseng, Y.-Ch.; Chen, Y.-S.; Sheu, J.-P. (1999). The Broadcast Storm Problem in a Mobile Ad Hoc Network, *Proceedings of the Annual International Conference on Mobile Computing and Networking*, pp. 151-161, ACM Press, Seattle (United States), August 1999.
Qayyum, A.; Viennot, L. & Laouiti, A. (2002). Multipoint Relaying for Flooding Broadcast Messages in Mobile Wireless Networks, *Proceedings of the 35th Hawaii International Conference on System Sciences*, IEEE ComSoc, Hawaii, HI (United States), January 2002.
Riley, G. F. (2003). The Georgia Tech Network Simulator, *Proceedings of the ACM SIGCOMM Workshop on Models, Methods and Tools for Reproducible Network Research*, pp. 5-12, ACM Press, Karlsruhe (Germany), August 2003.
Roy, A. (2005). *Adjacency Reduction in OSPF using SPT Reachability*, Internet-Draft `draft-roy-ospf-smart-peering-01.txt`, IETF, November 2005. *(obsolete)*
Roy, A. & Chandra, M. (2010). *Extensions to OSPF to Support Mobile Ad Hoc Networking*, RFC 5820, IETF, March 2010.
Toussaint, G. T. (1980). The Relative Neighborhood Graph on a Finite Planar Set, In: *Pattern Recognition*, No. 12, pp. 261-280.
Tseng, Y.-Ch.; Ni, S.-Y. & Shih, E.Y. (2003). Adaptive Approaches to Relieving Broadcast Storms in a Wireless Multihop Mobile Ad Hoc Network. *IEEE Transactions on Computers*, Vol. 52, No. 5, May 2003, pp. 545-557.

17

Multiple Multicast Tree Construction and Multiple Description Video Assignment Algorithms

Osamah Badarneh[1] and Michel Kadoch[2]
[1]*Yarmouk University*
[2]*Ecole de Technologie Superieure*
[1]*JORDAN*
[2]*CANADA*

1. Introduction

In this chapter, we introduce novel algorithms for constructing multiple multicast tree and assigning multiple description (MD) video to a group of heterogeneous multicast destinations. Our main objective is to increase the number of assigned MD video to each destination node. In order to achieve our objective, we propose to employ the independent-description property of MDC (multiple description coding) along with multiple multicast tree. We mean by independent-description property of MDC the following. If there are three video descriptions, for example, then receiving any subset of video descriptions, i.e., $(\{VD_1, VD_2, VD_3\}, \{VD_1, VD_2\}, \{VD_1, VD_3\}, \{VD_2, VD_3\}, \{VD_1\}, \{VD_2\}, \{VD_3\})$ of the video descriptions will reproduce the original video in different qualities depending on the number of video descriptions received.
A main issue of video multicasting for heterogeneous destinations is the assignment of video descriptions and the construction of multicast trees. However, the assignment of MD video and the construction of multicast tree can greatly affect the user satisfaction (i.e., affect the number of assigned video description to each destination and hence affect the quality of the received video. However, many questions are raised: How multiple multicast tree should be constructed? And how MD video should be assigned? Is it better to construct multiple multicast tree first and then assign the video descriptions? Or is it better to assign the video descriptions first and we then construct multiple multicast tree? Should we perform that in a distributed manner or in a centralized one? Does the independent-description of MDC increase the user satisfaction?
To answer these questions, we propose different algorithms to construct multiple multicast tree and to assign MD video. The proposed algorithms are: Serial MDC, Distributed MDC, Centralized MDC, and sequential MDC. Serial MDC algorithm constructs multiple paths, to each destination, and assigns a different video description to each of them. After that, it constructs multiple multicast tree based on the assignment of MD video. Distributed MDC algorithm assigns MD video and constructs multiple multicast tree in parallel and in distributed fashion. In Centralized MDC, the assignment of MD video and the construction of multiple multicast tree are performed in a centralized manner. However, Centralized MDC first constructs multiple multicast tree and then assigns different video description to each multicast tree. Finally, Sequential MDC sequntially assigns MD video to each multicast tree.This means that all destinations should be assigned the first description. Then,

destinations that need another description should be assigned the second video description, and so on. The main difference between Sequential MDC and Centralized MDC algorithms is that the former does not employ the independent-property of MDC.
We evaluate and compare our proposed algorithms under different network conditions. For example, network size, and multicast group size. Simulation results demonstrate that, indeed, the way of multicast trees construction and the assignment of MD video can greatly affect the user satisfaction. In addition, simulation results show that MDC can achieve higher user satisfaction compared to Layered Coding (LC) with a small cost in terms of number of pure forwarders nodes, bandwidth utilization, and aggregate tree delay. Furthermore, simulation results show that the independent-description property of MDC can increase the user satisfaction.
The rest of this chapter is organized as follows. In the next section, we present the related work. In section 3, we present our network model and problem formulation of video multicasting. In Section 4, we describe our proposed algorithms for constructing multiple node-disjoint multicast trees and assigning MD video. In Section 5, we evaluate our proposed algorithms. The complexity analysis of the protocols is presented in Section 6. Finally, our conclusions are presented in Section 7.

2. Related work

An ad hoc network is a multihop wireless network without a preinstalled infrastructure or centralized administration. It can be deployed in situations where infrastructure is unavailable or where temporary network is needed. In this network, nodes are free to move randomly anytime, anywhere, and arrange themselves as required. Since nodes are often not within the radio transmission range of each other, each node operates not only as a host but also as a router, forwarding packets for other mobile nodes. In a typical ad hoc environment, mobile nodes work as a group to accomplish a certain task. Hence, multicast is very useful and efficient means of supporting group-oriented applications. Multicast is an essential technology for many applications such as video distribution and group video conferencing, and results in bandwidth and power savings as compared to multiple unicast sessions.
Many researches over the last several years have focused on unicast and multicast video transmission over wireless ad hoc networks (Wei & Zakhor, 2007; Mao, Cheng, Hou & Sherali, 2006; Agrawal et al., 2006; Chow & Ishii, 2008; Mao, Hou, Cheng, Sherali, Midkiff & Zhang, 2006; Mao et al., 2003). The main objective of these researches is to improve the quality of the received video by exploiting the error resilience properties of MDC along with multiple paths. In other words, MD video are encoded and transmitted over different paths to each destination node. If only any path is broken, packets corresponding to the other descriptions on the other paths can still arrive at the destination node on time.
MDC has been proposed as an alternative of the LC (Layered Coding) technique. In contrast to LC, MDC is a coding technique which fragments a single media stream into independent bit-streams, where the multiple bit-streams are referred to as multiple descriptions. In order to decode the media stream, any description can be used (we referred to as "independent-description" property (Badarneh et al., 2008)); however, the quality improves with the number of descriptions received in parallel. The idea of MDC is to provide error resilience to media streams. Since an arbitrary subset of descriptions can be used to decode the original stream, network congestion or packet loss, which is common in best-effort networks

such as the Internet, will not interrupt the stream but only cause a temporary loss of quality. The quality of a stream can be expected to be roughly proportional to data rate sustained by the receiver (Goyal, 2001; Puri & Ramchandran, 1999).
Video multicast over wireless ad hoc networks with path diversity has been studied in Wei & Zakhor (2007); Mao, Cheng, Hou & Sherali (2006); Agrawal et al. (2006); Chow & Ishii (2008). Chow and Ishii have proposed a multicast protocol for video transmission called MT-MAODV (Multiple Trees Multicast Ad Hoc On-demand Distance Vector) (Chow & Ishii, 2008). An extension to the well-knownMAODV to construct two optimally disjoint multicast trees in a single routine for video multicast was proposed. MDC scheme is used to split the video into several independent and equally important video descriptions. Each description is transmitted over different tree. In (Mao, Cheng, Hou & Sherali, 2006), the authors introduced a multicast approach for multiple description video over ad hoc networks. An application-centric, cross-layer routing approach with the objective of minimizing the over all video distortion was proposed. In this approach multiple source trees for MD video multicast are used. Furthermore, each description is coded into a base layer and number of enhancement layers. Packets belonging to the same description from both the base layer and enhancement layers are transmitted on the same tree. The authors showed that this approach can effectively deal with frequent link failures and diverse link qualities in wireless ad hoc networks. Agrawal et al. have presented a multiple tree protocol called Robust Demand-driven Video Multicast Routing (RDVMR) (Agrawal et al., 2006). RDVMR explores the path diversity and error resilience properties of MDC. RDVMR deploys a novel path based Steiner tree heuristic to reduce the number of forwarding nodes in each tree, and constructs multiple trees in parallel with a reduced number of common nodes among them to provide robustness against path breaks and to reduces the total data overhead. Two multiple tree multicast routing protocols were presented in (Wei & Zakhor, 2007). Serial MDTMR protocol (Multiple Disjoint Trees Multicast Routing) constructs two disjoint multicast trees in a serial fashion. However, in order to reduce routing overhead and construction delay of serial MDTMR, parallel MNTMR (Multiple Nearly-disjoint Trees Multicast Routing) was suggested. This protocol constructs two nearly-disjoint multicast trees in a single routine by dividing the network virtually into two parts and tree construction is carried out simultaneously at both virtual topologies. Both serial MDTMR and parallel MNTMR protocols explore MDC to provide robustness for video multicast applications. In order to improve the quality of the received video, the video was split into two descriptions and each description was transmitted over a different tree.

3. Network model and problem formulation

3.1 Network model for multicasting

We consider a multi-hop wireless ad hoc network with $\mathcal{V}$ nodes. The nodes communicate with each other via wireless links. Each node in the network can communicate directly with a subset of the other nodes in a network. A node v can transmit directly to node u if the both nodes are within the transmission range of each other. We modeled a wireless ad hoc network as weighted $\mathcal{G} = (\mathcal{V}, \mathcal{E})$, where $\mathcal{V}$ is a set of wireless nodes each with random location and $\mathcal{E}$ is a set of wireless communication links between the nodes. A link between node pair $\{v, u\}$ indicates that both nodes v and u are within each other's transmission range. The nodes in set $\mathcal{V}$ can be of the following three types:

- **Multicast source node**: The node that sends out the multicast video packets. We denote it by $\mathcal{S}$.

- **Destination node**: A node that receives the multicast video packets. The set of destination nodes in a multicast tree is denoted by $\mathcal{Y} \subseteq \mathcal{V} - \mathcal{S}$
- **Forwarder node**: A node that is an intermediate hop in the path from the source $\mathcal{S}$ to a destination node in $\mathcal{Y}$. It is denoted by $\mathcal{F}$.

Two positive real-valued functions are defined on a link $e = \{v, u\} \in \mathcal{E}$, namely:

- **Link Delay**: $d(e) \in \Re^{+}$.
- **Link Bandwidth**: $Bw(e) \in \Re^{+}$.

In this work, we focus on the network layer, i.e., the construction of multiple multicast trees and the assignment of MD video. We assume that the physical and MAC layers dynamics, such as the link delay and bandwidth, are translated into the network layer parameters. These parameters can be measured at every node and distributed through the network using LSAs (Link State Advertisements) (Clausen & Jacquet, year 2003).

DEFINITION 1: *A path p from the multicast source $\mathcal{S}$ to a destination node in $\mathcal{G}$ is defined as a list of nodes $(v_1, v_2, \cdots, v_k)$ such that $\forall\, j, 1 \leq j \leq k, e_j = (v_j, v_{j+1}) \in \mathcal{E}$ and no node appears more than once.*

The delay of the path p is the sum of all link delays, that is,

$$d(p) = \sum_{j=1}^{k-1} d\left(e_j\right) \tag{1}$$

The bandwidth of the path p is the minimum available bandwidth of all links, which is defined as

$$Bw(p) = \min_{e_i \in p} \{Bw(e_i)\} \tag{2}$$

In case of $\mathcal{K}$ node-disjoint paths, $P = \{p_1, p_2, \cdots, p_{\mathcal{K}}\}$, then the delay of the $\mathcal{K}$ paths for a destination node is:

$$d(P) = \max_{p_j \in P} \left\{d\left(p_j\right)\right\} \tag{3}$$

Let L be the number of the multicast trees constrcuted to meet the destinations' requirments, then the delay of the tree-aggregate $T = t_1 \cup t_2 \cup \cdots t_L$ is defined as:

$$d(T) = \max_{l \in |1 \ldots L|} d(t_l) \tag{4}$$

where $d(t_l)$ is the delay of a multicast tree t_l, which is defined as the longest delay from the source $\mathcal{S}$ to the destinations on t_l, that is:

$$d(t_l) = \max_{p_i \in t_l} \{d(p_i)\}, \quad i = |1 \ldots m| \tag{5}$$

where m is the number of destinations on t_l.

3.2 Problem formulation

Our problem of MD video assignment can be formulated as follows: Given a wireless ad hoc network $\mathcal{G} = (\mathcal{V}, \mathcal{E})$, with $\mathcal{N}$ number of MD video, a link delay, a link bandwidth, a source $\mathcal{S}$, and a set of destinations $\mathcal{Y} = \{R_1, R_2, \cdots, R_m\}$ such that each destination node $R_i \in \mathcal{Y}$ requires a preference number of MD video, then construct multiple node-disjoint multicast tree spanning $\mathcal{Y} \cup \mathcal{S}$ such that the total number of the assigned video descriptions to each destination is maximized. That is:

$$maximize\,\{\mathcal{N}_{asg}(R_i)\} \tag{6}$$

where $\mathcal{N}_{asg}(R_i)$ is the number of the assigned video descriptions to the destination R_i.
To minimize the delay of every path from the source $\mathcal{S}$ to each destination $R_i \in \mathcal{Y}$, the shortest path tree algorithm is deployed.

4. Multiple multicast tree construction and multiple description video assignment algorithms

4.1 Serial MDC algorithm

The MD video assignment and multiple multicast trees construction algorithms are shown in algorithms $1-4$. At the beginning, let the multicast source has a partial topology that contains multiple paths to each destination, as shown in Fig. 1(a). Following, it arranges the destinations that require one and two video descriptions in a descending order according to their number of node-disjoint paths in the sets x and y, respectively. After that, it checks the destinations in the set y if any of them has only one path, if yes, it adds it to the set x. At the end of these steps, the sets x and y contain the destinations arranged in a descending order according to their number of paths. After that, the source node runs the algorithms $1-4$. We use the two colors: red and green to refer to the first and second descriptions, respectively. The multicast source starts with the set y and constructs its red (R) and green (G) paths for each destination if possible. To find the R-path, the green nodes (G-nodes) should be removed because they already have been assigned a description and they cannot be on another tree. However, to find the G-path, the red nodes (R-nodes) should be removed. The R and G paths are constructed using shortest path algorithm (in terms of delay).
When the set y is empty, the source node starts with the set x. Since any description can reproduce the original video signal, this, what we referred to as independent-description property of MDC, therefore the multicast source will assign any color (R or G) to each destination in the set x.
Based on the sets of multiple paths $\mathcal{K}_{R_i}$ (the R and G paths) for every destination R_i, then the multicast source S constructs multiple multicast trees for the video transmission according to algorithm 4. That is, all nodes that have been assigned the same color are attached to the same tree. For example, the nodes that have been assigned the R-color are attached to the first tree (R-tree) and the nodes that have been assigned the G-color are attached to the second tree (G-tree). Fig. 1 is an illustrative example.

4.2 Distributed MDC algorithm

In this algorithm the assignment of MD video and the construction of multiple multicast trees are performed in a distributed manner. Each node in the network will only select one video description to transmit it to its neighbor nodes. This condition is to ensure disjointness between multicast trees. Destination nodes are responsible to construct multiple node-disjoint paths to the multicast source, node S . Each destination node will select a number of disjoint

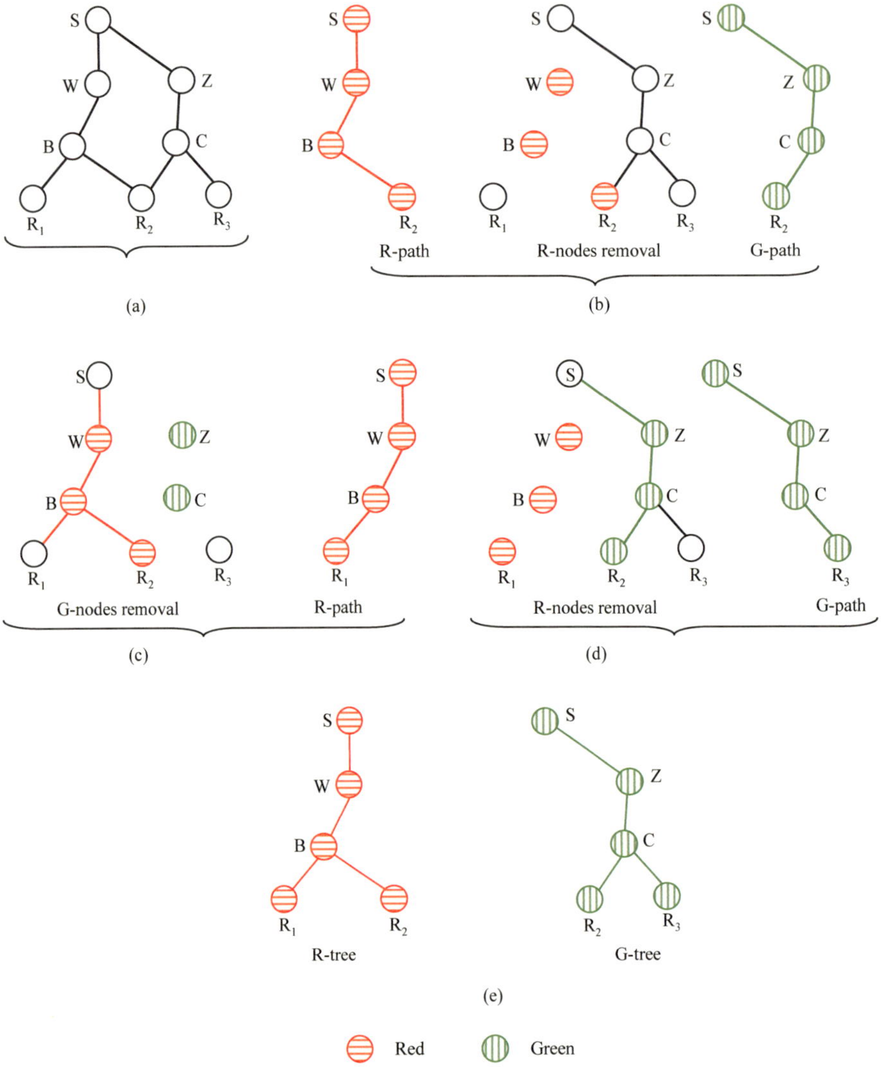

Fig. 1. Serial MDC: An illustrative example: (a) Partial topology. (b) Multiple paths construction and nodes removal for destination R_2. (c) Multiple paths construction and nodes removal for destination R_1. (d) Multiple paths construction and nodes removal for destination R_3. (e) Multiple multicast trees construction.

paths equal to its preference number of MD video. If there are two paths have the same video description, the one with shortest delay will be chosen.
The source node S will broadcasts the information of the available MD video and the bandwidth requirements for each description to its neighbor nodes. Neighbor nodes that have enough bandwidth will randomly choose one description and rebroadcasts it along with its bandwidth requirement to its neighbor nodes. As we mentioned previously, each node will only choose one description to transmit it to its neighbor nodes to maintain disjointness between multicast trees. This process will continue to reach a destination node.

Algorithm 1 Serial MDC

Given: $\mathcal{G} = (\mathcal{V},\mathcal{E})$, set x, and set y
for $\forall\, i \in$ set y **do**
  Z = set y
  Construct a R_{path} using algorithm 2
  Construct a G_{path} using algorithm 3
end for
for $\forall\, k \in$ set x **do**
  Z = set $x \cup$ set y
  Construct a R_{path} using algorithm 2
  if the $R_{path} = \phi$ **then**
    Construct a G_{path} using algorithm 3
  end if
end for

When a destination node receives information about a video description, it will rebroadcast this information to its neighbor nodes. This means also that a destination node could be a forwarder node. If this destination node has enough bandwidth it will select another description to receive. After a destination node selects its proper paths it will send this information to the source node.
After the multicast source $\mathcal{S}$ receives the paths for each destination node, it constructs multiple node-disjoint multicast trees. To do so, nodes that have the same video description should be added to the same tree. Algorithm 5 describes the construction of multiple multicast trees.
Fig. 2 shows an example of MD video assignment and construction of multiple multicast trees. The multicast source S broadcasts information about two video descriptions (VD_1, and VD_2) to its neighbor nodes, nodes W , and Z. Each node will randomly select one video description to rebroadcast. Therefore, node W selects VD_1 and node Z selects VD_1. After that, nodes W and Z will rebroadcast this information to their neighbors nodes, nodes B, and C. This process will continue until this information reached the destination nodes, nodes R_1, R_2, and R_3. Destination nodes R_1, and R_3 will select the paths $S \rightarrow W \rightarrow B \rightarrow R_1$, and $S \rightarrow Z \rightarrow C \rightarrow R_3$, respectively, to receive VD_1. The destination node R_2 has two paths with the same description, description VD_1. Therefore, it will select the path with minimum delay. Assume the path $S \rightarrow W \rightarrow B \rightarrow R_2$ is selected. Note that destination node R_2 receives the same video description through different paths. This can be related to the randomness of choosing a video description. Finally, the multicast source S will construct only one multicast tree using algorithm 5. Fig. 2(c) shows multicast tree t_1.

Algorithm 2 R_{path} Construction

for $\forall\, j\; G_{path} \in$ Z **do**
  $\mathcal{P}$ = Parents of G_{nodes}
  $\mathcal{V} \leftarrow \mathcal{V}$ - $\mathcal{P}$
end for
Construct a R_{path} using the shortest path (in terms of delay) algorithm

Algorithm 3 G_{path} Construction

for $\forall\ j\ R_{path} \in Z$ **do**
  $\mathcal{P}$ = Parents of R_{nodes}
  $\mathcal{V} \leftarrow \mathcal{V}$ - $\mathcal{P}$
end for
Construct a G_{path} using the shortest path (in terms of delay) algorithm

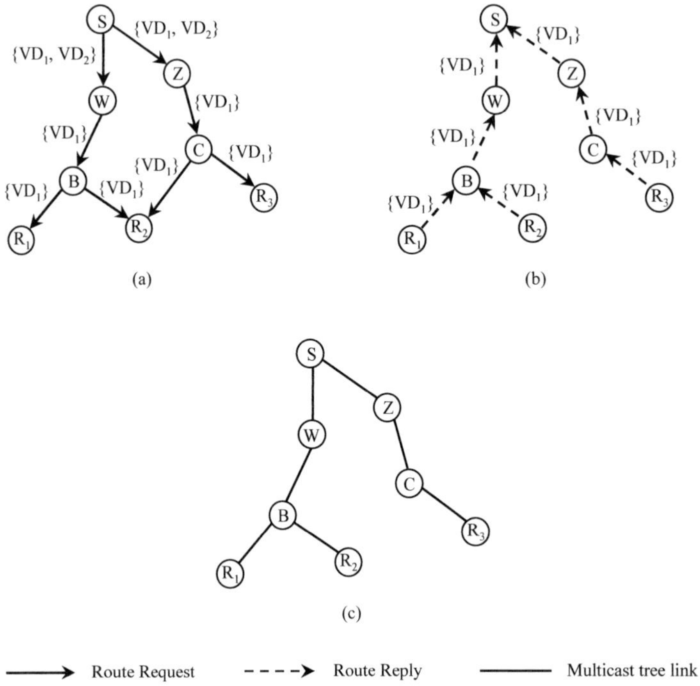

Fig. 2. Distributed MDC algorithm: (a) Route Request broadcasts. (b) Route Reply unicast. (c) Multicast tree construction.

Algorithm 4 Serial MDC: Multiple Multicast Tree Construction

Given: Z = set $x \cup$ set y
for $\forall\ i \in Z$ **do**
  if i has R_{color} **then**
    Add i to R_{tree}
  else if i has G_{color} **then**
    Add i to G_{tree}
  end if
end for

Algorithm 5 Distributed MDC: Multiple Multicast Tree Construction

for $i = 1$ *to* $\mathcal{V}$ **do**
  if node i has the 1st video description **then**
    Add node i to tree t_1
  else
    Add node i to tree t_2
  end if
end for

4.3 Centralized MDC algorithm

Before the construction of multiple node-disjoint multicast trees and the assignment of MD video, the multicast source S starts with constructing individual Multiple Node-Disjoint Paths (MNDP), with minimum delay, to each destination in the multicast group to meet the number of video descriptions required.

DEFINITION 1: MNDP problem: *consider a network represented by a graph $\mathcal{G} = (\mathcal{V},\mathcal{E})$ and a bandwidth constraint $\mathcal{W}$, find a MNDP, set P_i, from the multicast source node $\mathcal{S}$ to the destination node R_i such that:*

1. $d\left(p_{ij}\right)$ is minimized, $\forall\ p_{ij} \in P_i$
2. $Bw\left(p_{ij}\right) \geq \mathcal{W}, \ \forall\ p_{ij} \in P_i$

Algorithm 6 describes how MNDP are constructed. Before constructing multiple nodedisjoint paths to each destination, we first remove all links with capacity less than the bandwidth requirement, and then we construct multiple shortest paths (in terms of delay) on the residual network. Based on the sets of MNDP constructed, then multicast heuristic algorithm constructs Multiple Node-Disjoint Multicast Trees (MNDMT) for the video transmission, as shown in Algorithm 7.

Algorithm 6 Multiple Node-Disjoint Paths

$P_i = \phi$ /* MNDP set */
For each destination R_i do
Let $\mathcal{G}^*$ be equal to $\mathcal{G}$
repeat
  Find a shortest path p_{ij} to R_i (in terms of delay) in $\mathcal{G}^*$ such that $Bw\left(p_{ij}\right) \geq \mathcal{W}$
  Add p_{ij} to P_i
  Remove all forwarding nodes of p_{ij} in $\mathcal{G}^*$
until
The number of paths in P_i equal to the number of video descriptions required

As a simple example, we consider the partial network topology in Fig. 3(a), with a requirement of two descriptions for destination R_2 and one description for both destinations R_1 and R_3, to demonstrate the construction of multiple multicast trees. According to Algorithm 6, there are three path sets (Fig. 3(b)) P_1, P_2, and P_3 from the source S to the destinations R_1, R_2, and R_3, where $P_1 = \{p_{11}\} = \{S \to W \to B \to R_1\}$, $P_2 = \{p_{21}, p_{22}\} = \{S \to W \to B \to R_2, S \to Z \to C \to R_2\}$, and $P_3 = \{p_{31}\} = \{S \to Z \to C \to R_3\}$.

In Fig. 3(c)-(e), we show an example of multiple multicast trees construction using MNDMT. According to Algorithm 7, Step 4, the destination R_2 has the maximum number of paths, which is set P_2 , (two paths); therefore we have two multicast trees according to step 5, namely, $t_1 = p_{21}$ and $t_2 = p_{22}$ as seen in Fig. 3(c). The path p_{11} of the destination R_1 will be added to t_1 (Fig. 3(d)), according to Step 8, since it intersects t_1 with the most links. Because $P_1 = \phi$, then the algorithm picks up the next destination, R_3, and adds its path p_{31} to tree t_2 (Fig. 3(e)) according to Step 8. Since all the paths of each destination have been added, then the algorithm ends.
After constructing multiple multicast trees, Algorithm 8 assigns different video description to each tree. Therefore, trees t_1 and t_2 are assigned the first and second descriptions, respectively. Since any description can reproduce the original video signal, this we referred to as independent-description property of MDC, therefore the destination R_3 will be able to reproduce the original video signal. It is worth noting that if LC technique is used instead of MDC and according to Chen-LC algorithm, only one multicast tree will be constructed. Thus, they will be only assigned the basic layer.

4.4 Sequential MDC algorithm

Sequential algorithm constructs multiple disjoint multicast trees and assigns MD video to the destination nodes in a centralized fashion. However, the main difference between sequential MDC and centralized MDC algorithms is that the assignment of MD video is executed in a sequential way. This means that all the destination nodes should be first assigned the first video description (VD_1), then the destination nodes that require a second description they will be assigned the second video description (VD_2) and the destination nodes that require a third description they be assigned the third video description (VD_2) and so on. Therefore, to perform the assignment of MD video in a sequential way, the destination nodes on each multicast tree should be superset of the later, i.e., $t_L \subseteq t_{L-1} \cdots \subseteq t_2 \subseteq t_1$. Algorithm 7 is deployed to construct multiple disjoint multicast trees, and then algorithm 9 is executed to form the final version of the multiple multicast trees. After that, the trees $t_1, t_2, \cdots, t_L$ will be assigned the first, the second and the Lth description, respectively. It is worth pointing out that Sequential MDC algorithm does not employ the independent-property of MDC.

Algorithm 7 Multiple Node-Disjoint Multicast Trees

1: **for** $i = 1$ *to* m **do**
2: Find the set of MNDP P_i by algorithm 6
3: **end for**
4: Find a set P_i that has the maximum number of paths
5: initially, Let $T = P_i$, i.e., $t_1 = p_{i1}, t_2 = p_{i2}, \ldots, t_L = p_{iL}$
6: **for** $i = 2$ *to* m **do**
7: Add each path in P_i to T as follows:
8: Find a path $p_{ij} \in P_i$ such that it intersects a tree $t_k \subset T$ not covering R_i with the most links, and add p_{ij} to t_k
$$t_k \leftarrow t_k + p_{ij}$$
9: Remove p_{ij} from P_i
$$P_i \leftarrow P_i - p_{ij}$$
10: Repeat Steps (8) and (9) until $P_i = \phi$
11: **end for**

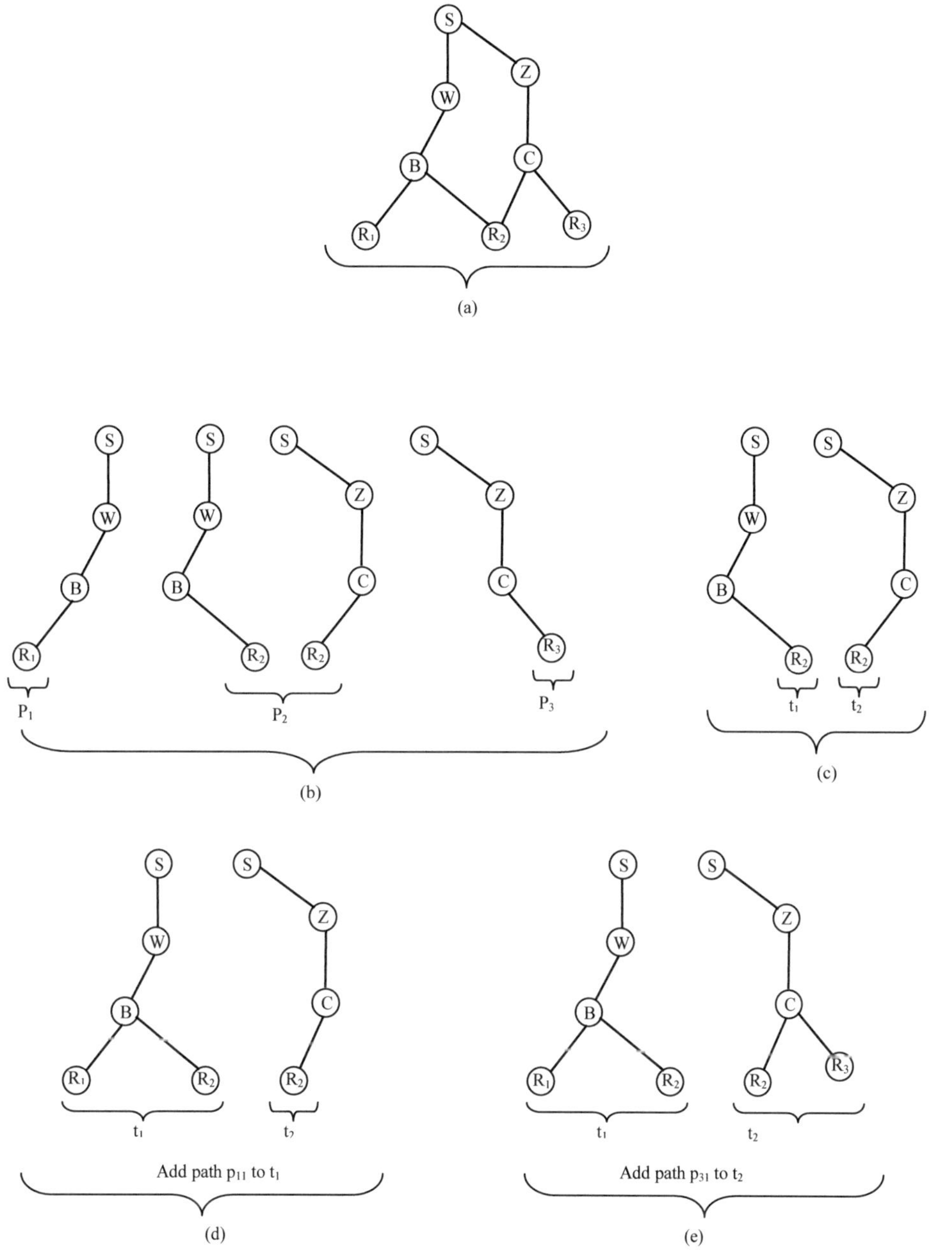

Fig. 3. Centralized MDC algorithm

We use Fig. 3(a), to explain how sequential MDC algorithm constructs multiple disjoint multicast trees. At the end of algorithm 7, two disjoint multicast trees are constructed, namely, t_1 and t_2 as seen in Fig. 3(e). However, in order to perform sequential assignment of MD video, R_3 should be connected to t_1. And because Sequential MDC algorithm maintains totally

Algorithm 8 MD video assignment

1: For $i = 1$ *to* L /*L is the number of the multicast trees constructed*/
2: For $j = 1$ *to* n /*n is the number of MD video, $(L \leq n)$*/
3: If $Bw(t_i) \geq Bw(VD_i)$ then
$VD_i \rightarrow t_i$

disjoint multicast trees, therefore, only one multicast tree, t_1 , is constructed as shown in Fig. 4.

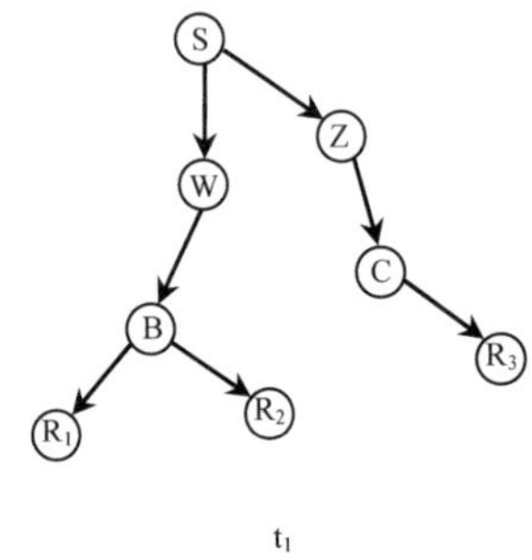

Fig. 4. Demonstration of sequential MDC algorithm

5. Performance evaluation

This section deals with the performance evaluation of our developed algorithms. In particular, we evaluate the performance of our proposed algorithms, namely, Serial MDC, Distributed MDC, Centralized MDC, and Sequential MDC algorithms and compare them with the algorithm proposed by Chen et al. in (Chen et al., 2004). Chen et al. proposed this algorithm for assigning a number of video layers that are encoded using LC technique; we referred to as Chen-LC. To make a fair comparison, we modified Chen-LC algorithm to construct node disjoint multicast trees. Moreover, in order to take the bandwidth requirements for MDC and LC into consideration, we consider the video sequences reported in (Gogate et al., 2002). Since all the video sequences have roughly the same bit rate, we consider the video sequence of "Football". The average video source rate is 1.5 Mbps for each description, whereas the average video source rate for the layered coder is 1.57 Mbps for the base layer and 1.45 Mbps for the enhancement layer.

We generate a wireless ad hoc network by placing a number of nodes at random locations in a square area of 1000×1000 m^2. The radio transmission range is 250 m and the number of video descriptions required by each destination is uniformly distributed to be $\in \{1,2\}$. The residual bandwidth of each link is randomly chosen from [2,10] Mbps. The delay in each link is randomly chosen from [1,20] ms. Moreover, the multicast source S and a set of destinations $\mathcal{Y}$ are randomly chosen from the network graph to form a multicast session. Any

Algorithm 9 Sequential algorithm

1: **for** $i = 1$ *to* L **do**
2: $t_L \subseteq t_{L-1} \cdots \subseteq t_2 \subseteq t_1$
3: **end for**

destination node is at least 2-hop away from the multicast source. For each simulation, several experiments have been run to ensure 95% confidence interval. The 95% confidence intervals are always plotted, when they are not visible it means that they are smaller than the curve markers.

To show the significance of our developed algorithms, we evaluate and compare its performance with the well-known multicast algorithm, Chen-LC, using the following metrics:

- **User satisfaction:** This metric is defined as the total number of the assigned video descriptions to all destinations divided by the total number of requested video descriptions by all destinations. This metric presents the effectiveness of a protocol.

$$User\ satisfaction = \frac{\sum_{i=1}^{n} \mathcal{N}_{asg}(R_i)}{\sum_{i=1}^{n} \mathcal{N}_{req}(R_i)} \tag{7}$$

where $\mathcal{N}_{asg}(R_i)$, and $\mathcal{N}_{req}(R_i)$, are the number of the assigned and requested video descriptions of the destination R_i respectively, and n is the number of destinations.

- **Number of pure forwarders (PF):** It is defined as the number of pure forwarders nodes on the aggregate multicast tree T that are not destinations. This measures the efficiency in terms of minimizing the number of pure forwarding nodes.

$$PF = \sum_{i=1}^{\mathcal{V}} I(v_i) \tag{8}$$

where $\mathcal{V}$ is the network size and $I(v_i)$ is defined as:

$$I(v_i) = \begin{cases} 1 & \text{for } v_i \in T - \{S, \mathcal{Y}\} \\ 0 & \text{otherwise} \end{cases}$$

- **Bandwidth utilization:** This metrics defined as the total used bandwidth for the video distribution tree(s).

$$Bandwidth\ utilization = \sum_{e \in u} Bw\,(e) \tag{9}$$

where e denotes a link, u is the set of used links, and $Bw\,(e)$ denotes the bandwidth devoted to video distribution in link e.

- **Aggregate tree delay:** It represents the longest delay from the multicast source s to a destination node R_i on the aggregate tree T, as seen in Equ.(4).

5.1 Varying number of multicast destinations

Fig. 5 to Fig. 8 illustrate Serial MDC, Distributed MDC, Centralized MDC, Sequential MDC and Chen-LC algorithms performance with varying number of multicast destination nodes while the network size is set to 50 nodes. Fig. 5 shows that Serial MDC, Distributed MDC, and Centralized MDC algorithms achieve higher user satisfaction compared to Chen-LC algorithm. This can be related to the independent-description property of MDC. Sequential MDC and Chen-LC algorithms have the same user satisfaction. This is because Sequential MDC algorithm does not employ the independent-property of MDC. In this case it is similar to Chen-LC. In other words, VD_1 is equivalent to the basic layer and VD_2 is equivalent to the enhancement layer. Serial MDC, Distributed MDC, and Centralized MDC algorithms are well scalable in term of number of destinations. Centralized MDC algorithm achieves a higher user satisfaction compared to Serial MDC and Distributed MDC algorithms. As the number of destinations increases, user satisfaction decreases gradually in Serial MDC, Distributed MDC, and Centralized MDC algorithms. However, the user satisfaction of Chen-LC and Sequential algorithms decreases sharply as the number of destinations increases. That is, as a result of the dependent-layer property of LC (for Chen-LC algorithm) and because Sequential MDC algorithm does not employ the independent-description property of MDC.

Fig. 6 depicts the number of pure forwarders nodes as a function of number of destination nodes. It can be seen that Centralized MDC has slightly higher number of pure forwarders nodes compared to the other algorithms. However, Distributed MDC has a lowest number of pure forwarders nodes. In Fig. 7, we plot the average bandwidth utilization. Clearly, the bandwidth utilization of Centralized MDC is slightly higher than the bandwidth utilization of the other algorithms. This is because Centralized MDC requires more number of pure forwarders, compared to the other algorithms; to constructs multiple node-disjoint trees (see Fig. 6). However, Distributed MDC requires a minimum bandwidth for video distribution trees. This is because Distributed MDC has a minimum number of pure forwarders nodes. We show in Fig. 8 the aggregate tree delay as a function of number of destinations. All algorithms achieve a comparable delay as compared to each other. As the number of destinations increases the aggregate tree delay increases. This is because more paths are constructed to build multiple multicast tree.

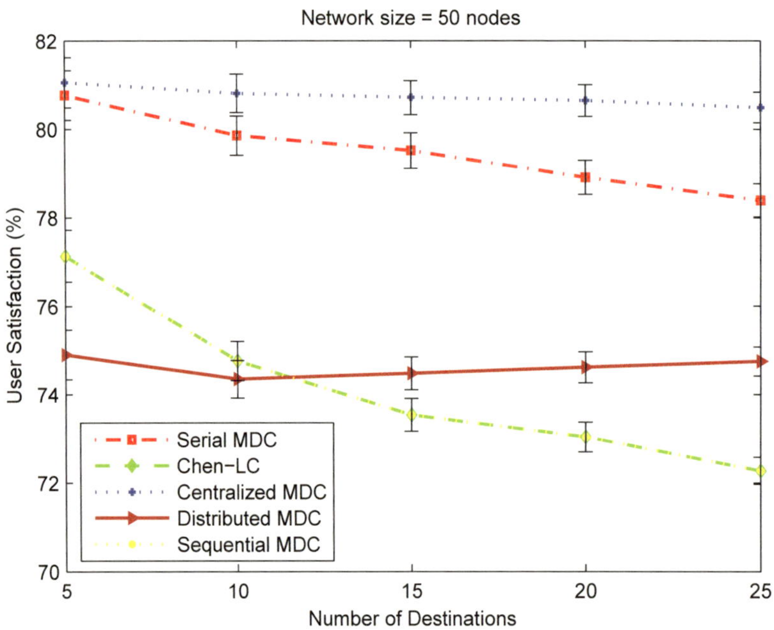

Fig. 5. User satisfaction versus number of destinations. Network size = 50 nodes

Fig. 9 to Fig.12 illustrate Serial MDC, Distributed MDC, Centralized MDC, Sequential MDC, and Chen-LC performance with varying number of multicast destination nodes while the network size is set to 100 nodes. As the network size increase from 50 nodes (Fig. 5) to 100 nodes (Fig. 9), the users satisfaction for all algorithms increases. This is because the number of nodes in the network increases. As a result, the number of paths to each destination is

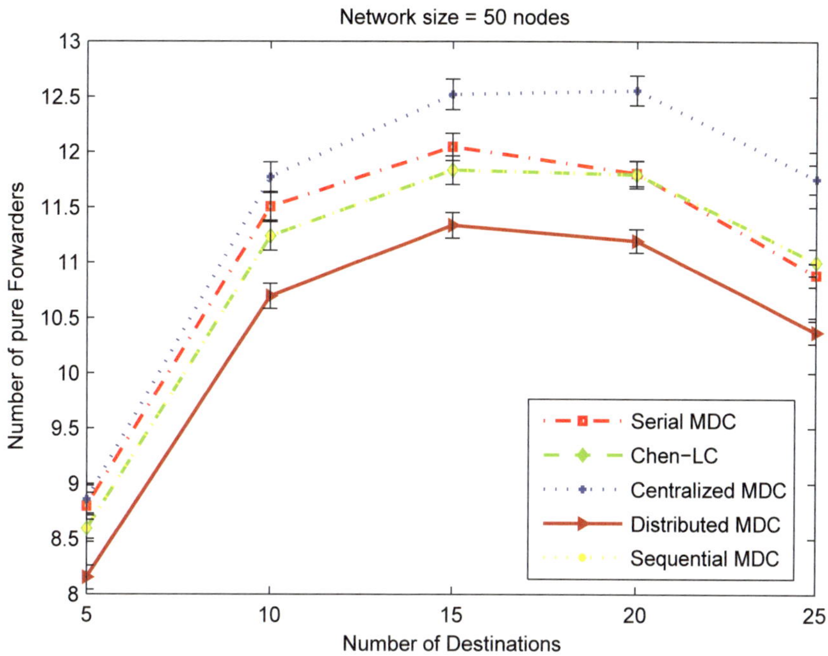

Fig. 6. Number of pure forwarders versus number of destinations. Network size = 50 nodes

increased. Again, Serial MDC, Distributed MDC, and Centralized MDC show good scalability as the number of destinations increases. As the number of nodes in the network increases from 50 to 100 nodes, the number of pure forwarders nodes, the bandwidth utilization, and the aggregate tree delay (in Fig. 10, Fig. 11, and Fig. 12) are increased compared to Fig. 6, Fig. 7, and Fig. 8, respectively.

5.2 Varying network size

Fig. 13 to Fig. 20 compare Serial MDC, Distributed MDC, Centralized MDC, Sequential MDC, and Chen-LC performance, with varying number of nodes in the network (network size) from 50 to 100 nodes, in terms of user satisfaction, number of pure forwarders, bandwidth utilization, and aggregate tree delay. The number of destinations is set to 10 and 30 nodes. Centralized MDC achieve a higher user satisfaction (see Fig. 13 and Fig. 17) compared to the other algorithms. The cost of that is the increase in the number of pure forwarders nodes (see Fig. 14 and Fig. 18), the bandwidth utilization (see Fig. 15 and Fig. 19), and the aggregate tree delay (Fig. 16 and Fig. 20). However this cost is still comparable. As the network size increases, the user satisfaction for all algorithms increases. We related that to the increase in the number of resources in the network, i.e., number of nodes, and bandwidth. Fig. 16 and Fig. 20 show that as the network size increases the aggregate tree delay decreases. This is because more alternate paths (with minimum delay) may exist.

Comparing Fig. 13 with Fig. 17, we can note that the user satisfaction, for all algorithms, decreases as the number of destinations increases from 10 to 30 nodes. This because the number of node-disjoint paths to the destination nodes decreases. Thus, the number of assigned MD video to each destination decreases.

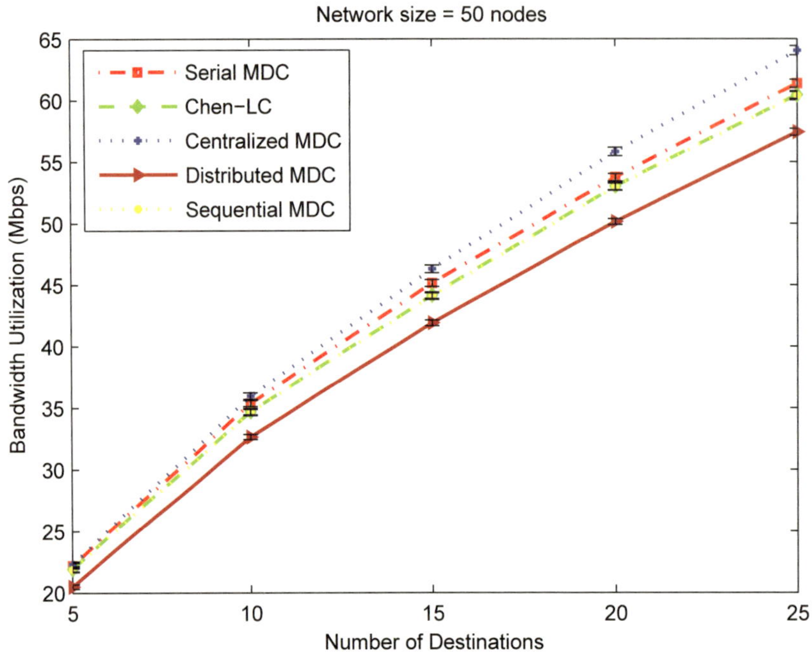

Fig. 7. Bandwidth utilization versus number of destinations. Network size = 50 nodes

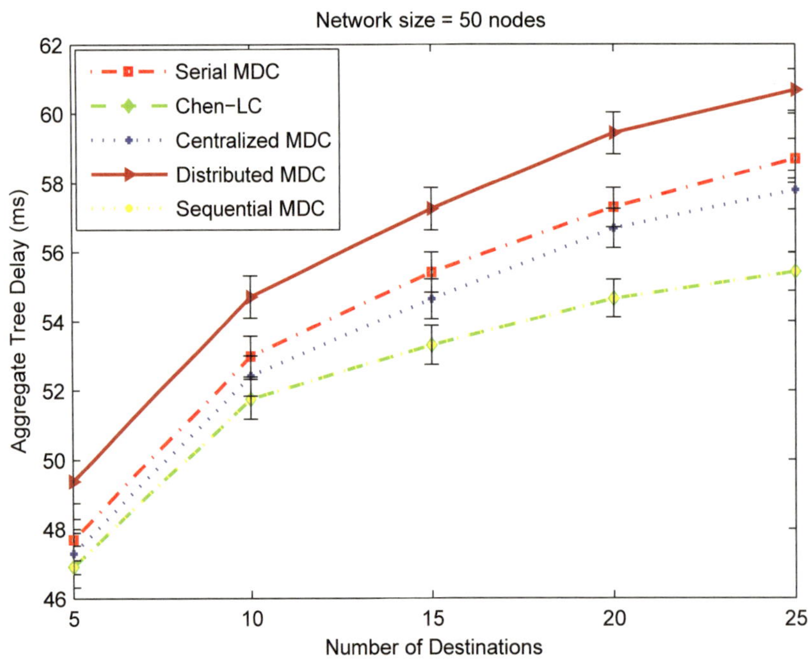

Fig. 8. Aggregate tree delay versus number of destinations. Network size = 50 nodes

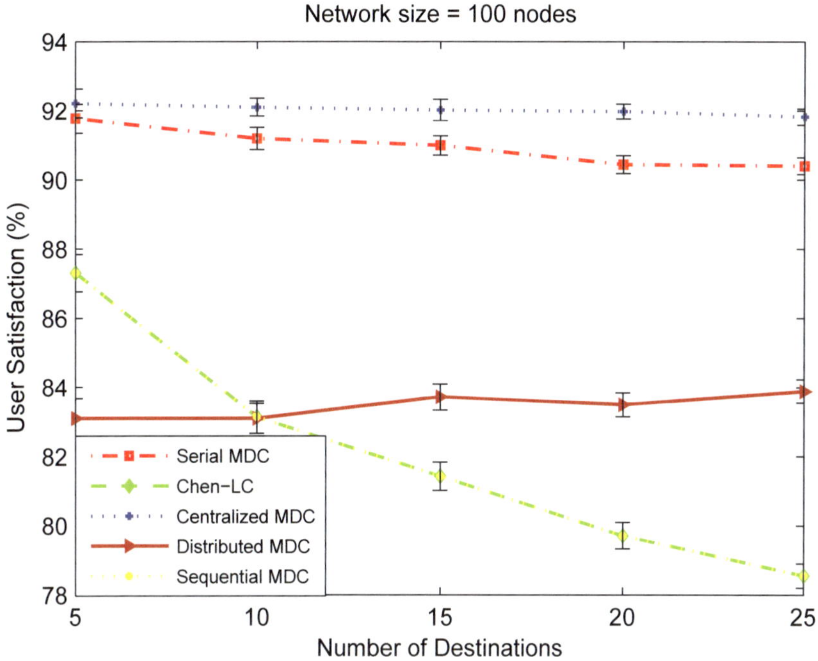

Fig. 9. User satisfaction versus number of destinations. Network size = 100 nodes

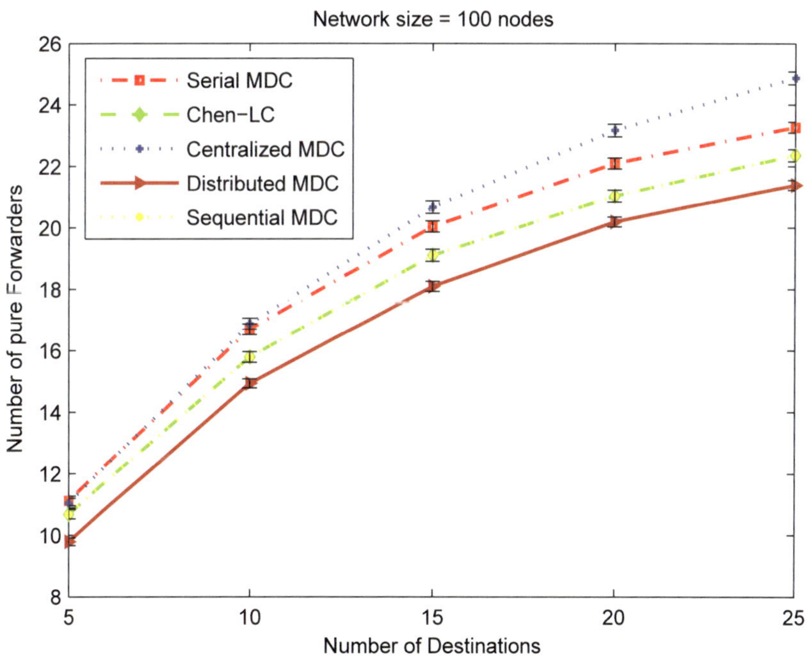

Fig. 10. Number of pure forwarders versus number of destinations. Network size = 100 nodes

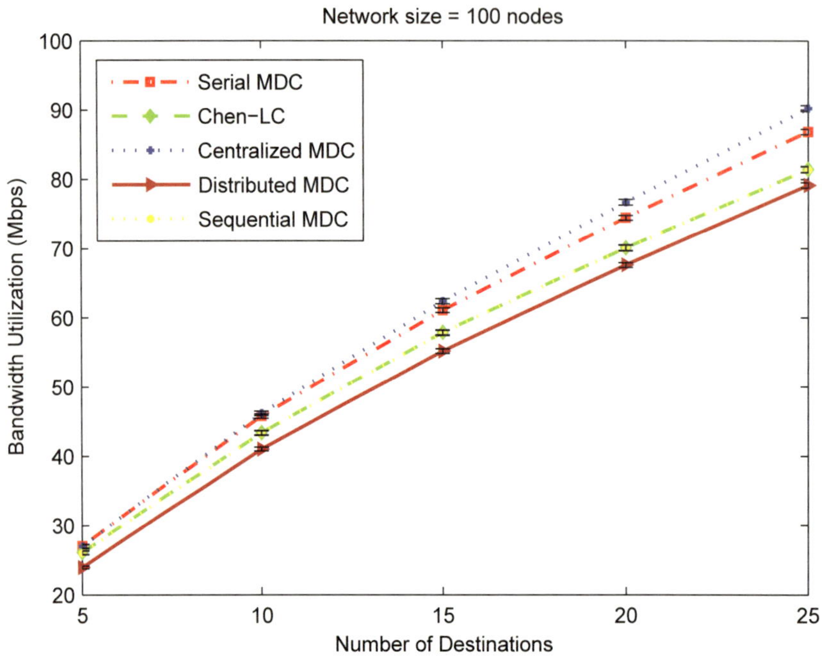

Fig. 11. Bandwidth utilization versus number of destinations. Network size = 100 nodes

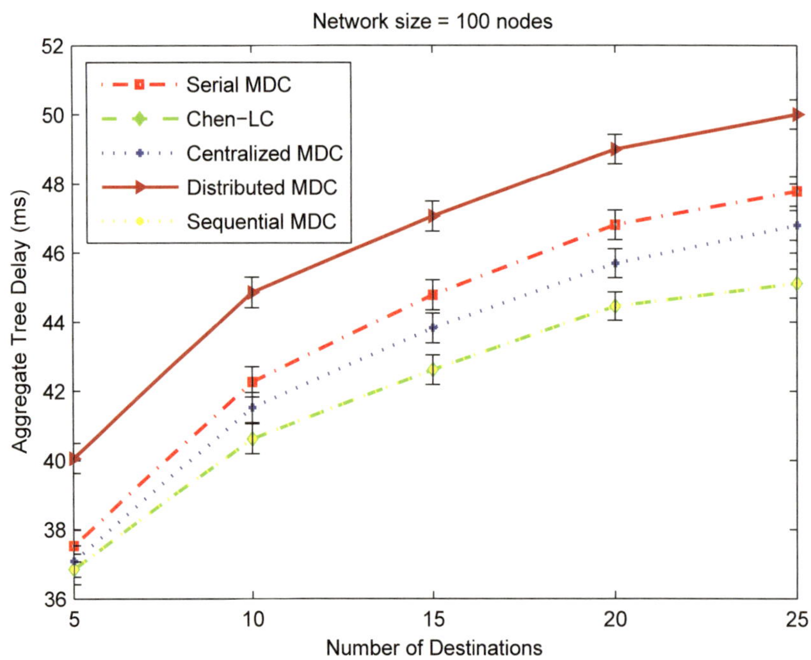

Fig. 12. Aggregate tree delay versus number of destinations. Network size = 100 nodes

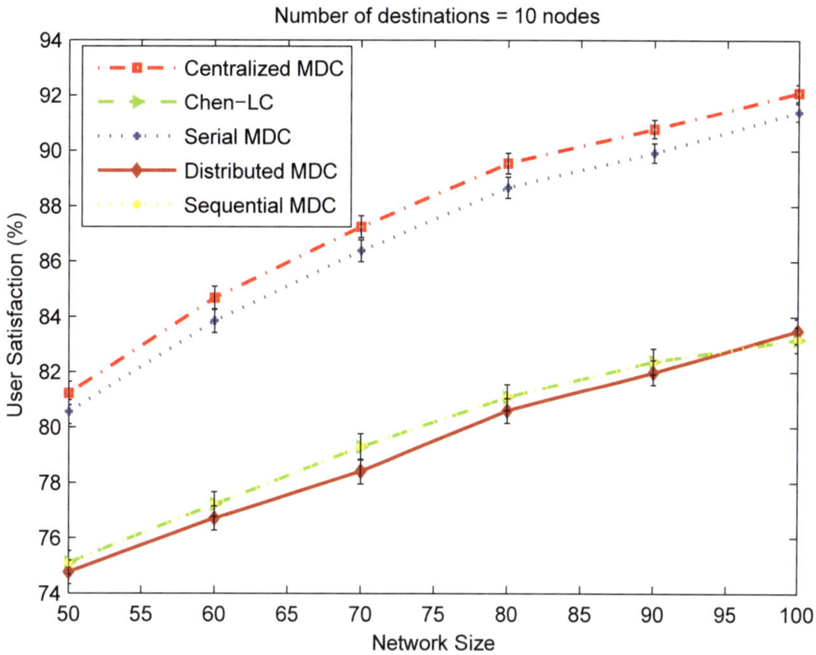

Fig. 13. User satisfaction versus network size. Number of destinations = 10 nodes

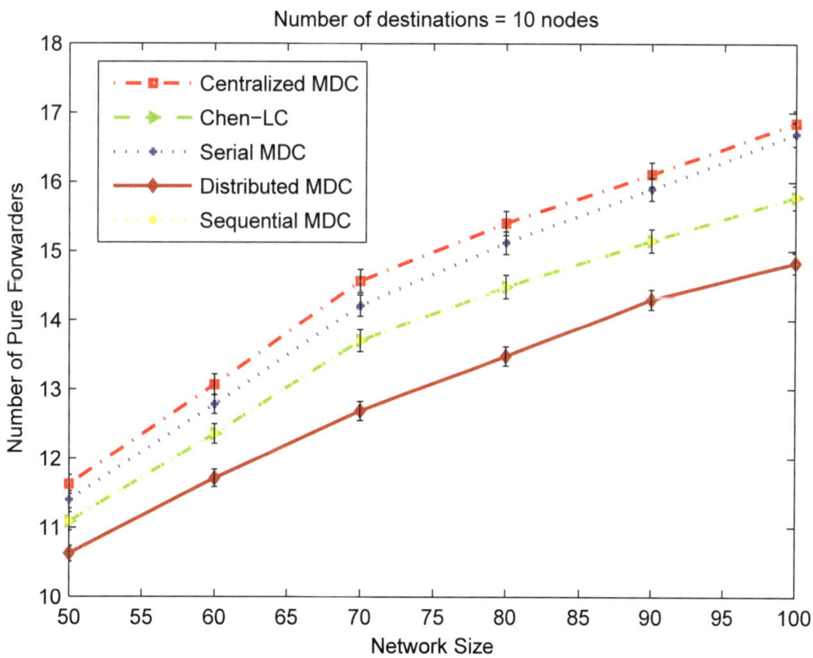

Fig. 14. Number of pure forwarders versus network size. Number of destinations = 10 nodes

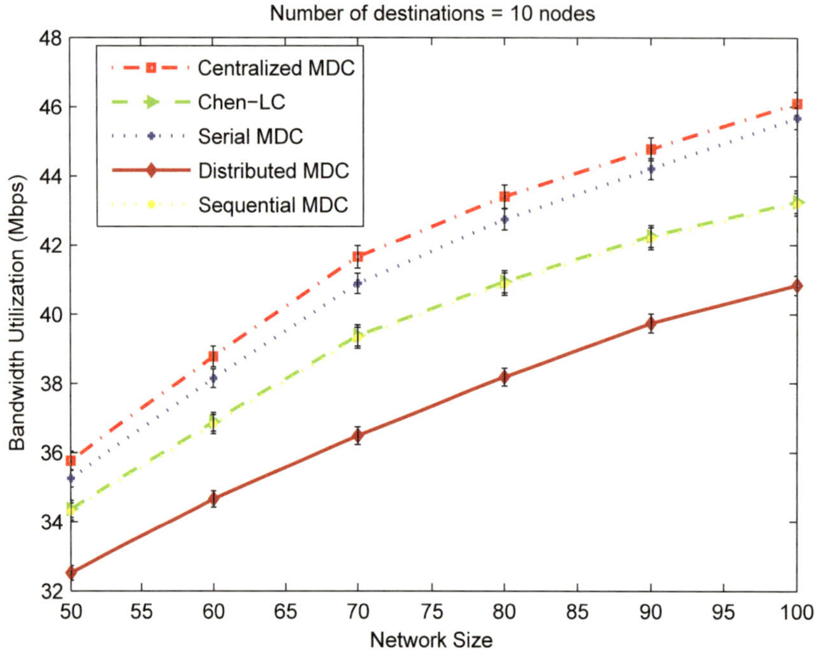

Fig. 15. Bandwidth utilization versus network size. Number of destinations = 10 nodes

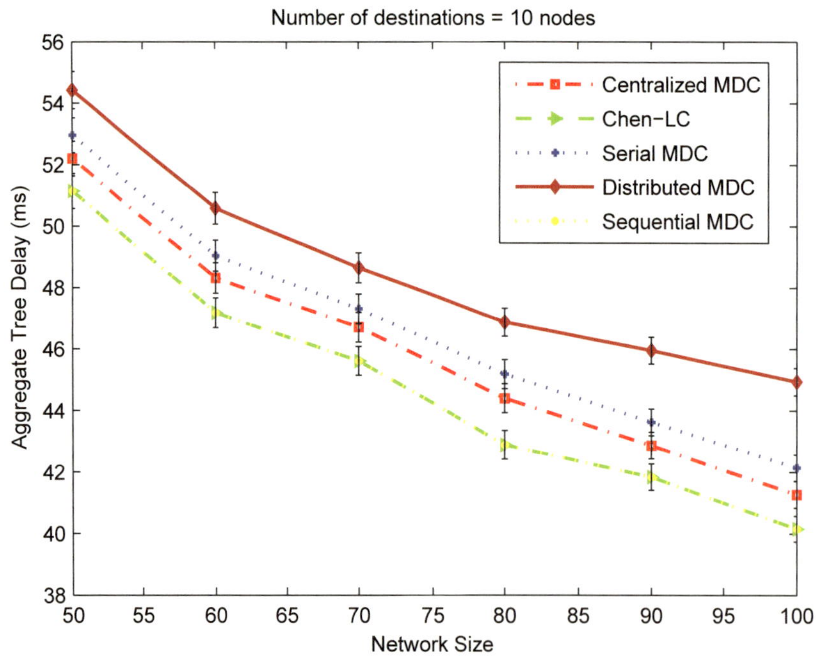

Fig. 16. Aggregate tree delay versus network size. Number of destinations = 10 nodes

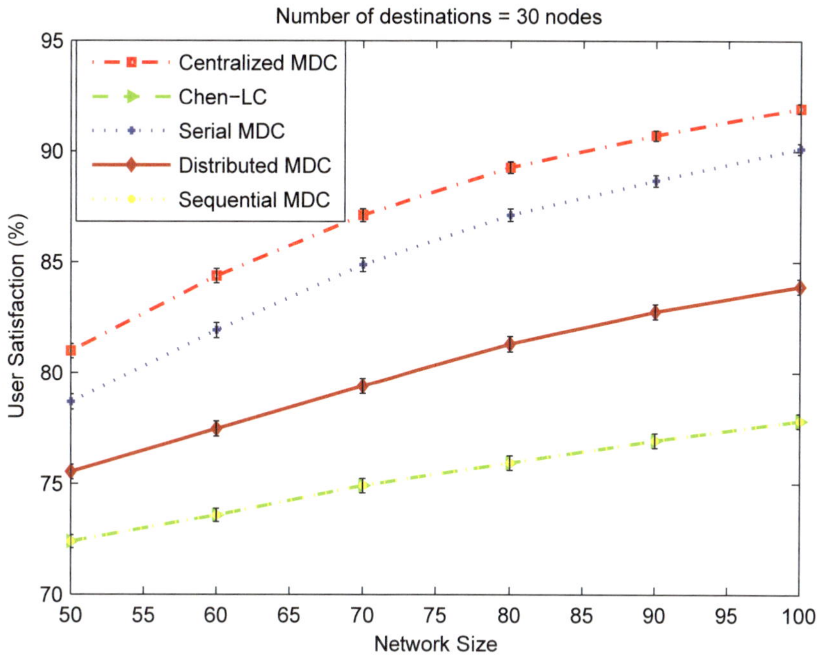

Fig. 17. User satisfaction versus network size. Number of destinations = 10 nodes

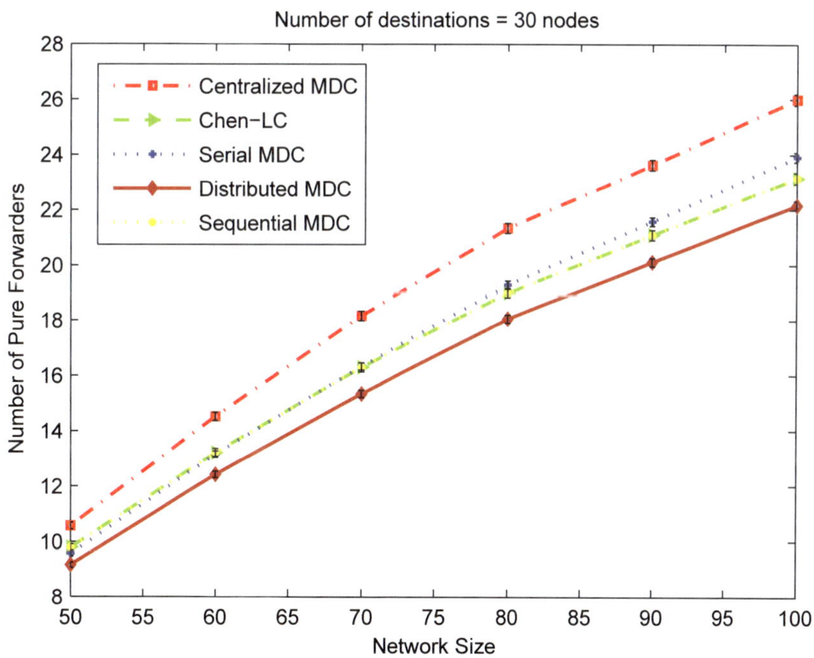

Fig. 18. Number of pure forwarders versus network size. Number of destinations = 10 nodes

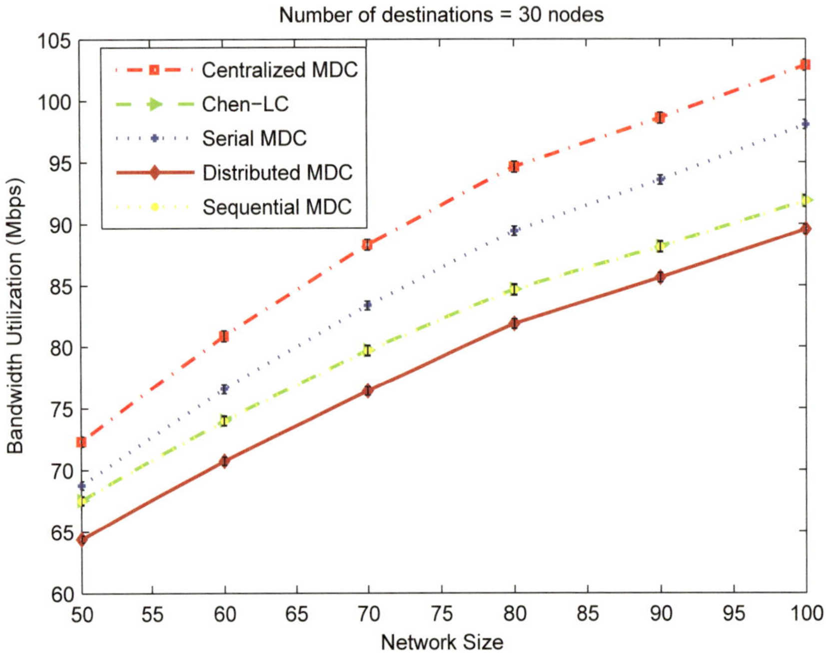

Fig. 19. Bandwidth utilization versus network size. Number of destinations = 10 nodes

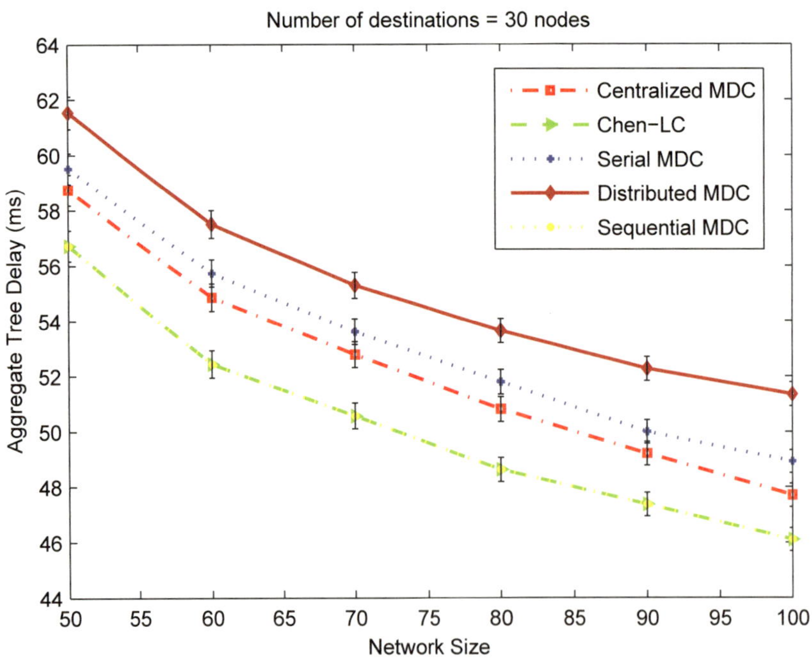

Fig. 20. Aggregate tree delay versus network size. Number of destinations = 10 nodes

6. Complexity analysis of the algorithms

We analyze the complexity of our proposed algorithms as follows. For Serial MDC, the shortest path algorithm (Dijkstra's algorithm) is of complexity $O(|\mathcal{V}|\log(|\mathcal{V}|) + |\mathcal{E}|) \leq O(\mathcal{V}^2)$ where $|\mathcal{V}|$ and $|\mathcal{E}|$ are the number of nodes and number of wireless communication links in the partial topology, respectively. Since it iterates $\mathcal{Y}$ times, where $\mathcal{Y}$ is the number of destination nodes. Therefore the complexity is $O(\mathcal{V}^2 \times \mathcal{Y})$ and finally the algorithm iterates $|\mathcal{N}_{req}(VD)|$ times, where $|\mathcal{N}_{req}(VD)|$ is the total number of required video descriptions for all destinations. As a result, the complexity of Serial MDC is given by $O(\mathcal{V}^2 \times \mathcal{Y} \times \mathcal{N}_{req}(VD))$. For Distributed MDC the complexity is given by $O(\mathcal{V})$. Finally, Centralized MDC, Sequential MDC, and Chen-LC algorithms have the same complexity of Serial MDC algorithm.

7. Conclusion

In this chapter we study the problem of multiple multicast trees construction and the assignment of MD video. Different algorithms are proposed for that purpose. These algorithms are: Serial MDC, Distributed MDC, Centralized MDC, and Sequential algorithms. Serial MDC, Distributed MDC, and Centralized MDC algorithms deploy the independent-description property of MDC, whereas Sequential MDC algorithm does not take this property into consideration. Simulation results demonstrate that deploying this property of MDC along with multiple multicast tree can greatly improve the user satisfaction. Furthermore, simulation results demonstrate that the way of multiple multicast tree construction and the assignment of MD video can affect the user satisfaction. In addition, simulation results show that MDC can achieve higher user satisfaction compared to Layered Coding (LC) with a small cost in terms of number of pure forwarders nodes, bandwidth utilization, and aggregate tree delay.

8. References

Agrawal, D., Reddy, T. B. & Murthy, C. S. R. (2006). Robust demand-driven video multicast over ad hoc wireless networks, *3rd International Conference on Broadband Communications, Networks and Systems* pp. 1–10.

Badarneh, O., Kadoch, M. & Elhakeem, A. (2008). A new approach for the construction of multiple multicast trees using multiple description video for wireless ad hoc networks, *IEEE Conference on Local Computer Networks* pp. 152–159.

Chen, J., Chan, S.-H. G. & Li, V. O. (2004). Multipath routing for video delivery over bandwidth-limited networks, *IEEE Journal on Selected Areas in Communications* **22**(10): 1920–1932.

Chow, C.-O. & Ishii, H. (2008). Multiple tree multicast ad hoc on-demand distance vector (mt-maodv) routing protocol for video multicast over mobile ad hoc networks, *IEICE-Transactions on Communications* **E91-B**: 428–436.

Clausen, T. & Jacquet, P. (year 2003). Optimized link state routing protocol (olsr). www.ietf.org/rfc/rfc3626.txt.

Gogate, N., Chung, D.-M., Panwar, S. & Wang, Y. (2002). Supporting image and video applications in a multihop radio environment using path diversity and multiple description coding, *IEEE transactions on circuits and systems for video technology* **12**(9): 777–792.

Goyal, V. K. (2001). Multiple description coding: Compression meets the network, *IEEE Signal Processing Magazine* **18**: 74–94.

Mao, S., Cheng, X., Hou, Y. T. & Sherali, H. D. (2006). Multiple description video multicast in wireless ad hoc networks, *Mobile Networks and Applications* **11**: 63–73.

Mao, S., Hou, Y., Cheng, X., Sherali, H., Midkiff, S. & Zhang, Y.-Q. (2006). On routing for multiple description video over wireless ad hoc networks, *IEEE Transactions on Multimedia* **8**(5): 1063–1074.

Mao, S., Lin, S., Panwar, S., Wang, Y. & Celebi, E. (2003). Video transport over ad hoc networks: multistream coding with multipath transport, *IEEE Journal on Selected Areas in Communications* **21**(10): 1721–1737.

Puri, R. & Ramchandran, K. (1999). Multiple description source coding through forward error correction codes, *IEEE Proceedings Asilomar Conference on Signals, Systems, and Computers* pp. 342–346.

Wei, W. & Zakhor, A. (2007). Multiple tree video multicast over wireless ad hoc networks, *IEEE Transactions on Circuits and Systems for Video Technology* **17**: 2–15.

Part 4

TCP in Ad Hoc Networks

18

TCP-MAC Interaction in Multi-hop Ad-hoc Networks

Farzaneh R. Armaghani[1] and Sudhanshu S. Jamuar[2]
[1]Monash University, GSIT,
[2]Department of Electrical Engineering, University of Malaya,
[1]Australia
[2]Malaysia

1. Introduction

Recent demands on affordable, portable wireless communication and computation devices have resulted in exponential growth of wireless networks ranging from Wireless Local Area Networks (WLAN) and Wireless Wide Area Networks (WWAN) to Ad-Hoc and Sensor networks. The major goal of wireless communication is to allow users to communicate together and to have access to global network anytime anywhere. This has led to wide acceptance of infrastructure based cellular networks (WWANs) where mobile stations communicate with a centralized controller, often referred as Access Point (AP) that is connected to the wired networks. On the other hand, WLANs have appeared as dominant popular technologies in many venues including a local area such as an academic campus or an airport terminal. These wireless networks mostly rely on IEEE 802.11 Wi-Fi (Wireless Fidelity) technology and its various derived versions (i.e. 802.11a,b,g).

IEEE 802.11 standard supports two operational modes: The infrastructure-based Wireless Local Area Networks (WLANs) and an infrastructure-less Ad-Hoc Networks. A WLAN (Conti, 2003) typically imposes the existence of an AP and normally is connected to the wired networks to provide internet access for mobile devices. Obviously, only one hop link is needed to communicate between mobile devices and AP. In contrast, there is no AP or infrastructure in Ad-Hoc networks. Any two stations can communicate directly when they are in the range of reception of each other. To this end, the stations may use multi-hop routing to deliver their packets to destinations. The ad-hoc protocols (Conti, 2003; Mohapatra & Krishnamurthy, 2005) are self-configured for address and routing in the face of mobility and the network topology may change in each configuration. The multi-hop wireless ad-hoc networks, or multi-hop wireless networks enable wireless networking in the environments where the wired or cellular connections are impossible, inadequate, or cost effective (e.g. battle field, disaster recovery, etc.).

The popularity of internet over the last decades has resulted in rapid advancement of demanding applications. The Transmission Control Protocol/Internet Protocol (TCP/IP) (Stevens, 1994) is a well-known de facto protocol in developing today's internet. Basically, TCP provides a connection-oriented and reliable end-to-end data delivery between two hosts in traditional wired networks. Since TCP is well tuned and due to its wide acceptance in internet, it is desirable to extend and adopt its functionality to wireless networks. On the

other hand, unique characteristics and usage of multi-hop wireless networks require robust, reliable and adaptive designs. This may be achieved by considering the interaction of different layers to meet the increasing demands of these networks.
The reliability in TCP is achieved by retransmitting lost packets and acknowledgment (ACK) confirmation. If the sender does not receive any acknowledgment within a timeout interval or receives duplicate ACKs in the case of out-of-order packets, the packet will be retransmitted. Any packet loss is assumed as congestion in wired networks. When a packet loss is detected, TCP invokes its congestion control mechanism to slow down the sending rate to reduce the congestion. However, packet losses are not mainly due to congestion in wireless networks. It might be due to some wireless specific properties such as high medium access contention, route breakage and high bit error rate in radio channels (Hanbali, Altman, & Nain, 2005; Xiang, Hongqiang, & Jiangfeng, 2005).
The key challenge of TCP protocol is its poor bandwidth utilization and performance when it runs over 802.11 multi-hop wireless networks. The reason can be explained due to the extensive number of medium access carried out by TCP. Basically, TCP sender will be informed of successful transmission by receiving the acknowledgment from the other end host. The MAC overhead can be caused by generating redundant ACK packets that compete in the same route with data packets for the media. Although the TCP-ACK packets are small, they may cause the same overhead as data packets in MAC layer resulting in wastage of wireless resources (Altman & Jimenez, 2003; de Oliveira & Braun, 2007). In fact, the short RTS/CTS control frames to provide the data delivery implemented by 802.11 MAC protocol, cannot eliminate the interference in large topologies (Xu, Gerla, & Bae, 2002). As load increases, the well-known hidden terminal effects caused by interference between ACK and data packets can impact TCP performance dramatically in long paths if TCP acknowledges every incoming data packets. One way to improve the TCP performance over 802.11 in multi-hop ad-hoc networks is to alleviate the medium access contention by reducing the number of generated ACKs, simply called as *delayed ACKs*. This can be done by merging several ACKs in one ACK which is possible due to cumulative ACK scheme used in TCP. Referring to already proposed approaches to reduce the number of induced ACKs, the TCP performance is still affected by a limitation of a method which dynamically selects the number of delayed ACKs based on the channel condition (Altman & Jimenez, 2003; de Oliveira & Braun, 2007). This motivates us to study the performance of TCP-ACKs in interaction with 802.11 over the multi-hop ad-hoc networks and develop a dynamic delayed ACK strategy to adjust TCP to these kinds of networks.

2. IEEE 802.11 challenges

The unique characteristics of ad-hoc designs impose several challenges in comparison with single hop networks such as cellular networks or WLANs, when they run over 802.11 MAC protocol.
The most serious challenge is the RTS/CTS handshaking implemented in 802.11, which is not efficient enough to prevent collisions due to large distribution of mobile nodes and multi-hop function in ad-hoc networks. It has been proved through analytical model and simulation experiments (Fu, et al., 2005; Xu, et al., 2002) that RTS/CTS cannot function well in topologies more than three hops (3 hop scenario) between sender and receiver. For larger number of hops, the RTS/CTS exchange cannot prevent the existence of famous hidden node problem which is discussed later.

Mobile nature of ad-hoc networks, where each node may experience different degree of channel contention and collision, is another problem, (Zhai et al., 2006). The interaction between the MAC and higher layers has a significant effect on the network performance. The interaction between MAC and TCP layer is investigated throughout this chapter.

2.1 Medium contention and spatial reuse

The Hidden and Exposed terminals are defined based on transmission range and sensing range of the nodes (Xu, et al., 2002). Transmission range represents the range within which a packet is successfully received if there is no interference from other radios. Sensing range is the range within which a transmitter triggers carrier sense detection to sense an ongoing signal. Spatial reuse facilitates maximum possible non-conflicting simultaneous transmissions in MAC layer. A hidden terminal is the one that can neither sense the transmission of a transmitter nor correctly receive the reservation packet (i.e. CTS control frame) from its corresponding receiver (Zhai, et al., 2006). In other words, a hidden terminal is a node that is within the transmission range of a receiver but out of the sensing range of an intended transmitter. Therefore, it can interfere with an ongoing transmission at the receiver by transmitting at the same time. Consider the scenario illustrated in Fig. 1a to see the cause of hidden nodes. Here node D is a hidden terminal while B is transmitting to C because it is out of B's sensing range. Therefore, D's transmission collides with RTS reception in C. After seven attempts, both B and D assume that C is unreachable and the packets will be dropped. This leads to bandwidth wastage as both data transmissions are destroyed.

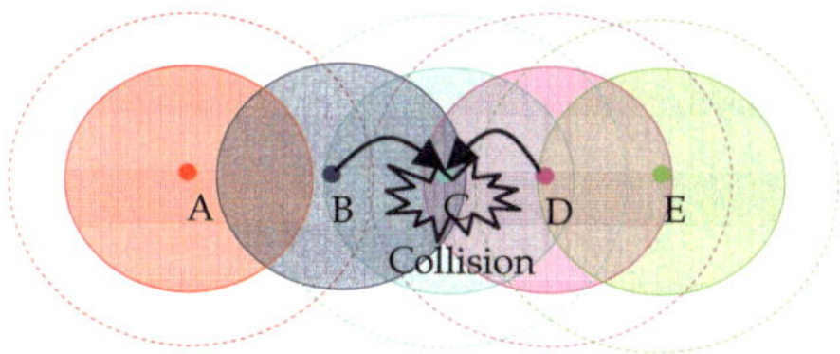

(a) Hidden terminals

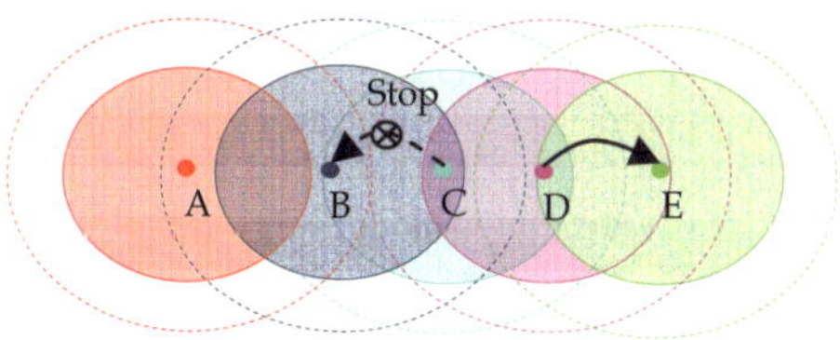

(b) Exposed terminals

Fig. 1. Contention and spatial reuse

To achieve high channel utilization, MAC needs to maximize the spatial reuse (Zhai, et al., 2006). Exposed terminal problem is a factor in 802.11 influencing the spatial reuse and is a problem which is caused by a terminal that is within the sensing range of a transmitter and can not interfere with the reception of the receiver; but it would not be able to start a transmission because it senses a busy media. As depicted in Fig. 1b, node C is considered as an exposed terminal when D is transmitting to E. C senses the medium as busy and it has to keep silence, even though it can transmit to B which is out of D's sensing range. In short, the hidden terminals reduce the network capacity due to increase in number of collisions, while exposed terminals affect the spatial reuse due to unnecessary deferring nodes from transmitting.

3. TCP-MAC interaction in multi-hop ad-hoc networks

TCP interaction with lower layers includes the proposals that address the inability of TCP to distinguish between losses due to route failures and network congestion. These proposals involving the network layer suggest notifying the TCP sender about routing failure, when the routing layer detects a route failure (Dongkyun, Toh, & Yanghee, 2000; Holland & Vaidya, 2002; Liu & Singh, 2001; Wang & Zhang, 2002; Yu, 2004). On the other hand, considerable research (Fu, et al., 2005; Gerla, Bagrodia, Lixia, Tang, & Lan, 1999; Gerla, Tang, & Bagrodia, 1999; Khalife & Malouch, 2006; S. Xu & T. Saadawi, 2001) has been carried out to show how TCP performance is significantly affected by MAC protocols in multi-hop ad-hoc networks. We discuss these proposals that address the interaction between TCP and MAC in this chapter.

The fundamental problem in interaction between 802.11 MAC and TCP arises from the impact of hidden and exposed terminals on TCP congestion control mechanism and the impact of TCP transmission rate and ACKs' overhead on existence of hidden and exposed terminals.

3.1 Impact of hidden terminal and exposed terminal problem

The RTS/CTS control frames implemented in 802.11 MAC protocol can not prevent the hidden and exposed terminal problems in the scenarios with more than three numbers of hops.

Normally, when there is an unsuccessful RTS transmission either due to collision or due to an unnecessary deferred transmission; the sender in MAC layer enters a backoff period and it reschedules its RTS transmission when its backoff timer expires. After seven successive unsuccessful attempts, it is assumed that the route has failed and the packet is dropped. The effect of this wrong route failure report on TCP operation is not negligible. In this case, the sender tries to find a new route to destination. If the route takes some time to restore, TCP enters its backoff state and probes for a restored route at increasingly longer time intervals. Hence, the route might be restored for quite some time but TCP remains idle until it retransmits the packet after its time out expiration. Same condition happens when a route failure is reported after four unsuccessful attempts to transmit a data packet in MAC layer.

Therefore, hidden and exposed terminals may cause a lack of ACKs at TCP sender, leading it to retransmit by timeout (de Oliveira & Braun, 2007). As a consequence, TCP invokes its slow start mechanism to slow down its transmission rate to the lowest level instead of fast retransmit. In other words, TCP may not receive three duplicate ACKs due to its small Congestion Window size (*cwnd)* at most of the times. *Cwnd* is defined as maximum number of data packets a TCP sender may inject into the network at anytime without waiting for an ACK from the receiver. Considering this problem, TCP end-to-end throughput may decreases significantly as the number of the hops grow due to considerable delay of waiting for the transmit timer to expire.

3.2 Impact of TCP transmission rate

Another major problem in interaction between TCP and MAC is based on the probability of packet dropping due to increase in link contention as the TCP offered load increases (Fu, et al., 2005). In fact, TCP keeps sending more packets during congestion avoidance phase while a node is trying to access the medium within the MAC retry limits. The increment of sending rate continues until TCP perceives any packet loss indication. As a result, the channel condition may be aggravated because there are more outstanding packets intended to obtain the channel simultaneously. Consequently, TCP experiences an incredible throughput decrease due to the limited spatial reuse imposed by 802.11 MAC protocol.

3.3 TCP redundant ACKs

The issue of spatial contention in multi-hop wireless networks, which is aggravated by hidden and exposed terminals, can be caused by generating redundant TCP acknowledgments. It is noted that although TCP-ACKs are much smaller than TCP data packets, their transmission requires the same signalling overhead of the 802.11 MAC protocol (Altman & Jimenez, 2003). In fact, the receiver must content for the medium using RTS/CTS frames for ACK transmissions exactly as the sender does for data transmissions. The problem can be explained through Fig. 2 in which node D is a hidden terminal when B is transmitting data packets to C.

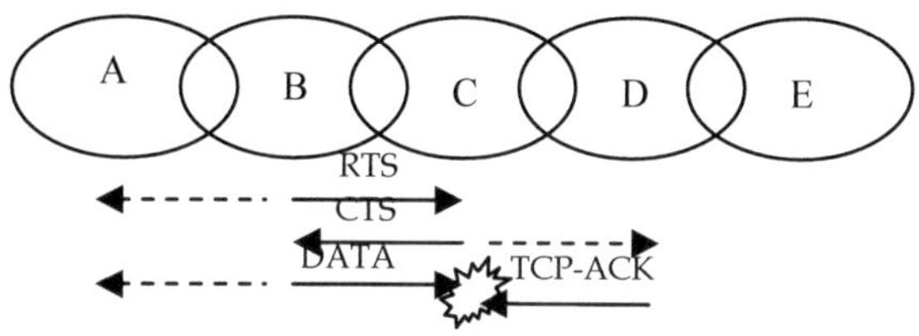

Fig. 2. Collision between DATA and TCP-ACK

After a successful RTS and CTS handshaking, B starts to send the data. Meanwhile, a TCP acknowledgment is flowing back from D to C. As it is shown in the Fig. 2, D's ACK transmission will interfere with C's data reception and TCP acknowledgment will be dropped. When TCP does not receive any ACK in a time interval, it assumes a data packet is lost and invokes its congestion control mechanism. When it comes down to TCP throughput, serious degradation will be observed.

4. TCP modifications over MAC layer in ad-hoc networks

In recent years, there has been an increasing amount of literature available on optimizing TCP performance in interaction with MAC layer over multi-hop wireless networks at both TCP sender and receiver side (De Oliveira & Braun, 2005, 2007; Fu, et al., 2005; Hamadani & Rakocevic, 2005) in which all of them try to minimize the effect of MAC overhead caused by spurious TCP retransmissions, TCP sending rate, and ACK packets. In order to mitigate the spatial contention, the proposal is to apply changes to the TCP *cwnd* size at the sender end. In contrast, proposals at receiver side aim to reduce the MAC overhead and ACK traffic by alternations on TCP acknowledgment strategy or advertised window (*rwin*).

4.1 Limiting TCP's packet output

The effect of 802.11 MAC protocol interferences on TCP is extensively studied (Fu, et al., 2005). They have suggested that for a given specific network topology there is an optimum *cwnd* size, which maximizes TCP throughput by improving the spatial reuse property of wireless networks. It is stated that in a chain topology consisting of h number of hops, the maximum number of simultaneous transmission is upper bounded by $\frac{h}{4}$, at which maximum spatial channel reuse is achieved (Fu, et al., 2005). Therefore, TCP achieves the highest throughput with its *cwnd* size being $\frac{h}{4}$ in an h-hop chain topology to limit the number of outstanding packets over the entire forwarding path.

TCP continuously increases its *cwnd* size until a packet loss is detected. It typically operates at an average window size that is larger than optimal, thereby leading to dropped packets caused by link layer contention. Therefore, two link layer modifications named *Link RED* and *adaptive pacing* is proposed to maintain the optimum *cwnd* size at TCP sender and to reduce the medium contention. The Link RED algorithm aims to reduce the contention on the wireless channel by monitoring the average number of retransmissions. The probability of dropping a packet is computed when this average number becomes greater than a threshold. The goal of adaptive pacing is to improve the spatial channel reuse when the *cwnd* size exceeds the optimum value. This mechanism is enabled within the Link RED and a node increases its backoff timer in MAC layer when it notices that the threshold has reached.

In another major study, Chen et al. (K. Chen, Xue, & Nahrstedt, 2003) have investigated the impact of the TCP's Congestion Window Limit (CWL) on TCP throughput by taking up the Fu et al.'s observations. Based on this claim that the impact of return path is not considered in (Fu, et al., 2005), a dynamic mechanism known as dynamic CWL is proposed in which the CWL depends on the Bandwidth Delay Product (BDP) of the connection. It is discussed that regardless of the MAC protocol, the BDP cannot exceed the Round-Trip Hop-Count (RTHC) in a wireless multi-hop network. In case of 802.11 MAC protocol, authors report that the BDP is less than $\frac{1}{4}$ of the RHTC as only four hops nodes away can transmit concurrently without collisions. As a result, when the *cwnd* exceeds $\frac{RHTC}{5}$, the TCP throughput decreases substantially. However, one major drawback is that the maximum retransmission timeout in TCP is set to 2s as opposed to the 240s which is given in the standard. This might affect the simulation result.

The maximum retransmission timeout in dynamic CWL has improved in (Papanastasiou & Ould-Khaoua, 2004), where it is proposed to throttle the sending rate increase during the congestion avoidance phase to a level below the standard of one segment per RTT. In this way, no upper bound is forced to *cwnd* as it is done in CWL and dynamic CWL and a significant improvement above dynamic CWL has been achieved by realistic setting of the maximum RTO. One can see that limiting the *cwnd* to slow down the transmission rate has been conclusively shown as an effective solution for spatial contention in MAC layer and consequent TCP throughput optimization.

4.2 Managing a shared medium

As discussed earlier, TCP performance suffers from an inefficient spatial channel usage when multiple flows are trying to access the shared radio channel leading to severe unfairness problem (Boggia et al., 2005). It is mentioned that, during the probing phase for available network bandwidth, TCP allows a large number of segments to be outstanding which in turn generates a high collision probability when TCP flows go through an 802.11 ad-hoc networks. A receiver-side approach has been then proposed by Boggia et el. to exploit the *rwin* filed of TCP segments to limit the number of in-flight packets in the network based on the medium condition. In this paper, a cross-layer algorithm which collects the total frame collision probability in MAC layer along the path has been proposed. The measured probability is then communicated to the TCP receiver to properly set the *rwin*. When there is not much collision in MAC layer, the total collision probability value is less than a threshold, *rwin* is increased exponentially by one segment. However, in a high congested media, the increment is much slower. Upon this strategy, a reasonable tradeoff has been achieved between TCP throughput and fairness. But the ratio of retransmitted segments is strongly reduced and the throughput does not improve significantly and stays similar to that obtained using NewReno. These findings suggest that by monitoring the channel condition, we can control the MAC overhead caused by spurious TCP retransmissions and redundant ACKs, resulting in a better TCP performance.

4.3 ACK thinning techniques

The sender's transmission of TCP-DATA segments and the receiver's ACK response contribute to spatial contention while TCP pair are communicating. The mechanisms dealing with the reciprocal ACK response aim to reduce the amount of ACK traffic for an optimized spatial contention. The first optimization of this nature has been introduced through standard TCP with delayed ACK option by delaying ACKs upon receiving two in-order data packets. It has been proved through extensive simulations (Lilakiatsakun & Seneviratne, 2003; S Xu & T Saadawi, 2001) that well-known TCP variants including Reno, New Reno, SACK and Vegas can perform better in case of throughput, bandwidth and energy consumption by employing the delayed ACKs. It has also been shown that in special cases the improvement in TCP throughput is in the range of 15-32% by deploying the optional delayed ACK mechanism (S Xu & T Saadawi, 2001). Such findings motivate the recent studies on more investigation over the degree of improvement that can be achievable with ACK thinning mechanisms.

The study of the behavior of delaying more than two ACKs over 802.11 MAC protocol was first carried out by Altman et al (Altman & Jimenez, 2003) in which the idea of standard delayed ACKs, which is combination of only two consecutive ACKs, has been extended. In the proposed scheme called TCP-LDA (Large Delayed Acknowledgment), an acknowledgment is sent only after a given number d of segments or after a certain fixed timeout. The dynamic aspect of TCP-LDA operates with d growing with an increasing packet sequence number from one up to $d = 4$. Once this limit is reached, the delay window (*dwin*) size, i.e. the number of the ACKs to be combined, is considered as fixed at 4 data packets even though a timer expires at TCP sender side. This may lead to the shortage of ACK phenomenon which is a condition with a bigger *dwin* size than a *cwnd* size. In this condition, a sender does not receive any ACK in a time interval and is just able to transmit upon a timeout. Moreover, TCP-LDA is not adaptable to different channel conditions and out-of-order packets are not taken into account. This means that when any indication of an out-of-order packet or dropped packet is received, *dwin* is never decremented again and still is fixed at 4 packets leading to poor TCP performance.

Singh et al (Singh & Kankipati, 2004) have shown enhancement in TCP performance and have derived the relation between throughput and number of data packets covered in one ACK. Based on the analysis, they have proposed TCP-ADA (Adaptive Delayed Acknowledgment) scheme, which tries to decrease the number of ACKs to one per congestion window. However, employing a large *dwin* size equal to *cwnd* size is not an efficient solution in all scenarios resulting in the burstiness of the forwarding packets in long paths. In this case, too many data packets are queued at TCP sender side, waiting for an acknowledgment to be received inducing packet drops in the router's buffer. Also they have not addressed the effect of packet loss event and out-of-order packets in the *dwin* size adjustment.

TCP-DCA (Delayed Cumulative Acknowledgment) (J. Chen et al., 2008) has been proposed to decrease the number of ACKs based on the path length. This study reveals that for a given topology and flow pattern, there exists an optimal delay window size at receiver that produces best TCP throughput. It is shown that path length is an important factor in choosing the *dwin* size because when travelling a longer path, a packet is more likely to suffer interference. For a 3 hop scenario in hidden terminal problem, it may rarely happen that the *dwin* size limit is equal to *cwnd* size. However, for long paths a high *dwin* limit aggravates the channel contention. If data packets arrive in order, the receiver generates one cumulative ACK for every *d* data packets. In case of an out of order packet, the receiver acknowledges immediately without any delay. To get the delay timer period of ACK timeout, the receiver monitors the packet inter-arrival interval and computes a smooth inter-arrival. Moreover, in TCP-DCA the sender reuses the advertised window (*rwin*) field in data packet header for advertising its *cwnd* size to the receiver to prevent the shortage of ACKs phenomenon. The modifications are at receiver side and the mechanism is simulated in scenarios with medium traffics while the higher loaded traffics are not taken into consideration. Table 1 shows these findings on the optimized *dwin* size based on the path length.

Path Length (h)	*dwin* limit
$h \leq 3$	*Cwnd*
$3 < h \leq 9$	5
$h \geq 10$	3

Table 1. Optimized delay window size in different path length

Olivera et al (De Oliveira & Braun, 2005, 2007) draws our attention to some drawbacks and limitations in the previous literature. The authors have improved the fix number of 4 packets handled with a single ACK after the start-up phase and the fixed ACK timeout in Altman and Jimenez's scheme.

The proposed dynamic delayed acknowledgment scheme called TCP-DAA (Dynamic Adaptive Acknowledgment) of Altman and Jimenez's applies the concept proposed in RFC 2581 by sending an immediate acknowledgment upon out-of-order packets or packets filling a gap at the receiver. This means that TCP-DAA is adaptive in the term of packet losses in the channel. The receiver maintains a dynamic delaying window (*dwin*) with size ranging from 2 to 4 full sized segments in networks with medium traffics which determines when an ACK will be produced. When a packet loss is observed; the *dwin* breaks down to two ACK packets. When there is no more losses reported; TCP enlarges *dwin* by one up to 4 packets. That means TCP receiver waits for 4 data packets before generating the ACK. To this end, the receiver implements an *ack-count* variable which increases by one until it reaches to the current value of *dwin* whenever a consecutive data packet is received.

To deal with the high delay variations in wireless ad-hoc environments, an effective mechanism has been employed to set the ACK timeout based on the packet inter-arrival times at the receiver. The rational is, in case of a single dropped packet, the next DATA packet will arrive out-of-order, thus triggerring immediate transmission of an ACK. However, if it was only a delay variation and the DATA packet arrives before the expected time for the subsequent packet; no timeout is triggered and the receiver avoids sending an extra and unnecessary ACK packet into the network (De Oliveira & Braun, 2005). The DAA method has been evaluated on string and mesh topologies of varying lengths and different number of flows. The results show an optimization in TCP throughput up to 50% over plain TCP NewReno on string topologies up to 8 hops and 20 flows (De Oliveira & Braun, 2005).

Later in (de Oliveira & Braun, 2007), an extension has been proposed to TCP-DAA concerning the robustness in high traffic environments with considerable packet losses. It has been pointed out that TCP-DAA may be inefficient in such environments. Based on these observations, an enhanced mechanism is suggested which is more adaptable in high traffic channels. In proposed mechanism called TCP-DAAp (TCP-DAA plus), *dwin* enlarges more gradually by a factor between zero and one to its limit of 4 packets to provide enough ACKs to the TCP sender. Additionally, *dwin* is reduced to one packet instead of two as a reaction to packet losses. This behavior is more conservative in comparison with TCP-DAA to prevent from transmitting a burst of data and the shortage of ACK phenomenon due to the small size of *cwnd* in such high loaded lossy environments. The achieved results obtained on a chain topology confirm that TCP-DAAp is as robust as the regular TCP mechanism under heavily constraint environments; however it does not provide the same improvements of TCP-DAA.

The main drawback of this work is that receiver is not dynamically notified to use TCP-DAA or TCP-DAAp in the scenarios mix of moderate and high traffic. Therefore, although the results have shown a significant and robust improvement, respectively in moderate and high traffics, the algorithm is not applicable in the real world where the traffic may change from time to time. To this end, it has been suggested to have an additional monitoring mechanism at the receiver to adjust the TCP-DAA strategy on the basis of channel condition (de Oliveira & Braun, 2007). Thus an adaptive receiver mechanism to switch between DAA and DAAp strategies in scenarios susceptible with high losses; is considered as future work by de Oliver and Braun (de Oliveira & Braun, 2007).

5. Monitoring Delayed Acknowledgment (TCP-MDA)

TCP-MDA dynamically reacts to the existing traffic in the network unlike the TCP-DAA. It can delay more ACKs in low load channels and less in the high traffics. Basically, in a high traffic network, it's more conservative to provide enough ACKs to TCP sender as there are more data packets intended to achieve the channel simultaneously. This is prominent in a long path topology where the spatial reuse property is limited due to hidden terminals and packet loss is more common to happen.

5.1 MAC collision probability measurement

The method implemented in TCP-MDA to measure the collision probability in 802.11 is based on the procedure used in (Boggia, et al., 2005) in which the measured collision probability has been employed to set the advertised window field of TCP to slow down the transmission rate. The same idea has been used in TCP-MDA to properly set the number of delayed ACKs as it is shown in Fig. 3.

A field called as $non\_collision\_prob_i^{tot}$ in the MAC protocol header has been added to meet our requirements. This field is set to 1 in the first hop of the path where no collision is detected. It should be recalled that TCP's functionality is based on the end hosts and it does not need any support from the intermediate nodes. Hence, the first hop is the hop which holds the source of the flow and the last hop is considered as TCP destination node. On the other hand, 802.11 is involved with the intermediate nodes as well as the source and destination nodes due to its responsibility for node-to-node delivery.

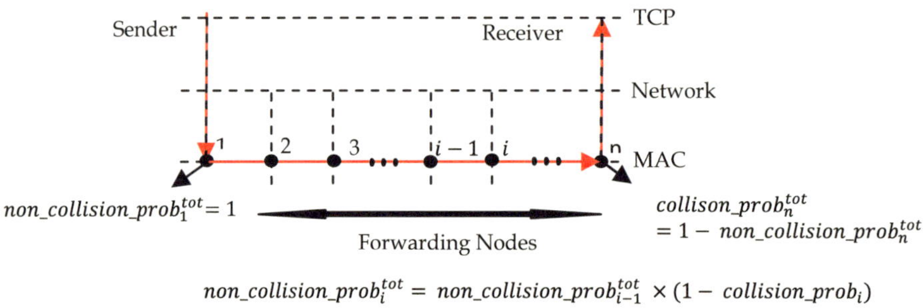

Fig. 3. MAC collision probability measurement

In TCP-MDA, TCP layer sends the SYN segment when a connection initiates. The MAC layer is conscious about flow start by receiving SYN at the first node and sets the $non\_collision\_prob_1^{tot} = 1$. For the same reason, TCP can interact with MAC layer at destination node, allowing the evaluation of the final collision probability and sends it to the TCP layer for the delayed ACK adjustment.

As the packet is traversing the path from sender to receiver in 802.11, the $non\_collision\_prob_i^{tot}$ field is calculated by taking into account the collision probability at each intermediate node (Boggia, et al., 2005).

We define $collision\_prob_i$ as the local collision probability at the ith forwarding node and it is given by the ratio between the number of retransmitted data packets and the total number of transmitted data packets.

The $non\_collision\_prob_i^{tot}$ is not a local value. It is defined as the product of collision probability at each node from sender to i^{th} node in a given path. Hence, in order to obtain the $non\_collision\_prob_i^{tot}$, it is needed to multiply the local non collision probability of the ith forwarding node which is $(1 - collision\_prob_i)$, with the overall non collision probability written in the header of the packet that is being forwarded from the $(i - 1)^{th}$ node. To this end, when a packet is received at node i, the overall $non\_collision\_prob_{i-1}^{tot}$ is read from the packet header. Then for each transmitted frame, the local non collision probability is estimated and the computed overall $non\_collision\_prob_i^{tot}$ is written back into the packet header. This procedure continues until the packet reaches to the last node and the $non\_collision\_prob_i^{tot}$ is given as:

$$non\_collision\_prob_i^{tot} = non\_collision\_prob_{i-1}^{tot} \times (1 - collision\_prob_i) \quad (1)$$

Finally, the total collision probability in the path at the last node is computed by using the value of total non collision probability which is read from the packet header. This is given as:

$$collision\_prob_n^{tot} = 1 - non\_collision\_prob_n^{tot} \quad (2)$$

It should be noted that all the calculated values are considered for the data packets and the transmission of RTC/CTS control frames is not taken into account. The pseudo-code depicted in Fig. 4 describes the whole process at one node.

```
collision_prob = 0;
transmitted_pkts = 0;
retransmitted_pkts = 0;
for (each data packet)
    // Packet is received at a node
    if (1st node of the path)
        non_collision_prob_1^tot = 1;
    else
        non_collision_prob_{i-1}^tot is read from the packet header;
        collision_prob_i = retransmitted_pkts_i / transmitted_pkts_i;
        non_collision_prob_i^tot = non_collision_prob_{i-1}^tot * (1-collision_prob_i);
    end if
    if (last node of the path)
        collision_prob_n^tot = 1 - non_collision_prob_n^tot;
    end if
    if retransmitted packet
        retransmitted_pkts_i + +;
    end if
    transmitted_pkts_i + +;
end for
```

Fig. 4. Packet processing at a single node to collect the collision probability

5.2 Delaying window strategy

The ACK processing in TCP-MDA is dependent on the calculated collision probability, *total_collision_prob*, in different channel traffics. Withholding ACK responses is done by maintaining a dynamic delaying window (*dwin*) at TCP receiver to define the number of data packets that would arrive before generating an ACK.

Like TCP-DAA, *dwin* size is initialized to one and it is gradually enlarged to its limit of 4 data packets. When the achieved *total_collision_prob* from MAC layer is less than a threshold (*collision_ thresh*), the channel is considered in the good condition and *dwin* is incremented by one for every received data packets. This means that *dwin* would become 4 faster and the receiver would generate less ACKs. It would be advantageous then to keep *dwin* at 4 as long as the channel is stable. When facing losses, however, *dwin* should be reduced due to the fact that during these periods the channel may have less packets than 4 in flight to trigger the fast retransmit mechanism at the sender. As a result, the channel may timeout if the receiver ACKs are not obtained quickly.

When receiver gets any indication of packet loss or the packet is overly delayed during transit, *dwin* reduces to two packets and again enlarges by one packet into its limit in low traffic channels. The reason to resume *dwin* growth from two instead of one is to go back to a behavior similar to that of the standard delayed acknowledgment (DA) in such situations, which performs better than configurations without it (de Oliveira & Braun, 2007). Figure 5 depicts the pseudo-code of TCP-MDA when a packet arrives at receiver.

To track the number of the delayed ACKs, TCP receiver maintains an *ack_count* variable ranging from one to the current value of *dwin*. Whenever a consecutive data packet is received, *ack_count* variable is increased by one. In this way, when *ack_count* = *dwin*, an ACK response is, immediately produced and *ack_count* is reset to one. It signifies the beginning of the next group of data packets for which the corresponding ACKs will be delayed. In fact, *ack_count* differentiates between each group of data packets.

It is also desirable to produce quick ACK responses so as to allow an increase of sending rate during the slow start phase at the sender. If ACKs are delayed too much during this phase, the sender would not receive enough ACKs to increase its sending rate efficiently due to the ACK requirements of TCP sender to clock out the data. A speeding factor μ, with $0 < \mu < 1$ is considered to enlarge the *dwin* in the startup phase instead of a fixed value of one. Additionally, *maxdwin* is considered as an indicator which turns true when the slow start phase is over and *dwin* reaches its maximum value of 4. Once the *maxdwin* is reached, then this mechanism is not activated again for the same connection. Hence this facility is for short life flows (de Oliveira & Braun, 2007).

```
dwin =1;
ack_count =1;
for (each data packet)
    // TCP sender
    Send SYN segment;
    // MAC layer
    Set non_collision_prob_1^tot=1 at the first node;
    Set non_collision_prob_i^tot as the packet is
    traveling hop by hop;
    Set collision_prob_n^tot at the last hop;
    // TCP Receiver
    Function DelayACK()
    if (data arrived in an interval)
        if (ack_count < (int)dwin)
            if (out of order packet?)
                OutofOrderPkt();
                break;
            else
                ack_count++;
                delay the ACK;
            end if
        else
            ack_count = 1;
            SendAck();
            if (collision_prob_n^tot < collision_thresh)
                if (dwin<4)
                    dwin +=1;
                end if
            else
                if (dwin<4)
                    dwin += μ' ;
                end if
            end if
        end if
    else
        TimeOut();
        break;
    end if
    end Function DelayAck
    Function OutofOrderPkt() & TimeOut()
    if (collision_prob_n^tot < collision_thresh)
        dwin = 2;
    else
        dwin = 1;
    end if
    ack_count = 1;
    SendAck();
    end Function OutofOrderPkt() & TimeOut()
end for
```

Fig. 5. TCP-MDA pseudo-code

The mechanism described above works well in moderate traffics; however, when the loss rates are considerable, it is desirable to enlarge *dwin* slowly to provide enough ACKs to TCP sender as there are more packets intended to achieve the channel. To meet this design, when *total_collision_prob* exceeds the *collision_thresh*, *dwin* is incremented by a factor μ' between zero and one. This is more aggressive in conditions with considerable losses due to small *cwnd* size in most of the times. In fact, *cwnd* size will be cut when a packet loss is perceived by a TCP sender. Thus, we need to provide enough ACKs to the corresponding sender to prevent from a transmission upon the timeout and to prevent from a bigger *dwin* size than *cwnd* size. For the same reason, it is more appropriate to reduce *dwin* to one instead of two as a reaction to packet loss. The optimized value for the *collision_thresh* is obtained through different simulation results which show the best value among all the indexes in (Armaghani et al., 2008). Figure 5 illustrates the pseudo code of the whole algorithm.

5.3 ACK timeout computation

For every successful delivered data and ACK packets, MDA method allows 4 data packets to produce one ACK response. However, it is desirable to trigger an immediate ACK without waiting an *ack_count* to reach the current *dwin* when a data packet is overly delayed during transmission. The ACK timeout is computed by the means of packets' inter-arrival time (Fig. 6).

That is, an ACK is generated when no data packets arrive within an average inter-arrival time since the last unacknowledged data packet. Therefore, an inter-arrival time gap between each received data packet which an ACK is to be delayed, say $i - 1, i, i + 1, \ldots$, and the previous data reception is recorded as $\delta_{i-1}, \delta_i, \delta_{i+1}, \ldots$.

It should be noted that the inter-arrival time between each data group is not taken into account. These collected inter-arrival time periods are used to calculate a smoothed average to estimate an expected inter-arrival time, $\bar{\delta}_i$ as given by following equation:

$$\bar{\delta}_i = \alpha \times \bar{\delta}_{i-1} + (1-\alpha) \times \delta_i \tag{3}$$

$\bar{\delta}_{i-1}$ is the last calculated average, δ_i is the data packet inter-arrival time sampled and $0 < \alpha < 1$ is an inter-arrival smoothing factor.

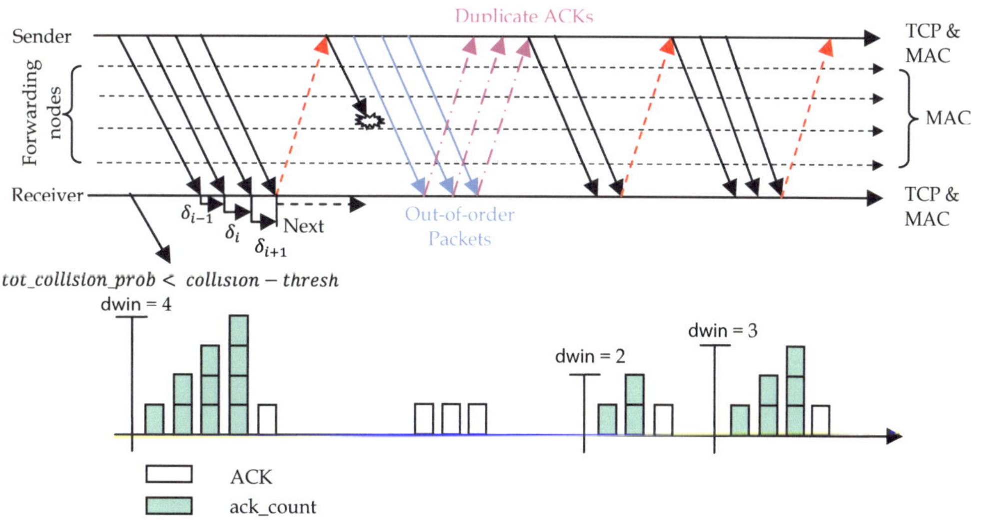

Fig. 6. An example of how TCP-MDA works in the moderate traffic

In case of out-of-order packets, an ACK is immediately prompted; otherwise the receiver waits for the period τ_i before responding. This effective timeout interval is calculated using a timeout tolerance factor k, with $k > 0$ as given in equation (4).

$$\tau_i = (2+k)\times\overline{\delta_i} \tag{4}$$

The rational here is that due to high delay variations in such environments, it is reasonable to wait for the time the second packet is expected. So, unnecessary timeouts are avoided to be triggered.

5.4 Sender side's modifications

The only requirement at the sender in TCP-MDA is to restrict its *cwnd* size to the maximum of 4 packets. This means that, TCP allows keeping 2, 3 or 4 packets outstanding in the network at any given time. This small size of *cwnd* has been reported in the literature (De Oliveira & Braun, 2005; Fu, et al., 2005) as an efficient size in the short range scenarios and has been thoroughly discussed in Section 4.1. It has been suggested that TCP sender can overcome the spatial contention property by confining the number of the packets in flight in the network (Fu, et al., 2005). So that a limit of $\frac{h}{4}$ for *cwnd* has been reported as an optimal setting in a chain topology; where h is the number of hops between sender and receiver. This setting has been followed in all the methodology's steps described in last sections to confirm the above conclusions and to make TCP-MDA more comparable with TCP-DAA.

5.5 Optimized numbers of delayed ACKs

Different simulations have been run to find the optimal number of in-order data packets to be waited before generating an ACK in different path lengths. In fact, delaying more ACKs in short range scenarios with less than three numbers of hops are found to be more effective as opposed to the upper bounded of four ACKs in scenarios dealing with moderate traffic. However, a large *dwin* over a long path can aggregate the situation by inducing a large burst of data into the network leading to more packet losses (J. Chen, et al., 2008). We take these considerations into account in scenarios, which is mix of low and high traffic/loss rates, An optimal *dwin* size, which acts best in comparison with the other sizes, have been obtained. TCP-MDA with optimal *dwin* size has been compared with TCP-MDA with *cwnd* limit setting and the results are discussed later.

5.6 Performance evaluation

To validate the proposed strategy various simulations representing the derived TCP-MDA scheme under different parameters is presented in this section. The system performance in term of throughput has been studied and the effects of different parameters have been investigated. The evaluation of TCP-MDA has been conducted with the Network Simulator-2 (ns-2) (Fall & Varadhan, 2008).

5.6.1 Simulation area setup

Two scenarios namely chain topology and grid topology as depicted in Fig. 7 have been considered throughout our experiment. The chain topology consist of n nodes with number of nodes *(n)* varying from 2 to 20 and number of concurrent flows varying from 1 to 20 in each simulation. For each simulation, TCP connection is sourced at the first node (node 0) and packets travel hop by hop over the chain to the end node ($1 \leq$ end node ≤ 19). Simulation has been done for a 5×5 grid topology with three and six TCP flows, respectively. In case of six TCP flows, half of the flows go horizontally and the other half go vertically, spaced evenly.

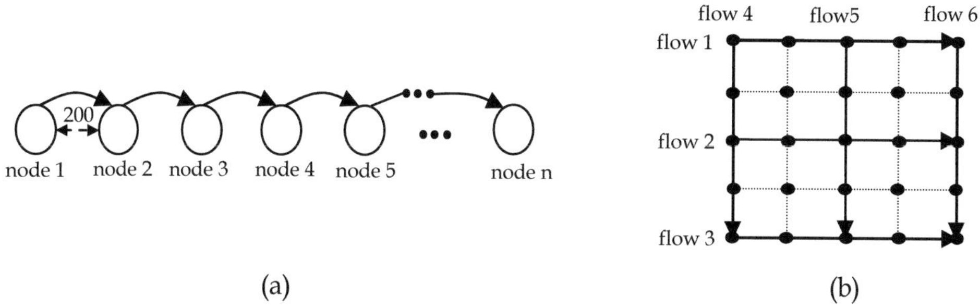

Fig. 7. Simulation scenarios: (a) Chain topology, (b) Grid topology

The nodes are considered as static to minimize the impact of routing dynamics and concentrate on the interaction between TCP and MAC protocol as it is widely followed in the previous researches (K. Chen, et al., 2003; De Oliveira & Braun, 2005; Lilakiatsakun & Seneviratne, 2003; Papanastasiou & Ould-Khaoua, 2004; S Xu & T Saadawi, 2001). In fact, the target is to investigate the dropped packets resulting from channel spatial reuse and contention rather than the dropped packets induced by the route failure which belongs to the mobility fact of network layer. IEEE 802.11 MAC protocol has been considered as widely studied underlying protocol in wireless networks along with Ad-hoc On-Demand Distance Vector Routing (AODV) protocol as a very popular routing protocol in ad-hoc networks. Moreover, nodes access the radio channel at the data rate of 2 Mbps with transmission range set to 250 m and interference range of 550 m.

A TCP-NewReno variant is used which starts transmitting FTP traffic along the chain topology and the packet size is set as 1,460 bytes. Most of the parameters are chosen as given in (de Oliveira & Braun, 2007). These parameters include the value of 0.75 for a as an inter-arrival smoothing factor and 0.2 for k as a tolerance factor tailored to compute the ACK timeout in sending the acknowledgments.We also set the startup parameter μ as 0.3 which provides the best result among the other indexes in (de Oliveira & Braun, 2007). End-to-end TCP throughput has been evaluated and has been defined as total bits transmitted and acknowledged over the simulation time (5).

$$Throughput(kbps) = \frac{ReceivedPackets(Packetsize)*8}{STime} \tag{5}$$

Scenarios starting from a moderate traffic/loss rate and ending to a noisy channel with extensive packet losses has been assumed in simulation. A Four State Markov Chain error model has been considered to model this environment as depicted in Fig. 8.

The error rate is changed from 0 in good state to 0.2 in the worst state. There will be more packet losses as the probability of error increases. Here, the packet drops are not only losses due to MAC collisions but also losses induced due to permanent external disturbance. In our proposed strategy, we account the packet losses due to the medium contention and external disturbance is not taken into account. The multi-state error model implements time based error state transitions. Transitions to the next error state occur at the end of the duration of the current state. The next error state is then selected using the transition state matrix (Fall & Varadhan, 2008).

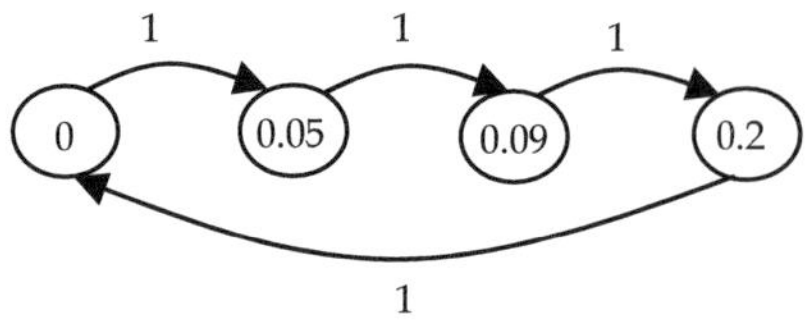

Fig. 8. Four state markov chain error model

The justification for employing this typical error model is its compatibility of introducing different state of collisions through the simulation time which has been achieved by monitoring the simulation trace file. An error rate more than 0.2 might lead to a high collision probability especially in larger ranges more than four hops. This would prevent the proposed strategy to properly show its functionality in low collision probability conditions. All simulation parameters are listed in Table 2. Each data point represents an average of 5 simulation runs with different random seed numbers and each run lasts for 1,000 s. We choose 1,000 s as we target the constrained scenarios ending to high packet loss. So that, it takes a longer time to reach a stable simulation condition.

Parameter	**Value**
Channel bandwidth	2 Mbps
Channel delay	25 μs
Transmission range	250 m
Interference range	550 m
Packet size	1460 bytes
Window limit	4 packets
Regular TCP	NewReno
Routing protocol	AODV
Traffic type	FTP
α	0.75
κ	0.2
μ	0.3
μ'	0.28
collision_thresh	0.3

Table 2. Simulation parameters

5.6.2 Throughput in the chain topology

As discussed earlier, we know that when the channel is in good condition, *dwin* is incremented by one to its limit of 4 to generate less ACKs. However, when the loss rates are considerable, it is more proper to enlarge *dwin* slowly by a factor μ' which has value between zero and one to provide enough ACKs to TCP sender. Monitoring the channel condition is done by comparing the achieved *total_collision_prob* from MAC layer with the *collision_ thresh*.

The optimized value for μ' has been obtained by the analytical evaluation given in (de Oliveira & Braun, 2007). It has been proved that following a very conservative procedure, *dwin* should be increased by about 0.28 for each in-order data packet received, and is same as simulation results in (de Oliveira & Braun, 2007). The same value of μ' has been used here in all the simulations based on the observations in (de Oliveira & Braun, 2007). In addition to μ', the optimized value of 0.3 for the *collision_thresh* has been obtained through different simulation results which show the best value among all the indexes. All the simulation results are presented in (Armaghani, et al. 2008) for different values of *collision_thresh.*

To evaluate the effectiveness of the TCP-MDA strategy against TCP-DAA and TCP-DAAp (De Oliveira & Braun, 2005, 2007), further simulations have been conducted for 2, 4, 9 and 16 hop scenarios. The rest of simulation parameters and experimental setup were identical to ones selected in Section 5.6.1. It can be concluded that in a 4 hop scenario with up to 5 concurrent flows, TCP-MDA performs similar to TCP-DAA (Fig. 9a). However, for concurrent flows more than 5, results show the improvement of TCP-MDA over the other protocols.

This behavior of TCP-MDA could be due to the setting of *collision_thresh* in our experiments. In fact, the total collision probability (*total_collision_prob)* measured in the MAC layer has been observed to be less than 0.3 in the scenarios up to 5 numbers of flows. Therefore, TCP-MDA throughput is same as TCP-DAA by enlarging *dwin* by one to its limit of 4 packets.

As the number of flows increases, the *total_collision_prob* has a value more than 0.3. Thus, TCP-MDA reacts under this condition by enlarging *dwin* more gradually in order to avoid timeout at the receiver and a bigger *dwin* size than *cwnd* size, i.e. shortage of ACK phenomenon. These results also show that the efficiency of TCP-DAA goes down to the level of TCP-DAAp when we have several concurrent flows running in the network. This behavior closely confirms the rationale of enlarging *dwin* more gradually in a higher loaded channel.

For 2 hop scenario simulation results do not show considerable improvements over the TCP-DAA. A possible explanation for this might be due to the limited spatial reuse property imposed by MAC layer. Spatial contention is negligible in a small network with a short path and we still have a steady state condition where less packet loss may occur with increased load. So that, the calculated collision probability in MAC layer is less than the assumed threshold in these scenarios and *dwin* enlarges by one to meet the need of combining 4 ACKs in one ACK.

The results of the evaluation with 9 hops are shown in Fig. 9b where we observed that TCP-MDA strategy again proves superior to all other protocols. The improvement ranges between 4 to 13% over TCP-DAAp and 10 to 30% over TCP-NewReno+DA+WL and even higher over TCP-DAA.

The throughput results for a 16 hop scenario depicted in Fig. 9c is not very encouraging. Although, TCP-MDA still seems to slightly outperform others, but TCP instability problem is observed in this experiment. This instability may be explained due to the fact that there are more hidden and exposed terminals that cannot sense each other for transmissions in longer path. In fact, there would be more timeout reports and retransmission efforts in the MAC layer. After several unsuccessful retransmission efforts, the MAC layer would report a link breakage and a route discovery would be triggered immediately after the route failure has been reported. In this way the source would have to wait for the duration until the new route has been established. This is likely to affect the throughput.

Another reason that could be attributed may be due to the consequence of high interference on TCP sender RTT estimation. This implies that longer end-to-end connection would result in higher amount of contention among nodes because all of them try to access the channel at

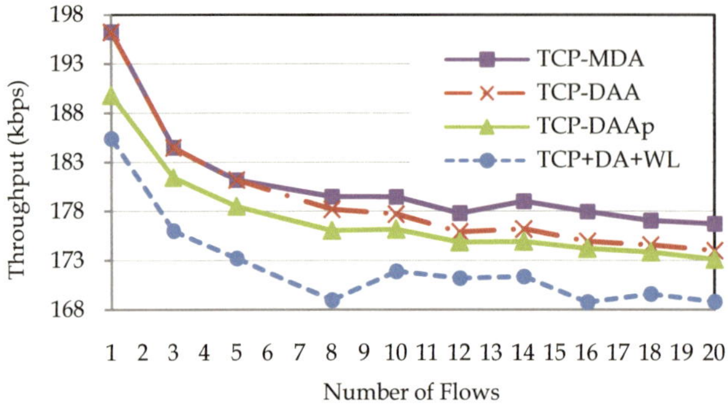

(a) 4 hop

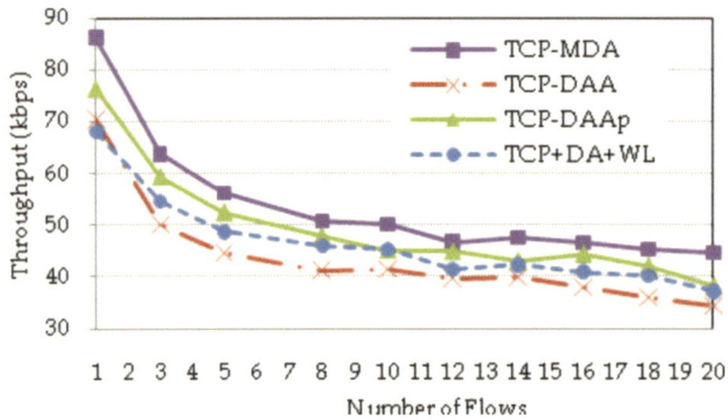

(b) 9 hop

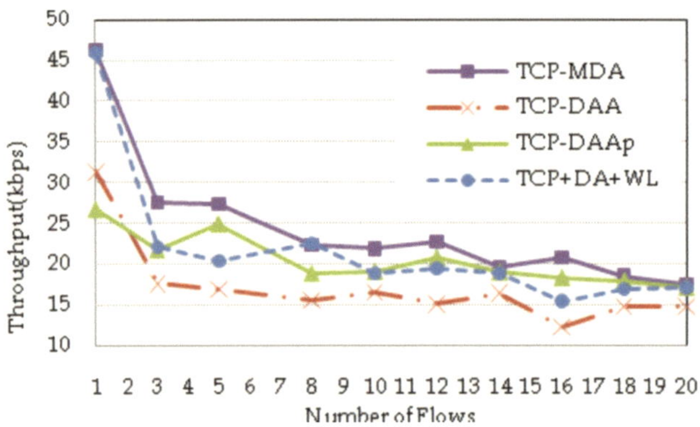

(c) 16 hop

Fig. 9. TCP throughput vs. number of flows in a 4 hop chain topology

the same time, and time for the TCP sender to detect lost packets would be longer. Figure 10 depicts a simple scenario where all nodes have at least one packet to send in the forward direction. We assume that node B and D initially have the channel access and they start to transmit at the same time. Soon after the transmission, there would be collision in packet from B to C with the packet from D to E. Meanwhile, A has been waiting to start transmitting several packets to B before releasing the channel.

However, B would be still unable to access the channel and buffers the new packets in addition to packet(s) already in its buffer and would start building up its queue. Therefore, a bottleneck may occur at node B of the path resulting to an artificial increase of the RTT delay measured by the sender. As a result, TCP would overestimate the available bandwidth and enlarges its *cwnd* size leading to the network overload in the next RTT. This procedure would continue until a packet drop would be reported within a MAC retry limits specified by 802.11 MAC standard.

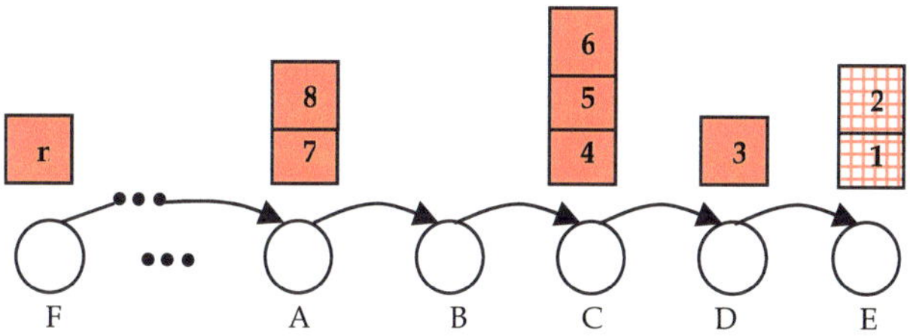

Fig. 10. Network overload scenario

It has been observed that TCP-MDA shows improvements in comparison with both TCP-DAA and TCP-DAAp in short range networks (up to 10 hops). This is basically because TCP-DAA and TCP-DAAp have not been designed for the scenarios facing the tradeoff between moderate and high loss rates, so they are more adaptable to the environment when they come together as TCP-MDA with a channel monitoring mechanism.

The drawback of the proposed strategy is that, TCP-MDA does not estimate the internal network state. However, in a channel with high loss rate, packet drops are not only due to the MAC collision. Packet loss might be due to the high medium induced errors and external disturbance. Since TCP-MDA is not tailored to monitor the channel state, so it is unable to demonstrate the level of medium errors.

5.6.3 Throughput in grid topology

Grid topology is a more complex scenario with various interactions among the nodes. Extensive channel contention exists and so more packet drops are expected as a consequence.
Grid topology is commonly used in literature to evaluate the effect of multiple interfering flows on TCP performance (Boggia, et al., 2005; J. Chen, et al., 2008; De Oliveira & Braun, 2005). Figure 11 compares the performance of TCP-MDA, TCP-DAA, TP-DAAp as well as standard TCP with DA extension.

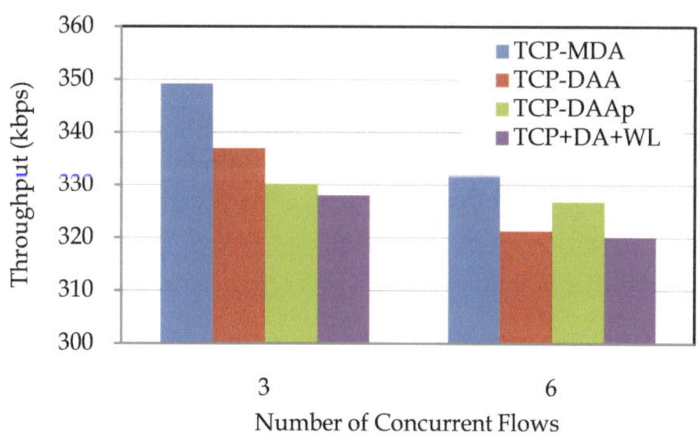

Fig. 11. TCP throughput over a 5×5 grid topology

Here, the flows do not share the same path but still interfere due to the hidden terminals and interference between nodes' transmission ranges. The results depicted in Fig. 11 again mirror the optimized throughput of TCP-MDA over the other protocols in scenarios with dynamic traffic. There are fewer contentions in the case of three flows and so TCP-MDA maintains the traffic by enlarging *dwin* rapidly up to four delayed ACKs. As the level of contention upsurges, TCP-MDA turns to perform more moderately by a gradual *dwin* enlargement, i.e. in the case of six cross flows.

TCP-DAAp provides a better throughput over TCP-DAA and TCP+DA+WL in the case of six flows which again prove the need of providing more ACKs in high traffic channels. In general, the same observation as chain topology holds true for grid topology. It can be deducted that in chain and grid scenarios, TCP-MDA benefits by delaying more ACKs in low traffic and less in high traffic channels.

5.6.4 Impact of congestion window limit

It is reported in earlier studies that limiting *cwnd* size improves TCP performance by maximizing the spatial reuse. So, in this study a limit of up to 4 packets has been considered for *cwnd* in scenarios with not more than 19 hops to make our work more comparable with the ones presented in (de Oliveira & Braun, 2007).

It would be noted that TCP-MDA may not provide the same improvement in some scenarios and the performance may degrade to the level of standard TCP that uses DA and window limit (WL). This behavior can be explained as following: first, limiting *cwnd* by itself would decrease the channel interference and maximize the spatial reuse. On the other hand, delaying ACKs helps TCP sender to slowdown its transmission rate by triggering the *cwnd* growth to its limit in a longer interval. In this way, the total number of induced data packets in the network might be affected by a slow transmission rate and the receiver delaying window adaption provides little extra improvement.

The above discussion has motivated to do more investigation on the impact of *cwnd* limit along with the *dwin* limit. To this end, we have run different simulations in which the *cwnd* has been unbounded and *dwin* size has been varied with different values. All the simulation parameters are same as in earlier simulations. Our objective has been to identify the relationship between TCP throughput and optimized *dwin* size in different path lengths. The results are presented in Fig. 12.

The above observations determine that *dwin* size in TCP-MDA is based on the path length of a TCP connection. We have observed that for a short path (hops ≤ 3); the ACK can be delayed up to a large value. The reason lies on the 802.11 capability to transmit the packets without collision in short ranges no matter what the burst size is.

However, employing a large *dwin* size is not an efficient solution in all scenarios resulting in the burstiness of the forwarding packets in long paths. In this case, too many data packets are queued at the TCP sender side, waiting for an acknowledgment to be received inducing packet drops in the router's buffer. Since there are more interfering nodes, there might be more packet losses because the packet has more chances to be interfered in a long path. The proper values for TCP-MDA *dwin* size according to our observations in different path length are listed in Table 3.

Although, there are more unsuccessful packet transmissions caused by interference in the chains between 4 and 6 hop counts, TCP-MDA still could maintain performance gain by delaying ACK for more data packets since a TCP sender is able to recover packet loss rather rapidly due to the small RTT.

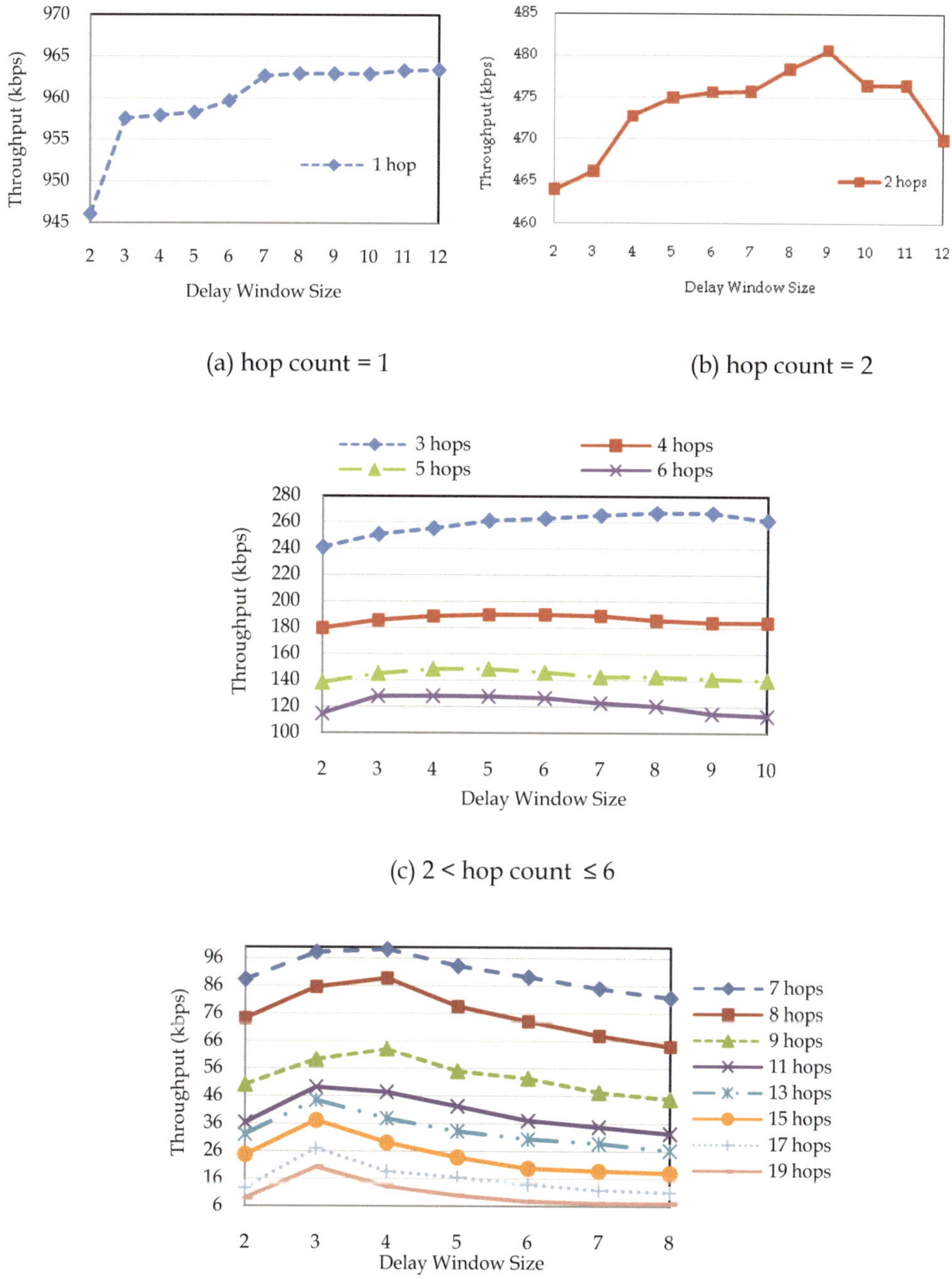

(a) hop count = 1

(b) hop count = 2

(c) 2 < hop count ≤ 6

(d) 6< hop count ≤ 19

Fig. 12. TCP throughput vs. delay window size in chain topology

Path length (No. of hops)	*dwin* Limit
$h \leq 3$	9
$3 < h \leq 5$	5
$5 < h \leq 9$	4
$9 < h \leq 19$	3

Table 3. Optimized *dwin* size in different path lengths

Similar trend also exists for paths longer than 6 hops, where TCP-MDA achieves throughput gain only when the delay window size is equal to 4. For larger topologies than 10 hops, large delay window size may not maintain throughput gain due to excessive data packet losses. Further, TCP-MDA spends more time detecting packet loss due to the larger RTT. Therefore, for long paths, large delay window is not preferred.

Next experiment is tailored to evaluate the performance of TCP-MDA - WL over TCP-MDA. Figure 13 shows the performance of proposed strategy with and without *cwnd* limit. It has been shown in previous simulations that TCP-MDA outperforms the other protocols in a short chain of hops. Here, the impact of *cwnd* limit has been investigated in our assumed scenario with medium and high loss rates. The number of the acknowledgments to be delayed in TCP-MDA - WL has been accordingly based on observation in each path length as listed in Table 3.

In this experiment, the two graphs in Fig. 13a and Fig. 13b show that *cwnd* limit on TCP-MDA does not bring considerable benefit on a short path with less number of flows. In fact, bounding the *cwnd* along with *dwin* may restrict the TCP performance by confining the total number of packets in flight in the network in small topologies, i.e. small burst size. Therefore, a large *dwin* solely might be enough effective on throughput improvement in these kinds of scenarios. For a longer path in Fig. 13c *cwnd* limit provides more throughput gain which is prominent in less number of flows.

6. Conclusion

In this chapter poor bandwidth utilization and performance of TCP, when it runs over 802.11 MAC protocol in multi-hop ad-hoc networks, has been addressed. This problem can be due to the extensive number of medium access carried out by TCP by generating redundant ACK packets that compete in the same route with data packets for the media. First, the reasons of TCP performance degradation in ad-hoc networks have been studied. Then the impact of delay acknowledgments has been studied which helps to improve TCP performance by reducing the number of generated ACK. Taking into account the importance of delay ACKs on TCP performance enhancement, a dynamic TCP-MAC interaction strategy has been proposed to reduce ACK induced overhead and consequently collisions.

The results have shown that the proposed dynamic TCP-MAC interaction approach reduces the number of ACKs transmitted by a TCP receiver by monitoring the medium collision probability and reacting to packet losses. The results comparison has shown improvement in short path lengths between 4 to 9 hops in a chain topology.

The impact of TCP *cwnd* size along with delayed ACK on MAC spatial reuse have also been studied. The findings show that with an unbounded *cwnd*, for each topology there exists an

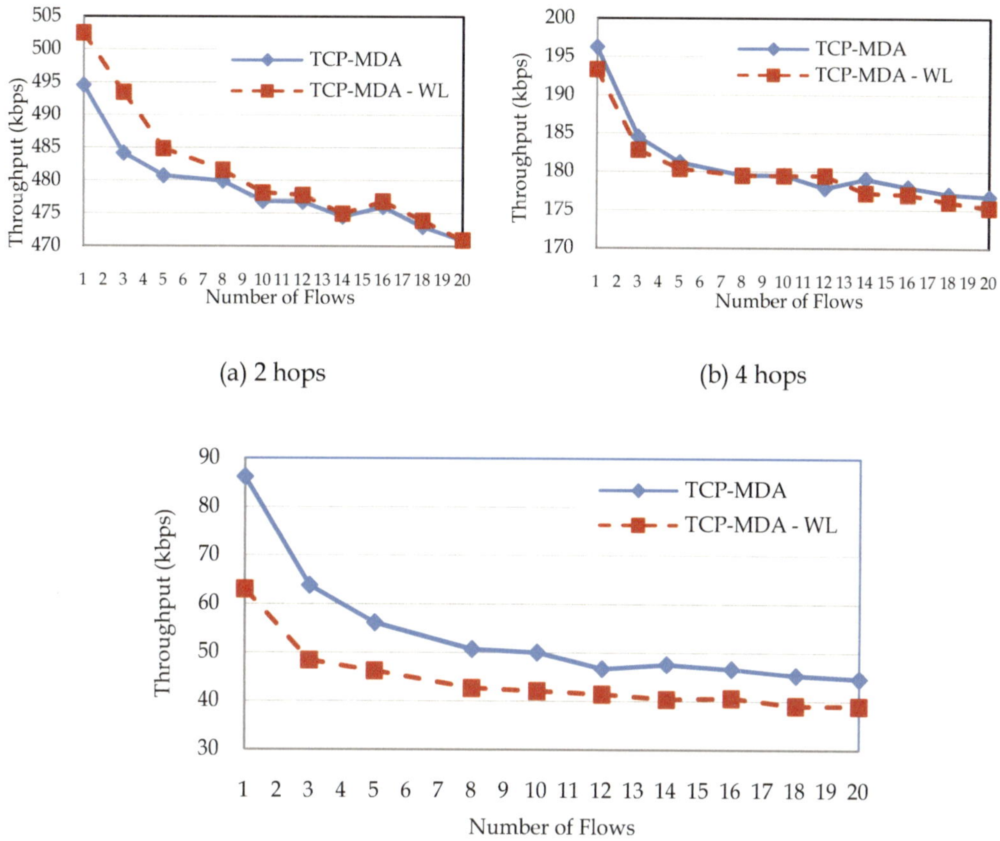

(a) 2 hops

(b) 4 hops

(c) 9 hops

Fig. 13. Comparison of TCP-MDA with and without *cwnd* limit

optimal delay window size in which TCP throughput is maximized. Armed with the optimized numbers of delayed ACKs, the proposed strategy has been evaluated with two different adjustments: with a limited *cwnd* and maximum delayed ACKs of 4 packets in all the topologies; and an unbounded *cwnd* along with a various *dwin* based on the path length of the topology. The related results draw to the conclusion that limiting *cwnd* is not beneficial in all the scenarios and a large *dwin* may solely help to alleviate the spatial reuse contention in short range topologies with less number of flows.

7. Directions for future work

Based on the achieved results, following problems may be subject for further study:

- Error detection mechanism: TCP-MDA is basically based on the consideration that having a dynamic loss rate, the packets suffer from the channel interference and MAC collision and accordingly the channel collision probability is taken into account. Hence,

TCP-MDA does not detect the exact internal state of the network. However, multi-hop networks are prone to much higher bit error rates in a lossy channel leading to very complex conditions. As a consequence, TCP-MDA lacks robustness in detecting what is exactly going on within the network so that it can't take proper action upon error based losses. So, it would be an interesting prospect to develop an end-to-end basis error detection mechanism to inform TCP about the actual cause of any packet loss so the TCP recovery mechanism can take the most appropriate action. This error model can be designed using more heuristic methods and fuzzy logic to consider more realistic transitions among the various states of the network.

- TCP sender adoption: TCP-MDA focuses more on the TCP receiver side and the only investigation on the sender side is over the impact of congestion window limitation. It is a basic TCP functionality that the sender relies on ACKs for computing its timeout interval and transmits new data packets. Moreover, TCP RTT computation can be affected by high/low delay variance. Hence, TCP performance can be disturbed by unnecessary delaying ACKs. As a result, a comprehensive study might reveal issues which are not captured in present research.

8. References

Altman, E., & Jimenez, T. (2003). Novel delayed ACK techniques for improving TCP performance in multihop wireless networks. *Lecture Notes in Computer Science, 2775*, 237-250.

Armaghani, F., Jamuar, S., Khatun, S., & Rasid, M. (2008). Performance Analysis of TCP with Delayed Acknowledgments in Multi-hop Ad-hoc Networks. To appear in *Wireless Personal Communications (on line available).*

Boggia, G., Camarda, P., Grieco, L. A., Mastrocristino, T., & Tesoriere (2005, 7-7 Sept. 2005). *A Cross-layer Approach to Enhance TCP Fairness in Wireless Ad-hoc Networks.* Paper presented at the Wireless Communication Systems, 2005. 2nd International Symposium on.

Chen, J., Gerla, M., Lee, Y., & Sanadidi, M. (2008). TCP with delayed ack for wireless networks. *Ad Hoc Networks, 6*(7), 1098-1116.

Chen, K., Xue, Y., & Nahrstedt, K. (2003). On setting TCP's congestion window limit in mobile ad hoc networks. *Communications, 201*, 03.

Conti, M. (2003). Body, personal, and local ad hoc wireless networks *The Handbook of Ad Hoc Wireless Networks* (pp. 3-24). Florida: CRC Press, Inc.

De Oliveira, R., & Braun, T. (2005). *A dynamic adaptive acknowledgment strategy for TCP over multihop wireless networks.* Proc. IEEE INFOCOM'05, March 2005, 1863 - 1874, Miami (USA)

de Oliveira, R., & Braun, T. (2007). A smart TCP acknowledgment approach for multihop wireless networks. *IEEE Transactions on Mobile Computing, 6*(2), 192-205.

Dongkyun, K., Toh, C. K., & Yanghee, C. (2000, 2000). *TCP-BuS: improving TCP performance in wireless ad hoc networks.* Paper presented at the Communications, 2000. ICC 2000. 2000 IEEE International Conference on.

Fall, K., & Varadhan, K. (2008). The ns manual. from The VINT Project: http://www.isi.edu/nsnam/ns/ns-documentation.html

Fu, Z., Luo, H., Zerfos, P., Lu, S., Zhang, L., & Gerla, M. (2005). The impact of multihop wireless channel on TCP performance. *IEEE Transactions on Mobile Computing, 4*(2), 209-221.

Gerla, M., Bagrodia, R., Lixia, Z., Tang, K., & Lan, W. (1999, 1999). *TCP over wireless multi-hop protocols: simulation and experiments*. Paper presented at the Communications, 1999. ICC '99. 1999 IEEE International Conference on.

Gerla, M., Tang, K., & Bagrodia, R. (1999, 25-26 Feb 1999). *TCP performance in wireless multi-hop networks*. Paper presented at the Mobile Computing Systems and Applications, 1999. Proceedings. WMCSA '99. Second IEEE Workshop on.

Hamadani, E., & Rakocevic, V. (2005). *Evaluating and Improving TCP Performance Against Contention Losses in Multihop Ad Hoc Networks*.

Hanbali, A., Altman, E., & Nain, P. (2005). A survey of TCP over ad hoc networks. *IEEE Communications Surveys & Tutorials, 7*(3), 22-36.

Holland, G., & Vaidya, N. (2002). Analysis of TCP performance over mobile ad hoc networks. *Wirel. Netw., 8*(2/3), 275-288.

Khalife, H., & Malouch, N. (2006, June 2006). *Interaction Between Hidden Node Collisions and Congestions in Multihop Wireless Ad-hoc Networks*. Paper presented at the Communications, 2006. ICC '06. IEEE International Conference on.

Lilakiatsakun, W., & Seneviratne, A. (2003). *TCP performances over wireless link deploying delayed ACK*. Proc. VTC '03, Vol. 3, April 2003, 1715 - 1719.

Liu, J., & Singh, S. (2001). ATCP: TCP for mobile ad hoc networks. *Selected Areas in Communications, IEEE Journal on, 19*(7), 1300-1315.

Mohapatra, P., & Krishnamurthy, S. (2005). *AD HOC NETWORKS: technologies and protocols*: Springer.

Papanastasiou, S., & Ould-Khaoua, M. (2004). TCP congestion window evolution and spatial reuse in MANETs. *Wireless Communications and Mobile Computing, 4*(6).

Singh, A., & Kankipati, K. (2004). *TCP-ADA: TCP with adaptive delayed acknowledgement for mobile ad hoc networks*. Proc. Wireless Communication and Networking Conference, March 2004, 1685 - 1690.

Stevens, W. (1994). *TCP/IP illustrated, vol. 1*: Addison-Wesley.

Wang, F., & Zhang, Y. (2002). *Improving TCP performance over mobile ad-hoc networks with out-of-order detection and response*. Paper presented at the Proceedings of the 3rd ACM international symposium on Mobile ad hoc networking \& computing.

Xiang, C., Hongqiang, Z., & Jiangfeng, W. (2005). A survey on improving TCP performance over wireless networks: Network Theory and Applications, voll6.[S. 1.]: Springer.

Xu, K., Gerla, M., & Bae, S. (2002). *How effective is the IEEE 802.11 RTS/CTS handshake in ad hoc networks*. Proc. IEEE GLOBECOM Vol. 1, Nov. 2002, 71 - 76.

Xu, S., & Saadawi, T. (2001). Does the IEEE 802.11 MAC protocol work well in multihop wireless ad hoc networks? *Communications Magazine, IEEE, 39*(6), 130-137.

Xu, S., & Saadawi, T. (2001). *Evaluation for TCP with Delayed ACK Option in Wireless multi-hop Networks*. Proc. IEEE VTC'01, Fall 2001, 267 - 271.

Yu, X. (2004). *Improving TCP performance over mobile ad hoc networks by exploiting cross-layer information awareness*. Paper presented at the Proceedings of the 10th annual international conference on Mobile computing and networking.

Zhai, H., Wang, J., Chen, X., & Fang, Y. (2006). Medium access control in mobile ad hoc networks: challenges and solutions. *Wireless Communications and Mobile Computing, 6*(2), 151-170.

19

The Effect of Packet Losses and Delay on TCP Traffic over Wireless Ad Hoc Networks

May Zin Oo and Mazliza Othman
University of Malaya
Kuala Lumpur,
Malaysia

1. Introduction

The popularity of wireless network has been growing steadily. Wireless ad hoc networks have been popular because they are very easy to implement without using base stations. The wireless ad hoc networks are complex distributed systems that consist of wireless mobile or static nodes that can freely and dynamically self-organize. The ad hoc networks allow nodes to seamlessly communicate in an area with no pre-existing infrastructure. Future advanced technology of ad hoc network will allow the forming of small ad hoc networks on campuses, during conferences and even in homes. Furthermore, there is an increasing need for easily portable ad hoc networks in rescue mission, especially for accessing rough terrains. However, the quick adaptation and ease of configuration of ad hoc networks come at a price.

In wireless ad hoc networks, route changes and network partitions occur frequently due to the unconstrained network topology changes. Moreover, this kind of network inherits the traditional problems of wireless communication, such as unprotected outside signals or interferences, unreliable wireless medium, asymmetric propagation properties of wireless channel, hidden and exposed terminal phenomena, transmission rate limitation and blindly invoking congestion control of transport layer. Although most of these limitations and complexities are due to the lack of fixed backbone or infrastructure, building ad hoc network temporarily is not only simple and easy to implement but also cost-effective and less time-consuming if compared to an infrastructure network that needs to establish a based station and fixed backbone. Among the above mentioned problems and limitations, the impact of transport layer limitations is analyzed across ad hoc routing protocols throughout the network topologies.

Transmission Control Protocol (TCP) (Postel, 1981) is the *de facto* standard designed to provide reliable end-to-end delivery of data packet in the wired networks. Normally, TCP is an independent protocol that is not related to the underlying network technology. However, some assumptions of TCP, such as consideration of only static node, packet losses due to congestion or buffer overflows are inspired from the features of wired networks. In the wireless network, these assumptions may not be correct all the time due to the rapid network topology changes, node movements and limited battery power. In order to apply TCP to an ad hoc environment, TCP has to overcome many problems, such as packet losses due to congestion, high bit errors, node mobility, longer delay and so on. The following TCP versions, Tahoe (Stevens, 1997), Reno (Allman, 1999), NewReno (Floyd & Henderson, 1999), Vegas (Brakno et al., 1994) and Westwood (Gerla et al., 2002), are enhanced versions of TCP and perform differently depending on how the routing protocols can quickly adapt route changes due to link breaks in an ad hoc network environment.

For wireless ad hoc networks, the issue of routing packets between any pair of nodes becomes a challenging task because the nodes can move randomly within the network. A path that is considered optimal at a given point in time might not work at all a few moments later. Traditional routing protocols such as DSDV (Perkins & Watson, 1994) are proactive in that they maintain routes to all nodes. They react to any change in the topology even if no traffic is affected by the change and they require periodic control messages to maintain routes to every node in the network. As mobility increases, more of scarce resources, such as bandwidth and power, will be used. Alternative reactive routing protocols, i.e. DSR (Johnson et al., 2007) and AODV (Perkins et al., 2003), determine the route when they explicitly need to route packets, thus avoiding nodes from updating every possible route in the network. However, these protocols tend to cause the broadcast storm problem (Tseng et al., 2002) due to the broadcast nature of the route discovery procedure. To avoid the discovery of a new route whenever a route fails, multipath routing protocols, i.e. AOMDV (Marina & Das, 2006) and OLSR (Clausen & Jacquet, 2003), were proposed which involve either on-demand or the usage of multiple relay points according to the link state information.

In the wireless ad hoc network, the behavior of protocols always vary depending on the core mechanisms of other protocols and factors such as node speeds, node movement patterns and background traffic. Almost all previous studies consider the importance of routing protocols over the performance of TCP (Ahuja at al., 2000; Dyer & Boppana, 2001; Gupta et al., 2004; El-Sayed, 2005; Kawadia & Kumar, 2005; Osipov & Tschudin, 2006; Mondal & Laqman, 2007; Anastasi et al., 2007; Sakib, 2009). Ahuja at al., (2000) considered four routing protocols: AODV, DSR, DSDV and SSA (Signal Stability-based Adaptive (Dube et al., 1997)) protocols and analyzed the performance of TCP. Dyer & Boppana (2001) also considered two on demand routing protocols, DSR and AODV, and proposed an adaptive proactive protocol (ADV) to enhance the TCP performance under a variety of conditions.

On the other hand, several papers (Ahuja at al., 2000; Chandran et al., 2001; Dyer & Boppana, 2001; Holland & Vaidya, 2002) discuss the effect of node mobility that may severely degrade the TCP performance due to the protocol's inability to manage efficiently mobility effects. As there are different versions of the TCP, many authors have compared the performance of different TCP versions by measuring throughput and fairness (Xu & Saadawi, 2000; Rakabawy et al. 2005, Kim et al., 2005). However, their analysis focus on the comparison of throughput and fairness, rarely considered packet loss rate depending on the increased number of connections. Some of them like, Kim et al. (2005), considered only TCP-NewReno and TCP-Vegas depending on AODV and OLSR routing protocols.

To the best of our knowledge, very few experimental analyses have been carried out so far (Lim et al., 2003; Oo & Othman, 2010) on the usage of multipath routing protocol. Their experiments are limited to using the ordinary TCP over multipath routing protocols. Therefore, this chapter discusses how the TCP variants interact to the use of routing protocols depending on the different topologies in the static and mobile ad hoc network environments.

The next section of this chapter is organized as follows. Section 2 briefly presents an overview of the ad hoc routing protocols and section 3 describes the variants of TCP that we have analyzed. Section 4 discusses the simulation methodology. Section 5 presents an analysis of the simulation results. Section 6 summarizes and concludes this chapter.

2. Overview of ad hoc routing protocols

2.1 Destination-Sequenced Distance Vector (DSDV)

DSDV (Perkins & Watson, 1994) is a proactive, hop-by-hop distance vector routing protocol. In DSDV, each node maintains a routing table of all possible destinations and the number of hops to each destination. Each node broadcasts its routing table information periodically throughout the network by using monotonically increased sequence numbers. The use of sequence number not only prevents the nodes from the occurence of stale routes but also avoids the formation of routing loops. If a node does not receive a periodic message from its neighbor for a while, it assumes that the link is broken. Moreover, its route update algorithm is very simple and guarantees loop free routes by transmitting a smaller update messages time to time. Therefore, the entire routing table need not be transmitted when the network topology changes occur.

2.2 Optimized Link State Routing Protocol (OLSR)

OLSR (Clausen & Jacquet, 2003) is a carefully designed protocol that works in a distributed manner and does not depend on any central entity. Each node chooses its neighbor nodes as multipoint relays (MPR) that are responsible for forwarding control traffic by flooding. MPRs provide the shortest path to a destination by declaring and exchanging the link information periodically for their MPR's selectors. By doing so, the nodes maintain the network topology information. The MPR is used to reduce the number of nodes that broadcasts the routing information throughout the network. To forward data traffic, a node selects its one hop symmetric neighbors, referred to as MPRset that covers all nodes that are two hops away.

The MPRset is calculated from information about the node's symmetric one hop and two hop neighbors. This information in turn is extracted from HELLO messages. Similar to the MPRset, a MPR Selectors set is maintained at each node. A MPR Selector set is the set of neighbors that have chosen the node as a MPR. Upon receiving a packet, a node checks its MPR Selector set to see if the sender has chosen the node as a MPR. If yes, the packet is forwarded, otherwise the packet is processed and discarded.

For route maintenance, Hello messages are broadcast periodically for link sensing, neighbor's detection and MPR selection process. The information contained in the HELLO message:

- how often the host sends Hello messages,
- willingness of a host to act as a Multipoint Relay, and
- information about its neighbor (i.e. interface address, link type and neighbor type)

The link type indicates that the link is symmetric, asymmetric or simply lost. The neighbor type is either symmetric, MPR or not a neighbor. If the link to the neighbor is symmetric, this node is chosen as MPR. After receiving a HELLO message information, a node builds its routing table. When a node receives a duplicate packet with the same sequence number, it discards the duplicate. A node updates its routing table either when a change in the neighbor is detected or a route to any destination has expired and a shorter route is detected for a destination.

2.3 Dynamic Source Routing (DSR)

In DSR (Johnson et al., 2007), each node is initialized by broadcasting a route request packet when it either needs a route to the destination or does not have a route in its route cache. On

receiving this request, each node broadcasts it by appending its address to the request packet until this packet reaches the destination. The destination node replies to the earliest request to the source node. This approach is known as source routing.
In DSR, each node not only quickly supports a route when a route break occurs but also tolerates the topological changes due to the monitoring of the operations of routes. Moreover, it is able to compute the correct routes in the presence of asymmetric link. It does not make use of the periodic routing, thereby saving bandwidth and reducing power consumption.

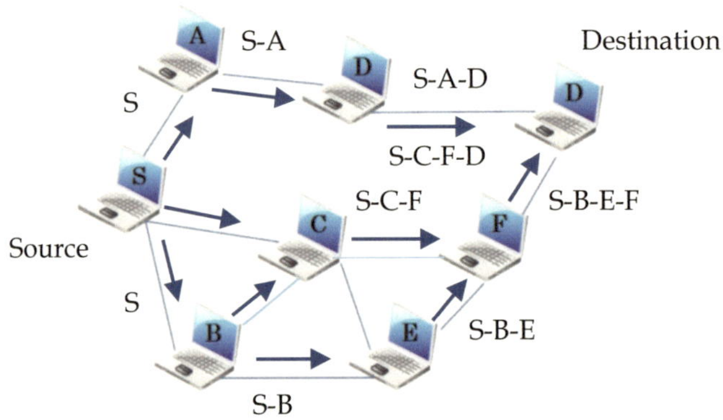

(a) Sending procedure of a request packet

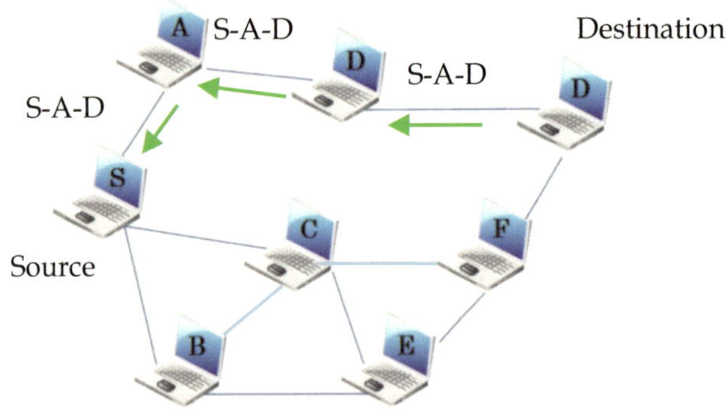

(b) Replying procedure of a reply packet

Fig. 1. Route discovery procedure of DSR

There are two main operations of DSR: route discovery and route maintenance. When a node wants to send a packet, and there is no route available to the destination, the node initiates a route discovery procedure. The source node broadcasts a route request to its neighbors by adding the destination address and route information that is recorded when the route request has passed. Upon receiving a route request, a node checks if it is the destination or if it knows a fresh route to the destination. If it is, the destination node has already found the complete route from the source and replied back to the source node. Otherwise, the node appends its address to the route information record and re-broadcasts the route request to its neighbors.To maintain the routes, each node constantly monitors the links it uses to forward the packets. If a node finds out that it cannot forward a packet, it sends a route error packet to its upstream nodes towards the source.

2.4 Ad-hoc On-demand Distance Vector (AODV)

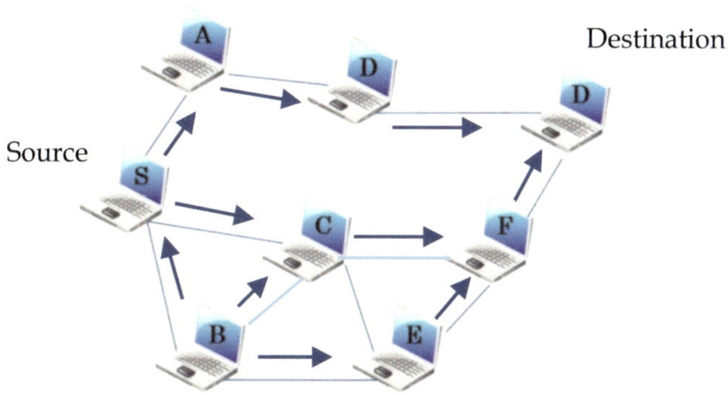

(a) Sending procedure of a request packet

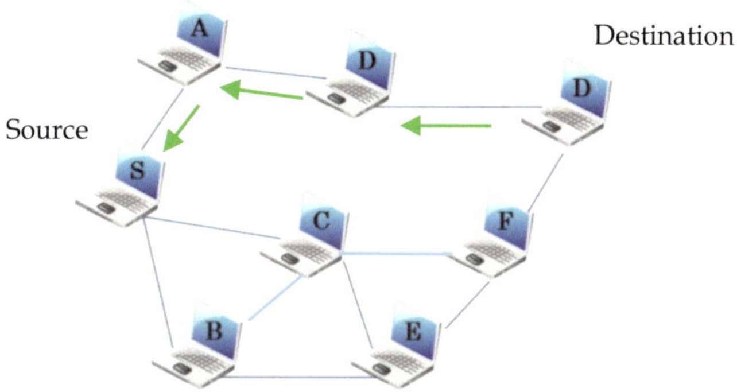

(b) Replying procedure of a reply packet

Fig. 2. Route discovery procedure of AODV

AODV (Perkins & Das, 2003) is based on DSDV and DSR routing protocols. In AODV, each node maintains a routing table, one entry per destination. Each entry records the next hop to the destination and its hop count (i.e. the distance from the current node to the destination node). AODV also uses a sequence number generated by a destination node to indicate the fresh-enough routes. Like DSR, AODV discovers a route through network-wide broadcasting. Unlike DSR, it does not record the nodes it has passed but only counts the number of hops. It builds the reversed routes to the source node by looking into the node that the route request has come. The responsibility of intermediate nodes is to check for fresh routes according to the hop count and destination sequence number and forwards the packets that they receive from their neighbors to the respective destinations.

AODV utilizes HELLO packets for route maintenance. If a node does not receive a HELLO packet within a certain time, or it receives a route break signal that is reported by the link layer, it sends a route error packet by either unicast or broadcast, depending on the precursor lists (i.e. active nodes towards the destination), in its routing table. It uses the periodic beaconing and sequence numbering procedures of DSDV and a similar route discovery procedure as in DSR. However, there are two major differences between DSR and AODV. The most distinguishing difference is that in DSR each packet carries full routing information, whereas in AODV the packets carry the destination address. This means that AODV is potentially less memory consuming than DSR. The other difference is that the route reply packets in DSR carry the address of every node along the route, whereas in AODV the route reply packets only carry only the destination IP address and sequence number. AODV avoids the stale route cache problem of DSR and it adapts the network topology changes quickly by resuming route discovery from the very beginning.

2.5 Ad-hoc On-demand Multipath Distance Vector (AOMDV)

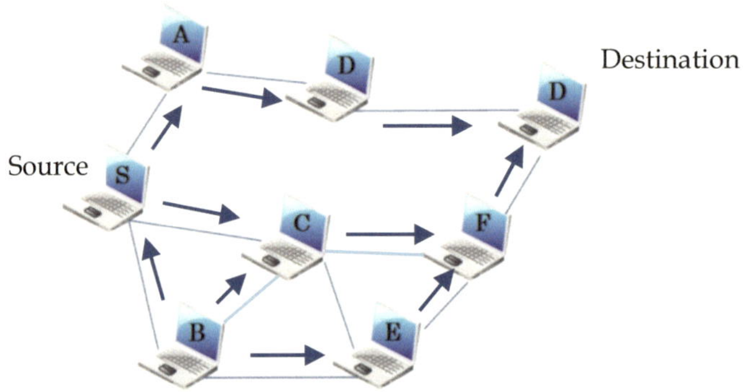

(a) Sending procedure of a request packet

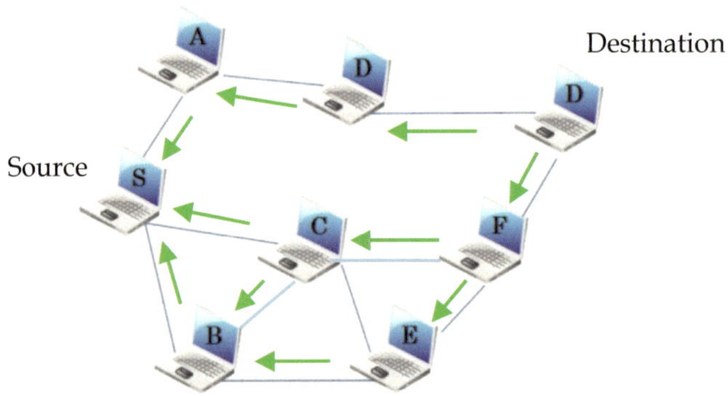

(b) Replying procedure of a reply packet

Fig. 3. Route discovery procedure of AOMDV

To overcome the invoking of a route discovery procedure whenever a route break occurs, Marina & Das (2006) proposed an AOMDV that allows each node to keep multiple paths to the destination. When a source node has data packets for a destination, it first checks its routing table to ascertain whether it already has a route to the destination node. If a route is available, it sends the data packets by utilizing its existing route. If not, it initiates a route discovery procedure by broadcasting RREQ to obtain a route to the intended destination.

AOMDV computes multiple paths and observes each route advertisement to define an alternate path to the source or the destination during a route discovery procedure. RREQ packets arriving at the nodes are copied and sent back to the source nodes. This approach may push the formation of loops due to accepting all copied routes. In order to eliminate any possibility of loops, it uses advertised hop count field in the route tables. The advertised hop count of a node S for a destination D is set the maximum hop count of the multiple paths for D at S.

The advertised hop count is initialized each time the sequence number is updated. By doing so, AOMDV only accepts alternative routes with lower hop counts. Each RREQ conveys an additional first hop field to indicate the first neighbor of the source node. The intermediate nodes do not discard duplicate copies of RREQ immediately as long as each RREQ provides a new node-disjoint path to the source. If an intermediate offers a new path, a reverse path is set up. It sends a RREP back to the source. At the destination, reverse routes are established like in the same situation of intermediate nodes. By computing multiple paths in a single route discovery attempt, a new route discovery is needed only when all paths fail.

3. Overview of transport protocols

3.1 TCP-Tahoe

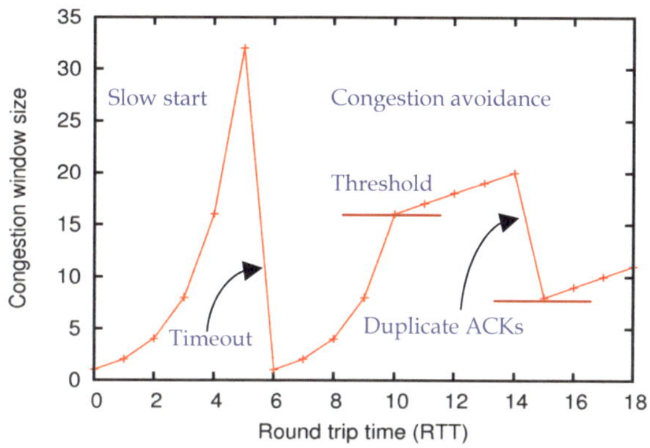

Fig. 1. Congestion control of TCP-Tahoe

The TCP protocol provides reliability, flow control, congestion avoidance, fairness, and in-order delivery. Originally, the protocol did not have congestion avoidance, causing the networks to become overloaded. TCP Tahoe introduced congestion avoidance, where dropped packets are used as an indication of congestion, and slow start, where the initial window size grows exponentially (i.e. a source node transmits one segment and wait for its ACK (acknowledgement). If the ACK is received, the congestion window is increased to transmit two segments. After receiving ACKs for those two segments, the congestion window is increased to four to transmit four segments) until either a congestion or timeout event is detected. In the congestion avoidance region, the initial window is increased linearly as shown in Fig. 1. In TCP-Tahoe (Stevens, 1997), there are two indications of packet losses: a timeout event and the receipt of duplicate ACKs. Whenever the timeout event occurs, Tahoe starts the slow start procedure by initiating congestion window size starting from one, whereas the congestion window (*cwnd*) is halved (i.e. *cwnd* = *cwnd*/2) when three duplicate ACKs are received.

3.2 TCP-Reno

Instead of starting transmission from a slow start after a relatively long idle period, Allman (1999) introduced TCP-Reno by adding fast retransmit and fast recovery algorithms. With fast retransmit, Reno attempts to retransmit packets before a timeout. However, a sender will initiate a slow-start procedure as if a timeout causes the retransmission. With fast recovery, Reno uses additive increase/multiplicative decrease at all the time, and only initiates the slow start when either a connection is established or a timeout occurs. In other words, Reno with fast recovery omits the slow start if no timeout occurs.

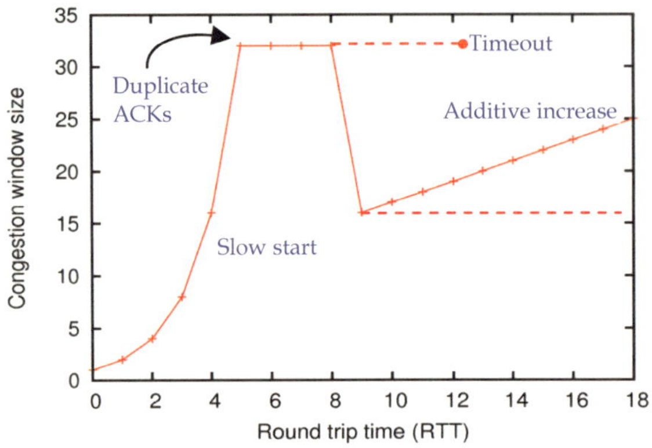

Fig. 2. Congestion control of TCP-Reno

3.3 TCP-NewReno

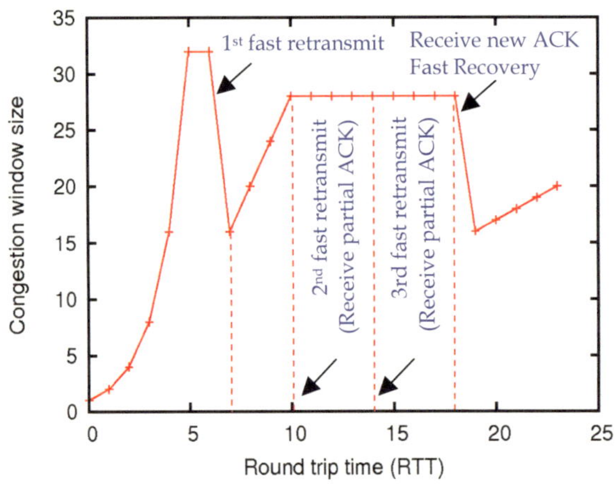

Fig. 3. Congestion control of TCP-NewReno

TCP-NewReno (Floyd & Henderson, 1999) is an improvement of Reno and it is an advanced fast transmit, where three duplicate acknowledgments signal a retransmission without a timeout with fast recovery. The fast recovery means that once a certain threshold of ACKs is received, the window size is decreased by half rather than starting over with slow start. Only during timeout does it go back into slowstart. NewReno increases the adoption of the TCP selective acknowledgements (SACK) (Mathis & Mahdavi, 1996) modification. TCP-NewReno possesses two kinds of ACKs: partial ACK and full ACK. The partial ACK acknowledges some segments at the fast recovery stage while the full ACK acknowledges all outstanding data. NewReno retransmits the segment based on the partial ACK. Upon receiving the full ACK, the sender sets the congestion window to slow start threshold and terminates the fast recovery. Then the congestion avoidance mechanism is resumed. In this way, the NewReno maintains a high throughput.

3.4 TCP-Vegas

Tahoe, Reno and NewReno variants are window-based transport protocols that adjust congestion window upon packet losses. On the other hand, Brakno et al., (1994) introduces a delay-based TCP, called TCP-Vegas, which does not violate the congestion avoidance paradigm of TCP. Instead of increasing the sending rate until a packet loss occurs, TCP Vegas prevents such losses by decreasing the sending rate when it senses incipient congestion even if there is no indication of packet loss. Vegas uses packet delay as an indication of congestion.

In a situation when a duplicate ACK is received, the timestamp for the ACK is compared to a timeout value. If the timestamp is greater than the timeout value, then Vegas will retransmit rather than wait for three duplicate ACKs. Vegas detects congestion at an incipient stage based on increasing Round Trip Time (RTT) values of the packets in the connection unlike other flavors, like NewReno, which detect a congestion only after it has actually happened via packet drops. TCP Vegas adopts a more sophisticated bandwidth estimation scheme. It uses the difference between expected and actual flow rates to estimate the available bandwidth in the network. When the network is not congested, the actual flow rate will be close to the expected flow rate. Otherwise, the actual flow rate will be smaller than the expected flow rate. So, TCP-Vegas can estimate the congestion level in the network and updates the window size accordingly. The difference between the flow rates can be easily calculated during the round trip time using the equation

$$Diff = (Expected \mid Actual)\, BaseRTT$$

where *Expected* is the expected rate, *Actual* is the actual rate, and *BaseRTT* is the minimum round trip time. Based on *Diff*, the source updates its window size as follows.

$$CWND = \begin{cases} CWND + 1 & if\ Diff < \alpha \\ CWND - 1 & if\ Diff > \beta \\ CWND & otherwise \end{cases}$$

3.5 TCP-Westwood

TCP-Westwood (Gerla et al., 2002) is a sender-side modification of the TCP congestion window algorithm. The key idea behind it is to estimate bandwidth to control the congestion window and the slow start threshold by monitoring the ACK packets.

A sender measures the rate of ACKs that it receives and estimates the data rate currently achieved by that connection. Whenever the packet losses occur (i.e. timeout or duplicate ACKs), the sender estimates the bandwidth to properly set the congestion window and slow start threshold. Instead of halving congestion window like Reno and NewReno, TCP Westwood backs off some value of cwnd and threshold based on the estimated value to ensure faster recovery. The improvement of Westwood is more significant in wireless networks with lossy links, since TCP Westwood relies on end-to-end bandwidth estimation to discriminate the cause of packet loss. Rather, it fully complies with the end-to-end TCP design principle.

4. Simulation methodology

We use simulations to study the variants of TCP over three ad hoc routing protocols. The simulation study is done using Network Simulator (NS-2) (McCanne & Floyd). NS-2 is a

discrete event simulator that was developed as part of the VINT project at the Lawrence Berkeley National University.
For the performance of TCP variants over routing protocols in a static environment we simulate a scenario of chain (6 nodes) and grid (25 nodes) in a rectangular topology of 1300m × 1000m, where each node has a transmission range of 200m. All nodes have a default bandwidth of 11Mbps and the simulation period is 360 seconds. We use an FTP (File Transfer Protocol) application with a packet size of 512 bytes. Each TCP variant is run over each routing protocol in static and mobile environments (Figs. 4 to 7).

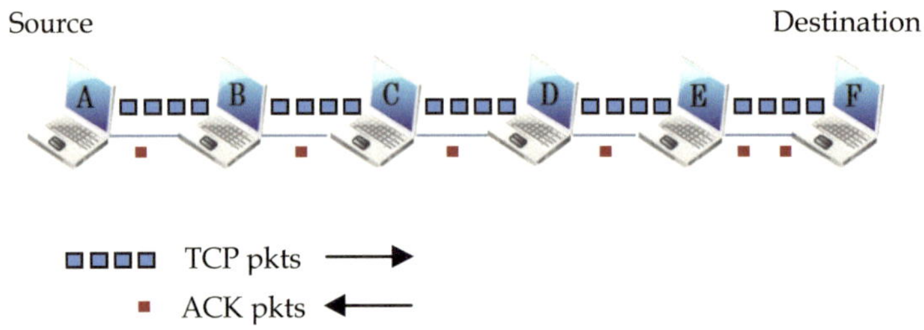

Fig. 4. Source node A connects to destination node F in a static ad hoc network

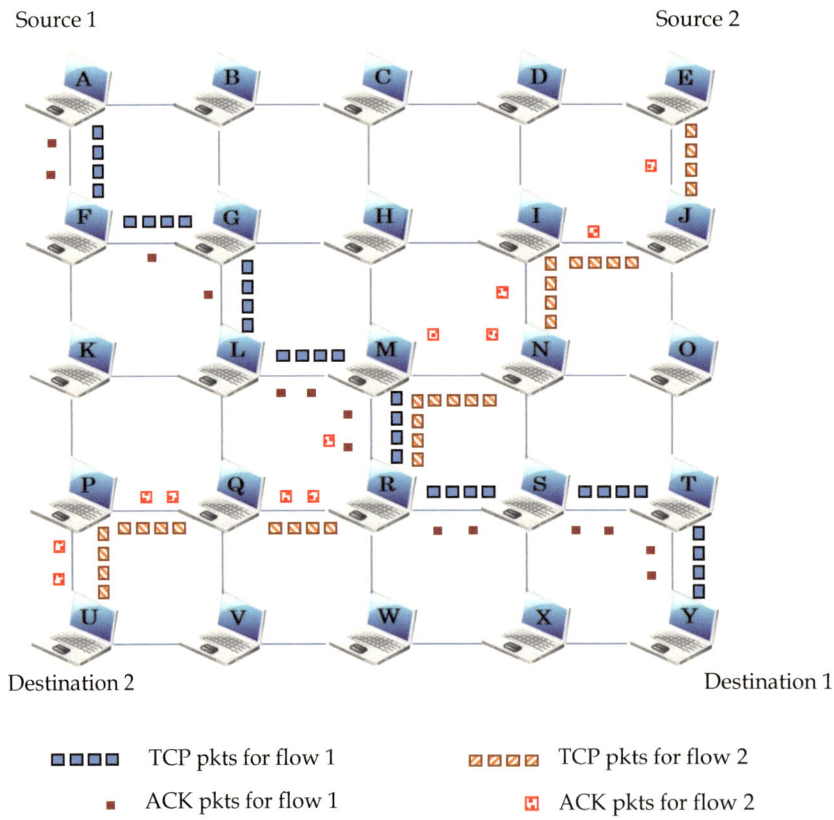

Fig. 5. Two TCP connections between a pair of nodes (A and Y) and node (E and U) in a static ad hoc network

In a mobile ad hoc environment, we manually change the network topology by adding movement for a few nodes. The "*setdest*" command under NS-2 directory (i.e. *ns-allinone-2.34\ns-2.34\indep-utils\cmu-scen-gen\setdest*) is used to generate the node movement. In Fig. 6, node C and node D start moving at 50 and 100 seconds of simulation time respectively. Both nodes turn back to its original position at 250 seconds, and they move at 10 m/s speed. Similarly, in Fig. 7, node I and node J start moving at 20 seconds and return to

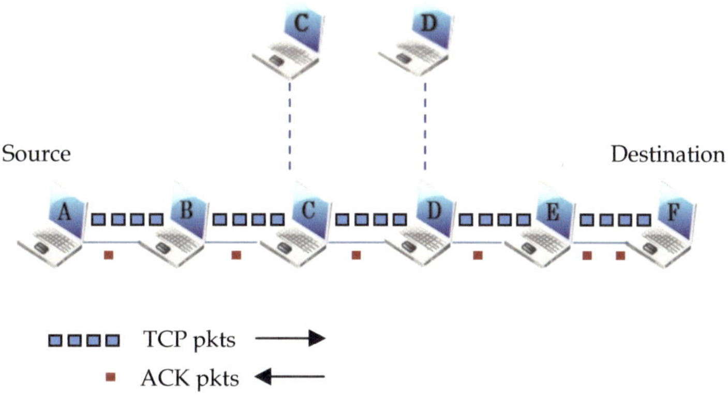

Fig. 6. Source node A connects to destination node F in mobile ad hoc network

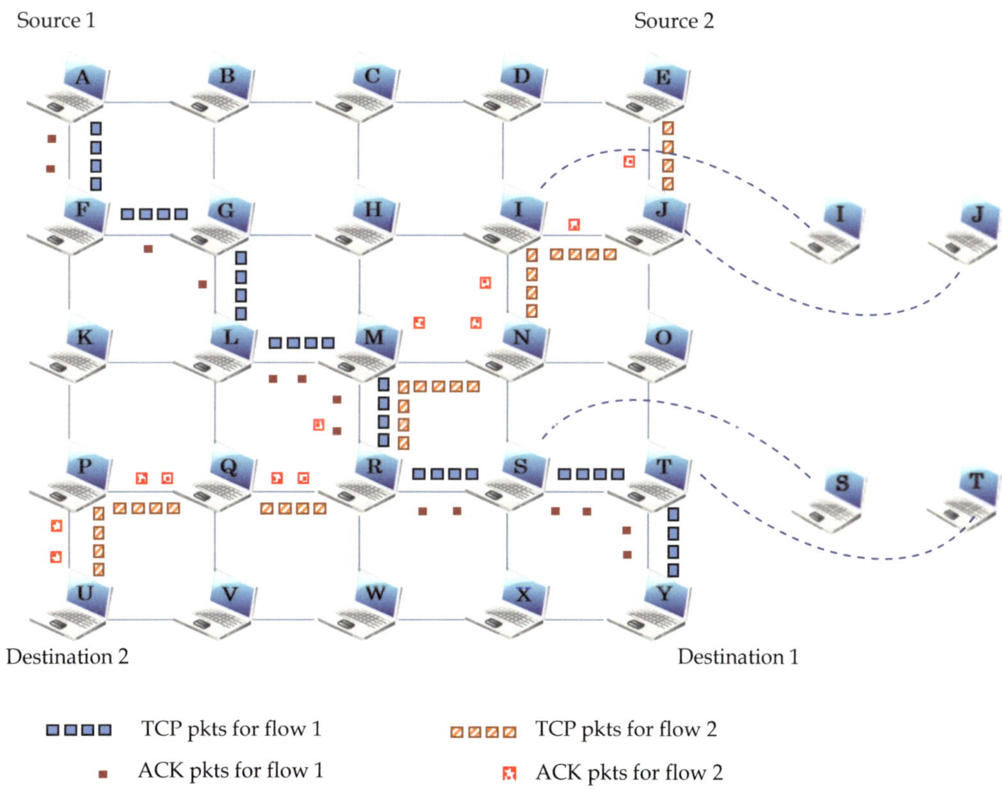

Fig. 7. Two TCP connections between a pair of nodes (A and Y) and node (E and U) in mobile ad hoc environment

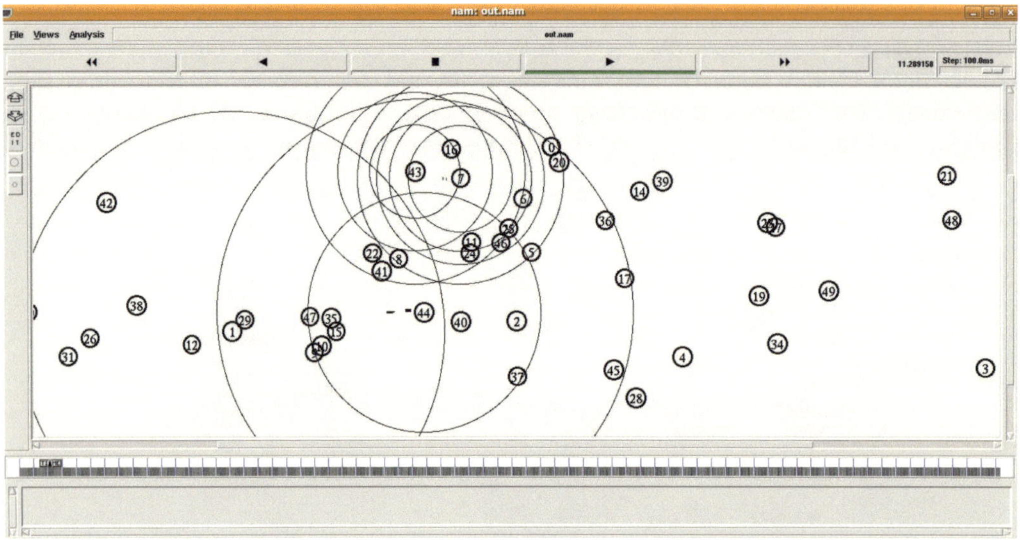

Fig. 8. The 50 pairs of random node movement in mobile environment

their original positions at 260 seconds. Also, node S and node T move to the left at 100 seconds and reach to their original positions at 300 seconds. In a chain topology, only one TCP connection is exchanged between a pair of source and destination while two TCP connections are transmitting in the grid topology for the static and mobile ad hoc environments.

To examine the TCP performance in a random topology, a moderate network of 50 nodes are randomly moved using Random Waypoint (RWP) mobility model (Camp et al., 2002) which is generated using the Bonnmotion v1.4 tool developed by the Communication Systems group at the Institute of Computer Science IV of the University of Bonn, Germany (*BonnMotion: a Mobility Scenario Generation and Analysis Tool 2009*). For example, the RWP mobility model can be generated by using the following command.

bm –f scenario1 RandomWaypoint –n 50 –d 900 –i 3600 –x 1600 –y 400 –h 10 –l 10 –p 0

n: the number of nodes that we wish to set
d: the simulation time
i: the cutting value that must be high default value because nodes have a higher probability of being near the center of the simulation area, while they are initially uniformly distributed over the simulation area in Random Waypoint model.
x: the coordinate of node position in x axis
y: the coordinate of node position in y axis
h: the maximum node speed
l: the minimum node speed
p: the maximum pause time

In the analyzed scenarios, the maximum pause time is set to zero for continuous movement, and nodes are allowed to move at 10 m/s speed. Simulations are run for 360 seconds simulation time. The number of tcp connections is varied between 10 to 50, and our performance analysis is examined by measuring the packet loss rate, delay and throughput.

5. Simulation results

In this section, we describe the results obtained from the simulation experiments in different scenarios. We simulate each variant of TCP (i.e. Tahoe, Reno, NewReno, Vegas and Westwood) over each routing protocol (i.e. DSDV, DSR, AODV, AOMDV and OLSR) in static and mobile ad hoc network environments. Then we measure how the topology changes affect the performance of TCP variants across each routing protocol in a 6-node chain and 25-node (5 x 5) grid topologies.

To examine the performance in the scenarios of random movement, the 50 pairs of nodes are simulated in 1600 x 400 simulation area for 360 seconds.

To analyze the network performance in those topologies, the packet loss rate (%), average end-to-end delay (msec) and throughput (kbps) are measured as performance metrics.

The packet loss rate (%) is the number of packet losses at the application layer while transferring data packets, i.e.

$$PLR = \frac{Dropped\ Packets}{Highest\ Packet\ ID + 1} * 100$$

The average end-to-end (EtE) delay (msec) is the transmission delay of data packets that are delivered to the intended destination successfully.

The throughput (kbps) is the rate of successfully delivered data per second to individual destinations during the network simulation.

5.1 Chain topology

Firstly, we analyze the packet loss rates of the TCP variants over the ad hoc routing protocols in static and mobile environments. The TCP variants over AODV incurs a lower packet loss rate than DSDV and DSR in static environment as shown in Table 1. Because DSDV sends periodic messages throughout the network, and DSR stores all route information in the control packets, the packet loss rates for both protocols increase due to the collision and congestion in the MAC layer. However, when the node movements are

Packet loss rate (%) / TCP variants	DSDV		DSR		AODV	
	Static	Mobile	Static	Mobile	Static	Mobile
Tahoe	0.09	0.75	0.09	0.14	0.05	0.45
Reno	0.09	0.75	0.09	0.14	0.05	0.45
NewReno	0.09	0.75	0.09	0.14	0.05	0.45
Vegas	0.00	0.46	0.00	0.11	0.00	0.33
Westwood	0.09	0.65	0.09	0.14	0.05	0.45

Table 1. The percentage of packet loss rate in chain topology

added, DSR achieves the lowest loss rate due to its route cache mechanism where all possible routes to the destination are kept. On the other hand, among the TCP variants, Vegas is the best protocol for all situations, and it achieves no losses over the routing protocols in the static environment.

Secondly, when we examine the average end-to-end delay, AODV incurs the lowest delay, whereas DSR incurs the highest delay over all TCP variants as shown in Table 2. AODV always keeps routes as a soft state, for example, routes expire after a timeout interval and a fresh route discovery is invoked. Accordingly, AODV is significantly better delay-wise and can possibly perform even better than others when node movements are added. Likewise, AODV has a special timer mechanism to detect route breaks and update fresh-enough routes whereas DSR does not contain any explicit mechanism to expire stale routes in the cache. The stale routes are later detected by route error packets, leading to performance degradation although it achieves a lower packet loss rate as shown in Table 1.

Delay (msec) / TCP variants	DSDV		DSR		AODV	
	Static	Mobile	Static	Mobile	Static	Mobile
Tahoe	432.4	604.7	646.6	639.7	418.5	415.0
Reno	432.4	604.7	646.6	639.7	418.5	415.0
NewReno	432.4	604.7	646.6	639.7	418.5	415.0
Vegas	72.9	333.4	71.3	70.5	67.79	67.4
Westwood	432.4	624.5	646.6	639.7	418.5	415.0

Table 2. The average end-to-end delay in chain topology

The delay-based Vegas achieves the lowest delay for the static and mobile ad hoc environments. The performance of Tahoe, Reno, NewReno and Westwood are not different enough to compare against each other in both environments. The delay difference of Vegas is lower than others by a factor of around 6 over DSDV and AODV, and by a factor of more than 9 over DSR for the static environment. However, the variants of TCP incur a lower delay over DSR and AODV whereas DSDV incurs a higher delay when the node movements are added. Especially, the delay of Vegas over DSDV suddenly increases once the nodes move as shown in Fig.1. No matter what variants of TCP are utilized, all of them achieve a lower delay over AODV routing protocol in both environments.

Throughput (kbps) / TCP variants	DSDV		DSR		AODV	
	Static	Mobile	Static	Mobile	Static	Mobile
Tahoe	97.5	27.8	97.4	39.9	101.4	30.6
Reno	97.5	27.8	97.4	39.9	101.4	30.6
NewReno	97.5	27.8	97.4	39.9	101.4	30.6
Vegas	74.4	17.5	100.4	33.9	104.6	39.8
Westwood	97.5	25.9	97.4	39.9	101.4	30.6

Table 3. The performance throughput in chain topology

Finally, Table 3 compares the throughput of TCP variants over each routing protocol. AODV supports higher throughput for all TCP variants, especially Vegas in both environments. However, the performance of Vegas is lower than others over DSDV in both environments. In DSR, Vegas achieves higher throughput than others in static environment, whereas its performance is lower than others in mobile environment. When the node movement is introduced, TCP variants over DSR achieve a higher throughput than DSDV and AODV because nodes with DSR always have backup routes in hand and keep them in their caches. As soon as a route break occurs due to congestion or collision, it can recover the route quickly before the TCP timeout. In this way, DSR attains higher throughput at moderate node movement in mobile environment as long as its route cache is not stale.

5.2 Grid topology

For the grid topology, one of the multipath routing protocols, AOMDV is considered to examine the performance of TCP variants. For the static environment, we encounter that TCP variants except Westwood have no packet loss over AODV as shown in Table 4. Another thing is that AOMDV also achieves a lower packet loss rate if compared to DSDV and DSR. However, finding multiple paths in a static environment is not effective if compared to the single path AODV, and even when a few node movement is added, AODV has a lower losses than AOMDV. AOMDV possibly performs better than AODV over the lossy links that occur due to the random node movement and increased traffic because it has fresh multiple alternative routes. Like in chain topology, Vegas upholds a lower packet loss than others, and there are no losses except over DSDV in the static environment.

Packet loss rate (%) / TCP variants	DSDV		DSR		AODV		AOMDV	
	Static	Mobile	Static	Mobile	Static	Mobile	Static	Mobile
Tahoe	0.29	1.73	0.15	0.88	0.00	0.33	0.06	0.57
Reno	0.29	1.73	0.15	0.86	0.00	0.33	0.06	0.57
NewReno	0.45	1.89	0.15	0.44	0.00	0.33	0.06	0.82
Vegas	0.17	0.20	0.00	0.14	0.00	0.07	0.00	0.18
Westwood	0.91	1.81	0.13	0.12	0.04	0.51	0.06	0.82

Table 4. The percentage of packet loss rate in grid topology

In Table 5, if we look at the end-to-end delay for all TCP variants in a static environment, DSDV has the lowest delay for both environments in the grid topology because the nodes in the grid topology are organized, therefore packet losses due to route break, congestion or collision of MAC layer could be recovered easily. The table-driven and periodic approach of DSDV, thus, suffers more losses possibly due to congestion, whereas it achieves the lowest delay compared to others. The delay of Westwood is the worst over DSR routing protocol in the static environment.

The delay-based protocol, Vegas always incurs a lower delay for all situations due to the consideration of actual and expected flow rates. On the other hand, Vegas obtains a lower delay over DSDV and DSR, and delay becomes higher over AODV and AOMDV protocols when the node movements are added. Although AODV achieves a lower delay in most situations, in the grid topology, it suffers a higher delay than others because the number of route discovery frequencies of AODV increases due to its flooding nature whenever a route break occurs due to network congestion.

Delay (msec) / TCP variants	DSDV		DSR		AODV		AOMDV	
	Static	Mobile	Static	Mobile	Static	Mobile	Static	Mobile
Tahoe	596.8	131.2	970.3	198.8	662.1	647.1	715.7	616.1
Reno	588.9	141.0	970.3	193.3	662.1	647.1	715.7	616.1
NewReno	614.5	134.7	970.3	183.3	662.1	647.1	715.7	662.1
Vegas	137.2	48.1	112.8	32.0	108.4	127.9	115.5	125.9
Westwood	574.4	150.6	1160.9	210.4	661.3	639.6	715.7	662.1

Table 5. The average end-to-end delay in grid topology

In Table 6, when throughput is compared, TCP variants over DSR perform better than others in both environments. In DSDV and AODV, Westwood is the best throughput in static environment, whereas it suffers the lowest throughput in mobile environment. In the grid topology, the possibility of congestion increases due to the channel contention. Whenever a packet loss occurs, Westwood attempts to select a slow start threshold and a congestion window depending on the effective bandwidth used at the time congestion is experienced, whereas Reno and NewReno blindly halves the congestion window after trying the fast retransmit and fast recovery procedures. Therefore, in grid topology, the performance of Westwood is significant if compared to others. On the other hand, the significance of performance throughput for all TCP variants can be seen over DSR routing in the mobile environment. DSR's route cache mechanism may not be effective enough to provide the routes that have been cached in high mobility and traffic scenarios, whereas in moderate situation, such as fewer node movement, DSR provides the highest throughput to all TCP variants.

Throughput (kbps) / TCP variants	DSDV		DSR		AODV		AOMDV	
	Static	Mobile	Static	Mobile	Static	Mobile	Static	Mobile
Tahoe	37.4	78.1	90.5	237.1	45.3	44.4	42.4	80.3
Reno	36.2	78.1	90.5	147.9	45.3	44.4	42.4	80.3
NewReno	37.4	76.4	90.5	161.1	45.3	44.4	42.4	39.2
Vegas	27.7	56.8	44.5	158.7	46.5	44.1	43.4	42.6
Westwood	56.2	36.2	60.7	123.4	54.4	52.9	42.4	39.2

Table 6. The performance throughput in grid topology

5.3 Random topology

As mentioned in section 4, we examine the performance of TCP variants and routing protocols in the random topology. All nodes move randomly across the RWP model. A node starts moving from a randomly chosen position and stays in one location for a certain period of time (i.e. a pause time). Once this time expires, the node chooses a destination and moves at a randomly chosen speed. This speed is selected from a uniformly distributed speed between minimum and maximum speed. Upon arrival at the destination, the above process is started over again.

5.3.1 Packet loss rate measurement

We vary the number of TCP connections from 10 to 50 connections and the network performance are measured in the 50 nodes random topology. TCP traffic is generated using a traffic generation tool under the NS-2 directory (i.e. *ns-allinone-2.34\ns-2.34\indep-utils\cmu-scen-gen*), for example, (*ns cbrgen.tcl -type* <*cbr | tcp*> *-nn* <*nodes*> *-seed* <*seed*> *-mc* <*connections*> *-rate* <*rate*>*)*. For the 20 number of connections and the 50 pair of nodes, the following command is used.

ns cbrgen.tcl –type cbr –nn 50 –seed 1 –mc 20 –rate 4

The percentage of packet loss rates for all TCP variants varies between 0.5 and 4 over all ad hoc routing protocols as shown in Fig. 9. The stability of TCP variants is encountered over DSR, and all losses vary between 0.5 and 1.2 (Fig. 9(b)). As the number of TCP connections increases, the packet loss rates of TCP variants decrease. In Fig. 9(c), the packet losses over AODV are the worst if we also look at the view of stability issue. The link failure detection mechanism of AODV based on HELLO messages generates frequent route failures with associated packet loss oscillation.

In the two multipath routing protocols AOMDV and OLSR, AOMDV encounters a greater packet loss rate than OLSR by a factor of up to 2 as shown in Fig. 9(d) and (e). Although AOMDV supports multiple paths between a source and destination, it is difficult to recover the packets during the time between the failure of a primary route and the finding of an alternative route. On the other hand, as OLSR nodes always have routes in hand due to its proactive nature, it reduces packet loss rates significantly (Oo & Othman, 2010).

Vegas is the best transport variant of TCP, and it is able to provide a lower packet loss rate in most situations. It is able to detect congestion in advance by estimating bandwidth before actual congestion happens. Other TCP variants like Tahoe, Reno, NewReno are not as good as Vegas when TCP connection flows grow, especially in DSDV and AODV.

5.3.2 Average end-to-end delay measurement

Although DSR has the lowest packet loss rate as mentioned in section 5.3.1, it is not good enough to apply in delay-sensitive applications. It suffers the highest delay especially for Tahoe, Reno and NewReno as shown in Fig. 10 (a). As the number of TCP connections increases, the average delay of TCP variants also increases. Vegas can transfer packets almost four times faster than over DSDV and OLSR, two times over AODV and DSR, five times over AOMDV than others as shown in Fig. 10.

The performances of window-based protocols such as Tahoe, Reno and NewReno are not very significant over each other, whereas delay-based Vegas protocol gains a significantly lower delay. On the other hand, when the route breaks occur, Tahoe, Reno and NewReno halves it congestion window and starts the slow start procedure after the TCP timeout expiry period, tending to the increased delay if compared to the Vegas. In Fig. 10 (a), TCP variants over DSDV are the best if compared to other routing protocols. DSDV starts discovering routes proactively, and it may increase the routing overhead, whereas it significantly reduces average end-to-end delay at the moderate network.

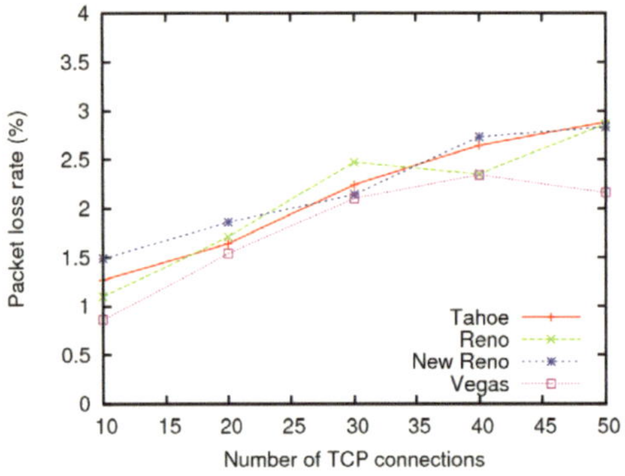

(a) DSDV

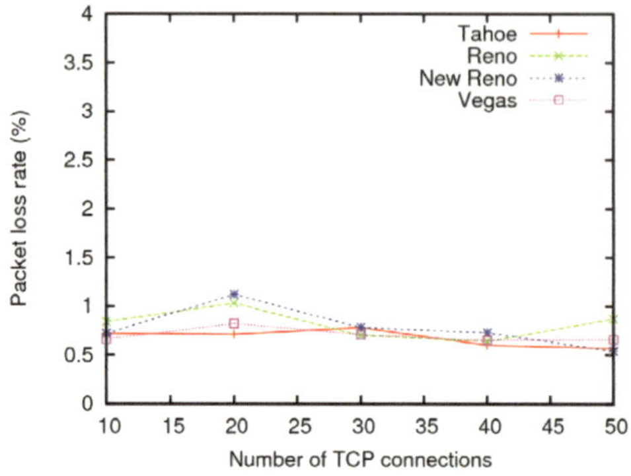

(b) DSR

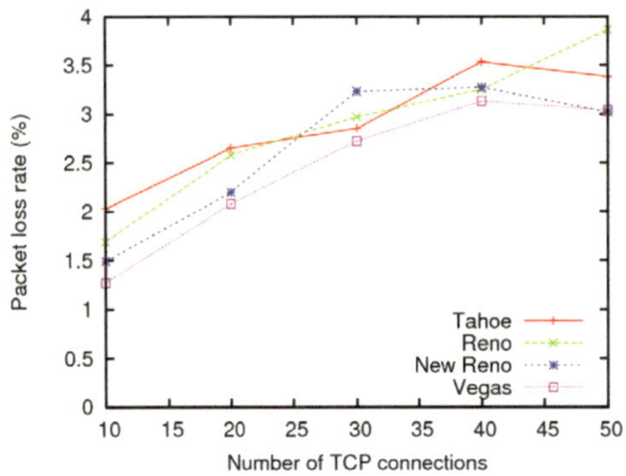

(c) AODV

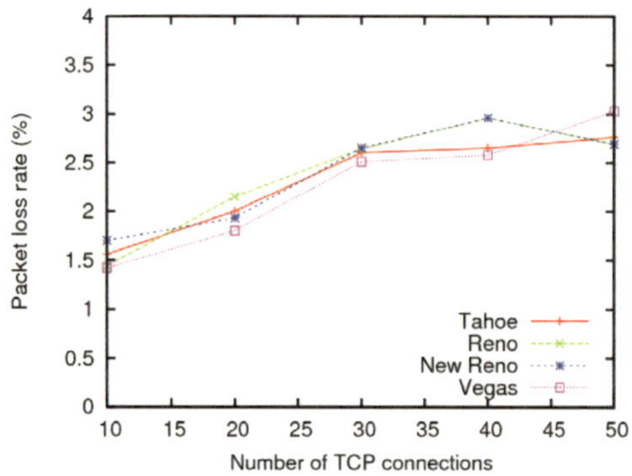

(d) AOMDV

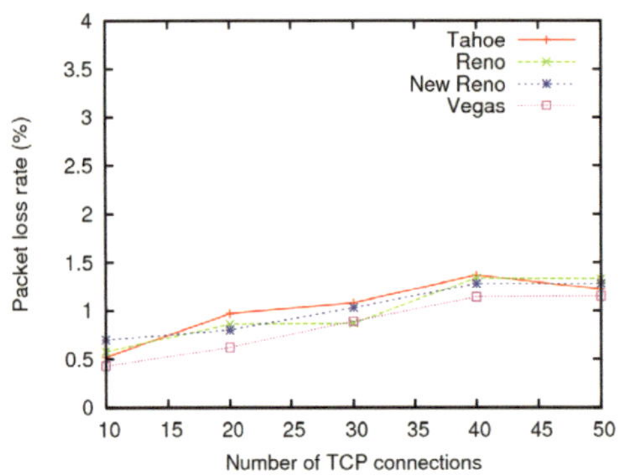

(e) OLSR

Fig. 9. Packet loss rates measurement in the random topology

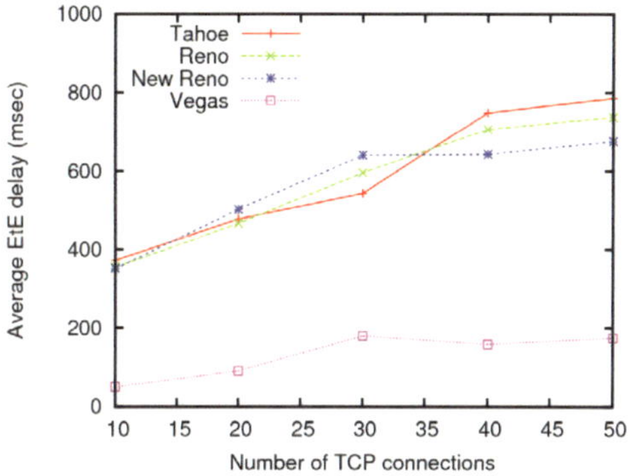

(a) DSDV

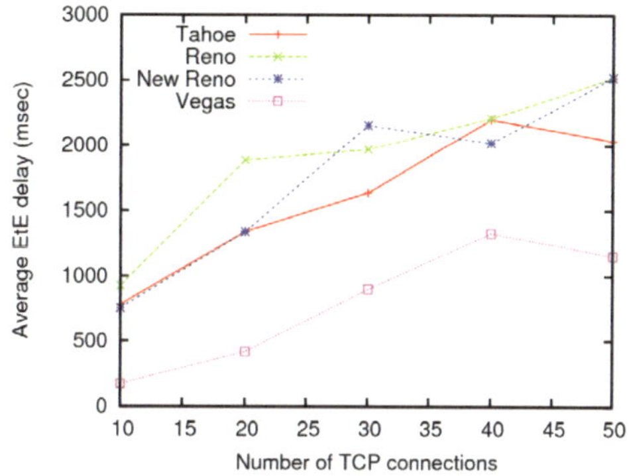

(b) DSR

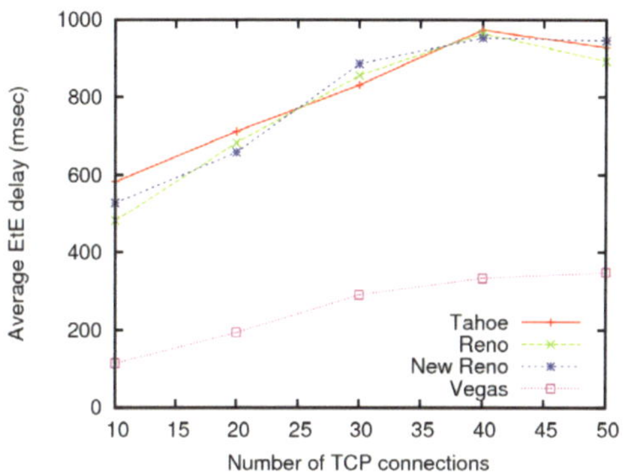

(c) AODV

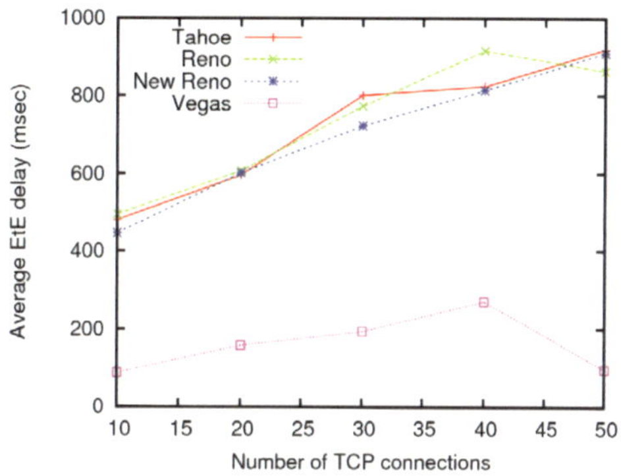

(d) AOMDV

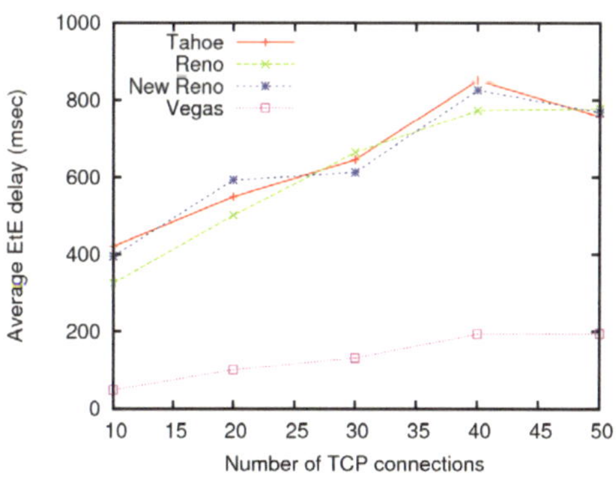

(e) OLSR

Fig. 10. Average end-to-end delay measurement in the random topology

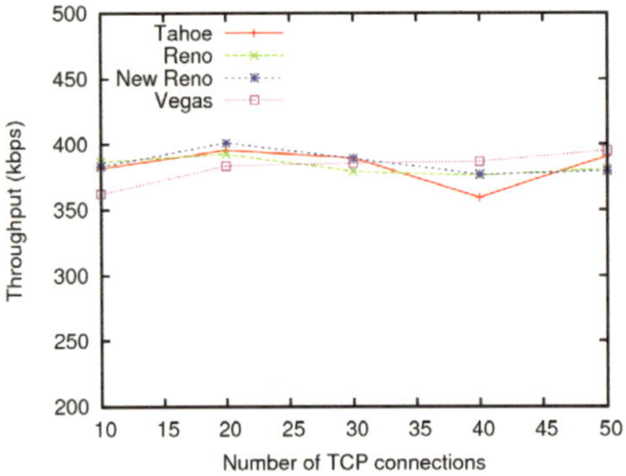

(a) DSDV

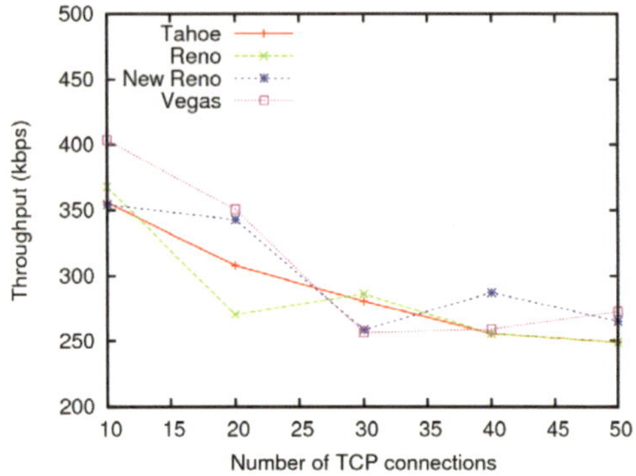

(b) DSR

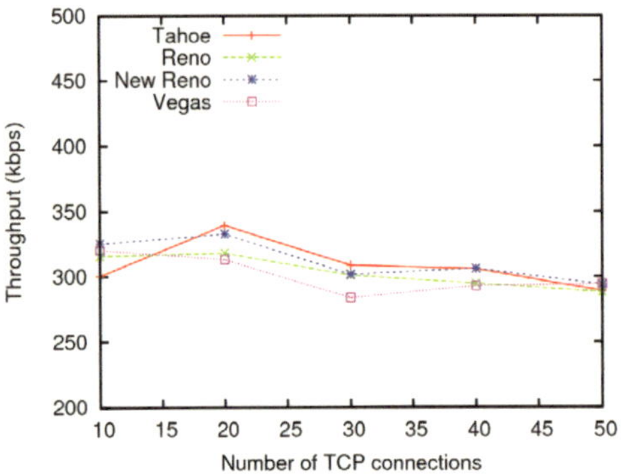

(c) AODV

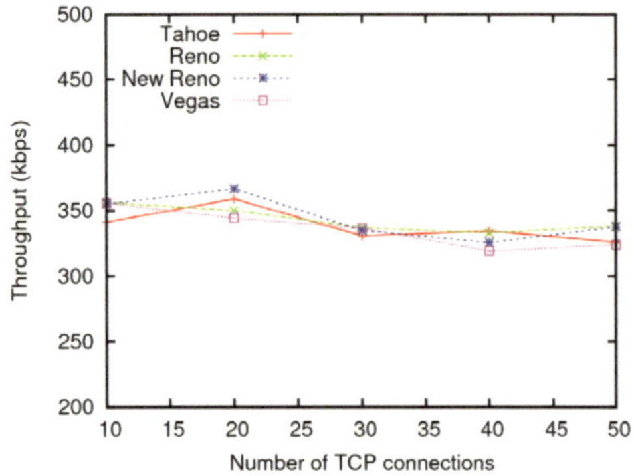

(d) AOMDV

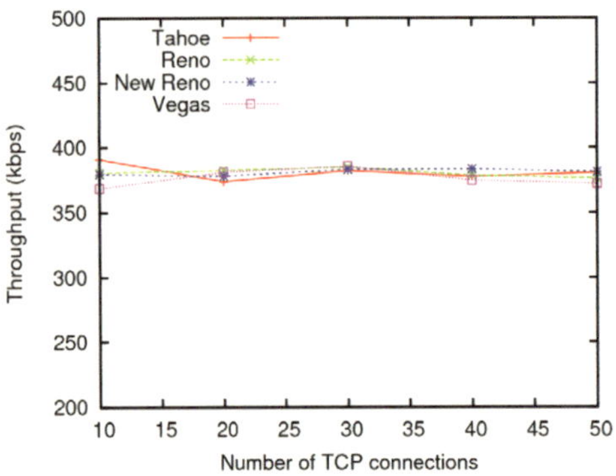

(e) OLSR

Fig. 11. Throughput measurement in the random topology

5.3.3 Throughput measurement

The TCP variants over DSDV achieve a higher throughput by a factor of almost 1.5 on average compared to others as shown in Fig. 11(a). The better stability of throughput for the TCP variants could be encountered in proactive routing protocols DSDV and OLSR (Fig. 11(e)). When the number of nodes increases, the possibility of congestion and the contention at the MAC layer increase in the network. However, when the routing layer protocols receive the collision reports from the link layer, they re-discover routes by sending the broadcast messages throughout the network. Therefore, in Fig. 11(c), AODV suffers a lower throughput if compared to others. Another thing is that DSR suffers the instability throughput for all TCP variants because when the node density and the number of connections increase, the stale route problem of DSR comes active and makes the performance worse (Fig. 11(b)).

6. Conclusion

In this chapter, we analyze the performance of TCP variants across ad hoc routing protocols in static and mobile ad hoc environments. The performance of TCP variants vary depending on the routing protocols, their core mechanisms and background changes, such as the node mobility, node speed, pause time and number of tcp connections and network topologies. In the chain topology, all of the TCP variants achieve a significantly lower delay over AODV routing protocol in both environments. Moreover, AODV provides a higher throughput for all TCP variants, especially for Vegas in both environments. One interesting thing is that AODV always achieves a lower delay, it suffers a higher delay than others in the grid topology. In the grid topology, although TCP variants have the lowest delay over DSDV in both environments, in the random topology, TCP variants incur a lower packet losses over DSR and OLSR, and encounter a lower delay over DSDV. On the other hand, DSDV and OLSR provide the highest data transfer rate (i.e. throughput) for all TCP variants in random topology. Among all TCP variants, Vegas is the best transport protocol and performs better than others in most situations.

7. Acknowledgement

This work is supported in part by University of Malaya Research Grand (UMRG) under grant RG024/09ICT.

8. References

BonnMotion: a Mobility Scenario Generation and Analysis Tool (2009). Available from: http://net.cs.unibonn.de/fileadmin/ag/martini/projekte/BonnMotion/src/BonnMotion_Docu.pdf.

Ahuja, A.; Agarwal, S.; Singh, J. P. & Shorey, R. (2000). Performance of TCP over Different Routing Protocols in Mobile Ad Hoc Networks, *IEEE 51st Vehicular Technology Conference*, pp. 2315-2319, 0-7803-5721-3, Tokyo, May 2000, Japan.

Allman, M. (1999). TCP Congestion Control, *Request for comment 2581*.

Anastasi, G.; Ancillotti, E.; Conti, M. & Passarella, A. (2007). Experimental Analysis of TCP Performance in Static Multi-hop Ad Hoc Networks, In: *Multi-hop Ad Hoc Networks from Theory to Reality*, Conti, M.; Crowcroft, J. & Passarella, A. (Ed.), page number (97-114), Nova Science, 1-60021-605-6, New York.

Boppana, R. & Konduru, S. (2001). An Adaptive Distance Vector Routing Algorithm for Mobile Ad Hoc Networks, *IEEE Infocom*, pp. 1753-1762, 0-7803-7016-3, Anchorage, April 2001, Alaska.

Brakno, L. S.; O'Malley, S. W. & Peterson, L. L. (1994). TCP Vegas: new techniques for congestion detection and avoidance, *ACM SIGCOMM Computer Communication Review*, Vol. 24, No. 4, (October 1994) page number (24-35), 0146-4833.

Camp, T., Boleng, J. & Davies, V. (2002). A survey of mobility models for ad hoc network research, *Wireless Communications and Mobile Computing Special Issue on Mobile Ad Hoc Networking: Research, Trends and Applications,* Vol. 2, No. 5, (August 2002) page number (483-502), 1530-8669.

Chandran, K.; Raghunathan, S.; Venkatesan, S. & Prakash, R. (2001). A Feedback Based Scheme for Improving TCP Performance in Ad Hoc Wireless Networks, *IEEE Personal Communication Magazine, Special Issue on Ad Hoc Networks*, Vol. 8, No. 6, (August 2001) page number (34-39), 1070-9916.

Clausen, T. & Jacquet, P. (2003). Optimized Link State Routing Protocol (OLSR), *Request for Comments 3626.*

Dube, R.; Rais, C. D.; Wang, K-Y. & Tripathi, S. K. (1997). Signal Stability-based Adaptive (SSA) Routing for Ad Hoc Mobile Networks. *IEEE Personal Communications Magazine*, Vol. 4, No. 1, (February 1997) page number (36-45), 1070-9916.

Dyer, T. D. & Boppana, R. V. (2001). A Comparison of TCP Performance over Three Routing Protocols for Mobile Ad Hoc Networks, *ACM Symposium on Mobile Ad Hoc Networking & Computing*, pp. 56-66, 1-58113-428-2, Long Beach, October 2001, ACM, California.

El-Sayed, H. M. (2005). Performance evaluation of TCP in mobile ad hoc networks, *The Second International Conference on Innovations in Information Technology,* September 2005.

Floyd, S. & Henderson, T. (1999). The NewReno Modification to TCP's Fast Recovery Algorithm, *Request for Comments 2582.*

Gerla, M.; Sanadidi, M. Y.; Zanella, R. W.; Casetti, A. & Mascolo, S. (2002). TCP Westwood: congestion window control using bandwidth estimation. *IEEE Global Telecommunications Conference*, pp. 1698-1702, 0-7803-7206-9, San Antonio, August 2002, IEEE Computer Society,TX.

Gupta, A.; Wormsbecker, I. & Williamson, C. (2004). Experimental Evaluation of TCP Performance in Multi-hop Wireless Ad Hoc Networks, *Proceedings of IEEE Annual Internation Symposium on Modeling, Analysis, and Simulation of Computer and Telecommunications Systems*, pp. 3-11, 0-7695-2251-3, Volendam, October 2004, IEEE Computer Society, The Netherlands.

Holland, G. & Vaidya, N. (2002). Analysis of TCP Performance over Mobile Ad Hoc Networks, *Wireless Networks,* Vol. 8, No. 2/3 (March 2002) page number (275-288), 1002-0038.

Johnson, D.; Hu, Y. & Maltz, D. (2007). The Dynamic Source Routing Protocol (DSR) for Mobile Ad Hoc Networks for IPv4, *Request for comment 4728*.

Kawadia, V. & Kumar, P. (2005). Experimental investigation into TCP Performance over Wireless Multihop Networks, *SIGCOMM Workshops*, pp. 22-25, 1-59593-026-4, Philadelphia, August 2005, ACM, USA.

Kim, D.; Bae, H. & Song, J. (2005). Analysis of the Interaction between TCP Variants and Routing Protocols in MANETs, *Proceedings of the IEEE International Conference on Parallel Processing Workshops*, pp. 380-386, 0-7695-2381-1, University of Oslo, June 2005, IEEE Computer Society, Norway.

Lim, H.; Xu, K. & Gerla, M. (2003). TCP performance over multipath routing in mobile ad hoc networks, *IEEE International Conference on Communication*, pp. 1064-1068, 0-7803-7802-4, Anchorage, May 2003, IEEE Computer Society, Alaska.

Marina, M. K. & Das, S. R. (2006). Ad hoc on-demand multipath distance vector routing, *Wireless Communications and Mobile Computing*, Vol. 6, No. 7, (November 2006) page number (969-988), 1530-8669.

Mathis, M. & Mahdavi, J. (1996). TCP Selective Acknowledgement Options, *Request for comment 2018*.

McCanne, S. & Floyd, S. *VINT Group, Network Simulator Ns-2*. Source code: http://www.isi.edu/nsnam/ns.

Mondal, S. A. & Luqman, F. B. (2007). Improving TCP performance over wired-wireless networks, *Computer Networks, Vol.* 51, No. 13, (September 2007), page number (3799-3811), 1389-1286.

Oo, M. Z. & Othman, M. (2010). Performance Comparisons of AOMDV and OLSR Routing Protocols for Mobile Ad Hoc Network, *2010 Second International Conference on Computer Engineering and Applications*, pp. 129-133, 978-0-7695-3982-9, Bali Island, March 2010, Indonesia.

Osipov, E. & Tschudin, C. (2006). Evaluating the Effect of Ad Hoc Routing on TCP Performance in IEEE 802.11 Based MANETs, In: *Next Generation Teletraffic and Wired/Wireless Advanced Networking*, Koucheryacy, Y.; Harju, J. & Lversen, V. B. (Ed.), page number (298-312), Springer Berlin, 978-3-540-34429-2, Heidelberg.

Perkins, C.; Belding-Royer, E. & Das., S. (2003) Ad hoc on-demand distance vector routing (AODV), *Request for Comments 3561*.

Perkins, C. E. & Watson, T. J. (1994). Highly dynamic destination sequenced distance vector routing (DSDV) for mobile computers, *ACM SIGCOMM Computer Communication Review*, Vol. 24, No. 4, (October 1994) page number (234-244), 0146-4833.

Postel, J. (1981). Transmission Control Protocol (TCP), *Request for comment 793*.

Rakabawy, E. S.; Lindemann, C. & Vernon, M. (2005). Improving TCP Performance for Multihop Wireless Networks, *IEEE International Conference on Dependable Systems and Networks*, pp. 684-693, 0-7695-2282-3, Yokohama, June 2005, IEEE Computer Society, Japan.

Sakib, A. M. (2009). Improving performance of TCP over mobile wireless networks, *Wireless Networks*, Vol. 15, No. 3, (April 2009) page number (331-340), 1002-0038.

Stevens, W. (1997). TCP Slow Start, Congestion Avoidance, Fast Retransmit, *Request for comment 2001*.

Tseng, Y.-C.; Ni, S.-Y.; Chen, Y.-S. & Sheu, J.-P. (2002). The broadcast storm problem in a mobile ad hoc network. *Wireless Networks,* Vol. 8, No. 2/3, (March-May 2002) page number (153-167), 1002-0038.

Xu, S. & Saadawi, T. (2000). Performance Evaluation of TCP Algorithms in Multi-hop Wireless Packet Networks, *Wireless Communications and Mobile Computing*, Vol. 2, No. 1, (December 2001) page number (85-100), 1530-8669.

Part 5

Other Topics

20

A Survey on the Characterization of the Capacity of Ad Hoc Wireless Networks

Paulo Cardieri[1] and Pedro Henrique Juliano Nardelli[2]
[1]*University of Campinas*
[2]*University of Oulu*
[1]*Brazil*
[2]*Finland*

1. Introduction

Even though the interest in ad hoc wireless networks has begun in the early 1970s, several technological difficulties, particularly those related to implementation, have postponed advances in this field until the 1990s, when important issues were investigated and solved, including medium access control, routing, energy consumption, among others. These advances have allowed for actual implementation and commercial deployment of wireless communication systems based on the ad hoc concept, including wireless sensor networks, Internet access in rural areas, etc. Despite the formidable advances in this field observed in the last two decades, one key problem remains open and is still subject to intense research effort: that of modeling and measuring the capacity of ad hoc networks (Andrews et al., 2008). The intrinsic characteristics of ad hoc networks, particularly the lack of a central coordination entity and its consequences, added to the peculiarities of the wireless communication channel, make the estimation of capacity of ad hoc networks a challenging task. Despite the mentioned difficulties, researchers have proposed a myriad of metrics for characterizing the capacity of ad hoc networks under different conditions and emphasizing different aspects of the network, as described throughout this chapter.

One of the first key results in this field was achieved by Kleinrock and Silvester (Kleinrock & Silvester, 1978) in late 1970's, when they investigated the relationship between capacity and transmission radius in a network of packet radios operating under ALOHA protocol. Takagi and Kleinrock further investigated this relationship in (Takagi & Kleinrock, 1984). Both works were based on the metric so called *expected forward progress*, defined in such way to capture the tradeoff relating the one-hop throughput and the average one-hop length. In fact, decreasing the one-hop length has conflicting effects on throughput: it may increase throughput due to the resulting link quality improvement, but it may also decrease throughput, due to a larger traffic and a higher contention level caused by the consequent larger number of hops between source and destination. Subbarao and Hughes (Subbarao & Hughes, 2000) improved the model previously proposed, by including the effects of the transmission system, and introduced the concept of *information efficiency*, defined as the product of the expected forward progress and the spectral efficiency of the transmission system. Nardelli and Cardieri extended the concept of information efficiency by taking into account the effects of channel reuse and multi-hop transmissions, leading to a new metric, named *aggregate multi-hop information efficiency* (Nardelli & Cardieri, 2008a; Nardelli et al.,

2009). Based on a similar concept as that of *information efficiency*, Weber et al. introduced the metric transmission capacity (Weber et al., 2005), which is related to the optimum density of concurrent transmissions that guarantees that outage constraints are met. Simply stated, transmission capacity is the area spectral efficiency of successful transmissions resulted from the optimal contention density. The capacity metrics cited above, to be described in Section 2, have in common their statistical basis, resulted from the statistical nature of several mechanisms related to wireless communications, such as the interaction among nodes sharing a given channel and the propagation effects.

Following a deterministic approach to characterizing capacity of ad hoc networks and focusing on the behavior of capacity scaling laws, Gupta and Kumar introduced the concept of *transport capacity* (Gupta & Kumar, 2000), which relates transmission rate and source-destination distance. Gupta and Kumar formulated the transport capacity from the perspective of the requirements for successful transmission, which were described according to two interference models: the Protocol Interference Model, which is geometric-based, and the Physical Interference Model, based on signal-to-interference ratio requirements. Gupta and Kumar investigated the behavior of the network capacity when the number of nodes grows (i.e., asymptotic capacity), to show that the per-node throughput decreases as $O(1/\sqrt{n})$, where n is the number of nodes in the network. This approach was followed by several authors to investigate the asymptotic capacity of wireless ad hoc networks in a variety of scenarios, such as different transmission constraints (Xie & Kumar, 2004; 2006), and with directional antennas (Sagduyu & Ephremides, 2004). Grossglauser and Tse presented an important extension of the work of Gupta and Kumar by considering the effects of mobility on the capacity (Grossglauser & Tse, 2002). They showed that, in a network with mobile nodes operating under a 2-hop relaying transmission scheme, the per-node throughput capacity may remain constant as the number of nodes in the network increases, at the cost of unbounded packet transmission delay. This important result motivated other researchers to further investigate the tradeoff between capacity and delay in mobile wireless networks (El Gamal et al., 2006), (Herdtner & Chong, 2005), (Neely & Modiano, 2005). In Section 3 we will discuss the main results on network capacity evaluation from the perspective of scaling laws.

The brief review presented above is an evidence of the complexity of the problem of characterizing capacity of ad hoc networks, leading to a number of different metrics, with different focuses and perspectives. While this large number of metrics is also an evidence of the importance of this field, it may also mislead researchers looking for appropriate models and metrics for a particular application or scenario. This chapter therefore aims at providing readers with an overview of capacity metrics for wireless ad hoc networks, emphasizing the rationale behind the metrics.

2. Statistical-based capacity metrics

The inherent random nature of ad hoc networks suggests a statistical approach to quantify capacity of such networks. Specifically, a statistical approach is very useful for the design of practical communication systems, when a set of quality requirements is imposed by the user application in mind. In this section we will discuss some statistical-based capacity metrics found in the literature, namely expected forward progress, information efficiency, transmission capacity and aggregate multi-hop information efficiency metrics. The specificities of each metric will be discussed and their application scenario will be pointed out.

2.1 Expected forward progress

As already mentioned, the work done by Kleinrock and Silvester (Kleinrock & Silvester, 1978) in the late 1970's was one of the first attempts to model capacity of ad hoc wireless networks (Kleinrock & Silvester, 1978). They proposed the metric *expected forward progress* (EFP), measured in meters and defined as the product of the distance traveled by a packet toward its destination and the probability that such packet is successfully received. Formally,

$$\text{EFP} = d \times (1 - P_{out}), \tag{1}$$

where d is the transmitter-receiver separation distance and P_{out} is the outage probability, i.e., the probability that the bit error rate (or other related metric) is higher than a given threshold. In (Kleinrock & Silvester, 1978) the authors introduced the idea of modeling network as a collection of nodes following a spatial point process, allowing for the use of tools and properties of Stochastic Geometry (Baddeley, 2007), making possible to derive analytical formulation relating several network parameters, such node density, propagation channel parameters, number of hops, packet error probability, etc. In fact, a plethora of analysis was performed based on the metric EFP (e.g. (Sousa & Silvester, 1990), (Sousa, 1990), (Zorzi & Pupolin, 1995)).

2.2 Information efficiency

Subbarao and Hughes (Subbarao & Hughes, 2000) extended the work done by Silvester and Kleinrock by including in the model the spectral efficiency of the transmission system, resulting in a new metric, named *information efficiency* (IE), which is formally defined as the product of EFP and the spectral efficiency η of the link connecting transmitter and receiver nodes, or

$$\text{IE} = \eta \times d \times (1 - P_{out}). \tag{2}$$

Roughly speaking, IE quantifies how efficiently the information bits can travel towards its destination.

In order to understand the tradeoff captured by the information efficiency, let us consider a transmission system in which modulation and error-correcting coding techniques should be selected to optimize the IE of the network. If a modulation technique with large cardinality is used, then the spectral efficiency of the system increases, at expenses of a higher minimum required signal-to-interference plus noise ratio (*SINR*) to achieve a given packet error probability. This higher required *SINR* clearly increases the outage probability P_{out}. Error correcting coding also plays an important role in this tradeoff, as it can reduce the minimum required *SINR*, at the expenses of a higher bandwidth, reducing therefore the spectral efficiency of the transmissions. These tradeoffs are captured by the information efficiency metric, allowing for a joint system design involving modulation, coding, transmission range, among other parameters. Following this approach, the performance of different transmission schemes was investigated, such as, discrete sequence spread spectrum (Subbarao & Hughes, 2000), frequency hopping (Liang & Stark, 2000), direct sequence mobile networks (Chandra & Hughes, 2003), direct sequence code-division multiple access with channel-adaptive routing (Souryal et al., 2005) and coded MIMO frequency hopping CDMA (Sui & Zeidler, 2009).

It should be noted that, from the perspective of the whole network, the information efficiency of a link does not tell us much about how efficiently the channel is being reused throughout the network area. We will return to this point when discussing the next two metrics.

2.3 Transmission capacity

Weber et al. proposed in (Weber et al., 2005) the transmission capacity (TmC) metric of single-hop ad hoc networks. TmC is defined as the product of the density of successful links and their communication rates, subject to a constraint on the outage probability. Formally,

$$\text{TmC} = \eta \times \lambda \times (1 - P_{out}), \quad (3)$$

where λ is the density of active links in the network. Therefore, TmC quantifies the spatial spectral efficiency of the network, capturing in its formulation the effects of active links density on the outage probability. In fact, with a high density of concurrent transmissions, information flow in the network is also higher, which is indicated by a high TmC. However, the downside of a high density of active links is an increase in the interference level, leading to a higher outage probability and, consequently, a lower transmission capacity. This tradeoff, together with the ones previously presented, are the basis of the TmC framework, which can be used to evaluate several transmission strategies with different focuses. For instance, TmC was used to study frequency hopping spread spectrum (Weber et al., 2005), interference cancelation (Weber, Andrews, Yang & de Veciana, 2007), threshold transmissions and channel inversion (Weber, Andrews & Jindal, 2007), power control (Jindal et al., 2008), among many others. In fact, TmC is one of the most flexible metrics to study single-hop ad hoc networks. However, in multi-hop links scenarios, TmC is not an appropriate metric, as it does not take into account the expected forward progress of packets, making this metric unsuitable to study, for instance, the effects of different routing strategies.

2.4 Aggregate multi-hop information efficiency

In (Mignaco & Cardieri, 2006), Mignaco and Cardieri extended the work done by Subbarao and Hughes by including the effects of spatial reuse in the definition of the IE, leading to a new metric named *aggregate information efficiency* (AIE). This new metric is defined as the sum of the IE of active links in the network per unit area. Nardelli and Cardieri further improved the network model used to define AIE, by including the effects of retransmissions (Nardelli & Cardieri, 2008a) and outage constraints (Nardelli & Cardieri, 2008b). Particularly, in (Nardelli & Cardieri, 2008b) the authors make the AIE an extension of the metric TmC, where the distance traveled by a packet is explicitly considered.

Nonetheless, the metric AIE does not yet take into account the effects of multi-hop communication links. In (Nardelli et al., 2009), Nardelli *et al.* addressed such limitation and proposed the metric *aggregate multi-hop information efficiency* (AMIE). The idea behind the evolution from AIE to AMIE is to abstract multi-hop links and evaluate the AMIE based on the end-to-end performance of multi-hop links. Formally, the aggregate multi-hop information efficiency is defined as

$$\text{AMIE} = d \times \eta \times \lambda \times (1 - P_{out})^h, \quad (4)$$

where h is the average number of hops between source and destination, and d, η, λ and P_{out} were already defined. The main advantage of the AMIE is to be more flexible and general than other similar metrics. Based on this metric, several transmission schemes and network scenarios have been investigated, such as M-QAM modulation with Reed-Solomon coding scheme and ARQ retransmissions (Nardelli et al., 2009), different access protocols with limited number of retransmissions and back-offs (Nardelli et al., 2010; Kaynia et al., 2010) and different hopping strategies (Nardelli & Cardieri, 2010).

3. Capacity scaling laws

In this section, we study the capacity of wireless networks from the perspective of scaling laws, that is, we are now interested in understanding how capacity scales as the number of nodes in the network grows. This is an important subject to be investigated, as it exposes how several intrinsic aspects of wireless communication, such as interference, channel reuse and resource limitation, affect the performance of a network. Throughput, measured in bit per second, is a typical metric of capacity of communication networks and, as such, is one of the quantities considered in this section. However, in ad hoc wireless networks, in their most general configuration, source and destination nodes may be far apart, such that direct communication (single hop) is not possible, requiring a multi hop connection, with neighboring nodes acting as relays. Clearly, multi hop connections leads to a traffic increase, as a given packet is transmitted several times before reaching its final destination. Therefore, source-destination separation distance must be taken into account when characterizing capacity in wireless ad hoc networks. In this sense, a very popular capacity metric for ad hoc networks is the *transport capacity*, measured in bit·meter per second. Consider a network with transport capacity of T bit·meter per second. This means that the rate between two nodes spaced one meter away from each other is T b/s. If the distance between the nodes is doubled, then the rate decreases to $T/2$ b/s.

Gupta and Kumar (Gupta & Kumar, 2000) investigated the transport capacity and the throughput capacity of wireless networks, and derived bounds that describe the behavior of the network capacity when the number of the nodes in the network increases. Several other authors extended the work done by Gupta and Kumar, by including other aspects in the models or improving the formulation. In this section we will review the main results from the work of Gupta and Kumar and some of the extensions, particularly those presented in (Xue & Kumar, 2006).

Before discussing the models and the results of capacity scaling law, we will review some auxiliary concepts and models. We will begin with a review of asymptotic notation, commonly used to describe the asymptotic behavior of capacity as the number of nodes in the network increases.

3.1 Some auxiliary definitions

3.1.1 Asymptotic notation

In the asymptotic analysis of capacity of wireless network, the results are often presented using the asymptotic notation (or *big O-notation*) (Bruijn, 2010). In this section we briefly review the definition of some of the notation commonly used. In the following, we will assume that $f(n)$ and $g(n)$ are functions that map positive integers to positive real numbers.

Definition 1 *We say that $f(n) = O(g(n))$ (or, more precisely, $f(n) \in O(g(n))$, or even $f(n)$ is $O(g(n))$)*[1]*, if there exists a constant c and there exists an integer $n_0 \geq 1$ such that $f(n) \leq c\,g(n)$ for $n \geq n_0$ (see Figure 1(a)).*

In other words, $f(n) = O(g(n))$ means that $g(n)$ grows at least as fast as $g(n)$.

[1]Formally, we should write $f(n) \in O(g(n))$, and the form $f(n) = O(g(n))$ is considered an abuse of notation. In fact, the symmetry that the equals sign implicitly suggests does not exist in the statements involving asymptotic notation.

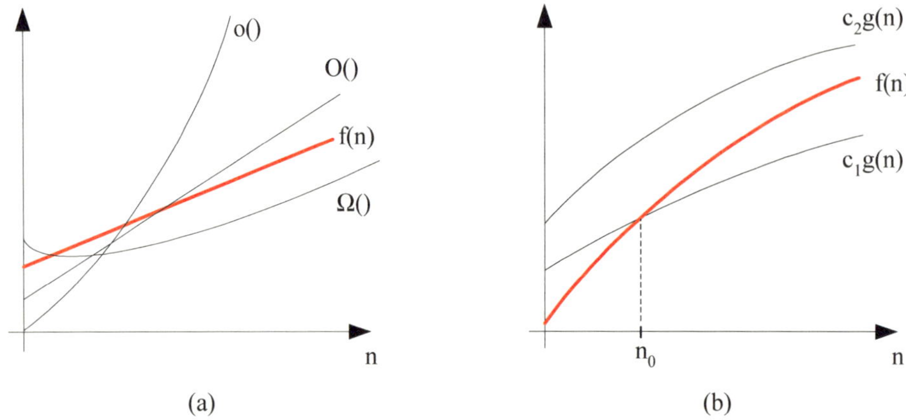

Fig. 1. (a) Interpretation of $O()$, $o()$ and $\Omega()$; (b) Interpretation of $f(n) = \Theta(g(n))$.

Definition 2 *We say that $f(n) = o(g(n))$ if for any positive constant c, there exists an integer $n_0 \geq 1$ such that $f(n) \leq c\, g(n)$ for $n \geq n_0$ (see Figure 1(a)).*

The difference between the definitions of $O()$ and $o()$ is that in the former there must exist at least one constant c such that $f(n) \leq c\, g(n)$, while in the latter the relation $f(n) \leq c\, g(n)$ must be true for any constant c. Therefore, $O()$ and $o()$ provide tight and loose upper bounds, respectively.

Definition 3 *We say that $f(n) = \Omega(g(n))$ if there exists a constant c and there exists an integer $n_0 \geq 1$ such that $f(n) \geq c\, g(n)$ for $n \geq n_0$ (see Figure 1(a)).*

Definition 4 *We say that $f(n) = \Theta(g(n))$ if there exist positive constants c_1 and c_2, and there exists $n_0 \geq 1$ such that $c_1\, g(n) \leq f(n) \leq c_2\, g(n)$, for $n \geq n_0$. Equivalently, $f(n) = \Theta(g(n))$ if $f(n) = O(g(n))$ and $f(n) = \Omega(g(n))$ (see Figure 1(b)).*

Note that $f(n) = \Theta(g(n))$ means that $g(n)$ is both a tight upper bound and a tight lower bound on $f(n)$.

3.1.2 Capacity metrics

Definition 5 (Transport capacity) *Let us suppose that node i successfully transmits to node j at rate λ_{ij} bits per second, and that the distance between i and j is d_{ij} meters. Therefore, we can say that the network transports $\lambda_{ij} \times d_{ij}$ bit·meter per second. Note that this metric expresses the difficulty of transmitting to a longer distances. Transport Capacity T of a network is evaluates as $\sum_{i \neq j} \lambda_{ij} d_{ij}$, where λ_{ij} is the feasible rate between nodes i and j.*

Definition 6 (Throughput capacity) *It is the guaranteed rate, measured in bits per second, that can be supported uniformly for all source-destination pairs.*

3.1.3 Interference models

Definition 7 (The protocol interference model) *Let $\{(X_i, X_{R(i)}) : k \in \mathcal{T}\}$ be the set of active transmitter-receiver pairs in the network. According to the* **protocol interference model***, this transmission is successfully received if the distance between nodes $X_{R(i)}$ (the intended receiver of node*

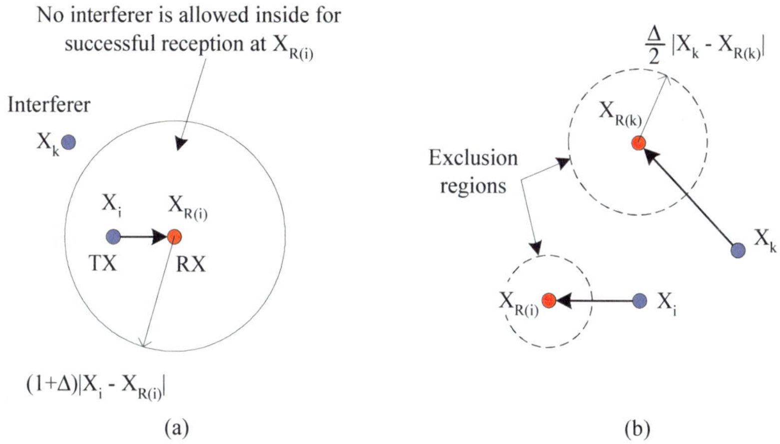

Fig. 2. The protocol model: (a) Disk around receiver $X_{R(i)}$ must be free of interfering nodes for correct reception at node $X_{R(i)}$; (b) Two links are successful if the corresponding exclusion regions are disjoint.

X_i transmission) and any other node X_k transmitting on the same channel is larger than the distance between X_i and $X_{R(i)}$, that is

$$|X_k - X_{R(i)}| \leq (1+\Delta)|X_i - X_{R(i)}|, \tag{5}$$

where $|X_k - X_{R(i)}|$ indicates the distance between nodes X_i and $X_{R(i)}$, and $\Delta > 0$ is the spatial protection margin. Figure 2(a) shows a geometric interpretation of this model. Now, let us consider two pairs of active nodes X_i and X_k, with X_i transmitting to $X_{R(i)}$ and X_k transmitting to $X_{R(k)}$, and with both pairs operating under the protocol model, represented by expression (5). We can show that, in order to have both transmissions successfully received, we must have

$$|X_{R(k)} - X_{R(i)}| \geq \frac{\Delta}{2}\left(|X_k - X_{R(k)}| + |X_i - X_{R(i)}|\right). \tag{6}$$

This result indicates that *circular exclusion regions* around the receivers $X_{R(j)}$ and $X_{R(k)}$, of radius $\Delta|X_i - X_{R(i)}|/2$ and $\Delta|X_k - X_{R(k)}|/2$, respectively, are disjoint, as shown Figure 2(b). Therefore, exclusion regions around receivers of each successful transmission are mutually disjoint, and consume a portion of the network area.

Definition 8 (The physical interference model) *Consider, as before, a set of active transmitter-receiver pairs $\{(X_i, X_{R(i)}) : i \in \mathcal{N}\}$, transmitting over the same channel, with a transmit power assignment $\{P_i\}$. According to the* **physical interference model**, *the transmission from node X_i is successfully received by node $X_{R(i)}$ if the signal-to-interference plus noise ratio* (SINR) *at $X_{R(i)}$ is equal to or larger than a given threshold β, that is*

$$\frac{\frac{P_i}{|X_i - X_{R(i)}|^{\eta}}}{\sigma^2 + \sum_{k \in \mathcal{N}, k \neq i} \frac{P_k}{|X_k - X_{R(i)}|^{\eta}}} \geq \beta, \tag{7}$$

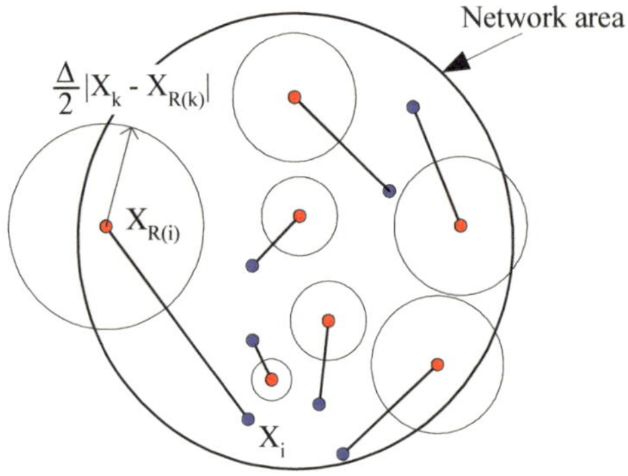

Fig. 3. Arbitrary network under the Protocol Interference model: successful links correspond to disjoint disks.

where σ^2 is the additive noise power. The threshold β depends on transmission parameters, such as modulation technique, error correcting coding and the minimum acceptable bit error rate.

3.2 Transport capacity in arbitrary networks with immobile nodes

We consider in this section a network of n immobile nodes, which can act simultaneously as source, relay or destination. These n nodes are arbitrarily located in a planar disk of unity area. This means that the positions of the nodes can be adjusted in order to satisfy the conditions for successful transmissions imposed by the interference model considered in the analysis. Every node selects randomly another node as the destination of its bits. The results of this analysis are presented in the sequel, for both the Protocol Interference model and the Physical Interference model.

3.2.1 Capacity under the protocol interference model

The authors of (Gupta & Kumar, 2000) showed that the transport capacity T_A of an arbitrary network with n nodes under the *Protocol Model* is

$$T_A = \Theta(W\sqrt{n}) \quad \text{bit} \cdot \text{meter/s}, \tag{8}$$

This means that the transport capacity *per node* is $\Theta(W\sqrt{1/n})$ bit·meter/s, and goes to zero as the number of nodes increases. Following (Xue & Kumar, 2006), this result can be proved using the fact that, under the Protocol Interference model, disks of radius equals to $\Delta|X_i - X_{R(i)}|/2$ centered at receiver nodes of successful links are disjoint (see Definition 7). Therefore, each successful link consumes a fraction of the network area and the sum of the area of disks of all successful links is upper limited by the network area (see Figure 3). Neglecting the border effects (i.e., when nodes are close to the boundary of the network area), we can write

$$\sum_{i \in \mathcal{T}(t)} \pi \left(\frac{\Delta}{2} d_i\right)^2 \leq 1 \;\rightarrow\; \sum_{i \in \mathcal{T}(t)} d_i^2 \leq \frac{4}{\pi \Delta^2}, \tag{9}$$

where d_i is the T-R separation distance $|X_i - X_{R(i)}|$ of the i-th T-R pair, and $\mathcal{T}(t)$ is the set of successful links at time t. This expression can be interpreted as follows: a set of n nodes is accommodated in such way[2] that condition (9) is satisfied. It should be noted that, at any

[2]Recall that we are dealing with the arbitrary network case.

given time t, at most $n/2$ nodes will be transmitting (the other $n/2$ nodes will be receiving). Now, we can use the Cauchy-Schwarz inequality to write

$$\sum_{i=1}^{n/2} d_i^2 \sum_{j=1}^{n/2} 1^2 \geq \left(\sum_{i=1}^{n/2} d_i \times 1 \right)^2 ,$$

or

$$\begin{aligned} \sum_{i=1}^{n/2} d_i &\leq \sqrt{\sum_{i=1}^{n/2} d_i^2 \frac{n}{2}} \\ &\leq \sqrt{\frac{2n}{\pi \Delta^2}}. \end{aligned}$$

Therefore, we have found an upper bound on the sum of the T-R separation distances of successful links. Now, if we assume that all sources transmit at rate W, then the transport capacity T_A of the network at a given time t is upper bounded as

$$T_A = W \sum_{i \in \mathcal{T}(t)} d_i \leq \sqrt{\frac{2}{\pi}} \frac{W}{\Delta} \sqrt{n},$$

or, $T_A = O(W\sqrt{n})$ bit-meter/s. Now, we can also show that a transport capacity of $\frac{W\sqrt{A}}{1+2\Delta} \frac{n}{\sqrt{n}+\sqrt{8\pi}}$ bit-meter/s is achievable under the Protocol Interference Model (see (Xue & Kumar, 2006) for details), completing the proof of (8).
Recalling that the network has n nodes, we can conclude that the transport capacity *per node* is $\Theta(W/\sqrt{n})$. This means that the transport capacity diminishes to zero as the number of users in the network increases. Note that we are assuming here that sources randomly select other nodes as their destinations and, therefore, the average source-destination separation distance does not depend on the number of nodes n. So, as n increases, we have more and more nodes willing to send their bits over paths with the same average length, but sharing the same available bandwidth.

3.2.2 Capacity under the physical interference model

Now, if the *Physical Interference model* is adopted, Kumar and Gupta (Gupta & Kumar, 2000) showed that the transport capacity is

$$T_A = O(W\, n^{\frac{\alpha-1}{\alpha}}) \;\; \text{bit} \cdot \text{meter/s}. \tag{10}$$

This upper bound can be proved recalling that, according to the Physical Interference model, a successful transmission requires that

$$\frac{P_i d_i^{-\alpha}}{N + \sum_{j \in \mathcal{T}, j \neq i} P_j d_j^{-\alpha}} \geq \beta. \tag{11}$$

If we include the desired signal power in the summation in denominator, and isolate the term d_i^{α}, we get

$$d_i^{\alpha} \leq \frac{(\beta+1) P_i}{\beta \left(N + \sum_{j \in \mathcal{T}} P_j d_j^{-\alpha} \right)}. \tag{12}$$

Noting that the T-R separation distance d_i is smaller than the diameter of the network area, i.e., $d_i \leq 2/\sqrt{\pi}$, then

$$d_i^\alpha \leq \frac{(\beta+1)\,P_i}{\beta\left[N+\left(\frac{\pi}{4}\right)^{\alpha/2}\sum_{j\in\mathcal{T}}P_j\right]} \leq \frac{(\beta+1)\,P_i}{\beta\left[\left(\frac{\pi}{4}\right)^{\alpha/2}\sum_{j\in\mathcal{T}}P_j\right]}. \tag{13}$$

Now, summing the quantities d_i^α of all active links, we get

$$\sum_{i\in\mathcal{T}} d_i^\alpha \leq \frac{(\beta+1)}{\beta}\left(\frac{4}{\pi}\right)^{\alpha/2}. \tag{14}$$

Next, we use the Holder's inequality, according to which, for $a,b>0$, $p,q\geq 1$ and $1/p+1/q=1$,

$$\sum ab \leq \left(\sum a^p\right)^{1/p}\left(\sum b^q\right)^{1/q}. \tag{15}$$

Therefore, recalling that there are at most $n/2$ links, then

$$\begin{aligned}
\sum_{i\in\mathcal{T}} d_i &\leq \left(\sum_{i\in\mathcal{T}} d_i^\alpha\right)^{1/\alpha}\left(\sum_{i\in\mathcal{T}} 1^{\frac{\alpha-1}{\alpha}}\right)^{\frac{\alpha-1}{\alpha}} \\
&\leq \left(\sum_{i\in\mathcal{T}} d_i^\alpha\right)^{1/\alpha}\left(\frac{n}{2}\right)^{\frac{\alpha-1}{\alpha}} \\
&\leq \left[\frac{(\beta+1)}{\beta}\left(\frac{4}{\pi}\right)^{\alpha/2}\right]^{1/\alpha}\left(\frac{n}{2}\right)^{\frac{\alpha-1}{\alpha}} \\
&\leq \frac{1}{\sqrt{\pi}}\left(\frac{2\beta+2}{\beta}\right)^{1/\alpha} n^{\frac{\alpha-1}{\alpha}}.
\end{aligned} \tag{16}$$

Finally, if all sources transmit at rate W, the transport capacity is upper bounded as

$$T_A = W\sum_{i\in\mathcal{T}} d_i \leq \frac{W}{\sqrt{\pi}}\left(\frac{2\beta+2}{\beta}\right)^{1/\alpha} n^{\frac{\alpha-1}{\alpha}}. \tag{17}$$

Note that if capacity is equitably shared among all sources, the transport capacity per node is $T_A = O(W/n^{1/\alpha})$, and goes to zero as n increases. Note also that this bound indicates that a larger path loss exponent α leads to a higher capacity. This can be explained by noting that larger α means stronger signal attenuation and, therefore, reduced interference. Consequently, concurrent links can be packed together, increasing capacity.

3.3 Throughput capacity in random networks with immobile nodes

3.3.1 Capacity under the protocol interference model

Gupta and Kumar also showed that the throughput capacity in bits per second of a random network under the Protocol Model is upper bounded by

$$\lambda(n) \leq \frac{c\,W}{\sqrt{n\log n}}. \tag{18}$$

This result can be proved using again the argument that successful transmissions consume portions of the network area. Let us consider a network with n nodes randomly placed on a

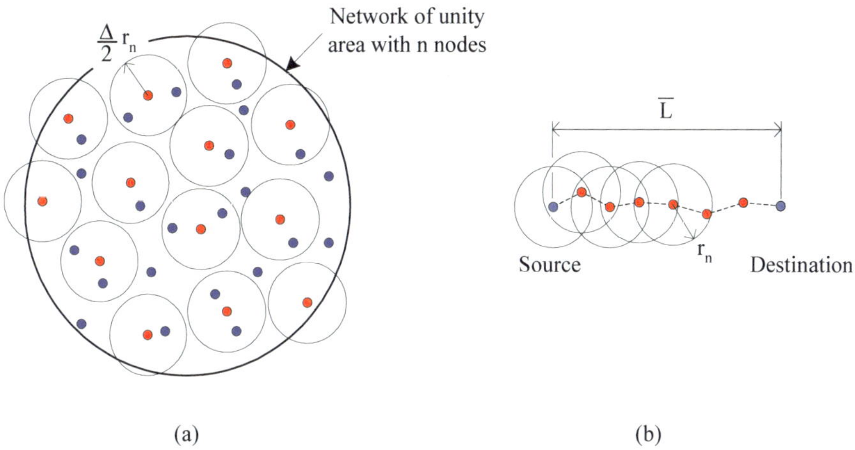

Fig. 4. The protocol model: (a) Disks around active receivers must be disjoint; (b) Average number of hops between source and destination.

disk of unity area. Let us also assume that all nodes transmit with a common transmission range r_n. In order to guarantee that no node is isolated in the network, it can be shown that r_n must be asymptotically larger than $\sqrt{\log n/\pi n}$ (Gupta & Kumar, 1998) (Penrose, 1997). Next, we recall that, under the Protocol Interference model, successful transmissions require that disks of radius $\Delta r_n/2$, centered at receivers, must be disjoint, as shown in Figure 4(a). Therefore, the number of successful transmissions N_S within a disk of unity area is upper bounded as

$$N_S < \frac{4}{\pi \Delta^2 r_n^2}. \tag{19}$$

Therefore, the aggregate number of bits transmitted per second in the network cannot be larger than

$$\frac{4W}{\pi \Delta^2 r_n^2},$$

where W is the common transmission rate of the individual transmissions.

Now, as before, let us consider that source nodes choose at random their destination nodes, and denote $\overline{L}$ the average source-destination separation distance. Note that $\overline{L}$ does not depend on the number of nodes in the network. Therefore, the average number of hops between source and destination is lower bounded by $\overline{L}/r_n$ (see Figure 4(b)). If each source generates bits at rate $\lambda(n)$, then the average number of bits transmitted by the whole network is given by $n\lambda(n)\overline{L}/r_n$ and must satisfy

$$n\lambda(n)\frac{\overline{L}}{r_n} \leq \frac{4W}{\pi \Delta^2 r_n^2}. \tag{20}$$

Finally, using $r_n > \sqrt{\log n/\pi n}$, we complete the proof of (18).

In this same context, i.e., random networks under the Protocol Interference model, Xue and Gupta presented in (Xue & Kumar, 2006) a transmission scheme that achieves a throughput

$$\lambda(n) \leq \frac{c\,W}{(1+\Delta)^2 \sqrt{n \log n}}. \tag{21}$$

To demonstrate that (21) is valid, n nodes are randomly placed in a square of unity area. This area is tessellated by cells of side $s_n = \sqrt{K \log n/n}$, as shown in Figure 5(a). We can

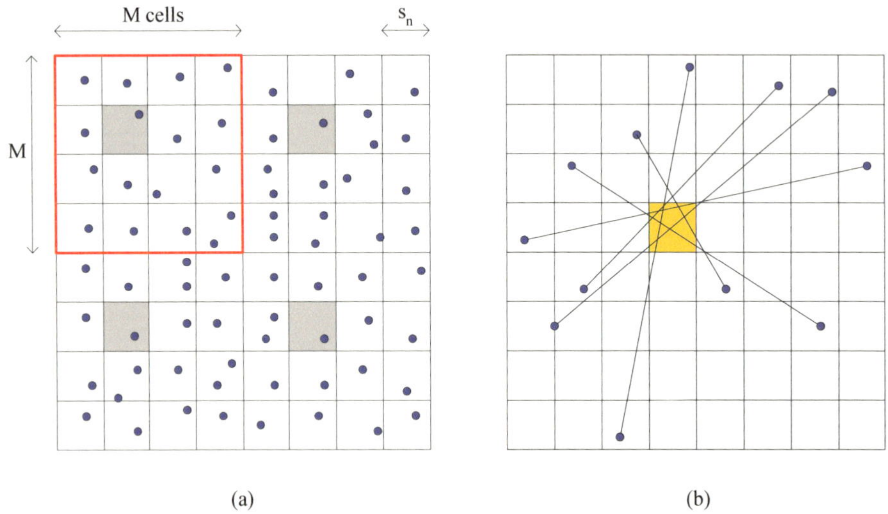

Fig. 5. The protocol model: (a) Tessellation of the unity square by cells of side s_n, with adjacent cells grouped in groups of M^2 cells ($M = 4$). Cells in blue are allowed to transmit concurrently; (b) Source-destination lines crossing a given cell (adapted from (El Gamal et al., 2006), copyright ©2006 IEEE).

show that, with probability approaching one, each cell has at least one but no more than $Ke \log n$ nodes (see (Xue & Kumar, 2006) for details). We suppose that nodes transmit with a common transmission range such that every node can transmit to any node located in its neighboring cells. In order to guarantee successful transmissions, by controlling interference, the following transmission scheme is used. We divide the cells into groups of M^2 adjacent cells (see Figure 5(a)). At each time-slot, one node from one cell of each group is allowed to transmit. Therefore, at each time-slot, there will be n/M^2 concurrent transmissions (or concurrent cells), as exemplified in Figure 5(b). Clearly, time is split into M^2 time-slots. Successful transmissions are guaranteed if concurrent cells are enough far apart, being the distance between concurrent cell controlled by the number M. Note that the required value of M for successful transmission does not depend on n, as only one node from each cell transmits at each time-slot. Therefore, under the Protocol Interference model, we can simply set $M = c(1 + \Delta)$ (Xue & Kumar, 2006). Since, as before, each source node chooses at random its destination node, bits reach their destination by means of multi-hop routes. Therefore, every node transmits not only its own bits, but also bits from other nodes. Therefore, the number of bits each node pumps to the network (its own bits and those from other nodes) is related to the number N_R of multi hop routes crossing the cell to which the node belongs (see Figure 5(b)). This number N_R, in turn, is related to the number of lines connecting a source and a destination that intersect a given cell. Xue and Gupta (Xue & Kumar, 2006) showed that, with probability approaching one, $N_R \le c\sqrt{n \log n}$. Therefore, the number of bits transmitted per second from a given cell is $\lambda(n)c'\sqrt{n \log n}$, where $\lambda(n)$ is the throughput per node. If W is the transmission rate in each time-slot, and recalling that there are $[c(1 + \Delta)]^2$ time-slots, then each cell transmits at rate $W/[c(1 + \Delta)]^2$. Therefore, the throughput per cell $\lambda(n)c'\sqrt{n \log n}$ is feasible if

$$\lambda(n)c'\sqrt{n \log n} \le \frac{W}{[c(1 + \Delta)]^2},$$

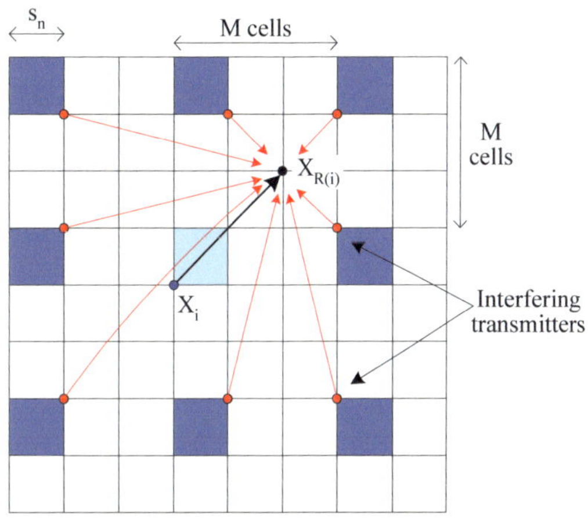

Fig. 6. Evaluation of the interference in a tesselated network under the Physical Interference model.

concluding the proof of (21). It should be noted that one node in each cell can be designated to handle all relay traffic, while all other nodes act as sources or destinations.
Note that while (18) gives an upper bound on the throughput per node, (21) gives a feasible throughput, and we say that *the order of the throughput* of random networks under the Protocol Interference model is

$$\lambda(n) = \Theta\left(\frac{W}{\sqrt{n \log n}}\right). \tag{22}$$

As noted in (Xue & Kumar, 2006), the result in (22) suggests that the throughput of random networks is almost that achieved in the best case scenario (arbitrary networks), in which throughput is $O(1/\sqrt{n})$, despite the fact that nodes are optimally located.

3.3.2 Capacity under the physical interference model

When the Physical Interference model is used, it can be shown that throughput per bits per second

$$\lambda(n) = \Theta\left(\frac{W}{\sqrt{n \log n}}\right) \tag{23}$$

is feasible. This result can be derived using the same transmission scheme used in Section 3.3 .1. We just need to show that M can be selected such that transmissions can achieve $SINR \geq \beta$, as required by the Physical Interference model for successful transmission (Xue & Kumar, 2006). In order to show that, let us consider the transmission from node X_i to receiver $X_{R(i)}$ in a network tesselated as before, as shown in Figure 6. This transmission is disturbed by transmissions from nodes located in the concurrent cells, which are arranged according to tiers of $8k$ cells, with $k = 1, 2, \cdots$. Using simple geometric arguments, we see that in the worst-case scenario, the distance between X_i and $X_{R(i)}$ is $2\sqrt{2}s_n$, and the distances between receiver $X_{R(i)}$ and interferers of the k-th tier are larger than $kMs_n - 2s_n$. The aggregate interference power can therefore be upper bounded as

$$\begin{aligned}\sum_{k\in\mathcal{N},k\neq i}\frac{P_k}{|X_k-X_{R(i)}|^\alpha} &\leq \sum_{k=1}^{\infty}8k\frac{P}{(kMs_n-2s_n)^\alpha}\\ &\leq \frac{8P}{(Ms_n)^\alpha}\sum_{k=1}^{\infty}\frac{k}{(k-2/M)^\alpha}.\end{aligned}$$

It can be shown that $\sum_{k=1}^{\infty}\frac{k}{(k-2/M)^\alpha}$ converges when $\alpha > 2$ (Xue & Kumar, 2006), and therefore there is a value of M sufficiently large that guarantees $SINR \geq \beta$ at the receiver. Therefore, the throughput

$$\lambda(n) = \frac{cW}{\sqrt{n\log n}}$$

is feasible in a random network under the Physical Interference model as well.

An upper bound on the throughput for random network under the Physical Interference model can be derived using the upper bound on the throughput for the case under the Protocol Interference model. In fact, successful links $(X_i, X_{R(i)})$ in a random network under the Physical Interference mode are also successful under the Protocol Model, for appropriate values of Δ and β. Therefore, an upper bound on the throughput for the Protocol Model also holds for the Physical Interference model. Therefore, for a random network under the Physical Interference model the throughput is upper bounded as

$$\lambda(n) < \frac{cW}{\sqrt{n}}. \tag{24}$$

3.4 Capacity with directional antennas

In the previous sections we assumed that transmitters and receivers are equipped with omnidirectional antennas. However, it is well known that directional antennas can reduce interference and, consequently, increase capacity. Yi et al. (Yi et al., 2007) extended the work done by Gupta and Kumar by including directional antennas in the model, and investigated the effects of directional antennas on the capacity scaling laws. The radiation pattern adopted by Yi et al. is modeled as a sector with beamwidth α, for the transmit antenna, and β, for the receive antenna. This is a rather optimistic model as it assumes that the energy irradiated outside the main bean is zero (i.e., sidelobes have zero gain). Following the same reasoning as in (Gupta & Kumar, 2000), the authors in (Yi et al., 2007) show that the throughput capacity per node for a arbitrary network under the Protocol Interference model scales as

$$\lambda(n) = O\left(\frac{1}{\sqrt{n\alpha\beta}}\right). \tag{25}$$

Therefore, capacity increases as beamwidth decreases, what can be explaining by the fact that directional antennas reduces the overall interference, and more concurrent transmissions can be accommodated at a given time. However, even though the use of directional antennas may increase capacity, it does not change the form of the scaling law of capacity. That would be possible if α and β decreased as fast as $1/\sqrt{n}$, leading to a constant throughput per node as the size n of the network increases.

Spyropoulos and Raghavendra (Spyropoulos & Raghavendra, 2003) also investigated the effects of directional antennas on the capacity scaling laws of ad hoc networks, but using more

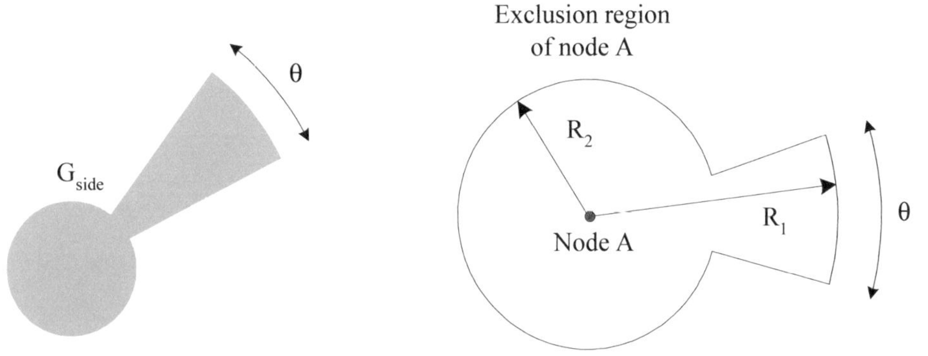

(a) (b)

Fig. 7. (a) Idealized radiation pattern; (b) Exclusion region created when the Protocol Interference model is used with the radiation pattern in (a): for successful reception at node A, no other receiver can be located inside such exclusion region.

general antenna models. First, they considered an idealized radiation pattern with beamwidth θ with unity gain, and constant sidelobe with gain $G_{side} < 1$, as shown in Figure 7(a). When this radiation pattern is assumed at both transmitters and receivers, the use of the Protocol Interference model results in an exclusion region as shown in Figure 7(b), in which R_1 and R_2 are given by

$$R_1 = \left[(P/P_{th})\, G_{side}\right]^{1/\alpha} \quad \text{and} \quad R_2 = \left[(P/P_{th})\, G_{side}^2\right]^{1/\alpha}. \tag{26}$$

Therefore, small gain G_{side} leads to small exclusion area, which, in turn, leads to a large number of concurrent transmissions. In fact, Spyropoulos and Raghavendra showed that the throughput capacity per node is upper bounded as

$$\lambda(n) \le \frac{cW}{\sqrt{n \log n}} \, \frac{1}{\theta G_{side} + (2\pi - \theta) G_{side}^2}. \tag{27}$$

In the directional antenna model adopted by Spyropoulos and Raghavendra, a narrow beam is steered towards the intended node, and out of the main beam, the antenna gain is constant. This, however, is not an appropriate model for the so called *smart antenna*, which are capable of not only steering a narrow beam towards a given direction, by also steering strong attenuation (nulls) towards some directions, in order to mitigate the signal from known interfering transmitters. In order to evaluate the effects of a smart antenna on the network capacity, Spyropoulos and Raghavendra considered that a smart antenna with N elements can steer a beam of gain $G_{max} = 1$ towards the desired direction, and gains $G_{null} \ll 1$ towards at most $N - 2$ different directions. Now, the use of this antenna model together with the Protocol Interference model allows for the accommodation of at most $N - 2$ receiving nodes within a circle of radius $R = (P/P_{th})^{1/\alpha}$, and the throughput capacity per bits/sec per node is upper bounded as

$$\lambda(n) \le \frac{cW(N-2)}{\sqrt{n \log n}}. \tag{28}$$

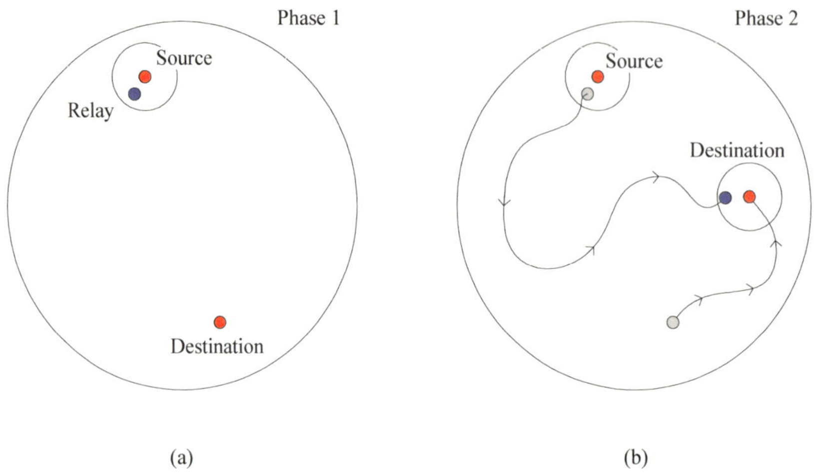

Fig. 8. The 2-hop relaying transmission scheme adopted by Tse and Grossglauser: (a) In Phase 1, source transmits its packet to a relay node within its transmission range; (b) in Phase 2, packet is sent to the destination when the relay node gets close enough to the destination node (adapted from (Grossglauser & Tse, 2002), copyright ©2002 IEEE).

3.5 Networks with mobile nodes

Grossglauser and Tse (Grossglauser & Tse, 2002) extended in another direction the work done by Gupta and Kumar, by introducing mobility in the model. As discussed in previous sections, throughput in a network with *immobile* nodes decays as $1/\sqrt{n}$ due to the traffic increase caused by multi hop connections between sources and destinations. Alternatively, one could use large transmission ranges in order to reduce the number of hops between source and destination. However, this strategy limits the number of concurrent transmissions, limiting the capacity of the network. Other alternative would be to restrict transmissions to neighbors. However, only a small fraction of sources are close enough to their destination nodes, limiting capacity as well. In the light of this observation, and considering a network of *mobile* nodes, Grossglauser and Tse (Grossglauser & Tse, 2002) developed a 2-hop relaying transmission scheme with two phases, described in the following as exemplificed in Figure 8:

- Phase 1: A packet generated by a node is either directly transmitted to the corresponding destination node, or relayed to a intermediate (relay) node. In the former case, the transmission session is concluded.
- Phase 2: If the packet is sent to a relay node, the packet is buffered until the relay node is close enough to the destination node, when the packet is eventually sent to its final destination.

Note that an essential aspect of this scheme is that , due to mobility, the relay node and the destination nodes will eventually be close enough to each other to allow communication between them. Based on this model, Grossglauser an Tse showed that the average long-term throughput per S-D pair remains constant as n increases, that is, throughput scales as $\Theta(1)$. An important aspect of this analysis is that the mobility model adopted assumes that, at a given time, a node is equally likely to be in any part of the network, meaning that the network topology completely changes over time. Clearly, this mobility model is an oversimplification of a real scenario, but the results obtained under this model can be viewed as upper bound on the performance.

Grossglaucer and Tse pointed out that throughput remains constant as n increases at the expenses of an increasing delay. This has motivated several studies of the tradeoff between delay and throughput in ad hoc networks (El Gamal et al., 2006), (Herdtner & Chong, 2005), (Lin et al., 2006), (Neely & Modiano, 2005), (Sharma et al., 2007). For instance, El Gamal et. al (El Gamal et al., 2006) investigated this tradeoff not only for mobile networks, but also for static networks. For mobile networks, they considered a network operating under the same 2-hop relaying transmission scheme adopted by Grossglauser an Tse, and assumed a mobility model named random-walk model, according to which nodes move a distance $1/\sqrt{n}$ per unit time. They then showed that the throughput scales as $\Theta(1)$, as in (Grossglauser & Tse, 2002), but the delay scales as $\Theta(n \log n)$. For static network, El Gamal et. al showed that, at throughput $\Theta(1/\sqrt{n \log n})$ (as in the work done by Gupta and Kumar), the average delay is $\Theta(\sqrt{n/\log n})$.
Another important extension of the work done by Grossglaucer and Tse is the one carried out by Herdtner and Chong (Herdtner & Chong, 2005) in which the authors showed that mobility alone does not increase capacity of ad hoc networks. Specifically, they showed that if the buffer size of nodes is finite and limited to $\Theta(1)$, i.e., it remains constant as n increase, then the throughput capacity is only $O(1/\sqrt{n})$, instead of $\Theta(1)$. Therefore, a scaling law for throughput in a mobile network in the form $\Theta(1)$ is only possible if the buffer size increases as n increases.
Lin et al. (Lin et al., 2006) investigated the tradeoff between capacity and delay in a mobile wireless network, assuming a Brownian motion model. A key parameter in this mobility model is the variance σ^2, which is related to the time required by a node to move to different parts of the network. Large σ^2 means that the node will take a short amount of time to move. The authors of (Lin et al., 2006) showed that, under the 2-hop relaying transmission scheme proposed by Grossgluaser and Tse, throughput of $\Theta(1)$ is achieved at the expenses of an average delay of $\Omega(\log n/\sigma^2)$, showing how the node speed affects the delay.

4. Summary

This chapter provided an overview of metrics for capacity evaluation of ad hoc wireless networks. The peculiarities of wireless ad hoc networks make the estimation of capacity of this kind of networks a complex task, which is evidenced by the variety of capacity metrics found in the literature.
The capacity metrics discussed in this chapter can be classified into two groups: metrics based on a statistical approach, and metrics focused on the network scalability. In the first group, discussed in Section 2, capacity metrics incorporate aspect from the physical layer (e.g. modulation parameters, spectral efficiency, etc.) and from the network layer (e.g. spatial reuse, number of hops, etc.). Therefore, these metrics are suitable for network design and parameter optimization.
The metrics in the second group, discussed in Section 3, essentially describe how network capacity behaves when the number of nodes in the network grows. As can be noted from the discussion presented in Section 3, the scaling laws derived are closely related to the particular network model and transmission scheme assumed. Therefore, even though the resulting scaling laws are rather pessimistic (per-node capacity vanishes as the size of the network increases), the results can be used as guideline for the design of more appropriate transmission schemes, that would hopefully result in non-vanishing capacity.

5. References

Andrews, J., Shakkottai, S., Heath, R., Jindal, N., Haenggi, M., Berry, R., Guo, D., Neely, M., Weber, S., Jafar, S. & Yener, A. (2008). Rethinking information theory for mobile ad hoc networks, *IEEE Communications Magazine* 46(12): 94–101.

Baddeley, A. (2007). Spatial point processes and their applications, *Stochastic Geometry*, Springer, pp. 1–75.

Bruijn, N. (2010). *Asymptotic Methods in Analysis*, Dover Publications.

Chandra, M. & Hughes, B. (2003). Optimizing information efficiency in a direct-sequence mobile packet radio network, *Communications, IEEE Transactions on* 51(1): 22–24. URL: *10.1109/TCOMM.2002.807607*

El Gamal, A., Mammen, J., Prabhakar, B. & Shah, D. (2006). Optimal throughput-delay scaling in wireless networks - part i: the fluid model, *Information Theory, IEEE Transactions on* 52(6): 2568 –2592.

Grossglauser, M. & Tse, D. (2002). Mobility increases the capacity of ad hoc wireless networks, *Networking, IEEE/ACM Transactions on* 10(4): 477–486. URL: *10.1109/TNET.2002.801403*

Gupta, P. & Kumar, P. (2000). The capacity of wireless networks, *IEEE Trans. on Information Theory* 46(2): 388–404.

Gupta, P. & Kumar, P. R. (1998). *Stochastic Analysis, Control, Optimization and Applications: A Volume in Honor of W.H. Fleming*, Birkhauser, chapter Critical power for asymptotic connectivity in wireless networks, pp. 547 – 560.

Herdtner, J. & Chong, E. (2005). Throughput-storage tradeoff in ad hoc networks, *INFOCOM 2005. 24th Annual Joint Conference of the IEEE Computer and Communications Societies. Proceedings IEEE*, Vol. 4, pp. 2536 – 2542 vol. 4.

Jindal, N., Weber, S. & Andrews, J. (2008). Fractional power control for decentralized wireless networks, *IEEE Trans. on Wireless Communications* 7(12): 5482–5492.

Kaynia, M., Nardelli, P., Cardieri, P. & Latva-aho, M. (2010). On the optimal design of MAC protocols in multi-hop ad hoc networks, *Sixth Workshop on Spatial Stochastic Models for Wireless Networks*.

Kleinrock, L. & Silvester, J. (1978). Optimum transmission radii for packet radio networks or why six is a magic number, *National Telecommunications Conference*.

Liang, P. & Stark, W. (2000). Transmission range control and information efficiency for FH packet radio networks, *MILCOM 2000. 21st Century Military Communications Conference Proceedings*, Vol. 2, pp. 861–865 vol.2. URL: *10.1109/MILCOM.2000.904053*

Lin, X., Sharma, G., Mazumdar, R. & Shroff, N. (2006). Degenerate delay-capacity tradeoffs in ad-hoc networks with brownian mobility, *Information Theory, IEEE Transactions on* 52(6): 2777 –2784.

Mignaco, A. & Cardieri, P. (2006). Total information efficiency in multihop wireless networks, *IEEE International Performance, Computing, and Communications Conference*.

Nardelli, P. & Cardieri, P. (2008a). Aggregate information efficiency and packet delay in wireless ad hoc networks, *IEEE Wireless Communications and Networking Conference*.

Nardelli, P. & Cardieri, P. (2008b). Aggregate information efficiency in wireless ad hoc networks with outage constraints, *IEEE International Workshop on Signal Processing Advances in Wireless Communications*.

Nardelli, P., de Abreu, G. & Cardieri, P. (2009). Multi-hop aggregate information efficiency in wireless ad hoc networks, *Communications, 2009. ICC '09. IEEE International Conference*

on, pp. 1–6.
URL: *10.1109/ICC.2009.5199200*

Nardelli, P. H. J. & Cardieri, P. (2010). Exploiting location information to improve the efficiency of wireless networks. Submitted, available at http://sites.google.com/site/phjnardelli.

Nardelli, P., Kaynia, M. & Latva-aho, M. (2010). Efficiency of the ALOHA protocol in multi-hop networks, *IEEE International Workshop on Signal Processing Advances in Wireless Communications*.

Neely, M. & Modiano, E. (2005). Capacity and delay tradeoffs for ad hoc mobile networks, *Information Theory, IEEE Transactions on* 51(6): 1917 – 1937.

Penrose, M. D. (1997). The longest edge of the random minimal spanning tree, *The Annals of Applied Probability* 7(2): 340 – 361.

Sagduyu, Y. E. & Ephremides, A. (2004). On the capacity bounds of wireless networks with directional antennas, *Proc. Conference on Information Sciences and Systems*.

Sharma, G., Mazumdar, R. & Shroff, B. (2007). Delay and capacity trade-offs in mobile ad hoc networks: A global perspective, *Networking, IEEE/ACM Transactions on* 15(5): 981 –992.

Souryal, M., Vojcic, B. & Pickholtz, R. (2005). Information efficiency of multihop packet radio networks with channel-adaptive routing, *Selected Areas in Communications, IEEE Journal on* 23(1): 40–50.
URL: *10.1109/JSAC.2004.837366(410) 23*

Sousa, E. (1990). Optimum transmission ranges in a frequency hopping multi-hop packet radio network, *Telecommunications Symposium, 1990. ITS '90 Symposium Record., SBT/IEEE International*, pp. 608–611.
URL: *10.1109/ITS.1990.175675*

Sousa, E. & Silvester, J. (1990). Optimum transmission ranges in a direct-sequence spread-spectrum multi-hop packet radio network, *IEEE Journal on Selected Areas in Communications* 8(5): 762–771.

Spyropoulos, A. & Raghavendra, C. (2003). Asympotic capacity bounds for ad-hoc networks revisited: the directional and smart antenna cases, *Global Telecommunications Conference, 2003. GLOBECOM '03. IEEE*, Vol. 3, pp. 1216 – 1220 vol.3.

Subbarao, M. & Hughes, B. (2000). Optimal transmission ranges and code rates for frequency-hop packet radio networks, *IEEE Transactions on Communications* 48(4): 670–678.

Sui, H. & Zeidler, J. (2009). Information efficiency and transmission range optimization for coded MIMO FH-CDMA ad hoc networks in time-varying environments, *Communications, IEEE Transactions on* 57(2): 481–491.
URL: *10.1109/TCOMM.2009.02.070076*

Takagi, H. & Kleinrock, L. (1984). Optimal transmission ranges for randomly distributed packet radio terminals, *Communications, IEEE Transactions on* 32(3): 246–257.

Weber, S., Andrews, J. & Jindal, N. (2007). The effect of fading, channel inversion, and threshold scheduling on ad hoc networks, *IEEE Trans. on Information Theory* 53(11): 4127–4149.

Weber, S., Andrews, J., Yang, X. & de Veciana, G. (2007). Transmission capacity of wireless ad hoc networks with successive interference cancellation, *IEEE Transactions on Information Theory* 53(8): 2799–2814.

Weber, S., Yang, X., Andrews, J. & de Veciana, G. (2005). Transmission capacity of wireless ad hoc networks with outage constraints, *IEEE Transactions on Information Theory* 51(12): 4091–4102.

Xie, L. & Kumar, P. (2004). A network information theory for wireless communication: scaling laws and optimal operation, *Information Theory, IEEE Transactions on* 50(5): 748–767. URL: *10.1109/TIT.2004.826631*

Xie, L. & Kumar, P. (2006). On the path-loss attenuation regime for positive cost and linear scaling of transport capacity in wireless networks, *IEEE Transactions on Information Theory* 52(6): 2313–2328.

Xue, F. & Kumar, P. R. (2006). Scaling laws for ad hoc wireless networks: An information theoretic approach, *Foundations and Trends in Networking* 1(2): 127–248.

Yi, S., Pei, Y., Kalyanaraman, S. & Azimi-Sadjadi, B. (2007). How is the capacity of ad hoc networks improved with directional antennas?, *Wireless Networks* 13(5): 635–648.

Zorzi, M. & Pupolin, S. (1995). Optimum transmission ranges in multihop packet radio networks in the presence of fading, *Communications, IEEE Transactions on* 43(7): 2201–2205.
URL: *10.1109/26.392962*

21

Design and Analysis of a Multi-level Location Information Based Routing Scheme for Mobile Ad hoc Networks

Koushik Majumder[1], Sudhabindu Ray[2] and Subir Kumar Sarkar[2]
[1]*Department of Computer Science and Engineering, West Bengal University of Technology, Kolkata*
[2]*Department of Electronics and Telecommunication Engineering, Jadavpur University, Kolkata*
India

1. Introduction

Classical routing algorithms for MANET are basically route based, i.e. nodes maintain routes to the other nodes in the network. Many existing routing protocols (DSDV (Perkins & Bhagwat, 1994), WRP (Murthy & Garcia-Luna-Aceves, 1996), FSR (Haas. & Pearlman, 1998), ANDMAR (Gerla et al., 2000), DSR (Johnson & Maltz, 1996), AODV (Perkins & Royer, 1999), TORA (Park & Corson, 1997)) proposed within the MANET working group of IETF, are designed. These algorithms are basically of two types - proactive and reactive.
In case of proactive protocols like DSDV (Perkins & Bhagwat, 1994) , CGSR (Chiang et al., 1997), STAR (Garcia-Luna-Aceves & Spohn, 1999), OLSR (Clausen et al., 2001), HSR (Iwata et al., 1999), GSR (Chen & Gerla, 1998) the nodes in the adhoc network must keep track of all the routes to all other nodes, so that, whenever a node wants to send a data packet to another destination node, it can do that without wasting any time for path setup. This necessitates periodic exchange of routing information between the nodes of the network. The immediate disadvantage of these schemes is that too much network traffic will be consumed when the size of the network or the mobility of nodes increases.
In case of reactive routing protocols such as DSR, AODV, ABR (Toh, 1997), SSA (Dube et al., 1997), FORP (Su & Gerla, 1999), PLBR (Sisodia et al., 2002) a lazy approach is applied. Here the nodes need not maintain the routes to all other nodes. Thus, there is no need of periodic exchange of routing information between nodes. Routes to the destinations are determined on demand by flooding the whole network with route query packets. The immediate disadvantage of this approach is - flooding becomes prohibitive as the size of the network grows.
Some proposed algorithms claim to have the best of these two classes. Protocols like CEDAR (Sinha et al., 1999), ZRP (Haas, 1997), and ZHLS (Joa-Ng & Lu, 1999) combine both a proactive and a reactive approach.
A new family of routing algorithms, which are known as position-based routing algorithms such as GLS (Li et al., 2000), SLURP (Seung-Chul. et al., 2001), SLALoM (Cheng et al., 2002), DLM (Xue et al., 2001), were, introduced which use information about the physical position

of the participating nodes. They eliminate some of the limitations of topology based routing algorithms by using this extra information.
Commonly, each node determines its own position using GPS or some other type of positioning service. Position-based routing protocols have certain advantages over Topology-based routing protocols.

1. The nodes have neither to store routing tables nor to transmit messages to keep routing tables up to date.
2. Reduced overhead, as the establishment and maintenance of routes is (usually) not required in a protocol that uses location information for routing.

2. Location services

A Location service is responsible for providing location information of nodes in the network. Mobile nodes register their current location with this service. When a node does not know the position of the destination node, it contacts the location server and requests that information.
Existing location services (Amouris et al., 1999) can be classified according to how many nodes host the service. This can be either some specific nodes or all nodes on the network. Furthermore, each location server may maintain the position of some specific or all nodes in the network. These can be abbreviated as:

- some-for-some
- some-for-all
- all-for-some
- all-for-all

3. Review of previous work

Several location service schemes have been proposed in the literature: GLS, SLURP, SLALoM and DLM are some representative examples.

3.1 GLS (Li et al., 2000)

Grid Location Service (GLS) divides an area containing the ad hoc network into a hierarchical grid of squares. The largest square is called the level-H square. The level-H square is then recursively divided into four level-(H-1) squares until level-0 squares are reached, forming a so-called quad-tree. In each level-i square (for i > 0), node A selects three location servers, one in each level-(i-1) square that A is not in. The structure of GLS is shown in Fig. 1.

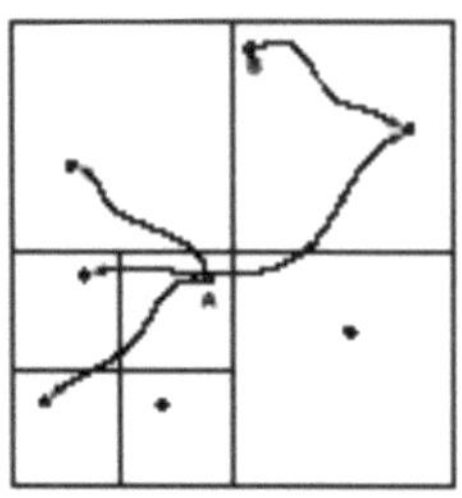

Fig. 1. Structure of GLS

GLS selects the location servers based on the node ID in each server's service area (i.e., a quadrant). For a node C to be node A's location server, C must have the smallest ID that is larger than A's ID in that quadrant, i.e., C = min {x | node x is in the quadrant, ID(x) > ID (A)} Each node updates its location servers with its *exact* location after it moves a threshold distance δ. To query for a particular node A, a node B sends the query to the node that is closest to A for which B has location information, and so on. Eventually the query would reach one of A's location servers.

3.2 SLURP (Seung-Chul. et al., 2001)

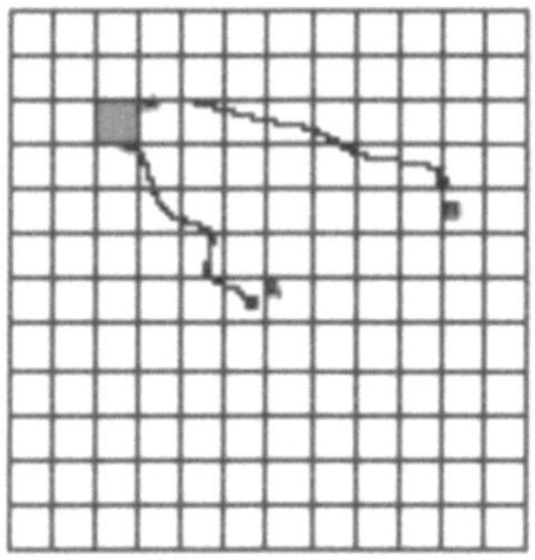

Fig. 2. Flat Grid of Squares used in SLURP

In SLURP, the entire network area (a square) is divided into a flat grid of squares. Node A selects its location servers by applying a hash function to A's ID and obtains the (x, y) coordinate of a point in the entire area. The square containing that point is called the home square for node A. All nodes in that square store A's exact location information. Every time node A moves to a different square, it updates its home square with new location information. For any node B, that wishes to communicate with node A, the same hash function is applied to node A's ID to obtain A's home square. A query packet is then forwarded to A's home square to retrieve A's location information. This is illustrated in Fig. 2.

3.3 SLALoM (Cheng et al., 2002)

SLALoM combines the strengths of SLURP and GLS. In this scheme, each node is assigned multiple home regions distributed uniformly over the area in which the nodes move about. (The nodes in these home regions act as location servers for the node.) It is assumed that the mobile nodes are capable of knowing their current location, using for example, the Global Positioning System (GPS), and are equipped with radios. It is also assumed that the nodes move about in a square region of area A. According to SLALoM, the square is divided into G unit regions called *order-1 squares*. It then combines K^2 of the order-1 squares to form *order-2* squares. A node's home region will consist of an order-1 square. With some exceptions, every node has a home region in each order-2 square. Hence, every node has O (A/K^2) home regions.

Maintaining location. Let v be a node in the network. Suppose it lies in the order-1 square R_i and R_i is inside order-2 square Q_j. We say that a home region of v is *near* v if the home region lies in Q_j or it lies in one of the eight order-2 squares that are neighbors of Q_j. Otherwise, a home region is *far* from v.

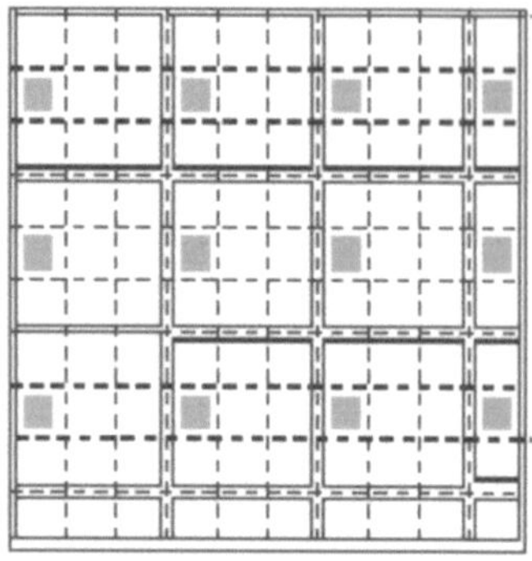

Fig. 3. Network Hierarchy of SLALoM

The following invariant always holds for *v*: *all home regions of v know v is in* Q_j. *In addition, all home regions near v know v is in* R_i.

Location Updation. Each time a node moves into a new order-1 square, it has to inform its 9 nearby home regions of its current exact location. This entails 9 broadcasts in a unit region. Furthermore, if such a move also causes the node to move into a new order-2 square, then it has to inform all its far home regions of its current approximate location. This requires O (A/K^2) broadcasts in a unit region.

Paging. If a node *u* wishes to find the location of another node *v*, it sends a unicast to a home region of *v* closest to it. If this home region is near *v* then *u* obtains the exact location of *v*. On the other hand, if the home region is far from *v* then *u* obtains an approximate location of *v*. Node *u* then routes its message to a home region near *v*, R_k. The node that receives the message at R_k then sends it to the exact location of *v*.

3.4 DLM (Xue et al., 2001)

DLM partitions the entire network much like GLS, i.e., there are H + 1 level of squares. The location servers are duplicated uniformly across the region, one server in every level-K square. Here K is a system parameter between 1 and H. The servers are chosen by hashing to a point in each level-K square; therefore, we say DLM also uses a *two-level server structure.*

DLM uses two addressing policies: complete and partial address. In complete address policy, all the location servers store the exact location of a node. In case of the partial address policy, each location server stores location information with different granularity. For i > K, if the location server of node A is located in the same level-i square in which A resides in, the servers store only which level-(i-1) square A is in. If the server is located in the same level-K square as A, the complete location information is stored.

The query operation is straightforward if the complete address policy is used. Node B simply queries the nearest location server of A to obtain A's location. If the partial address policy is used, node B simply queries the nearest location server of A. If the complete address of A is found, then the query is complete. Otherwise the server of A indicates which level-(i-1) square A is in, the query is then forwarded to A's location server in that level-(i-1) square. This process continues until A's complete information is found.

4. Proposed scheme

Position based routing protocols need not store the route information. Here the main component is the geographic location information of the nodes. In our proposed Layered

Square Location Management (LSLM) scheme, we have assumed that each node is equipped with GPS system through which the node can acquire its current geographic location. We also assume that each node has a transmission range of r_t.

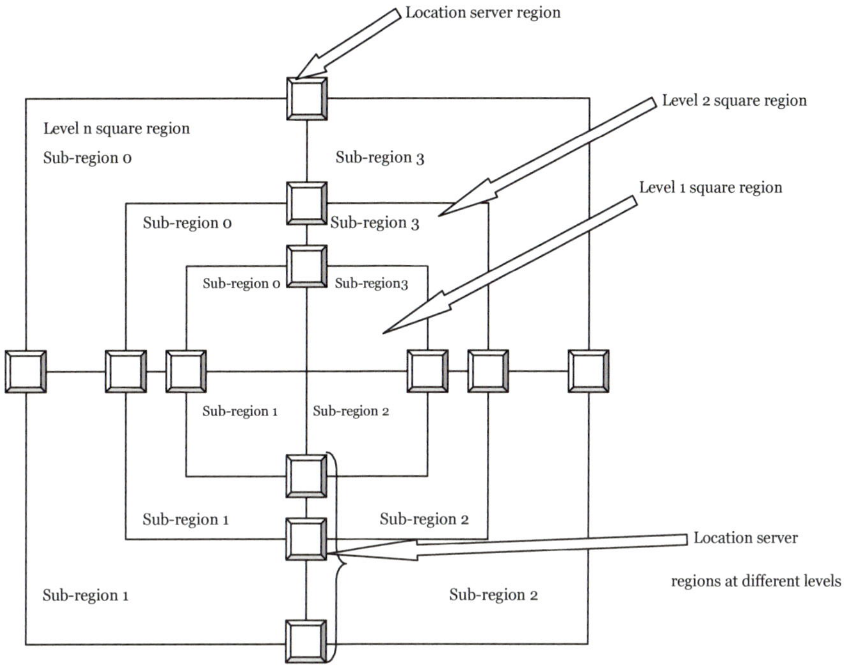

Fig. 4. Complete network structure of proposed LSLM

In our scheme, we have divided the entire network area into L level of square regions. The arrangement is such that each level i square region encapsulates the level (i-1) square region and is encapsulated by level (i+1) square region. Each square region has a side length of $2.2^l s$, where l denotes the level number and s depends on the node density. The innermost region is the level-1 square region and the outermost region is the level-L square region.

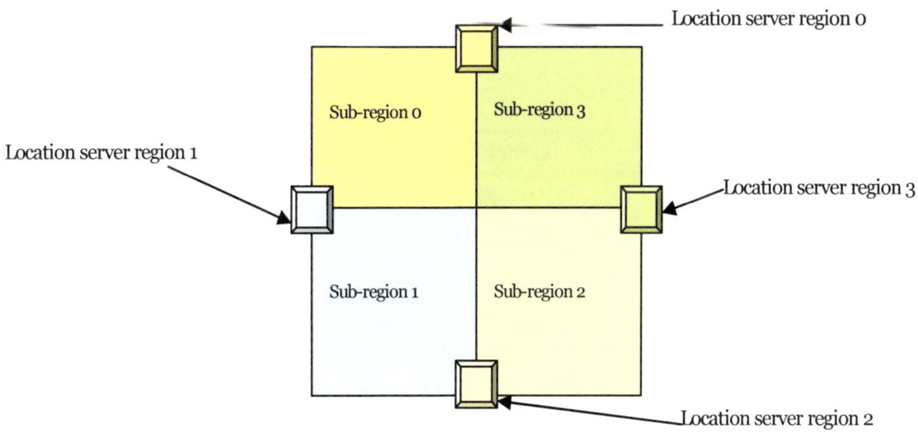

Fig. 5. Assignment of location server region

The square region at each level is further subdivided into four sub-regions: sub-region-0, sub-region-1, sub-region-2, and sub-region-3. In each level we have four location server regions, where each location server region is a square area having side length of r. All the nodes residing in the location server region act as location servers. These location servers are responsible for keeping track of the location information of the nodes. We have shown the arrangement of the location server regions within the square region at each level in Fig. 5. Each location server region has a fixed sub-region within the square region at each level assigned to it. The location server is responsible for keeping track of the location information of all the nodes within this sub-region. The Table:1 shows the assignment of the sub-regions within the square region at each level.

Location server region 0	Sub-region0
Location server region 1	Sub-region1
Location server region 2	Sub-region2
Location server region 3	Sub-region3

Table 1. Assignment of sub-regions to location server regions

In case of MANET, changes in network topology can be frequent and mobility of the nodes can be high. Therefore, cost for location update will be a major burden. If we keep track of only the exact location information of the nodes, then there is a possibility of this information becoming stale quickly as the mobile nodes frequently change their location. This will require frequent invocation of expensive location update routines. To address the issue, we have applied the concept of multi-level location information. We have assumed that the location information can be of two types - fully qualified location information and relative location information.

The fully qualified location information of a node includes the following components.

Node -id
x-coordinate
y- coordinate
Location server-id

Fig. 6. Fully qualified location information

Location server id has three components.

Level no.
Sub-region no.
Server-id

Fig. 7. Location server id

From Fig. 6 and Fig. 7, we can see that the fully qualified location information of a node A contains the current x and y coordinate position of A, the node id and the id of the location server that is currently keeping track of the location information of A. Location server id has three components embedded in it. The level no. of the square where the location server is currently in is indicated by the "level no". A location server is responsible for keeping track

of the fully qualified location information of the nodes within a sub-region and the "sub-region no." corresponds to this particular sub-region. "Server id" uniquely identifies the sever within a location server region.
On the other hand relative location information has two components.

Location server id
Node id

Fig. 8. Relative location information

The location servers within a location server region are responsible for keeping track of the fully qualified location information of only those nodes that are currently within its assigned sub-region within a particular level i. On the other hand, the location servers within a particular location server region also keep track of the relative location information of all the nodes that are currently within other sub-regions of the same level i square region. When a node moves within a particular sub-region, it needs to notify only the single location server region - that is currently in charge of that sub-region, regarding the change in its fully qualified location information. This reduces the location update cost.

4.1 Location update

We can divide the location update mechanism in three categories.

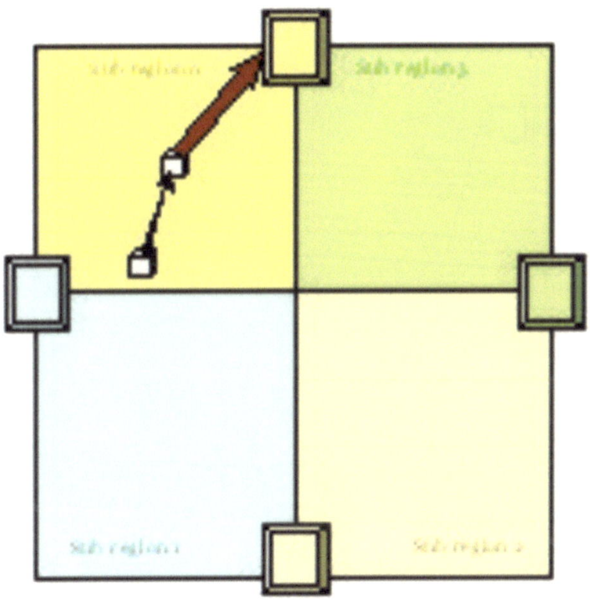

Fig. 9. Location update for node movement within sub-region

I. Location update for node movement within sub-region:
When a node A moves within its current sub-region, it needs to notify only those location servers that are currently in charge of this particular sub-region. This set of location servers are currently keeping track of the fully qualified location information of node A and any changes in the x and y coordinate positions of node A must be reflected to them. As A is

moving within its sub-region, there will be no change in its relative location information. Therefore, node A need not inform the location servers in other sub-regions of that particular level. Moreover, A does not need to notify the location server regions in other levels, as they contain neither the fully qualified location information nor the relative location information of node A.

II. Location update for node movement between sub-regions:

In this case, node A is moving from one sub-region to another within the same level i square region. After reaching the new sub-region, A probes its neighbors to get information about its new location server region. Once it gets this information, A sends its current x and y coordinate positions and the node id to the new location server region. Now node A is under the direct supervision of the new location server region. Therefore, this new location server region needs to update the location information regarding node A from relative to fully qualified one. The new location server region then needs to send the new relative location information of node A to other location server regions, which are within the same level i square region. These other location server regions now need to modify the location information about node A accordingly. If they contained the fully qualified location information, in that case, they need to update it with the new relative location information of node A. On the other hand, if they contained the old relative location information, in that case, they need to update the entry with the new relative location information of node A.

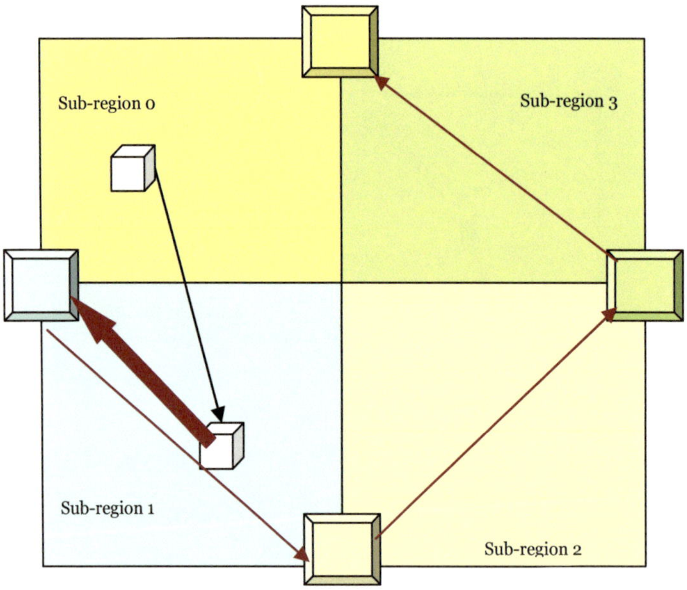

Fig. 10. Location update for node movement between sub-regions

III. Location update for node movement between square regions at different levels:

In this case, (illustrated in Fig. 11), node A moves from a square region at one level to a square region at another level. After reaching its new sub-region within its new square region, the node probes its neighbors to get information about its new location server region. Once A gets this information, it sends its previous fully qualified address and the current x and y coordinate positions to this new location server region. From node A's previous fully qualified address, the new location server region can know the previous level no. of the node. The previous level no. is required by the new location server region in sending the new relative address of A, (i.e., current location server id and node id) to a location server region in the previous level. This information is then relayed to all the other location server regions in the previous level. Those location server regions after analyzing the current relative address of the node, find that the level no. of node A has already changed, i.e., node A is no longer in the square region at their level. Therefore, they delete the entry corresponding to node A from their database.

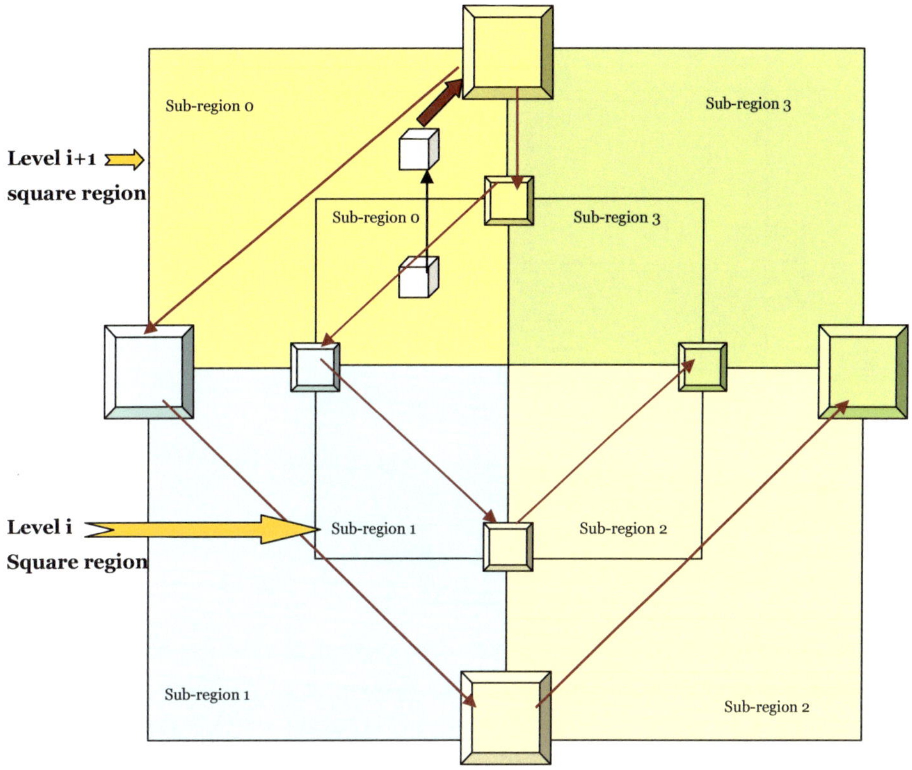

Fig. 11. Location update for node movement between square regions at different levels

The new location server region is in a square region, which is at a different level than the level of node A's previous square region. Therefore, the new location server region must make a new entry in its location information database about the new fully qualified location information of node A. This new location server region then needs to send the new relative location information of node A to other location server regions within the new square region. These other location servers previously had no location information about node A. Therefore, they need to make new entries in their location information database about the new relative location information of node A.

4.2 Location query

Suppose node S wants to send a data packet to a destination node D but the location information of node D is unknown to S. Corresponding to three location update scenarios three situations can evolve.

I. Destination D is within same sub-region at same level as of source S:
In this case the location server region that is in-charge of the sub-region contains the fully qualified address of node D. The source node S sends the data packet to the location server region. The location server region extracts the current x and y coordinate position of node D from its fully qualified address and sends the data packet to node D at that location.

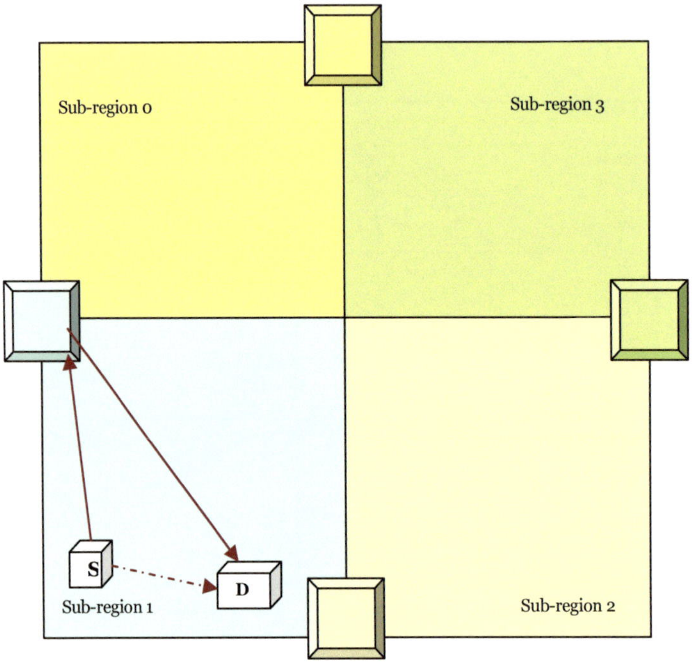

Fig. 12. Destination D is within same sub-region at same level as of source S

II. Destination D is within other sub-region at the same level as of source S:
In this case the source S sends the data packet to the assigned location server region of its sub-region. But as the destination D is within a different sub-region, therefore, the location server region of node S contains only the relative location information about destination D. From this information, the location server region of node S can find the location server region, which is currently containing the fully qualified address of node D. The location server region of node S then sends the data packet forwarded by S, to that particular location server region. This new location server region ultimately sends the data packet to the destination node D.

III. Destination D is within other square region at different level than that of source S:
The location server region now sends the data packet to the location server region of the square region that is encompassing the current level square region. It also forwards the packet to the location server region of the square region that is contained by the current level square region. The location server regions at other levels now follow the previously mentioned steps for location query. This process is continued until the destination node D is found or the network boundary is reached. Thus, if the destination node falls within the network boundary, the data packet is propagated from the source node S to the destination node D through the intermediate location server regions.

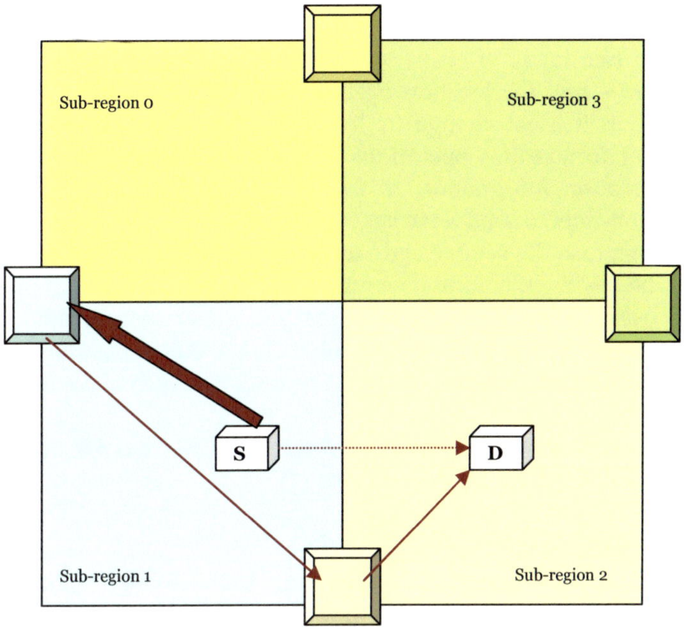

Fig. 13. Destination D is within other sub-region at the same level as of source S

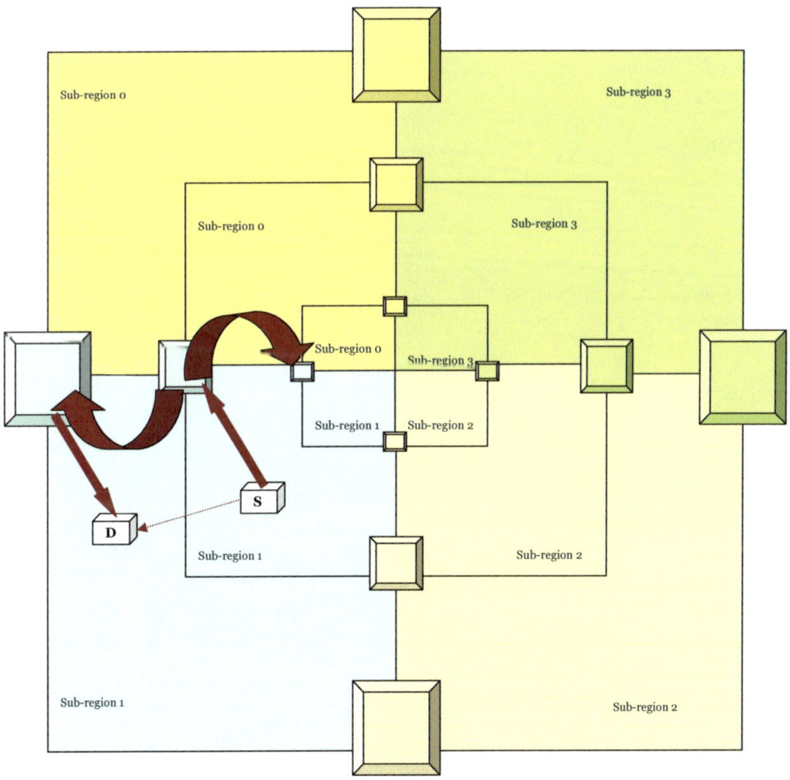

Fig. 14. Destination D is within other square region at different level than that of source S

5. Analysis of Layered Square Location Management (LSLM)

There are mainly two types of costs, which are important for any location management scheme. These are - cost for location update and cost for location query. When a node changes its position it must change its location information at the location server. The number of packet forwarding operations it needs to perform per second, in order to maintain fresh location information, is known as the location updation cost $Cost_{update}$. Similarly if a node wants to send a packet to a destination node whose location information is unknown, in that case the sender node must perform location query, to find the location information of the destination node. The number of packet forwarding operations that each node needs to perform for the purpose of location query defines the location query cost $Cost_{query}$. There is also a third type of cost, which is known as the storage cost. The storage cost $Cost_{storage}$ signifies the number of location records that each of the location servers needs to store.

In the following sections we analyze these three types of costs for our proposed Layered Square Location Management (LSLM) scheme.

5.1 Location updation cost [$Cost_{update}$]:

In our proposed scheme, location update has been divided into three parts. As a consequence, the cost for location update can also be divided into three parts - i>Cost for location update for node movement within sub-region ($Cost_{update-intra-subregion}$) ii>Cost for location update for node movement between sub-regions ($Cost_{update-inter-subregion}$) iii>Cost for location update for node movement between square regions at different levels ($Cost_{update-inter-level}$).

Thus we can write,

$$Cost_{update} = Cost_{update-intra-subregion} + Cost_{update-inter-subregion} + Cost_{update-inter-level}$$

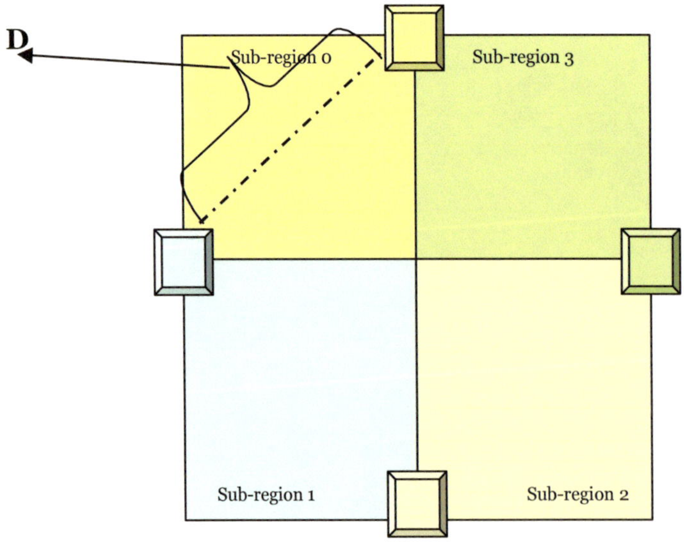

Fig. 15. Distance D

The cost for location update depends upon the amount of forwarding load, where forwarding load is determined by the number of hops traversed by a packet during location update operation. Thus the forwarding load, and as a consequence the cost will be greater for a packet traveling a greater distance. Cost for location update for node movement within sub-region ($Cost_{update\text{-}intra\text{-}subregion}$) is basically the product of updation frequency and the cost of updation of one location server region. The cost of updation of one location server region is proportional to the average number of hops an update packet takes to reach the assigned location server region. We denote this cost by Cost (1). We can approximate this cost by considering the distance $D=\sqrt{2.2^l.s}$; where l denotes level number (Fig. 15).
Let us denote z as the average progress for each forwarding hop, where z is a function of the radio transmission range r_t and the node density (γ) (Seung-Chul.et al., 2001). We assume both r_t and γ are constants. Therefore, z is also a constant. It is possible to derive the average number of hops an update packet takes by D/z. If we consider the average velocity of a node as v, and the transmission range of a node as r_t, then the updation frequency is v/r_t. Thus,

$$Cost_{update\text{-}intra\text{-}subregion} = v/\ r_t\,.\ Cost\ (1)$$

$$\text{And}\quad Cost\ (1) \propto \sum_{l=0}^{L} \sqrt{2.2^l.s}/z$$

$$\approx \sqrt{2}.s.L/z.$$

If we assume S as the side length of the square region at the maximum level, i.e. L^{th} level square region, then, $S \propto 2^L$. Thus, $L \propto \log S$. Since, $S \propto \sqrt{N}$, (N=Total Number of nodes in the network), we have $L \propto \log\sqrt{N}$. Thus,

$$Cost_{update\text{-}intra\text{-}subregion} = O\ (v.\log\sqrt{N}). \qquad (1)$$

Cost for location update for node movement between sub-regions ($Cost_{update\text{-}inter\text{-}subregion}$) is the product of the boundary crossing rate (Ω) and the cost for updating the four location server regions (Cost(4)). So,

$$Cost_{update\text{-}inter\text{-}subregion} = \Omega.\ Cost\ (4).$$

The boundary-crossing rate is proved (Yu et al., 2004) to be proportional to v. The cost of updating four location server regions can be approximated by $4(D_l)/z$. Thus

$$Cost\ (4) \propto \sum_{l=0}^{L} 4.\ \sqrt{2.2^l.s}/z$$

$$\approx 4\sqrt{2}.s.L/z.$$

Therefore,

$$Cost_{update\text{-}inter\text{-}subregion} = O\ (v.\log\sqrt{N}). \qquad (2)$$

Similarly we can formulate $Cost_{update\text{-}inter\text{-}level}$ as

$$Cost_{update\text{-}inter\text{-}level} = \Omega.Cost\ (8).$$

We can approximate the cost of updating eight location server regions by $4(D_l + D_{l-1})/z$. Thus

$$\text{Cost (8)} \propto \sum_{l=0}^{L} 6.\ \sqrt{2}.2^l.s/z$$
$$\approx 6\sqrt{2}.s.L/z.$$

Therefore,

$$\text{Cost}_{\text{update-inter-level}} = O\ (v.\log\sqrt{N}). \qquad (3)$$

Thus from "(1)", "(2)" and "(3)" we have

$$\text{Cost}_{\text{update}} = \text{Cost}_{\text{update-intra-subregion}} + \text{Cost}_{\text{update-inter-subregion}} + \text{Cost}_{\text{update-inter-level}} = O\ (v.\log\sqrt{N}).$$

5.2 Location query cost [$\text{Cost}_{\text{query}}$]:

If a source node has some data to send to a destination node, the source node must first query a location server region to get the current location information of the destination node. The cost for this activity of querying the location information is known as location query cost ($\text{Cost}_{\text{query}}$). In order to calculate $\text{Cost}_{\text{query}}$, we have to measure the expected number of forwarding hops traveled by a query packet from the source node to its assigned location server region, which can be approximated by D/z. Therefore, the expected query cost is,

$$\text{Cost}_{\text{query}} = \sum_{l=0}^{L} \sqrt{2}.2^l.s/z$$
$$\propto H$$
$$= O\ (\log\sqrt{N}).$$

5.3 Storage cost [$\text{Cost}_{\text{storage}}$]:

In order to calculate the expected storage cost we need to find the average number of records stored by a location server node in the network. Dividing the total number of records stored in the network by the total number of nodes acting as location servers gives us the average number of records. Each node in the network stores its address at the four location server regions of its current layer of existence. Earlier we have mentioned that each location server region is a square area having side length of r. Hence, the area covered by a location server region can be expressed by r^2. The average number of nodes (γ) is assumed to be constant. Thus the average number of nodes serving as location servers within a location server region is $r^2.\ \gamma$. Now, the expected storage cost can be expressed as

$$\text{Cost}_{\text{storage}} = (N.4.\ r^2.\ \gamma)/(L.\ 4.\ r^2.\ \gamma) = N/L,$$

where, N= Total number of nodes in the network; L= Maximum level number. Since $L \propto \log\sqrt{N}$; the expected storage cost, $\text{Cost}_{\text{storage}} = O\ (N)$.

6. Conclusion

In this paper, we have presented Layered Square Location Management (LSLM), a novel scheme for the management of location information of the nodes in mobile ad hoc network. The effectiveness of a location management scheme depends on reducing the costs associated with the major location management functions- location update and location query. In case of a location service scheme we can reduce the location query cost by employing various caching strategies which is not possible for location update cost. Keeping track of only the exact location information, makes location update highly expensive due to the high mobility of nodes. In our scheme by dividing the entire network area into L levels of square regions and using multi-level location information, we have been able to provide a unique way to reduce the cost associated with both location update and location query. Further investigation on performance analysis of this scheme in different network scenarios can be taken as extended work.

7. References

Amouris, K.N.; Papavassiliou, S. & Li, M. (1999). A position-based multi-zone routing protocol for wide area mobile ad-hoc networks. In *Proceedings of the IEEE Vehicular Technology Conference (VTC)*, pages 1365-1369

Chen, T. W. & Gerla, M. (June 1998). Global State Routing: A New Routing Scheme for Ad Hoc Wireless Networks, *Proceedings of IEEE ICC 1998*, pp. 171-175

Cheng, Christine. T.; Lemberg, H. L.; Philip, Sumesh. J.; Berg, E. van. den. & Zhang, T. (March 2002). SLALoM: A scalable location management scheme for large mobile ad-hoc networks. In *Proceedings of IEEE WCNC*

Chiang, C. C.; Wu, H. K.; Liu, W. & Gerla, M. (April 1997). Routing in Clustered Multi-Hop Mobile Wireless Networks with Fading Channel, *Proceedings of IEEE SICON 1997*, pp. 197-211

Clausen, T. H.; Hansen, G.; Christensen, L. & Behrmann, G. (September 2001). The Optimized Link State Routing Protocol, Evaluation Through Experiments and Simulation, *Proceedings of IEEE Symposium on Wireless Personal Mobile Communications 2001*

Dube, R.; Rais, C. D.; Wang, K. Y. & Tripathi, S. K. (February 1997). Signal Stability-Based Adaptive Routing for Ad Hoc Mobile Networks, *IEEE Personal Communications Magazine*, pp. 36-45

Garcia-Luna-Aceves, J. J. & Spohn, M. (October 1999). Source-Tree Routing in Wireless Networks, *Proceedings of IEEE ICNP 1999*, pp. 273-282

Gerla, M.; Pei, G. & Hong, X. (August 2000). Lanmar: Landmark routing for large scale wireless ad hoc networks with group mobility. In *Proceedings of the First IEEE/ACM Workshop on Mobile Ad Hoc Networking and Computing (MobiHOC)*

Haas, Z. J. (October 1997). The Routing Algorithm for the Reconfigurable Wireless Networks, *Proceedings of ICUPC 1997*, vol. 2, pp. 562-566

Haas, Z. J. & Pearlman, M. R. (August 1998). "The zone routing protocol (ZRP) for ad hoc networks (Internet Draft)"

Iwata, A.; Chiang, C. C.; Pei, G.; Gerla, M. & Chen, T. W. (August 1999). Scalable Routing Strategies for Ad Hoc Wireless Networks, *IEEE Journal on Selected Areas in Communications*, vol. 17, no. 8, pp. 1369-1379

Joa-Ng, M. & Lu, I. T. (August 1999). A Peer-to-Peer Zone-Based Two-Level Link State Routing for Mobile Ad Hoc Networks, *IEEE Journal on Selected Areas in Communications,* vol. 17, no. 8, pp. 1415- 1425

Johnson, David. B. & Maltz, David. A. (1996). (Dynamic source routing in ad hoc wireless networks. In Imielinski and Korth, editors, *Mobile Computing,* volume 353. Kluwer Academic Publishers

Li, J.; Jannotti, J.; Couto, D. De.; Karger, D. & Morris, R. (August 2000). A scalable location service for geographic ad-hoc routing. In *Proceedings of ACM MobiCom,* pages 120.130

Murthy, S. & Garcia-Luna-Aceves, J.J. (October 1996). An efficient Routing Protocol for Wireless Networks. *ACM Mobile Networks and App. Journal, Special Issue on Routing in Mobile Communication Networks*

Park, V.D. & Corson, M.S. (April 1997). A Highly Adaptive Distributed Routing Algorithm for Mobile Wireless Networks. *Proceedings of IEEE INFOCOM'97, Kobe, Japan*

Perkins, Charles. E. & Bhagwat, Pravin. (August 1994). Highly dynamic Destination-Sequenced Distance-Vector routing (DSDV) for mobile computers. In *Proceedings of the SIGCOMM '94 Conference on Communications Architectures, Protocols and Applications*, pages 234–244

Perkins, Charles. & Royer, Elizabeth. (1999). Ad-hoc on-demand distance vector routing. In Proceedings *of IEEE Workshop on Mobile Computing Systems and Applications*

Seung-Chul.; Woo, M. & Singh, Suresh. (2001). Scalable routing protocol for ad hoc networks. *Wireless Networks*, 7(5):513-529

Sinha, P.; Sivakumar, R. & Bharghavan, V. (August 1999). CEDAR: A Core Extraction Distributed Ad Hoc Routing Algorithm, *IEEE Journal on Selected Areas in Communications,* vol. 17, no. 8, pp. 1454-1466

Sisodia, R. S.; Manoj, B. S. & Murthy, C. Siva. Ram. (March 2002). A Preferred Link-Based Routing Protocol for Ad Hoc Wireless Networks, *Journal of Communications and Networks,* vol. 4, no. 1, pp. 14-21

Su, W. & Gerla, M. (December 1999). IPv6 Flow Handoff in Ad Hoc Wireless Networks Using Mobility Prediction, *Proceedings of IEEE GLOBECOM 1999,* pp. 271-275

Toh, C. K. (March 1997). Associativity-Based Routing for Ad Hoc Mobile Networks, *Wireless Personal Communications,* vol. 4, no. 2, pp. 1-36

Xue, Y.; Li, B. & Nahrstedt, K. (2001). A scalable location management scheme in mobile ad-hoc networks. In *Proceedings of the IEEE Conference on Local Computer Networks (LCN '01)*

Yu, Yinzhe.; Lu, Guor-Huar. & Zhang, Zhi-Li. (2004). "Enhancing location service scalability with HIGH-GRADE," Dept. of Comp. Sci. & Eng., University of Minnesota, Technical Report TR-04-002

22

Power Control in Ad Hoc Networks

Muhammad Mazhar Abbas and Hasan Mahmood
Quaid-i-Azam University
Pakistan

1. Introduction

In this chapter, we present the power control techniques used in ad hoc networks. Traditionally, the power control has been implemented and used effectively in cellular networks. While the use of transmission power control in infrastructure based networks has proven to work well and improve performance, the application of power control techniques to ad hoc networks has many challenges and implementation complexities (Chauh & Zhang, 2006) (Basagni et al., 2004). The power control is of great significance in ad hoc networks because of their organizational structure and lack of central management. With the implementation of effective power control techniques, the ad hoc network can improve their vital parameters, such as power consumption, interference distribution, throughput, routing, connectivity, clustering, backbone management, and organization (Basagni et al., 2004).
We discuss several power control algorithms commonly used in ad hoc networks to get insight of power control techniques and their effectiveness. Most of the algorithms are adapted from cellular networks, modified accordingly, and proposed for ad hoc networks. Moreover, we argue the enhancement in performance of ad hoc networks with the use of these power control algorithms.
The power control requirements vary depending on the physical and network layer implementation of ad hoc networks (Stüber, 2002). We show the application of the prevailing power control algorithms to different physical layer models and discuss their performance. The application to CDMA based networks is emphasized as these types of networks have strict power control requirements and the performance is severely degraded without appropriate power control. In cellular networks, the power control requirements are stringent, especially in multiple access technologies. The appropriate allocation of power to the transmitters facilitates interference control and saves energy.
The near-far effect starts to dominate as the transmission power levels are not properly managed. The advantage of cellular networks over ad hoc networks is the presence of central management, and as a consequence, the uplink power control can be achieved. This is in contrast to ad hoc networks, which lack central management and most of the nodes are in peer to peer configuration (Blogh & Hanzo, 2002).
In addition, transmit power control is a cross layer design problem affecting all layers of the OSI model from physical layer to transport layer (Jia et al., 2005). In general, power conservative protocols are divided into two main categories: transmitter power control protocols and power management algorithms. Second class can be further divided into MAC layer protocols and network layer protocols (Ilyas, 2003).

At the end of the chapter, we discuss the concept of joint power control and routing in ad hoc networks. Power can be controlled in ad hoc networks by choosing optimal routes. The existing routing protocols may be classified as, uniform, non-uniform, proactive, reactive, hybrid, source, and non-source routing protocols (?Chaudhuri & Johnson, 2002). To further explain joint power control and routing techniques, we discuss a Minimum Average Transmission Power Routing (MATPR) technique (Cai et al., 2002), which implements a power control routing protocol using the concept of blind multi-user detection to achieve the task of minimum power consumption. The Power Aware Routing Optimization (PARO) technique (Gomez et al., 2003), a protocol for the minimization of transmission power in ad hoc networks, is based on the concept of node to node power conservation using intermediate nodes, usually called redirectors. PARO is efficient in both static and dynamic environments and is based on three main operations: overhearing, redirecting, and route maintenance.

2. Cellular networks

The wireless cellular networks require a fixed and well defined infrastructure. This type of network infrastructure is suitable to efficiently manage the network operations. Generally the network can be managed and operated by a central operations point. In the field, the the physical parameters, such as transmission frequency, resource allocation, and power control parameters are monitored and controlled by base station which have fixed location. We focus on power control for these types of configurations in order to study and analyze implementation to ad hoc networks.

Power control is a necessary feature in cellular communication networks with multiple access technologies. Power control has many management features such as interference control, energy saving, and connectivity (Almgren et al., 2009). In power control mechanism each user transmits and receives at an appropriate energy level, i.e., the transmission powers are controlled in such a way that the interference is minimized, while achieving sufficient quality of service (Lee, 1991).

In the absence of power control, the near-far effect is introduced as all the mobile users transmit at same power level or at a level which is not suitable at receivers in the network. In other words, the transmitters close to the base station create interference to neighboring users which are in the vicinity. In the absence of power control, the system capacity degrades as compared to other wireless systems (Hanly & Tse, 1999). The power control also increases the battery life by using a minimum required transmission power and is equally important in both uplink and downlink transmissions. In uplink transmission, the near-far effect problem is created as the signals of mobile propagate through different channels before reaching their corresponding base station (Moradi et al., 2006). The purpose of power control is to allow all mobile signals to be received with same power at the base station. Uplink power control enhances capacity of networks (Gilhousen et al., 1991). On the other hand, in downlink transmission, the near-far effect problem is not as important, because signals from the base station reach the mobile station while propagating through same channel (Lee et al., 1995). Uplink power control algorithms achieve their functions through open loop and closed loop power control, which can be further divided into closed outer loop power control and closed inner loop power control. In open loop power control, the mobile user adjusts its transmission power based on the received signaling power from the base station (Chockalingam & Milstein, 1998). In closed-loop power control, based on the measurement of the link quality, the base

station sends a power control command instructing the mobile to increase or decrease its transmission power level and sets the target signal-to-interference ratio (SIR) to such a level that sufficient quality of service is guaranteed (Rintamäki, 2002).
Power can be controlled in a centralized or distributed fashion. In centralized form a controller manages the information of all the established connections and channel gains, and controls the transmission power level (Grandh et al., 1993). While in the distributed form a controller controls only one transmitter of a single connection. It controls transmission power based on local information such as the signal to interference ratio and channel gains of the specific connection. Distributed form of power control is easy to use in common practice because it does not require extensive computational work (Zender, 1993).
Although we aim to discuss power control techniques for wireless ad hoc networks, it is important to get insight for the similar techniques used in cellular networks. These techniques were initially applied to cellular networks, and with the advent of ad hoc network were adapted and modified to meet new requirements. Some of the basic power control algorithms are presented below which are related to wireless cellular networks and their implementations.

2.1 Power control as eigen value problem

In the era of 1980s the concept of Signal to Interference Ratio (SIR) balancing in power control algorithms for cellular networks based on Code Division Multiple Access (CDMA) and other technologies were used by researchers (Nettleton, 1980) (Nettleton & Alavi, 1983) (Alavi & Nettleton, 1982). Initially, the power control problem was focused and treated as an eigen value problem with a non negative matrix G and corresponding balance power vectors p_u and p_d which satisfy the eigen value problem as

$$Gp_u = [(1+\gamma_u)/\gamma_u]p_u \tag{1}$$

and

$$G^T p_d = [(1+\gamma_d)/\gamma_d]p_d \tag{2}$$

where γ_u and γ_d are desired uplink and downlink $SIRs$. By taking $\lambda(G)$ as eigen value of G a solution to the above problem is given as

$$[(1+\gamma_u)/\gamma_u] = [(1+\gamma_d)/\gamma_d] \in \lambda(G) \tag{3}$$

Another solution to SIR balancing problem is given as

$$\gamma_u = \gamma_d = 1/(\rho - 1) \tag{4}$$

where spectral radius ρ is such that $\rho > 1$.
Iterative methods are very effective in solving these type of problems. One approach (Foschini & Miljanic, 1993) to solve the above eigen value problem iteratively is by solving liner algebraic equations, represented as $AP = b$, where $P = [p_1, p_2, ...p_N]^T$, and

$$P(k+1) = (1-A)P(k) + b \tag{5}$$

This algorithm converges and the method use derivative named as surrogate derivative and concludes that their algorithm is converging synchronously.

A generalized frame work for convergence is given in (Yates, 1995). By using proper power control, the interference is eliminated and we get iteration as

$$p_i(k+1) = \gamma_i^{tar}(k)p_i(k)/\gamma_i(k) \tag{6}$$

where p_i is the power of ith user and γ_i^{tar} is the target SIR

2.2 Distributed power control techniques

The Distributed Power Control (DPC) algorithm is applied at individual nodes in the network and the objective is to converge system power allocations to a suitable level (Grandhi et al., 1994). This can be accomplished by using feedback power control (Ariyavisitakul, 1994). In this method the power is adjusted in steps which may have fixed or variable size. It is seen that the performance of a power control algorithm with fixed step size and variable step size is almost the same. In addition, the higher power control rate can accommodate the effect of fast fading.

With the implementation of distributed power control, the SIR of the system can be controlled and managed to some extent. As a result, the outage probability of an individual link or a set of links can be reduced or entirely eliminated. The implementation of this type of method requires a distributed power control algorithm which reduces the outage probability to zero by keeping SIR above threshold value (Zander, 1992).

In another approach, a smaller balancing systems can be constructed by turning the transmitter of cells off so the outage probability is minimized. In some scenarios, if the value of SIR for a mobile is less than threshold value then outage probability is reduced and mobile is dropped from network (Wu, 1999). This improves the remaining network SIR.

An optimal SIR based distributed power control technique can be used by unconstrained and constrained optimization (Qian & Gajic, 2003). The theme of this algorithm is to establish a proportionality between transmission power and the error between the actual SIR and the desired SIR. Difference of transmission power from time step k to $k+1$ is given as

$$\Delta P_i(k+1) = P_i(k+1) - P_i(k) \tag{7}$$

The error between desired SIR and actual SIR is given as

$$e_i(k) = \gamma_i^{des} - \gamma_i \tag{8}$$

Then the proposed algorithm is described as

$$\Delta P_i(k+1) = \alpha_i(k)e_i(k) \tag{9}$$

where $\alpha_i(k)$ is the gain. Thus power allocation is given as

$$P_i(k+1) = P_i(k) + \alpha_i(k)(\gamma_i^{des} - \gamma_i) \tag{10}$$

2.3 Discrete time dynamic optimal power control

In this method, the reverse link system information is used for power control. A cost function, consisting of weighted sum of powers and some additional parameters is defined. An optimal power control law is presented based on a cost function comprising of weighted sum of power, power update information, and SIR error. It is also assumed that there is no significant change in SIR from one step to the next. For this purpose, a technique named as discrete time dynamic optical control is implemented (Koskie & Gajic, 2003). The general cost function and sufficient conditions for optimality are defined as

$$J[N] = g(x[N]) + \sum_{K=0}^{N-1} L(x[k], u[k], k) \tag{11}$$

and

$$\begin{pmatrix} H_{xx} & H_{xu} \\ H_{xu}^T & H_{uu} \end{pmatrix} > 0, H_{uu} > 0 \tag{12}$$

where J is the controller, L is the cost function and H is the hamiltonian. Some of the different optimal controllers for three cost functions are

$$J_I = (1/2) \sum_{k=0}^{K} (qe^2[k] + su^2[k]) \text{ for cheap power cost} \tag{13}$$

$$J_{II} = (1/2) \sum_{k=0}^{K} (qe^2[k] + 2rp[k] + su^2[k]) \text{ for linear power cost} \tag{14}$$

$$J_{III} = (1/2) \sum_{k=0}^{K} (qe^2[k] + rp^2[k] + su^2[k]) \text{ for quadratic power cost} \tag{15}$$

This method considerably saves power and improve quality of service.

2.4 Linear and bilinear power control techniques

The optimization of power conservation results in improved SIR distribution for the entire network. Although these optimizations are based on some estimates, as a consequence, errors are introduced in the actual results (Gajic et al., 2004).

The power control techniques named as linear and additive power updates algorithm and bilinear control algorithm are based on optimization of SIR error. It can be seen that mobile power is updated by using a distributive linear control law, given as

$$P_i(k+1) = P_i(k) + U_i(k) \tag{16}$$

where $i = 1, 2, \ldots\ldots n$. By minimizing SIR error and after other calculations the optimized power updates can be obtained as

$$P_i^*(k+1) = P_i^*(k) + U_i^*(k) \tag{17}$$

$$P_i^*(k) = \gamma_i^{tar} I_i[P\_i^*(k)] / g_{ii} \tag{18}$$

Where P_i^* is the optimized power. In the second algorithm, bilinear control law is used for update of power as

$$P_i(k+1) = P_i(k) U_i(k) \tag{19}$$

where $i = 1, 2, \ldots\ldots n$, and corresponding optimized power is same as in above case.

2.5 Power control technique based on relaxation method

This method is particularly useful in networks with multiple access technology, such as CDMA. A relaxation method can be used in solving iterative power control techniques. Two common techniques for iterative solution of power control problems can be used effectively with relaxation method. Application to Jacobi iteration method and Gauss Siddel iteration method for solution of power control problem, by introducing a relaxation parameter in these techniques, is presented as a modified Jacobi iteration

$$P_i(k+1) = [1 - \beta + \beta \frac{\gamma_i^{tar}}{\gamma_i(k)}] P_i(k) \tag{20}$$

and modified Gauss Siddle iteration

$$P_i(k+1) = [1 - \beta + \beta \frac{\gamma_i^{tar}}{\gamma_i(k,k+1)}] P_i(k) \tag{21}$$

The Gauss Siddle iteration with relaxation parameter β is more efficient than Jacobi iteration technique for solution of power control problem. The algorithms implemented by relaxation method converge faster than simple distributed power control algorithm (Siddiqua et al., 2007).

2.6 Distance based power control technique

The distance between transmitters and receivers can be estimated in a wireless networks. The attenuation of the signals is proportional to the distance which they travel. Therefore, if the information about the distances is know in real time or a prior, the power can be adjusted efficiently (Nuaymi et al., 2001). If a base station is present, the transmit power of each mobile station can be controlled by using distance information between base station and mobile stations. This algorithm computes the transmitted power P_m of a mobile node m as

$$P_m = k x_{a_m m}^n \tag{22}$$

where

$$x_{a_m m} = \begin{pmatrix} \frac{d_{a_m m}}{rR}, & \text{if} \quad \mathrm{d_{a_m m} > d_{min}} \\ \frac{d_{min}}{R}, & \text{if} \quad \mathrm{d_{a_m m} \leq d_{min}} \end{pmatrix} \tag{23}$$

where R is the base to mobile maximum distance and $d_{a_m m}$ is the distance between mobile and assigned base station.

2.7 Kalman filter based power control technique

In an uplink closed loop power control algorithm based on Kalman filter technique, the controller or a base station estimates SIR in a closed loop system (Rohi et al., 2007). The SIR can be estimated by any suitable method. The outage probability calculated by this method is smaller as compared to others. According to algorithm details, the base station estimates the SIR for a user and provide as input to Kalman predictor. Its output is compared with the desired SIR and the difference is quantized by a PCM. The transmitted power of user is then updated and SIR estimation is given as

$$SIR_{n+1}^* = SIR_n^* + \alpha \Delta SIR_n^* + \omega_n \tag{24}$$

and

$$\Delta SIR_n^* = SIR_n^* - SIR_{n-1}^* \tag{25}$$

The outage probability is given as $P_0 = P_r(SIR_r < SIR_0)$. Where SIR_r is measured at base station and SIR_0 is the minimum value of SIR for achieving desired BER.

2.8 Power control technique based on linear quadratic control theory

The state-space formulation and linear quadratic control technique can be used to solve the problem of power control by considering each mobile to base station link as an independent subsystem described as

$$S_i(n+1) = S_i(n) + V_i(n) \tag{26}$$

where

$$S_i(n) = P_i(n)/I_i(n) \tag{27}$$

and

$$V_i(n) = U_i(n)/I_i(n) \tag{28}$$

and

$$I_i(n) = \sum_{j\neq i}^{Q} P_i W_{ij} + n_i/G_{ki} \tag{29}$$

The input to each subsystem $U_i(n)$ depends on the total interference produced by other users plus the noise in the system and each $S_i(n)$ track is made equal to the threshold value of SIR (Osery & Abdallah, 2000). For the discrete case the new state is given by

$$\varsigma_i(n+1) = \varsigma_i(n) + e_i(n) \tag{30}$$

where error $e_i(n) = S_i(n) - \gamma^*$. Each subsystem can now be expressed as a second-order linear state-space system as

$$X_i(n+1) = \begin{pmatrix} e_i(n+1) \\ s_i(n+1) \end{pmatrix} = \begin{pmatrix} 1 & 1 \\ 0 & 1 \end{pmatrix} X_i(n) + \begin{pmatrix} 0 \\ 1 \end{pmatrix} V_i(n) \tag{31}$$

The feedback controller $V_i(n) = -[k_\varsigma k_s]x_i(n) + k_s\gamma^*$, where $[k_\varsigma k_s]$ is the gain matrix which are found by solving the Riccati equation. If the right feedback gains $[k_\varsigma k_s]$ is chosen, the steady-state $S_i(n)$ will go to the threshold SIR. To find the optimum feedback control for the state-space representation given above, the Linear Quadratic Control theory is used. After the gain matrix $[k_\varsigma k_s]$ is found, the power control can be expressed as

$$P_i(n+1) = min[P_i, S_i(n+1)I_i(n)] \tag{32}$$

The method assures that the maximum transmission power of the mobile i will not be exceeded. This method reaches a zero outage probability with less iterations than other distributed power control methods. This approach was also found to be more effective in handling a large number of mobile stations in the system.

2.9 Power control technique based on utility and pricing

The power control algorithm can be implemented in a distributed fashion based on utility and pricing concepts (Shah et al., 1998). The efficiency of this protocol can be improved in low BER and high SIR conditions. The formula of SIR of user j at base station k is given as

$$\gamma = \frac{W}{R} \frac{h_{jk}p_j}{\sum_{\forall i \neq j} h_{ik}p_i + \sigma_k^2} \tag{33}$$

In this method, by introducing a pricing factor the utility is maximized and as a result helps in power control problem. A general utility function which is a monotonically increasing function of SIR is given as

$$u_j = \frac{E}{p_j} Rf(\gamma) \tag{34}$$

Where $f(\gamma)$ is a measure of efficiency of protocol. The power control problem is considered as a cooperative power control game. The user maximizes its utility at equilibrium point with maximum SIR value as $Max \quad u_i(p_1, p_2, ...p_N), \forall i = 1, 2, ...N$, and $f(\gamma^*) = \gamma^* f\prime(\gamma^*)$. We can also consider a monotonically increasing pricing function, $F = \beta p_j$, which is assumed to depend upon a cost function, given as,

$$c_{ij} = -\frac{\partial u_i}{\partial p_j} p_j \tag{35}$$

and some inequalities,

$$p_1^* < p_2^* < .... < p_N^* \tag{36}$$

$$u_1^* > u_2^* > .... > u_N^* \tag{37}$$

$$c_1^* < c_2^* < .... < c_N^* \tag{38}$$

A distributed power control algorithm for a wireless cellular system based on sigmoid like utility function can also be implemented (Xiao et al., 2003). In this algorithm, the power control problem is considered as a multi player non-cooperative game. This algorithm is valid for both voice and data users. The value of SIR for user i with transmission power P can be written as

$$SIR_i = \frac{G_{ii}P_i}{\sum_{j \neq i} G_{ij}P_j + \sigma_i} \tag{39}$$

The main goal of this algorithm is to maximize the net utility by transmission power adjustment and softening the hard SIR requirements as

$$NU_i(SIR_i, P_i) = U_i(SIR_i) - C_i(P_i) \tag{40}$$

where $C_i(P_i) = \alpha_i P_i$ is assumed cost function of power for the user i. The power control problem is then defined as $max_{P \geq i} NU_i$. By solving above equation the optimal power for user i is

$$\hat{P}_i = \frac{R_i}{G_{ii}} f_i^{-1}(\alpha \frac{R_i}{G_{ii}}) = \frac{R_i}{G_{ii}} \widehat{SIR_i} \tag{41}$$

After introducing the iteration factor the above equation can be written as

$$\hat{P}_i(K+1) = \frac{R_i(K)}{G_{ii}(K)} \widehat{SIR_i}(K) = P_i(K) \frac{\widehat{SIR_i}(K)}{SIR_i(K)} \tag{42}$$

Thus, by using utility based power control protocol, a user can control its power by decreasing its SIR and even turn off transmission during heavily loaded network.

2.10 Opportunistic power control technique

In this distributed opportunistic power control algorithm, the transmission power depends on channel gain by observing feedback from the receiver. The transmission rate is managed by SIR at the receiver (Leung & Sung, 2006). The SIR of a terminal i in a cellular system comprising of N mobile terminals can be written as $\gamma_i = \frac{P_i}{R_i}$, where R_i is the effective interference to terminal i, and is given as

$$R_i = \frac{\sum_{j \neq i} G_{ij} + \sigma_i}{G_{ii}} \tag{43}$$

and the power of the terminal i is upgraded as

$$P_i^{n+1} = I_i(R_i^n, P_i^n) \tag{44}$$

The proposed opportunistic power control algorithm is given as

$$I_i^{opp}(R_i^n) = \frac{\varsigma_i}{R_i^n} \tag{45}$$

This algorithm converges and equation $P_i^n R_i^n = \varsigma_i$ is satisfied. The transmission power of terminal i varies directly with ς_i.

2.11 Power control technique based on simple prediction Method

A simple prediction is sometimes useful for power control in wireless networks (Neto et al., 2004). This approach can be used to implement a distributed power control algorithm, based on simple prediction method, and by considering both path gain and SIR as time varying functions using Taylor series. Discrete-time $SINR$ is given as

$$\gamma_i(k) = \frac{g_i(k) p_i(k)}{I_i(k)} \tag{46}$$

where $p_i(k) = \frac{I_i(k)\gamma_i(k)}{g_i(k)}$ is known as necessary transmission power. The transmission power at instant $k+1$ is given by using Taylor series as

$$p_i(k+1) = \gamma_t \frac{\hat{I}_i(k+1)}{\hat{g}_i(k+1)} = \gamma_t \frac{2I_i(k) - I_i(k-1)}{2g_i(k) - g_i(k-1)} \tag{47}$$

3. Ad hoc networks

Wireless networks without any fixed infrastructure are called ad hoc networks, also often called as infrastructure less networks. Generally, ad hoc wireless networks are self-creating, self-organizing, and self-administrating networks (Cayirci & Rong, 2009). Ad hoc network consists of mobile nodes which communicate with each other through wireless medium without any fixed infrastructure. Nodes in mobile ad-hoc network are free to move and organize themselves in an arbitrary fashion. The nodes in a mobile ad hoc network (MANET) must collaborate amongst themselves and each node may acts as a relay when required. Mobile ad hoc networks have a fully decentralized topology and they are dynamically changing (Jindal et al., 2004). Ad hoc networks are very popular in military applications for many years. The concept of ad hoc networks was first used in commercial area in 1990s, at the same time, the idea of a collection of mobile nodes was originated. In the mid of 1990s some routing protocols were standardized by a commission known as Internet Engineering Task Force (IETF). First standard IEEE802.11 for wireless network was introduced in 1997. The latest standard is faster and applied for longer communication. Today, ad hoc networks are attractive and challenging topic of research due to its tremendous applications (Perkins, 2001). The most popular applications of ad hoc networks are temporary communication networks, relief operations, operations in congested and small areas (Ramanathan & Redi, 2002).

Ad hoc networks have many challenges which includes high error probability of transmission, limited capacity, hidden and exposed terminals problem, interference, mobility, node failures, topology maintenance, self healing, node search, synchronization, transmission reliability, and congestion control etc. (Goldsmith & Wicker, 2002).

3.1 Importance of power control in ad hoc networks

Unlike cellular networks, in ad hoc networks the power control is not trivial and is usually managed in a distributed fashion. The nodes in the ad hoc network communicate with all other nodes by sending packets to the neighboring nodes. The choice of an appropriate power level for packets at a particular node is very crucial matter, as it indirectly effects the physical layer, network layer, and transport layer of the system by determining the quality of received signal, range of transmission and magnitude of interference respectively (Kawadia & Kumar, 2005). The nodes in ad hoc networks use different modes of operation such as transmit mode, receive mode, idle mode and sleep mode. As a result, these different types of nodes have different power consumption requirements (?).

Power consumption of ad hoc networks can be controlled either by controlling transmission power or by choosing optimal routs for transmission. Transmit power control is a cross layer design problem affecting all layers of the OSI model from physical layer to transport layer (Jia et al., 2005). In general power conservative protocols can be divided into two main categories as transmitter power control protocols and power management algorithms. Second class can be further divided into MAC layer protocols and Network layer protocols (Ilyas, 2003). In the subsequent sections, we present the details of some of these protocols.

4. Power control techniques in ad hoc networks

The power control issue is one of the major challenges prevailing in ad hoc networks. There are many power control algorithms presented by various authors and researchers. Some of these algorithms are discussed in this chapter. We begin by presenting algorithms which are based on 802.11 medium access layer, then discuss some of the challenges faced by CDMA networks as in these types of networks, power control is an essential component. In addition, power control techniques at network layer are also presented. These power control methods are jointly implemented with routing protocols and clustering configurations.

4.1 Simple modifications to 802.11

The 802.11 MAC standard is slightly modified and adapted in order to obtain a distributed power control loop based algorithm which results in lower energy consumption and higher throughput. In this algorithm, in contrast to the original IEEE 802.11 MAC protocol, the transmissions occur at different power levels which are chosen by the algorithm (Agarwal et al., 2001). The main purpose of the algorithm is to calculate a minimum transmit power level for each node to successfully transmit to neighboring nodes. During transmission, a ratio of the signal strength of the last received message to the minimum acceptable signal strength at the node currently transmitting the message is included in the CTS and DATA message headers. The receiver encodes the ratio of received signal strength of RTS message to minimum acceptable signal strength in the header of CTS reply message.

The transmitter will also encode into it the ratio with respect to CTS upon transmitting the DATA message. In this way RTS-CTS-DATA-ACK exchange provides an opportunity to both receiver and transmitter to inform each other not only about their signals strength but also about their transmit power levels. Each node maintains a small table with fields cf-pwr, dr-pwr and a count down timer field. The field cf-pwr maintains an exponential weighted average history of the received signal strength ratio received from each neighbor and the dr-pwr field maintains an exponential weighted average (EWA) history of the cf-pwr field at instances when packet loss occurred. Upon receiving a CTS or DATA message from a node its cf-pwr field in the table is updated by decreasing the transmit power level by one, unless the countdown timer shows zero value. The dr-pwr field in the table is updated by increasing transmit power level by one for the timeout during wait for CTS, DATA or ACK message.

A power control scheme based on modifications to BASIC power control scheme (CTS-RTS hand shake), can be implemented which saves power without degrading throughput (Jung & Vaidya, 2002). In this algorithm just like BASIC power control scheme RTS and CTS, messages are sent at maximum power level P_{max} while DATA and ACK messages are sent with minimum power level. The novelty of this protocol is that ACK-DATA collision avoidance can be made possible by transmitting DATA with maximum power level for a short period so that nodes in CS zone can sense it. P_{max} is achieved periodically during transmission of DATA. The nodes which may interfere with ACK reception stops their transmission by observing that system is busy and power is saved.

In another modification (Lin & Lau, 2003), a protocol for power control named as PCMAC is implemented. This protocol overcomes the problems created by asymmetrical links efficiently. The IEEE 802.11 standard protocol is modified by introducing an extra channel for power control. In contrast to the BASIC scheme for power control, in PCMAC protocol RTS, CTS, DATA, and ACK transmission occur at minimal necessary power level while the broadcast packets at maximal power level. During reception of DATA, the receiver calculates the noise power level by estimating the noise and signal strength. It then informs the neighboring terminals with this information by using power control channel. Keeping in view this information the neighboring terminals take any suitable action. This algorithm replaces the four way handshake by three way hand shake. Also, each terminal manages the three tables named as sent table, receive table for data packet transmission, and a power history table maintaining the record of necessary power level to reach other terminals. This necessary power level is calculated as $P_{nec} = Rx_{th}P_T/E$, Where E is the received signal strength, P_T, the power level at which a packet is transmitted, included in the RTS, CTS and broadcast packet head.

4.2 Link collision avoidance technique

In Asymmetric Link Collision Avoidance (ALCA) based power control protocol the power levels are managed by the announcement of the Current Transmission Duration Information (CTDI) through N different carrier durations (CD) (Pires et al., 2005). This protocol overcomes the problem of DATA, ACK frames collision in BASIC power control scheme due to asymmetric links. ALCA is based on two major steps. Firstly the transmitting node computes the CTDI and then allows the nodes in CS-Zone (CSZ) to recover required CTDI by choosing an appropriate CD from N different CDs. Secondly, the terminal in the CSZ finds a suitable extended inter frame space (EIFS) value based on CD extracted by the DATA carrier. This protocol saves power considerably.

4.3 Power control dual channel protocol

The power control dual channel (PCDC) protocol permits simultaneous interference limited transmissions in the neighboring area of receiver by modifying typical RTS-CTS handshake process in mobile ad hoc networks (Muqattash & Krunz, 2004). This protocol gives much importance to the network layer and MAC layer interaction as power control issue is considered a joint MAC and network layer problem. The MAC layer controls the transmission power of route request (RREQ) packets and affects the network layer. As it is evident from the name, there are two main channels in protocol - data and control channel. The control channel has further two sub channels named as RTS-CTS channel and ACK channel. This protocol is based on the following assumptions:

i. Channel gain remains stationary during transmission of DATA packets.
ii. The gain between two nodes remain same in both sides.
iii. Data and control packets observe same gain between a pair of nodes.

This protocol is distributed in nature and has advantages over other protocols because of the availability of reserved channels. Although functioning of the protocol has relatively stringent assumptions, a relatively smooth and better performance can be achieved under normal channel conditions.

4.4 Power control MAC protocol

This protocol uses an access window and improves the network throughput at low energy consumption by allowing multiple transmissions (Muqattash & Krunz, 2005). It is a distributed, asynchronous and adaptive power control protocol and is named as POWMAC which is based on single channel and single transceiver design. The novel features of POWMAC are described as:

i. Collision avoidance information (CAI) is included in control packets instead of simple RTS/CTS control packets.
ii. Required transmission power is calculated at the intended receiver.
iii. Some CTS and Decide-to-Send (DTS) packets are transmitted towards the potentially interfering terminals.
iv. An access window (AW) is introduced which stops the transmission of DATA packets for a short period and reduces the collisions between control and data packets by informing the transmitters about ensuing transmission.

4.5 Distributed correlative power control technique

The correlative power control can be described as a transmitting node that predicts the interference by using a prediction filter after observing the interference around it. It also includes this information with RTS message. The receiving node assigns a power to CTS by observing this included predicted interference. The receiver repeats the whole procedure and then sends CTS message along with predicted interference to transmitter. The transmitter then assigns power to DATA by observing this predicted value of interference included in CTS message. The same procedure is adopted before sending DATA and ACK messages by transmitter and receiver (Alawieh et al., 2007). The minimum transmission power can be calculated as $P_{min} = k/Gain$, where the channel loss $Gain$ can be measured as a ratio of the received and transmitted powers P_r and P_t. The received signal power can be given as

$$P_r = P_t r^{-4} G^2 h^2 10^{\Im/10} \tag{48}$$

Where r is the distance between two nodes, h is the height of the antenna, G is the antenna gain and $\Im$ is shadowing component. The transmission power of CTS is given as

$$P_{CTS} = max(P_{min}, \zeta \times I/Gain) \tag{49}$$

The transmission power of DATA is given as

$$P_{DATA} = max(P_{min}, \zeta \times I_{+}/Gain) \tag{50}$$

The transmission power of ACK is given as

$$P_{ACK} = max(P_{min}, \zeta \times I_{++}/Gain) \tag{51}$$

Where I is the predicted interference plus noise power.

4.6 Adaptive power control techniques

The adaptive power control technique uses a two ray ground propagation model (Zhang et al., 2005a). This adaptive power control algorithm is based on the relationship between transmission powers of RTS-CTS, CTS-DATA, and DATA-ACK pairs using single channel setup. The relationship between transmit power P_t and receive power P_r can be written as

$$P_r = P_t * G_t * G_r * h_t^2 * h_r^2 / d^4 \tag{52}$$

where h_t and h_r are the heights, and G_t and G_r are the gains of transmitter and reviver's antenna respectively. The relationship between the transmissions powers of RTS/CTS/DATA can be given as

$$P_{RTS,r} = P_{RTS,t} * G_t * G_r * h^4 / d^4 \quad (53)$$

$$P_{DATA,r} = P_{DATA,t} * G_t * G_r * h^4 / d^4 \quad (54)$$

$$P_{CTS,r} = P_{CTS,t} * G_t * G_r * h^4 / d^4 \quad (55)$$

where d is the distance between two nodes, h is the antenna height and P is the power. A successful transmission between receiver and transmitter can take place by satisfying the following necessary conditions:

$$P_{i,t} * P_{j,t} \geq P_w(i,j) \quad (56)$$

and

$$P_{i,t} \geq k / g_{(t,r)} \quad (57)$$

where $g_{(t,r)}$ is the ratio of attenuation gains between transmitter $P_{i,t}$ and receiver $P_{i,r}$, P_w is the cross coefficient of i and j.
In a similar type of protocol the delivery of packets is based on a delivery curve function, which shows a relationship between successfully delivered packet and total transmitted packets (Zhang et al., 2005b). This is also an adaptive power control protocol assuming the successive correlations between the transmission powers of four way hand shake frame for improvement of system throughput and energy saving. It helps the protocol to choose the best working profile and packet correlations are considered in protocol operation. In this protocol transmission power of RTS and CTS are considered same while those of DATA and ACK are similar. The main goal of this protocol is to find minimum powers P_{RTS} and P_{DATA} for successive communication.

4.7 Neighbor detection power control technique

According to this protocol a node initially increases its transmission power until it detects some neighbors around it and again adjusts its power according to the node degree (Abasgholi et al., 2008). After this step, any increase in transmission power decreases the number of one hop neighbors. The umber of neighbors increases with the decrease in transmission power which ultimately enhance the network throughput. The transmission power is varied between a minimum and a maximum value. The change in power can be calculated as

$$P_t = P_c - 5\, log(d_t / d_c) \quad (58)$$

where P_t and P_c are the targeted and current transmission powers. d_t and d_c are targeted and current node degrees which can be calculate as $d_c = D.\pi.r_c^2$ and $d_t = D.\pi.r_t^2$

4.8 Decoupled adaptive power control technique

The objective of these class of protocols is to strictly prohibit the hidden terminals creation (Ho & Liew, 2006). The two protocols named as Decoupled Adaptive Power Control (DAPC) and Progressive Uniformly Scaled Power Control (PUSPC), focus on the hidden terminal avoidance and minimizing mutual interference for enhancing overall network capacity. In the first DAPC protocol each node continuously monitors its surrounding, adjust their powers in a disturbed manner through various iterations by collecting information from neighboring nodes and create hindrance to new hidden terminals and interfering links. The second PUSPC protocol overcomes the deficiencies in DAPC. This protocol deals with two sets named as power control set and finished set. In the beginning of operations all nodes lie in the power control set with same power and they start reducing power through each iteration and after some time few nodes shift to finished set with different powers.

4.9 Autonomous power control technique

This protocol allows nodes to send DATA/ACK packets with power level calculated by keeping in view the distance between transmitter and receiver and RTS/CTS packets with an adjustable power (Chen et al., 2006). This protocol is based on autonomous power control MAC protocol (APCMP). A dynamic network structure is proposed where the main goal of protocol is to reduce energy consumption and improve network efficiency. The protocol describes the initial adjustment of a power level for DATA/ACK messages transmission depending upon the average distance between a transmitter and its neighbors at that time. The power level for RTS/CTS messages is adjusted in proportionality to the above adjusted power for DATA/ACK messages. Usually transmission power level for RTS/CTS is taken a little grater than transmission power for DATA/ACK. The distance between a transmitter and receiver can be estimated as

$$d = k\sqrt{\frac{p^{*}_{RTS/CTS^{\alpha}}}{p_{rec}}} \tag{59}$$

where k is the coefficient, $p_{RTS/CTS}$ is the transmitting power level for the RTS/CTS packet, p_{rec} is the received signal power level, and α is a constant which depends on the antenna gain, system loss, and wavelength. The average estimated distance from transmitter to n neighboring nodes is calculated as

$$\bar{d} = 1/n\sum_{i=1}^{n} d_i \tag{60}$$

where d_i is the estimated distance from the transmitter to the ith neighbor. The transmission power level for DATA/ACK can be calculated as

$$p_{DATA/ACK} = \bar{d}^{k} \times Rx_{thresh} \tag{61}$$

where Rx_{thresh} is the minimum necessary received signal strength. The transmission power level for RTS/CTS can be calculated as

$$p_{RTS/CTS} = p_{DATA/ACK} \times \alpha \tag{62}$$

Where α is a proportionality parameter such that $\alpha > 1$.

4.10 Load sensitive power control technique

In these family of protocols, the power is optimized by keeping in view the load, number of stations and grid area of the network (Park & Sivakumar, 2002). The algorithm denies the concept that throughput can always be maximized with minimum transmission power. We present two of these types of transmission control protocols, namely - the Common Power Control (CPC) and Independent Power Control (IPC).

In CPC all nodes prefer to use the same transmission power while in IPC the nodes are independent to use different transmission powers. The operation of CPC and IPC is initially based on the continuous monitoring of contention time (CT). Each node in the network maintains two threshold values for CT, upper threshold and lower threshold by observing its CT values continuously. A node increase its transmission power if its measured CT lies above the upper bound and decrease its transmission power if its measured CT lies below the lower bound, while it maintains transmission power if its measured CT lies between two bounds. In all cases the main purpose of protocol is to maximize throughput per low energy consumption.

5. Power control for CDMA networks

The ad hoc networks which employ CDMA technology benefit the most from power control. While the use of transmission power control in these types of networks benefit in saving overall consumption for the entire network, the power control algorithms substantially increase the throughput of CDMA networks. In the presence of uncontrolled interference, the performance of CDMA networks degrades considerably. The peer to peer nature of ad hoc networks prohibits the nodes to achieve perfect power control, therefore, CDMA networks with all their benefits fail to perform well. In this section we discuss power control algorithms used in CDMA based ad hoc networks.

5.1 Single busy tone power control technique

This protocol utilizes three channels named as Data Channel (DCH), Control Channel (CCH), and a Busy Tone (BT) separated by use of frequency (Zhou et al., 2005). This protocol, known as single busy Tone CDMA (SBTCDMA), is based on the combined action of RTS/CTS hand shake, single busy tone, and power control utilization. This protocol achieves better channel gain at the cost of less energy consumption after successful solution of hidden node problem. All RTS/CTS packets are transmitted through CCH with common code for all nodes and DATA packets are transmitted in DCH with separate code for each node.

In this protocol initially each node maintains network allocation vector (NAV) and a CTS table. The neighbors of a transmitter update their NAV regularly. Initially if CCH is idle for a short period, a node *i* check its NAV and finding it zero starts operation by a sending an RTS packet to another node *j*. After receiving RTS successfully the receiver *j* waits for a short period and then sends the CTS packet to transmitter.

After sending RTS, it immediately turn on busy tone signal and wait for DATA packet. The data packet will be sent by node *i* after successful completion of RTS/CTS packets exchange using DCH and in the mean time it also updates its NAV also. All other neighbors of node *i* except *j* remain silent by updating their NAV only and save power.

5.2 Dual reservation power control technique

The dual reservation power control technique a CDMA based multi channel MAC protocol which utilizes three common code channels (Min et al., 2007). The system configuration consists of a broadcast channel and data channel which considerably improves the network throughput and reduces near far interference. The main features of the protocol are described as,

i. Code synchronization is done through RTS/CTS handshake by using common code channel.
ii. Dual reservation scheme is presented through ACK piggybacking using data channel.
iii. Near-Far interference is reduced dynamically.
iv. Broadcast messages are supported with busy tone.

According to the protocol each node in the network maintains three lists regularly, available code list (ACL), occupied code list (OCL) and forbidden code list (FCL). Initially a node j transmit RTS message along with ACL on common code channel with P_{max} to node j. Upon receiving RTS the node i, after comparing node its ACL, and calculating P_{min} sends a CTS message on common code channel including selected data channel and P_{min}. Then node i sends data on P_{min} and after receiving data successfully the receiving node sends an ACK message on data channel otherwise a piggybacking ACK is used. When a node is busy in transmission it broadcast a message as a busy tone on channel by just switching on its transceiver. This process decreases the collision probability and data is transmitted with less power.

A similar protocol (Muqattash & Krunz, 2003), also based on CDMA, efficiently solves the near-far effect problem and allow simultaneous transmission in the vicinity of receiver. This Protocol operates at two frequency channels namely Data and Control channels. Available bandwidth, split into two frequency bands, is for simultaneous transmission to take place. All the nodes on control channel use the common code while all nodes on data channel use different codes. The RTS/CTS hand shake takes place through control channel and all interfering nodes are allowed to transmit concurrently. In addition, the transmitter and receiver must agree on spreading code and transmission power. The minimum required transmit power can be given as

$$P_{CDMA} = \frac{\xi_{max}\mu^* P_{thermal} d^n}{k} \tag{63}$$

where $P_{thermal}$ is the thermal noise power, ξ_{max} is maximum planned noise rise, and μ^* is the ratio needed to achieve the target bit error rate at that receiver. The minimum transmission power at which the data is transmitted and decoded correctly, is given as

$$P_{min} = \frac{\mu^*(P_{thermal} + P_{MAI})}{G} \tag{64}$$

where G is the channel gain.

5.3 Power control technique using channel access method

In this algorithm a distributed power control algorithm along with channel access protocol for CDMA based ad hoc networks to maintain the quality of service is used (Sun et al., 2003). Dynamic range of power for all terminals is considered as the ratio of maximum transmission power and minimum transmission power. The ith link's transmission power in a distributed power control algorithm proposed with adaptive protection margin can be calculated as

$$P_i(k+1) = \frac{\delta\gamma_i}{SIR_i(k)} \times P_i(k) = \delta \times [\gamma_i \times (\sum_{j\neq i} G_{ij}P_j(k) + \eta_i)/G_{ii}] \quad (65)$$

Where protection factor δ, that provides a protection margin for active links, should be greater than one as $\delta > 1$. This increases overall network performance.

5.4 Joint distributed power control and routing protocol

In joint distributed power control and routing protocol, a joint distributed power control and routing protocol for CDMA based ad hoc networks keeping in view the quality of service aspect obtained under low energy and acceptable BER constraints is used (Comaniciu & Poor, 2003). All retransmission are statistically independent of one another and a packet transmission from a node wait for the successful reception from previous transmitter. Probability of correct packet reception depends on SIR, described as

$$P(\gamma) = (1 - BER)^M \quad (66)$$

where M is the packet length. A link can operate on minimum power if received SIR is equal to the optimal SIR γ^*, that can be achieved by the solution of following equation

$$\gamma\frac{d\tilde{P}_c(\gamma)}{d\gamma} - \tilde{P}_c(\gamma) = 0 \quad (67)$$

As minimum $SIR = \gamma^*$, this gives for a link (i, j) the value of P_i as

$$\underset{r(i,j)}{\mathbf{min}} \frac{h_{ij}P_i}{\frac{1}{L}\sum_{k\neq i,k\neq j}^{N} h_{(k,j)}P_k + \sigma^2} = \gamma^* \quad (68)$$

The solution is possible by using iterative power control algorithm as

$$P_i(n+1) = T(\mathbf{P}(\mathbf{n})) \quad (69)$$

If $SIR < \gamma^*$, then the system consume more energy as many retransmission occur and if $SIR > \gamma^*$, then the energy is consumed to overcome the surplus gain. Therefore, a better quality of service the necessary condition achieved by power control is given as

$$SIR_{(i,j)} \geq \gamma^*, \forall(i,j) \quad (70)$$

where $(i, j) = 1, 2, ...N$. All new entries will follow the above necessary condition for active transmission and the power vectors converge to minimum power solution. The algorithm operation stops if further decrease in transmission power is not possible.

6. Joint power control, routing, and clustering

In joint power control, routing, and clustering, the algorithm is implemented at network layer. The routes are carefully chosen such that the impact of interference or power consumption is minimum. The information at the network layer is accessed from the physical layer parameters, thus making it a cross layer system. Information exchange between layers contribute in making routing decisions.

We present some protocols used for power control with joint power control and routing or joint power control and clustering techniques.

6.1 Dynamic forwarding nodes

We can jointly address the transmission power assignment problem by using the concept of power control, routing and clustering. In this protocol we propose a mechanism which is based on careful selection of dynamic forwarding nodes for the enhancement of system throughput while keeping the energy consumption low (Yener & Kishore, 2004). A node i with an intended receiver j update its transmission power with the knowledge of received SIR and channel gain as

$$P_i(n+1) = \frac{\gamma^*}{h_{ij}}(\gamma_j(n) - \frac{P_i(n)h_{ij}}{N}) \tag{71}$$

where γ_j is the received interference, and γ^* is the target SIR. Now two clusters of nodes are considered say C_1 has at least $L+2$ nodes and C_2 having at least one node. If a node in C_1, say A wants to communicate with a node in C_2 say X and in the mean time a node B in C_1 also wants to communicate with another node D in the same cluster C_1, then there are two possibilities. In the first option node A transmit to node X directly and transmission between other two nodes of C_1 can also take place while in the second option the node A transmit to X by using $L-1$ hops in C_1 to another node E in C_1 and then E will transmit to X. During the transmission of nodes E and X the transmission between B and D nodes can also take place. Here node E is considered as forwarding node. The necessary transmission powers levels for node A and node B in the first option are calculated as

$$P_{AX} = \frac{N\sigma^2(d_{AX}^4(K-1) + d_{BD}^4)}{K-2} \tag{72}$$

$$P_{BD} = \frac{N\sigma^2(d_{BD}^4(K-1) + d_{AX}^4)}{K-2} \tag{73}$$

where d_{AX} is the distance between node A and X, d_{BD} is the distance between node B and D, and $K = N/\gamma^* + 1$.
The total transmission power for first option is therefore calculated as

$$P_1 = P_{AX} + P_{BD} = \frac{N\sigma^2}{K-2}(d_{AX}^4 + d_{BD}^4) \tag{74}$$

Similarly the total power for second option is calculated as

$$P_2 = \frac{N\sigma^2}{K-2}(d_{EX}^4 + d_{BD}^4 + \frac{K-2}{K-1}\sum_{i=E_1,E_2,\ldots,E_{L-1}} d_{i,i+1}^4) \tag{75}$$

6.2 Power control by clustering in CDMA ad hoc networks

While clustering in ad hoc networks has many benefits, this approach can significantly improve power consumption and performance (Hasan et al., 2003). A clustered system for ad hoc networks based on combination of a broadcast channel CSMA and two CDMA uplink and downlink channels using the joint concept of successive interference cancellation (SIC), user ordering, and open loop power control can improve network throughput. CSMA also helps in the cluster management, routing and mobility control. All nodes in the network communicate through cluster heads using above described three channels. The received power P_r at the cluster head can be written as

$$P_r = \rho^2 * 10^{-\zeta/10} \bar{P} \frac{P_t}{\gamma^{\alpha}} \quad (76)$$

where $\bar{P}$ is the constant revived power at a distance of one meter, ζ is the shadowing factor, α is the path loss exponent, and ρ is the fading amplitude.

The system describes two types of communications, intra cluster and inter cluster. In first type, all communication is done through cluster heads which are usually one hop or two hops away from its member nodes and maintain three tables namely routing table, membership table and a forwarding table. The communications from nodes to cluster head take place by RTS/CTS/ACK handshake on CSMA channel.

In the second type of communication the nodes transmit packets to their respective cluster head and then it is responsible for successful transmission to nearby cluster head using gateway nodes. Ultimately the packet reached the cluster head of the destination node.

6.3 Common power control technique

In this protocol a distributed, asynchronous and adaptive method named as COMPOW is based on the concept, that all homogeneously dispersed nodes in the network use a common necessary power level (Narayanaswamy et al., 2000). It is important to note that link between transmitter and receiver should be bidirectional. As all receivers, even using common power level may not have common $SINR$, so the transmission powers of all nodes should be kept small so that nodes can have nearly equal $SINR$ value and also create less interference to others. Moreover it is also proved by the proposed protocol, that power per route can be saved and MAC contention is minimized on using low power levels at all nodes. A routing table is maintained for each power level by sending and receiving HELLO messages at that power level. The number of entries in the table for a power level describe the number of nodes that can access the specific power level in minimum hops. The minimum power level whose routing table has the same number of entries as the routing table with maximum power level, is known as optimum power level. The routing table of this optimum power level is considered as master routing table. This protocol can easily be implemented with OSI model.

6.4 Joint power control and clustering techniques

According to this protocol the node clusters are made on the basis of transmit power level without taking into consideration their physical location (Kawadia & Kumar, 2003). The joint problem of power control and clustering can be effectively used for a non homogeneously build network. There are three protocols namely CLUSTERPOW for network capacity enhancement, tunneled CLUSTERPOW for optimization achievement, and implementation of MINPOW for optimal routing.

A high transmit power level is required for communication between two separate clusters but intra cluster communication can be done at low transmission power level with multiple hops. Along with source node the CLUSTERPOW protocol is implemented at each node from source to destination route. The dynamic routing daemons are proposed for each power level which maintains their own routing table by communicating with their daemons of same power level on other nodes by sending HELLO messages at specific power level. The choice of next hop depends upon the minimum necessary power routing table. This protocol provides loop free routes for minimum power assignment.

The second protocol, tunneled CLUSTERPOW is an advance form of CLUSTERPOW protocol which is responsible for the successful transmission of packets to its destination by transmitting packet at small power level hop by hop instead of direct transmission to destination. In this protocol a dynamic tunneling mechanism is required for each packet delivery, which makes it more complicated than CLUSTERPOW protocol.

These protocols play an important role in network capacity maximization. To minimize power consumption during communication MINPOW protocol based on the concept of link cost is proposed and implemented at network layer by sending HELLO packets with each power level. The HELLO packets with maximum power level have routing information while others, considered as beacons, have knowledge of total consumed power, packet transmission power and sequence number of the HELLO packets. Transmission power level can be calculated by knowing receiver power level and distance between receiver and transmitter. Total consumed power can be calculated as

$$P_{total} = P_{rec} + P_{trans}(p) \tag{77}$$

where p is the transmission power level of a beacon packet. Link cost is calculated as

$$LinkCost = min_{beacons}(P_{total}) + P_{rec} \tag{78}$$

6.5 Joint power control and routing in ad hoc networks

Power can be controlled in ad hoc networks by choosing optimal routes. The choice of existing routes depends on a complicated cost metric. This cost metric takes in to account the minimum energy requirements of a particular route, while the routing protocol manages the power consumption for the entire network. The existing routing protocols may be classified as, uniform, non uniform, proactive, reactive, hybrid, source, and non source routing protocols (Kadu & Chaudhari, 2002).

6.6 Minimum average transmission power routing technique

In minimum average transmission power routing technique, a power control routing protocol named as minimum average transmission power routing (MATPR) is used (Cai et al., 2002). This protocol is used for CDMA based ad hoc networks using the concept of blind multi-user detection to achieve the task of minimum power consumption. It is applicable to both real time services and data services. Multi-user interference, appropriate coding scheme and acknowledgment scheme is jointly considered for the protocol design.

In normal data service, if a node say B receives a packet with errors from another node say A, it will transmit the same packet with error to A and then A will retransmit the packet to B. Now upon successful reception of error free packet B will send a correct received message to A to inform other party. This explains that packet transmission latency from source to destination does not matter in normal data service. The average power for successful transmission of packet can be calculated as $P_{av} = \frac{\alpha P_i}{(1-P_e)}$. In real time service it has great importance as BER is very low. Thus the error correction takes place through error correcting codes rather than retransmission process. MATPR protocol acts as proactive, where as the blind detector is installed at receiving node.

6.7 Power aware routing optimization technique

Power aware routing optimization techniques are implemented at network layer. The Power Aware Routing Protocol (PARO) uses the same principle of power aware routing optimization (Gomez et al., 2003). This protocol minimizes transmission power in ad hoc networks and is based on the concept of node to node power conservation using intermediate nodes usually called as redirectors. The redirectors play their role in transmission even if the source and destination pairs are in direct transmission range. The increasing number of redirectors between source and destination lowers the transmission power of packets. According to the PARO a node keeps its transmitter on for L/C seconds during transmission where L and C are length of frame and speed of channel respectively. PARO is efficient in both static and dynamic environments. This protocol is based on three main operations namely overhearing, redirecting, and route maintenance.

7. Conclusion

In this chapter we presented the power control techniques used in ad hoc networks. Traditionally, the power control has been implemented and used effectively in cellular networks. While the use of transmission power control in infrastructure based networks has proven to work well and improve performance, the application of power control techniques to ad hoc networks has many challenges and implementation complexities. We presented power control algorithms which are applicable to ad hoc networks. The power control is of great significance in ad hoc networks because of their organizational structure and lack of central management. It is seen that with the implementation of efficient power control techniques, ad hoc networks can improve their vital parameters, such as power consumption, interference distribution, throughput, routing, connectivity, clustering, backbone management, and organization.

8. References

Chauh, M. C. & Zhang, Q. (2006). *Design and Performance of 3G wireless Networks and Wireless Lans*, Springer, ISBN 0-387-24153-1, USA.

Basagni, S.; Conti, M.; Giordano, S. & Stojmenovic, I. (2004). *Mobile Ad Hoc Networking*, Wiley Interscience, ISBN 0-471-37313-3, USA.

Stüber, G. (2002), *Principles of mobile communication*, Kluwer Academic Publishers, ISBN 0-306-47315-1, USA.

Blogh, J. & Hanzo, L. (2002), *Third generation system and intelligent wireless networking*, John Wiley and Sons, ISBN 0-470-84519-8, USA.

Almgren, M.; Andersson, H. & Wallstedt, K. (1994), Power Control in Cellular Systems, *Proceedings of IEEE Veh. Tech. Conf., VTC '94*, pp. 833-837, ISBN o-7803-1927-3, July 1994, Stockholm .

Lee, W. C. Y. (1991), Power control in CDMA (cellular radio), *Proc. IEEE Veh. Tech. Conf.(VTC)*, PP. CH2944, ISBN 0-87942-5822, St Louis, May 1991, Missouri.

Hanly, S. V.& Tse, D. (1999), Power Control and Capacity of Spread Spectrum Wireless Networks, *Automatica, Vol., 35, No., 12*, December 1999, pp. 1987-2012, ISSN. 0005-1098.

Moradi, H.; Samie, M.& Fallah, M. (2006), An Overview of MAI Effect on the Uplink

Performance of the UMTS Air Interface, *Wireless Personal Communication, Vol., 36, No., 3,* February 2006, pp. 277-289, ISSN. 1572-834X.

Gilhousen, K. S.; Jacobs, I. M.; Viteri, A. J.; Weaver, L. A. & Wheatly, C. E. (1991), On the Capacity of a Cellular CDMA System, *IEEE Trans. Veh. Technology*, Vol., 40, No., 2, (May 1991), pp. 303-312, ISSN. 0018-9545.

Lee, T. H.; Lin, J. C. & Su, Y. T. (1995), Down link Power Control Algorithm for Cellular Radio Systems, *IEEE Trans. Veh. Technology*, Vol., 44, No., 1, (March 1995), pp. 89-94, ISSN. 0018-9545.

Chockalingam, A. & Milstein, L. B. (1998), Open Loop Power Control Performance in DS-CDMA Networks with Frequency selective Fading and Non Stationary Base Stations, *Wireless Networks*, Vol., 4, No., 4, (June 1998), pp. 249-261, ISSN. 1572-8196.

Rintamäki, M. (2002), Power control in CDMA cellular communication systems, *Encyclopedia of Telecommunications*,(2002), Wiley Inter Science, ISBN 0471-36972-1.

Grandhi, S. A.; Vijayan, R.; Goodman, D. J.& Zender, J. (1993), Centralized Power Control in Cellular Radio Systems, *IEEE Trans. Veh. Technology*, Vol., 42, No., 4, (December 1993), pp. 466-468, ISSN. 0018-9545.

Zender, J. (1993), Transmitter Power Control for Co-Channel Interference Management in Cellular Radio Systems, *Proceedings of 4th WINLAB Workshop*, New Brunswick, 1993, New Jersey.

Nettleton, R. W. (1980), Traffic Theory and Interference Management for A Spread Spectrum Cellular Mobile Radio System, *Proc. IEEE ICC-80*, pp. 24.5.1-24.5.5, 1980, Seattle WA.

Nettleton, R. W. & Alavi, H. (1983), Power control for a spread spectrum cellular mobile radio system, *IEEE Veh. Technology Conference*, pp. 242-246, May 1983, Toranto.

Alavi, H. & Nettleton, R. W. (1982), Downstream Power Control for a Spread-Spectrum Cellular Mobile Radio System, *Proceedings of IEEE Global Telecomm. Conference*, pp. 84-88, November 1982, Miami Florida.

Foschini, G. J. & Miljanic, Z. (1993), A Simple Distributed Autonomous Power Control Algorithm and Its Convergence, *IEEE Trans. Veh. Technology*, Vol., 42, No., 4, (December 1993), pp. 641-646, ISSN 0018-9545.

Yates, R. D. (1995), A Framework for Uplink Power Control in Cellular Radio Systems, *IEEE Journal on Selected Areas in Communication*, Vol., 13, No., 7, (October 1995), pp. 1341-1347, ISSN. 0733-8716.

Grandhi, S. A.; Vijayan, R. & Goodman, D. J. (1994), Distributed Power Control in Cellular Radio Systems, *IEEE Trans. Communication*, Vol., 42, No., 2, (February 1994), pp. 226-228, ISSN. 0090-6778.

Ariyavisitakul, S. (1994), Signal and Interference Statistics of a CDMA System With Feedback Power Control, *IEEE Trans. Communication*, Vol., 42, No., 2, (February 1994), pp. 1626-1634, ISSN. 0090-6778.

Zander, J. (1992), Distributed Cochannel Interference Control in Cellular Radio Systems, *IEEE Trans. Veh. Technology*, Vol., 41, No., 3, (December 1992), pp. 305-311, ISSN. 0018-9545.

Wu, Q. (1999), Performance of Optimum Transmitter Power Control in CDMA Cellular Mobile Systems, *IEEE Trans. Veh. Technology*, Vol., 48, No., 2, (April 1999), pp. 571-575, ISSN. 0018-9545.

Qian, L. & Gajic, Z. (2003), Optimal Distributed Power Control in Cellular Wireless systems, *Dynamics of Continuous, Discrete and Impulsive Systems*, Vol., 10 (2003), pp. 537-559,

ISSN. 1201-3390.

Koskie, S. & Gajic, Z. (2003), Optimal SIR based Power Control in 3G Wireless CDMA Networks, *American Control Conference*, pp. 57-62, ISBN 0-7803-7896-2, Denver, June 4- 6, 2003, Colorado.

Gajic, Z.; Skataric, D. & Koskie, S. (2004), Optimal SIR based Power Updates in Wireless CDMA Communication Systems, *43rd IEEE Conference on Decision and Control*, pp. 5146-5151, Paradise Island, December 14-17, 2004, The Bahamas.

Siddiqua, P.; Ahmed, S.; Hasan, M. S. & Aditya, S. K. (2007), Power Control Algorithm for WCDMA, *National Conference on Communication and Information Security*, November 2007, Daffodil.

Nuaymi, L.; Godlewski, P. & Lagrange, X. (2001), Power Allocation and Control for the Downlink in Cellular CDMA Networks, *Proceedings of the 12th IEEE International Symposium on Personal, Indoor and Mobile Radio Communications*, pp. C29-C31, ISBN 2001-965456, October 2001, San Diago.

Rohi, H.; Hajghassem, H. & Armaki, S. H. M. (2007), Kalman-Based Power Control for DS-CDMA Cellular Mobile Systems, *Proc. of the 4th Int. Conf. on Mobile Technology*, Applications and Systems, ISBN 978-1-59593-819-0, September 2007, Singapore.

Osery, A. E. & Abdallah, C. (2000), Distributed Power Control in CDMA Cellular Systems, *IEEE Antennas and Propagation Magazine*, Vol. 42, No. 4, August 2000, pp. 152-159, ISSN 1045-9243.

Shah, V.; Mandayam, N. B. & Goodman, D. J. (1998), Power Control for Wireless Data Based on Utility and Pricing, *Proceedings of Personal Indoor Mobile Radio Communications Conference, PIMRC'98*, pp. 1427-1432, Boston, September 1998, Cambridge.

Xiao, M.; Shroff, N. B. & Chong, E. K. P. (2003), A Utility-Based Power Control Scheme in Wireless Cellular systems, *IEEE/ACM Transactions on Networking*, Vol. 11, No. 2 (April 2003).

Leung, K. K. & Sung, C. W. (2006), An Opportunistic Power Control Algorithm for Cellular Network, *IEEE/ACM Transactions on Networking*, Vol. 14, No. 3 (June 2006), pp. 470-478, ISSN 1063-6692.

Neto, R. A. O.; Chaves, F. S, Cavalcanti, F. R. P. & Maciel, T. F. (2004), A New Distributed Power Control Algorithm Based on a Simple Prediction Method, *ICT' 04, LNCS 3124*, pp. 431-436, 2004.

Cayirci, E. & Rong, C. (2009), *Security in Wireless Ad Hoc and Sensor Networks*, John Wiley and Sons, ISBN 978-0-470-02748-6, USA.

Jindal, N.; Mitra, U. & Goldsmith, A. (2004), Capacity of Ad Hoc Networks with Node Cooperation, *Proc. IEEE ISIT*, pp. 269, July 2004, Chicago.

Perkins, C. E. (2001), *Ad Hoc Networking*, Addision Wesley, ISBN 03215-79070, USA.

Ramanathan, R. & Redi, J. (2002), A Brief Overview of Ad Hoc Networks, *IEEE Communications Magazine*, Vol. 40, No. 5, (May 2002), pp. 20-22, ISSN 0163-6804.

Goldsmith, A. J. & Wicker, S. B. (2002), Design Challenges for Energy Constrained Ad Hoc Wireless Networks, *IEEE Wireless Communications*, Vol. 9, No. 4, (2002), pp. 8-27, ISSN 11536-1284.

Kawadia, V. & Kumar, P. R. (2005), Principles and Protocols for Power Control in Wireless Ad Hoc Networks, *IEEE Journal on Selected Areas in Communication*, Vol. 23, No. 1, (January 2005), pp. 1-13, ISSN 0733-8716.

Chaudhuri, S. P. & Johnson, D. B. (2002), Power Mode Scheduling in Ad Hoc Networks, *Proceedings of the 10th IEEE International Conference on Network Protocols (ICNP'02)*, pp.192-193, November 2002. Paris.

Jia, L.; Liu, X.; Noubir, G. and Rajaaraman, R. (2005), Transmission power control for ad hoc wireless networks: throughput, energy and fairness, *Proceedings of IEEE WCNC*, pp. 619-625, 0-7803-8967-0, March 2005. New Oreleans, La USA.

Ilyas, M. (2003), *The Hand Book of Ad Hoc Wireless Networks*, CRC Press, 0-8493-1332-5, USA.

Agarwal, S.; Krishnamurthy, s.; Katz, R. H. & Dao, S. K. (2001), Distributed Power Control in Ad-hoc Wireless Networks, *Proc. PIMRC01*, Sheraton hotel San Diego, September-October 2001, California.

Jung, E. & Vaidya, N. H. (2002), A Power Control MAC Protocol for Ad Hoc Networks, *MOBICOM'02*, pp. 36-47, Atlanta, September 2002,Georgia.

Lin, X. H. & Lau, V. K. N. (2003), Power Control for IEEE 802.11 Ad Hoc Networks: Issues and a New Algorithm, *International Conference on Parallel Processing, ICPP'03*, Kaohsiung, October 2003, Taiwan.

Pires, A. A.; Rezende, J. F. & Cordeiro, C. (2005), ALCA: A new scheme for power control on 802.11 Ad Hoc networks, *IEEE symposium on a WOWMOM'05*, 2005.

Muqattash, A. & Krunz, M. M. (2004), A Distributed transmission Power Control Protocol for Mobile Ad Hoc Networks, IEEE Transactions on mobile computing, Vol. 3, No. 2 , April 2004, pp. 113-128, 1536-1233.

Muqattash, A. & Krunz, M. M. (2005), POWMAC: A Single-Channel Power Control Protocol for Throughput Enhancement in Wireless Ad Hoc Networks, *IEEE Journal on Selected Areas in Communication*, Vol. 23, No. 5, May 2005, pp. 1067-1084, 0733-8716.

Alawieh, B.; Assi, C. & Ajib, W. (2007), A Distributed Correlative Power Control scheme for Mobile Ad Hoc Networks using Prediction Filters, *21st International Conference on Advance Networking and Applications(AINA'07)*,Niagara Falls, May 2007, Canada.

Zhang, J.; Fang, Z. & Bensaou, B. (2005), Adaptive Power Control for Single Channel Ad hoc Network, *IEEE ICC'05*, Vol., 5, pp. 3156-3160, COEX convention center, May 2005, Seoul.

Zhang, J.; Fang, Z. & Bensaou, B. (2005), Adaptive Power Control Algorithm for Ad Hoc Networks with Short and Long Term Packet Correlations, *Proc. IEEE Conference on Local Computer Networks LCN'05*, 2005.

Abasgholi, B.; Kazemi, R.; Arezoomand, M. & Enayati, A. R. (2008), Neighbor Detection Power Control MAC protocol in Mobile Ad Hoc Networks, *3rd International Symposium on Wireless Pervasive Computing*, Petros M. Nomikos Conference centre, May 2008, Santorini Greece.

Ho, W. H. & Liew, S. C. (2006), Distributed Adaptive Power Control in IEEE 802.11 Wireless Networks, *IEEE international Conference on MASS*, pp. 170-179, October 2006,Vancouver, Canada.

Chen, H. H.; Fan, Z. & Li, J. (2006), Autonomous Power Control MAC Protocol for Mobile Ad Hoc Networks, *EURASIP Journal on Wireless Communications and Networking*, Vol. 2006 , Article ID 36040 , (2006), pp. 1-10, ISSN1460-3705.

Park, S. J. & Sivakumar, R. (2002), Load Sensitive Transmission Power Control in Wireless Ad Hoc Networks, *Proc. of IEEE GLOBECOM'02*,Taipei, Vol. 01, pp. 42-46, 0-7803-7632-3, November 2002, Taiwan.

Zhou, X.; Li, J. & Yang, J. (2005), A Novel Power Control Algorithm and MAC Protocol for CDMA Based Mobile Ad Hoc Networks, *IEEE MILCOM 05*, Atlantic City, Vol 2, pp. 754-760, Oct 2005, New Jersey.

Min, J.; Huimin, C. & Yuhua, Y. (2007), A Dual Reservation CDMA Based MAC Protocol with Power Control for Ad Hoc Networks, *Journal of Electronics(China)*, Vol. 24, No. 1, January 2007, 1993-0615.

Muqattash, A. & Krunz, M. (2003), CDMA Based MAC Protocol for Wireless Ad Hoc Networks, *MobiHoc'03*, Annapolis, pp. 153-163, June 1-3, 2003, Maryland USA.

Sun, J.; Hu, Y.; Wang, W. & Liu, Y. (2003), Channel Access and Power Control Algorithem with QOS for CDMA based Wireless Ad Hoc Networks, *14th IEEE International Symposium on Personal Indoor and Mobile Radio Communication Proceedings*, September 2003.

Comaniciu, C. & Poor, H. V. (2003), QoS Provisioning for Wireless Ad Hoc Data Networks, *Proceedings of the 42nd IEEE Conference on Decision and Contro*l, vol. 1, pp. 92-97, Hawaii, December 2003, USA.

Yener, A. & Kishore, S. (2004), Distributed power control and routing for clustered CDMA wireless ad hoc networks, *Proceedings of the 60th IEEE Vehicular Technology Conference (VTC'04)*, vol. 4, pp. 2951-2955, Los Angeles, September 2004, USA.

Hasan, A.; Yang, K. & Andrews, J. G. (2003), Clustered CDMA AdHoc Networks Without Closed Loop Power Control, *IEEE MILCOM03*, Boston, October 2003, USA.

Narayanaswamy, S.; Kawadia, V.; Sreenivas, R. S. & Kumar, P. R. (2000), Power control in ad-hoc networks: Theory, architecture, algorithm and implementation of the COMPOW protocol, *European Wireless Conference*, 2000.

Kawadia, V. & Kumar, P. R. (2003), Power Control and Clustering in Ad Hoc Networks, *IEEE INFOCOM'03*, San Francisco, pp. 459-469, April 2003. California.

Kadu, R. K. & Chaudhari, N. V. (2002), A Study of Power Saving Techniques in Mobile Ad Hoc Network, *2008 International Conference on Computer Science and Information Technology*, August 2002.

Cai, Z.; Lu, M. & Wang, X. (2002), Minimum Average Transmission Power Routing in CDMA AD Hoc Networks Utilizing the Blind Multiuser Detection, *Proc. of IPDPS'02*, 2002.

Gomez, J.; Campbell, A. T.; Naghshineh, M. & Bisdikian, C. (2003), PARO: Supporting Dynamic Power Controlled Routing in Wireless Ad Hoc Networks, *Wireless Networks*, Vol. 9, No. 5, September 2003, pp. 443-460, 1022-0038.